CUMMINGS PUBLISHING COMPANY
Menlo Park, California
Reading, Massachusetts • London • Amsterdam • Don Mills, Ontario • Sydney

ALTERNATE EDITION

Physics for College Students

with Applications to the Life Sciences

DONALD E. TILLEY

Collège militaire royal de Saint-Jean

Saint-Jean, Quebec

WALTER THUMM

Queen's University

Kingston, Ontario

Prints on pages x, 40, 96, 206, 242, 296, 326, 388, 432, 466, 516, 552, and 614 are courtesy of The Library of Congress; on pages 152, 268, and 356, of PHYSICS TODAY, and on page 670, of Floyd Clark, Cal Tech.

Alternate Edition
Physics for College Students

This work is an alternate edition of Physics for College Students.
Copyright © 1974 and 1976 Cummings Publishing Company, Inc.
Philippines Copyright 1974 and 1976.

All rights reserved. No part of this publication may be reproduced, stored in a retrieval system, or transmitted, in any form or by any means, electronic, mechanical, photocopying, recording, or otherwise, without the prior written permission of the publisher.
Printed in the United States of America.
Published simultaneously in Canada.
Library of Congress Catalog Card Number: 75-30169
ISBN: 0-8465-7534-5
BCDEFGHIJKL-HA-79876

Cummings Publishing Company
2727 Sand Hill Road
Menlo Park, California 94025

Preface

This book is an alternate version of *Physics for College Students with Applications to the Life Sciences*. In response to suggestions from instructors, topics have been rearranged so that the first semester can end with the subject of heat, and the second semester can include all the material on wave motion, together with electricity and optics, and conclude with the chapter on quantum physics.

Our main intent is to provide the student with a grasp of fundamental physical principles in a modern context. The mathematics used is no more advanced than high school algebra and trigonometry. And we give recognition to physics as a human endeavor by a historical introduction to each chapter as well as by references throughout to contemporary applications.

There are three major reasons for our emphasis on the applications of physical principles to the life sciences, and to medicine in particular. First, we wish to show that physics is not an isolated intellectual pursuit and that, in fact, boundaries between "disciplines" are arbitrary and are adopted primarily for convenience. Second, we have found that medical examples are of interest to students taking a physics course as part of a general education. And finally, we wish to make this book suitable for many students heading for careers in the life sciences, including medicine.

For these latter students in particular we have provided some additional examples from the life sciences as a postscript to each chapter after the end of the questions. This does not imply, however, that *Physics for College Students* would be unsuitable for general physics courses preceding the calculus-based physics of the physical science and engineering students. Above all, we have given priority to presenting the principles of physics with clarity and accuracy.

The questions for each chapter are grouped into a main set and a supplementary set. The order of the questions in the main set corresponds rather closely with the order in which subject matter is presented in the text. These questions are arranged in order of increasing difficulty within major topic groups, but there are no exceptionally demanding problems in the main set.

The most difficult problems appear in the supplementary questions, along with some general review questions, and several questions intended to supplement topics that have been merely introduced in text. The supplementary questions are also concerned on occasion with the derivations of some equations that were simply quoted in text.

The student who attempts roughly half the questions in the main sets will have allotted reasonable effort to this phase of the learning process. Answers to the odd-numbered questions in both the main sets and the supplementary questions appear in Appendix F.

Although this alternate edition of *Physics for College Students* is intended for a two-semester course, it is nevertheless quite feasible for the book to be used, with appropriate adaptation, in a two-quarter or a three-quarter course.

```
1.1 Some Elementary Particles
     ↓
1.2 Particle Interactions,
    Nuclei, Atoms
     ↓
2 Motion and Force
     ↓
3 Mechanics
     ↓
4 Energy
     ├──────────────────────────── 5 Frames of Reference
     │                               and Relativity
6 Fluid Mechanics ────┤
     ↓
7 Disordered Energy
     ↓
8 Thermal Processes
     ↓
9 Disorder and         ├──────────── 10 Mechanical Waves:
  Its Increase         │                 Sound
                       ↓
11 Electric Potential
   and Circuits
     ↓
12 Electromagnetism
     ├──────────────────────────── 13 Alternating Currents
     ↓
14 Electromagnetic Waves
     ├──────────────────────────── 15 Optics
     ↓
16 Quantum Physics
```

A word about our rationale regarding Chapter 1: By presenting some facts about atoms and their constituents in the first chapter, we are able to treat interesting modern examples throughout the book. Somewhere, sometime, everyone has heard of such objects as electrons and atoms. We feel that having this knowledge summarized at the outset facilitates the ability to use these ideas effectively. Nonetheless it is possible to omit, or leave as optional reading, most of Chapter 1 and start the course with Chapter 2 on Newtonian mechanics. In this case the first two sections of Chapter 1 would become collateral reading for Section 2.7 and the remainder of Chapter 1 would serve to augment Section 4.4.

The introduction to mechanics in Chapter 2 and in the first sections of Chapter 3 is developed in such a way as to avoid making heavy demands on a student's analytical ability. Systematic analytical procedures are then emphasized in Sections 3.5 and 3.6.

Rearrangements and possible omissions are indicated in the chart on the top of the page. For example, Chapter 5 on relativity could be skipped, in part or entirely, with minimal effect on most of the remaining material. The same could be said of the chapters in which we deal with "practical" details to a greater extent than in the remainder of the book. We refer here particularly to Chapter 6 on fluid mechanics, Chapter 10 on sound, Chapter 13 on alternating currents, and Chapter 15 on optics. With the exception of the first section in Chapter 10 (Wave Phenomena), in these chapters most of the underlying physical principles have been encountered in other contexts so that omission of these chapters would cause a minimum of disruption in a course whose thrust was unequivocably *principles of physics*. On the other hand, for instructors presenting a course with a more practical bias, the chapters on sound, alternating currents, and optics, as well as the one on fluid mechanics, will provide a helpful extension into some of the areas of applied physics.

In the historical introductions we frequently use certain scientific terms before their precise definitions are given. Such terms are, however, discussed carefully in the main body of the chapter in question.

The suggested reading lists are provided for the convenience of the student and the instructor. It is certainly not intended that any student attempt all, or perhaps any, of this additional reading, but it is there with sufficient description to facilitate selection by those inclined to pursue certain topics further.

Our thanks for valuable suggestions are due many colleagues, students, and friends. In particular we wish to thank Dr. John F. Morgan, Department of Ophthalmology, Hotel Dieu Hospital and Queen's University, Kingston, Ontario; Professor David Atherton of the Physics Department of Queen's University; and two former students, Steve Wight and Ken Johns. For typing the manuscript as well as for making many constructive comments we are indebted to our wives.

Finally, our thanks to the Cummings editorial staff.

DONALD E. TILLEY

WALTER THUMM

*Saint-Jean, Quebec,
and Kingston, Ontario*

January 1976

Contents

1. **The Atom and Its Nucleus, 1**

    1.1 Some elementary particles, 3
    1.2 Particle interactions, nuclei, atoms, 5
    1.3 Ions and particle detection, 8
    1.4 Discovery of the nucleus, 11
    1.5 Radioactivity and charge conservation, 13
    1.6 Nuclear reactions, 20
    1.7 Applications of radioisotopes, 23
    Questions, 28
    Supplementary questions, 30
    Additional applications to medicine and the life sciences, 30

2. **Motion and Force, 41**

    2.1 Motion along a straight line, 43
    2.2 Vectors, 50
    2.3 Motion in three dimensions, 52
    2.4 Force, mass, and acceleration, 55
    2.5 Gravitational forces, 59
    2.6 Motion of a particle in a gravitational field, 63
    2.7 Electric forces, 70
    2.8 Charges in electric field, 75
    2.9 Electric forces in atoms, 79
    Questions, 81
    Supplementary questions, 88
    Additional applications to medicine and the life sciences, 90

3. **Mechanics, 97**

    3.1 Newton's third law, 99
    3.2 Momentum and Newton's second law, 100
    3.3 Conservation of momentum, 102
    3.4 Center of mass, 106
    3.5 Analysis using components of vectors, 108
    3.6 Using Newton's laws, 110
    3.7 Friction, 115
    3.8 Statics of rigid bodies, 119
    3.9 Rotation, 126
    3.10 Simple harmonic motion, 130
    Questions, 137
    Supplementary questions, 144
    Additional applications to medicine and the life sciences, 150

4. **Energy, 153**

    4.1 Energy in Newtonian mechanics, 155
    4.2 Thermal energy, 162
    4.3 Simple machines, 166

4.4 Energy and mass, *170*
4.5 Nuclear fission and fusion, *181*
4.6 Remarks on high-energy physics, *187*
Questions, 192
Supplementary questions, 198
Additional applications to medicine and the life sciences, 202

5. Frames of Reference and Relativity, 207

5.1 Frames of reference in Newtonian mechanics, *209*
5.2 Accelerated frames, *212*
5.3 Relativity of time, *218*
5.4 Lorentz transformations, *222*
5.5 Consequences of the Lorentz transformations, *225*
5.6 Relativistic mechanics, *228*
5.7 Remarks on special and general relativity, *230*
Questions, 232
Supplementary questions, 236
An example from the medical laboratory, 238

6. Fluid Mechanics, 243

6.1 Fluid statics, *245*
6.2 Pascal's principle, *251*
6.3 Archimedes' principle, *251*
6.4 Fluid dynamics, *253*
Questions, 260
Supplementary questions, 262
Additional applications to medicine and the life sciences, 265

7. Disordered Energy, 269

7.1 Gases, *271*
7.2 Kinetic theory interpretation of temperature, *274*
7.3 Thermal energy and temperature, *278*
7.4 Calculation and measurement of thermal energy, *280*
7.5 Heat, *283*
7.6 Calorimeters and specific heats, *286*
Questions, 290
Supplementary questions, 292
Additional applications to medicine and the life sciences, 294

8. Thermal Processes, 297

8.1 Thermal expansion, *299*
8.2 Heat transfer, *302*
8.3 Thermometers, *304*
8.4 Changes of phase and phase equilibrium, *310*
8.5 Water vapor in air, *317*
8.6 Equilibrium states; microscopic and macroscopic viewpoints, *319*
Questions, 321
Supplementary questions, 323
Additional applications to medicine and the life sciences, 324

9. Disorder and Its Increase, 327

9.1 Calculation of entropy changes, *329*
9.2 The second law of thermodynamics, *332*
9.3 Heat engines and refrigerators, *337*

9.4 Degradation of energy, *342*
9.5 Entropy in statistical mechanics, *343*
9.6 Low-temperature phenomena, *348*
   Questions, *349*
   Supplementary questions, *351*
   Additional applications to medicine and the life sciences, *352*

10. Mechanical Waves: Sound, *357*

10.1 Wave phenomena, *359*
10.2 Sound, *361*
10.3 Vibrating systems and resonance, *366*
10.4 Characteristics of sound, *368*
10.5 The ear and hearing, *370*
10.6 Doppler effect, *371*
10.7 Supersonic speeds and shock waves, *374*
10.8 Sonic spectrum, *375*
   Questions, *379*
   Supplementary questions, *382*
   Additional applications to medicine and the life sciences, *383*

11. Electric Potential and Circuits, *389*

11.1 Electric potential, *391*
11.2 Current electricity, *395*
11.3 The complete circuit, *399*
11.4 Ohm's law, *406*
11.5 Power and energy in electric circuits, *408*
11.6 Series and parallel circuits, *410*
11.7 Multiloop circuits, *419*
   Questions, *420*
   Supplementary questions, *425*
   Additional applications to medicine and the life sciences, *429*

12. Electromagnetism, *433*

12.1 Sources of a magnetic field $\vec{B}$, *435*
12.2 The force exerted by $\vec{B}$ on a moving charge, *438*
12.3 Circular motion of a charge in a uniform magnetic field, *440*
12.4 Magnetic force and electric current, *443*
12.5 Faraday's law of electromagnetic induction, *449*
   Questions, *455*
   Supplementary questions, *458*
   Additional applications to medicine and the life sciences, *462*

13. Alternating Currents, *467*

13.1 Alternating voltages and currents, *470*
13.2 Resistance, reactance, and impedance, *475*
13.3 Power in AC circuits, *484*
13.4 Electronics, *486*
13.5 Electricity and safety, *494*
13.6 Supplementary topic; phasors and the addition of voltages in series circuits, *496*
   Questions, *499*
   Supplementary questions, *503*
   Additional applications to medicine and the life sciences, *509*

14. Electromagnetic Waves, *517*

14.1 Characteristics of electromagnetic waves, *520*
14.2 Interference, *528*
14.3 Reflection, refraction, and dispersion, *539*
14.4 Doppler effect in astronomy, *542*
   Questions, *546*
   Supplementary questions, *548*
   Additional applications to medicine and the life sciences, *550*

15. Optics, *553*

15.1 Straight lines, *556*
15.2 Reflection, *557*
15.3 Refraction, *563*
15.4 Lenses, *568*
15.5 The eye and seeing, *580*
15.6 Optical instruments, *586*
   Questions, *599*
   Supplementary questions, *603*
   Additional applications to medicine and the life sciences, *608*

16. Quantum Physics, *615*

16.1 Properties of photons, *617*
16.2 Experimental evidence of photons, *620*
16.3 Energy levels, *624*
16.4 Some quantum mechanics, *627*
16.5 The hydrogen atom, *634*
16.6 The electronic structure of atoms, *639*
16.7 Energy bands in solids, *642*
16.8 Semiconductors, *644*
16.9 The laser, *647*
16.10 Interaction of electromagnetic radiation with matter, *650*
16.11 Quantum electrodynamics, *652*
   Questions, *657*
   Supplementary questions, *660*
   Additional applications to medicine and the life sciences, *663*

Epilogue, *671*

Appendix A: Mathematical Review, *673*

A.1 Some basics of algebra, *673*
A.2 Algebraic sums, *675*
A.3 Exponents, *675*
A.4 Significant figures, *677*
A.5 Equations, *679*
A.6 Proportionality, *681*
A.7 Geometry, *683*
A.8 Trigonometry, *684*

Appendix B: Various Physical Constants and Other Physical Data, *690*

Appendix C: The Metric System and Some Equivalents, *691*

Appendix D: Alphabetical List of the Elements, *692*

Appendix E: Natural Trigonometric Functions, *693*

Appendix F: Answers and Hints for Odd-Numbered Questions, *694*

Index, *731*

Ernest Rutherford

1 The Atom and Its Nucleus

Physicists attempt to understand nature. What little understanding has been achieved is one of mankind's most valuable possessions. Physics is of enormous practical value; indeed much of our modern technology arose rather directly from the work of physicists. But physics has many faces. In addition to the evident utilitarian aspect there is also to be found intellectual adventure together with a most significant portion of natural philosophy.

In this book, although basic physical principles and some modern applications are our main concern, we feel that a historical perspective can only enhance one's appreciation of the ideas in physics and of their impact on our culture. For this reason we start each chapter with brief historical comments.

The reader will find that some of our greatest advances in understanding have been made by men with the good taste to examine the simplest things: a falling stone, a planetary orbit, a hydrogen atom.

In this first chapter, we assemble some of the facts of present day physics about atoms and the atomic nuclei.

An outstanding pioneer of nuclear physics and possibly the greatest experimental physicist of our century was a New Zealander, Ernest Rutherford. His research career began at an exciting time. Marvelously penetrating x-rays were discovered by W. C. Röntgen (1845–1923) in Germany in 1895. Several weeks later in France, Henri Becquerel (1852–1908), searching for x-ray emission from phosphorescent substances, discovered instead strange radiations emitted from a compound of uranium. This radiating activity, or *radioactivity* as it is now known, was soon shown to involve the emission of three distinct types of rays: alpha particles, beta particles, and gamma rays (about which we will have more to say later).

At McGill University in Montreal, Rutherford's research turned to the study of radioactivity. By 1902, working with Frederick Soddy (1877–1956), he had discovered the then astounding fact that radioactivity involved the sudden and spontaneous transformation of one atom into a chemically different atom and formulated the *Rutherford-Soddy law* of radioactive decay. The alpha particle, in a beautifully

"It is not in the nature of things for any one man to make a sudden violent discovery; science goes step by step, and every man depends on the work of his predecessors. . . . Scientists are not dependent on the ideas of a single man, but on the combined wisdom of thousands of men."

Ernest Rutherford
(1871–1937)

Physics Today

Glenn T. Seaborg
(1912-   )

simple series of experiments, was shown to carry a positive electrical charge and, in fact, to be a helium ion (i.e., a helium nucleus—that is a helium atom stripped of its two electrons).

Later, at Manchester University in England, high-speed alpha particles were the key to Rutherford's greatest work: the study of the scattering of these particles as they passed through matter, an experimental and theoretical investigation which, in 1911, revealed the existence of the *atomic nucleus* as a minute concentration of mass and electric charge. Again using alpha particles, this time as projectiles to bombard nitrogen nuclei, Rutherford, in 1919, showed that a nuclear reaction was obtained which produced nuclei of hydrogen and oxygen. This was the first laboratory controlled change of one chemical element into another, a transmutation of matter, the alchemists' dream come true.

And the applications of this fundamental knowledge of the atom and its nucleus have become legion. As Nobel prize winner Dr. Glenn T. Seaborg reportedly stated at a symposium on biomedical research, "We are already experiencing its (atomic and nuclear science) benefits every day—every day it is helping to save human lives and relieve human suffering in this country and throughout the world. Such benefits reveal a pay-off on our investment in the atom that few people realize."

1.1 SOME ELEMENTARY PARTICLES

The bewildering variety of the objects that we see around us and the events that we witness daily all arise from the interaction of relatively few different types of tiny *identical particles*. This idea is one of the most important and remarkable conclusions of science. We therefore start this course with a description of these particles, the basic building blocks of all matter.

The forthcoming facts are the end product of thousands of experiments and rest on an imposing interwoven structure of experimental evidence. Although only a very small portion of this vital evidence can be presented in this chapter, much of the background upon which this knowledge rests will emerge in the remainder of the book.

Electrons and Electric Charge

The *electron* is the particle of primary interest in much of physics, chemistry, and electrical technology. It possesses *mass* and *electric charge*, two fundamental physical quantities which will be discussed at some length in subsequent chapters.

Objects with a small mass are agile in the sense that the smaller the mass of an object, the greater the speed the object will acquire when a given push is applied for a specified length of time. And the electron's mass is indeed small, the smallest ever discovered, except for those particles with no mass at all. By comparison, the mass of a mere thimbleful of water is tremendous, a billion billion billion (10^{27}) times greater than the mass of an electron.

An object endowed with electric charge experiences a force, a push or a pull, in the presence of a second object which also possesses an electric charge (see Fig. 1-1). The size of this "electrical" force is proportional to the amount of electric charge on the first object. That is, doubling the amount of charge doubles the electric force, tripling the charge triples the force, etc. Experiments have revealed that there are two different types of electric charges: one type is named *negative charge* and the other, *positive charge*. Two different charges of the same type, both positive or both negative, exert repulsive forces on each other. But a negative charge in the presence of a positive charge experiences a pull toward the positive charge, an *attractive* force. The positive charge experiences a force of the same size directed toward the negative charge. In summary: *Like charges repel, unlike attract.*

An electron's charge is of the type called negative. The symbol $-e$ is conventionally used to denote the charge on an electron. This particular amount of electric charge is apparently of fundamental significance because every sufficiently accurate measurement of electrical charge indicates that the only quantities of charge that occur in nature are integral multiples of the charge, e. For instance, in a gas through

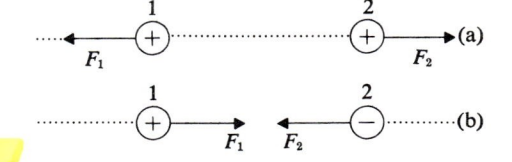

Fig. 1-1 Electrical forces.
(a) Like electrical charges repel each other.
(b) Unlike charges attract. In each case F_1 is the force experienced by the *electrical charge* 1 due to the presence of *charge* 2, and F_2 is the force experienced by *charge* 2 due to the presence of *charge* 1.

*See, however, the remarks on quarks in Section 4.6.

which an electric spark is passing, one finds particles with charges $+3e$, $+2e$, $+e$, 0 (uncharged or electrically neutral), and $-e$, but never anything with a charge of something like $(3/11)e$ or $2.7e$.*

In early studies of radioactivity, one type of unidentified particle emerging with high speed from certain types of radioactive material was christened a *beta particle* (β-particle). Subsequently the β-particle was identified to be an *electron*, but the old name has survived to the extent that electrons originating in a radioactive decay are often called β-particles.

Protons

An important constituent of all the materials that we encounter in our daily life, and which accounts for nearly half of their mass, is the second particle on our list, the *proton*. The proton is massive compared to an electron. Precise measurements yield the result that a proton's mass is nearly 1836 times the mass of an electron.

A proton possesses a positive electric charge, $+e$. That is, the positive charge on a proton is of exactly the same size as the negative charge on an electron.

Neutrons

More than half the mass of most of the objects around us arises from particles called *neutrons*. As its name suggests, the neutron, unlike the proton, carries no electric charge, that is, the neutron is electrically neutral. A neutron has a slightly greater mass than a proton, the neutron mass being nearly 1839 times the mass of an electron.

Photons

Particles of light are called *photons*. A photon always travels at a speed of 3.0×10^8 meters per second (that is, about 186,000 miles per second). A photon has no mass at all and no electric charge.*

A photon does possess *energy*. A discussion of this most important physical quantity, energy, possessed in fact by every particle, we postpone to later chapters. Now we merely point out that photons can be classified according to their energy. For example, a particle of red light is a photon with a certain definite amount of energy. A stream of such identical photons constitutes a beam of red light. A photon which has nearly twice the energy of a red light photon is a particle of violet light.

Photons of any energy are possible; at least, no upper limit to the energy of photons has yet been ascertained. An x-ray beam is composed of photons, the energy of which is about ten thousand times greater than that of a red light photon. The radiation called *gamma rays* (γ-rays) discovered in radioactivity experiments, consists of photons whose energy is typically about a million times that of a red light photon.

*We follow the convenient practice of most research physicists and use the word mass in only one sense. In many introductory textbooks the reader will find two sorts of masses: mass as we use the term and also another mass called "relativistic mass." Relativistic mass is never used in this book.

4   THE ATOM AND ITS NUCLEUS

Any beam of photons, no matter what the photon energy, can be referred to by the general term, *electromagnetic radiation*.* Some different names that are used for electromagnetic radiation, arranged in order of increasing photon energy, are as follows: infrared; visible light of colors red, orange, yellow, green, blue, and violet; ultraviolet; x-rays; and γ-rays. This orderly arrangement is known as the *electromagnetic spectrum*, of which visible light forms just a very, very small part. In other words, the human eye is sensitive to only a small range of electromagnetic radiations.

*To be discussed in some detail in Chapter 14.

1.2 PARTICLE INTERACTIONS, NUCLEI, ATOMS

Different particles interact with each other, exerting forces whose nature depends on the properties of the particles involved. One consequence of certain types of particle interactions is the formation of stable structures composed of several different particles.

Nuclei

The strongest forces that are known are the "nuclear forces" that attract a neutron to another neutron, or a neutron to a proton, or a proton to a proton. This nuclear force has a very *short range*, about 10^{-15} meter (one million billionth of a meter). This means that if two particles are separated by a distance greater than about 10^{-15} meter, they experience no nuclear force. [Of course, if the two particles in question are both charged particles, protons, then in addition to the attractive short range nuclear force, there will be also an *electric* (repulsive) force.]

This strong nuclear force allows an assembly of neutrons and protons to cluster together and form a structure called a *nucleus*. For example, 92 protons and 146 neutrons, a total of 238 particles, can form a tiny compact nucleus whose radius is merely 10^{-14} meter. This particular nucleus is called the uranium-238 nucleus and is denoted by the symbol $_{92}U^{238}$. In this convention the number 238 in the superscript position denotes the total of the number of neutrons and protons in the nucleus and is called the *mass number* of the nucleus. The number in the subscript position specifies what is termed the *atomic number* of the nucleus, the number of protons in the nucleus. Since each proton carries a positive electric charge $+e$, the nucleus carries a positive electric charge given by

$$(\text{atomic number}) \times e.$$

The particle that Rutherford used with such skill, the *alpha particle* (α-particle), is the helium nucleus, $_2He^4$, a particularly stable structure composed of two protons and two neutrons.

About 1500 different nuclei have been discovered. The nucleus $_{92}U^{238}$ is the most massive nucleus which has been found remaining in the material from which our planet was formed.

Atoms

The electric force, exerted by one electrically charged particle on another, leads to interactions which are weaker than the nuclear interactions but are not restricted to a short range.

An electron, because it carries a negative electric charge, experiences an electric attractive force in the presence of the positive charge on a nucleus. This electric force makes possible the formation of a rather loose structure consisting of electrons whirling about a tiny but massive central nucleus. When the number of orbiting electrons is equal to the number of protons in the nucleus (the atomic number), the structure contains as much positive charge as negative charge; it is electrically neutral. Such a structure is called an *atom*.

For example, a uranium atom consists of a uranium nucleus (containing 92 protons) surrounded by 92 orbiting electrons. The simplest atom is a hydrogen atom: it consists of one proton as the nucleus and one orbiting electron.

The average distance from the nucleus to the outermost electron in any atom is roughly 10^{-10} meter. Evidently atoms are very small. Some hundred million atoms, side by side, are required to give a line one centimeter long (approximately 0.4 inch). Although the head of a pin has a mass of only about 8 milligrams, and its volume is roughly 1 cubic millimeter, this pinhead contains around 10^{20}, that is 100,000,000,000,000,-000,000, iron atoms!

Nevertheless, the nucleus of the atom is minute compared to the size of the atom. Even for uranium, the nuclear radius is merely one ten thousandth of the atomic radius. But over 99.9% of the mass of the atom is concentrated in its tiny nucleus. We see that an atom is mostly empty space!

All these facts quite naturally suggest that atoms be pictured as shown in Fig. 1-2. But nature has turned out to be more subtle than such a picture suggests! We will return to the question of pictorial representations of atoms after encountering some experimental evidence that is relevant to the subtleties involved.

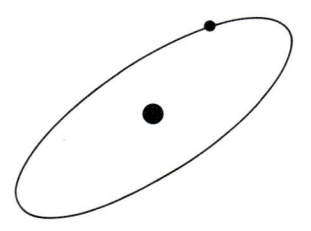

Fig. 1-2 Hydrogen, a simple atom.

Elements and Isotopes

The interaction of a certain atom with other atoms, in other words, the *chemical behavior* of the atom, is determined by the atom's *electron configuration* or the distribution of electrons throughout the atom. This electron configuration is ultimately governed simply by the amount of positive charge in the nucleus. And the nuclear charge is determined by the number of protons in the nucleus (the atomic number).

Atoms with the same number of protons in the nucleus thus have the same chemical behavior; they are said to be atoms of the same *element*. We find occurring naturally on our planet some 90 different elements ranging from the element hydrogen with a nucleus containing just one proton, up to uranium with 92 protons in its nucleus.

Atoms of the same element can have a different number of neutrons in their nucleus. Such atoms, which have the same atomic number but different mass numbers, are called *isotopes* of the element in question.

Hydrogen, for example, has three known isotopes: *common hydrogen*, H-1, with a proton as its nucleus ($_1H^1$); *deuterium*, H-2, with a deuteron as its nucleus ($_1H^2$); and *tritium*, H-3, with a triton as its nucleus ($_1H^3$).

In uranium ore we find three uranium isotopes: U-238, U-235, and U-234. Separation of the different isotopes cannot be effected by chemical processes because the different isotopes have the same chemical properties. Effective methods of separation must exploit the difference in mass of the isotopes. A major factor inhibiting the rapid proliferation of A-bombs throughout the world is this technological difficulty of extracting the isotope U-235 from natural uranium metal which is more than 99% composed of the isotope U-238.

Molecules

Atoms interact with other atoms (of the same or of a different kind). In certain circumstances the result is the formation of a larger structure called a *molecule*.* As an example, two hydrogen atoms can interact to form a hydrogen molecule which consists of two well-separated protons (their separation is some hundred thousand times the range of nuclear force) and two electrons orbiting between the protons.

While many molecules are relatively simple structures consisting of only a few atoms, there are also many other kinds of molecules, each composed of numerous atoms—sometimes thousands. For example, myoglobin, a substance which is found in muscle tissue and which resembles the hemoglobin found in red blood cells, is composed of molecules of some 2500 atoms each.

Different types of molecules interact in what are called *chemical reactions* to produce new types of molecules. *All the familiar objects around us are aggregates of atoms or molecules.*

As a review, let us look at some water in a glass and think of its constituents. The water is an aggregate of water molecules. Each water molecule is composed of atoms: two atoms of hydrogen and one of oxygen. Each hydrogen atom consists of an electron and a nucleus containing a proton and sometimes one or even two neutrons. An oxygen atom contains eight electrons and a nucleus with eight protons and eight, nine, or ten neutrons depending on just which oxygen isotope it is. Thus, water is built up from electrons, protons, and neutrons.

*In the inert gases (*helium, neon, argon, krypton, xenon, *and* radon*), the atoms do not combine to form a larger stable structure. In this case these atoms themselves are called molecules. However, atoms of some of these "inert" elements in some cases can be made to combine with atoms of other elements. The first such compound involving one of these inert elements was xenon hexafluoroplatinate prepared by Neil Bartlett at the University of British Columbia in 1962.*

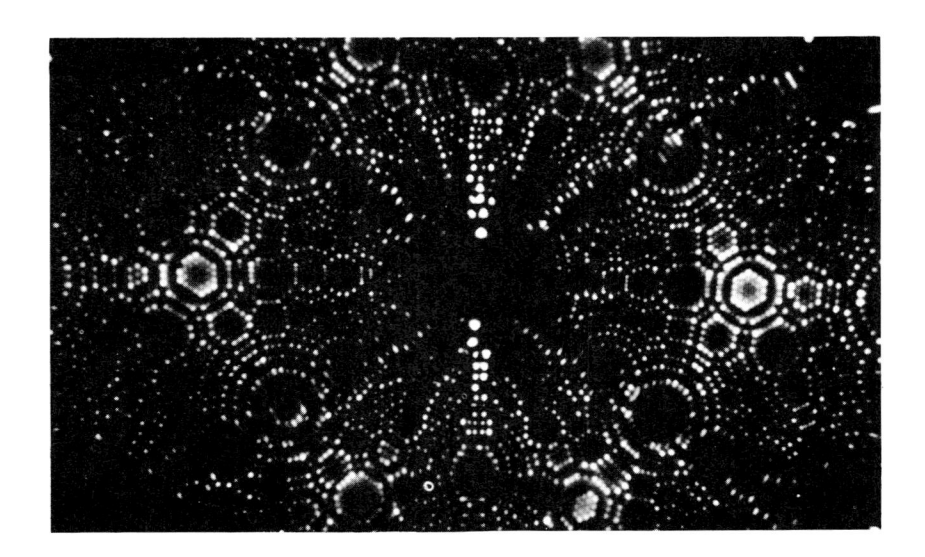

Fig. 1-3 Field ion microscope image of a tungsten needle tip, that is, a hemispherical tungsten crystal approximately 420 angstroms in radius. (Courtesy Dr. E. W. Müller, The Pennsylvania State University.)

Up to this point you have seen none of the evidence supporting this view of matter. Some will come later. But at this point it is reassuring to examine the remarkable photograph (Fig. 1-3) taken by Dr. Erwin W. Müller of The Pennsylvania State University with his field ion microscope. The smallest of the bright dots come from single atoms in the crystal of tungsten, while the larger dots arise from clusters of atoms. Evidently, we are looking rather directly at the arrangement of atoms in a metal, thanks to a magnification of nearly two million diameters!

1.3 IONS AND PARTICLE DETECTION

Ions

An atom or molecule which is not electrically neutral is called an *ion*. The process of *ionization* generally consists of the removal of one or more electrons from a neutral atom or molecule which then becomes a positive ion. Thus if an electron is knocked off a neutral helium atom, the remaining structure, the nucleus ($_2$He^4) plus one orbital electron, is called a helium ion or singly ionized helium. Now if a second electron is ripped away, doubly ionized helium remains. This is just the bare nucleus, $_2$He^4, which we have previously identified as an α-particle.

Now consider what happens when an α-particle is projected through air at a high speed, say 1/20 of the speed of light. As the α-particle passes close by or through the oxygen and nitrogen molecules that comprise the air, this α-particle exerts a strong electric force on electrons that it passes—strong enough to pull some electrons right out of these molecules. Consequently, the moving α-particle leaves in its wake a *trail*

8 THE ATOM AND ITS NUCLEUS

of ions. The α-particle is slowed down by these encounters and is eventually stopped after traveling about 3 centimeters through the air.

Any high-speed charged particle leaves a trail of ions along its path as it passes through matter. A β-particle (electron), projected at 95% of the speed of light, travels some 3 meters through air, leaving an ion trail which is much less dense than that of the doubly charged massive α-particle.

A photon, although it carries no electric charge, does interact with electrically charged particles. A γ-ray (high-energy photon) passing through matter will occasionally cause an electron to be ejected from an atom, leaving an ion to commemorate the event. But compared to α-particles, γ-rays are ineffective ionizing agents, to say the least. The relative ionization effectiveness of typical α-, β-, and γ-radiation is about 10,000/100/1 respectively, a fact which has some considerable bearing on the stopping and detection of these various radiations.

Track Recorders

The path through matter of a high-speed charged particle is strewn with ions. Detection of these ions is therefore a key to the detection of these interesting fast particles. Many ingenious devices accomplish this task, and thus have led to much of our knowledge of the subatomic world.

The entire trail of ions is rendered visible (see Fig. 1-4) in the *cloud chamber*, invented by C. T. R. Wilson (1869–1959) in England in 1907. Air, saturated with water vapor, becomes supersaturated (in other words, is more than ready to condense) as it is cooled by a rapid expansion of the chamber. At this instant any ions in the chamber serve as centers for condensation of the water vapor. A small, but visible, water drop grows about the ion. The ion trail that had been made by a fast charged particle suddenly becomes visible as a line of small drops of water in the air of the cloud chamber!

The *bubble chamber*, a track recorder more suitable for modern high-energy physics, was invented by the American physicist, D. A. Glaser (1926–), in 1952. Here a trail of ions in a superheated liquid becomes visible as a trail of bubbles, each ion serving as a center of boiling.

Tracks in a *spark chamber* are visible because of short electric sparks initiated by ions.

A thick, very sensitive, photographic film called a *nuclear emulsion* is a light-weight continuously sensitive track recorder that has won its share of glory in mountain-top expeditions, balloon ascents, and rocket flights, bringing back records of new particles and new reactions in the cosmic radiation that bombards our planet. A trail of ions and the associated disruption within the emulsion constitutes a latent image which, after a photographic development process, is a track whose

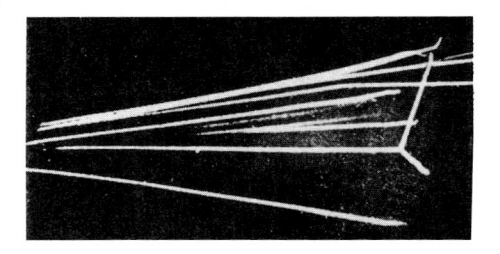

Fig. 1-4 Cloud-chamber photograph of the collision between an electrically charged particle and the nucleus of an atom. (From A. B. Arons, *Development of Concepts of Physics*, Addison-Wesley, 1965.)

Physics Today

D. A. Glaser (1926–)

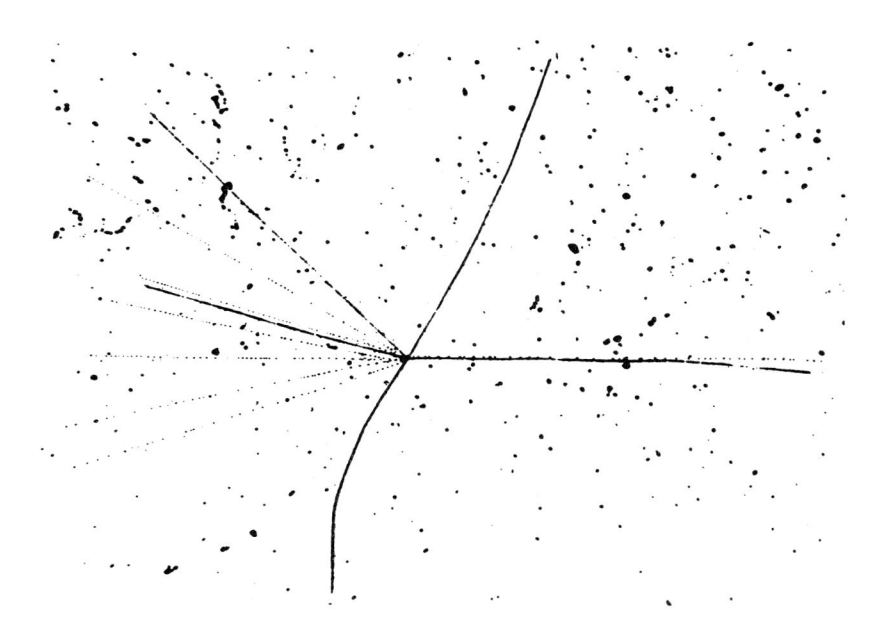

Fig. 1-5 Tracks produced in a nuclear emulsion plate by the interaction of high-energy heavy cosmic ray particles with emulsion grains. The photographic plate was carried to a high altitude by a balloon. (Courtesy of Brookhaven National Laboratory.)

details are clearly visible under examination with a microscope (Fig. 1-5). The ion density along the track yields information about the charge and mass of the particle that made the track.

Particle Detectors

Many devices simply detect that ionizing or associated events have occurred, without making visible an entire track. One example is the fluorescent coating on the inside surface at the front of a television picture tube which emits light when struck by high-speed electrons. The pattern of electron impacts is thus detected and constitutes the picture on the screen.

In many of Rutherford's experiments, the α-particle detector was a zinc sulphide screen. A tiny flash of green light (a scintillation), visible under a microscope, is emitted from a point on the screen which has been struck by an α-particle. Modern *scintillation detectors* often employ a large crystal of sodium iodide (for γ-ray detection) or anthracene (for electron detection). The flash of light, produced when the γ-ray or β-particle disrupts atoms of the crystal, is led to a photomultiplier tube (which multiplies, or amplifies, the signal). The final result is that a sizable electric signal is obtained, sufficient to be recorded in an electronic counting device. Such an assembly is called a *scintillation counter*.

The famous *Geiger counter* had its beginning in 1908 when Rutherford and his assistant, Hans Geiger (1882–1945) designed a counter which could detect α-particles. This device was ultimately improved by

Geiger and his research assistant W. Müller. Essentially the Geiger (or G. M. for Geiger-Müller) counter is a gas-filled tube with a central wire and an outer metal cylinder (Fig. 1-6), between which is connected a high-voltage battery (about 1000 volts). When an ionizing event occurs in a molecule of the gas within the tube, the electron which was ejected from the molecule is pulled toward the positively charged central wire with such a force that it acquires sufficient speed between collisions with the gas molecules to knock electrons out of these molecules. An avalanche of electrons develops. The arrival of these electrons at the central wire produces an electric signal sufficient to activate an electronic counter or to produce a click in a loudspeaker.

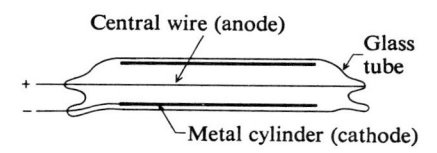

Fig. 1-6 Construction of a simple Geiger-Müller tube.

1.4 DISCOVERY OF THE NUCLEUS

Rutherford's investigation of α-particle scattering in passage through a thin foil (Fig. 1-7) is one of the most significant experiments in the history of physics. The consequent discovery that an atom's mass was almost entirely contained in a highly charged central nucleus proved to be an essential first step in understanding atomic structure. Subsequent progress in atomic physics, led by Niels Bohr (1885–1962) of Denmark, was rapid, and the study of the nucleus itself, *nuclear physics*, could begin in a systematic way. Rutherford's idea of using a scattering

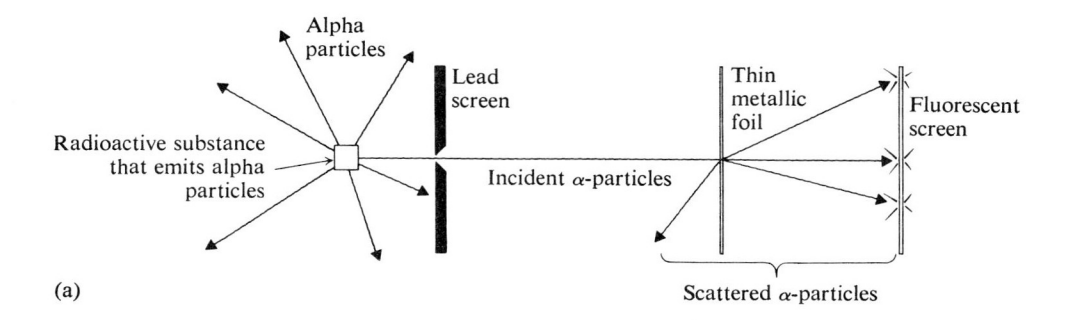

Fig. 1-7 Rutherford scattering (a) The experimental setup. (b) The scattering of alpha particles by the nuclei of the metallic foil. The fraction scattered through sizable angles has been greatly exaggerated.

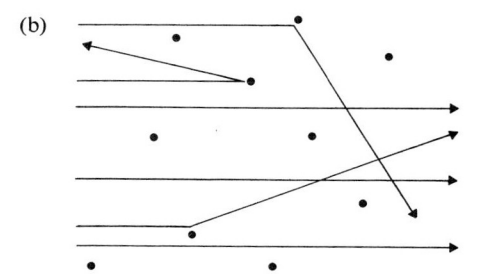

experiment to investigate interactions at small distances is now the standard approach in modern high-energy physics.

Rutherford's experimental setup, indicated in Fig. 1-7, shows a narrow beam of α-particles incident on a thin gold foil. Now it must not be imagined that this beam can be aimed accurately enough to hit an area, on the foil, of nuclear or atomic dimensions. Rather, the places on the foil struck by the different α-particles of the beam are sprinkled at random over an area of the surface containing millions of gold atoms. (It is as if nuclear physicists always use a wobbly machine gun instead of a steady rifle, and fire at an assembly of millions of tiny identical targets, not caring which particular target is struck.)

The object of the experiment is to find out something about the contents of a gold atom by observing the deflection (or scattering) that various α-particles experience in their encounters with the gold atoms in the foil.

To understand the implications of different scattering events, consider a collision between a moving golf ball and a stationary ping-pong ball. The massive golf ball will experience only a small deflection from such an encounter. But if, on the other hand, the ping-pong ball is projected at the golf ball, large deflections of the projectile can occur. And sometimes the ping-pong ball even bounces straight backward.

Rutherford observed that almost all α-particles went through the foil without appreciable deviation. All but one in 10,000 α-particles were deviated by less than 10°. From this he could conclude that most of the α-particles met (or interacted strongly with) only objects much less massive than themselves (golf ball hits ping-pong ball). Therefore most of the gold atom was devoid of massive objects.

However, there were a few α-particles deflected through large angles. Some even came almost straight back! This is now like the ping-pong ball hitting the golf ball. *Clearly, on rare occasions, the α-particle encountered something much more massive than itself.* Evidently this massive object had exerted a strong force on the α-particle in these rare encounters. Was it the familiar electric force of repulsion between two charged objects?

At this point Rutherford made the assumption that an atom had a nucleus and that this minute massive nucleus was indeed electrically charged and contained nearly all the atom's mass. He then went to work with paper and pencil and calculated just what fraction of α-particles would then be deflected, by the electrical push exerted by the nucleus, through angles greater than 10°, 20°, . . . , etc. There was detailed numerical agreement between these calculations and the experimental results amassed by his assistants, H. Geiger and E. Marsden (1889–). Thus was discovered the presently held concept of the nucleus. It is this indirect way that is now used throughout nuclear and particle physics to "see" what is going on.

To determine the size of the nucleus it is necessary to increase the speed of the incident α-particle until it can penetrate to the nucleus. When the nuclear forces come into play, the scattering changes. Such scattering experiments, using very high-speed particles, yield the nuclear size quoted in Section 1.2.

1.5 RADIOACTIVITY AND CHARGE CONSERVATION

Radioactivity

Most of the some 1500 different nuclei that have been discovered are unstable. In fact, only 280 stable nuclear isotopes are presently known. An unstable nucleus suddenly and spontaneously changes. This change is accompanied by the emission of one or more rapidly moving particles. Such a nucleus is said to be radioactive and the change is called *radioactive decay*.

The original nucleus is named the *parent* nucleus and the residual nucleus, the *daughter*. A radioactive decay process is classified as α-, β-, or γ-decay, according to the type of particle emitted.

α-decay

A typical example of α-decay is the sudden breakup of a uranium nucleus, $_{92}U^{238}$, as a high-speed α-particle ($_2He^4$) is emitted, leaving a residual nucleus with atomic number reduced by 2 and mass number reduced by 4. This is a nucleus $_{90}Th^{234}$ of the element thorium. In symbols this α-decay is

$$_{92}U^{238} \longrightarrow {}_{90}Th^{234} + {}_2He^4. \tag{1.1}$$

Another example of α-decay involves radon (Rn)

$$_{86}Rn^{222} \longrightarrow {}_{84}Po^{218} + {}_2He^4$$

This is a decay process which provided Rutherford with energetic α-particles.

β-decay

In a β-decay process such as

$$_{82}Pb^{214} \longrightarrow {}_{83}Bi^{214} + {}_{-1}\beta^0, \tag{1.2}$$

the lead nucleus $_{82}Pb^{214}$ suddenly transforms into the nucleus $_{83}Bi^{214}$ of the element bismuth, which has a greater nuclear charge (83) but the same mass number (214). This transformation is accompanied by the creation of an electron ($_{-1}\beta^0$) which generally moves off with a speed close to that of light. But this electron did not exist in the nucleus before the transformation process, rather it was created at the instant of the radioactive decay. (We shall return to the subject of β-decay in Chapter 4 to give a more complete account of radioactive decay products.)

γ-decay

In γ-decay, the parent and daughter have the same charge and the same mass number. A nucleus evidently changes from an unstable to a different (but not at all necessarily completely stable) configuration as it emits the γ-ray. An example is

$$_{86}Rn^{222*} \longrightarrow {_{86}Rn^{222}} + {_{0}\gamma^{0}}. \tag{1.3}$$

The unstable parent nucleus $_{86}Rn^{222*}$ (the asterisk is used to indicate the nucleus is not in its most stable state), is itself the daughter nucleus of a preceding α-decay.

Radioactive Series

The radioactive elements in the preceding examples are all members of what is called a radioactive series, the uranium series shown in Fig. 1-8. The series starts with the uranium nucleus $_{92}U^{238}$. This decays by α-emission into $_{90}Th^{234}$, but this daughter is itself unstable and undergoes β-decay into the palladium nucleus $_{91}Pa^{234}$. This nucleus is also unstable and decays, and so on. Various radioactive transformations of the original nucleus continue until it finally achieves a *stable* configuration as a nucleus of the isotope of lead, $_{82}Pb^{206}$.

The scientific information that is so undramatically but succinctly represented in tables of data or charts like those of Fig. 1-8 is usually acquired only after intense human effort. This is particularly true of two of the entries in Fig. 1-8, the polonium nucleus $_{84}Po^{210}$ and the radium nucleus $_{88}Ra^{226}$. The discovery of the element polonium was made in France in 1898 by Marie Curie (1867–1934) and her husband Pierre Curie (1859–1906) and named after her native country, Poland. The Curies, attempting to isolate the radioactive substance in the mineral pitchblende, went on to also discover the element radium. The efforts required to achieve the chemical separation of minute amounts of radium from masses of pitchblende, largely carried out in an abandoned wooden shed, were nothing short of heroic. (The story of the Curies is chronicled by Mme. Curie's daughter, Eve. See Suggested Reading at the end of this Chapter.)

The uranium series of Fig. 1-8 is only one of three different radioactive series found in nature. These series are named the uranium, actinium, and thorium series respectively, after elements at or near the beginning of the series. A fourth radioactive series is also known. This series was first discovered starting from artificially produced plutonium-241; it is named the neptunium series after the longest-lived member of the chain.

Physics Today

Marie Curie
(1867–1934)

14   THE ATOM AND ITS NUCLEUS

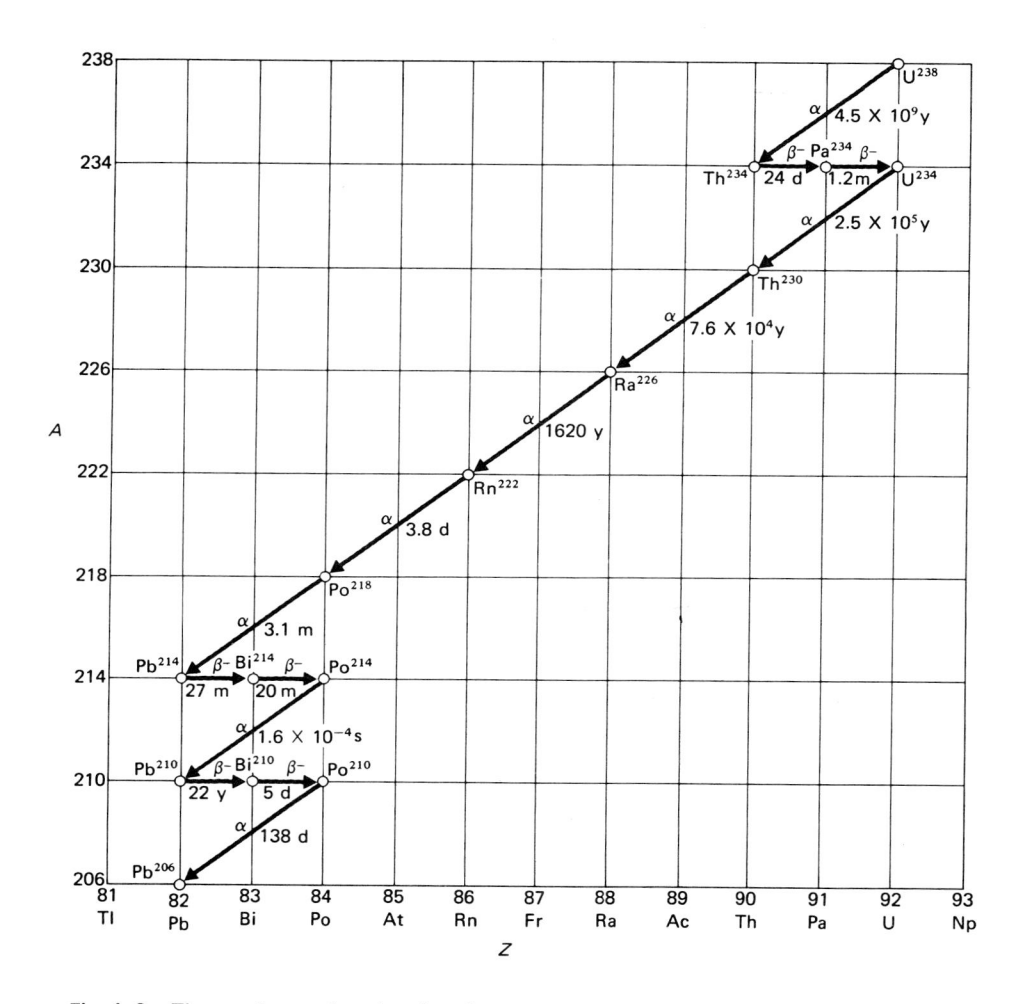

Fig. 1-8 The uranium series, showing the main decay modes leading to lead-206. The respective half-lives are shown in seconds (s), minutes (m), days (d), or years (y).

Law of Conservation of Electric Charge

The radioactive decay of a nucleus is an example of a violent and dramatic *change*. The nucleus itself changes, and in β- or γ-decay new particles are created so that even the number of particles changes. In any such apparently chaotic situation it is most interesting to look for quantities that are unchanged or *conserved*.

Examination of the subscripts appearing in the Eqs. (1.1), (1.2), and (1.3), which summarize the experimental observations on these α-, β-, and γ-decays, reveals that the electric charge before the decay is exactly the same as the net charge after the decay has occurred. Thus for the β-decay of Eq. (1.2), the lead nucleus initially present has a charge of $+82e$. Change occurs. An electron ($_{-1}\beta^0$) is created with its *negative* charge of $-e$, and the residual bismuth nucleus has a charge of $+83e$. Since

$$+82e = +83e + (-e),$$

we have the result that the charge before the process equals the algebraic sum of the charges after the process. (Notice that attention must be paid to algebraic signs ($\pm$) which are always to be included in an algebraic sum.) Evidently in this β-decay, electric charge is conserved.

Now in all experiments that have kept track of electric charge during various processes, a similar result has been obtained. No one has yet discovered a process in which the net charge (algebraic sum) increases or decreases.

We summarize these experimental observations in what we call a *law of nature* or a *physical law*, the *law of conservation of electric charge*:

> *Electric charge is conserved.* The algebraic sum of the electric charges present before any process equals the algebraic sum of the charges existing after the process.

This law of conservation of electric charge is a basic physical law whose correctness we assume in constructing physical theories.

There is evidence in the decays given by Eqs. (1.1), (1.2), (1.3), that the *mass number* given as the superscript, that is, the *total* of the number of protons and the number of neutrons, is another conserved quantity. But, since a variety of reactions discovered since 1950 reveal that protons and neutrons can be changed into other particles, the law of conservation of mass number must be modified in a fashion that we will take up in Chapter 4.

The statements which are labeled physical laws are a major part of what scientists call an understanding of nature. Having just met a physical law, the law of conservation of electric charge, it is well to emphasize the nature of any physical law:

1. A physical law is a generalization obtained, directly or indirectly, from *experiments*. New experiments may show that the "law" has to be modified or its range of application restricted.
2. The physical law usually can most fruitfully be given as a mathematical statement. One can then bring to bear mathematical methods to deduce the implications of the law in different situations.
3. The question of the validity of the underlying physical laws or physical assumptions in any theory is a matter which rests on experimental evidence. There is nothing in mathematics alone that can tell us whether or not a physical law is true.

All this implies that experimental facts are vital. But, as in the Rutherford scattering experiment, the meaningful interpretation of an experiment requires a theory. Theory and experiment become interwoven.

Exponential Decay and Half-life

In 1902 Rutherford and Soddy found that when a sample of a given type of radioactive substance was separated from its parent, so that no new nuclei of the substance were being created, then the number of radioactive nuclei remaining as time elapsed decreased in the very special way given by the graph in Fig. 1-9. This is named the graph of an "exponential decay." Each type of radioactive nuclei is characterized by a quantity called its *half-life*, which has the following significance. At the end of *any* time interval equal to the half-life of the nuclei in question, the number of parent nuclei remaining is half the number present at the beginning of the interval. In other words, the half-life is the time it takes for half of a given radioactive sample to decay.

The half-lives of the different types of nuclei encountered in the uranium series are given in Fig. 1-8: 4.5×10^9 years for the uranium α-decay of Eq. (1.1), 27 minutes for the lead β-decay of Eq. (1.2). (Other half-lives are given in Table 1 of Section 1.7.)

Example A chemical separation of the lead isotope Pb-214 from its radioactive parent is completed at 10:00 A.M. This lead isotope decays with a half-life of 27 minutes. At 11:00 A.M. there are 8.0×10^6 nuclei of Pb-214 remaining. How many will remain at 11:27 A.M.? At 11:54 A.M.?

Solution At 11:27 A.M., one half-life later than 11:00 A.M., the number remaining will be

$\frac{1}{2} \times$ (number of nuclei at 11:00 A.M.)
$= \frac{1}{2} \times 8.0 \times 10^6$ nuclei $= 4.0 \times 10^6$ nuclei.

At 11:54 A.M., there will remain 1/2 the number at 11:27 A.M.:

$\frac{1}{2} \times 4.0 \times 10^6$ nuclei $= 2.0 \times 10^6$ nuclei.

No matter *how* old the sample is at a given moment, one half-life later we will find that 1/2 of the sample will have survived.

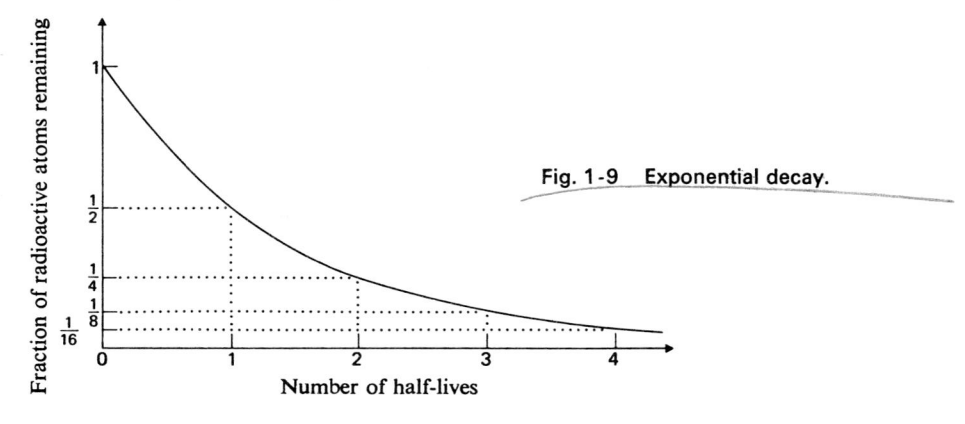

Fig. 1-9 Exponential decay.

The decay events take place completely at *random*. There has been found no way of predicting the precise instant at which a particular nucleus will decay. However, the shorter the half-life, the greater the chances of decay of any particular nucleus during any forthcoming time interval.

From the mathematical properties of the exponential decay one can show, with some labor that we will omit, that at an instant when there are N identical radioactive nuclei present, the number which will decay in the following short time interval is given by

$$\text{Number of decays} = 0.69 \times \left(\frac{\text{time interval}}{\text{half-life}}\right) \times N. \qquad (1.4)$$

This result is just an approximation, but the approximation is good if the time interval is not greater than 1% of the half-life of the nuclei in question. This equation we shall call the *Rutherford-Soddy law of radioactive decay*.

As an example, consider a sample containing 2.7×10^{21} atoms of radium-226 (this is about one gram of radium). The half-life of this radium isotope is 5.1×10^{10} sec (over 1600 years). Eq. (1.4) gives the number of disintegrations that take place in 1.0 sec.

$$\text{Number of decays} = 0.69 \times \left(\frac{1.0 \text{ sec}}{5.1 \times 10^{10} \text{ sec}}\right) \times 2.7 \times 10^{21}$$

$$= 3.7 \times 10^{10}.$$

A source with this number, 3.7×10^{10}, of decays per second, is said to have an *activity* of 1 *curie*. Cancer therapy units very often use the radioactive cobalt isotope (Co-60) with source activities ranging from 2000 to 5000 curies. Student laboratory sources and luminous watch dials have activities of about one millionth of a curie, a microcurie.

Probability and Quantum Mechanics

The Rutherford-Soddy law of radioactive decay does not tell us when a particular nucleus will decay. Just when this event occurs appears to be a matter of chance. But, like a bookie before a horse race, we can compute odds or, more formally, probabilities.

For instance, from the preceding example, we can predict that in the next second, in a sample containing 2.7×10^{21} radium atoms, 3.7×10^{10} will disintegrate. The fraction is

$$\frac{3.7 \times 10^{10}}{2.7 \times 10^{21}} = \frac{1}{73 \times 10^{9}}.$$

Thus we can say that there is one chance in 73 billion that a particular radium nucleus will decay within the next second. Equivalently, we say that the *probability* of a particular radium nucleus decaying in the next second is

$$\frac{1}{73 \times 10^9}.$$

As far as the decay of one particular nucleus is concerned, we have now seen that the Rutherford-Soddy law allows us to compute only *probabilities*. Probabilities, not certainties. This feature of radioactive decay suggests that perhaps events in this submicroscopic domain happen by *chance*. Maybe some of the most fundamental laws of nature are, like the one of Rutherford and Soddy, laws which determine merely probabilities.

And indeed it seems that this is the way things are. With the theory known as *quantum mechanics*, largely developed between 1924 and 1928, one determines only the probability that a given event will occur. Moreover, the internal consistency of this theory demands that in our description of nature there be an element of uncertainty which is irreducible.

This theory, quantum mechanics, applied to the hydrogen atom, does *not* lead to a picture like that of Fig. 1-2, in which the electron has a precisely defined orbit and is shown at some instant at a very definite position. In quantum mechanics, there is always an uncertainty associated with the electron's position. At any instant, the most one can determine is that there is a certain chance or probability of this electron being in a given region.

The relative probabilities of the electron being in different regions of space can be represented by a *cloud* (Fig. 1-10) which is dense where the electron is most likely to be and thinner in regions where the electron is less likely to be found.

But do not think that the electron itself is as big as this cloud. The electron has always been found to interact with other particles as if it were very small—smaller than a dot in the cloud of Fig. 1-10. And just how much smaller is one of the things physicists are trying to discover.

More will be said about quantum mechanics in later chapters, particularly in Chapter 16. It is *the* successful theory of the structure and behavior of atoms and molecules. Theories of the nucleus are far from perfected, and our understanding of the elementary particles is, at best, rudimentary. But whatever theoretical success has been achieved in these domains has rested on quantum mechanics. It appears we live in a quantum mechanical world, a world in which chance plays a fundamental role.

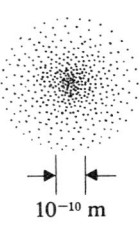

Fig. 1-10 Quantum mechanical representation of a hydrogen atom. The cloud is most dense where the electron is most likely to be found. The nucleus is at the center of the cloud.

1.6 NUCLEAR REACTIONS

Spontaneous change of a nucleus, that is, radioactive decay, occurs for relatively few of the nuclei that are found on our planet. But any nucleus can be made to participate in a nuclear reaction and to undergo an essentially immediate and perhaps radical change.

Nuclear reactions are produced by bombarding matter with a beam of particles, often high-speed protons or α-particles. If such a particle has enough initial speed to penetrate to a nucleus, in spite of the electrical repulsive force, strong nuclear forces influence the interaction. Different things can happen.

Sometimes the nucleus is left unchanged and a particle emerges in a direction different from that of the incident beam. This is called *elastic scattering*.

A second possibility is that the incident particle be captured by the target nucleus forming a new system which is an unstable structure called a *compound nucleus*. This unstable compound nucleus undergoes a decay, often simply emitting a neutron or a γ-ray, to form another nucleus. The half-life of this decay is extremely short, typically 10^{-14} sec or less.

It is also possible that the incident particle simply knock out of the target nucleus some neutrons and protons which may emerge in different directions or be found together as a deuteron, an α-particle, or even something larger.

The greater the speed of the incident particle, the greater the number of different possible outcomes. As we will discuss in Chapter 4, a specific process becomes possible only if the incident particle has sufficient *energy*.

Nuclear reactions are symbolized by showing first the incident particle and the target nucleus and then the final products. Any possible intermediate state of affairs, such as a compound nucleus, is usually omitted. (The lapse of time involved in creating the final products is entirely negligible for almost all practical purposes, typically 10^{-23} sec to 10^{-14} sec.) Thus, for Rutherford's historic first laboratory-induced transmutation, which involved bombarding nitrogen ($_7N^{14}$) with α-particles ($_2He^4$), one writes first simply

$$_2He^4 + {_7N^{14}} \longrightarrow ?$$

Rutherford found that charged particles were produced which traveled further through the gas than the α-particles. By a variety of techniques, including observations of the tracks in a cloud chamber, the emitted particle was identified as a proton ($_1H^1$). Later it was correctly inferred that a nucleus of an oxygen isotope, oxygen-17, had been created. The entire reaction is symbolized by

$$_2He^4 + {_7N^{14}} \longrightarrow {_8O^{17}} + {_1H^1}.$$

After this discovery, attention turned to the construction of particle accelerators to provide a beam of high-speed particles capable of producing a variety of nuclear reactions. In the early 1930s such machines were built: for example, the Cockcroft-Walton accelerator. J. D. Cockcroft and E. T. S. Walton in England built a machine by means of which they were able to engender nuclear reactions.

In the United States Ernest Lawrence (1901–1958) and M. S. Livingston (1905–) built the first cyclotron in Berkeley, California, spawning generations of such machines of ever-increasing size and capability. In later chapters we shall discuss these accelerators and the many new types of particles that are created in reactions produced by the beams of high-speed particles emerging from these giant machines.

Another experimental approach to the study of nuclear reactions is also very fruitful. When the neutron was discovered in 1932 by James Chadwick in England, the great Italian theoretical physicist Enrico Fermi (1901–1954) recognized that this electrically neutral particle would be a particularly suitable object with which to probe the nucleus. An incident neutron feels no electrical repulsion and interacts strongly with the protons and neutrons inside the nucleus. Fermi turned to experimental physics, and, with his colleagues, bombarded targets of various elements with neutrons. They soon discovered a host of new radioisotopes.

A typical "neutron capture" reaction,

$$_0n^1 + {_{27}}Co^{59} \longrightarrow {_{27}}Co^{60} + {_0}\gamma^0,$$

produces the useful radioisotope, cobalt-60. Today the copious flux of neutrons in nuclear reactors is used to produce, in this way, considerable quantities of cobalt-60 that are employed throughout the world.

After some four decades of research, using particle accelerators and nuclear reactors, over a thousand radioisotopes have been discovered. These radioisotopes, as well as the stable isotopes, are generally listed in an orderly fashion in what is known as an *isotope chart* (or chart of the nuclides). A part of such a chart is shown in Fig. 1-11 on the following page.

The radioactivity of these newly discovered nuclei is of the same nature as that displayed by the naturally occurring radioisotopes, such as those in the uranium series (Fig. 1-8). Presumably these artificially produced radioisotopes are not now found in nature because neither they nor their parents have sufficiently long half-lives.

One new feature that the reader will notice in the isotope chart is the occurrence of decays labeled β^+. In such a decay a *positron* (β^+) is emitted. This particle has the same mass as an electron (β^-) but an opposite electric charge. More will be said about this interesting particle in Chapter 4.

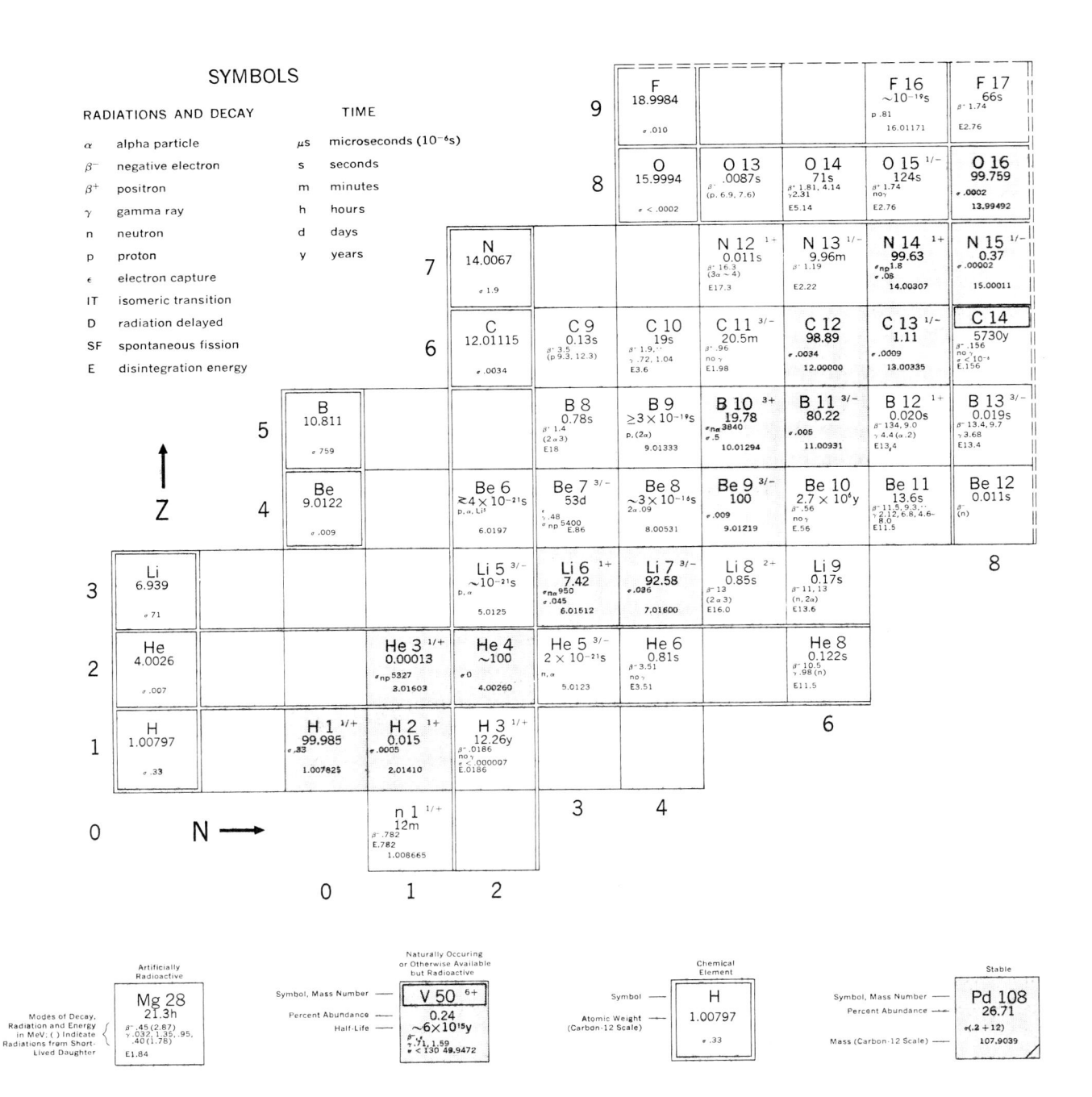

Fig. 1-11 Part of an isotope chart adapted from Chart of the Nuclides, ninth edition. The horizontal number *N* refers to the number of neutrons, and the vertical number *Z* to the number of protons. (Courtesy Knolls Atomic Power Laboratory, Schenectady, New York. Operated by General Electric Company for the United States Energy Research and Development Administration Naval Reactors Branch.)

22 THE ATOM AND ITS NUCLEUS

1.7 APPLICATIONS OF RADIOISOTOPES

It is always interesting to see how newly acquired knowledge can be used. But the consequences of the discoveries encountered in this chapter are so intricately woven through the fabric of physics and chemistry that any cursory listing of applications would be misleading. Instead we therefore quite arbitrarily select for discussion only a few more or less direct applications of radioisotopes. The wide variety in the constructive use of radioisotopes is implicit in Fig. 1-12.

Radioisotopes as Radiation Sources

As sources of high-energy γ-rays or β-particles, radioisotopes are much more suitable in some applications than elaborate x-ray machines or particle accelerators. A radioisotope source is compact and since the radioactive decays are spontaneous, there is no way the source can get out of order and stop radiating.

Some of the radioisotopes in frequent use are listed in Table 1-1.

Table 1-1. HALF-LIVES OF SOME RADIOISOTOPES

Radioisotope	Half-life	Radiation emitted
Phosphorus-32	14.3 days	β^-
Cobalt-60	5.24 years	β^-, γ
Iodine-131	8.05 days	β^-, γ
Cesium-137	30 years	β^-, γ

For a given application, one selects a source which emits the desired type of radiation. If a source emits both β-particles and γ-rays, and the β-particles are not wanted, it is a simple matter to stop these electrons in a metal sheet which is thin enough to have little effect on the γ-ray intensity.

Fig. 1-12 Radioisotopes in industry and medicine. (a) The Gamma-cell is a portable irradiation unit designed for industrial and other research work. It is used to study the effects of γ-rays on various materials.
(b) Cobalt-60 used in cancer treatments. In agriculture a mobile Cobalt-60 Irradiator is used to irradiate various foods with γ-rays to increase their shelf life or, in the case of potatoes and other vegetables, to inhibit sprouting. (Courtesy of Atomic Energy of Canada Limited.)

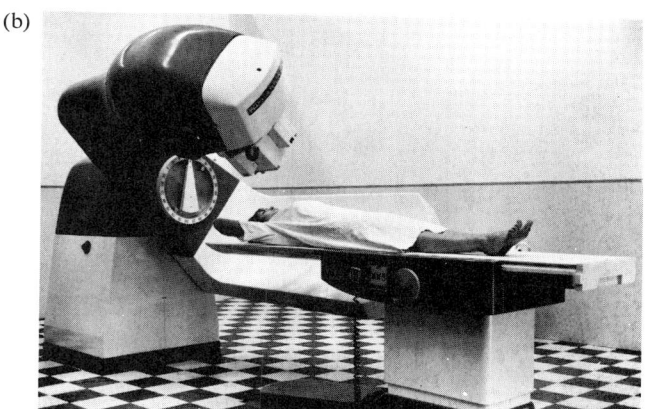

Only a few radioisotopes have half-lives suitable for a practical source. The half-life must be long enough so that the source activity will not decrease with inconvenient rapidity. Yet to have a large source *activity* (a source which emits many particles per second is said to have a large activity) the half-life must not be too long.

To understand this latter point, assume we have produced a given number of atoms of two different radioisotopes, one with a half-life of 1 year and the other with a half-life of 1000 years. Then, as you are asked to verify in Question S-1, the Rutherford-Soddy law, Eq. (1.4), shows that the number of decays per second of the isotope with the shorter half-life will be 1000 times greater than that of the other isotope.

The radioisotope probably most frequently used at present as a radiation source for a great variety of purposes (Fig. 1-12) is cobalt-60. Strong cobalt sources have acquired the somewhat alarming and quite misleading name "cobalt bomb." Such cobalt sources have replaced high-voltage x-ray units in cancer clinics. It is estimated that Canadian-produced cobalt units alone provide nearly 1.5 million treatments per year, in some 40 different countries.

In industry, cobalt sources are used extensively to inspect metal structures in order to detect flaws before these flaws become serious enough to precipitate structural failure. Major airlines "x-ray" their aircraft periodically, often using γ-ray units employing Cs-137 or Co-60.

In agriculture, γ-ray irradiation is used to increase the shelf-life of certain vegetables and to inhibit the sprouting of stored potatoes. Research involving the development of new plant species employs radioisotopes to engender genetic mutations.

Radioisotope sources are used to sterilize surgical instruments and dressings, especially those made of non-heat-resistant materials. This method of sterilization is based on the lethal effect of radiation on bacteria.

When mobile, compact, and rugged radiation sources are required, radioisotopes are without a competitor. Minute sources are often useful in medicine. For instance, instead of surgical removal of the pituitary gland, one can implant tiny yttrium-90 oxide beads (about 1 mm in diameter), glasslike in nature, directly into the gland. The β-particles, emitted in the decay of Y-90, destroy the gland. Since the range of penetration of the β-particles is short, surrounding tissue is not significantly damaged.

Small radioisotope sources aid in the exploration of oil-bearing geologic formations. Radioactive needles (γ-ray emitters) are lowered into the ground, in so-called test wells, and then the intensity of the radiation is measured at several nearby places. From this information a geologist is able to draw a "profile" of the underlying earth layers. Study of this profile may then reveal whether or not the expensive operation of major drilling would be justified.

Identification of a Radioisotope

The detection of the presence of traces of some element in an object is a demanding task using traditional chemical procedures. However, the most minute quantities, as little as a few million molecules, of a *radioisotope* can be detected if its half-life is not too long.

One simply uses a particle detector, often a scintillation counter, to detect the radiations emitted. Identification of the type of radioisotope responsible for the radiation is accomplished by determining the characteristics of the emitted radiation, and where possible, the half-life of the decay. The half-life can be determined from a graph such as Fig. 1-9, if there is a detectable decrease in the counting rate (and consequently a decrease in the number of radioactive nuclei) during the time interval over which observations are made.

The actual number of radioactive nuclei in the sample can be calculated, using the Rutherford-Soddy law, Eq. (1.4), from the measurement of the number of particles emitted in a measured time interval.

The comparative ease of identification of radioisotopes has led to the use of a *neutron activation analysis* to determine which elements are contained in a sample. For detection of traces, the neutron activation method is often considerably superior to the usual chemical analysis. The sample is bombarded by neutrons. Nuclear reactions occur and radioisotopes are created. The subsequent identification of a particular radioisotope will determine that a certain element was present as a target in the bombarded sample. Such analysis by neutron activation is coming into widespread use in criminology.

Tracers

Radioisotopes enter into chemical reactions and form molecules in the same way as the normal stable isotopes of a given element. A small percentage of molecules can thus be tagged with a radioisotope and the progress of an assembly of the molecules can be *traced* as they move about in some chemical, biological or industrial process.

In medicine, iodine-131 finds extensive application, often as a tracer. Sodium iodide is readily taken up by the thyroid. By detecting the rate at which iodine-131 accumulates in the thyroid, the functioning of this gland can be monitored. Also by scanning the region of the thyroid with a suitable detector, abnormalities in size, shape, and tissue can be detected. Moreover, when irradiation of the thyroid is desired as a treatment of hyperthyroidism, iodine-131 is used as a mobile source which nature takes right to the target. In this connection, it is interesting to note that now after some 20 years of experience with this isotope in the medical field, I-131 continues to hold its place as a most effective tool for the investigation and treatment of thyroid disorders. Another radioisotope which is popular for scanning the thyroid and other organs is technitium-99 in the so-called excited state. Tc-99 is formed in the decay of molybdenum-99. Upon its "creation" in this decay process the Tc-99 is not in its most stable state. Then as the Tc-99 goes to its stable state it gives off energy (called excitation energy) by emitting a useful low-energy gamma ray.

Medical radioisotope diagnostic procedures were greatly improved by the introduction of the *gamma camera*. This device is, in general, replacing smaller detectors (*rectilinear scanners*) which must move back and forth across the region of interest. For example, the lung scan procedure shown in Fig. 1-13 takes about five to ten minutes with the gamma camera shown. It takes well over half an hour to achieve somewhat similar results with a rectilinear scanner.

In a typical industrial use of tracers, manufacturers of soap and detergents test the effectiveness of their products in washing dirt from clothes by adding radioactive tracers to the dirt and measuring the proportion of the tracers that is carried off in the dirty water.

Studies of friction between surfaces and the wear of moving parts are now performed by detecting the transfer of radioisotopes from one surface to another. For investigations with piston rings, radioisotopes are created throughout the volume of the piston ring by irradiating the ring in a nuclear reactor.

Ionizing Radiation and Biological Tissue

As the reader is no doubt well aware, and as can be construed by implication from our discussion of the use of radiation to destroy malignant tissues, ionizing radiation is harmful. Biologists agree that the ionization of the complex molecules of living matter causes damage. Why then does man employ ionizing radiation at all? And how big a radiation dose is really harmful?

Before answering the first question it is necessary to realize that mankind, as well as other life forms, has successfully evolved in the very face of a steady stream of ionizing radiation: cosmic rays, radioactive material in the ground and in the air, and even radioactive isotopes in the living tissues themselves. But this *background radiation* is small, being in a year about the same amount that a person is subjected to in a single chest x-ray.

So what is the answer to the question? Small enough dosages of radiation generally seem to cause no irreparable damage to an individual (although the hazard of gene mutation which might have a significant effect on offspring is ever present). With this fact in mind, exposure to radiation (whether as a hospital patient, or as a physicist working with an accelerator, or even as a mountain climber who subjects himself to the more intense cosmic radiation experienced at high altitudes) is essentially a matter of calculated risk. In the medical situation,

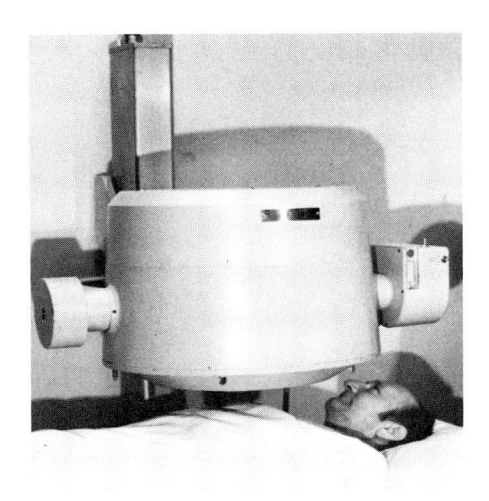

Fig. 1-13 Lung scan, carried out with a gamma camera. In this procedure the patient receives an intravenous injection of 300 microcuries of I-131 immediately prior to the scan. (Photo by Alan Kidd, taken in Kingston General Hospital, through the courtesy of T. W. Challis, M.D.).

for instance, the advantages of early diagnosis are presumed to outweigh possible damage due to radiation. And on these grounds diagnostic x-rays are taken. There is no doubt but that in competent hands these ionizing radiations are exceedingly powerful and useful tools to the great advantage of mankind.

Well then, how much radiation is considered "safe"? There is no precise threshold; that is, one cannot say a certain dose is absolutely safe and one slightly larger is dangerous. However, international scientific authorities have from time to time attempted to grapple with this problem to provide some guidelines for the general public as well as for radiation workers. For the general public, it is recommended that the total radiation dosage (exclusive of medical dosages) for an individual by age 30 not exceed a specified dosage which is about 50 times the annual natural background dosage. For radiation workers, who form only a small part of the world's genetic pool, the guideline is, of course, higher: about 100 times natural background dosage per year.

Dating with Radioisotopes

A radioactive nucleus, provided it is prevented from participating in a nuclear reaction with other nuclei by its strong electric charge and its blanket of orbital electrons, has a half-life which is essentially unchanged by the state of chemical combination of the atom in which it resides. A collection of radioisotopes therefore forms an excellent clock. It is thus possible to establish dates which are of considerable interest in archaeology, history, and geology.

The American, W. F. Libby, was awarded the 1960 Nobel prize in chemistry for his work in *radiocarbon dating.* His method makes ingenious use of a radioisotope, carbon-14, which has a half-life of about 5730 years. This isotope is continually created in our atmosphere by a nuclear reaction as neutrons produced by cosmic rays bombard nitrogen. The carbon-14 atoms combine chemically with oxygen molecules to form carbon dioxide. The result is that a very small portion, about 1 part in 10^8, of the carbon taken in by living plants, and hence by the creatures that eat the plants, is radioactive.

Now as soon as an organism dies, its intake of carbon-14 stops. This carbon-14 then decays without replenishment. The ratio of the number of carbon-14 atoms to the number of stable carbon-12 atoms in the object, decreases as time goes on. The time elapsed since death is inferred from measurement of this ratio. In this manner, reaching back as much as 50,000 years, geological events can be dated and the age of archaeological artifacts established. The uncertainties of these measurements in the case of specimens like the Dead Sea scrolls (about 2000 years old) and Egyptian mummies (about 5000 years old) are estimated to be only about 100 years.

Physics Today

W. F. Libby
(1908–   )

The age of the earth, some 4.5 billion years, has been determined by measuring the relative amounts of uranium-238 and lead-206 contained in geological specimens. These isotopes are the first and the last in the uranium series shown in Fig. 1-8. U-238 decays with a half-life of 4.5 billion years and, after going through all the steps in the uranium series, the residual nucleus finally transforms itself into lead-206. The amount of Pb-206 present in the rock therefore increases as time elapses. By determining how much lead-206 has accumulated in this way, the age of the rock can be calculated.

QUESTIONS

1. Two separated objects, A and B, each possess a positive charge.
   (a) Is the force F_A exerted on A by B an attractive force?
   (b) Draw a diagram showing objects A and B and represent the force F_A by an arrow which has its tail at A and points in the direction of the force acting on A.
2. Draw a diagram analogous to Fig. 1-1 for the case when objects 1 and 2 are both electrons. Does one electron repel another electron?
3. State, in terms of the amount of electric charge denoted by the symbol e (a positive number), the electric charge possessed by each of the following particles: electron, proton, neutron, photon.
4. Classify as attractive, repulsive, or zero, the electric force exerted by each of the following particles on each of the other particles: proton, electron, neutron.
5. Arrange in order of increasing photon energy the following types of electromagnetic radiation: visible light, x-rays, infrared, ultraviolet.
6. What is the mass and speed of a photon?
7. Arrange the following particles in order of increasing mass: proton, neutron, electron, photon, β-particle.
8. Define the terms *mass number* and *atomic number*.
9. For a nucleus with a mass number of 27 and an atomic number of 13, find:
   (a) the number of protons;
   (b) the total of the number of protons and neutrons;
   (c) the number of neutrons.
10. Find the number of neutrons and protons in each of the following nuclei: $_4\text{Be}^9$, $_8\text{O}^{17}$, $_{53}\text{I}^{131}$.
11. (a) What is an α-particle?
    (b) What is the electric charge possessed by an α-particle (in terms of the charge e)?
    (c) What is the atomic number of an α-particle?
    (d) What is the mass number of an α-particle?
12. Suppose that a pin head were shaped like a flat disk and that some 10^{12} iron atoms, arranged "cheek to jowl" so to speak, were required to cover the circular area of the pin head. Assume the pin head to be 10^7 atoms deep. How many atoms would there be in this pin head?
13. What is the meaning of the statement that an atom is an electrically neutral system of particles?
14. (a) What is the same about two isotopes of a certain element?
    (b) What is different?
    (c) Explain their similar behavior in a chemical reaction.
15. Name all the constituents of deuterium and compare the forces involved in the formation of its nucleus with those involved in the formation of the atom.
16. What is an ion?
17. (a) Describe the events that occur as a high-speed α-particle passes through air.
    (b) What is one actually seeing when looking at an α-particle track in a cloud chamber (Fig. 1-4)?
18. In the Rutherford scattering experiment, almost all α-particles go through the gold foil without appreciable deviation. Why?
19. In the Rutherford scattering experiment, some α-particles are scattered so that they come almost straight back from the foil. Explain this observation.

20. Suppose there were no nucleus and instead atoms were constructed of isolated protons and electrons distributed uniformly throughout the entire volume of the atom. Contrast the results that would be obtained in scattering protons from such matter with Rutherford's observations on α-particle scattering from real matter.

21. What is meant by the term "radioactive series"? Name one of these series.

22. Evaluate the net electric charge (algebraic sum) and the sum of the mass numbers before and after the decays given by Eqs. (1.1), (1.2), and (1.3).

23. "Iodine-131 is a radioactive isotope having a half-life of about eight days. It emits β-particles and γ-rays." Explain the statement in the quotation marks with particular emphasis on the terms "radioactive," "isotope," "131," "half-life," "β-particles," and "γ-rays."

24. Suppose a radioactive isotope had a half-life of 10 days. How much of this isotope would be left after 20 days if the original amount were 4 billion atoms?

25. Radium ($_{88}Ra^{226}$) is used as a therapeutic agent in cancer treatment. But it is not actually the radium itself which is the source of the useful radiation for this purpose. It is the gamma rays from the lead ($_{82}Pb^{214}$) and the bismuth ($_{83}Bi^{214}$), which appear in the decay chain of radium, that are used therapeutically.
   (a) Refer to Fig. 1-8 and trace the decay chain from radium-226 to bismuth-214.
   (b) Write the appropriate decay processes leading from radium-226 to bismuth-214 using notation such as shown in Eq. (1.1).
   (c) In addition to the therapeutically useful γ-ray photons, what other particles are emitted in the decay of lead-214 and of bismuth-214?
   (d) Would you expect radon-222 to be useful in cancer therapy in place of radium-226? Explain.

26. What is the unit of activity of a radioactive material?

27. If a particular sample of I-131 has an activity of 4 millicuries at 1 P.M. on January 1, what will be its activity at 1 P.M. on January 17 of the same year? (1 millicurie = 10^{-3} curie.)

28. If the activity of a certain radioactive sample decreases to 1/8 of its original value in 50 days, what is the half-life of the sample?

29. How many α-particles are emitted per second from a milligram of polonium-218 which has a half-life of 3.05 minutes? (The number of atoms in one gram of polonium-218 is approximately 2.7×10^{21}.)

30. Approximately how many counts/sec would a radiation detector indicate when held near a 20-millicurie radiocobalt needle? (Assume that one particle is emitted in each disintegration and that 15% of these particles are detected by the radiation counter.)

31. The half-life of the uranium isotope U-235 is about 7×10^8 years while the half-life of the iodine isotope I-131 is about 8 days. Suppose one had an equal number of atoms of each of these two isotopes. Which one would yield the greater number of disintegrations per second? Which one is said to have the higher activity?

32. The neutrons that Chadwick discovered were emitted when the beryllium nucleus, $_4Be^9$, was bombarded with α-particles.
   (a) What is the most massive possible residual nucleus after this reaction?
   (b) Write down the equation symbolizing this nuclear reaction.

33. In experimental studies of nuclear reactions, what advantages are there in using neutrons as the bombarding particles?

34. Give the nuclear reaction involved in making I-131 from I-130 by neutron bombardment.

35. Give the nuclear reaction involved in making phosphorous-32 from phosphorous-31 by neutron bombardment.

36. What is the relationship between the number of nuclei that undergo β-decay within a thin source in a given time interval and the number of β-particles emitted by the source, assuming that a negligible number of β-particles are stopped within the source?

37. The *effective half-life* of I-131 in the body is roughly half of its radioactive half-life of about 8 days. This is so because, in addition to radioactive decay, there is also biological elimination of this isotope from the body. Keeping this fact in mind, how long would it take before the activity of the I-131 accumulated in a person's thyroid would be reduced to 1/1024 of its initial value? [*Hint:* $2^{10} = 1024$.]

38. Give a brief outline of the use of radioisotopes in any one of these four fields:

   (a) medicine, (b) industry,
   (c) agriculture, (d) archaeology.

39. (a) What basic assumption regarding the ratio of C-14 to C-12 has to be accepted in order that there be any validity in the carbon dating process?
    (b) Why is radiocarbon dating unable to assign dates to bones that are about a million years old?

SUPPLEMENTARY QUESTIONS

S-1. Source A contains N radioactive atoms which have a half-life of T_A sec. Source B contains the same number of atoms, but their half-life is T_B sec. Using the Rutherford-Soddy law, Eq. (1.4), show that

$$\frac{\text{decays per sec in source } A}{\text{decays per sec in source } B} = \frac{T_B}{T_A}.$$

S-2. Discuss briefly the role of these scientists in the development of nuclear science: Madame Curie, Ernest Rutherford, Hans Geiger, James Chadwick, Henri Becquerel.

S-3. In Section 1.7 it was indicated that Tc-99 is a useful, shortlived radioisotope obtained from the decay of Mo-99. Using the information available in Appendix D, write the nuclear reaction showing this decay process.

S-4. A patient has been given a tracer dose of I-131. Daily measurements over the thyroid gland reveal that the quantity of I-131 therein is decreasing at such a rate that the effective half-life is 5.0 days. What is the biological half-life of iodine in this patient? To determine the answer, use an expression derived from the assumption that physiological elimination follows a law similar in form to the Rutherford-Soddy radioactive decay law:

$$\frac{1}{T} = \frac{1}{T_b} + \frac{1}{T_p}.$$

In this expression, T is the effective half-life, T_b the biological half-life, and T_p the radioactive, or physical, half-life discussed in Section 1.5.

S-5. The unit of radiation exposure is the *roentgen* (defined in terms of the ionizing effect of radiation on a standard mass of air). A common regulation is that devices which may produce x-rays, such as a TV set, should not exceed an exposure rate of 0.5 milliroentgen per hour (mR/hr) at a distance of two inches from the outer surface of the device. The exposure a person experiences due to natural causes (the so-called background radiation) is about 120 mR in a year. TV sets today emit virtually no x-rays. But suppose you had one which just met the standard requirement (0.5 mR/hr). How many hours would you have to sit just two inches from the set to be subjected to an exposure comparable to background exposure experienced in a month?

ADDITIONAL APPLICATIONS TO MEDICINE AND THE LIFE SCIENCES

Physics in Medicine

The applications of physics in medicine are manifold. From the early beginnings of western civilization there has been a close relationship between medicine and the development of physics. To students of either the history of medicine or the history of physics these things should become evident: (1) how such early physical observations as the electricity of Galvani and Volta were tied in with physiological observations, (2) how often significant advances in physics were made by physicians, and (3) how often significant medical advances were made possible by physicists. Probably x-rays remain the outstanding example of a new discovery in physics that to some extent revolutionized medical diagnostic procedures.

New possibilities for medicine will emerge from today's physics and technology, and a comprehension of basic physical principles will be invaluable to anyone in the medical field. To emphasize this assertion, here is a random sampling from an almost endless list of medical

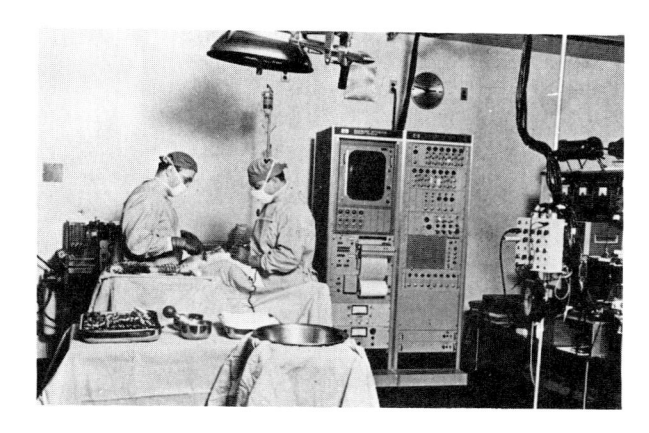

Fig. 1-14 Modern OR (operating room). (Courtesy *Hewlett-Packard Journal*.)

areas related to physics: prosthetic devices, centrifuge, circulation of blood, body fluid drainage, thermometers, blood clotting, antiseptics and surface tension, viscosity of blood, elasticity of the blood vessels, patient monitoring instruments, cobalt therapy, removal of foreign bodies, operating room precautions, electrosurgery, bioelectric currents, ECG, EEG, EMG, microscopy, vision, ophthalmoscopy, infrared radiation, audiometry, basal metabolism, body temperature, blood pressure, air and water mattresses, oxygen therapy.

The technological complexity of modern medicine is readily seen in Fig. 1-14, which shows an operating room. The chief basis of this technology is physics.

Particle Detectors: Solid State Detectors

Some solids are able to store ionizing radiation energy. Certain phosphors after irradiation emit light when heated. This effect is the basis of *thermoluminescent dosimetry* (TLD), since the amount of emitted visible light is dependent on the amount of radiation energy first absorbed by the phosphor. TLD-type dosimeters which are capable of measuring radiation exposure with a precision of ± 3%, and yet which are only about 2 or 3 mm square, are used in cancer radiation therapy.

Another application of such solid luminescent dosimeters is in the determination of the natural ionizing "background radiation" environment as related to plant growth. In this kind of ecological study the obvious advantages of solid detectors are their relative resistance to weathering, to small temperature changes, and to humidity changes.

As an example, in one study a number of such dosimeters were imbedded in the bark of oak trees. Readings were made every two weeks, and during the onset of foliation in the spring, a marked increase in ionizing radiation activity was noted.

Question

Considering the example just cited, how might one distinguish between the general background radiation environment, which consists largely of gamma rays, and the beta radiation due to radioactive materials in the tree sap?

Answer

Station similar detectors near the tree; or, as was actually done, wrap certain detectors in aluminum foil to filter out the beta rays due to radioactive trace elements in the sap.

Typical Values of Activities of Radioisotopes Used in Medicine

Thyroid scan (I-131):   50 - 200 μCi
             (Tc-99):   1 mCi

Yttrium-90 pellets (or seeds) implanted for destruction of pituitary gland (see Fig. 1-15): 2-4 pellets of 1-2 mCi each.

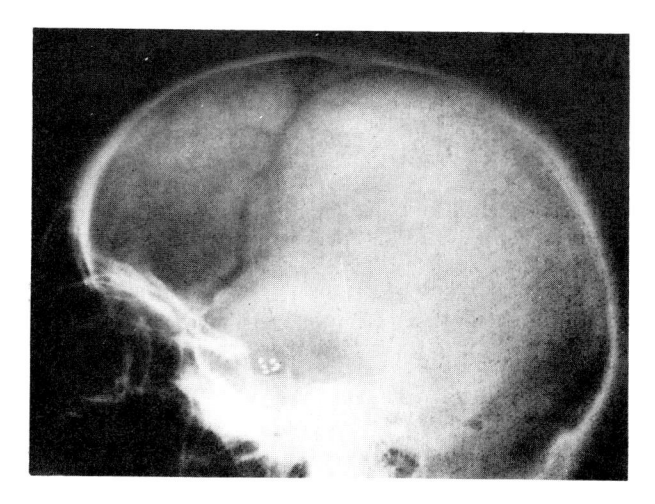

Fig. 1-15 Radiograph (or "x-ray") showing ceramic beads of yttrium-90 oxide implanted in the region of the pituitary gland. (Courtesy U.S. Energy Research and Development Administration.)

ADDITIONAL APPLICATIONS

Intraocular tumor location (P-32): 500 μCi.
Intraperitonal administration of radiogold (Au-198) in treatment of ovarian cancer: 150 μCi.
Radiocobalt (Co-60) needles (or wires) for interstitial implants in tumors: 50 - 100 mCi.

Question

About how many counts/sec would a radiation counter indicate when held near a 50-mCi radiocobalt needle? (Assume that one particle is emitted in each disintegration and that 10% of these particles are detected by the radiation counter.)

Answer

$$\frac{10}{100} \times \frac{50 \times 10^{-3} \text{Ci}}{1 \text{Ci}} \times 3.7 \times 10^{10} \text{ counts/sec}$$

$$= 19 \times 10^7 \text{ counts/sec.}$$

Further Reading

Suggested references relating to this and subsequent examples in nuclear medicine are:
(a) For the medical viewpoint, particularly specific clinical aspects: Blahd, W.H., ed., *Nuclear Medicine*, McGraw-Hill, New York, 1965, or Wagner, H.N., ed., *Principles of Nuclear Medicine*, Saunders, Philadelphia, 1968.
(b) For emphasis on the physical aspects: Quimby, E. H., S. Feitelberg, and W. Gross, *Radioactive Nuclides in Medicine and Biology*, Lea and Febiger, Philadelphia, 1970.

Radiography: Radioisotopes as Radiation Sources

Diagnostic radiography employing x-ray machines involving complex electric circuitry is one of the main medical diagnostic procedures today. For certain purposes it would be desirable to have a light and truly portable machine which is simple and independent of electrical requirements. A suitable radioisotope for medical radiography should give off gamma rays in the 30-130 keV region and should have a reasonably long half-life and high enough activity to permit short exposure times without requiring unduly large amounts of the isotope. That is, the isotope should have a high *specific activity*, usually expressed in mCi/gm. One isotope which would be very suitable if produced economically enough is gadolinium-153. It has a half-life of 240 days and gives off gamma rays of 40 and 90 keV.

Neutron Activation Analysis

Neutron activation analysis shows promise of becoming another significant component of the arsenal of nuclear medicine.

Although the neutron-activation technique is used almost exclusively to detect various chemical elements and measure their quantity, it is basically an isotopic method. It is the particular isotope nucleus that interacts with the neutron to produce a given radioisotope of an element. For quantitative analyses the technique is applicable to elements only when the isotopic abundances of a sample are known. This is the case in most samples, because the elements are present in their natural isotopic abundance.

The neutron activation process is now being used with increasing frequency in studies of human metabolism. Because of its exceptionally high detection sensitivity (from 0.5 part per million down to as little as 0.001 part per million, depending on the element) neutron activation analysis has helped to extend the study of metabolic behavior even to trace elements in the body.

Radioisotopes produced in body tissue samples by neutron activation are also routinely used in diagnostic studies of the changes in normal body functions caused by disease. For example, neutron bombardment of small samples of fingernail clippings containing a trace of sodium will produce a radioisotope of sodium. The subsequent identification of this radioisotope, sodium-24, which has a 15-hour half-life, will determine that sodium-23 was present as a target in the bombarded fingernails. And a certain presence of sodium in the nails appears to be associated with cystic fibrosis.

Biological Half-Life (see also Question S-4)

When dealing with radioactive tracers in biology and medicine, one needs to know something of the rate at which the organism involved expels the material containing the radioisotope during normal physiological processes. Analogous to the physical half-life (T_p) of p. 17 is the so-called biological half-life (T_b). If, as seems to be the case quite commonly, the biological elimination follows an exponential curve such as shown in Fig. 1-9, then a little work with exponents will show that the effective half-life (T) is given by

$$\frac{1}{T} = \frac{1}{T_b} + \frac{1}{T_p} .$$

In calculating dosages of radioisotopes in the case of animals and humans, the effective half-life T is the crucial parameter.

Technetium-99

The radioisotope which may well be replacing I-131 as the most commonly employed one in medical diagnosis is technetium-99. This isotope has a short-lived excited (metastable) state (in this state its symbol is Tc-99* or Tc-99m). This short half-life (about 6 hours) is good from the medical point of view because it means that the radioactivity within the patient does not persist very much longer than the time required for the necessary counting and scanning procedures.

Of course, for a hospital to have such a short-lived isotope readily on hand poses a problem. The difficulty is resolved by having available the somewhat longer-lived parent, molybdenum-99 (Mo-99), with a half-life of 66.6 hours. When Tc-99* is required it is eluted (washed out) from the container (the *generator*) holding Mo-99. (Hospitals generally acquire one or more 100-mCi generators weekly.)

In view of this milking process the Tc source is called a *technetium cow* (see Fig. 1-16). The gamma ray that is useful for the detection apparatus, usually a scintillation detector, is the 0.140-MeV gamma ray emitted as the excited technetium nucleus goes to the ground state.

Question

Write the nuclear reaction showing the decay of Mo-99 to Tc-99.

Answer

$$_{42}Mo^{99} \rightarrow {}_{43}Tc^{99*} + {}_{-1}\beta^{0}$$

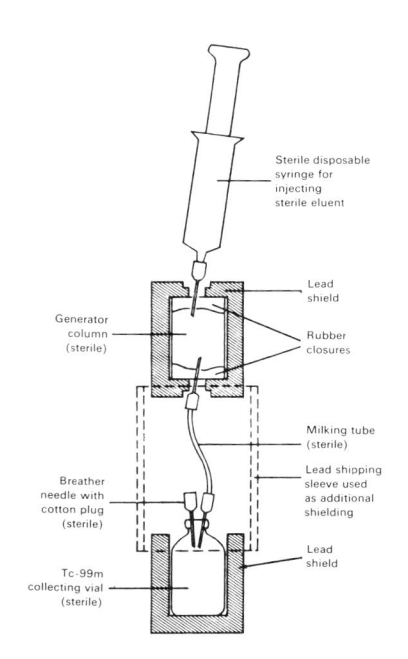

Fig. 1-16 Technetium cow.

Gamma Camera

Moving scanners (see Fig. 1-17), that is detectors, that move back and forth across the region of interest are, as stated in this chapter, being replaced in many medical diagnostic procedures by stationary detectors which yield an "instant scan." For example, a renal (kidney) scan which takes a half hour or so with a rectilinear scanner can be done in several minutes with a gamma camera. A widely used device of this kind is the *Anger camera* (see Fig. 1-13 and Fig. 1-18) developed by Hal Anger of the Donner Laboratories at the University of California. This detector uses a thin sodium-iodide crystal up to perhaps 25 cm in diameter behind which there are 19 photomultiplier tubes. Gamma rays from the radioisotope in the patient

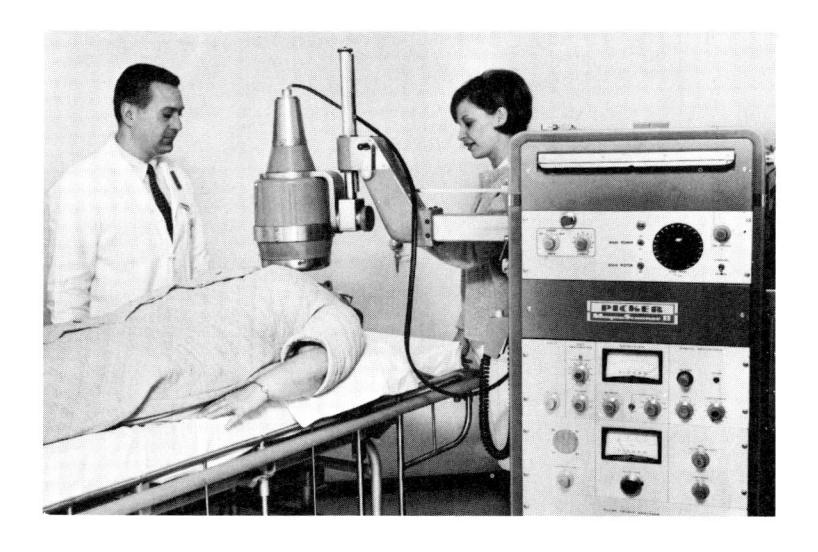

Fig. 1-17 A patient undergoing a thyroid scan. The patient has been given 50 microcuries of I-131 the previous day. (Courtesy Herbert A. Selenkow, M.D., Thyroid Laboratory, Peter Bent Brigham Hospital, Boston, Mass.)

ADDITIONAL APPLICATIONS 33

pass through collimators to the scintillator; then the distribution of the light among the various phototubes indicates the locale of origin of the rays. A picture of about 100,000 such separate signals originating at different points may be displayed on an oscilloscope and/or recorded with a Polaroid camera.

Such detectors, in general called *gamma cameras*, are particularly useful in brain scans.

Nonradioactive Isotope Tracers

The use of isotopes as tracers is not the prerogative solely of radioisotopes, although for many procedures they are indispensable. Nonradioactive isotopes can also be used, and indeed they have certain advantages. Enriching the iron in blood with a nonradioactive isotope of iron has already been used for studies of the function of iron in blood plasma. And a stable isotope of calcium has been used to determine the metabolic pathway of calcium in children. In general, the method involves the administration of a nonradioactive salt enriched in a specific isotope. Measurement of the change in the concentration of the administered isotope over a period of time provides useful information about the metabolic role of the element in question.

Question

What is the advantage of employing nonradioactive isotopes in tracer studies when such studies lend themselves to the use of stable isotopes?

Answer

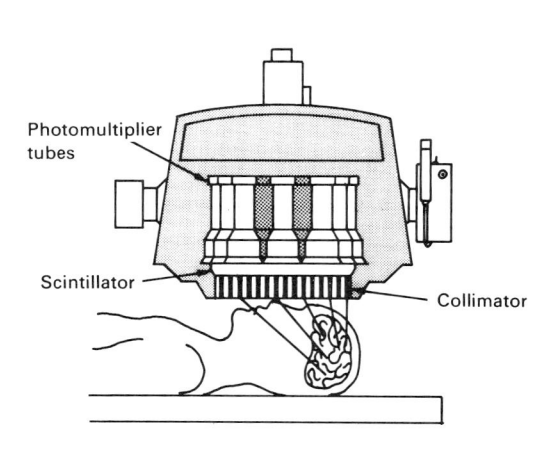

The obvious major advantage of using a nonradioactive isotope as a tracer is that the patient is not subjected to ionizing radiation, a particularly important consideration in studies involving children and pregnant women.

Radiation Exposure and Radiation Guides: Ionizing Radiation and Biological Tissue

Some *exposures and their effects on humans* are shown in Table 1-2. Some *typical values* of radiation exposures in roentgens (R) or milliroentgens (mR) are shown in Table 1-3.

Figure 1-19 shows the comparison of the sensitivities of various organs and organ systems to x- or γ-rays.
Radiation guides: suggested limits on *absorbed dose* (see Fig. 1-20) due to external radiation as proposed by the ICRP (International Commission on Radiological Protection).

Fig. 1-18 Anger camera. The gamma rays from the injected radioisotope pass through collimators immediately above the patient.

Table 1-2

Exposure, roentgens	Effect
25-50 R whole body	Possible blood changes (e.g., blood-cell destruction)
$\sim$ 100 R whole body	Possible radiation sickness
$\sim$ 250 R to gonads	Temporary sterility ($\sim$ 1 year)
$\sim$ 300 R localized	Loss of hair, reddening of skin
$\sim$ 400 R whole body	Death to 50% of exposed human population, or 50% chance of death in average individual
600 R or more, whole body	Death

Fig. 1-19 Comparison of sensitivities of various organ systems to single exposures of x- or γ-radiation. Solid lines represent the range of exposures over which effects have been observed. Dotted extensions to the left of the solid lines indicate that the effects might be detectable at lower exposure values. The symbol LD_{50} means that such an exposure is lethal to 50% of a large group; that is, mammalian LD_{50} is the exposure that would cause 50% of the mammals so exposed to die of the immediate effects of the radiation. (From J. F. Thomson and R. L. Straube, "Organ and Organism Response," in *Radiation Biology and Medicine*, edited by W. D. Claus, Reading, Mass.: Addison-Wesley, 1958.)

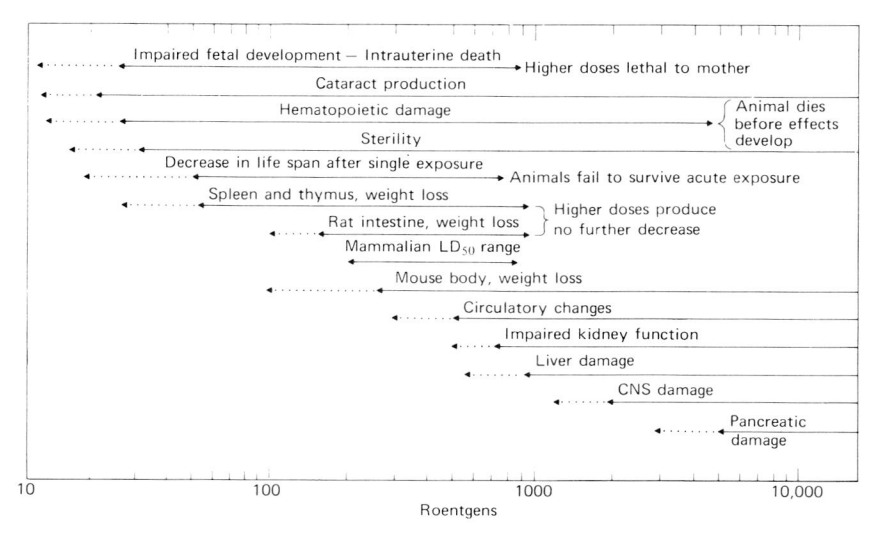

Table 1-3

Radiation source	Location	Exposure
Wrist watch (1) (1 μCi of Ra per watch)	Central body, including sex organs, at average distance of 1 ft	40 mR/yr
Luminous dials in airplane cabin (100 dials with 3 μCi of Ra each) (1)	Pilot at an average distance of 1 yd from the dials	1300 mR/yr
Phosphate rock (commercial fertilizer 0.01 to 0.025% U)(1)	Flat surface ground	280-700 mR/yr
People (1)	Packed in crowd	2 mR/yr
X-rays (2)	Lumbar spine, AP	1080 mR
	Lumbar spine, Lat.	4450 mR
	Chest, PA	10-65 mR
	Pelvis, AP	395 mR
X-rays, dental	Single film	0.05-5R
Fluoroscopy without image amplifier (3)		3 R/min
with image amplifier (3)		0.3 to 0.6 R/min

(1) W.F. Libby, *Science*, 122, 3158, 57 (1955).
(2) H.E. Duggan, *Canadian Medical Association Journal*, 91, 894 (1964).
(3) G.M. Ardran and H.E. Crooks, *British Journal of Radiology*, 26, 352 (1953).

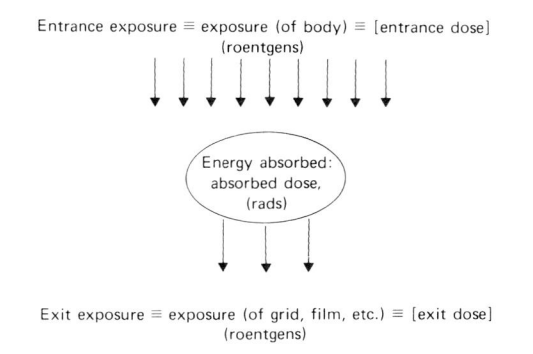

Fig. 1-20 Exposure (measured in R) and absorbed dose (measured in rads). Variations of terminology found in different publications are given.

Table 1-4

Occupational	Nonoccupational	General population
Critical organs (includes the whole body, head and trunk, active blood-forming organs, gonads)	Individuals in vicinity of controlled areas	Average exposure, exclusive of background radiation
3 rems/13 weeks; Cumulative for life: [5(Age − 18)] rems	1.5 rem/yr (adult workers occasionally entering radiation areas); 0.5 rem/yr (general public, including children, living in the neighborhood of controlled areas)	5 rems by age 30

Table 1-5

Dose to the Gonads per Year Resulting from Background Radiation*

External Radiation	
Cosmic rays	24
Local gamma rays (strongly dependent on locale)	50
Radon in the air (3×10^{-13} curies/liter)	1
Internal radiation	
Potassium 40	21
Carbon 14, radon, and its decay products	4
Total gonadal dose per year	100

* Approximate values in mrads/year per person in the U.K. (Varies considerably with locale.) Based on data in *The Hazards to Man of Nuclear and Allied Radiations*, London: 1956 and 1960. Report to the Medical Research Council. Crown copyright.

The guides are expressed in rem. (In case of x-rays such as used in medical diagnosis the exposure in roentgens is numerically about equal to the absorbed dose in rems. For details on the relation between *roentgen, rad* and *rem*, the reader is referred to Chapter 11 of the book by A. Ridgway and W. Thumm, cited in the Suggested Reading at the end of Chapter 1. (See Table 1-4.) Dose from *background radiation*: (See Table 1-5.)

Medical Dosages: The NCRP. (National Committee on Radiation Protection and Measurements) in Report 39 estimates that in the United States the average dosage (averaged over the entire population) due to diagnostic procedures is about 55 mrem/yr, while that due to therapeutic radiation is about 10 mrem/yr. The diagnostic figure is rather high compared to those reported by other advanced countries (United Kingdom: 14 mrem/yr). Yet according to present U.S. Public Health Service estimates, the U.S. average may go up over 90 mrem/yr before the end of the 1970s. At the same time some authorities in this field suggest the average dosage could be considerably reduced without lowering medical benefits.*

Question

What is the maximum accumulated permissible radiation dose to the whole body suggested for a 22-year old person employed in work connected with radiation? (Refer to Table 1-4.)

Answer

$$\text{Dose} = [5(22 - 18)] \text{ rems}$$
$$= 20 \text{ rems, exclusive of medically received radiation.}$$

* See the article by Karl Z. Morgan, *Comments on Radiation Hazards and Risks*, in the October 1971 issue of *The Physics Teacher*.

SUGGESTED READING

ANDRADE, E. N. DA C., *Rutherford and the Nuclear Atom*. Garden City, N.Y.: Doubleday, 1964. This 218-page paperback is one of the Doubleday Anchor Science Study Series written for beginning students in physics (as well as for the general public) by distinguished authors intimately involved in the topics of which they write. This particular book, completely nonmathematical in treatment, gives a genuine insight into the developments leading to the concept of the nuclear atom, as well as a lively picture of all the personalities involved. For example, here you will meet Rutherford's assistants, Geiger and Marsden, portrayed in the actual atmosphere of their time by an author who was present.

BOOTH, VERNE H., *The Structure of Atoms*. New York: Macmillan, 1964 (first edition). Very readable; essentially descriptive. Recommended for those who wish a relatively nonmathematical but considerably fuller discussion of the matters treated in our Chapter 1, especially as regards atomic physics rather than nuclear physics.

CARO, D. E., J. A. McDONELL, B. M. SPICER, *Modern Physics*. London: Edward Arnold (Publishers) Ltd., 1962. Here is offered excellent coverage, in some 200 pages, of the field of "modern," i.e., atomic and nuclear, physics. The material is presented in a historical context and, although calculus is used occasionally, any reader with high school algebra could benefit from an acquaintance with this book.

CLINE, BARBARA L., *The Questioners*. New York: Thomas Y. Crowell Co., 1965. A lively account of the work and thoughts of the men who pioneered nuclear and atomic physics. Written for all to enjoy, by a woman who is herself not a professional scientist but who received aid from some of the best physicists of our day, this work has been very well received by the scientific community. The book concludes with a brief excursion into the roles of some physicists in recent developments in both physics and world affairs. Any reader curious to know more about modern physics and its creators will find reading this book a rewarding experience in that it vividly illuminates science as a human enterprise. (This book is now available in a paperback edition by the Signet Science Library under the title, *Men Who Made a New Physics*.)

CURIE, EVE, *Madame Curie*. Garden City, N.Y.: Doubleday, 1939. This is the biography of one of the most famous experimental scientists of all time. Reading this fascinating historical account of Madame Curie's achievements should (or at least may) dispel some of those peculiar notions that intellect as applied to science is a prerogative of the male sex. Written by Mme. Curie's daughter, Eve, this book combines a warm, human insight with incisive scientific facts.

DAVIS, HELEN MILES, *The Chemical Elements*. Washington, D.C.: Science Service, Inc., 1961 (third edition). This paperback, written by a person dedicated to the popularization of science, is a veritable font of information on the elements from their discovery, to properties, characteristic compounds, practical applications, etc. The author, Miss Helen Miles Davis, died in 1957 before a second edition was fully revised. The revision was carried on by G. T.

Seaborg, who shared the 1951 Nobel prize for chemistry with E. M. McMillan as the result of their discoveries in the field of transuranic elements.

ERWALL, L. G., H. G. FORSBERG, and K. LJUNGGREN, *Industrial Isotope Techniques*. New York: John Wiley and Sons, Inc., 1964. For those students who wish to know more about the industrial application of radioisotopes. This book, like the one by Johns to be mentioned shortly, is written for those with a professional interest in the field concerned.

JOHNS, H. E. and J. R. CUNNINGHAM, *The Physics of Radiology*. Springfield, Ill.: Charles C. Thomas, 1974. Chapter XV offers an excellent summary of the clinical use of radioisotopes. Very profitable reading for those students interested in medical matters and who find themselves dissatisfied with the relatively short shrift this topic received in our Chapter 1.

LAPP, R. E., *Matter*. New York: Time, Inc., 1963. Written by a physicist who has spent much time and energy on interpreting physics, particularly atomic, nuclear, and radiation physics, to the layman, this book is almost a must. Nowhere else under a single cover will the reader find such a wealth of relevant pictures, some of them quite rare, from people (e.g., W. D. Coolidge showing Thomas Edison the process of x-ray tube filament manufacture) to things (e.g., the outlandish behavior of liquid helium, which creeps up, over, and down the outside of a container intended to hold it). One of the outstanding features of the book is the section on the chemical elements. Not only are the elements' characteristics described but pictures of samples of the elements are shown.

LAPP, R. E., and H. L. ANDREWS, *Nuclear Radiation Physics*. Englewood Cliffs, N.J.: Prentice-Hall, 1966 (third edition). For those readers with some physics background who wish a more thorough treatment of radioactivity. Knowledge of elementary calculus is presumed.

MANDELKERN, LEO, *An Introduction to Macromolecules*. New York: Springer-Verlag New York Inc., 1972. Students beginning the study of physics, chemistry, or biology will be interested in the importance of macromolecules. This book provides a non-mathematical discussion of the structures and properties of all classes of macromolecules, from proteins and nucleic acids to synthetic polymers.

PATON, A. R., *Science for the Non-Scientist*. Minneapolis: Burgess, 1962. This 125-page paperback gives, among other things, the chemist's viewpoint on the structure of matter. Highly recommended for those readers who wish to read something on atomic and molecular binding, all at a nonmathematical level.

PAULING, LINUS, and ROGER HAYWARD, *The Architecture of Molecules.* San Francisco: W. H. Freeman, 1964. Beautifully illustrated introduction to the subject of how atoms are arranged and interconnected in molecules. The book's brief but lucid explanations, coupled with the concrete representations of molecular structures in the form of illustrations, clarify such diverse molecular properties as hardness, color, density, nutrient, and toxic qualities. The explanations of Hayward's illustrations are by the Nobel prize chemist, Pauling, who authored the definitive text on the nature of the chemical bond. Written particularly for the beginner in atomic and molecular science.

RIDGWAY, A., and W. THUMM, *The Physics of Medical Radiography* (2nd revised printing). Reading, Mass.: Addison-Wesley, 1973. Chapter 11 of this book is recommended for those readers who have had their appetites regarding radiation, its dosimetry and effect on biological tissues whetted by our short excursion into this field. Subjects such as exposure and absorbed dose, the detection of radiation, as well as radiation hazards (particularly with respect to x-rays), and protective measures are treated in a descriptive, relatively nonmathematical fashion.

WEBER, R. L. (ed.), *A Random Walk in Science.* New York: Crane, Russak and Co. Ltd., 1973. The student will enjoy this anthology which has been specifically arranged for casual reading, each article being only loosely related to the others near it. There are anecdotes about noted scientists, items of historical interest, and articles showing the often bizarre progress of scientific theories.

Galileo Galilei

2 Motion and Force

Having seen something of the nature of physics and made an acquaintance with facts of the submicroscopic world in Chapter 1, we now find it necessary to backtrack in time. Indeed we shall go far back to bring the reader to grips with some of the key physical principles which in retrospect will make many of the phenomena of the first chapter more comprehensible.

Efforts to understand the simplest motions of ordinary objects yielded a golden harvest. Major contributions to this study were made by the Italian, Galileo Galilei, who is widely regarded as the father of physics. From his time, physics has been stamped with two characteristics whose importance appears to increase as the decades go by. First, he demonstrated the value of expressing theories and experimental results in the concise language of mathematics. Next, he insisted on the supremacy of experimental results, saying that "one sole experiment, or concludent demonstration, produced on the contrary part, sufficeth to batter to the ground . . . a thousand probable arguments."

In mechanics, he was the first to show the importance of the concept of *acceleration*. He found that the motion of all falling bodies, in an idealized "free fall," could be described with utmost simplicity: all such bodies have the same constant acceleration. In this connection there is the story familiar to most people but the historical accuracy of which is questionable: that Galileo actually climbed the leaning Tower of Pisa and publicly demonstrated that different weights, dropped simultaneously, will hit the ground simultaneously, thus disproving Aristotle's (384–322 B.C.) contention that heavy objects fall faster than light ones.

The starting point in understanding motion is the *law of inertia*, which we shall discuss in this chapter. This simple law, a radical departure from the Aristotelian notions accepted by his contemporaries, was inferred by Galileo after a series of experiments with rolling balls.

Galileo designed many ingenious instruments, such as the first pendulum clock. He made the first thermometer, as well as one of the first microscopes and telescopes. The telescope led him to several

. . . an object "continues in its motion in a straight line until something intervenes to halt or slacken or deflect it."

Galileo Galilei
(1564–1642)

capital discoveries in astronomy which, as a dangerous side effect, gave publicity to his support of the hated sun-centered picture of our solar system that had been advanced by Copernicus (1473–1543). This, coupled with his sustained attacks on Aristotelian philosophy, led to a series of encounters between Galileo and the learned men of the Church. This clash of ideas culminated in the tragic scene in which the aging scientist, facing the Inquisition, pledged to "abjure, curse, and detest the said heresies and errors . . . ." Sentenced to house imprisonment for the rest of his life, he passed his last years enfeebled and finally blind, dying at the age of 78.

In the year that Galileo died, there was born a young Englishman who was to become the dominant figure in the history of mechanics and probably in all science. A biographical sketch of this man, Sir Isaac Newton (1642–1727), follows in the next chapter. Newton, benefitting from Galileo's experiments and ideas concerning inertia and acceleration, formulated a few succinct laws of motion and showed that these were at once a vehicle for thought as well as a method of precise calculation, which together were adequate for the understanding of a tremendous range of natural phenomena. This mechanics is still the basic subject for physics students. It still gives a simple and accurate description of much that concerns the modern engineer, the scientist, as well as the philosopher.

Of course, in this century we have learned that the great French mathematician, Lagrange (1736–1813), was wrong when he lamented that the law of nature can only be discovered once. There have been deep-seated changes in our ideas about the most fundamental things, and the domain of applicability of Newton's mechanics has been staked out. In 1905, Albert Einstein (1879–1955) showed how mechanics must be modified for speeds near the speed of light. During the next two decades the realization grew that, in the realm of atoms and their nuclei, the "classical mechanics" of Newton failed and profound changes in viewpoint were introduced. This "revolution" in physics culminated, between 1924 and 1928, in the formulation of the quantum mechanics we mentioned earlier in connection with atomic structure. But even for those who are interested only in seeing the view from the latest peak that has been scaled, Newton's mechanics is the necessary path.

Our plan of attack is this. First we learn how to describe motion, and this involves terms like *velocity* and *acceleration*. Whereas this beginning is inherently mathematical, in this chapter emphasis is placed on giving a picture and an appreciation of how motion progresses. Mathematical entities called vectors have been used primarily to help you picture what is going on. After these preliminaries we plunge into the heart of mechanics: force, mass, and acceleration.

2.1 MOTION ALONG A STRAIGHT LINE

Rate of Change

Science is very much concerned with a description of changes occurring in the world, and a key term in this description is the phrase *rate of change*. For example, the motion of a particle is described in terms of velocity and acceleration, and velocity is defined as the rate of change of position, while acceleration is defined as the rate of change of velocity. (If you have the good fortune to have money in the bank, you are concerned with its rate of change which the bank determines by specifying an interest rate.) Newton, before he could develop his marvelously successful mechanics, had to invent mathematical methods for calculating rates of change; the branch of mathematics he thus created is known as the calculus. Without getting into mathematical details or worrying about mathematical rigor, we can still gain some appreciation of the meaning and use of rates of change.

Let us develop our general ideas with a specific example in mind: the height of a plant which grows in an irregular manner. As time goes on the height increases. The average rate of change of the height over a certain interval of time is defined to be the change in height divided by the time interval. Thus, if the plant grew 5 inches in the third and fourth days, its average rate of change of height during these days is

$$\frac{5 \text{ inches}}{2 \text{ days}} = 2.5 \text{ inches per day.}$$

We see thus that no subtleties are involved in the calculation of average rates of change. But now let us ask: What is the rate of change of height at a certain instant of time, say at 12 noon on the fourth day? Again, to find the rate of change, we wish to divide the change by the time interval during which the change took place, and if the plant grew 0.60 inch between 9:00 A.M. and 3:00 P.M., we can compute the average rate of change for this interval to be

$$\frac{0.60 \text{ inch}}{6.0 \text{ hours}} = \frac{0.60 \text{ inch}}{0.25 \text{ day}} = 2.4 \text{ inches per day.}$$

This is a good approximation to the rate of change of height at 12:00 noon, but it is not exact if the plant has been growing irregularly during the six-hour interval. What we need to do to find the rate of change of height at the instant of 12:00 noon is to consider ever-decreasing time intervals, each of which includes 12:00 noon, and divide the ever-decreasing growth by the corresponding time interval. The sequence of average rates of change, obtained as the time interval becomes arbitrarily small, converges to the number which is defined as the rate of change at the particular instant in question. At this point we hasten to reassure the reader who is perplexed that students of calculus spend many hours, and examine many special cases, before they master the ideas involved in the definition of this instantaneous rate of change. Our object here is to merely introduce the idea and to prepare you for the extremely important application which follows.

If we know, either by measurement or theory, the rate of change at different times, we can then compute changes, using the equation:

Change occurring in a time interval
= (rate of change at any given instant within the interval)
× (time interval).   (2.1)

This equation is exact if the rate of change is constant. Otherwise, the equation is not exact but is a good approximation for time intervals so short that the rate of change itself does not alter significantly during the interval. This equation is usually a very bad approximation if the time interval is so long that the rate of change alters appreciably during the time interval in question. In the latter case we must chop the long time interval into several short intervals, and apply Eq. (2.1) to compute the change for each short interval, and then add up all these changes to compute the total change. This procedure will be illustrated in the forthcoming section on velocity.

It is worthwhile learning the notation used for the factors in Eq. (2.1). If we are concerned with some quantity like the height of a plant which we can denote by the letter y, then we use the symbol Δy to denote "the change in y."* That is,

Δy = (new value of y) − (old value of y).

Similarly, if the symbol t denotes the reading of our clock, the symbol Δt is used to denote the interval of time during which the change Δy occurred. With this understanding we can rewrite Eq. (2.1) in the form

Δy = (rate of change of y at any given instant
during the interval Δt) × (Δt).   (2.2)

(Notice that the symbol Δ is always to be followed by the symbol representing the changing quantity. Remember that Δ alone does not represent a number; it is quantities like Δy or Δt that are numbers. Therefore, in an equation like Eq. (2.2) *the symbol Δ cannot be canceled out*.)

Frame of Reference

Even though we live in three-dimensional space, many motions take place in just one dimension and, since motions in one dimension are the simplest, we will study these first.

The symbol Δ is the Greek capital letter, delta. The expression Δy is read as "delta y."

44 MOTION AND FORCE

Consider a car traveling along a straight section of a highway. To describe motion we must first select a *frame of reference*, which in this case could be furnished by a line painted on the highway along the path of the car and extending indefinitely in both directions away from a certain blob of paint. This frame of reference is represented in Fig. 2-1 by the line *OX*, with the point *O* representing the blob. To determine the position of the car, it is merely necessary to give its distance *x* from the point *O* with the convention that distances to the right of *O* are positive numbers and distances to the left of *O* are negative. Speaking mathematically, the line *OX* is called the *X-axis* with *origin* at point *O*, and *x* is named the *position coordinate* of point *C* which represents the car. *A mathematical description of the motion consists of giving the value of the position coordinate, x, corresponding to each instant of time, t.* A meter stick and a stopwatch suffice to make the measurements. A portion of such a description is pictured in Fig. 2-2.

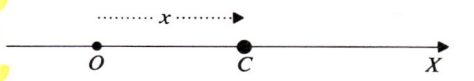

Fig. 2-1 The line *OX* is the *X*-axis with origin at the point *O* and *x* is the position coordinate of the point *C* which represents the car. When *C* is to the right of *O*, the position coordinate *x* is a positive number.

Fig. 2-2 Values of the position coordinate *x* at various instants of time *t*.

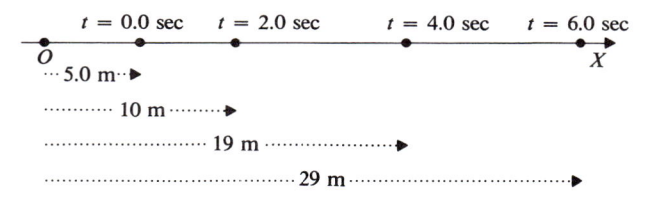

Velocity

What is velocity? Let us use the concept of rate of change discussed earlier. *Velocity is the rate of change of the position coordinate.* For example, the needle of a car's speedometer indicates, at any instant, the size (or magnitude) of the velocity of the car. In North America, speedometers are, of course, calibrated in miles per hour, rather than the meters per second which are used in this book. (You can convert from one system to the other using the fact that 60 mph equals 26.8 meters/sec.)

Applying the general equation, Eq. (2.1), to this situation of an automobile speedometer, we can relate the change in the position coordinate Δx which occurs during a time interval Δt to the velocity v at any given instant within this interval by the approximate expression

$$\Delta x = v \Delta t. \quad (2.3)$$

Example 1 Consider the motion pictured in Fig. 2-2. If the speedometer reads 8.0 meters/sec at the instant $t = 2.0$ sec and the position coordinate is 10.0 meters, can we estimate the position of the car 0.20 sec later?

2.1 MOTION ALONG A STRAIGHT LINE

Solution Yes. Using Eq. (2.3) we calculate

$$\Delta x = (8.0 \text{ meters/sec}) \times (0.20 \text{ sec}) = 1.6 \text{ meters.}$$

Thus the new position coordinate is

$$10.0 \text{ meters} + 1.6 \text{ meters} = 11.6 \text{ meters.}$$

However, we realize that the above answer is probably slightly in error if the velocity has been changing during the 0.20-sec interval. In fact, if within this interval the driver slammed on the brakes, causing a relatively large change in the velocity, then the error in the calculation of Δx would be appreciable. On the other hand, if the velocity were held constant at 8.0 meters/sec during the 0.20-sec interval, our evaluation of Δx would be exact.

Example 2 Now suppose we have a record, as tabulated below, of the velocities and the times at which they occurred, and from this data we wish to find the change in the position coordinate for the large time interval extending from $t = 8.0$ sec to $t = 14.0$ sec.

Time (seconds)	8.0	9.0	10.0	11.0	12.0	13.0	14.0
Velocity (meters/sec)	10.0	11.0	13.0	9.0	4.0	2.0	1.0

Solution We notice that the velocity varies from 13.0 to 1.0 meter/sec during this large time interval, so that direct application of Eq. (2.3) would give an answer that would probably be very misleading. A much more reliable estimate is obtained by subdividing the large time interval into many small intervals and applying Eq. (2.3) to each small interval. One such subdivision is to consider intervals 2.0 sec long, and use for each interval the velocity which occurs at the middle of the interval. Thus for the first such interval stretching from $t = 8.0$ sec to $t = 10.0$ sec we get

$$\Delta x = (11.0 \text{ meters/sec}) \times (2.0 \text{ sec}) = 22 \text{ meters.}$$

For the next interval we have

$$\Delta x = (9.0 \text{ meters/sec}) \times (2.0 \text{ sec}) = 18 \text{ meters,}$$

while in the last interval

$$\Delta x = (2.0 \text{ meters/sec}) \times (2.0 \text{ sec}) = 4.0 \text{ meters.}$$

The estimate for the change in the position coordinate for the original large time interval is the sum:

$$22 \text{ meters} + 18 \text{ meters} + 4.0 \text{ meters} = 44 \text{ meters.}$$

At this juncture, the student may be framing several questions.

First: Is this trivial little example worth bothering about? The answer is that some of our grandest laws of nature simply determine rates of change. So it is vital to understand how the rate of change determines the change for a short time interval and consequently allows one to piece together the changes for a succession of time intervals so that the evolution in time can be followed step by step.

Next: Observing that our method gave an answer that was only an approximation to the actual change, one might ask whether or not it is possible to solve all such problems exactly. The answer is that, although many special problems of this type can be solved exactly using the integral calculus, the general problem must be approached in the relatively cumbersome way we have indicated. Moreover, by subdividing the large time interval into very many very small time intervals, the approximation can be made very good indeed. Computers are just the things to handle the large number of routine calculations required.

The student should notice the significance of a positive or negative velocity v in the equation $\Delta x = v \Delta t$. If v is positive, then Δx is positive and x has increased during the time interval Δt so that the motion is in the direction of increasing x, that is, toward the right in Fig. 2-1 and Fig. 2-2. However, if v is negative, Δx is negative and the motion is toward the left.

A positive number like 30 meters/sec, which just gives the magnitude or size of the velocity and not its direction is called the *speed*. One must know both the speed and the direction of motion to completely know the velocity. Speedometers are correctly named because they do measure speed, not velocity.

We close this section on velocity by drawing the reader's attention to the fact that Eq. (2.3) could, of course, be written

$$v = \frac{\Delta x}{\Delta t},$$

which gives us an approximate expression for the velocity at any given instant within the time interval, Δt, during which the position coordinate changed by an amount Δx. The approximation will be good if Δt is small.

Acceleration

Again we employ the concept of the rate of change. This time we are interested in the rate of change of velocity. *The rate of change of velocity is named the acceleration.* A car is accelerated when one steps on the gas and also, according to the usage of the word in physics, when one applies the brakes. In each case the velocity is changing. The unit for measurement of acceleration is a velocity unit divided by a time unit, that is, meters per second per second, which is written as *meter/sec/sec* or, as is more usually done, as m/sec^2.

The change in velocity Δv which occurs during a time interval Δt is, according to Eq. (2.1), determined by the acceleration (denoted by the letter a) at any given instant during this interval using the approximate relationship

$$\Delta v = a\Delta t. \qquad (2.4)$$

Again, it is to be emphasized that this equation is

1. Exact if the acceleration is constant,
2. A good approximation if Δt is so short an interval that the acceleration does not alter appreciably,
3. Usually a very poor approximation if Δt is so long that large changes in the acceleration occur. (To obviate such a poor approximation, we should subdivide the long interval into several short ones and apply $\Delta v = a\Delta t$ to each short interval.)

Motion with *constant* acceleration merits special attention since not only does it occur in many situations but also it is historically important. It was Galileo's studies of motion involving uniform or constant acceleration which focused scientists' thoughts upon the subject of acceleration. And acceleration was revealed by Newton to be a key ingredient in the formulation of the laws of mechanics.

For motion with *constant* acceleration, the change in velocity Δv in any time interval Δt is given exactly by

$$\Delta v = a\Delta t. \qquad (2.4)$$

In order to find the change in the position coordinate Δx in this time interval Δt we use Eq. (2.3), the approximate relationship

$$\Delta x = v\Delta t. \qquad (2.3)$$

The velocity during the time interval Δt changes from some initial value v_0 to the value $v_0 + a\Delta t$ at the end of the interval, and the question arises as to which value of the velocity should be used in Eq. (2.3) to obtain the best estimate for Δx. If we select the velocity that is attained at the midpoint of the time interval, that is, the velocity $v_0 + \frac{1}{2}a\Delta t$, and substitute this value for v in Eq. (2.3) we obtain

$$\Delta x = (v_0 + \tfrac{1}{2}a\Delta t)\Delta t$$

which yields

$$\Delta x = v_0\Delta t + \tfrac{1}{2}a(\Delta t)^2. \qquad (2.5)$$

Now here is some good news. Eq. (2.5) is not just an approximation; it turns out (we omit the proof) that this equation gives Δx exactly, no matter how long the time interval, provided that the acceleration does not change.

Galileo's experiments showed that for a ball starting from rest ($v_0 = 0$) and rolling down an inclined plane, the distance Δx was proportional to the square of the time interval, in agreement with the prediction of Eq. (2.5) which, for this case, is simply

$$\Delta x = \tfrac{1}{2} a (\Delta t)^2. \tag{2.6}$$

Galileo concluded that the ball moved with constant acceleration down the plane, and correctly inferred that a freely falling body moves with constant acceleration. Comparing the falling motions of bodies of different weights, it was learned that in free fall all bodies have almost the same constant downward acceleration of 9.8 meters/sec², which is named the *acceleration due to gravity*.

At this point we would make an aside to draw the reader's attention to two relevant and very significant attitudes adopted by Galileo. First, although the accelerations of objects falling through the air depend upon the size and the shape of the object as well as upon the atmospheric conditions and even upon the precise location on the earth, Galileo realized that the *essential* point was a constant acceleration downward with deviations caused by perhaps quite complicated factors. His very instructive simple description of the motion was obtained only by ignoring these complications. Galileo thus demonstrated the great value of *idealizations* of reality.

Next we note that Galileo was not daunted by the fact that his experimental results were in conflict with Aristotle's widely accepted teaching that heavier bodies should fall faster. Such *acceptance of experimental results*, even though they disagree with the perhaps plausible ideas of the most renowned men, has been the crucial factor in the development of science since Galileo's time.

Here is a typical problem that will help to clarify ideas on motion with constant acceleration.

Example 3 A stone is dropped from rest down a deep well (see Fig. 2-3). How far has it fallen after 1.60 sec, and what is its velocity at this instant?

Solution We know the acceleration is 9.8 meters/sec² downward, and we select our axis OX to point downwards as in Fig. 2-3, with the origin O at the top of the well. Using Eq. (2.6) with $a = 9.8$ meters/sec² and $\Delta t = 1.60$ sec, we obtain

$$\Delta x = \tfrac{1}{2} \times (9.8 \text{ meters/sec}^2) \times (1.60 \text{ sec})^2 = 12.5 \text{ meters}.$$

Since the stone has never backtracked, the change in its position coordinate Δx is the same as the distance through which it has fallen.

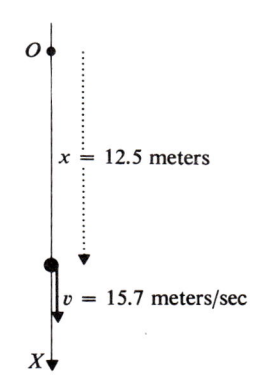

Fig. 2-3 The position coordinate ($x = 12.5$ meters) and the velocity ($v = 15.7$ meters/sec) of a stone 1.6 seconds after it was released at the origin.

2.1 MOTION ALONG A STRAIGHT LINE   49

Incidentally, since the particle started at the origin, its position coordinate x at any instant is the same as its change Δx which has occurred since it was at the origin (where $x = 0$).

The change in velocity is found from Eq. (2.4) to be

$$\Delta v = (9.8 \text{ meters/sec}^2) \times (1.60 \text{ sec}) = 15.7 \text{ meters/sec}.$$

Since $\Delta v = v - v_0 = v - 0 = v$, this change in velocity is just the velocity itself, i.e., $v = 15.7$ meters/sec.

Before leaving our discussion of motion with constant acceleration, we will derive a useful result that will be needed later. It is this: We can relate position, velocity, and acceleration without referring to the time interval. If we consider only the case with zero initial velocity, then after a time interval Δt, $\Delta v = v$, and thus Eq. (2.4) gives

$$\Delta t = \frac{v}{a}.$$

Substituting this value of Δt in Eq. (2.6) yields

$$\Delta x = \tfrac{1}{2} a \left(\frac{v}{a}\right)^2.$$

Rearranging, we obtain the desired relationship,

$$v^2 = 2a\Delta x. \qquad (2.7)$$

Eq. (2.6) and Eq. (2.7) are frequently of use in solving practical problems of motion.*

2.2 VECTORS

Vector Quantities

Velocity has a *direction* as well as a *magnitude* or size; 60 miles/hr north is a different velocity from 60 miles/hr east. Acceleration also has both direction and magnitude, as is the case for several other physical quantities which will soon be encountered, such as force and momentum. Physical quantities that have both direction and magnitude (and add in a special way to be described shortly) are called *vector quantities*, and are represented by mathematical entities called *vectors*.

A brief dip into the mathematics of vectors is necessary to be able to describe three-dimensional motion with some measure of ease and clarity. A vector can be represented by an arrow, with the length of the arrow proportional to the magnitude of the vector and with the direction of the arrow giving the vector's direction in space. A vector is symbolized by a letter with an arrow above it such as $\vec{V}$ (although in many other books the reader will find the practice of simply using bold-faced letters to denote vectors). The magnitude of the vector $\vec{V}$

---

*The velocity v defined in Section 2.1 is the x-component (v_x) of the velocity vector $\vec{v}$ which will be defined in Section 2.3. And in Section 2.1, the acceleration a denotes the x-component (a_x) of the acceleration vector $\vec{a}$.

is a positive number denoted simply by V, the same symbol but with the arrow omitted.

In view of the above, vector equations make statements about directions as well as magnitudes. The equation

$$\vec{c} = \vec{d}$$

means not only that the vectors have the same magnitude ($c = d$), but also that they point in the same direction. All parallel vectors of the same magnitude, as in Fig. 2-4, are regarded as equivalent in the following vector algebra.

Addition and Multiplication

The sum of two vectors, $\vec{a}$ and $\vec{b}$, is defined by the triangle rule shown in Fig. 2-5 as the vector denoted by $\vec{a} + \vec{b}$. In words, the rule for computation of a vector sum is as follows: Place the tail of the second vector at the tip of the first; then the sum is the vector drawn from the tail of the first to the tip of the last. The rule can be extended to define the sum of any number of vectors, as illustrated for three vectors in Fig. 2-6. A little experimentation by the reader will reveal that the order of addition doesn't matter. In the parlance of "modern" mathematics, we say the addition of vectors is commutative. In other words,

$$\vec{a} + \vec{b} = \vec{b} + \vec{a}.$$

The rule for multiplication of a vector by a positive or negative number is given in Fig. 2-7. Notice that this rule implies that the negative of a given vector is a vector which has the same magnitude as this given vector but points in the opposite direction.

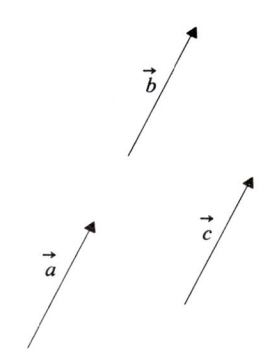

Fig. 2-4 Parallel vectors of the same magnitude are equal.
$\vec{a} = \vec{b} = \vec{c}$

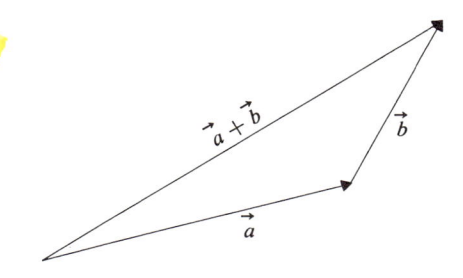

Fig. 2-5 The definition of the vector sum $\vec{a} + \vec{b}$ of two vectors $\vec{a}$ and $\vec{b}$.

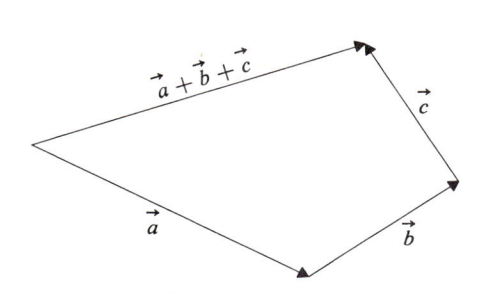

Fig. 2-6 The vector $\vec{a} + \vec{b} + \vec{c}$ is the vector sum of the vectors $\vec{a}$, $\vec{b}$, and $\vec{c}$.

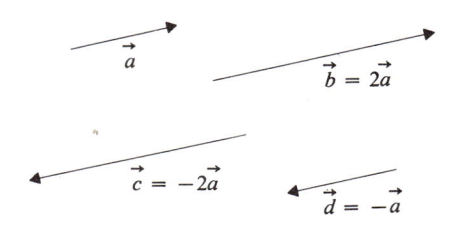

Fig. 2-7 Multiplication of a vector $\vec{a}$ by the positive number 2 gives a vector $\vec{b}$ in the same direction with a magnitude $2a$. Multiplication of the vector $\vec{a}$ by the negative number -2 gives a vector $\vec{c}$ in the opposite direction with a magnitude $2a$. Multiplication of $\vec{a}$ by the negative number -1 gives a vector $\vec{d}$ in the opposite direction with the same magnitude.

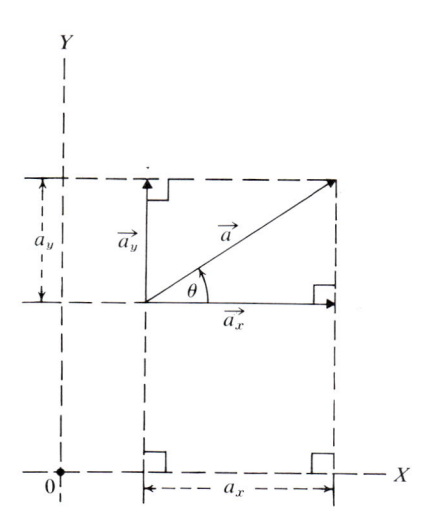

Fig. 2-8 The vector components of $\vec{a}$ resolved along *OX* and *OY* are the vectors $\vec{a}_x$ and $\vec{a}_y$. The x-component of $\vec{a}$ is $a_x = a \cos \theta$ and the y-component of $\vec{a}$ is $a_y = a \sin \theta$. When the components are known, the magnitude of the vector can be determined from $a = \sqrt{a_x^2 + a_y^2}$ and the angle from $\tan \theta = a_y/a_x$.

Components of a Vector

It is often convenient to express a given vector in a plane as a sum of two perpendicular vectors which are then called *vector components* of the given vector. The vector is said to be *resolved into components* along the two perpendicular directions of these vector components.

If we introduce in the plane two perpendicular coordinate axes, *OX* and *OY*, any vector $\vec{a}$ in the plane can be resolved into vector components $\vec{a}_x$ and $\vec{a}_y$, parallel to *OX* and *OY* respectively, such that

$$\vec{a} = \vec{a}_x + \vec{a}_y,$$

as illustrated in Fig. 2-8.

The *positive or negative number* a_x, called the *x-component* of $\vec{a}$, has a magnitude which is just the projection of $\vec{a}$ along *OX*. From the triangle in Fig. 2-8, we see that

$$a_x = a \cos \theta.$$

The number a_x is positive if the vector component, $\vec{a}_x$, points in the positive x-direction. Negative a_x corresponds to the vector component $\vec{a}_x$ in the negative x-direction.

The y-component of $\vec{a}$, the number a_y, is related in a similar way to the direction of the *Y*-axis:

$$a_y = a \sin \theta.$$

2.3 MOTION IN THREE DIMENSIONS

Position Vector

Since space as we conceive it has three dimensions (i.e., length, width, and height), it now behooves us to generalize our discussion of one-dimensional motion to describe three-dimensional motions such as those executed by a bird. Again, the first requirement is a *frame of reference*, something to which we can relate to make position measurements to specify the location of the bird. This frame of reference, represented in Fig. 2-10 by the axes *OX*, *OY*, *OZ*, could be provided at a

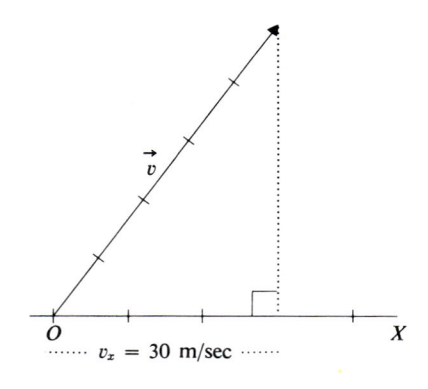

Fig. 2-9 The velocity vector $\vec{v}$ has a magnitude (named the speed) of 50 meters/sec. The x-component of this vector is $v_x = 30$ meters/sec.

Fig. 2-10 Frame of reference with axes *OX*, *OY*, *OZ*. The vector $\vec{r}$ is the position vector of the bird.

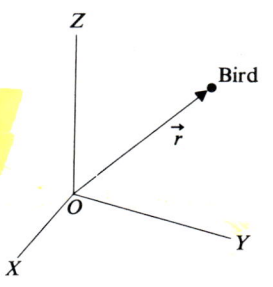

52 MOTION AND FORCE

street corner by perpendicular lines on the intersecting sidewalks, with a telephone pole conveniently located at the intersection. A vector $\vec{r}$, with its tail at the origin (the intersection of the perpendicular lines on the sidewalk and the foot of the telephone pole) and its tip at the bird, determines the position at any instant. And, as time goes on, the tip of this vector traces out the trajectory of the bird. A description of the motion consists of a specification of this *position vector* $\vec{r}$ for each instant of time.

Velocity Vector

In Fig. 2-11 we show the position vector $\vec{r}$ at a certain instant t and also the new position vector, $\vec{r} + \Delta\vec{r}$, after the interval Δt. The change in the position vector is $\Delta\vec{r}$. *The velocity $\vec{v}$ at the instant when the particle is at P is defined as the rate of change of the position vector $\vec{r}$ at this instant.* From Eq. (2.1), the change in the position vector $\Delta\vec{r}$ which occurs during the time interval Δt is related to the velocity vector $\vec{v}$ at the point P by the approximate relationship

$$\Delta\vec{r} = \vec{v}\Delta t. \tag{2.8}$$

The direction of the velocity vector $\vec{v}$ at P is such that it just touches the path at P, that is, $\vec{v}$ is in the direction of the *tangent* to the path. This direction is approximately that of the vector $\Delta\vec{r}$ if Δt is small. The magnitude of the velocity vector is called the *speed*.

Often we are presented with a knowledge of an object's velocity at different instants of time and its position at some instant. From this data one can compute the path of a particle using the approximate relationship, $\Delta\vec{r} = \vec{v}\Delta t$, to determine $\Delta\vec{r}$ for a short time interval, which includes the instant at which the velocity $\vec{v}$ occurred. Thus in Fig. 2-12 the particle is initially at A with a velocity $\vec{v}_1$. During Δt_1 it moves a distance which is approximately $\Delta\vec{r}_1 = \vec{v}_1\Delta t_1$ to reach the point B. To find out where the object goes in the next short interval Δt_2, we should use a velocity $\vec{v}_2$ that occurs within this interval. This step-by-step procedure gives us the approximate path $A\ B\ C\ D$. If we now repeat the procedure, using very short time intervals and consequently many more steps, we can trace out the path with greater accuracy.

Acceleration Vector

As an object moves along a path, its velocity can change in both *magnitude and direction*. In Fig. 2-13 we show the velocity vector $\vec{v}$ at the point P and a different velocity vector, $\vec{v} + \Delta\vec{v}$, at the point Q which is reached after a time interval Δt. The change in the velocity vector, denoted by $\Delta\vec{v}$, is found by the usual triangle rule noting that $\Delta\vec{v}$ is the vector

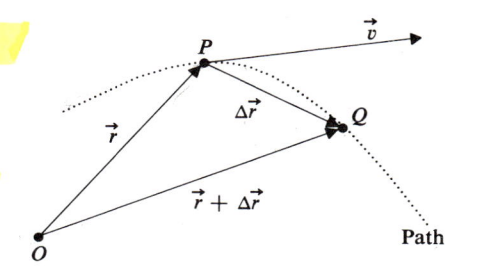

Fig. 2-11 Bird moves from point P to Q in Δt seconds. The change in the position vector is $\Delta\vec{r}$.

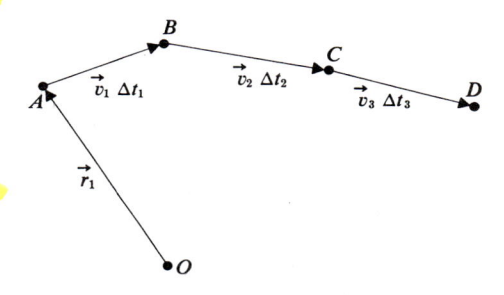

Fig. 2-12 Approximate path *ABCD*.

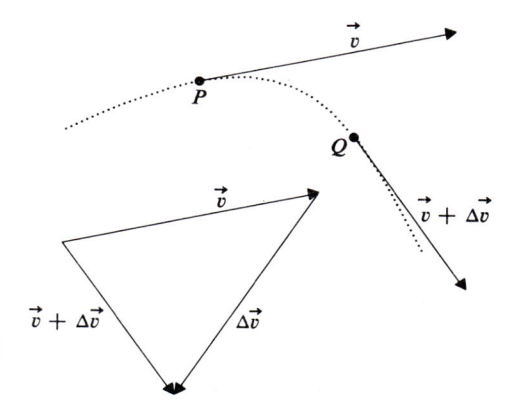

Fig. 2-13 $\Delta\vec{v}$ is the change in the velocity vector that occurs in the time interval Δt as the particle moves from P to Q.

which must be added to $\vec{v}$ to produce the new vector $\vec{v} + \Delta\vec{v}$. The *acceleration vector* at the instant when the particle is at P is defined as the *rate of change of the velocity vector* at this instant. The change in the velocity vector $\vec{v}$ which occurs during the time interval Δt is, according to Eq. (2.1), related to the acceleration vector at the point P by the approximate relationship

$$\Delta\vec{v} = \vec{a}\Delta t. \tag{2.9}$$

The approximation is good if the acceleration vector $\vec{a}$ does not change appreciably during the time interval Δt; this usually requires that the time interval be short so that the point Q in Fig. 2-13 is close to the point P. It is important to notice that the acceleration vector $\vec{a}$ points in the direction of the *change in the velocity* vector (the direction of $\Delta\vec{v}$) for Q very close to P. From Eq. (2.9) one can deduce a multitude of important results which are displayed in Fig. 2-14. These diagrammatically represented results are worthy of study and full understanding as the reader peruses the next few paragraphs.

Centripetal Acceleration

Notice that for a curved path $\vec{a}$ always points inward, and that if the speed is constant, $\vec{a}$ points towards the center of the arc of the circle. The fact that there is an *acceleration* for circular motion with *constant* speed often seems surprising to students. The point is that the velocity vector changes *direction* even though its magnitude, the speed, is unchanged. If you examine the direction of this change $\Delta\vec{v}$, in Fig. 2-15, you will find that when Q is very close to P, $\Delta\vec{v}$ points toward the center of the circular arc; therefore this is the direction of the acceleration vector. This particular acceleration, directed toward the center of the circle, has acquired a name of its own: *the centripetal acceleration.* The interesting result, that this centripetal acceleration has a magnitude $a = v^2/r$, is a consequence of the geometry of the situation. (We will omit the proof. The interested reader is referred to page 97 of the book by Holton and Roller listed in the Suggested Reading at the end of this chapter.)

If the object is speeding up as it goes along a curved path, it has, in addition to a centripetal acceleration, v^2/r, an acceleration in the direction of the velocity. And its total acceleration is the vector sum of these two accelerations.

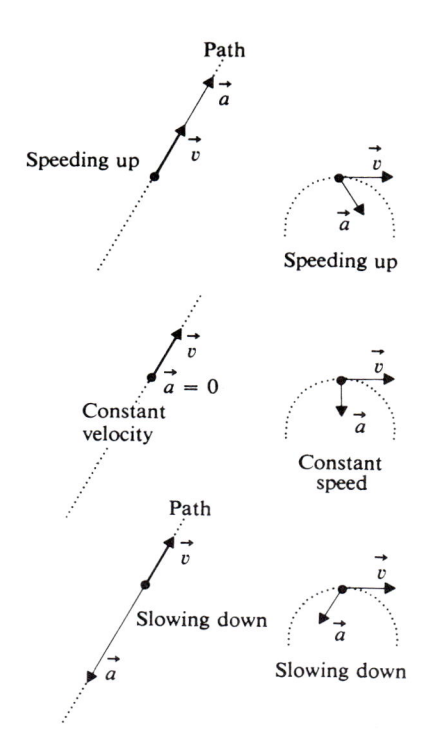

Fig. 2-14 The direction of the acceleration vector for motion along a straight line and along a curved path.

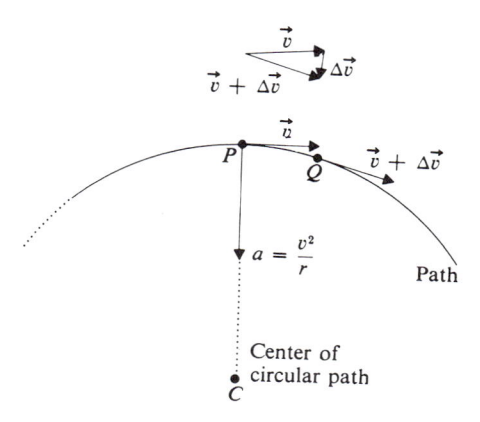

Fig. 2-15 Circular motion with constant speed v. For point Q very close to P, the vector $\Delta\vec{v}$ is in direction PC.

Determination of Motion

To describe motion we are equipped with the position vector $\vec{r}$, the velocity vector $\vec{v}$ which is the rate of change of the position vector, and the acceleration vector $\vec{a}$ which is the rate of change of the velocity

54 MOTION AND FORCE

vector. You will be relieved to learn that we do not have to continue in this vein and consider the rate of change of the acceleration! The acceleration is the quantity that Newton found to be so important.

When we take up the study of Newton's second law of motion, it will become apparent that our laws of nature usually specify the acceleration at each instant of time or the acceleration that would be encountered at each point of space. However, something more must be known before we can say what motion will occur, for if the object be started differently, different motions will result. But if we know the initial position vector $\vec{r}_1$ and the initial velocity vector $\vec{v}_1$, and also have complete information about the acceleration vector $\vec{a}$, then the entire motion is determined and we can compute all the velocities that will be encountered, and trace out the path taken, by repeated use of the approximate relations

$$\Delta \vec{r} = \vec{v} \Delta t$$

and

$$\Delta \vec{v} = \vec{a} \Delta t.$$

It is worthwhile thinking this through for the motion pictured in Fig. 2-12. The object starts at point A with a velocity $\vec{v}_1$, and in a short time interval, Δt_1, its position vector changes by an amount $\Delta \vec{r}_1 = \vec{v}_1 \Delta t_1$, which brings the particle to the point B. The change in the velocity vector during this interval is given by $\Delta \vec{v}_1 = \vec{a}_1 \Delta t_1$, where $\vec{a}_1$ is an acceleration occurring for this portion of the motion. For the next time interval, Δt_2, we use the new velocity $\vec{v}_1 + \Delta \vec{v}_1$ in the calculation of $\Delta \vec{r}_2$ and an appropriate acceleration for this interval for the calculation of $\Delta \vec{v}_2$. In this way, step by step, we discover the velocity at different points and trace out the path. In practice, a computer would be given the task of performing the large number of calculations required to achieve reasonable accuracy, or, better still, for many types of motion the entire problem could be solved in a few lines using a mathematical operation known as integration, which was developed in the calculus of Newton and Leibnitz (German philosopher and mathematician, 1646–1716).

2.4 FORCE, MASS, AND ACCELERATION

So far we have looked at motion without concerning ourselves about how this motion, and changes in it, were brought about. This study of motion is usually termed *kinematics*. We now concern ourselves with *dynamics*, the study of motion including the interactions which bring about changes in motion.

The central law of mechanics, the key to Newton's synthesis* of much of the natural philosophy of his time, the law governing a large realm

---

*Because of Newton's achievement in the formulation of the subject of mechanics, other contributors to this field are often overlooked. Names such as Christopher Wren and Christiaan Huygens should be included among the predecessors of Newton. Leonhard Euler (1707–1783) should also be remembered; some 60 years after Newton's Principia, Euler was apparently the first to enunciate Newton's famous second law in essentially the form in which we recognize it today. Readers interested in these historical developments are referred to an illuminating article in the March 1972 issue of The Physics Teacher: "Where Credit is Due: The Second Law of Motion and Newton's Equations," by V. V. Raman.

of our present engineering and science, is the relationship *force equals mass times acceleration* which, expressed in symbols, is

$$\vec{F} = m\vec{a}. \tag{2.10}$$

This law, which is commonly referred to as *Newton's second law of motion*, involves two physical quantities, *force* and *mass*, which must be carefully discussed as we examine the meaning of the law.

Force

In physics a force is a *push* or a *pull* (any other meanings that this word has in daily life are excluded). We sharpen up this idea of a push or a pull by specifying how to measure force. A definition that describes a sequence of operations to measure some physical quantity is called an *operational definition* of the quantity in question. An operational definition of force can be given in terms of spring balances, with the force under investigation applied to a hook as shown in Fig. 2-16. The force which produces a certain definite extension we shall call a *force of unit magnitude*. Naming this force unit the *newton* (the abbreviation is N), we say that the force applied to the hook in Fig. 2-16 is 1 newton. Two such forces acting in the same direction furnish a force of 2 newtons; two equal forces in the same direction which combine to give 1 newton are each 0.5 newton, etc. This method is adequate in principle to define the magnitude and direction of force but in practice, for modern scientific work, spring balances are replaced by more accurate detectors.

Just how strong a force is a force of 1 newton? Well, the gravitational pull (weight) on 100 cubic centimeters (a teacupful) of water is nearly 1 newton. And a large man whose weight is 1000 N weighs 225 pounds.

A force has magnitude and direction. It is a vital experimental fact, presented by Simon Stevin (Belgium) in 1605, that forces add according to the triangle rule for addition of vectors. This is verified by applying simultaneously two forces to the same spring balance and discovering the single force to which they are equivalent. Forces are thus *vector quantities*. As we shall stress in later applications, the force $\vec{F}$ that appears in Newton's second law,

$$\vec{F} = m\vec{a}, \tag{2.10}$$

is the vector sum of all the forces acting on the object whose acceleration is $\vec{a}$. This sum of forces we shall call the *resultant force*. So to understand motion it is necessary to think about a vector sum of forces and, in particular, it must be realized that when several forces act on a body their vector sum may be zero, and then the motion proceeds as if there were no forces at all!

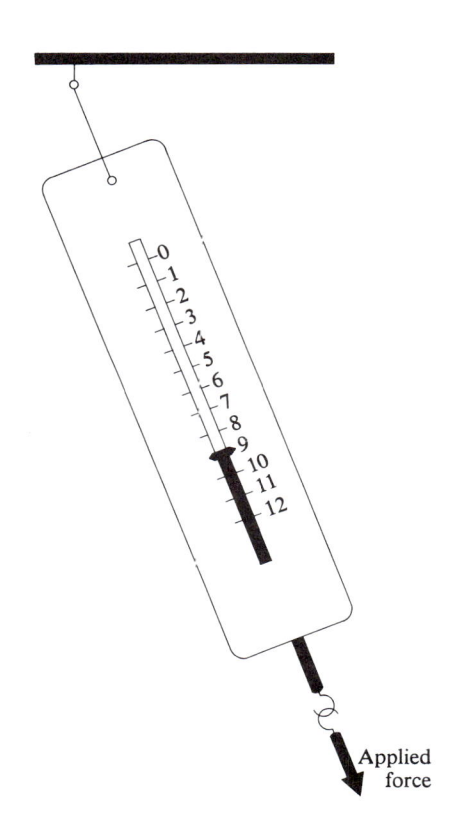

Fig. 2-16 A spring balance which can be used to define the magnitude and direction of force.

56 MOTION AND FORCE

The Law of Inertia

Profiting from Galileo's experiments and reflections, Newton took as his starting point what is called *Newton's first law* or the *law of inertia: Every body continues in its state of rest or of uniform motion in a straight line, except insofar as it is compelled by forces to change that state.* We now avail ourselves of some of the concepts and terms we have belabored earlier in this chapter. We rephrase this law simply as follows: Zero resultant force implies that the velocity vector is constant and, conversely, if the velocity vector is constant the resultant force is zero.

This innocent "law" is by no means obvious! In fact, it seems to fly in the face of everyday experience with objects that sooner or later stop when the push or pull is removed. Without careful analysis one tends to think, as Aristotle did, that rest is the "natural state" and that force is required to maintain motion. Galileo and Newton asserted that *constant velocity* is the natural state of affairs and that only *changes in velocity* have to be explained by the presence of a (resultant) force. "Rest" is just one special kind of constant velocity. A body has an inherent tendency to keep its velocity vector constant: we say the body has inertia.

Of course, direct experimental studies of motion without force are frustrated by ever-present frictional effects. However, the tendency of a body to keep moving with a constant velocity does become apparent in situations, like that of a puck sliding along the ice, where frictional forces are small. Nevertheless, one has to idealize, as Galileo did, to arrive at this law of inertia. Keeping this fact in mind should permit us to be quite charitable toward Aristotle and his misconception!

Motion with constant velocity, when there are forces acting, occurs frequently. Before you can accept the idea that cars, airplanes, and track stars, when moving with constant velocity, do indeed have *zero resultant force* acting upon them, you must learn to think of *all the forces acting on the body* in question. A runner has a natural inclination to concentrate on the force exerted by his feet on the track. But in fact his acceleration will be determined by the horizontal forces exerted by the track on his feet and by the air on his body, and if his velocity is constant these two force vectors do add up to zero resultant force.

We will come back to the law of inertia when we examine the laws of physics in different frames of reference. For the moment, let us remark that the law of inertia is not true if you use a car as the frame of reference while the brakes are being applied. Just think of how free objects are accelerated forward into the dashboard! In all our work, until you are notified to the contrary, we will assume that our frame of reference for the measurement of position, velocity, and acceleration is a frame in which the law of inertia is true: such a frame is called an *inertial frame of reference.*

Newton's Second Law and Mass

Newton's second law, $\vec{F} = m\vec{a}$ [Eq. (2.10)], focuses attention on the *acceleration vector*, the rate of *change* of velocity rather than the velocity. It states that for a given object the resultant force vector and the acceleration vector are parallel, and one determines the other. (In some situations you may find it convenient to speak of the resultant force as the cause of the acceleration, while in other cases it may be natural to think of the acceleration causing the resultant force. But really, force and acceleration are related by $\vec{F} = m\vec{a}$, and that is that.)

A remarkable feature of Newton's second law is that to any object we can assign one single number called *mass*, and that this number completely characterizes the relationship between the resultant force and the acceleration for that object. For a given object, by measuring the resultant force (in newtons) and the associated acceleration (in m/sec²), it is found that the ratio F/a is constant, and this constant is defined as the mass m of the object in kilograms.

Example 4 What is the mass of a car which has an acceleration of 2.0 meters/sec² when the resultant force on the car is 2.6×10^3 N?

Solution The mass equals F/a, hence

$$m = (2.6 \times 10^3 \text{ N})/(2.0 \text{ meters/sec}^2)$$
$$= 1.3 \times 10^3 \text{ kg.}$$

Now, knowing the mass of an object, one can find the acceleration produced by a given resultant force.

Example 5 Suppose the car of Example 4 were subjected to a resultant force of 3.9×10^3 N. What would be the car's acceleration?

Solution The magnitude of the acceleration vector $\vec{a}$ can be determined by rearranging the equation $F = ma$ and solving for a:

$$a = \frac{F}{m} = \frac{3.9 \times 10^3 \text{ N}}{1.3 \times 10^3 \text{ kg}} = 3.0 \text{ meters/sec}^2.$$

At this point we would ask the reader to reconsider these two examples from the standpoint of the units involved. As we have defined it, mass is determined by a ratio F/a and the unit of mass is the N/(m/sec²). It is convenient to give this cumbersome term a new name, the *kilogram*.* That is, by definition,

$$1 \text{ kg} = 1 \text{ N/(m/sec}^2).$$

Therefore

$$1 \text{ N} = 1 \text{ kg m/sec}^2,$$

and

$$1 \text{ N/kg} = 1 \text{ m/sec}^2.$$

---

*In actual scientific practice, the standard of mass, designated as one kilogram, is the mass of a cylinder of platinum-iridium which is kept at the International Bureau of Weights and Measures at Sèvres in France. The newton is then defined by $1 \text{ N} = 1 \text{ kg m/sec}^2$. We have chosen to introduce force (and the newton) before mass (and the kilogram) for pedagogical reasons.

MOTION AND FORCE

The rules for the manipulation of the units in Examples 4 and 5 are evidently a simple consequence of the definition of the kilogram. This practice of introducing a new name for an awkward conglomeration of familiar units will be repeated as we develop the subject.

One final point regarding the equation $F = ma$: If we apply forces of the same magnitude F to different masses, the larger mass will have the smaller acceleration. For instance, a 50-N force gives a 5.0-kg mass an acceleration of 10 m/sec^2, while the same 50-N force would give a larger mass of 25 kg an acceleration of only 2.0 m/sec^2. We say that the larger mass has greater *inertia*, and that mass is a *quantitative measure of inertia*. The rather vague term *inertia* refers to the tendency of an object to continue to move at the same speed in the same direction, that is, to maintain a constant velocity vector.

The mks System

As a postscript to our discussion of force, mass, and acceleration, we point out that the units used throughout this book are those of the *mks system*,* so named because the fundamental units employed are the *meter* for length, the *kilogram* for mass, and the *second* for time. Among the units of this system are the *newton* as the unit of force, the *joule* as the unit of work, the *watt* as the unit of power (see Table 4-2) and the *coulomb* as the unit of electric charge.

The mks system is widely employed in the world of science and also of electrical engineering. Of course there are other systems of units also in common use: for example, the engineering system in which force is expressed in *pounds* and length in *feet* (Table 4-2). To preclude such man-made complications diverting the reader from concentration on the principles of physics, we will employ only one system of units, the mks system.

*What we have called here the mks system is essentially similar to the Système International d'Unités (SI) adopted at the Eleventh General Conference of Weights and Measures in 1960. However, we have retained the American spelling of meter and kilogram instead of the recommended metre and kilogramme. For further details on this subject of units the reader is referred to the book by Theodore Wildi cited in the Suggested Reading at the end of this chapter.

2.5 GRAVITATIONAL FORCES

Many of the implications of Newton's second law, $\vec{F} = m\vec{a}$, are brought out by examining the motion of an object that is subjected to just one force which then is itself the resultant force $\vec{F}$. Motion determined by a gravitational force is as interesting a starting point in this space age as it was in Newton's time.

Gravitational Field $\vec{g}$

In the presence of any massive object like our planet earth, any other mass m, say a marble, experiences an attractive force called a gravitational force. Measurement of the gravitational force exerted on different masses m at a given location shows that this gravitational force is strictly proportional to this mass m; that is, we can write for the gravitational force $\vec{F}$,

$$\vec{F} = m\vec{g}, \qquad (2.11)$$

where the proportionality factor is denoted by a vector $\vec{g}$ in the direction of $\vec{F}$. *This vector $\vec{g}$ is called the gravitational field at the point in question and we say the gravitational field $\vec{g}$ exerts a force $\vec{F} = m\vec{g}$ on the mass m.* This gravitational force $\vec{F} = m\vec{g}$ has a name that is very familiar: it is called the *weight* of the object.

At different points on the earth or in "space," $\vec{g}$ generally has a different magnitude and a different direction, as shown in Fig. 2-17. This figure gives us an idea of the gravitational field that will be encountered at different points in the vicinity of our planet. Notice that $\vec{g}$ always points to the center of the earth and that the magnitude of $\vec{g}$ decreases as one moves away from the earth. Near the earth's surface the magnitude of $\vec{g}$ is approximately 9.8 N/kg, varying slightly from place to place. At an altitude of 4000 miles, $g = 2.45$ N/kg, merely one fourth of its value at the earth's surface. Thus, an astronaut whose mass is 68 kg will weigh (68 kg) × (9.8 N/kg) = 6.7 × 10² N (this is about 150 pounds force) at the earth's surface, and at an altitude of 4000 miles his weight will have decreased to (68 kg) × (2.45 N/kg) = 1.67 × 10² N (37.5 pounds force).

The important point to notice is that the mass m of an object does not depend on where the object is located, but its weight $m\vec{g}$, being just the gravitational force exerted by the gravitational field $\vec{g}$ on the object, decreases if the object is moved to a position where the gravitational field $\vec{g}$ is smaller. By moving to a high altitude someone who is overweight can indeed decrease his weight, but this probably is not really what he is interested in; his mass will not be changed by the trip, and there will be just as much of him as before.

Sources of Gravitational Fields

We have learned what a gravitational field $\vec{g}$ at a certain point in space does. It exerts a force $m\vec{g}$ if an object of mass m is placed at that point. Now what is the source of the gravitational field? The answer is that other objects produce this field $\vec{g}$ according to the following rule:

A particle of mass M produces at a distance r a gravitational field $\vec{g}$ of magnitude

$$g = \frac{GM}{r^2}. \qquad (2.12)$$

The vector $\vec{g}$ points toward the particle that produced it, as shown in Fig. 2-18. G is called the *gravitational constant*. Its value has been measured to be 6.67×10^{-11} N m²/kg² in very difficult and delicate

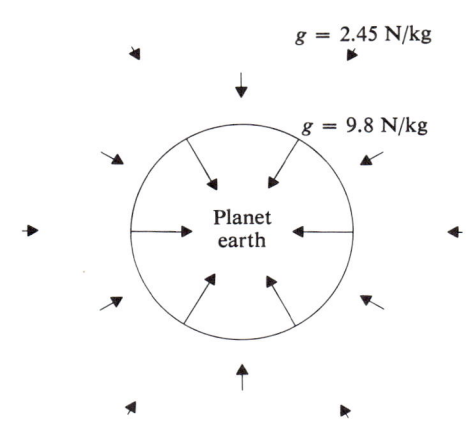

Fig. 2-17 The gravitational field produced by the earth at several points in space.

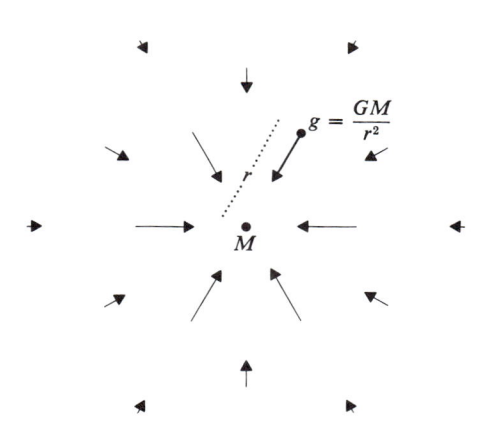

Fig. 2-18 The gravitational field vector at several points in space due to a particle of mass M. (Refer also to Fig. 2-24 showing an electric field.)

60 MOTION AND FORCE

experiments. This experiment was first performed by Lord Cavendish (1731–1810) in 1798: he measured the very feeble gravitational force exerted by one laboratory-size object on another. (The experimental arrangement involved is quite similar to that employed by Coulomb in determining the force between electrical charges as described later in this chapter.)

Notice that, according to Eq. (2.12), the field $\vec{g}$ produced by a particle of mass M is large (physicists often say strong) at points near M, and small (or weak) at points far from M. Also at a given distance, the larger the mass M the stronger the field $\vec{g}$.

We also need to know, of course, how to find the field at a given point when there are many sources contributing. The rule is simple: just compute the field that each particle would produce at the point in question as if that particle were the only source; then the field $\vec{g}$ at this point is the vector sum of such contributions from all the particles. This rule is known as the *superposition principle* for gravitational fields, and one says that the field $\vec{g}$ is the superposition of the contributions produced by the individual particles. This idea is demonstrated in Fig. 2-19. Superposition principles similar to this are found in many other branches of physics; for example, we will encounter such a principle again in our study of waves in Chapter 10. (For gravitational fields, the superposition principle is a consequence of the law of vector addition of gravitational forces.)

Armed with the superposition principle, we have a method of predicting the gravitational field that will be produced by any distribution of particles; for instance, all the particles that comprise the planet earth. Newton, after considerable labor on a problem which calculus students now find rather routine, proved that the gravitational field produced at external points by mass distributed uniformly over a *spherical* shell is the same as if the entire *mass* of this shell were located at its center. Therefore, to the approximation that the earth can be regarded as an assembly of concentric uniform spherical shells, the gravitational field outside the earth can be computed as if its entire mass M_e were located at its center. With this knowledge, we can use the expression for particle sources, Eq. (2.12), and so say that the earth's field at a distance r from the earth's center is

$$g = \frac{GM_e}{r^2}$$

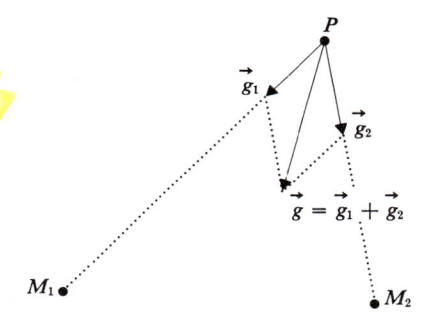

Fig. 2-19 The superposition principle. Mass M_1 alone would produce field $\vec{g}_1$, at point P; M_2 alone would produce $\vec{g}_2$. When both are present the field is the vector sum of $\vec{g}_1$ and $\vec{g}_2$.

for r not less than the radius r_e of the earth. This equation describes the gravitational field depicted in Fig. 2-17. Denoting by g_s, the gravitational field at the earth's surface where $r = r_e$, we obtain

$$g_s = \frac{GM_e}{r_e^2} \qquad (2.13)$$

Now we notice a remarkable thing. We have before our eyes a method of determining the mass of the earth! In Eq. (2.11) we know, from measuring the force on unit mass at the surface of the earth, that $g_s = 9.8$ N/kg; the radius of the earth r_e can be found, using surveying data and the geometry of a sphere, to be 6.37×10^6 m (approximately 4000 miles); the Cavendish experiment and later improvements give $G = 6.67 \times 10^{-11}$ N m²/kg². Inserting these numbers in Eq. (2.13) and solving for M_e, we find the mass of the earth is 5.97×10^{24} kg.

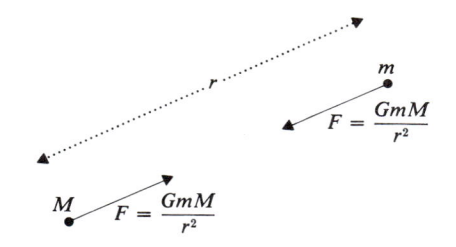

Newton's Law of Gravitation

To determine the gravitational force exerted by a particle of mass M on another particle of mass m we think about the problem in two distinct stages. First, we think of M as a source of a gravitational field at all points in space and determine the field that M produces at the location of the other mass m which is a distance r away. According to Eq. (2.12) this field has a magnitude

$$g = \frac{GM}{r^2} \qquad (2.12)$$

and a direction towards the source M. This field $\vec{g}$ exerts a force of magnitude

$$F = mg$$

on the mass m. Substituting for g the value given by Eq. (2.12) we get

$$F = G\frac{Mm}{r^2}. \qquad (2.14)$$

This is the magnitude of the force exerted by M on m; it acts on m and is directed toward M so it is an attractive force. To get the force exerted by m on M we compute the field produced by m at the location of M as $g = Gm/r^2$, and this field exerts a force $\vec{F} = M\vec{g}$ on M so that the magnitude of the force on M is $F = GMm/r^2$; the direction is toward the source m. Notice that the force exerted by m on M is equal in magnitude and opposite in direction to the force exerted by M on m (see Fig. 2-20). Forces with this property are said to obey *Newton's third law*, a topic to which we shall return later. The rule for the gravitational force exerted by one particle on another, expressed by Eq. (2.14), is called *Newton's law of gravitation*, since this was the gravitational force law assumed by Newton in his successful attempt to understand planetary motion.

The law need not be restricted to particles. It turns out to be true if M and m are the masses of any two spherically symmetrical distributions of mass with a distance r between their centers. The gravitational force between two objects is extremely weak unless at least one of the objects has a huge mass, like the mass of our planet.

Fig. 2-20 Newton's law of gravitation.

62 MOTION AND FORCE

Example 6 What is the gravitational force exerted by a 0.0100-kg marble on a 10.0-kg cannon ball which is 0.100 m away?

Solution We use Eq. (2.14) and substitute the appropriate numbers:

$$F = \frac{(6.67 \times 10^{-11} \text{ N m}^2/\text{kg}^2) \times (10.0 \text{ kg}) \times (0.0100 \text{ kg})}{(0.100 \text{ meter})^2}$$
$$= 6.67 \times 10^{-10} \text{ N}.$$

As mentioned earlier, using a very sensitive torsion balance, Cavendish managed to measure such a minute force in a similar situation and then, knowing F, m, M, and r, he could compute the gravitational constant G from Newton's law of gravitation, $F = GMm/r^2$ [Eq. (2.14)].

2.6 MOTION OF A PARTICLE IN A GRAVITATIONAL FIELD

We are now in a position to treat, at one fell swoop, the motion of a planet about the sun, the orbit of an earth satellite, or simply the familiar trajectory of a baseball as it leaves a bat and arcs toward the outfield. Let us consider a particle of mass m which is acted upon only by a gravitational field $\vec{g}$ and so experiences a resultant force $m\vec{g}$.

According to Newton's second law, this force equals the mass times the acceleration vector $\vec{a}$:

$$m\vec{g} = m\vec{a}.$$

We have the simple result that

$$\vec{a} = \vec{g}.$$

This tells us that at any point along the path of the particle, its acceleration is exactly given by the value of $\vec{g}$ at that point so that if we know $\vec{g}$ everywhere we also know the acceleration. Besides being called the gravitational field, we see that $\vec{g}$ can also be called the *acceleration due to gravity*.

Various Trajectories

Now let us consider the motion of this particle of mass m in a gravitational field produced by a fixed particle of mass M (or a spherically symmetric distribution of mass M). This field, as we already stated in Eq. (2.12), is given by

$$g = \frac{GM}{r^2}.$$

The source M, its field $\vec{g}$ at various points in space, and the particle m are shown in Fig. 2-21. Now what will be the motion of m? As we have stressed, the force $m\vec{g}$ acting on m determines the acceleration $\vec{a} = \vec{g}$ but this, by itself, is not enough to determine the motion. We must also know the velocity vector of m at some position; this information is usually called the starting or initial conditions which comprise the initial velocity $\vec{v}_1$ and the initial position $\vec{r}_1$. Knowing the initial conditions we can trace out the trajectory (as in Fig. 2-12) by finding the change in position vector, $\Delta \vec{r}_1 = \vec{v}_1 \Delta t_1$, in the first small time interval Δt_1 during which the change in velocity is $\Delta \vec{v} = \vec{a} \Delta t_1$. For the next small time interval the change in position vector is $\Delta \vec{r}_2 = (\vec{v}_1 + \Delta \vec{v}) \Delta t_2 = (\vec{v} + \vec{a} \Delta t_1) \Delta t_2$. Knowing the initial conditions $\vec{r}_1$ and $\vec{v}_1$ and knowing $\vec{a}$ everywhere, we can trace out the path. In Fig. 2-21 you can see the orbits that result from different starting speeds v_1. For the first path, the starting speed is zero, that is, the particle is released at rest. It falls in a straight line in the direction of $\vec{g}$ with increasing acceleration as it encounters stronger fields, and the velocity therefore increases rapidly until impact with M.

The second trajectory corresponds to projection with a low speed in a direction perpendicular to $\vec{g}$; the path is curved and the speed increases until impact. The curved path can be shown to be a portion of a curve known as an ellipse, one of the famous conic sections studied by Greek mathematicians. The three basically different ways to slice a cone (Fig. 2-22) produce three curves: the ellipse, the parabola, and the hyperbola.

The third orbit results from larger initial speed. The orbit is an ellipse, and the particle goes around this orbit forever (assuming, of course, no frictional retarding force). The point named *perigee* is the point of closest approach and here the orbital speed is a maximum. Minimum speed is attained at *apogee*, the starting point for this orbit where the distance from M is the greatest. The velocity and acceleration vectors are shown at a typical point Q_1. At this point the acceleration vector has a component in the direction of $\vec{v}$, the velocity vector, so the speed is increasing. However, at point Q_2, the acceleration vector has a component in the direction opposite to the velocity, so the speed is decreasing.

For a moment, let us think of M as the earth and m as a satellite. A question often asked is "What keeps the satellite up?" The question seems to assume that there is a law of nature that things naturally move down unless there is some force holding them up. But this is not so! Here the key physical law, $\vec{F} = m\vec{a}$, tells us that the *acceleration* is always directed downward (toward the earth's center) but the direction of the acceleration is not necessarily the direction in which the satellite goes; instead, the acceleration determines the direction of the *change in velocity*, $\Delta \vec{v}$. And the satellite moves not in the direction of $\Delta \vec{v}$, but in the usually quite different direction of the velocity vector $\vec{v}$. The path is indeed deflected in the direction of the force, but unless $\vec{v}$ is initially in the same or opposite direction to the force, the velocity vector does not become aligned with the force: *the satellite does not simply follow a path in the direction of the force.* Instead of erroneously thinking that

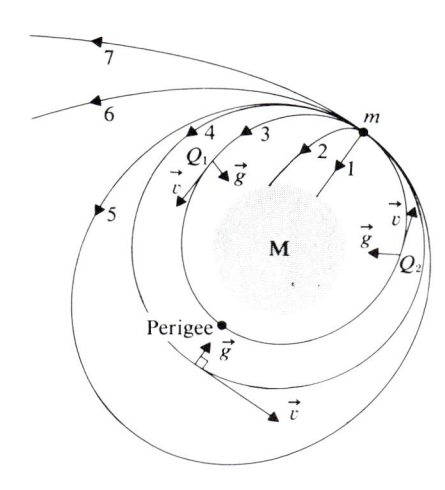

Fig. 2-21 Orbits of the mass m when projected perpendicular to $\vec{g}$ with different speeds.

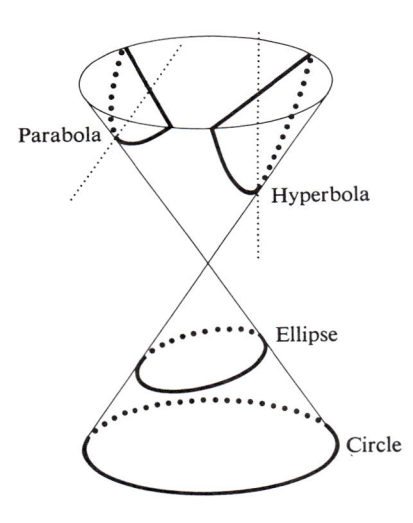

Fig. 2-22 The conic sections. The ellipse, the hyperbola, and the parabola are the curves formed by the intersections of planes at various angles with a cone. The circle and a segment of a straight line are special cases of the ellipse.

the natural thing for any such object is to "fall down to the ground," one should realize that all such objects are, in fact, "in orbit," and just what shape an orbit has will depend on the initial conditions. Of course, some narrow orbits like paths 1 and 2 of Fig. 2-21 happen to be intercepted by the attracting mass M (and that is indeed very convenient for tennis players and baseball players, to say the least!).

Other sources of misconception are popular science accounts which speak of putting orbiting H-bombs into outer space, ready to drop on signal! Well, suppose an H-bomb were strapped to a satellite and then released at some point in its orbit. Then the position and velocity of the H-bomb are just right for it to continue in the same orbit as the satellite, and so it will! To make it drop from rest straight down, it would have to be ejected backwards with a speed, relative to the satellite, equal to the satellite's speed in orbit. Then the H-bomb's initial velocity would be zero and its trajectory would be that of path 1 in Fig. 2-21.

Confusion also exists about the interesting phenomenon called "weightlessness" which is displayed by astronauts floating along, without any support, in the middle of their capsule. There is still a gravitational force $m\vec{g}$ acting on an astronaut of mass m and so he still has *weight*, as we have defined the term. But with no other forces acting on him, he experiences the same acceleration $\vec{g}$ that the capsule experiences, and with the same velocity as the capsule, he follows the same orbit. No forces arising from contact with the capsule are required to keep the astonaut moving in the same way the capsule moves. (We shall return to this subject and discuss it from another point of view when we treat accelerated frames of reference.)

Back to our general problem: the orbits illustrated in Fig. 2-21. These orbits are obtained with increasing initial speeds: orbit 4 is a circle, which is a special case of an ellipse; orbit 5 is an ellipse with the initial position as perigee; trajectory 6 is the parabolic path followed by m when it is projected with just enough speed (the so-called escape velocity) to continue moving away from M never to return; trajectory 7 is a hyperbolic escape path resulting from projection with a speed greater than the escape velocity. Later, when we discuss the law of conservation of energy, we will show how the escape velocity can be found.

Circular Orbits

The motion in the circular orbit (path 4 in Fig. 2-21) is particularly simple. Because the acceleration vector is always perpendicular to the velocity vector, the velocity vector changes in direction but not in magnitude, which is another way of saying the speed v is constant. The acceleration vector has only a centripetal component and, recalling the relation for the centripetal acceleration given earlier, $a = v^2/r$, we can deduce an expression for the speed in a circular orbit. Let us go back to Newton's second law, $\vec{F} = m\vec{a}$, applied to the motion of the mass m. In this equation, the magnitude of the resultant force $\vec{F}$ is given by Newton's law of gravitation as

$$F = G\frac{mM}{r^2}, \qquad (2.14)$$

and the magnitude of the acceleration by $a = v^2/r$, so Newton's second law yields

$$G\frac{mM}{r^2} = m\frac{v^2}{r}.$$

Note this equation is simply of the form $F = ma$.* Now multiplying both sides of the equation by r and then taking the square root of each side, we find

$$v = \sqrt{\frac{GM}{r}}. \qquad (2.15)$$

*The term centripetal force *is a name given to whatever force produces the centripetal acceleration. In* this example the centripetal force is the gravitational force acting on the mass m.

Example 7 What is the orbital speed of a satellite, at an altitude of three hundred miles?

Solution We know that $r = 4.0 \times 10^3$ miles $+ 0.3 \times 10^3$ miles $= 4.3 \times 10^3$ miles $= 7.0 \times 10^6$ meters, and that the mass of the earth $M = 6.0 \times 10^{24}$ kg. Using Eq. (2.15), we compute the speed in orbit to be

$$v = \sqrt{\frac{(6.67 \times 10^{-11} \text{ N m}^2/\text{kg}^2) \times (6.0 \times 10^{24} \text{ kg})}{7.0 \times 10^6 \text{ meters}}}$$

$$= 7.5 \times 10^3 \text{ meters/sec}.$$

This is approximately 17,000 miles per hour.

Inspection of Eq. (2.15) reveals that the larger the radius of the orbit, the smaller the orbital speed v. The moon, at a radius of 2.4×10^5 miles, has an orbital speed less than 1/7 of the speed of the satellite we have considered.

The calculation of the time T required to go once around a circular orbit (whose circumference is $2\pi r$) is straightforward because the speed v is constant.

Example 8 What is the period T of the satellite of Example 7, with a speed of 7.5×10^3 meters/sec?

Solution The circumference is 4.4×10^7 meters; hence

$$T = (4.4 \times 10^7 \text{ meters}/7.5 \times 10^3 \text{ meters/sec})$$
$$= 5.9 \times 10^3 \text{ sec}$$
$$= 1.64 \text{ hour}.$$

For the moon, this period T is nearly one month. And, in general, one can write

$$T = \frac{\text{distance}}{\text{speed}}$$

$$= \frac{2\pi r}{v}$$

$$= \frac{2\pi r}{\sqrt{\frac{GM}{r}}}$$

where we have substituted for v the expression given by Eq. (2.15). By squaring both sides of the equation, our last result can be rewritten in the form

$$T^2 = \left(\frac{4\pi^2}{GM}\right) r^3 \tag{2.16}$$

which shows that the "square of the period T is proportional to the cube of orbit radius." This result suggests a method of determining the mass M of any planet which has a moon. It is only necessary to determine, by astronomical observation, the moon's period T and the radius of its orbit r, and then we can calculate the planet's mass M from Eq. (2.16).

All of our discussion of the situation depicted in Fig. 2-21 can be applied to many different physical situations. But two remarks are in order. First, we have assumed a frame of reference in which Newton's second law is valid; such a frame can be pictured to have its axes always pointing toward certain distant stars and its origin such that these stars have no acceleration relative to this frame of reference. So in looking at Fig. 2-21, when thinking of M as the earth and m as an earth satellite, you would see the earth spinning and completing one revolution per day. The second assumption we have made is that M is fixed in our frame of reference. This is a good approximation only if M is much greater than m so that M experiences only a very small acceleration because of the force exerted on it by m.

Kepler's Laws

The tale for the case in which M represents the sun and m a planet is vital for modern space science and also was certainly of great historical importance. Why? Observations of the stately planetary motions had been made for thousands of years. The description of a planet's trajectory relative to an earth-centered frame is complex. A simple description of roughly circular orbits about a fixed sun, known in ancient times, was rediscovered by Copernicus. This heretical idea of demoting mankind's

Courtesy of Burndy Library

Tycho Brahe
(1546–1601)

*The mathematical support Kepler provided for the Copernican idea of the heliocentric solar system places Kepler in the front ranks of the great scientists. The stature of his work may be indicated by the fact that Dec. 27, 1971, the quarcentenary of his birth was declared a "UNESCO event." More on Kepler appears in Chapter 15.

home from the center of the universe to just one of many minor planets led to fierce disputes which were ultimately settled, not by the decisions of authorities but by the lonely efforts of the few who were willing to make precise experimental measurements and apply unfettered human reason to the best of their ability, all this in the climate of political, religious, and social hostility. No Nobel prizes for them (even had these been instituted in those days)! Years of precise observations of planetary positions were tabulated by Tycho Brahe (1546–1601), and this data was given a neat mathematical description by Johannes Kepler (1571–1630) in the form of three laws.*

1. The planets move around the sun in orbits which are ellipses, with the sun at one focus.
2. The radius vector (from sun to planet) sweeps over equal areas in equal times.
3. The squares of the periods of revolution of the planets around the sun are proportional to the cubes of the (average) radii of their respective orbits.

Newton tested his new mechanics by showing that Kepler's empirical laws were all a consequence of $F = ma$ with $F = GmM/r^2$. You have seen the derivation of Kepler's third law, at least for circular orbits. In the discussion concerning Fig. 2-21 we stated (and Newton proved) that an orbit is either an ellipse, a parabola, or a hyperbola, and that the latter two orbits imply escape. The remaining law of Kepler, the second in our list, is a consequence of the fact that the gravitational force is directed along the line joining m and M. This leads to what is known as *the conservation of angular momentum*, which we shall discuss in Section 3.9.

Projectiles Near the Earth

An important part of Newton's idea was that the story for earth satellites like the moon was not different in principle from that for falling apples near the earth's surface—the apple just manages to travel over a very short portion of a very narrow orbit (in vertical fall the apple's orbit has shrunk to a straight line) before it bangs into the earth, or Newton's head, as the case may be. In Fig. 2-21, with M as the earth, if we take a tiny portion near the earth's surface and enlarge it greatly, we get a picture like Fig. 2-23 suitable for discussing the trajectories of apples, baseballs, and bullets. And happily, a simplification occurs since the gravitational field $\vec{g}$ does not change appreciably in magnitude over the entire region of interest. This is so because the percentage change in distance from the center of the earth is so small (0.1% for a 4-mile change in altitude). Over such a limited region we can consider the earth's surface as flat with $\vec{g}$ vertical with a constant value of approximately 9.8 N/kg. If the only force acting on a mass m is the gravitational force $m\vec{g}$ (its weight), then Newton's second law gives

$$m\vec{g} = m\vec{a}$$

so that the acceleration $\vec{a} = \vec{g}$, as we have seen before. The acceleration $\vec{a}$ is therefore always vertically downward and constant ($a = 9.8$ meters/sec²). Therefore, the change in the velocity vector $\Delta \vec{v}$ is always vertically downward, and this means that there is no change in the horizontal component of the velocity vector. So the horizontal motion is extremely simple; it is a motion with constant speed, and hence,

Horizontal distance traveled
$$= \text{(horizontal component of velocity)} \times \text{(time)}.$$

While the horizontal motion is proceeding, there is also a vertical motion with a changing vertical component of velocity but with a constant vertical acceleration of 9.8 meters/sec² downward. This vertical motion has been completely solved already—we have Eq. (2.4) and Eq. (2.5) or Eq. (2.6) at our disposal. To see how this works we consider the following example:

Example 9 Consider a baseball projected horizontally with a speed of 30 m/sec as in Fig. 2-23. Where will the ball be after 1.60 sec?

Solution Initially the horizontal component of the velocity is 30 m/sec. This component does not change; there is no horizontal acceleration (assuming that the air friction on the ball is negligible—in real life this is, of course, not so—nevertheless the frictional forces are relatively small and therefore our simplified analysis is not far from the truth). Therefore, the horizontal distance equals (30 meters/sec) × (1.6 sec) = 48 meters.

Now for vertical motion, we note that the vertical component of the velocity is initially zero since the ball was projected horizontally. The vertical acceleration is 9.8 m/sec² downward. Using Eq. (2.6), we find the vertical distance the ball has fallen is $\frac{1}{2} \times (9.8 \text{ meters/sec}^2) \times (1.60 \text{ sec})^2 = 12.5$ meters.

Knowing both the horizontal and vertical distances, we can locate the ball as shown in Fig. 2-23.

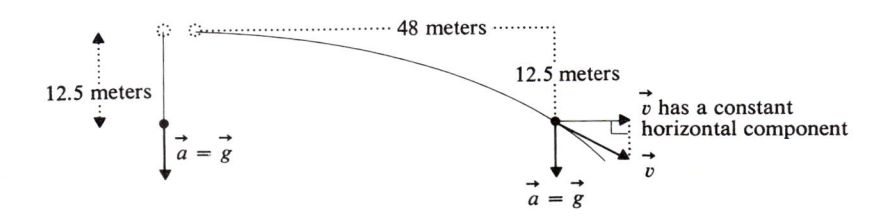

Fig. 2-23 The parabolic trajectory of a baseball with a constant acceleration $\vec{a} = \vec{g}$.

2.6 MOTION OF A PARTICLE IN A GRAVITATIONAL FIELD

It is clear that the method of Example 9 can be used to locate the ball at any time. And so one can trace out a projectile's path, which turns out to be a parabola. If a second ball, at the left of Fig. 2-23, were to be released from rest as the first is projected horizontally, the vertical motions of the two balls would be identical: they would both hit the ground at the same time.

2.7 ELECTRIC FORCES

In Chapter 1 we employed the fact that the attractive force exerted by the protons in the nucleus of an atom on the orbital electrons is an electric force. And the repulsive force exerted by one electron on another is also electrical. Therefore the configuration of electrons in an atom, and consequently the entire chemical behavior of atoms, is ultimately governed by electric forces.

In the study of electric forces we are aided by a very fortunate circumstance: There is a precise analogy between electric and gravitational forces. *The electric charge possessed by an object plays the same role in determining electric forces as does the object's mass in gravitational forces.* With the substitution of *charge* for *mass* one finds that there are really no new formulas to learn. All the understanding of gravitational matters can be readily carried over into the realm of electricity. That is not to say there is no difference between the gravitational and the electrical situation.

There is in fact a tremendous difference between electrical and gravitational phenomena. Two reasons for this are

1. The electric force exerted by one part of an atom on another part is much stronger than the feeble gravitational force with which that part presumably attracts the other.

2. There are *two* types of electric charge, *positive* and *negative*. Consequently, cancellation occurs and large objects can have a small net charge or be electrically neutral (zero net charge). Electric forces can be *repulsive* or *attractive*. On the other hand, only one type of mass occurs and gravitational forces are always attractive.

Electric Charge, the Coulomb

Recall that we denoted the charge on the proton by the symbol e and the electron's charge by $-e$. This quantity of charge would be the convenient unit of charge for calculations involving such particles. However, in the mks system a much larger amount of charge, named a *coulomb*, has been selected as the unit of electric charge. Just what laboratory procedures are specified to determine a charge in coulombs (coul) will be indicated later in this section.

Measurement of the charge e gives

$$e = 1.60 \times 10^{-19} \text{ coul},$$

so a collection of 6.2×10^{18} ($= 1/1.60 \times 10^{-19}$) protons has a total charge of 1.0 coul. The symbols Q or q will be used to denote amounts of electric charge.

Electric Field $\vec{E}$

For electric forces we use the same two-stage description that was employed for gravitational forces:

1. A source produces a field at all points in space.
2. The field at any location exerts a force on an electrically charged object at that location.

Analogous quantities are listed in Table 2-1.

With each point in space we associate a vector $\vec{E}$ which is called the *electric field* at that point. The source of an electric field is a (distant) electric charge. A stationary particle possessing a positive charge Q produces at a distance r an electric field $\vec{E}$ of magnitude, in mks units, given by

$$E = \frac{kQ}{r^2} \tag{2.17}$$

where the constant, $k = 9.0 \times 10^9$ N m²/coul², and the mks unit of E is the N/coul.

The electric field vector $\vec{E}$ points away from the positive charge that produced it (see Fig. 2-24). When the source is a negative charge, the field points toward the source.

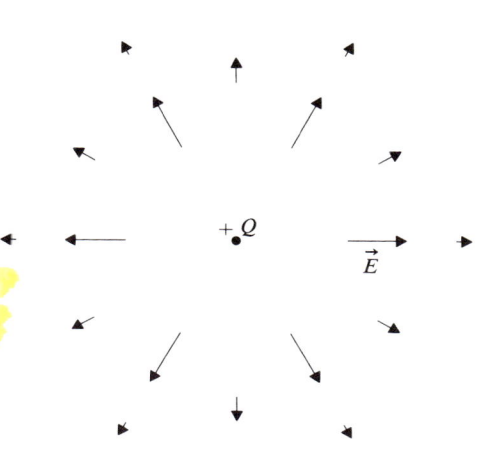

Fig. 2-24 Representation of the electric field due to a positive point charge. If the point charge were negative the arrows representing the electric field at the various points would be directed inward. (Compare this to Fig. 2-18, in which a gravitational field is shown.)

Table 2-1. ANALOGOUS QUANTITIES

Electrical	Gravitational
Electric charge q	Mass m
Electric field $\vec{E}$	Gravitational field $\vec{g}$
Electric force $\vec{F}$ exerted by $\vec{E}$ on q: $\vec{F} = q\vec{E}$	Gravitational force $\vec{F}$ exerted by $\vec{g}$ on m: $\vec{F} = m\vec{g}$
Source of electric field: charge Q	Source of gravitational field: mass M
Particle of charge Q produces at distance r an electric field of magnitude $E = \dfrac{kQ}{r^2}$	Particle of mass M produces at distance r a gravitational field of magnitude $g = \dfrac{GM}{r^2}$
Coulomb's law: $F = \dfrac{kQq}{r^2}$	Newton's law of gravitation: $F = \dfrac{GMm}{r^2}$

According to Eq. (2.17), the electric field created by the source charge Q is strong at points near the source (r small) and is weak far from the source. The value of the constant, $k(= 9.0 \times 10^9$ N m²/coul²) is determined by our choice of the size of the unit of charge, the coulomb, as we shall see later.

When many different charges contribute to the electric field at a given point in space, this electric field vector is simply the *vector sum* of the fields that each source would individually contribute, if each were the only source. This is the *superposition principle* for electric fields.

In electrical work one often has parallel plates (Fig. 2-25) with positive charge uniformly distributed over the surface of one plate and an equal amount of negative charge on the other plate. Computing the superposition of the contributions from each of these sources, we find that there turns out to be the same field $\vec{E}$ at all points between the plates; that is, $\vec{E}$ is uniform, except in the regions which are very close to the edges of the plates. Notice that $\vec{E}$ points away from the positive charges toward the negative charges.

Now suppose that at a certain point in space, we know the field $\vec{E}$ produced by some distant sources. What does $\vec{E}$ do? Well, if there happens to be a charge q at the point in question, then $\vec{E}$ exerts a force $\vec{F}$ on q given by

$$\vec{F} = q\vec{E}. \quad (2.18)$$

This force, named the *electric force*, is in the direction of $\vec{E}$ if the charge q is positive, but is in the opposite direction if q is a negative charge.

Example 10 The charged plates of Fig. 2-25 produce a uniform electric field between the plates. $\vec{E}$ points vertically upward and has a magnitude given by

$$E = 200 \text{ N/coul}.$$

(a) Find the electric force exerted on a proton placed in the region between the plates.
(b) Find the force exerted on an electron placed in the region between the plates.

Solution
(a) The proton has a charge

$$q = +e = +1.6 \times 10^{-19} \text{ coul}.$$

The electric force exerted by the electric field $\vec{E}$ on this charge q has a magnitude

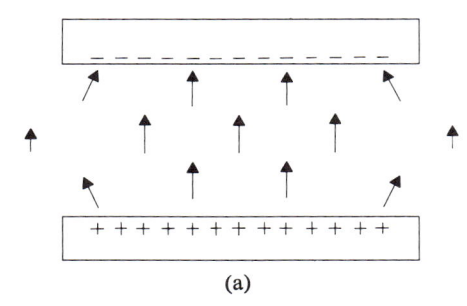

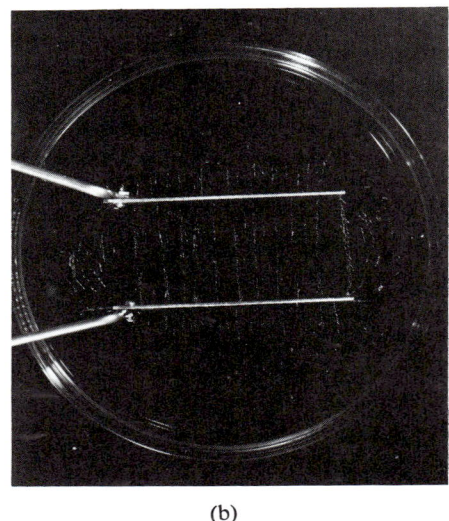

$$\begin{aligned} F &= qE \\ &= (1.6 \times 10^{-19} \text{ coul}) \times (200 \text{ N/coul}) \\ &= 3.2 \times 10^{-17} \text{ N}. \end{aligned}$$

Fig. 2-25 The uniform electric field $\vec{E}$ in the region between charged parallel plates.
(a) Schematic showing $\vec{E}$ vectors.
(b) Electric field displayed by means of grass seeds floating in oil. The magnitude of $\vec{E}$ is about 10^5 N/coul. The potential difference (Chapter 11) employed was 5 kilovolts. (Photo courtesy R. H. Parker, B. C. Institute of Technology.)

Since the proton's charge is positive, this electric force is in the same direction as the electric field.

(b) The electron has a charge

$$q = -e = -1.6 \times 10^{-19} \text{ coul.}$$

The electric force exerted by $\vec{E}$ on the electron has a *magnitude*, $F = 3.2 \times 10^{-17}$ N, as in the case of the proton. But because the electron's charge is negative, this electric force is in the opposite direction to that of the electric field. *Remark:* When we place a proton (or electron) between the plates, there is a new force exerted on the charges which reside on the plates. These charges may then alter their positions on the plates, and consequently the field $\vec{E}$ that they produce at the location of the proton will be a little different from the previous value, 200 N/coul. This effect is small unless the proton is very close to a plate.

A formal definition of the electric field $\vec{E}$ at a point in space is obtained from Eq. (2.18):

$$\vec{E} = \vec{F}/q$$

where $\vec{F}$ is the electric force that a charge q experiences when placed at the point in question. This "test charge" q should be small enough so that it has a negligible influence on the charges that are producing the field.

The key assertion of a *field* theory of gravitational or electric (or magnetic) forces is that the force acting on an object is determined by the field at the location of the object. Many different arrangements of sources could have produced this field. But, if the field is known, the discussion of the forces on an object can proceed without explicit reference to the field sources. For instance, the force on an electric charge in a television antenna is determined by the electric field at the location of the antenna. Knowing this field, it is not necessary to keep in mind the pertinent source charges in the TV broadcasting station and in the adjacent earth, trees, and roofs.

The field concept was introduced into studies of electricity and magnetism by Michael Faraday (1791–1867), who started his scientific career as a lowly, poorly paid laboratory assistant and ultimately became perhaps the greatest experimental scientist of all time, credited with fundamental discoveries ranging throughout chemistry and physics. The notion of electric and magnetic fields is central to the unification of theories of light, electricity, and magnetism and the consequent discovery of radio waves, the topics of Chapter 14. To this very day, the idea of a field is the fundamental primitive concept in the most successful description of nature that physicists have created—quantum electrodynamics. The field idea has staying power.

Coulomb's Law

Analogous to Newton's law of gravitation, $F = GMm/r^2$ [Eq. (2.14)], there is a law giving the electric force F exerted by a particle of charge Q on another particle with charge q. The charge Q sets up at all points in space an electric field. At a distance r from Q, this field has a magnitude

$$E = \frac{kQ}{r^2}. \tag{2.17}$$

A charge q at such a point (Fig. 2-26) experiences a force

$$\vec{F} = q\vec{E}, \tag{2.18}$$

where the magnitude of the force is, of course, given by

$$F = qE.$$

Substituting for E the value given by Eq. (2.17), we obtain

$$F = \frac{kQq}{r^2}. \tag{2.19}$$

This result is known as *Coulomb's law*.

If Q is positive, $\vec{E}$ points away from Q. Now if q is positive, the force exerted by $\vec{E}$ is in the direction of $\vec{E}$, that is, away from Q. In other words, a positive charge q is repelled by a positive charge Q. The reader can easily verify that the force directions are correctly summarized by the statement: *like charges repel and unlike attract*.

Like gravitational forces, Coulomb forces obey *Newton's third law* in that the force exerted on Q by q is equal in magnitude and opposite in direction to the force exerted on q by Q.

In Coulomb's law, we have facts that can be subjected to direct experimental test, in contrast with the field statements, Eqs. (2.17) and (2.18). And historically Coulomb's law came first; only later was the idea introduced that one charge produces a *field* which then exerts a force on the other charge. We have presented the fruitful field idea first because of the advantage of acquiring the habit of thinking about physical phenomena in terms of fields.

Coulomb's Experiments

Although the Greeks knew as long ago as 600 B.C. that amber, when rubbed, temporarily acquired some attracting force, a quantitative understanding of this force awaited the experiments of Charles A. Coulomb (1736–1806) in 1785.

Coulomb employed a torsion balance (see Fig. 2-27). The two pith balls, A and B, were initially in contact; then both were given an electrical charge of the same sign, whereupon the movable ball B was repelled by the fixed ball A.

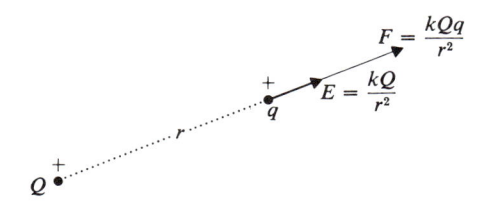

Fig. 2-26 The Coulomb force, $F = \frac{kQq}{r^2}$, exerted by Q on q.

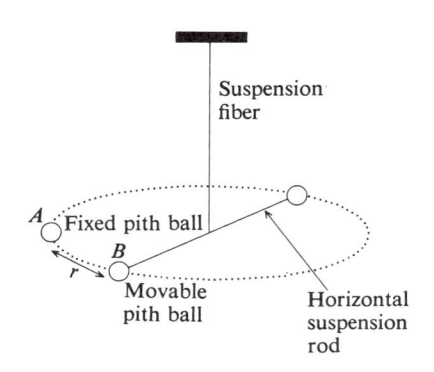

Fig. 2-27 Coulomb torsion balance.

The ball *B* came to rest when the twisting effect of the electrical force of repulsion between *B* and *A* was equal to the opposing twisting force exerted by the suspension fiber as it was disturbed from its normal equilibrium condition by the rotation of the horizontal suspension rod.

Coulomb had earlier carried out studies which showed that the angle through which such a suspension fiber is twisted (providing a certain limit is not exceeded) is directly proportional to the twisting force applied to the ball *B* on the end of the horizontal suspension rod. Consequently, he was able to relate the electrical repulsion force (which balanced the restoring force or twisting force exerted by the suspension fiber when the ball *B* was at rest) to the angle of twist, an angle which could be readily measured. He found that if he wanted the balls *A* and *B* closer together, he needed to twist the suspension fiber more. (With reference to Fig. 2-27, the twist applied at the top of the suspension fiber would have to be clockwise, as seen from the top, to bring *B* closer to *A*.) By means of this experiment, Coulomb was able to conclude that, with a given electrical charge on *A* and *B*, the electrical force of repulsion varied inversely as the square of the distance *r* between the centers of the balls. In other words, if *r* were reduced to half its original value, the force would become not twice as large, but four times as large.

This same torsion balance also lent itself to another investigation, one in which the relationship between the electrical force and the electric charge was revealed. As a consequence of this investigation, Coulomb concluded that the force varied directly as the product of the charges on *A* and *B*. He could alter the charge on a ball in a predictable way by touching it with an identical neutral ball because, if the material is such that charges are mobile, half the charge goes to the second ball.

Coulomb's experiments gave the result that the force exerted by charge *Q* on a charge *q* was proportional to Qq/r^2. The value of the constant of proportionality depends on the amount of charge which is arbitrarily selected as unit charge. The mks unit of charge, the coulomb, is defined in terms of something we have yet to discuss: electric current. With the size of the coulomb fixed, as described in Section 9.4, value of the proportionality constant, *k*, in Coulomb's law can be determined by measurement of the force and of the distance between two measured charges. The experimental result is 8.987×10^9 N m²/coul², which we have approximated by 9.0×10^9 N m²/coul².

2.8 CHARGES IN ELECTRIC FIELDS

Millikan's Measurement of *e*

As small drops of oil are squirted out of an atomizer, they acquire various electric charges by friction. Wandering around in air, even a neutral drop will occasionally encounter ions and acquire an electric charge. Consider such a charged oil drop, of mass M and net positive charge q, which has been squirted into the air in the region between charged plates (Fig. 2-25) where there is a uniform electric field $\vec{E}$ pointing vertically upward. The forces acting on the oil drop are then

1. An electric force
$$\vec{F}_{\text{electric}} = q\vec{E} \tag{2.18}$$
acting vertically upward.

2. A gravitational force (the weight of the drop)
$$\vec{F}_{\text{gravitational}} = M\vec{g} \tag{2.11}$$
acting vertically downward.

3. A viscous drag exerted by the air on the drop when the drop moves through the air.

By adjusting the charge on the plates, the electric field $\vec{E}$ can be varied until the electric force on the drop just balances the drop's weight, $qE = Mg$. The drop will then neither rise nor fall.

The mass M of the drop can be determined from a detailed study of its fall through the viscous air. The field $\vec{E}$ is determined by routine electrical measurements. Hence the charge on the drop can be calculated from $q = Mg/E$.

This is the basic idea behind a long series of careful experiments carried out by R. A. Millikan (1868–1953) during the first decades of this century. He found that the measured charge on the oil drop was always

$$\pm 1.60 \times 10^{-19} \text{ coul}, \pm 3.20 \times 10^{-19} \text{ coul}, \pm 4.80 \times 10^{-19} \text{ coul}, \ldots$$

That is,
$$q = (\text{integer}) \times e$$
where
$$e = 1.60 \times 10^{-19} \text{ coul}.$$

This experimental result, which was quoted in Section 1.1, has been confirmed in particle track recorder observations of thousands of nuclear reactions.

The presence of positive ions, that is, atoms or molecules from which an electron has been stripped, accounts for the positively charged oil drops. Drops with a negative charge have acquired an excess of electrons.

Motion of an Electron in a Uniform Electric Field

If we enclose the metal plates of Fig. 2-25 in a glass tube and pump out the air inside the tube we have what is called a *vacuum tube*. The metal plates are called *electrodes* and since there are just two of them, this

Physics Today

R. A. Millikan
(1868–1953)

vacuum tube is called a *diode*. If we heat the negative electrode (named the *cathode*), electrons will be "boiled off" from the hot surface.

Between the plates an electron (charge $q = -e$) experiences only an electric force

$$\vec{F} = q\vec{E} = -e\vec{E}.$$

The gravitational force on an electron is entirely negligible compared to even the most feeble electric forces encountered in practical work. If a good vacuum is maintained, an electron usually gets a free run from the cathode to the positive electrode (the *anode*), with little chance of a collision with the remaining air molecules. Newton's second law, $\vec{F} = m\vec{a}$, applied to this electron gives

$$eE = ma.$$

The electron's acceleration consequently has a magnitude

$$a = (e/m)E. \qquad (2.20)$$

The electron is of course accelerated in the direction of the electric force, and this direction is opposite to that of $\vec{E}$.

Notice that in a given field $\vec{E}$, the acceleration of any charged particle is proportional to its "*charge-to-mass ratio*" (e/m for the electron). Measurement of a particle's acceleration in a known electric field suffices to determine its charge-to-mass ratio.

The electron has a charge-to-mass ratio which is the world's largest, 1836 times that of the proton. Agility, a large acceleration in a given field, is therefore a characteristic of an electron. It is this characteristic that is exploited in many electronic devices, and in fact led to the discovery of the electron in 1897 by J. J. Thomson (1856–1940). Using electric and magnetic forces (Section 12.2), he was able to measure the electron's charge-to-mass ratio. The modern value is

$$e/m = 1.76 \times 10^{11} \text{ coul/kg}.$$

From this knowledge and from Millikan's determination of e, the electron's mass can be computed:

$$m = 9.11 \times 10^{-31} \text{ kg}.$$

The motion of the electron in a uniform electric field is a motion with a constant acceleration. It is therefore mathematically identical to that of a baseball in a uniform gravitational field. If the electron starts from rest near the cathode it will travel in a straight line to the anode, speeding up as it goes, with its distance from the cathode after a time interval Δt given by

$$\Delta x = \tfrac{1}{2} a (\Delta t)^2 \qquad (2.6)$$

and its velocity, from Eq. (2.4), by

$$v = a \Delta t.$$

Courtesy of Cavendish Laboratory

J.J. Thomson
(1856–1940)

The only difference between the electron's motion and that of a falling baseball lies in the numerical value of the acceleration (and the subsequent speeds attained). Electric fields of 10^6 N/coul are common. The acceleration imparted by such a field to an electron is

$$a = (e/m)E \quad (2.20)$$
$$= (1.76 \times 10^{11} \text{ coul/kg})(10^6 \text{ N/coul})$$
$$= 1.76 \times 10^{17} \text{ m/sec}^2.$$

This is about 2×10^{16} times the baseball's acceleration (9.8 m/sec²)!

Clearly it does not take very long (about 10^{-8} sec) for this electron to be accelerated to a speed approaching the velocity of light (3.00×10^8 m/sec). As these high speeds are reached, our Newtonian mechanics no longer gives the right answers! The modifications required are given by Einstein's special theory of relativity (Chapter 5).

If an electron is projected horizontally and enters the region between the plates of Fig. 2-25, it will follow a parabolic trajectory just like that of the baseball depicted in Fig. 2-23. After it has emerged from the region between the plates and moves through the region where the electric field $\vec{E}$ is zero, its velocity vector will be constant. In other words, the electron, after leaving the region between the plates, will move in a straight line at a constant speed. The net result is that the electron's trajectory has been bent by passage between the plates. The electron has been deflected.

Such deflection of a beam of electrons by passage through "deflection plates" is the key to the functioning of a television picture tube* or a *cathode ray oscilloscope* (Fig. 2-28). A beam of electrons is fired from an electron gun, proceeds down the picture tube, and makes a bright spot on the fluorescent screen where it strikes. On route the beam passes between plates arranged to give a vertical deflection and then between plates which give a horizontal deflection. The charge on the deflection plates is periodically altered, the electric field then changes accordingly, and the deflection of the beam passing between the plates is changed. The spot on the screen where the beam strikes is thus moved

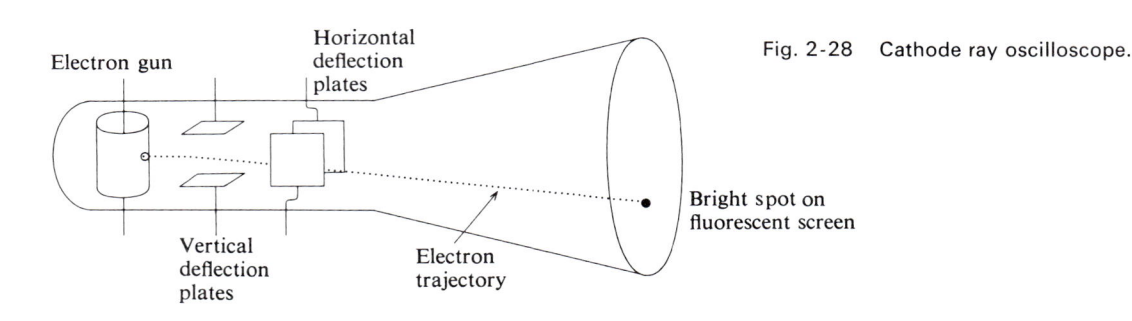

*In practice, modern TV picture tubes employ a magnetic deflecting force (Section 12.2) instead of the electric forces described in this chapter.

Fig. 2-28 Cathode ray oscilloscope.

78 MOTION AND FORCE

back and forth and up and down. By altering the beam intensity according to the incoming signal, the spot brightness changes as the spot moves and a picture is "painted" on the screen.

2.9 ELECTRIC FORCES IN ATOMS

Coulomb Forces

The nucleus of an atom with atomic number Z possesses a positive electric charge, $+Ze$. (Recall that there are Z protons in the nucleus and Z orbital electrons in the neutral atom). The force that deflects an incident α-particle in the Rutherford scattering experiment is the repulsive Coulomb force,

$$F = k\frac{(Ze)(2e)}{r^2},$$

exerted by the nucleus on the α-particle which has a charge $+2e$. A determination of the α-particle orbit in the presence of the nucleus proceeds in the same fashion as the analogous astronomical orbit problem and the answer is similar: the α-particle trajectory is a conic section, a hyperbola in this case.

The *attractive* coulomb force exerted by the nucleus on an electron, in precise analogy to the attractive gravitational force exerted by the earth on a satellite, gives rise to three different possible types of electron orbits (described by the three conic sections of Fig. 2-22). The hyperbolic and parabolic orbits describe an electron coming in from remote regions and moving off again in a different direction. These orbits correspond to electron scattering by a nucleus. The third possibility is an elliptic electron orbit about the nucleus.

In the elliptic orbit of an electron about a proton, since the electron does not travel indefinitely far from the proton, we say the electron is *bound* to the proton and that we have a *bound state* of the system comprising the proton and the electron, that is, a bound state of the hydrogen atom. Bohr showed that the light radiated by a hydrogen atom, together with this picture of electron orbits, implied that not all elliptic orbits were possible. When the electron is in the smallest of the possible orbits, it is most tightly bound to the proton. The atom is then said to be in its *ground state*. In this ground state the electron moves in a circular orbit of diameter about 1 angstrom unit (10^{-10} meter). Less tightly bound states, corresponding to larger possible orbits, are called *excited states* of the atom.

Quantum mechanics has blurred out (recall Section 1.5 and Fig. 1-10) this picture of electron orbits, but the idea remains of a ground state and excited states of any bound system of particles.

Hierarchy of Forces

The interactions of different chunks of matter arise from the interactions of the particles comprising the matter. In the submicroscopic world of elementary particles, it seems that there are just the four fundamental types of forces or interactions that are listed in Table 2-2 in order of increasing strength.

Table 2-2. ELEMENTARY INTERACTIONS

Name	State of knowledge
Gravitational force	Law known
Weak interaction	Knowledge incomplete
Electromagnetic interaction	Law known
Nuclear forces	Knowledge incomplete

The gravitational force is by far the weakest. To see how insignificant gravitational forces are in atomic physics, it is merely necessary to compute the ratio of the gravitational force, GMm/r^2, exerted by a proton (mass M, charge Q) on an electron (mass m, charge q), to the corresponding electric or Coulomb force, ke^2/r^2. The answer is

$$\frac{\text{gravitational force}}{\text{electric force}} = \frac{1}{2.3 \times 10^{39}}.$$

The exercise of expressing this number in words will give some idea of just how small it is.

The "weak interaction" is the name for the interaction which leads to β-decays.

An electric field exerts a force on a charged particle, whether or not it is moving. As we shall discuss in Chapter 12, there is an additional force acting on a *moving* charged particle: a magnetic force exerted by a magnetic field. The electric and magnetic interaction together comprise what is called the *electromagnetic interaction* that is listed above in Table 2-2.

By far the strongest force that has been discovered is the nuclear force binding together the constituents of a nucleus. Our knowledge of this force is still rudimentary.

The interaction of two macroscopic chunks of matter, say a brick sliding down a board, is conveniently described in terms of a "contact force" which generally involves a "frictional force." From a fundamental point of view, such forces ultimately arise from the electromagnetic interaction of elementary particles.

QUESTIONS

1. Between the ages of 14.0 and 14.5 years, George grows 2.0 inches taller.
   (a) What is his average rate of change of height for this interval?
   (b) When 14.0 years old, George was 5.0 ft tall. Assuming his rate of change of height (his growth rate) is constant between the ages of 14.0 and 17.0 years, find the change in his height in this interval.
   (c) What is his height when he is 17.0 years old?
   (d) From this data can you tell how tall he will be when he is 30 years old? If not, why not?

2. Define the term *velocity* for a one-dimensional motion.

3. Under what circumstances does the relationship $\Delta x = v \Delta t$ hold exactly?

4. A sprinter in a 100-meter race passes the 50-meter mark with a velocity at that instant of 10 m/sec.
   (a) Where would you expect him to be 2.0 sec later?
   (b) Would you expect his time for the race to be more than 10 sec? Why?

5. From the speedometer readings in Example 2, estimate the change in the car's position coordinate for the time interval extending from $t = 8.0$ sec to $t = 14.0$ sec, using $\Delta x = v \Delta t$ to calculate Δx for a succession of six time intervals, each of 1.0 sec duration. For each 1.0-sec interval, use the velocity which is half the sum of the velocity at the beginning of the interval and the velocity at the end of the interval.

6. When a stopwatch reads zero the position coordinate of a particle is x_0. The particle moves with a velocity v which is constant. Show that the equation $\Delta x = v \Delta t$ implies that the particle's position coordinate x at the instant t is given by
$$x = x_0 + vt.$$

7. Define the term *acceleration* for a one-dimensional motion.

8. Under what circumstances does the relationship $\Delta v = a \Delta t$ hold exactly?

9. A car is moving with a velocity of 10 m/sec in the positive x-direction when a stopwatch reads 8.0 sec. It has a constant acceleration of 1.5 m/sec² in the positive x-direction.
   (a) What change in velocity has occurred when the stopwatch reads 12 sec?
   (b) What is the velocity when the stopwatch reads 12 sec?

10. At the instant that a policeman's stopwatch reads 2.00 sec, a car is moving in the positive x-direction with a velocity of 22 m/sec. The car is slowing down (decelerating), the magnitude of its acceleration having the constant value of 2.0 m/sec²
    (a) What change in velocity has occurred when the stopwatch reads 5.00 sec?
    (b) What is the velocity when the stopwatch reads 5.00 sec?
    (c) What will the stopwatch read when the car is at rest?

11. A policeman starts a stopwatch when a car is 40 m beyond him and traveling at a speed of 30 m/sec. The car begins a constant deceleration of 3.0 m/sec².
    (a) How far is the car from the policeman at the instants $t = 2.00$ sec, 4.00 sec, 6.00 sec, 8.00 sec, 10.00 sec? Plot a graph showing the distance of the car from the policeman as a function of time for stopwatch readings from 0.00 to 10.00 sec.
    (b) Plot a graph showing speed versus time for the stopwatch readings considered in part (a).

12. A train starts at a station and maintains a constant acceleration for 12.0 kilometers to the next station where its speed is 36 m/sec. Find the acceleration of the train and the time taken.

13. A train starts from rest and, after a constant acceleration for 5.00 minutes, reaches a speed of 20 m/sec. How far did the train go in the fifth minute?

14. A rocket-driven sled, used in experiments on the physiological effects of large accelerations, runs on a straight horizontal track. Starting from rest, a speed of 450 m/sec (about 1000 miles per hour) is attained in 1.8 sec.
    (a) Calculate the acceleration, assuming it to be constant. Compare this value of acceleration with that of a freely falling body.
    (b) How far does this sled travel during the 1.8 sec interval?

15. A subway train travels a distance of 1000 m between stations, accelerating at a constant rate of 1.2 m/sec² for the first half of the trip and decelerating at this same rate for the last half. Find the maximum speed attained and the time required for the trip.

Fig. 2-29 The Leaning Tower of Pisa.

16. A 1.0-kg gold ball and a 10-kg lead ball are dropped at the same instant as shown in Fig. 2-29.
    (a) Which hits the ground first? Why?
    (b) What is the acceleration of the lead ball as it starts? What is the acceleration of this ball halfway down?
    (c) What is the velocity of the lead ball 2.0 sec after it is released?
    (d) How far does the lead ball fall in the first 2.0-sec interval?
    (e) What is the velocity of the lead ball after it has traveled 4.0 meters downward?
    (f) What assumption have you made in answering the foregoing questions?

17. A ball is dropped from the top of a building 39.2 m above the sidewalk.
    (a) How long will it take the ball to reach the sidewalk?
    (b) What is the ball's speed just before it hits the sidewalk?
    (c) What is the average velocity of the ball during the interval from $t = 1.00$ sec to $t = 2.00$ sec, assuming that the ball was dropped at $t = 0$? How far did the ball travel in this time interval?

18. If the ball in the preceding problem, instead of being released at rest, were thrown vertically downward with a speed of 19.6 m/sec, how long would it take to reach the sidewalk and what would be its speed just before impact?

19. (a) The vector $\vec{v}$ is defined as 40 m/sec, east. What is its magnitude (the speed)?
    (b) Another vector $\vec{b}$ is 40 m/sec, north. Can you say $\vec{b} = \vec{v}$?
    (c) Can you say $b = v$?

20. Sketch a diagram showing three vectors $\vec{a}, \vec{b}$ and $\vec{c}$ for which
    $$\vec{a} + \vec{b} + \vec{c} = 0$$
    in the case when
    (a) $\vec{a}$ and $\vec{c}$ have the same direction;
    (b) no two vectors have the same direction.

21. A velocity vector $\vec{v}$ is 40 m/sec, east. Another velocity vector is $\Delta\vec{v}$, 30 m/sec, north. Draw a vector diagram to scale representing these vectors. Now draw in and label the vector which is the sum of these two vectors, $\vec{v} + \Delta\vec{v}$. Determine the magnitude of this vector sum from your diagram. Use the theorem of Pythagoras to check your answer.

22. The vector $\vec{a}$ is 2.0 m/sec², north. The vector $\Delta\vec{v}$ is given by the equation $\Delta\vec{v} = \vec{a}\Delta t$. Find the magnitude and direction of $\Delta\vec{v}$ for $\Delta t = 3.0$ sec and for $\Delta t = 5.0$ sec.

23. The vector $\vec{P} = 30$ units north. The vector $\vec{Q} = -\vec{P}$. Give the magnitude and direction of $\vec{Q}$.

24. Assume the axis OX points east. A velocity $\vec{v} = 40$ m/sec, east.
    (a) Find its component v_x.
    (b) Another vector $\vec{u}$ points north and has a magnitude of 30 m/sec. Find u_x.
    (c) A third vector $\vec{w}$ points west and has a magnitude of 60 m/sec. Find w_x.

25. Assume that axis OX points east and axis OY points north. A certain velocity is 60 m/sec, 30° north of east. Find the x- and y-components of this velocity.

26. In Fig. 2-9, calculate the angle between $\vec{v}$ and the X-axis. Find v_y.

27. The position coordinates, x and y, of a particle in a plane are the x- and y-components of the particle's position vector. Find the position coordinates of a particle whose position vector has a magnitude of 9.0 m and is directed northwest. The axis OX points east, and OY points north.

28. A point Q is 4.00 m from the origin O, and the angle measured counterclockwise from the X-axis to OQ is 30°. Find the x- and y-components of the position vector of Q.

29. Find the x- and y-components of the acceleration vector of a falling stone in a coordinate system with OX horizontal and OY directed vertically upward.

30. From Fig. 2-8 show that
$$a = \sqrt{a_x^2 + a_y^2}$$
and that
$$\tan \theta = a_y/a_x .$$
These important relationships show how the magnitude and direction of a vector can be determined from knowledge of its components.

31. The position coordinates of a particle are $x = 8.0$ m, $y = 6.0$ m. Find the magnitude and direction of the particle's position vector $\vec{r}$.

32. As a baseball leaves a bat it has a velocity component of 30 m/sec in a vertically upward direction and a velocity component of 40 m/sec in a horizontal direction. Draw a diagram showing the velocity vector and its components. Find the speed of the baseball and the angle between its direction of travel and the horizontal.

33. Define position vector, velocity vector, and acceleration vector of a particle.

34. A car's position vector $\vec{r}$ is 30 meters east at $t = 10$ sec. It is traveling with a constant velocity vector $\vec{v}$ which is 10 m/sec east. Find the change in the position vector ($\Delta \vec{r}$) which occurs in the next 2.0 sec interval. Then find the magnitude and direction of the vector $\vec{r} + \Delta \vec{r}$. What is the position vector of the particle at $t = 12$ sec?

35. (a) A car's speed is increasing as it is driven straight north. Draw a diagram showing the directions of the velocity and acceleration vectors.
    (b) The driver of the car steps on the brakes. Draw the velocity and acceleration vectors.
    (c) Justify the relative directions of the acceleration and velocity vectors given above, using $\Delta \vec{v} = \vec{a} \Delta t$, and assuming for simplicity that the acceleration is constant during the time interval Δt.

36. Traveling with constant speed, a car goes first down a straight road, next around a gentle curve, and finally around a sharp turn. Compare the centripetal accelerations for these three cases and state which is greatest and which is least.

37. A car is speeding up as it goes around a curved portion of a racetrack. Draw the path, and then draw the velocity and acceleration vectors at one instant.

38. A car is going around a circular track of 100-meter radius with a constant speed of 20 m/sec. What is the centripetal acceleration at the instant when the car is due north from the center of the circle and traveling east? Draw a diagram showing the path, the velocity vector, and the acceleration vector.

39. Just after the instant mentioned in Question 38, the driver steps on the gas and acquires a tangential component of acceleration equal to 3.0 m/sec². Draw a diagram showing the path, the tangential and centripetal components of the acceleration. Draw in *the* acceleration vector (the vector sum of the tangential and centripetal components). Find the magnitude and direction of this acceleration vector.

40. A roller-coaster starts from rest and goes down a straight incline 60 m long, maintaining a constant acceleration of 1.5 m/sec². The roller-coaster then moves with constant speed along a circular section of horizontal track which has a radius of 30 m. Find the roller-coaster's acceleration when it is traveling along the curved horizontal track.

41. Compute the vector sum of the following forces:
    (a) 3 N east; 2 N east
    (b) 3 N east; 2 N west
    (c) 3 N east; 3 N west
    (d) 3 N east; 4 N north
    (e) 5 N north; 6 N east; 5 N south; 2 N west.

42. (a) While a car is traveling east, the horizontal force exerted on the car by the road is 300 N, east. The air exerts a force on the car of 200 N, west. Find the vector sum of these forces acting on the car.
    (b) The brakes of the car are applied. Then the horizontal force exerted on the car by the road is 1200 N, west. The air still exerts a force on the car of 200 N, west. What is the vector sum of these forces acting on the car?

QUESTIONS 83

43. Shortly after jumping from an aircraft a man experiences an upward force exerted by the air of 200 N. The downward gravitational force acting on the man is 800 N. What is the resultant force acting on the man?

44. You are pushing a wagon with a force of 200 N and it is traveling with a constant velocity of 3.0 m/sec. What is the total retarding force acting on the wagon?

45. The force acting down (weight, the gravitational force exerted by the earth on the object) on a paratrooper and his parachute is 1000 N. After a short time his velocity is constant. What can be said about the force exerted by the air on the parachute and on the man?

46. To measure the mass of a trunk, it is pushed along a sheet of ice by a measured horizontal force of 100 N. The acceleration is observed to be 2.0 m/sec². What is the mass of the trunk?

47. An empty box acquires an acceleration of 1.50 m/sec² when acted upon by a certain resultant force $\vec{F}$. When a brick is placed in the box, the acceleration, with the same resultant force acting on the box, is observed to be 0.50 m/sec². If four such bricks are placed in the box, what acceleration will be imparted to the box by the same resultant force $\vec{F}$?

48. The car in Question 42 has a mass of 1.40×10^3 kg. Find the magnitude and direction of its acceleration vector in Question 42(a) and also in Question 42(b).

49. A 20-kg box sliding across the floor slows down from 2.5 m/sec to 1.0 m/sec in 3.0 sec. Assuming the force (and therefore the acceleration) is constant, find the resultant force acting on the box. Give the magnitude and the direction of the force relative to the velocity vector of the box.

50. A 2.0-kg stone falls with an acceleration of 9.8 m/sec². What is the resultant force acting on the stone?

51. A 0.50-kg cardboard box falls with an acceleration of 8.0 m/sec². What is the resultant force acting on the box?

Fig. 2-30 Car coming to a sudden stop.

52. Refer to Fig. 2-30.
    (a) Explain what is shown there in terms of Newton's laws of motion.
    (b) Justify the use of seat belts in terms of your explanation in (a) above.

53. (a) Define weight.
    (b) How is the vector $\vec{g}$ (the gravitational field) at a point in space defined?

54. At the surface of a certain planet the gravitational field $\vec{g}$ has a magnitude of 2.0 N/kg. A 4.0-kg brass ball is transported to this planet. Give
    (a) the mass of the brass ball on the earth and on the planet;
    (b) the weight of the brass ball on the earth and on the planet.

55. (a) What is the gravitational field at a point where a 2.00-kg mass experiences a gravitational force of 18.0 N?
    (b) What gravitational force would be exerted on a 6.00-kg mass placed at this point?

56. (a) The gravitational field produced by a particle has a value $\vec{g}_1$ at a distance r_1 and a value $\vec{g}_2$ at a distance r_2. Express the ratio g_1/g_2 in terms of r_1 and r_2.
    (b) A particle produces a gravitational field of magnitude 1.0×10^{-11} N/kg at a distance of 2.0 m. What field does it produce at a distance of 6.0 m?

57. A 1.0-kg particle produces at a distance of 1.0 meter a gravitational field of what magnitude and direction? What force will this field exert on a 10-kg particle?

58. Using the data in Appendix B, calculate the gravitational field produced by the moon's mass
    (a) at the surface of the moon, and
    (b) at the location of the earth.

59. Using the data in Appendix B, calculate the gravitational field produced by the sun's mass
    (a) at the surface of the sun, and
    (b) at the location of the earth.

60. State the superposition principle for gravitational fields.

61. By considering the gravitational force, $\vec{F} = m\vec{g}$, exerted on a mass at a point where the gravitational field is $\vec{g}$, show that the superposition principle for gravitational fields ($\vec{g} = \vec{g}_1 + \vec{g}_2$ in Fig. 2-19) is a consequence of the superposition principle for the corresponding gravitational forces $\vec{F} = \vec{F}_1 + \vec{F}_2$.

62. At a given point P on the earth's surface, it sometimes happens that the gravitational fields contributed by the earth, the moon, and the sun are all in the same direction.
    (a) Sketch a diagram showing the relative positions of P, the earth's center, the moon, and the sun.
    (b) Using the results of Questions 58 and 59, calculate the superposition of these three fields at the point P.
    (c) Calculate the superposition of these three fields at a point Q on the earth's surface such that PQ is a diameter of the earth. (Neglect the change in magnitude of the individual fields due to the change PQ in the distance from the moon and the sun.)

63. The sources of a gravitational field are *two* particles, one of mass 1.0 kg and the other of mass 2.0 kg separated by 1.0 meter.
    (a) Find the gravitational field at the point midway between the particles.
    (b) Find the gravitational field at a point in space which is 2 meters from the 1.0-kg mass and 1.0 meter from the 2.0-kg mass.

64. Locate the point between the earth and the sun at which the superposition of their gravitational fields is zero.

65. What is the distance from the earth's center to a point outside the earth where the earth's gravitational field is 1/25 of its value at the earth's surface?

66. Show that Newton's law of gravitation, $F = GMm/r^2$, together with the definition of a gravitational field ($\vec{g} = \vec{F}/m$), imply that the field produced at a distance r from a source M is given by
$$g = GM/r^2,$$
as stated in Eq. (2.12).

67. Use Newton's law of gravitation to compute the force exerted by a 1.0-kg particle on a 10-kg particle which is 1.0 meter away. Check your answer by comparing with the result obtained by the two-stage process used in Question 57.

68. A 100-kg man and a 50-kg woman are approximately 20 m apart. Estimate the attractive force exerted on the man by the woman. Compare this force to the man's weight and decide whether or not the man would be aware of this attraction.

69. Calculate the magnitude of the force exerted on the earth by the sun. (Use the data in Appendix B.)

70. Explain how Cavendish was able to determine the mass of the earth.

71. What is the ratio of the force exerted on the moon by the sun to the force exerted on the moon by the earth? (Use the data in Appendix B.)

72. Show that $\vec{F} = m\vec{g}$ and Newton's second law imply that particles of different masses have the same acceleration in a gravitational field.

73. Do particles always move in the direction of the resultant force acting upon them? That is, in a short time interval Δt, is $\Delta \vec{r}$ always in the direction of $\vec{F}$. Is $\Delta \vec{v}$ in the direction of $\vec{F}$?

74. What happens to refuse that is released from an earth satellite? Does it fall straight down to earth? Explain.

75. An astronaut's weight changes as his altitude changes. Why? Is his weight really zero in orbit? Why?

76. What type of orbits can an escaping object take?

77. For the parabolic orbit of Fig. 2-21, draw in velocity and acceleration vectors at an instant long after projection. Will the escaping mass m be speeding up or slowing down? Why?

78. Show that Eq. (2.15), $v = \sqrt{GM/r}$, can be rewritten in the convenient form
$$v = \sqrt{gr},$$

and that for an earth satellite we have

$$v = \sqrt{g_s r_e (r_e/r)}$$

where $\vec{g}_s$ is the gravitational field at the earth's surface and r_e is the earth's radius.
(a) From this result, evaluate the speed of a satellite for an orbit near the earth's surface ($r = r_e$).
(b) Calculate the orbital speed of the moon ($r/r_e = 60$).
(c) Show that the moon's period is nearly one month.

79. At what distance from the earth's center will a satellite moving in the plane of the equator always be vertically above the same point on the earth's surface?

80. Explain how the mass of a planet with a moon can be determined.

81. A satellite of Jupiter has a period of 1.77 days and an orbital radius of 4.22×10^8 m. Find the mass of Jupiter.

82. A 20×10^3-kg spaceship with its rocket engines turned off is drifting in a remote region of space where the gravitational field due to other objects is very small. A small satellite, moving almost solely under the influence of the gravitational force exerted by the spaceship, has an orbit about the spaceship of radius 10^2 m. What is the period of revolution and the speed of this satellite?

83. Determine the mass of the sun from the value of G and the following data concerning the earth's orbital motion:

1 year = 3.15×10^7 sec,
radius of earth's orbit = 93×10^6 miles
$= 1.5 \times 10^{11}$ meters.

84. Calculate the mass of the earth from the data given in Appendix B concerning the moon's orbit.

85. The orbit of Mars about the sun has an average radius which is 1.52 times the average radius of the earth's orbit. What is the period (in years) of the orbital motion of Mars?

86. Halley's Comet has a period of about 76 years. Therefore from Kepler's third law you know something about the average radius of its orbit compared to the earth's orbit. Which is larger? Knowing that we do see the comet approximately every 76 years, sketch the appearance of its orbit.

87. Does the acceleration of a baseball, after it has been struck and is traveling toward the outfield, depend on who hit it? Does this acceleration depend on whether the baseball is going up or down?

88. A bullet leaves a rifle traveling horizontally with a speed of 1000 m/sec and hits a wall which is 50 meters away.
(a) How long does the bullet's flight last?
(b) How far will it have fallen by the time it hits the wall?

89. A baseball is projected horizontally with a speed of 30.0 m/sec from the top of a building which is 40.0 m high.
(a) Verify that after 1.60 sec the ball will have the position shown in Fig. 2-23.
(b) Find the magnitude of the velocity vector after 1.60 sec.
(c) How long will it take for the ball to hit the ground?

90. Show that if a second ball, at the left in Fig. 2-23, were to be released from rest as the first is projected horizontally, the vertical motions of the two balls would be identical: they would both hit the ground at the same time.

91. A car, traveling horizontally, drives over the edge of a cliff which is 19.6 m high and lands at a horizontal distance of 70 m from the base of the cliff. What was the speed of the car as it went over the edge of the cliff?

92. With a muzzle velocity of 500 m/sec, approximately how high above the bullseye must a rifle be aimed at a target which is 50.0 m away? Neglect air resistance.

93. (a) State two differences between electrical and gravitational forces.
(b) Give the electrical analogue of the expression for the gravitational field $\vec{g}$ produced by a mass M: $g = GM/r^2$.

94. (a) Define the term *electric field*.
(b) State the mks unit for electric field.

95. A positively charged oil drop is injected into the region between two oppositely charged, horizontally oriented plates like those of Fig. 2-25.
(a) If the charge on the drop is $+2e$, what is the magnitude of the electric field required that an electric force of 1.6×10^{-15} N be exerted on the drop?

86 MOTION AND FORCE

(b) Sketch the orientation of the plates and indicate the kind of charge on each in order that the electric field between them give rise to a force which might balance the gravitational force on the oil drop.

96. A charge of $+8.0 \times 10^{-8}$ coul is 0.40 m from a charge of -2.0×10^{-8} coul.
    (a) Determine the electric field produced by the negative charge at the location of the positive charge.
    (b) What force does this electric field exert on the positive charge?
    (c) Check this result by computing this force directly from Coulomb's law.

97. (a) Determine the electric field produced by the charges in the preceding question at the point P which is 0.30 meters from the positive charge and 0.10 meters from the negative charge.
    (b) What force would be exerted by this field on an electron placed at P?

98. What is the Coulomb attractive force exerted by a helium nucleus on a single orbital electron at a radius of 1 angstrom unit (10^{-10} m)?

99. Describe the essence of each of the following experiments:
    (a) Coulomb's experiment which led to what is now known as Coulomb's law.
    (b) The Millikan oil drop experiment which led to determination of the fundamental unit of charge.

100. In the Millikan oil drop experiment, an oil drop with a charge of -3.2×10^{-19} coul is held in equilibrium by an electric field of magnitude 8.0×10^3 N/coul directed downward. What is the mass of the oil drop?

101. Show that when the only force acting on an electron is the force exerted by an electric field $\vec{E}$, the electron's acceleration is given by $a = Ee/m$, where m is the electron's mass.

102. The plates of Fig. 2-25 are 4.0×10^{-2} m apart. The plates are charged as shown and produce a uniform field of 5.0×10^4 N/coul in the region between the plates. Find
    (a) the acceleration of an electron in this region;
    (b) the time required for an electron, which starts at rest at the upper plate, to reach the lower plate.

103. An electron emerges from the hot cathode of a vacuum tube with a speed of 6.0×10^5 m/sec. The electron then encounters an electric field of 20 N/coul in a direction opposite to the velocity vector of the electron.
    (a) Find the magnitude and direction of the force acting on the electron.
    (b) Determine the acceleration of the electron.
    (c) Find the distance the electron will travel in 3.0×10^{-8} sec and its speed after this time interval.

104. Describe the principal parts within the tube of a cathode ray oscilloscope and explain the influence of each part on an electron trajectory.

105. (a) Sketch the three different types of electron orbits that are possible (according to Newtonian mechanics) in the electric field produced by a nucleus.
    (b) Which orbit corresponds to a bound state?

106. Distinguish between the *ground state* and *excited state* of an atom.

107. Consider the gravitational force and the electric force exerted on an electron by a proton. Show that the ratio of these forces is given by

$$\frac{\text{gravitational force}}{\text{electric force}} = \frac{1}{2.3 \times 10^{39}}.$$

SUPPLEMENTARY QUESTIONS

S-1. (a) Denote the values of the velocity and position coordinate at the instant t by v_x and x, respectively. Their initial values, that is, their values at $t = 0$, are denoted by v_{x0} and x_0, respectively. Then
$$\Delta t = t - 0 = t,$$
$$\Delta x = x - x_0.$$
and
$$\Delta v_x = v_x - v_{x0}.$$
Making these substitutions, show that Eq. (2.4) yields
$$v_x = v_{x0} + a_x t$$
and that Eq. (2.5) gives
$$x = x_0 + v_{x0} t + \tfrac{1}{2} a_x t^2,$$
where a_x (the x-component of the acceleration vector) is assumed to have a constant value.
(b) Show that $x - x_0 = \tfrac{1}{2}(v_{x0} + v_x)t$.

S-2. When the acceleration is constant, $v_x^2 = v_{x0}^2 + 2a_x(x - x_0)$. Derive this equation by solving $v_x = v_{x0} + at$, for t, and substituting the resulting expression for t in the equation $x - x_0 = \tfrac{1}{2}(v_{x0} + v_x)t$.

S-3. *Amber Light Problem.* A motorist is approaching a traffic light with a speed v_0 when the light turns from green to amber. Suppose it takes a time t_0 for the motorist to make a decision and then to apply the brakes, and that the maximum braking deceleration has a magnitude a.
(a) Show that the motorist cannot stop without passing the light if his distance from the light is less than $v_0 t_0 + (v_0^2/2a)$.
(b) Suppose the amber light remains on for a time t_1. If the motorist wants to continue at speed v_0 and still pass the light before it turns red, what is the maximum distance he can be from the light as it turns from green to amber?
(c) Show that there is a zone in which the motorist will pass through a red light, whether he continues at constant speed or attempts to stop, if his speed v_0 is greater than $2a(t_1 - t_0)$.
(d) Evaluate this critical speed when $a = 5.0$ m/sec^2, $t_1 = 4$ sec and $t_0 = 1$ sec.

S-4. A rocket ascends vertically with an acceleration of 19.8 m/sec^2 for 80 sec. Then, with its fuel supply exhausted, it continues upward. What is the maximum height attained and the time taken to reach this height?

S-5. The pilot of a plane releases a flare from a certain height. The flare falls freely and, at exactly 3:00 p.m., passes an observer flying 240 meters above the ground. A second observer sees the flare strike the ground at 5.00 sec after 3:00 p.m. (Disregard air resistance.)
(a) At what height was the plane flying when the flare was released?
(b) At what time was the flare released?

S-6. In the electron gun of a television tube, an electron is accelerated through a distance of 1.20×10^{-2} m and emerges with a speed of 8.0×10^6 m/sec. The initial velocity is negligible compared to the final velocity. Find the electron's acceleration within the gun, assuming this acceleration to have been constant. For what time interval was the electron accelerated?

S-7. In an investigation of an accident that occurred in a 30-mile-per-hour (about 13 m/sec) zone, skid marks made by one automobile were 29 m long. If a maximum deceleration of 5 m/sec^2 is assumed, can it be concluded that the automobile was speeding?

S-8. A ball is thrown vertically upward with an initial speed of 19.6 m/sec.
(a) How long will the ball take to reach maximum height?
(b) What is the maximum height it will attain?
(c) At what times will the speed be 4.9 m/sec?
(d) At what times will the ball be 14.7 m above the point of projection?

S-9. A portion of a rocket is detached at a height of 4.9×10^2 m while the rocket is climbing vertically with a speed of 98 m/sec. Neglecting air resistance, estimate the time required for this portion to hit the ground.

S-10. Point P_1 has position vector $\vec{r}_1$ and point P_2 has position vector $\vec{r}_2$. Verify that the vector directed from P_1 to P_2 is $\vec{r}_2 - \vec{r}_1$.

S-11. Using the result of the preceding question show that the midpoint M of the line $P_1 P_2$ has the position vector given by $\vec{r}_M = \tfrac{1}{2}(\vec{r}_1 + \vec{r}_2)$.

88 MOTION AND FORCE

S-12. A particle experiences a constant acceleration $\vec{a}$ which is south at 10 m/sec². At $t = 0$ its position vector is $\vec{r}_0 = 1.0$ meter east; its velocity vector is $\vec{v}_0 = 30$ meters/sec east. Estimate where it is at $t = 1.0$ sec and at $t = 2.0$ sec.

S-13. Mars has a mass which is 0.11 of the earth's mass and a mean radius which is 0.53 of the earth's radius.
(a) Evaluate the gravitational field on the surface of Mars.
(b) What would be the weight of a 100-kg man on the earth and on Mars?

S-14. Figure 2-31 shows trajectories corresponding to three different speeds of projection from the earth's surface. (Disregard air resistance.)
(a) Identify the conic section corresponding to each trajectory.
(b) Describe the change in speed that occurs along each portion of the path and give the locations (if any) where maximum and minimum speeds are attained.
(c) Under what conditions will projection with a speed less than the escape speed result in a parabolic trajectory?

S-15. A planet is observed to have a moon which has a period T and a circular orbit of radius r.
(a) What is the mass of the planet?
(b) The radius of the planet is 1/50 of the radius of the satellite's orbit. Find the value of the gravitational field on the surface of the planet.

S-16. A stone is projected from the top of a building with a speed of 20.0 m/sec at an angle of 36.9° above the horizontal. The building is 50.0 m high. What horizontal distance will the stone travel before striking the ground?

S-17. A ball is thrown at a speed of 25.0 m/sec at an angle of 53.1° above the horizontal. At what height will it strike a vertical wall which is 30.0 m away?

S-18. A charge of $+8e$ is a distance D to the left of a charge $-2e$. Find the point where the electric field is zero.

Fig. 2-31 Trajectories corresponding to a speed of projection which is (a) less than the escape speed, (b) equal to the escape speed, (c) greater than the escape speed.

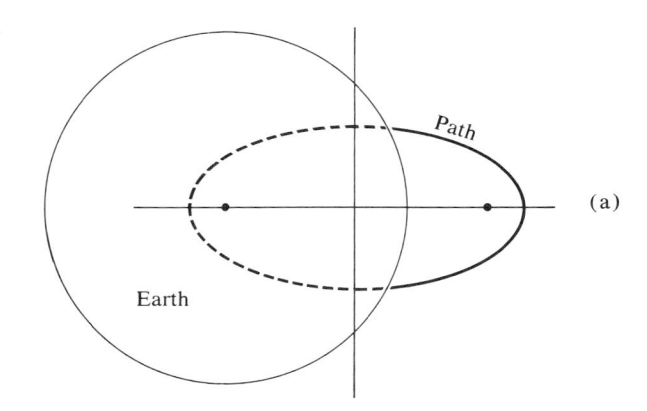

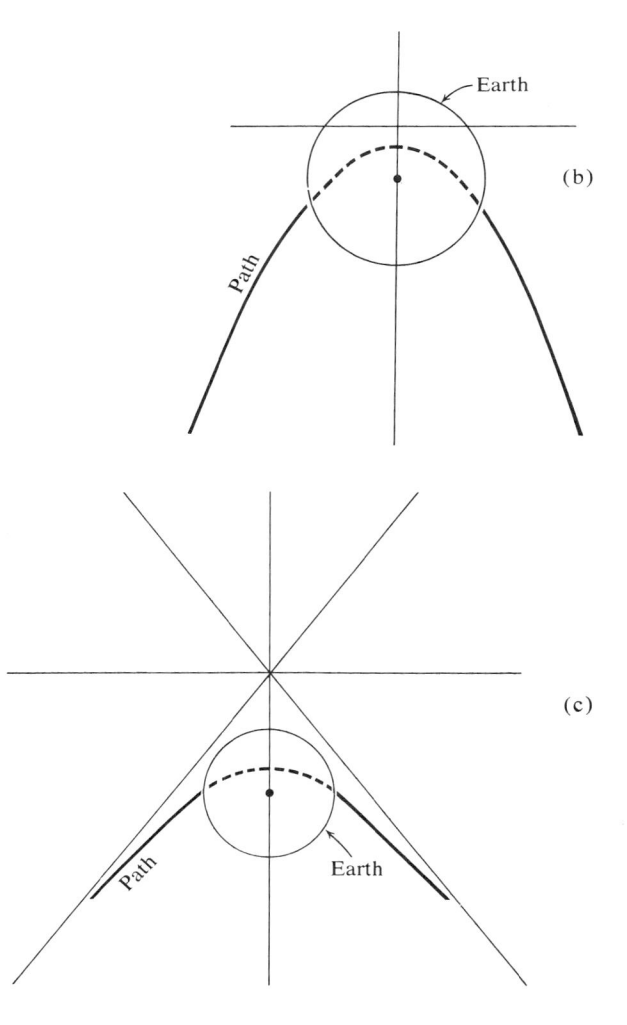

S-19. An electron emerges from the electron gun in a cathode ray oscilloscope at a speed of 2.0×10^7 m/sec and then travels horizontally with a constant velocity until, while passing between deflection plates, it encounters an electric field of 2.0×10^4 N/coul directed vertically downward. This deflecting field is established over a region which has a horizontal length of 2.0×10^{-2} m.

(a) For how long a time interval will the electron be subjected to a deflecting force?

(b) What is the vertical acceleration of the electron during the time interval determined above?

(c) What is the vertical component of velocity and the vertical displacement at the end of this time interval?

ADDITIONAL APPLICATIONS TO MEDICINE AND THE LIFE SCIENCES

Hypodermic Jet Injection

A "needle-free" hypodermic injection procedure makes use of the fact that a sufficiently high-speed small-diameter liquid jet can penetrate several layers of human tissue. Velocities higher than the velocity of sound are obtained by some types of *hypodermic jet injection* devices, although for injection the velocity need not necessarily be that high.

Question

Suppose the time required for a 0.10 cm³ injection is 0.20 sec, what is the average velocity of the jet? Assume the cross-sectional area of the jet is 1.5×10^{-8} m².

Answer

Since the volume is 0.10 cm³ (= 1.0×10^{-7} m³) and the cross-sectional area of the jet is 1.5×10^{-8} m², the "length" of the fluid stream is given by

$$\Delta x \text{ (length)} = \frac{\text{volume}}{\text{cross-sectional area}}$$

$$= \frac{1.0 \times 10^{-7} \text{ m}^3}{1.5 \times 10^{-8} \text{ m}^2}$$

$$= 6.7 \text{ m}.$$

This length is ejected in 0.20 sec so the velocity is

$$v = \frac{\Delta x}{\Delta t} = \frac{6.7 \text{ m}}{0.20 \text{ sec}}$$

$$= 340 \text{ m/sec}.$$

Gravity in Medical Practice

A wide variety of both simple and more complex clinical procedures depends upon the force of gravity. For example, blood circulation is affected by gravity; any change in a person's position has an effect on blood pressure.

The more or less horizontal position achieved when a person faints and falls is in a sense a protective mechanism in that in this position blood can more readily flow into the head.

Drainage and irrigation of certain body parts is also achieved by taking appropriate advantage of the force of gravity.

Intravenous infusions are usually dependent on gravity (see Fig. 2-32).

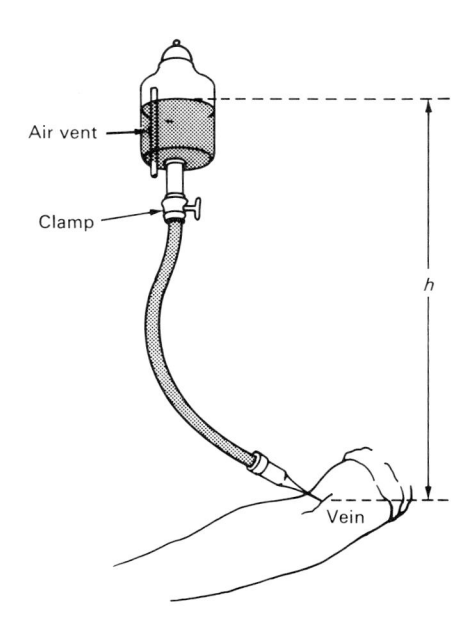

Fig. 2-32 One way of increasing the rate of flow in a transfusion set such as that shown is by increasing the height, *h*.

The *rocking bed* used in the case of patients who have some paralysis of the diaphragm, and those who suffer from certain circulatory diseases, is a simple mechanical device depending on the force of gravity. For example, in the case of the patient with partial paralysis of the diaphragm (polio patients), the rocking bed causes the feet to be elevated and the head to be lowered. Now the contents of the abdominal cavity pressing upon the diaphragm cause the diaphragm to rise and the lungs to be compressed. Exhalation results. Next the bed rocks into its alternate position, raising the head and thus reducing the pressure on the diaphragm. Air is now drawn into the lungs. This procedure is known as *mechanical respiration.*

Further Reading

For many specific applications of physical principles in nursing practice see Jensen, J. T., *Physics for the Health Professions,* Lippincott, Philadelphia, 1975.

String Electrometer: Electric Forces

A *string electrometer* is a device which depends on the fact that unlike electric charges attract each other. This electrometer is a key component of a variety of radiation meters such as the ionization chambers used for making measurements of medical x-ray exposures. Such radiation measurement devices are often called *condenser R-meters.*

The "string" may be a platinum-coated quartz fiber, which is hung like a loose jump rope from a fiber support in such a manner that the fiber can swing toward a so-called deflection electrode (see Fig. 2-33). Opposite charges are now placed on the fiber and the deflection electrode. The Coulomb force of the fixed deflection electrode now causes the oppositely charged quartz fiber to move toward the deflection electrode, against the restoring force (the restoring force being the spring action of the quartz fiber as it becomes twisted from its normal equilibrium position). At some pre-designed maximum condition of charge, the quartz fiber has its maximum deflection. A light in the electrometer casts a shadow of the fiber on a scale, which in this condition of maximum charge reads zero (for zero radiation exposure).

Now as the electrometer and the ionization chamber (see Fig. 2-34) to which it is attached are exposed to radiation, ionization of the air provides a mechanism for charge transfer, and the charge leaks off the electrodes. As the charge is reduced, the Coulomb attractive force between the fiber and the deflection electrode becomes less, so that the fiber under its elastic restoring force returns toward its equilibrium. The shadow of the fiber on the scale shows higher and higher exposure readings as the fiber moves toward its equilibrium position. The more radiation, the more ionization, the more charge leaks off, the more the fiber returns to its equilibrium position, and the higher the exposure reading indicated on the scale by the shadow of the fiber.

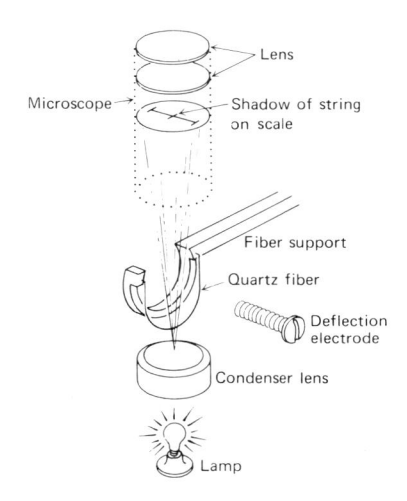

Fig. 2-33 Quartz-fiber or "string" electrometer, as used in the Victoreen Model 570 R-meter. (Courtesy Victoreen Instrumentation Divison, Sheller-Globe Corporation.)

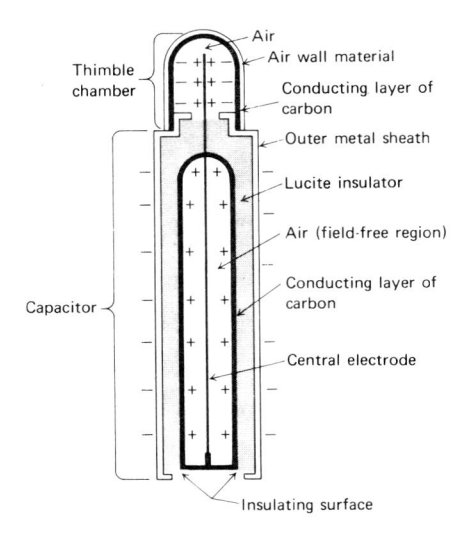

Fig. 2-34 Diagram of a Victoreen condenser chamber, showing the thimble ionization chamber affixed to a cylindrical capacitor.

ADDITIONAL APPLICATIONS 91

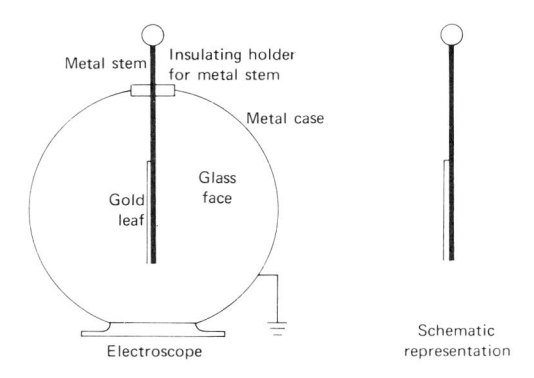

Fig. 2-35 Leaf electroscope. The leaf in this case, as indicated, is a piece of thin gold foil.

Question

Can you suggest an even simpler device which depends upon the Coulomb force and would indicate ionizing radiation?

Answer

The elementary leaf electroscope (see Fig. 2-35). The essentials of this device consist simply of a vertical stem made of a conductor along which hangs a thin metal leaf (say of aluminum), which is connected at its upper end to the upper end of the stem. When a charge is placed on the stem, the charge distributes itself over the stem and leaf; thus the bottom of the leaf moves away from the stem. If ionizing radiation is present, the ionized air will permit the charge to leak off this electroscope, and under the influence of gravity the leaf will fall back to hang parallel to the stem.

Electrophoresis

Electrophoresis involves the separation of particles having different migration velocities under the influence of an electric field. When suspended in an aqueous solvent, nearly all particles will become either negatively or positively charged. The sign and magnitude of the charge depends on the nature of the particles and the solvent used. When an electric field is applied across a trough or tube containing the particles of interest, migration takes place to the appropriate electrode. If different types of particles are present, they are likely to exhibit different migration velocities, depending on the sign and magnitudes of their charge. This method thus permits separation of different types of particles.

Probably the most exciting application of electrophoresis in medical research was the work by two-time Nobel prize winner, Linus Pauling (chemistry prize and peace prize) who, using electrophoresis, found that the hemoglobin of patients with sickle cell anemia has a different electrophoretic mobility than that of normal hemoglobin. Since this abnormality is hereditary, these electrophoretic studies have become of great interest to geneticists.

Question

What would you expect to happen to the speed of migration of particles as the electric field is increased?

Answer

The speed would increase since $F = qE$ and hence, if q is constant, a larger E means a larger F which in turn means a larger acceleration.

Physiological Monitor

Since electrical changes produced throughout the various constituents of animals and humans are what trigger reactions and motions of all kinds, it is not surprising that the *oscilloscope* has become a standard tool in basic biology and in both research and clinical procedures in medicine. In the latter area considerable use is made of various types of *physiological monitors* (see Fig. 2-36). Essentially these monitors are oscilloscopes which respond directly either to body electrical changes or to some other body change, such as blood pressure, by means of a mechanism (transducer) which transforms the relevant body changes into electrical changes.

A typical example of such a physiological monitor is a battery-operated oscilloscope designed for monitoring ECG (electrocardiogram), and EEG (electroencephalogram), and pulse-waveform. In the first two modes (ECG and EEG), the body electrical signals are used directly. The signals are amplified and applied on the vertical deflection plates (see Fig. 2-28). In the last mode (pulse-waveform) it is not a body electrical change which is employed directly; rather the *photoelectric effect* is used to produce an electrical signal which the oscilloscope "paints" on its screen. One uses a *pulse sensor* placed on some body extremity such as a finger tip. This photoelectric sensor operates on the variation of light transmission and reflection of the flesh which results from the slight swelling and contraction of the capillaries during each blood pressure cycle.

Question

If an oscilloscope is to be used to show an ECG trace, how frequently should the electric charge on the horizontal deflection plates be reversed to give a useful horizontal sweep of the electron beam across the oscilloscope screen?

Answer

The answer depends on the heart rate. Assuming a heart rate of 60 beats per minute (for simplicity in arithmetic), say that we want to see two complete heart beat displays on the screen. The horizontal sweep then needs to be of 2 seconds duration. That means the charge on the horizontal deflection plates has to be reversed at the end of 2 seconds to bring the electron beam back to its starting point. This return of the electron beam is done as quickly as possible. This completes one cycle of the horizontal sweep. The same process is repeated during subsequent cycles. (One says the period or duration of each cycle is 2.0 sec and that the sweep speed is 0.5 cycle/sec, or 0.5 hertz.)

Further Reading

For a popularized summary of many of the technological advancements of medicine, including patient monitoring devices, the reader is referred to Carlisle, N., and J. Carlisle, *Marvels of Medical Engineering*, Sterling, London, 1967. For a more rigorous treatment of such developments see Segal, B. L., and D. G. Kilpatrick, ed., *Engineering in the Practice of Medicine*, Williams and Wilkins, Baltimore, Md., 1967.

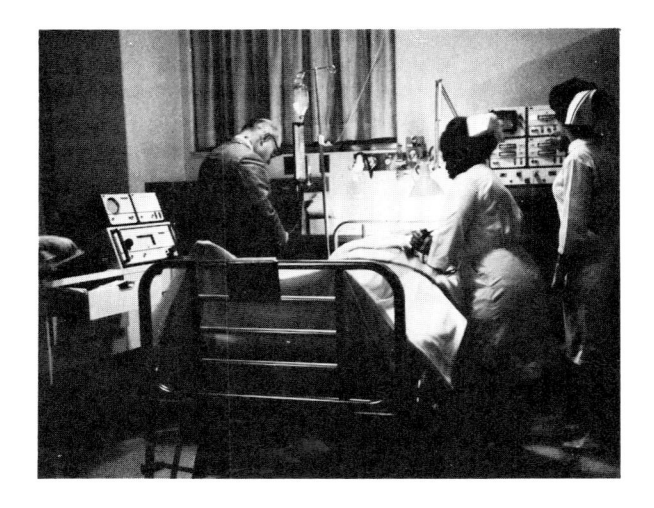

Fig. 2-36 Various physiological monitors are seen in this I.C.U. (intensive care unit). (Courtesy *Hewlett-Packard Journal*.)

SUGGESTED READING

BELL, E. T., *Men of Mathematics*. New York: Simon and Schuster, 1937. Interesting bibliographical chapters are devoted to Newton and to Lagrange. Bell's book may be highly recommended to any reader interested in the historical developments of mathematics, traced through the lives of the great mathematicians. Much of the book reads like an exciting novel.

BRONOWSKI, J., *The Ascent of Man*. Boston/Toronto: Little, Brown and Co., 1973. At this stage in the study of physics the student might profitably delve into one of the most exciting works of this decade. This book is nothing less than a full-scale history of science developed from the acclaimed thirteen part BBC television series written by Jacob Bronowski, mathematician, as well as statistician, poet, historian, and teacher. Bronowski died in La Jolla, California in 1974, shortly after completing this monumental work.

EINSTEIN, A. and L. INFELD, *The Evolution of Physics*. New York: Simon and Schuster, 1950. The first 38 pages are concerned with the topics of this chapter. Gravitational fields are introduced on pp. 129-132. This charming and very instructive book is written by the masters for those without any knowledge of mathematics or physics.

FEYNMAN, R. P., R. B. LEIGHTON, and M. SANDS, *The Feynman Lectures on Physics*. Reading, Mass.: Addison-Wesley, 1963. This famous book based on the lectures of one of the world's leading physicists will be enjoyed by all readers. The level, often very high, nevertheless is sometimes appropriate for the beginning physics student. Chapters 7 to 9 will be particularly helpful and stimulating.

HOLTON, G. and D. H. D. ROLLER, *Foundations of Modern Physical Science*. Reading, Mass.: Addison-Wesley, 1958. An excellent elementary account, with an accent on the history of physics.

NEWMAN, J. R., *The World of Mathematics*. New York: Simon and Schuster, 1956. The lives and contributions of Kepler, Galileo, and Newton are among the many topics of interest in this four-volume work which is both historical and mathematical.

WILDI, T., *Units*. Quebec City: Volta, Inc., 1971. This short book (about 130 pages) restricts itself, as the title implies, to a discussion of systems of units. It includes a short history of the various systems, a description of the International System (SI), and also the current standards by which fundamental units are defined. A good portion of the book is devoted to conversions from one system to another, a matter which may be of considerable interest to those readers planning to pursue studies in engineering.

Sir Isaac Newton

3 Mechanics

In the first part of this chapter we continue with the development of Newtonian mechanics. Newton's third law is stated, and momentum is introduced. Also we discover how to apply the laws of motion to describe the average motion of a large body or of any system of particles. The important idea of a conservation law is discussed in connection with the law of conservation of momentum.

The reader who has achieved some familiarity with the contents of the previous chapter will be beginning to appreciate the scope and impact of the scientific work of Sir Isaac Newton. After the passage of 300 years, we can still be as filled with awe at Newton's achievements as were his contemporaries such as Alexander Pope who wrote,

Nature and Nature's Laws lay hid in night;
God said, Let Newton Be!—And all was light.

and Edmund Halley who felt,

Nearer to the gods no mortal may approach.

Isaac Newton was born in the hamlet of Woolsthorpe, England, on Christmas Day in the year of Galileo's death, 1642. His father, a farmer, described as a wild, extravagant, weak man, was already dead. His mother, reputed to be thrifty and industrious, remarried when her son was three years old and then left his upbringing to the care of his grandmother. (Newton's far from extraordinary genealogy should give pause to advocates of any rigid eugenics program!)

Young Isaac, a puny, premature baby, showed in childhood early signs of talent in the design and construction of many gadgets: a water clock, water wheels, sundials, and a working model of a windmill. As a teenager he rose to the top of his class and, after he was discovered shirking farm chores to study mathematics, his mother wisely encouraged him to continue at school and, in 1661, to go to Cambridge University.

Newton took his degree at Cambridge in 1665 and then, with the university closed because of the great plague that was raging in London, spent two golden years of full leisure and quiet contemplation in his isolated little home in Woolsthorpe. At the end of this interval, although not yet 25 years old, he had laid the foundations for his

"I keep the subject constantly before me and wait till the first dawnings open little by little into the full light."

Sir Isaac Newton
(1642–1727)

greatest work: the laws of motion and universal gravitation, the calculus, and the nature of white light.

Returning to Cambridge, Newton soon became Lucasian Professor of Mathematics, a chair which he held for nearly 30 years. His discoveries in mathematics and his brilliant experimental work in optics brought him fame, but his main contributions to mechanics were unknown to the world until, skillfully coaxed by the astronomer, Edmund Halley, Newton consented to publish his findings. Two years of prodigious efforts culminated in the publication in 1687 of the most celebrated of scientific works, his *Principia*, or, in full, *Philosophiae Naturalis Principia Mathematica*. Not until 1704 did his beautifully presented *Opticks* appear. This latter work, which attests to his skill in experimental physics, alone would have earned him a formidable reputation.

Mathematicians and physicists regret the many diversions that interrupted Newton's scientific work. In 1696 he was made Warden of the Mint, and three years later became Master of the Mint. Contrary to popular superstition which portrays such men of abstract thought as being impractical in business and finance, Newton proved to be an able administrator and one of the best Masters that the Mint ever had. The portrait suggested by his brilliant labors in mathematics and physics and useful toil at the Mint is incomplete. Newton also had a strong mystical bent which led him to devote a large fraction of his time to alchemy and theology.

Knighted in 1705, honored by his countrymen and renowned abroad, Newton maintained good health and a peerless intellect until over 80, dying in 1727 in his 85th year.

The most illustrious scientists who came after Newton have been unanimous in their assessment of the magnitude of his contributions. Gottfried Wilhelm Leibnitz, the great German mathematician and philosopher who independently discovered calculus, wrote: "Taking mathematics from the beginning of the world to the time when Newton lived, what he has done is much the better part." The famous French mathematician and theoretical physicist, Pierre-Simon Laplace (1749–1827) states, "The *Principia* is pre-eminent above any other production of human genius." More recently, Sir Arthur S. Eddington (1882–1944) points out, "To suppose that Newton's great scientific reputation is tossing up and down on these latter-day revolutions is to confuse science with omniscience."

As a closing thought on the life of this great man, we select Newton's own evaluation of his efforts, which has a characteristic charm: "I do not know what I may appear to the world; but to myself I seem to have been only like a boy playing on the seashore, and diverting myself in now and then finding a smoother pebble or a prettier shell than ordinary, whilst the great ocean of truth lay all undiscovered before me."

3.1 NEWTON'S THIRD LAW

A very general statement (which we have in fact briefly mentioned in the previous chapter) about the nature of forces is known as Newton's third law: *The force exerted by an object A on another object B is equal in magnitude and opposite in direction to the force exerted by object B on object A.* The situation is illustrated in Fig. 3-1, and can be written as the vector equation

$$\vec{F}_{\text{on } A \text{ by } B} = -\vec{F}_{\text{on } B \text{ by } A}$$

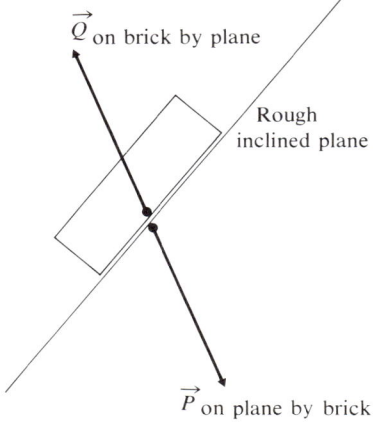

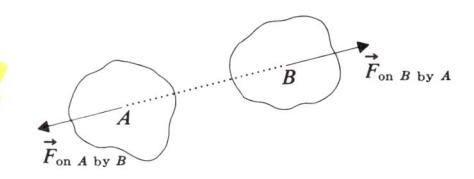

Fig. 3-1 Newton's third law.

(The significance of the minus sign is, as the reader may recall from our discussion of vectors, that the two forces are parallel but act in opposite directions.) It is essential to understand that the two forces referred to in the third law always act on *different* objects. Complete understanding of this latter statement is of the utmost importance. Whatever difficulties arise from matters in which Newton's third law is involved almost invariably may be traced to inadequate realization that each of the pair of forces involved acts on a different body.

We have already encountered an example of a pair of forces which satisfy Newton's third law: the gravitational force exerted by particle m on particle M is equal and opposite to that exerted by particle M on particle m (Fig. 2-20). At the surface of contact between two objects we meet other such pairs of forces. Consider the brick which is sliding down the rough inclined plane of Fig. 3-2. The brick exerts a force $\vec{P}$ on the plane and the plane exerts a force $\vec{Q}$ on the brick. Newton's third law states $\vec{P} = -\vec{Q}$. Suppose we wish to determine the acceleration of the brick. According to Newton's second law, $\vec{F} = m\vec{a}$, we must determine the resultant of all the forces acting *on* the brick; the force diagram Fig. 3-3 must therefore include the force $\vec{Q}$ because this

Fig. 3-2 Two forces which are related by Newton's third law.

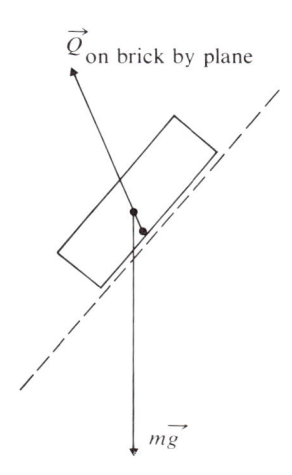

Fig. 3-3 The forces which determine the acceleration of the brick. Resultant force = $\vec{Q} + m\vec{g}$.

acts on the brick, but, and this is the point where it is easy to err, it must not include the force $\vec{P}$ because this force does *not* act on the brick; instead it acts on the plane.

Example 1 A man (Fig. 3-4), in throwing a 5.00-kg stone vertically upward, exerts an upward force of 200 N on the stone.
(a) How large is the downward force that the stone exerts on the man's hand?
(b) What is the acceleration, if any, of the stone?

Solution
(a) According to Newton's third law, the stone exerts a downward force on the man's hand of exactly 200 N.
(b) The acceleration of the stone is determined by the forces acting *on* it, which are (1) its weight, (5.00 kg) × (9.8 N/kg) = 49 N, acting downward and (2) the upward force of 200 N exerted by the man *on* the stone. Since these forces are not equal there is an acceleration. The resultant force acting on the stone is 200 N − 49 N = 151 N upward. The acceleration is determined from Newton's second law:

$$151 \text{ N} = (5.00 \text{ kg}) \times a.$$

Therefore,

$$\vec{a} = 30.2 \text{ meters/sec}^2 \text{ upward}.$$

Two points are to be noted: (1) The force exerted by the stone on the man is *not* determined by the *weight* of the stone. *It is always exactly the force exerted by the man on the stone*, whatever this force happens to be. (2) The force exerted *by* the stone is *not* one of the forces to be considered in determining the acceleration of the stone.

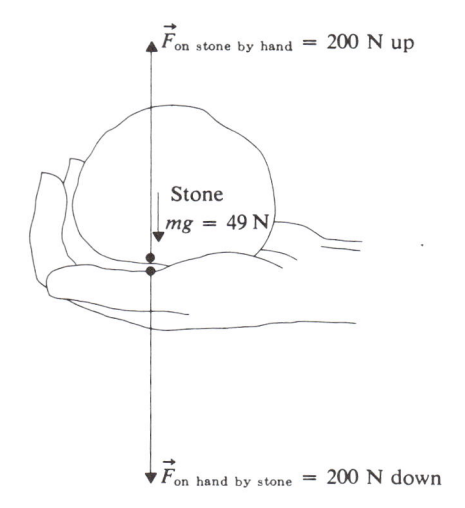

Fig. 3-4 The forces referred to in Example 1.

3.2 MOMENTUM AND NEWTON'S SECOND LAW

The product $m\vec{v}$ of the mass of a particle and its velocity vector is defined to be the *momentum* vector $\vec{p}$ of the particle:

$$\vec{p} = m\vec{v} \tag{3.1}$$

You will soon see that this quantity merits attention because it enters into the formulation of interesting physical laws. From the definition it is apparent that a 15×10^3-kg freight car moving south at 3.0 meters/sec has the same momentum as an automobile of mass 1.5×10^3 kg moving south at 30 meters/sec.

From the definition of momentum, $\vec{p} = m\vec{v}$, it follows that when the velocity changes by an amount $\Delta \vec{v}$, the associated change in momentum $\Delta \vec{p}$ is given by

$$\Delta \vec{p} = m \Delta \vec{v},$$

for a particle of constant mass m. Now if we consider a time interval

100   MECHANICS

Δt during which there is no appreciable change in the acceleration $\vec{a}$, then, from the definition of acceleration we have

$$\Delta \vec{v} = \vec{a}\Delta t.$$

Therefore,

$$\Delta \vec{p} = m\vec{a}\Delta t.$$

Since $\vec{F} = m\vec{a}$, we now have the useful result

$$\Delta \vec{p} = \vec{F}\Delta t,$$

which states that the change in momentum is equal to the product of the force and the time for which this force acts.

We can divide both sides of the last equation by the time interval Δt and get

$$(\Delta \vec{p}/\Delta t) = \vec{F}, \tag{3.2}$$

which states that the rate of change of momentum equals the resultant force. Newton actually formulated the second law this way instead of saying, as we have, that $\vec{F} = m\vec{a}$. The two statements are equivalent if the mass m is constant, as has been assumed in all the algebraic manipulations in this section. However, as we shall see in due course, at particle velocities near the velocity of light we must modify some of our mechanics. It turns out that the formulation of Newton's law given by Eq. (3.2), that is, in words,

Rate of change of momentum = resultant force,

is correct even for very large velocities, but that $\vec{F} = m\vec{a}$ gives the wrong answer in the realm of velocities near the velocity of light. In this sense, Newton's formulation, $\Delta \vec{p}/\Delta t = \vec{F}$, is preferable to $\vec{F} = m\vec{a}$.

Momentum of a System

Suppose we select for our consideration a system which consists of many particles. The system need not be a rigid body or even just one large body. In the analysis of a certain collision it might be convenient to define our system as the combination of a Ford traveling north and a Chevrolet traveling east. It is rather surprising that even systems like this obey simple mechanical laws.

A most important quantity is the total momentum vector $\vec{P}$ of the system which is defined to be the vector sum of the momentum vectors of the particles belonging to the system (that is, $\vec{P} = \vec{p}_1 + \vec{p}_2 + \ldots$). To discover how $\vec{P}$ changes as time goes on, we proceed as follows. For each individual particle the rate of change of its momentum vector equals the resultant force acting on it. The rate of change of the total momentum vector, $\Delta \vec{P}/\Delta t$, is the sum of the rates of change of each individual momentum vector and so equals the sum of all the forces acting on all the particles. Fortunately, in summing all these forces, the forces exerted by one particle of the system on another particle of the system cancel out in pairs according to Newton's third law. There remains only the vector sum of the forces exerted by particles outside the system on the particles of the system: such forces are called *external forces*. Denoting the vector sum of these external forces by $\vec{F}_{\text{ext}}$ we obtain the result

$$\frac{\Delta \vec{P}}{\Delta t} = \vec{F}_{\text{ext}}, \tag{3.3}$$

which states that the rate of change of momentum of a system of particles equals the vector sum of the *external forces* acting on the system. This result has the same form as Newton's second law for a particle, but now we know that it is true for a system of many particles. A key point to grasp is that the *internal forces*, forces exerted by one particle of the system on other particles of the system, do not change the total momentum of the system. The momentum changes of a pair of particles produced by their interaction are always vectors of equal magnitude and opposite direction, and hence these changes cancel out.

3.3 CONSERVATION OF MOMENTUM

A most important consequence of the equation

$$\frac{\Delta \vec{P}}{\Delta t} = \vec{F}_{\text{ext}}, \tag{3.3}$$

is that if $\vec{F}_{\text{ext}}$ is zero, there is no change in the total momentum vector $\vec{P}$. We have been led from Newton's second and third laws to assert the *law of conservation of momentum*, which can be expressed as follows: *When the vector sum of the external forces on a system of particles is zero, the total momentum $\vec{P}$ of the system remains constant.*

As has been pointed out in Chapter 1, in the terminology of physics a quantity whose total amount does not change while individual amounts alter is said to be conserved, and we speak of a law of conservation of this quantity. Conservation laws are regarded by physicists as particularly important for several reasons. For one thing, it is natural to suspect that a quantity which remains the same throughout complex and perhaps violent changes is somehow of fundamental significance. Certainly a conservation law has the great merit of being easy to apply; in any process in which a certain quantity is conserved one has merely to equate the total of the conserved quantity at the beginning of the process to its total at the end. It is, moreover, interesting to note that in the history of physics conservation laws, which arose within the context of a particular theory have, in several instances, turned out to be valid even in domains where the original theory was no longer successful. Thus in the physics of elementary particles, although Newtonian mechanics fails, we find that momentum and energy (and several other quantities, electric charge for one) are conserved.

One system with no external forces is the system comprising all the particles of the universe. The law of conservation of momentum then states that the momentum of the universe is constant as time goes on, any change in momentum of one part being canceled out by an equal and opposite change in another part.

Momentum conservation can sometimes be applied to systems even when $\vec{F}_{ext}$ is not zero. Such a system is a shell exploding in midair. This system is, of course, being continually acted on by the gravitational attraction of the earth. The forces which are exerted by particles that are outside the system and which do not participate in a violent process such as a collision or explosion, will change the total momentum of the system only a very little during the short time interval in which the violent process takes place. For such systems, to a very good approximation, we can assume momentum conservation and state

$$\begin{bmatrix}\text{Total momentum vector just}\\ \text{before collision or explosion}\end{bmatrix} = \begin{bmatrix}\text{total momentum vector just}\\ \text{after collision or explosion.}\end{bmatrix} \quad (3.4)$$

To see how to apply this law, let us consider a few examples.

Example 2 A cannon (Fig. 3-5) fires a 3.0-kg cannon ball which emerges from the cannon traveling horizontally with a speed of 100 meters/sec. The mass of the cannon is 2.0×10^3 kg. What are the various velocities of interest (cannon ball's, cannon's) just as the ball emerges?

Solution We all know that the cannon will recoil, and the reader will also now recognize the physical law governing recoil: the law of conservation of momentum. The recoil velocity is easily computed. Before firing, the total momentum vector of the system comprising the cannon ball and the cannon was evidently zero and, therefore, from Eq. (3.4) we conclude the total momentum vector of the same system must still be zero just after the cannon is fired. So the cannon must recoil with a momentum equal and opposite to that of the cannon ball's momentum in order that the vector sum of these two momentum vectors be zero. Equality of magnitudes of these momenta requires

$$(2.0 \times 10^3 \text{ kg}) \times V = (3.0 \text{ kg}) \times (100 \text{ meters/sec}),$$

where V is the speed of the cannon. Therefore,

$$V = 0.15 \text{ meter/sec.}$$

The required velocities can thus be summarized as follows: The cannon ball goes *one way at 100 meters/sec* and the cannon goes in the *opposite direction at 0.15 meter/sec.*

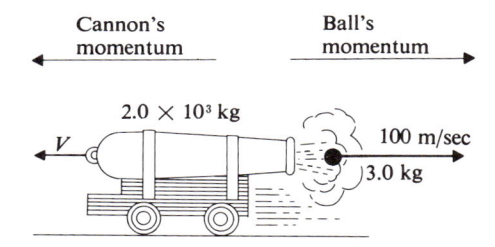

Fig. 3-5 The cannon recoils with a momentum equal and opposite to that of the cannon ball.

Example 3 Now let us suppose that the 3.0-kg cannon ball of Fig. 3-5, while traveling horizontally at 100 m/sec, strikes a 57-kg block of wood and becomes embedded in it. What is the speed of the wood with the embedded cannon ball at this instant?

Solution Before collision, the system of the block and the cannon ball has the momentum only of the ball, 3.0×100 kg m/sec. Just after collision, the block with the cannon ball embedded in it (total mass = 57 kg + 3 kg) must move in the original direction of motion of the cannon ball with a speed v such that momentum is conserved. Eq. (3.4) for this collision therefore gives

$$(3.0 \text{ kg}) \times (100 \text{ m/sec}) = (60 \text{ kg}) \times v.$$

Hence,
$$v = 5.0 \text{ m/sec}.$$

Before leaving this situation, it is interesting to consider a different choice of system: that comprising the cannon, the cannon ball, and the block. It is easy to verify that, for this system, the total momentum vector is zero at all instants considered. (We have neglected the effect of gravitational forces which would give the cannon ball a vertical component of momentum.)

We emphasize that momentum conservation is a *vector* law by examining a collision in which several different directions are involved.

Example 4 A 1.20×10^3-kg car is traveling north at 30.0 meters/sec and collides with a 3.60×10^3-kg truck traveling east at 10.0 meters/sec. The vehicles interlock and move off together. Find their common velocity $\vec{v}$.

Solution The total momentum before collision, as shown in Fig. 3-6, is directed northeast and has a magnitude

$$P = \sqrt{(1.20 \times 10^3 \text{ kg})^2 \times (30.0 \text{ m/sec})^2 + (3.60 \times 10^3 \text{ kg})^2 \times (10.0 \text{ m/sec})^2}$$
$$= 3.60 \times 10^4 \times \sqrt{2} \text{ kg m/sec}$$
$$= 5.09 \times 10^4 \text{ kg m/sec}.$$

After collision, momentum conservation requires that the interlocked vehicles go northeast with a speed v determined by Eq. (3.4):

$$(3.60 \times 10^3 \text{ kg} + 1.20 \times 10^3 \text{ kg})v = 5.09 \times 10^4 \text{ kg m/sec}.$$

Therefore
$$v = 10.6 \text{ m/sec, and}$$
$$\vec{v} = 10.6 \text{ m/sec northeast}.$$

Valuable and profound as the law of conservation of momentum

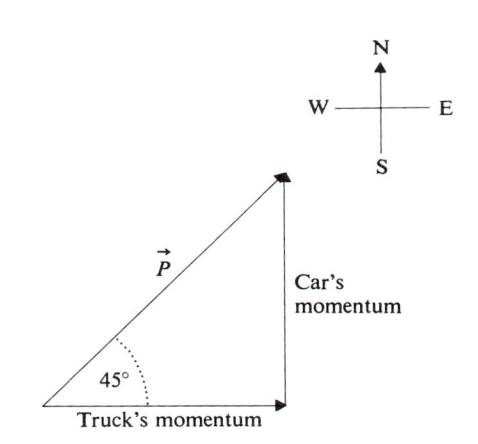

Fig. 3-6 The total momentum $\vec{P}$ is the *vector* sum of the car's and truck's momentum *vectors*.

is, one should not get the impression that this law alone is sufficient to determine what happens after a collision. Detailed prediction requires a knowledge of the forces acting during the collision. However, whatever does happen must conserve momentum. Thus the car and truck might very well bounce off one another and go off in different directions. Conservation of momentum merely tells us that the vector sum of the momentum of the truck and the momentum of the car must be the vector $\vec{P}$ determined above.

Rocket Propulsion

Momentum conservation permits us to understand rocket propulsion. A rocket increases its forward momentum simply by ejecting fuel backwards with as large a momentum as possible. Let us analyze the motion of a rocket in remote space so that gravitational forces can be ignored. We take as a convenient frame of reference a space ship traveling with a constant velocity which happens to be equal to the velocity of the accelerating rocket at a certain instant. The rocket ejects fuel backward at a rate R (in kilograms per second) with a velocity V relative to the rocket. Consider the system of mass m comprising the rocket and all the fuel it has at this instant. In the next interval Δt a mass $R\Delta t$ of fuel has acquired a momentum $(R\Delta t)V$ backwards, and, since momentum is conserved, we know that during the fuel ejection the fuel must have exerted a force forward on the rocket (and its remaining fuel) sufficient to impart forward momentum of equal magnitude. Denoting the change in the rocket's speed by Δv, the forward momentum increase is $(m - R\Delta t)\Delta v$ which is approximated by simply $m\Delta v$, if Δt is small. Momentum conservation for our system therefore requires

$$(R\Delta t)V = m\Delta v,$$

so

$$RV = m(\Delta v/\Delta t),$$

or

$$RV = ma,$$

where a $(= \Delta v/\Delta t)$ is the rocket's acceleration. This equation shows that RV plays the role of the force in $F = ma$; it is called the rocket *thrust*.

$$\text{Thrust} = RV.$$

Evidently to design a rocket engine with a large thrust, one must employ a fuel ejection velocity as large as possible and squirt out large quantities of the fuel rapidly.

The Present Status of the Momentum Conservation Law

In atomic physics and in the world of the elementary particles, the law of conservation of momentum has survived every experimental test and therefore plays a vital role in our analysis of events. In the field of quantum physics, with which we shall concern ourselves in Chapter 16, Newton's second and third laws are no longer successfully employed, but the momentum conservation law survives. So in our derivation of the law of conservation of momentum from Newton's third law, we built better than we knew; we constructed something of greater significance than appeared on the surface. The law of conservation of momentum seems to be one of nature's most fundamental laws.

3.4 CENTER OF MASS

An examination of the motion of any system of particles shows that, even though the system undergoes wild gyrations and perhaps explosions, there is a representative point C called the center of mass of the system whose motion gives a simple description of the average motion of the entire system. The mass center C of a system is a point moving with a velocity $\vec{V}_c$ such that an (imaginary) particle with a mass M equal to the total mass of the system, located at C, would have a momentum $M\vec{V}_c$ equal to the total momentum $\vec{P}$ of the system. This requires

$$M\vec{V}_c = m_1\vec{v}_1 + m_2\vec{v}_2 + \ldots \tag{3.5}$$

where the sum extends over all the momentum vectors $m_1\vec{v}_1$, $m_2\vec{v}_2$, etc., of the particles of the system. It is plausible and easy to prove that Eq. (3.5) will be satisfied by defining the center of mass C to be the point with a position vector $\vec{r}_c$ given by

$$M\vec{r}_c = m_1\vec{r}_1 + m_2\vec{r}_2 + \ldots \tag{3.6}$$

An investigation of the consequences of the equation defining the location of the center of mass, Eq. (3.6), shows that a uniform regular solid has its center of mass at its geometrical center. (It can be shown that the center of mass of a system coincides with what is called the center of gravity. In a uniform gravitational field the center of gravity is the balance point where the system in any orientation can be supported in equilibrium by the application of a single force.) Let us now consider an example.

Example 5 Locate the center of mass of the system of two particles (Fig. 3-7) of mass 10 kg and 5.0 kg with position vectors 3.0 meters and 6.0 meters in the direction OX.

Solution With all vectors in the same direction, Eq. (3.6) gives

$$(15 \text{ kg}) \times r_c = (10 \text{ kg}) \times (3.0 \text{ meters}) + (5.0 \text{ kg}) \times (6.0 \text{ meters}).$$

Hence,

$$r_c = 4.0 \text{ meters, and the vector } \vec{r}_c \text{ is in the direction } OX.$$

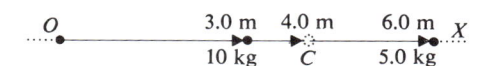

Fig. 3-7 The mass center C of a system of two particles.

106 MECHANICS

As seems intuitively plausible, we see that the mass center lies on the line joining the particles and is closer to the more massive particle. Notice that there does not have to be any actual matter at the location of the center of mass of the system.

The motion of the mass center of a system can be found by just thinking of an imaginary particle of mass M at the mass center. We know that the total momentum $\vec{P}$ of the system is equal to $M\vec{V}_c$. We have shown [Eq. (3.3)] that the rate of change of the total momentum $\vec{P}$ equals the vector sum of the external forces acting on the system, $\vec{F}_{\text{ext}}$. Therefore, $\vec{F}_{\text{ext}}$ equals the rate of change of $M\vec{V}_c$ which, for constant mass M, is $M\vec{a}_c$, where $\vec{a}_c$ is the acceleration of the center of mass. This remarkable result,

$$\vec{F}_{\text{ext}} = M\vec{a}_c, \qquad (3.7)$$

is called the *law of motion of the center of mass,* which states: *the center of mass of a system moves like a particle of mass equal to the total mass of the system, acted upon by the vector sum of the external forces.* Evidently, the problem of describing the average motion of an entire complex system is simply the problem of determining the motion of a single particle of mass M at the center of mass.* No matter what the point of application of an external force, the effect of the force on the motion of the center of mass is the same as if the force were *applied at the center of mass.* Some examples follow.

*When $\vec{F}_{\text{ext}}$ is zero, and consequently $\vec{P}$ remains constant, the motion of the center of mass C is extremely simple. No matter how the parts of the system fly about, the center of mass moves with a constant velocity $\vec{V}_c$.

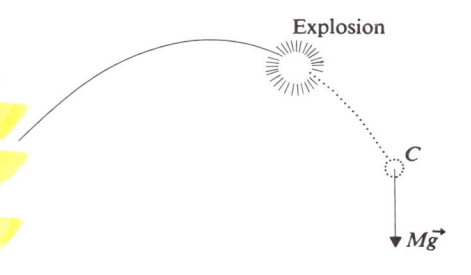

Fig. 3-8 The internal forces involved in an explosion of a shell do not influence the motion of its mass center C.

Example 6 A shell blows up in midair (Fig. 3-8). Describe the motion of the mass center.

Solution Neglecting air resistance, the vector sum of the external forces is just $M\vec{g}$ (the explosion involves only *internal forces*). Thus the law of motion of the mass center gives

$$M\vec{g} = M\vec{a}_c.$$

Hence, $\vec{a}_c = \vec{g}$ before and after explosion. The mass center moves on its parabolic trajectory as if nothing had happened. Note, however, that we have not said anything about the individual motions of the various shell fragments.

Example 7 A 10-kg stick which is 5 meters long lies in a north-south direction on a sheet of ice (Fig. 3-9). It is pushed by a 300-N force (about 68-lb force) applied at one end and directed east. Find the acceleration of the center of the stick.

Solution The geometrical center of the stick is also its mass center because the stick is uniform. The mass center moves like a 10-kg particle acted upon by a force of 300 N east. (The fact that the force is applied

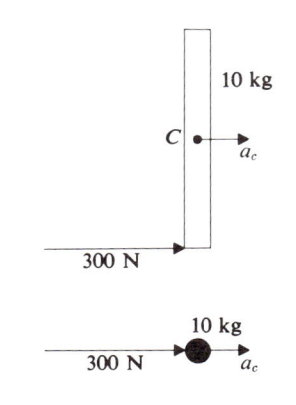

Fig. 3-9 The acceleration of the mass center of the stick is the same as that of a 10-kg particle acted upon by a 300-N force.

3.4 CENTER OF MASS

at the end of the stick does not affect the acceleration of the mass center.) Therefore,
$$300 \text{ N} = (10 \text{ kg}) \times a_c;$$
so $a_c = 30$ m/sec² and the direction of $\vec{a}_c$ is east.

3.5 ANALYSIS USING COMPONENTS OF VECTORS

When we wish to apply the equation, $\vec{F}_{\text{ext}} = M\vec{a}_c$, to situations involving several non-parallel forces, it is most convenient to introduce components of vectors. It is therefore necessary to return to the subject of vectors in order to acquire facility in working with components.

A first point to notice is that two vectors $\vec{a}$ and $\vec{b}$ are equal *if and only if* their corresponding components are equal. Consequently, for vectors in the *XY*-plane, the one vector equation
$$\vec{a} = \vec{b}$$
is equivalent to two equations relating components:
$$a_x = b_x, \qquad a_y = b_y.$$

Addition of Vectors by Adding Components

One major reason for the great usefulness of components is that the component along a given direction of the vector sum of any number of vectors is just the algebraic sum of the components of these vectors in the specified direction. This is illustrated in Fig. 3-10 for the vector addition of two vectors $\vec{a}$ and $\vec{b}$ to form a sum $\vec{s} = \vec{a} + \vec{b}$. Along the *X*-axis, the *x*-component of the vector sum s_x is the algebraic sum $a_x + b_x$ of the *x*-components of the two vectors that have been added; that is
$$s_x = a_x + b_x.$$
Similarly, along the *Y*-axis we have
$$s_y = a_y + b_y.$$

Knowledge of the components (s_x, s_y) completely determines the vector $\vec{s}$. If we wish to find the numerical value of s and the angle θ that $\vec{s}$ makes with the *X*-axis, we can use
$$s = \sqrt{s_x^2 + s_y^2}$$
and
$$\tan \theta = \frac{s_y}{s_x}.$$

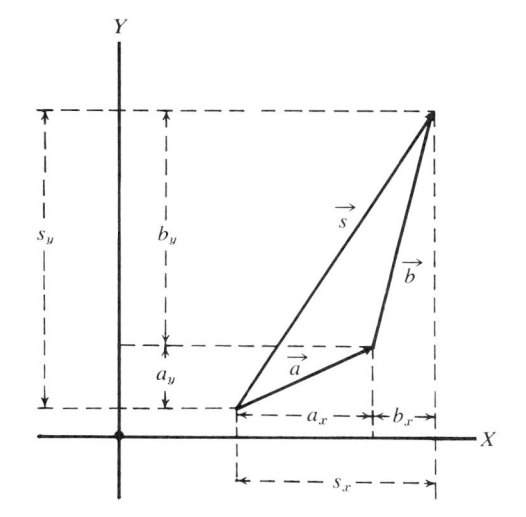

Fig. 3-10 Geometric proof that $s_x = a_x + b_x$ and $s_y = a_y + b_y$.

108   MECHANICS

It is evident that, by working with components, the addition of any number of vectors can be performed analytically simply by computing algebraic sums. Because the problem of adding vectors arises in nearly every branch of physics, the topic merits special attention. The following is a general systematic procedure for adding vectors analytically:

1. Introduce rectangular coordinate axes. Try to select an orientation of the axes that will minimize the labor necessary to compute the components along the axes.
2. Resolve each vector into its x- and y-components.
3. Calculate the algebraic sum of the x-components of the different vectors to determine the x-component of their vector sum. The y-components are treated similarly.
4. When necessary, the magnitude s and the direction of the vector sum $\vec{s}$ can be obtained from (s_x, s_y).
5. A check that the evaluation of $\vec{s}$ is at least approximately correct should be made by sketching a rough geometric head-to-tail addition of the vectors according to the triangle rule or its extension (Figs. 2-5 and 2-6).

Example 8 Figure 3-11 shows two forces acting on a particle. The force $\vec{F}_1$ is directed vertically downward and has a magnitude of 20.0 N. The force $\vec{F}_2$, of magnitude 50.0 N, makes an angle of 36.9° above a horizontal direction. Find the single force $\vec{F}$ equivalent to this combination of forces.

Solution From the vector addition rule for forces, the forces $\vec{F}_1$ and $\vec{F}_2$ are together equivalent to the single force $\vec{F}$, the *resultant force*, which is given by
$$\vec{F} = \vec{F}_1 + \vec{F}_2.$$

We can evaluate this vector sum by adding corresponding components. With axis OX horizontal and OY vertically upward we find

$$\begin{aligned}
F_x &= F_{1x} + F_{2x} \\
&= 0 + (50.0 \text{ N}) \cos (36.9°) \\
&= (50.0 \text{ N})(0.800) \\
&= 40.0 \text{ N}. \\
F_y &= F_{1y} + F_{2y} \\
&= -20.0 \text{ N} + (50.0 \text{ N}) \cos (53.1°)^* \\
&= -20.0 \text{ N} + (50.0 \text{ N})(0.600) \\
&= -20.0 \text{ N} + 30.0 \text{ N} \\
&= 10.0 \text{ N}.
\end{aligned}$$

Therefore
$$\begin{aligned}
F &= \sqrt{F_x^2 + F_y^2} \\
&= \sqrt{(40.0 \text{ N})^2 + (10.0 \text{ N})^2} = 41.2 \text{ N},
\end{aligned}$$

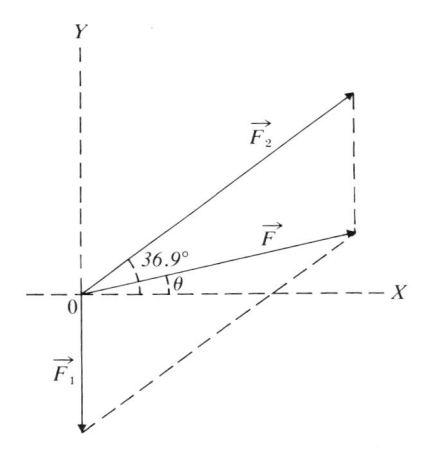

Fig. 3-11 Vector addition of the two forces of Example 8. $\vec{F}$ is the resultant of $\vec{F}_1$ and $\vec{F}_2$.

*The magnitude of the component of a vector along any axis is the product of the magnitude of the vector and the cosine of the acute angle between the vector and the axis. This component is a negative number if the corresponding vector component points in the negative direction for the axis in question.

3.5 ANALYSIS USING COMPONENTS OF VECTORS

and

$$\tan \theta = \frac{F_y}{F_x} = \frac{10.0 \text{ N}}{40.0 \text{ N}} = 0.250,$$

$$\theta = 14.0°.$$

The resultant force, the single force equivalent to the combination of $\vec{F}_1$ and $\vec{F}_2$, is a force of magnitude 41.2 N at an angle 14.0° above the horizontal X-axis.

3.6 USING NEWTON'S LAWS

It is simple enough to remember the formula $\vec{F} = m\vec{a}$ for a particle, or $\vec{F}_{\text{ext}} = M\vec{a}_c$ for the acceleration of the center of mass C of a system of particles; but a student must also learn and practice a systematic procedure for applying these laws in order to analyze with confidence a great variety of problems. The procedure given below will, if mastered, assist the student in solving many different types of problems not only in mechanics but also in all the various branches of physics.

1. *Preliminary sketch.* First draw a rough schematic diagram of the objects of interest. Visualize what will happen.
2. *System selection.* Select the object or objects which will comprise the system whose motion you are going to study. This can often be done in different ways, so for clarity of thought it is vital that you specify the system.
3. *Coordinate axes.* Choose perpendicular coordinate axes fixed in an inertial frame of reference. Orient the axes in such a way as to minimize future labor. When the acceleration vector has a known constant direction, it is usually convenient to have a coordinate axis in this direction.
4. *Free body diagram.* Draw a new schematic diagram showing *only the system* you have selected. Draw in vectors for all the *external forces acting on all the objects of the system*. Do *not* include forces exerted *by* the objects of the system.
5. *Finding an unknown acceleration or an unknown force.* Resolve into components along the coordinate axes each force appearing on the free body diagram. Then use these components to write down the following equations:

   Algebraic sum of x-components of external forces $= Ma_x$, (3.8a)
   Algebraic sum of y-components of external forces $= Ma_y$, (3.8b)

   where a_x and a_y are the components of the acceleration of the system's center of mass. For motion in the XY-plane, these equations are equivalent to the law of motion of the center of mass, $\vec{F}_{\text{ext}} = M\vec{a}_c$.

Finally, solve Eqs. (3.8) to find the unknown force or the acceleration.
6. *Check*. Always, in any physics problem, think about the answer. If it is a numerical answer, is it a reasonable size? Are the units correct? Procedures for checking a formula will be illustrated after Example 10.
7. *Shortcuts*. Experienced and confident students may skip through steps 1 and 2, with the picture in their mind's eye, but one should always be sure to draw the *free body diagram* and ensure that it is correct.

Here, then, is an example with some of the above steps indicated in the solution.

Example 9 A train consists of a locomotive and three identical passenger cars, each of mass 10×10^3 kg. The train has a forward acceleration of 0.15 meter/sec². Find the force $\vec{F}_1$ exerted on the second passenger car by the first.

Solution
(Step 1) Fig. 3-12 shows the train and the masses of the cars. Each car has the same acceleration. The second car is pulled forward by the first and backward by the third car.
(Step 2) The system selected has been circled in Fig. 3-12. It comprises the second and the tnird cars. For this system $M = 10 \times 10^3$ kg $+ 10 \times 10^3$ kg $= 20 \times 10^3$ kg.
(Step 3) Any frame of reference fixed in the earth is an appropriate inertial frame. Since the acceleration is obviously horizontal and to the right in Fig. 3-12, we pick an axis OX in this direction, with OY vertical.
(Step 4) The free body diagram for the system we have selected is shown in Fig. 3-13. The external forces acting on this system are: The weight Mg of the two cars, the force N exerted on the cars by the rails, the force F_1 exerted on the second car by the

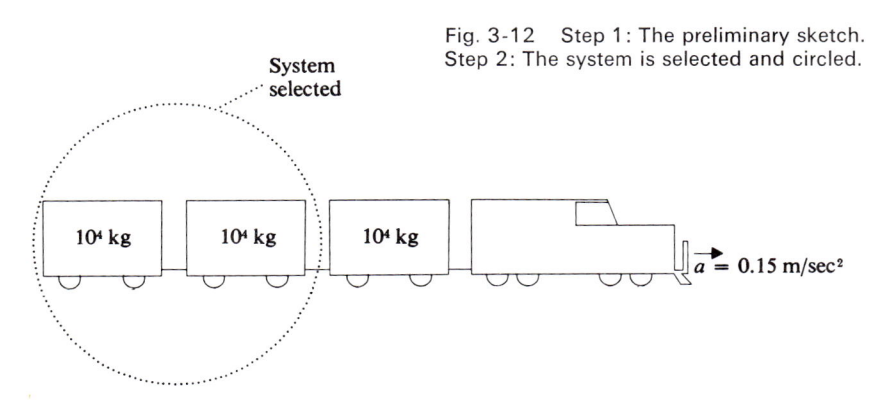

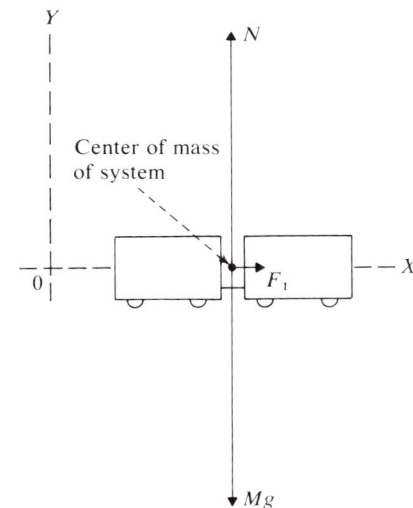

Fig. 3-12 Step 1: The preliminary sketch. Step 2: The system is selected and circled.

Fig. 3-13 Step 4: The free body diagram for the system circled in Fig. 3-12.

3.6 USING NEWTON'S LAWS 111

first. (For the purpose of determining the motion of the center of mass C of this system, all these forces could be represented as if they were applied at C.) Notice that the forces acting between the second and third cars do not appear on this free body diagram because they are *internal* forces for the system we have selected.

(Step 5) The acceleration of the center of mass C has the components $a_x = 0.15$ m/sec² and $a_y = 0$. Eq. (3.8a), the law of motion of the center of mass for *x*-components, yields

$$F_1 = Ma_x$$
$$= (20 \times 10^3 \text{ kg})(0.15 \text{ m/sec}^2)$$
$$= 3.0 \times 10^3 \text{ N},$$

so the force $\vec{F}_1$ exerted on the second car by the first is in the positive *x*-direction and has a magnitude of 3.0×10^3 N. (The equation for *y*-components gives only $N - Mg = Ma_y = 0$ which implies $N = Mg$.)

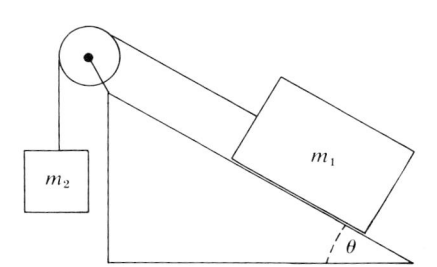

Fig. 3-14 Schematic diagram for Example 10.

Example 10 A mass m_1, on a smooth (frictionless) plane inclined at an angle θ with the horizontal, is attached to a light rope which passes around a light frictionless pulley and suspends a second mass m_2. Find the tension in the rope and the acceleration of the two masses.

Solution

(Step 1) The schematic diagram is shown in Fig. 3-14. We assume that the rope does not stretch. Then the two masses move with the same speed at every instant and consequently they have accelerations of the same magnitude. Whether m_2 is accelerated up or down, will depend upon the values m_1, m_2 and the angle θ.

To understand the forces involved in this problem, we must know something about ropes and pulleys. The tension in the rope at any cross-section is the force T exerted by the portion of the rope on one side of that cross-section upon the portion of the rope on the other side (Fig. 3-15). The key fact about a "light" rope and a "light frictionless" pulley is that the tension has the same value T throughout all segments of the rope; that is, the tension does not change when the rope passes around such an idealized pulley. (The reasons are examined in Questions 42 and S-36.)

(Step 2) First we select the mass m_1 as the system and study its motion.

(Step 3) We choose an axis OX fixed on the inclined plane and pointing down the plane.

(Step 4) Figure 3-16 is the free body diagram showing the external forces acting on m_1. The rope exerts a force on m_1 that has a magnitude

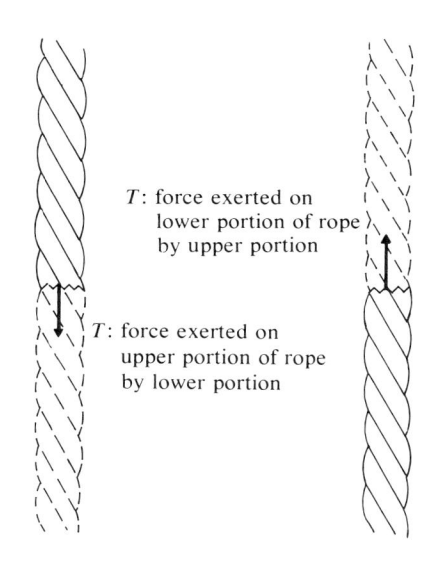

Fig. 3-15 Tension in the rope is a force of magnitude T. Newton's third law requires that the two forces shown above have equal magnitudes and opposite directions.

112 MECHANICS

which we have called the tension T in the rope. This force is a pull (you cannot push anything with a rope), so it acts up the plane along the line of the rope. The force N exerted on m_1 by the smooth plane is in a direction *normal* to the plane. An idealized, perfectly smooth plane exerts no frictional force; that is, the force exerted by a smooth plane has no component parallel to the surface. Therefore this force must be normal to the surface.

(Step 5) We resolve the weight $m_1\vec{g}$ into an x-component $m_1 g \sin \theta$ and a y-component $m_1 g \cos \theta$. The fact that $m_1\vec{g}$ is *replaced* by its components, and therefore is to be deleted from the figure, is indicated by the two short parallel lines drawn through this vector in Fig. 3-16. The law of motion of the center of mass for x-components [Eq. (3.8a)] yields

$$m_1 g \sin \theta - T = m_1 a \qquad (3.9)$$

where a denotes the x-component of the acceleration of m_1. Then the acceleration of m_1 is down or up the plane according to whether a is positive or negative.

We now must repeat the entire procedure for the system consisting of the mass m_2. For a coordinate axis directed vertically upward and the free body diagram as in Fig. 3-17, the law of motion of the mass center for vertical components yields

$$T - m_2 g = m_2 a. \qquad (3.10)$$

Notice that the tension is *not* equal to $m_2 g$ unless the acceleration of the suspended mass is zero.

Equations (3.9) and (3.10) are simultaneous equations in two unknowns, T and a. We can eliminate T by adding these equations. This yields

$$m_1 g \sin \theta - m_2 g = m_1 a + m_2 a,$$

which can be solved for a, giving

$$a = \frac{m_1 g \sin \theta - m_2 g}{m_1 + m_2}. \qquad (3.11)$$

Substituting this value for a in Eq. (3.10) and solving for T we obtain

$$T = \frac{m_1 m_2 \, g(1 + \sin \theta)}{m_1 + m_2} \qquad (3.12)$$

(Step 6) With the problem solved using symbols instead of numerical values, there are several checks that can be made. For instance, if the plane is horizontal corresponding to $\theta = 0$, and if the

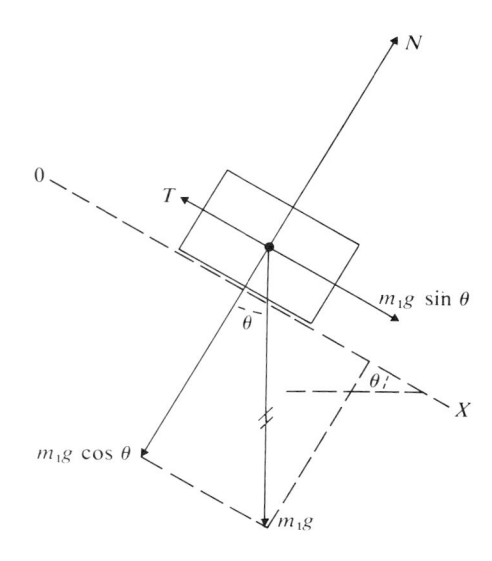

Fig. 3-16 Free body diagram for mass m_1 of Example 10.

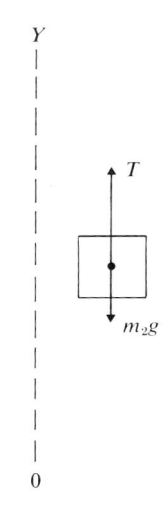

Fig. 3-17 Free body diagram for the block of mass m_2 of Example 10.

3.6 USING NEWTON'S LAWS

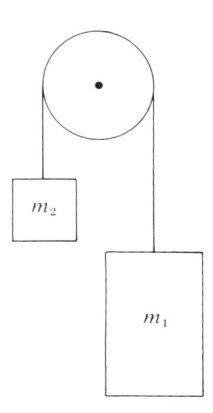

Fig. 3-18 Atwood's machine. Compare with Fig. 3-14 for the case when $\theta = 90°$.

suspended mass m_2 is zero, then it is obvious that the mass m_1 will remain at rest on the horizontal board and that the tension in the rope will be zero. And when we substitute $\theta = 0$, $m_2 = 0$ into Eqs. (3.11) and (3.12), using the fact that $\sin \theta = 0$, we find that these equations do correctly predict that $a = 0$ and $T = 0$. And if $m_1 = 0$, then m_2 will fall freely with an acceleration $\vec{g}$, and the tension will be zero. Substituting the value $m_1 = 0$ into Eqs. (3.11) and (3.12), we find that these equations do give the right answers, $a = -g$ and $T = 0$. If we give θ the value 90°, we achieve the situation in Fig. 3-18 in which the masses are suspended at the ends of a rope which passes over a pulley. In this device, called Atwood's machine, the acceleration of the masses, from Eq. (3.11) with $\sin \theta = \sin 90° = 1$, is given by

$$a = \left(\frac{m_1 - m_2}{m_1 + m_2}\right) g. \tag{3.13}$$

Now if these suspended masses are equal we know that their acceleration must be zero, and this is easily seen to be the prediction of Eq. (3.13) when $m_1 = m_2$.

Dimensions

Another way to spot some obvious errors in an equation is to check what are called the physical *dimensions* of each term in the equation. All the quantities that are introduced in physics can be expressed in terms of products or quotients of certain fundamental physical quantities such as length (L), mass (M), time (T), and electric charge (Q). For instance, a *velocity* is a *length* divided by a *time*. We say velocity has the physical dimensions $[L/T]$. A dimensional formula for a physical quantity is an expression, such as $[L/T]$ for velocity, which shows how length, mass, time, and charge enter into any formula for the physical quantity. The square brackets indicate that *only* the dimensions are represented, and that numerical values have been disregarded. Thus both 60 miles per hour and 2 centimeters per second have the dimensions $[L/T]$ since each is a length divided by a time.

The dimensions of any area are $[L^2]$ and of any volume $[L^3]$. Since an acceleration is a length divided by a time squared, the dimensional formula for an acceleration is $[L/T^2]$ which can be written $[LT^{-2}]$. The dimensions of force are the dimensions of mass times acceleration, or $[MLT^{-2}]$.

In science we start any mathematical work with equations in which all the terms have the same dimensions. After any legitimate manipulations of these equations to deduce new equations, it must remain true that every term in a given equation has the same dimensions. A mathematical error can lead to a proposed equation with terms of different dimensions. Such errors can easily be detected by checking the dimensions of all terms in an equation.

For example, in the equation

$$x = x_0 + v_{x0}t + \tfrac{1}{2}a_x t^2,$$

the dimensions are

$$[L] = [L] + \left[\frac{L}{T} \times T\right] + \left[\frac{L}{T^2} \times T^2\right].$$

Each term has the dimensions of a length, so the equation is dimensionally correct. A number like the quantity $\tfrac{1}{2}$ in this equation has no physical dimensions—it is called a dimensionless number. Notice that the equation would still have been dimensionally correct even if the number $\tfrac{1}{2}$ had been erroneously omitted. Obviously dimensional checks will not reveal all types of errors, but certain errors can easily be spotted. Suppose that after some manipulation we had found the acceleration of Example 10 to be

$$a = \frac{m_1 m_2 g}{m_1 + m_2}.$$

Here the right-hand side has the dimensions

$$\left[\frac{M^2}{M}\frac{L}{T^2}\right] = \left[\frac{ML}{T^2}\right]$$

which is not the dimension $[L/T^2]$ of acceleration. Because the proposed equation is not dimensionally correct, we could conclude that an error has been made. To check the dimensions for the expression for the acceleration actually found in Example 10, we first note that $\sin \theta$ is a dimensionless quantity since it is the ratio of two lengths. Then we obtain

$$\left[\frac{(m_1 \sin \theta - m_2)g}{m_1 + m_2}\right] = \frac{[M][LT^{-2}]}{[M]} = [LT^{-2}]$$

which are the dimensions of acceleration. Therefore our expression for acceleration is at least dimensionally correct.

3.7 FRICTION

We are often concerned in daily experience not only with gravitational forces, but also with the force exerted on one object by another when two solid objects are in contact, such as the block and the inclined plane in Fig. 3-19. The force exerted on the block by the plane can be resolved into two vector components: a vector component $\vec{N}$ perpendicular to the surfaces in contact, called the *normal component*; and a vector component parallel to the surfaces in contact, called the *force of friction* $\vec{f}$.

The direction of the friction force $\vec{f}$ is such that it always opposes the slipping or the tendency to slip of the surfaces in contact. If the block were sliding down the plane, the frictional force $\vec{f}$ would have the direction shown in Fig. 3-19.

It is found that the magnitude f of the frictional force is proportional to the magnitude N of the normal component. Experimental results are most conveniently summarized in the following empirical relations called the laws of friction:

1. While sliding occurs

$$f = \mu_k N \qquad (3.14)$$

where f is called the force of *sliding friction*, and the constant of proportionality μ_k, is called the *coefficient of kinetic (or sliding) friction*. This coefficient has a value which depends on the nature of the surfaces in contact, being small if the surfaces are smooth and large if the surfaces are rough. Although the coefficient of kinetic friction for two given surfaces has slightly different values at different relative speeds, in our work we shall assume for simplicity that μ_k is independent of speed.

2. If the surfaces in contact are at rest relative to each other, the frictional force $\vec{f}$, now named a force of *static friction*, has a magnitude which is not greater than

$$f_{\max} = \mu_s N, \qquad (3.15)$$

where the constant of proportionality μ_s, called the *coefficient of static friction*, is characteristic of the surface in contact and has a value greater than μ_k. When the surfaces in contact are at the point of slipping, $f = f_{\max} = \mu_s N$. Otherwise, the force of static friction is less than this value:

$$f < \mu_s N. \qquad (3.16)$$

Experiments show that μ_k and μ_s are nearly independent of the apparent area of contact between the two surfaces.

Microscopic examination reveals that even apparently smooth surfaces actually appear as shown in Fig. 3-20. Since the surfaces touch only at isolated high spots, the actual contact area is very much smaller than the apparent contact area. If the apparent contact area is decreased while N is kept constant, plastic deformation of high spots occurs in such a way that the actual contact area remains the same. At the contact points, molecules of the different bodies are close enough together to exert strong attractive intermolecular forces on one another, leading to the formation of "cold welds." The force of kinetic friction is associated

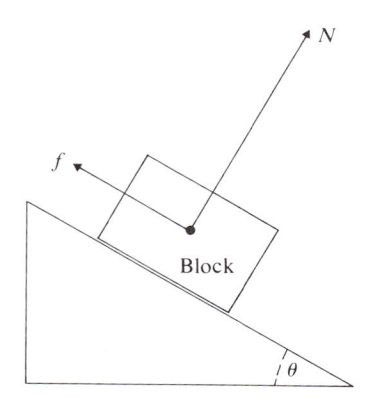

Fig. 3-19 Block sliding or tending to slide down the plane. The force exerted on the block by the plane is resolved into a component $\vec{N}$ normal to the surfaces in contact and a component parallel to these surfaces called the force of friction $\vec{f}$.

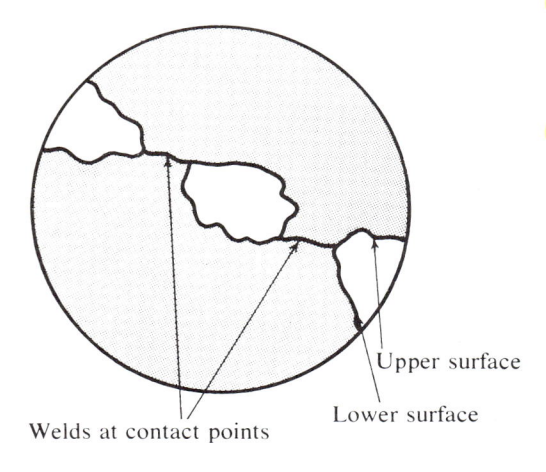

Fig. 3-20 Microscopic view of apparently smooth surfaces.

with the continual rupture and formation of thousands of such welds as one surface slides over the other.

Typical numerical values of coefficients of friction are given in Table 3-1.

Table 3-1. COEFFICIENTS OF FRICTION

Surfaces in Contact	Static, μ_s	Kinetic, μ_k
Steel on steel	0.74	0.57
Copper on steel	0.53	0.36
Glass on glass	0.94	0.40
Teflon on steel	0.04	0.04

These tabulated values are obtained for surfaces exposed to air and therefore coated with a film of oxide. When metal surfaces are carefully cleaned in a vacuum and then placed in contact, the surfaces become welded together and the corresponding value of the coefficient of friction is enormous.

Example 11 A block of mass m slides down a plane inclined at an angle θ with the horizontal. The coefficient of kinetic friction between the block and the plane is μ_k. Find the acceleration of the block.

Solution
(Step 1) Figure 3-19 is the schematic diagram.
(Step 2) The block is selected as the system.
(Step 3) We choose coordinate axes fixed in the plane with the X-axis directed down the plane.
(Step 4) In the free body diagram of Fig. 3-21, the frictional force $\vec{f}$ exerted on the block by the plane is parallel to the plane and in the direction opposite to the direction of motion of the block. The force $\vec{N}$ is the normal component of the force exerted on the block by the plane. The only other external force acting on the block is its weight $m\vec{g}$.
(Step 5) The weight $m\vec{g}$ is resolved into an x-component, $mg \sin \theta$, and a y-component of magnitude $mg \cos \theta$. The vector $m\vec{g}$ is replaced by its components. Then the law of motion of the center of mass for the x-components [Eq. (3.8a)] gives

$$mg \sin \theta - f = ma_x.$$

For the y-components, Eq. (3.8b) gives

$$N - mg \cos \theta = ma_y = 0$$

where $a_y = 0$ because the acceleration $\vec{a}$ is parallel to the plane.

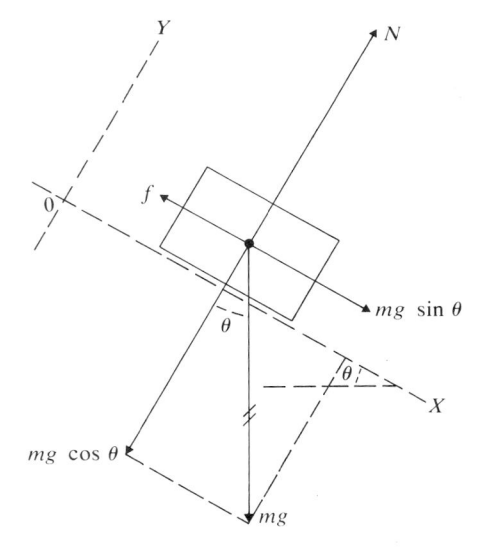

Fig. 3-21 Free body diagram for Example 11.

Consequently,
$$N = mg \cos \theta.$$

The law of sliding friction relates f and N:
$$f = \mu_k N.$$

Substituting $\mu_k N$ for f in the equation for a_x and then using $N = mg \cos \theta$ we obtain
$$a_x = g \sin \theta - \mu_k g \cos \theta.$$

(Step 6) To check dimensions, first note that $\sin \theta$, $\cos \theta$ and μ_k are ratios which are dimensionless. Then it is apparent that each term in this equation has the dimensions of acceleration. The equation is therefore dimensionally correct.

If the plane were vertical, the block would fall with an acceleration $\vec{g}$. And our equation for the acceleration does correctly predict that $a = g$ when $\theta = 90°$ ($\sin 90° = 1$, $\cos 90° = 0$).

(A word of caution is necessary. Our result was derived under the assumption that the frictional force has a negative x-component of magnitude $f = \mu_k N$. Since this is only true when the block is sliding in the positive x-direction, our result can be used only for this case.)

Example 12 The coefficient of static friction between the block and the inclined plane in Fig. 3-19 is given by $\mu_s = 0.75$. With the block at rest on the plane and θ initially zero, the left-hand end of the inclined plane is raised very slowly, and the angle θ increases. At what value of θ will the block start to slip?

Solution
(Step 1) Figure 3-19 is the schematic diagram. We are concerned here with a body at rest. Such a problem in *statics* can be treated by the methods used for dynamics, using zero for the value of the acceleration. (Problems in which we must also consider the conditions for rotational equilibrium will be treated in the following section.) While the block is at rest on the inclined plane, the force of friction acting on the block is a force of *static friction*.
(Step 2) We select the block as the system.
(Step 3) The coordinate axis OX is fixed on the inclined plane and points down the plane.
(Step 4) Figure 3-22 shows the free body diagram. The force of static friction exerted on the block by the plane is directed up the plane, opposing the tendency of the block to slip.

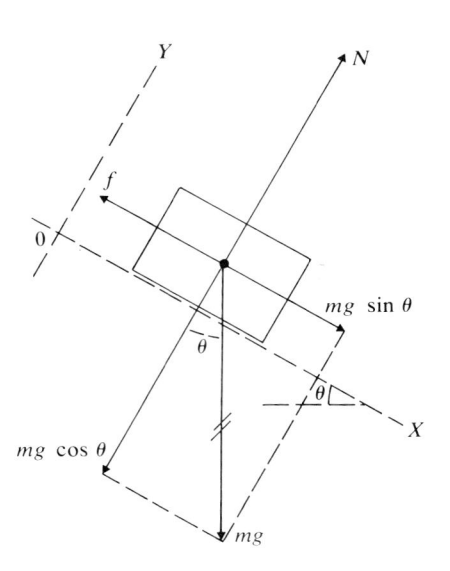

Fig. 3-22 Free body diagram for Example 12.

118 MECHANICS

(Step 5) We resolve the weight $m\vec{g}$ into x- and y-components and replace the vector $m\vec{g}$ by these components. Then the law of motion of the center of mass (for the special case when the acceleration is zero) gives for the x-components,

$$mg \sin \theta - f = ma_x = 0,$$

and for the y-components,

$$N - mg \cos \theta = ma_y = 0.$$

Consequently,

$$f = mg \sin \theta,$$
$$N = mg \cos \theta.$$

Dividing the first of these equations by the second gives

$$\frac{f}{N} = \frac{\sin \theta}{\cos \theta} = \tan \theta.$$

Suppose the block slips if θ is increased above the value θ_{max}. Then when the angle of inclination is θ_{max} the block is on the point of slipping, and the force of friction attains its maximum value given by

$$f_{max} = \mu_s N.$$

Then

$$\mu_s = \frac{f_{max}}{N} = \tan \theta_{max},$$
$$\tan \theta_{max} = \mu_s = 0.75,$$
$$\theta_{max} = 36.9°.$$

(At angles of inclination less than θ_{max}, the force of static friction will be less than $\mu_s N$. From the analysis above we see that f then will simply assume a value that cancels the component of the weight along the plane: that is, $f = mg \sin \theta$.)

3.8 STATICS OF RIGID BODIES

Torque About an Axis

The effect of a force on the rotational motion of an extended object depends not only upon the magnitude and direction of the force but also upon its line of action (see Fig. 3-23). That quantity which is the significant measure of the turning effect of a force about an axis is the *torque*, denoted by τ (Greek letter tau), which we now define.

Consider a force $\vec{F}$ lying in a plane and an axis perpendicular to the plane through the point A in that plane (Fig. 3-23). The perpendicular distance from the line of action of $\vec{F}$ to the axis is called the *moment arm*

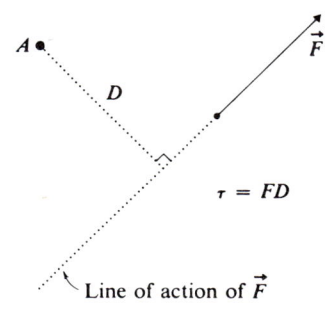

Fig. 3-23 The torque τ of a force F about an axis through A and perpendicular to the plane of the paper is equal to FD. The moment arm is D.

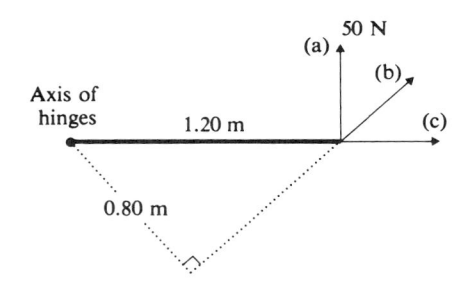

Fig. 3-24 The torque changes when the direction of the force changes because then the moment arm is different. The view is that seen looking down at the top of the door. The plane of the paper is horizontal.

*Torque and work or energy are very different physical quantities. To keep this distinction in mind the torque unit, the newton-meter, is not called by the same name as the energy unit, the joule.

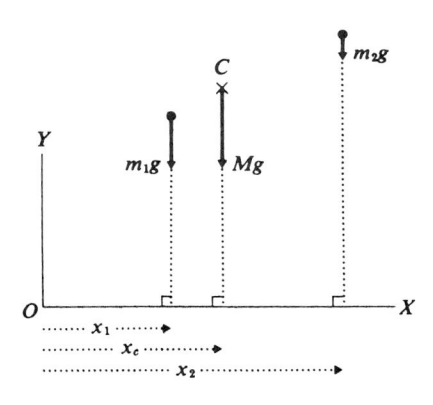

Fig. 3-25 The sum of the torques, m_1gx_1 and m_2gx_2, about an axis through the origin and perpendicular to the plane of the paper equals the torque Mgx_c.

D. The magnitude of the torque τ of $\vec{F}$ about the axis is defined to be the *product of the magnitude of the force and the moment arm*: that is

$$\tau = FD. \tag{3.17}$$

Torque can have either a clockwise or a counterclockwise sense.

Example 13 A 50-N horizontal force is applied to the outer edge of a door (Fig. 3-24) which is 1.20 meters wide. Find the torque of this force about the vertical axis of the hinges if

(a) The force is directed perpendicular to the door;
(b) The force is directed in such a way that the moment arm is 0.80 meter;
(c) The force is directed so that its line of action passes through the axis of the hinges.

Solution
(a) Here the moment arm D is simply the width of the door, so

$$\tau = (50 \text{ N})(1.20 \text{ meters})$$
$$= 60 \text{ newton-meters}*;$$

(b) The torque is now given by

$$\tau = (50 \text{ N})(0.80 \text{ meter})$$
$$= 40 \text{ newton-meters};$$

(c) In this circumstance the moment arm D is 0. Therefore

$$\tau = (50 \text{ N})(0) = 0.$$

We can see from the example above that the torque of the 50-N force applied at the edge of the door can be varied from a maximum value when the moment arm is the door width to a minimum value of zero when the moment arm is zero, simply by varying the direction of the force.

Center of Gravity

Consider a system (Fig. 3-25) of two particles m_1 and m_2 in a uniform gravitational field $\vec{g}$. Now recall that the definition of the mass center of this system is, according to Eq. (3.6)

$$M\vec{r}_c = m_1\vec{r}_1 + m_2\vec{r}_2. \tag{3.6}$$

We may resolve $\vec{r}_c$, $\vec{r}_1$, and $\vec{r}_2$ into x- and y-components and equate the x-components and the y-components separately. For the x-components we have

$$Mx_c = m_1x_1 + m_2x_2. \tag{3.18}$$

120 MECHANICS

Multiplying each side of this equation by g we obtain

$$Mgx_c = m_1gx_1 + m_2gx_2. \quad (3.19)$$

This equation has an interesting interpretation. Each term is a torque of a gravitational force mg with a moment arm x about an axis through the origin. Therefore Eq. (3.19) states that the sum of the torques of the gravitational forces acting on the system is the same as if the entire weight Mg were applied at the mass center C. The point of application C of this equivalent resultant force Mg is called the *center of gravity*. Our result is that in a uniform gravitational field, the mass center is the center of gravity.

This proof for a two-particle system can be extended without difficulty to systems comprising any number of particles. Therefore in any problem involving a system of particles in the uniform gravitational field near the earth's surface, we can boldly place the entire weight Mg at the mass center and know that this will give the right answer as far as gravitational torques are concerned.

Example 14 Consider the system of Example 5 which comprises a 10.0-kg particle and a 5.0-kg particle separated by 3.00 meters with a mass center 1.00 meter from the 10.0-kg particle (Fig. 3-26). Find the sum of the gravitational torques of this system about a perpendicular axis through the point A which is a horizontal distance of 8.0 meters from the mass center.

Solution With the entire weight considered as concentrated at the mass center, we obtain for this torque

$$(15.0 \text{ kg})(9.8 \text{ m/sec}^2)(8.0 \text{ meters}) = 1.18 \times 10^3 \text{ newton-meters}.$$

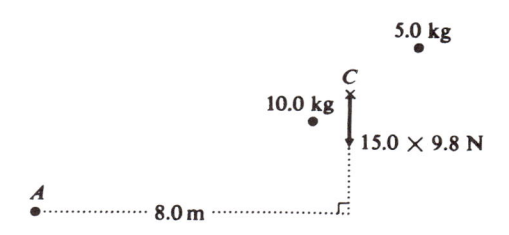

Fig. 3-26 The gravitational torque about A can be computed by treating the system as if the entire weight of the system were applied at the center of gravity C.

Conditions of Equilibrium for a Rigid Body

In the construction of bridges, buildings, and most experimental apparatus, we are concerned with bodies at rest. Often these bodies do not bend or vibrate appreciably, so that it is a useful approximation to consider that they are rigid bodies. That is, their parts all maintain a fixed location with respect to each other when external forces are applied.

The forces acting on a rigid body which remains at rest (or, more generally, has no translational or rotational acceleration) must satisfy two conditions of equilibrium. First, from the law of motion of the mass center,

$$\vec{F}_{\text{ext}} = M\vec{a}_c, \quad (3.7)$$

we conclude, since $\vec{a}_c$ is zero, that

$$\vec{F}_{\text{ext}} = 0. \quad (3.20)$$

3.8 STATICS OF RIGID BODIES

In words, the first condition of equilibrium is this: *The vector sum of all the external forces acting on a body in equilibrium is zero.* When all the forces acting on the body lie in the *XY*-plane, Eq. (3.20) is equivalent to two equations:

Algebraic sum of x-components of external forces = 0   (3.21)

and

Algebraic sum of y-components of external forces = 0   (3.22)

Even when the resultant of the external forces is zero, these external forces may cause a rigid body to spin more rapidly or more slowly. It is the second condition of equilibrium that guarantees rotational equilibrium. And, although we have not studied the dynamics of rotation, the discussion at the beginning of this section suggests that torques are involved. It turns out that, for any system of forces, the second condition of equilibrium may be stated as follows: For *any* axis

Sum of clockwise torques = *sum of counterclockwise torques.*   (3.23)

In particular, for any system of forces lying in a plane, the most convenient axis will be an axis perpendicular to the plane of the forces.

Example 15 A 10.00-meter uniform log which weighs 1200 N is supported at a point *A* which is 2.00 meters from its left end and at a point *B*, 6.00 meters from the same end. Find the forces F_A and F_B exerted by the supports on the log.

Solution In any statics problem, we follow the procedure stressed in Section 3.6 for using Newton's laws. The only new aspect comes at the end, when we apply the conditions of equilibrium, Eqs. (3.21), (3.22), and (3.23).

(Step 1) We draw a schematic diagram such as is shown in Fig. 3-27. Clearly the supports must push up on the log.
(Step 2) We select the log as the system in equilibrium.
(Step 3) The *Y*-axis is chosen to be directed vertically upward.
(Step 4) We draw a free body diagram as shown in Fig. 3-28. This includes the unknown forces F_A and F_B pushing up on the log. From the discussion of the center of gravity we learned that the 1200-N weight can be represented as acting at the mass center, which, for a uniform log, will be at its geometrical center.
(Step 5) We are now in a position to apply the conditions of equilibrium. The algebraic manipulations are often simplified by applying the torque equation first. *If one selects the axis about which the torques are to be taken so that this axis passes through the line of action of an unknown force, this force will have zero moment arm and therefore zero torque, and so will not appear in the torque*

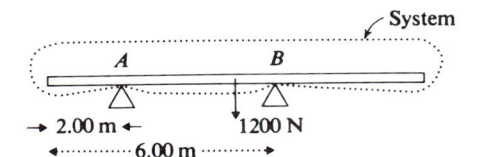

Fig. 3-27 Schematic diagram for Example 15. The system which has been selected is enclosed by the dotted line.

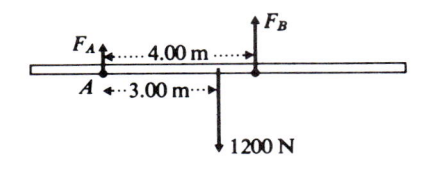

Fig. 3-28 Free body diagram for system shown in Fig. 3-27.

122   MECHANICS

equation. Thus if we take torques about a horizontal axis perpendicular to the log and passing through the point A, the torque of F_A about this axis is zero, and Eq. (3.23) yields

$$(1200 \text{ N})(3.00 \text{ meters}) = F_B \times (4.00 \text{ meters}).$$

This gives $F_B = 900$ N.

The first condition of equilibrium, Eq. (3.22) requires

$$F_A + 900 \text{ N} - 1200 \text{ N} = 0$$

Therefore

$$F_A = 300 \text{ N}.$$

(Step 6) The answers make sense: A total force of 300 N + 900 N = 1200 N is exerted by the supports in an upward direction. The answers can be checked by examining the torques about a different axis, say an axis through the center of gravity of the log. Then the counterclockwise torque is

$$(900 \text{ N})(1.00 \text{ meter}) = 900 \text{ N-meters}.$$

The clockwise torque is

$$(300 \text{ N})(3.00 \text{ meters}) = 900 \text{ N-meters}.$$

The second condition of equilibrium expressed by Eq. (3.23) is therefore satisfied.

Example 16 A uniform beam, which weighs 400 N and is 5.00 meters long, is hinged at a wall at its lower end by a frictionless hinge. A horizontal rope, 3.00 meters long is fastened between the upper end of the beam and the wall. Find the force T exerted by the rope on the beam and also find the horizontal and vertical components, F_h and F_v, of the force exerted by the hinge on the beam.

Solution

(Step 1) The schematic diagram is shown in Fig. 3-29.
(Step 2) The system in equilibrium that we select is the beam.
(Step 3) We choose the X-axis to be directed horizontally to the right and the Y-axis to be directed vertically upward.
(Step 4) The free body diagram for the beam is given in Fig. 3-30. The force exerted by the hinge on the beam has been replaced by its horizontal and vertical components. If you imagine removing the hinge and making your hand supply the force required to produce equilibrium, you will discover the directions of these components. However, even if you place on the free body diagram a component in the direction opposite to the actual force, no harm is done. When you complete the solution, this component will appear as a negative number. The negative sign indicates that the assumed direction of this component

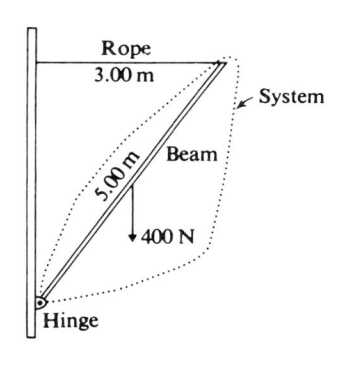

Fig. 3-29 Schematic diagram for Example 16. The system selected has been enclosed by the dotted line.

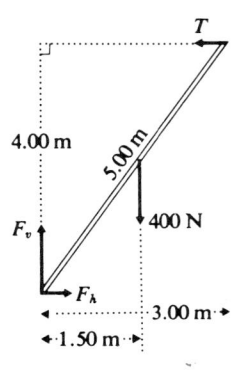

Fig. 3-30 Free body diagram for system of Fig. 3-29.

3.8 STATICS OF RIGID BODIES

must be reversed. Again note that the weight (400 N) is represented as acting at the mass center of the beam. The rope pulls on the beam and therefore exerts a force on the beam which acts to the left.

(Step 5) We select an axis perpendicular to the beam and passing through the hinge. Then, since the moment arms of F_v and F_h are zero, these forces have zero torque about this axis. The 400-N weight has a moment arm of 1.50 meters and therefore exerts a clockwise torque of

$$(400 \text{ N})(1.50 \text{ meters}) = 600 \text{ N-meters}.$$

The moment arm for the force T is evidently the distance along the wall from the hinge to the rope which is, by the Pythagorean theorem,

$$\sqrt{(5.00 \text{ meters})^2 - (3.00 \text{ meters})^2} = 4.00 \text{ meters}.$$

Now Eq. (3.23) gives

$$600 \text{ N-meters} = T \times (4.00 \text{ meters}).$$

Therefore,

$$T = 150 \text{ N}.$$

The first condition of equilibrium, in particular Eq. (3.21), yields

$$F_h - T = 0.$$

Therefore,

$$F_h = 150 \text{ N}.$$

Also, Eq. (3.22) gives

$$F_v = 400 \text{ N}.$$

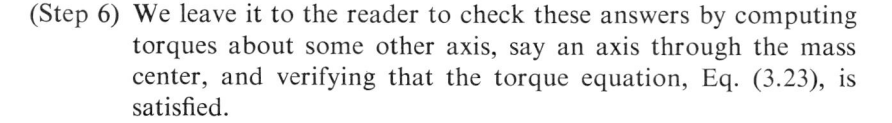

(Step 6) We leave it to the reader to check these answers by computing torques about some other axis, say an axis through the mass center, and verifying that the torque equation, Eq. (3.23), is satisfied.

Example 17 Rather complicated systems of pulleys and weights are often used in hospitals. In Fig. 3-31 we show an arrangement known as *Russel traction*, set up for the treatment of a fractured femur. Traction is used to overcome muscle contraction, which would produce overriding and misalignment at the fracture. The object is to exert forces which will keep the two sections of the bone in alignment and just touching. We wish to find the magnitude of the traction force at the fracture.

To understand the forces involved in Russel traction, we must recall some facts about ropes and pulleys. The tension in a rope at any cross section is the force exerted by the portion of the rope on one side of that cross section upon the portion of the rope on the other side. Since the suspended mass m (Fig. 3-31) is in equilibrium, the tension T is equal to

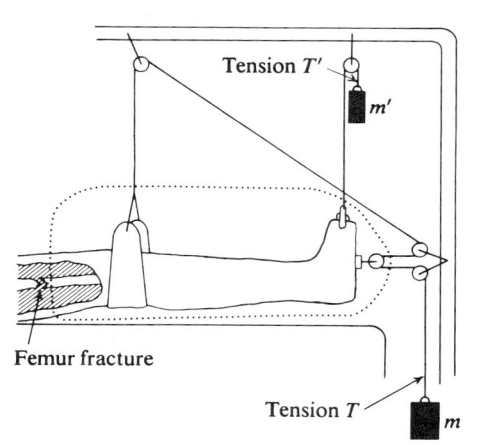

Fig 3-31 Russel traction for fracture of a femur. The system to which equilibrium conditions will be applied has been outlined.

124 MECHANICS

mg. Similarly, the tension T' equals $m'g$. The key fact about pulleys in statics is that, if pulley friction is negligible, the tension in the rope does not change when the rope passes around a pulley. Therefore, the tension has the same value $T(=mg)$ throughout the longer rope that passes around four different pulleys. We can now proceed with the analysis.

Solution
(Step 1) Fig. 3-31 is the schematic diagram.
(Step 2) The *system* to be selected can be chosen in many different ways. We choose the system to be everything within the dashed line shown in Fig. 3-31.
(Step 3) The X-axis is chosen to be directed horizontally to the right and the Y-axis vertically upward.
(Step 4) The free body diagram for the system is shown in Fig. 3-32.

At each cross section of a "cut" rope there is a pull in the direction of the rope and of magnitude equal to the tension in the rope. The resultant of the forces exerted by the upper part of the leg *on* the system is denoted by the force R. Most of this force R is furnished by muscles. The weight (Mg) of the system has been applied at the mass center of the system.

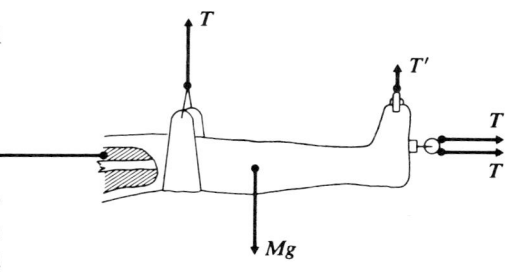

Fig. 3-32 Free body diagram for system outlined in Fig. 3-31.

(Step 5) The first condition of equilibrium, expressed in part by Eq. (3.21), is sufficient to determine the traction force R:

$$T + T - R = 0.$$

Therefore

$$R = 2T = 2\,mg.$$

A 10-lb weight, hung from the longer rope thus produces a traction force of 20 lb.

Stability

When the conditions of equilibrium are satisfied, we know that if a rigid body is undisturbed it will not have any translational acceleration, nor will its rotational motion change. However, if a rigid body at rest is slightly displaced from equilibrium, three different types of behavior are possible. We distinguish between these three possibilities:

1. *Stable equilibrium*: Forces come into play which tend to restore the body to its equilibrium position.
2. *Unstable equilibrium*: Forces arise which push or pull the object even further away from its equilibrium position.
3. *Neutral equilibrium*: Any slightly displaced position is still a position of equilibrium.

Let us consider some examples. A cube resting on a side is in stable equilibrium. This same cube balanced on a corner is in unstable equilibrium. A sphere resting on a horizontal plane is in neutral equilibrium. Clearly, in the design of most structures, we wish to achieve not only equilibrium but *stable equilibrium*.

3.9 ROTATION

Most machinery has rotating parts which are almost rigid. A mechanical engineer is therefore particularly interested in the application of Newtonian mechanics to the rotational motion of rigid bodies. This study involves ideas that are important in many fields of physics. The first step is to identify quantities that are appropriate for the description of rigid body rotation.

Angular Velocity ω

Fig. 3-33(a) shows a rigid body rotating about a fixed axis OZ which is perpendicular to the plane of the diagram (a plane parallel to the XY-plane) and which passes outward through the point A. The orientation of the rigid body at any instant is given by specifying the angle θ between a fixed direction, say the direction of the X-axis, and a line AP painted on the body. *The angular velocity ω of the rigid body is defined to be the rate of change of the angle θ.* This definition, together with Eq. (2.1), implies that the angle $\Delta\theta$ (Fig. 3-33b) through which the line AP rotates in a time interval Δt is given by the approximate relationship

$$\Delta\theta = \omega \Delta t. \qquad (3.24)$$

This expression is a good approximation for a time interval Δt so short that ω does not alter appreciably during this time interval. If ω is constant, $\Delta\theta$ is precisely equal to $\omega \Delta t$.

It is often convenient to measure angles in radians instead of degrees. The *radian measure* of an angle is defined* to be *the ratio of the length of the subtending arc to the radius of the arc.* For the angle $\Delta\theta$ of Fig. 3-33(b), a subtending arc has a length Δs and radius R, so this angle is given in radians by

$$\Delta\theta = \Delta s/R \quad (\Delta\theta \text{ in radians}). \qquad (3.25)$$

If the angle $\Delta\theta$ is increased to 360° or 1 revolution, the subtending arc Δs becomes a circumference of length $2\pi R$ and Eq. (3.25) gives $\Delta\theta = 2\pi R/R = 2\pi$ radians. Consequently,

$$2\pi \text{ radians} = 360 \text{ degrees} = 1 \text{ revolution}.$$

With angles measured in radians, the corresponding unit for angular

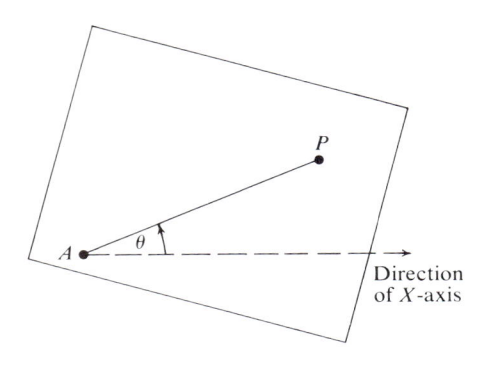

Fig. 3-33(a) Orientation of rigid body at any instant is given by specifying the angle θ.

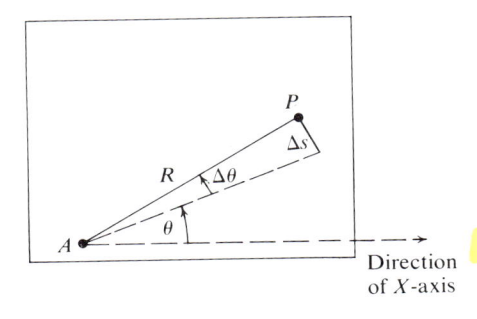

Fig. 3-33(b) Rigid body experiences an angular displacement $\Delta\theta$ in a time interval Δt, and the particle P moves a distance Δs.

*Radian measure is reviewed in Appendix A.

velocity ω is the radian/sec. Often the number f of revolutions per second is used. Then

$$\omega = 2\pi f. \quad (\omega \text{ radians/sec}, f \text{ revolutions/sec}). \quad (3.26)$$

Since the body is rigid, all lines painted on it rotate through the same angle in a given time interval. The angle $\Delta\theta$ and the angular velocity ω are therefore characteristic of the body as a whole, and this is the reason for focusing attention on $\Delta\theta$ and ω rather than on the familiar linear displacement and linear speed of some particle P of the rotating body. A particle's speed v is proportional to its distance R from the axis of rotation: that is,

$$v = R\omega \quad (\omega \text{ in radians per unit time}). \quad (3.27)$$

To understand this, notice that as the body rotates during a small time interval Δt, particle P moves in a circular arc of length Δs and radius R so its speed is given by

$$v = \Delta s/\Delta t = R\Delta\theta/\Delta t = R\omega,$$

where we have used Eqs. (3.25) and (3.24).

Because the particle P is traveling in a curved path of radius R, the particle must have a centripetal acceleration given by (Section 2.3)

$$a = v^2/R.$$

Using $v = R\omega$, this can be written

$$a = R\omega^2. \quad (3.28)$$

Example 18 A propeller, 2.20 m long, rotates about a fixed axis through its center. What is the speed of the propeller tip when the angular velocity of the propeller is 18×10^3 revolutions per minute?

Solution To use Eq. (3.27) we must express ω in radians per unit time. From Eq. (3.26) the propeller's angular velocity is

$$\omega = 2\pi(18 \times 10^3/60) \text{ radians/sec}$$
$$= 1.88 \times 10^3 \text{ radians/sec}.$$

The speed of the tip, moving in a circle of 1.10 m radius, is therefore

$$v = R\omega \quad (3.27)$$
$$= 1.10 \times 1.88 \times 10^3 \text{ m/sec}$$
$$= 2.07 \times 10^3 \text{ m/sec}$$

Angular Acceleration and Torque

The only type of rotational motion that must be considered to understand later sections of this book is the special case in which the angular velocity ω is *constant*. Therefore, rather than treating in detail the kinematics and dynamics of rotation with varying angular velocity, we will merely mention some of the principal quantities involved.

The angular acceleration α of the rotating rigid body represented in Fig. 3-33 is the rate of change of its angular velocity ω. The resultant external torque τ (the difference between the counterclockwise and the clockwise torques of the external forces about the axis of rotation) determines the angular acceleration of a rigid body rotating about a fixed axis, in accordance with the equation

$$\tau = I\alpha, \qquad (3.29)$$

where I is a quantity which depends on the distribution of mass about the axis of rotation, and which is called the *moment of inertia* of the rigid body about the axis.* This simple and fundamental equation of rotational dynamics can be deduced by applying Newton's laws of motion to each particle of the body.

The dynamical equation for rotational motion, $\tau = I\alpha$, should be compared to the dynamical equation which determines the linear acceleration a of a particle of mass m, $F = ma$. Analogous quantities for rotational and linear motion are seen to be *angular acceleration* and *linear acceleration*, *torque* and *force*, and *moment of inertia* and *mass*. We see that the moment of inertia I is a measure of rotational inertia. The larger the value of I, the greater the torque required to produce a given angular acceleration.

It should be noted that in the special case when the angular acceleration is zero, the equation $\tau = I\alpha$ reduces to the *condition of rotational equilibrium* given by Eq. (3.23), a fundamental relationship of the statics of rigid bodies.

Remarks on Angular Momentum

Although we will not continue the discussion of the dynamics of rotating systems, we would nevertheless like the reader to be aware of another significant dynamical quantity associated with rotation, *angular momentum*. The importance of angular momentum as a quantity that is conserved for any isolated system is demonstrated in situations ranging from the rotation of galaxies to the most recent experiments with atoms, nuclei, and elementary particles.

A particle with a momentum vector $\vec{p}$ has an angular momentum vector $\vec{J}_A$ about a point A (Fig. 3-34) which is defined to be a vector of magnitude

$$J_A = pD,$$

where D is the distance from A to the line through the particle in the

---

*The definition of the moment of intertia I of a rigid body about a given axis is

$$I = m_1R_1^2 + m_2R_2^2 + \ldots + m_nR_n^2,$$

where the particles of which the body is composed have masses $m_1, m_2, \ldots, m_n$, and are located at distances from the axis $R_1, R_2, \ldots, R_n$, respectively. Notice that the further a particle is from the axis, the greater is its contribution to the moment of inertia. A simple formula expressing the moment of inertia of a body about an axis can usually be found if the body is geometrically simple. For instance, a uniform cylinder of mass M and radius R has a moment of inertia about its axis given by $I = \frac{1}{2}MR^2$. For most objects, integral calculus is required to evaluate the sum which gives the moment of inertia.

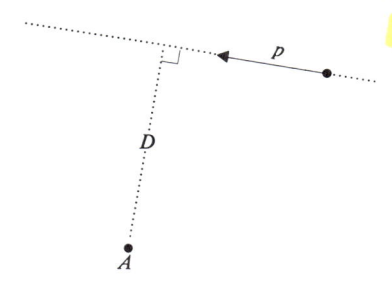

Fig. 3-34 The particle with linear momentum vector $\vec{p}$ has an angular momentum vector $\vec{J}_A$ about the point A of magnitude given by $J = pD$. The vector $\vec{J}$ is directed out of the paper towards the reader.

128 MECHANICS

direction of $\vec{p}$. The direction of $\vec{J}$ is perpendicular to the plane containing A and this line through the particle in the direction of $\vec{p}$. In Fig. 3-34, $\vec{J}$ is directed outward from the paper.

Now from Newton's second law it can be shown that if there is only one force acting on an orbiting particle and this force is always directed toward A, then

$$J_A = \text{constant.}$$

Kepler's second law, describing how a planet speeds up as it approaches perihelion and slows down as it reaches aphelion, can be rephrased as the statement that a planet moves along its orbit in such a way that its angular momentum about the sun is constant.

The angular momentum of a system of many particles is the vector sum of the angular momentum of each of the particles. There are many familiar motions in which the angular momentum of a large body about its *mass center* is almost constant: the spinning of the earth as it moves along in its orbit; the motion of a gyroscope; and all the gyrations performed by a diver during his trajectory from diving board to water; and even the seemingly impossible antics of a cat which in falling any distance at all can manage to land on its feet.

All the experimental evidence in every realm of physics is in accord with the *law of conservation of angular momentum*: the total angular momentum vector of an isolated system about any point in an inertial frame is constant. This law, which can be derived from Newton's second and third laws, apparently is valid even in atomic physics where Newtonian mechanics is inadequate.

One of the real surprises that emerges from a study of the behavior of atoms is that the angular momentum of a system cannot assume any one of a continuous range of values. Instead, the situation is analogous to that encountered with electric charge. There is a natural unit of angular momentum, universally denoted by $\hbar$, which is very small:

$$\hbar = 1.06 \times 10^{-34} \text{ joule-sec.}$$

The component of the angular momentum of any system along a given direction is restricted to be an integral or half-integral multiple of $\hbar$! This restriction, a typical quantum mechanical phenomenon, is inexplicable from the point of view of Newtonian mechanics.

Each type of particle possesses a characteristic angular momentum about its mass center. This intrinsic angular momentum ($S\hbar$) called *spin*, is conventionally given by specifying the value of the so-called *spin quantum number*, S. Electrons, protons, and neutrons all have "half-integral spin." That is, $S = \frac{1}{2}$ for these particles. Photons have $S = 1$. Certain particles, such as the π-meson, to be mentioned in the next chapter, have no spin at all ($S = 0$).

3.10 SIMPLE HARMONIC MOTION

The bobbing motion of a mass suspended from a spring, or the small oscillations of a pendulum, are examples of what is called *simple harmonic motion*. An understanding of simple harmonic motion is particularly important not only because this motion occurs in a variety of circumstances, but also because the study of this type of motion involves mathematical terms and relationships that are necessary for the analysis of sound, light, alternating electric current, and phenomena of quantum physics.

Any motion that is repeated in equal time intervals of duration T is called periodic; T is called the *period of the motion*. The motion is oscillatory if it takes place back and forth over the same path. One complete execution of the motion is called a *cycle*. For example, the mass m in Fig. 3-35, completes one cycle in any round trip, say from the point O to $x = A$, back through O to $x = -A$, and then back again to O. Since one cycle is executed in a time interval of one period T, the number of cycles per unit time is $1/T$ which is called the *frequency f*:

$$f = 1/T. \tag{3.30}$$

The mks unit for frequency is the cycle/sec, named the *hertz* (Hz).

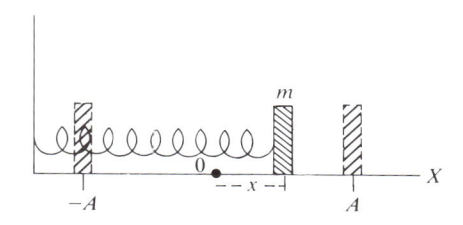

Fig. 3-35 Simple Harmonic Motion. The mass m oscillates back and forth on a smooth table along the axis OX between the points $x = A$ and $x = -A$. The amplitude of this simple harmonic motion is A.

The Position Coordinate in Simple Harmonic Motion

Simple harmonic motion is the particular type of oscillatory motion executed by the point P in Fig. 3-36(a) as the arrow of constant length A rotates about the origin with a constant angular velocity ω. The point P is always at the foot of the perpendicular from the tip of the arrow to the X-axis. At the instant t, the arrow makes an angle

$$\theta = \omega t + \theta_0 \tag{3.31}$$

with the X-axis. As time elapses, θ changes and the arrow rotates about a perpendicular axis through its tail, which is fixed at the point O. The point P moves along the X-axis, and its position coordinate x has the time dependence characteristic of simple harmonic motion:

$$x = \text{projection of } A \text{ on } X\text{-axis} \tag{3.32a}$$
$$= A \cos \theta \tag{3.32b}$$
$$= A \cos (\omega t + \theta_0), \tag{3.32c}$$

where A, θ_0, and ω are constants. Fig. 3-36(b) shows the corresponding graph of x as a function of time.

The positive constant A is named the *amplitude* of the simple harmonic motion executed by the point P. As the arrow of length A rotates, point P moves back and forth along the X-axis between the

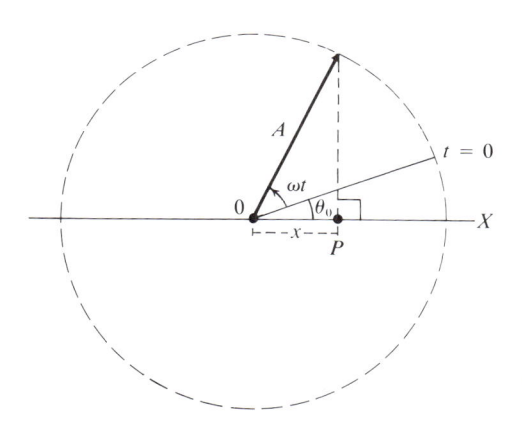

Fig. 3-36(a) Point P executes simple harmonic motion along OX as the arrow tip moves in a circle of radius A with a constant angular velocity ω.

130 MECHANICS

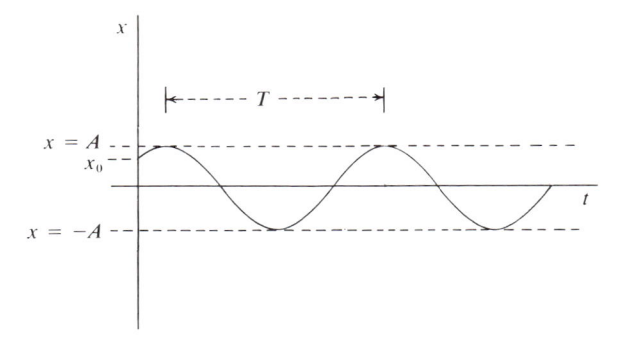

Fig. 3-36(b) The graph shows the coordinate x as a function of time for a particle oscillating with simple harmonic motion of amplitude A, with the initial position given by $x_0 = A \cos \theta_0$ and the period by $T = 2\pi/\omega$. Such a graph of a cosine or sine function is called a *sine curve*, and we say that x is a *sinusoidal function* of time.

points $x = A$ and $x = -A$. Evidently the *amplitude* is the maximum value attained by the position coordinate x. The total range of the motion is the distance 2A. The point O, at the center of the range of the motion, is called the equilibrium position (for a reason that will become clear as we investigate the dynamics of this motion).

The angle θ is called the *phase angle* or, briefly, the *phase* of the coordinate x. The phase angle changes at the constant rate ω. The value of θ at $t = 0$ is θ_0, which is called the *initial phase*.

The point P completes one cycle of its simple harmonic motion as the rotating vector makes exactly one complete revolution. Therefore the period T of the simple harmonic motion, being the time required for one cycle, is also the time required for one revolution (2π radians) of the arrow, which is rotating with constant angular velocity ω. The frequency f of the simple harmonic motion of the point P is the same as the number of revolutions per unit time of the rotating arrow. We have

$$f = 1/T, \qquad (3.30)$$

and, from Eq. (3.26),

$$\omega = 2\pi f, \qquad (3.33)$$

which gives ω in *radians per unit time*. This constant, ω, the angular velocity of the arrow, is called the *angular frequency* of the simple harmonic motion executed by P.

In summary, *a simple harmonic motion of angular frequency ω, amplitude A, and initial phase θ_0 is a motion that can be considered as the projection on a diameter* (say the X-axis) *of the tip of an arrow. The arrow has a length A, makes an angle* (the phase angle) $\theta = \omega t + \theta_0$ *with the X-axis, and rotates about its fixed tail with a constant angular velocity ω.*

Example 19 An arrow of length 2.00 m makes 4.00 revolutions per second. The initial phase angle is $-\pi/2$ radians. Describe the simple harmonic motion of the point P which is the projection of the arrow tip on the X-axis (Fig. 3-37). Locate P, first at the initial instant ($t = 0$) and next at an instant one-quarter of a period later.

3.10 SIMPLE HARMONIC MOTION

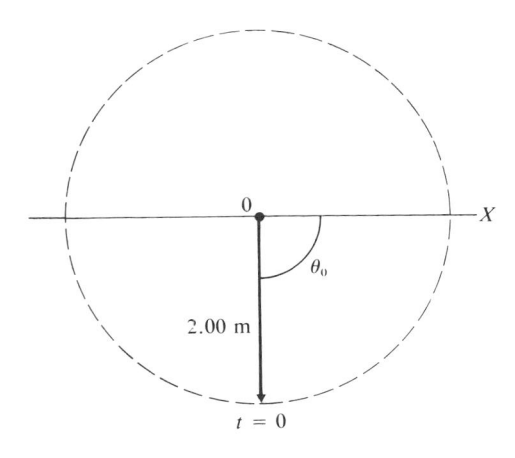

Fig. 3-37 The initial phase angle is $-\pi/2$ radians, and the amplitude is 2.00 m.

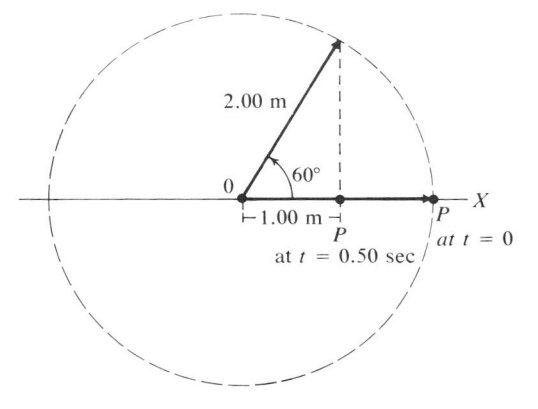

Fig. 3-38 Two positions of a particle P executing simple harmonic motion with an amplitude of 2.00 m and an initial phase angle of zero.

Solution The initial position of the arrow is shown on the diagram in Fig. 3-37. The number of revolutions per second of the arrow is the number of cycles per second (f) of the associated harmonic oscillation. The simple harmonic motion executed by P therefore has a frequency $f = 4.00$ hertz, a period $T = 1/f = 0.250$ sec, and an angular frequency $\omega = (2\pi \text{ radians})(4.00 \text{ hertz}) = 25.1$ radians/sec. The amplitude of this simple harmonic motion is the length of the arrow: $A = 2.00$ m. From the initial position of the arrow in Fig. 3-37, we see that its projection on the X-axis is zero, so initially the point P is at the origin O. One-quarter of a period later the arrow will have completed one-quarter of a revolution and rotated to the position of the axis OX. Projection of the arrow on the X-axis gives the position of P as

$$x = 2.0 \text{ m}.$$

At this instant P reaches maximum displacement from the equilibrium position O.

In summary, P executes simple harmonic motion with period 0.250 sec, amplitude 2.00 m, and initial phase $-\pi/2$ radians.

Example 20 A particle executes simple harmonic motion with a period of 3.00 sec. The total range of its motion is 4.00 m. Initially the particle is at rest to the right of its equilibrium position. Where will the particle be 0.50 sec later?

Solution First we construct the diagram in Fig. 3-38 to represent this simple harmonic motion. Here $2A = 4.00$ m, so the length of the arrow in Fig. 3-38 is $A = 2.00$ m. Since the particle starts at rest, the arrow tip must initially be traveling vertically. Consequently the arrow is horizontal, and since the particle is to the right of the equilibrium position, the initial arrow orientation must be as shown in Fig. 3-38, with the initial phase $\theta_0 = 0$. The angular frequency is $\omega = 2\pi/T = (2\pi/3.00)$ radians/sec. At $t = 0.50$ sec, the phase angle is

$$\begin{aligned} \theta &= \omega t + \theta_0 \\ &= (2\pi \text{ radians}/3.00 \text{ sec})(0.50 \text{ sec}) + 0 \\ &= \pi/3 \text{ radians,} \end{aligned}$$

which is 60°. Therefore, at $t = 0.50$ sec, the arrow makes a 60° angle with the X-axis, and its projection is

$$\begin{aligned} x &= A \cos \theta \\ &= (2.00 \text{ m}) \cos 60° \\ &= (2.00 \text{ m})(\tfrac{1}{2}) = 1.00 \text{ m}. \end{aligned} \quad (3.32b)$$

The particle is 1.00 m to the right of its equilibrium position.

Velocity and Acceleration in Simple Harmonic Motion

Since the arrow tip moves in a circle of radius A, with an angular velocity ω, its velocity vector $\vec{v}$ has a magnitude, from Eq. (3.27), of

$$v = \omega A,$$

and a direction tangent to its circular path (Fig. 3-39). The angle between $\vec{v}$ and the vertical is the phase angle θ. The point P moves along the X-axis with a velocity v_x which is just the x-component of the velocity $\vec{v}$ of the arrow tip:

$$v_x = -v \sin \theta \qquad (3.34a)$$
$$= -\omega A \sin (\omega t + \theta_0). \qquad (3.34b)$$

The negative sign in this expression corresponds to the fact that for θ between 0° and 180°, where $\sin \theta$ is positive, the point P moves in the negative x-direction; and for θ between 180° and 360°, the point P moves in the positive x-direction.

Our result shows that the point P moves back and forth with varying velocity, reaching a maximum speed ωA as it passes through the equilibrium position ($\theta = \pi/2$ radians or $-\pi/2$ radians), and coming to rest at $x = A$ (where $\theta = 0$) and at $x = -A$ (where $\theta = \pi$ radians).

The acceleration of the arrow tip (Fig. 3-40) is just a centripetal acceleration directed toward O with a magnitude, from Eq. (3.28), given by

$$a = \omega^2 A.$$

The x-component of the acceleration of the arrow tip is the acceleration a_x of the point P as it moves along the X-axis:

$$a_x = -\omega^2 A \cos \theta \qquad (3.35a)$$
$$= -\omega^2 A \cos (\omega t + \theta_0). \qquad (3.35b)$$

Now since $x = A \cos \theta$, we reach the important result

$$a_x = -\omega^2 x. \qquad (3.36)$$

The minus sign corresponds to the fact that the acceleration of P is always directed toward O, so that a_x is negative when x is positive and vice versa.

The equation $a_x = -\omega^2 x$ states that *in simple harmonic motion, a point P moves back and forth over the same path in such a way that its acceleration is proportional to its distance from a fixed point O and is always directed toward O*. This statement is an alternative definition of simple harmonic motion. The magnitude of the acceleration of the point P is greatest when x reaches its maximum magnitude, that is, $x = A$ or

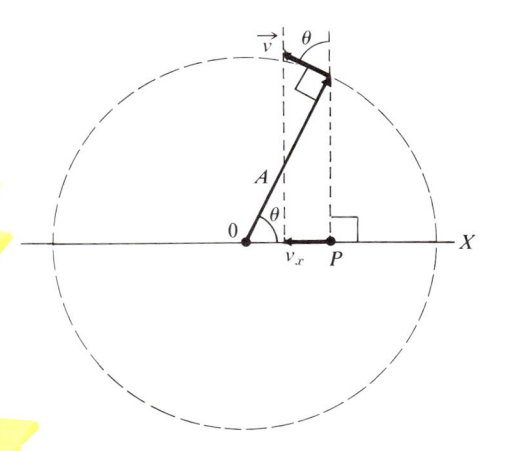

Fig. 3-39 Point P moves with a velocity which is the x-component of the velocity of the arrow tip.

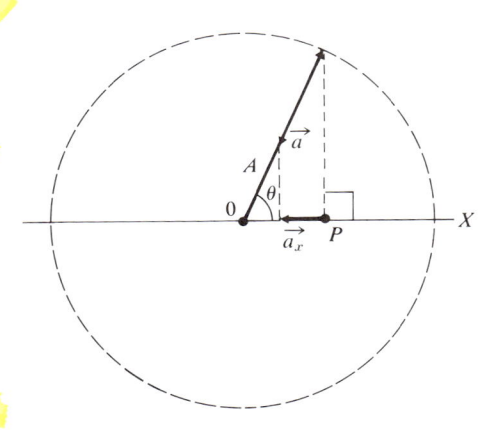

Fig. 3-40 The arrow tip has centripetal acceleration $\vec{a}$. Point P has an acceleration which is the x-component of $\vec{a}$. The acceleration of P is always directed toward the equilibrium position O.

3.10 SIMPLE HARMONIC MOTION 133

$x = -A$. As the point P sweeps through the equilibrium position, its acceleration is zero.

The relationship between a_x and x is to be contrasted with the relationship between a_x and v_x. At the equilibrium position where x and a_x are zero, the speed is a maximum. Where x and a_x reach their maximum magnitudes, the speed is zero.

Dynamics of Simple Harmonic Motion

Up to this point we have been concerned only with the description of simple harmonic motion. We now ask what are the circumstances under which a particle will execute such a motion? What are the characteristics of the resultant force acting on a particle that cause it to oscillate in this particular way? In a sense we have a solution in search of a problem.

Now, when a particle P of mass m executes simple harmonic motion along the X-axis, its acceleration is

$$a_x = -\omega^2 x. \tag{3.36}$$

Then Newton's second law, $F_x = ma_x$, implies that the resultant force F_x acting on this particle must be

$$F_x = -m\omega^2 x. \tag{3.37}$$

This answers our question. The resultant force that causes simple harmonic motion is a restoring force always directed toward O (as is the acceleration), and this force is proportional to the particle's distance from O. The point O where this resultant force is zero is therefore a position of *stable equilibrium*.

The magnitude of the constant of proportionality in Eq. (3.37) is called the force constant k; then

$$F_x = -kx \tag{3.38}$$

and

$$\omega^2 = k/m. \tag{3.39}$$

The period T of the motion is, from Eqs. (3.30) and (3.33)

$$T = 2\pi/\omega$$

$$T = 2\pi \sqrt{m/k}. \tag{3.40}$$

This shows that the period is determined entirely by the force constant and the particle's mass. In any simple harmonic motion the period is independent of the amplitude A and of the initial phase θ_0.

The constants A and θ_0 determine the particle's initial position x_0 and its initial velocity v_{x0}. Conversely, the amplitude and the initial phase angle are determined by the initial position and the initial velocity. For example, if $v_{x0} = 0$ and $x_0 = 2.00$ m, then, as we found in Example 20, $A = 2.00$ m and $\theta_0 = 0$.

The force law

$$F_x = -kx, \qquad (3.38)$$

is known as *Hooke's law*. Whenever the forces acting on a particle are such that the particle has a position of stable equilibrium, then, at least for small displacements x from this equilibrium position, the resultant force F_x acting on the particle must obey Hooke's law. To understand the reasons for this, consider the graph (Fig. 3-41) of a possible resultant force F_x plotted as a function of x when the point O is assumed to be a position of stable equilibrium. This assumption implies that, at least for small displacements, F_x must be a *restoring* force always directed toward O. Then, when x is positive, $\vec{F}$ must be in the negative x-direction, which implies that F_x is negative. Also, when x is negative, F_x must be positive. Therefore, the graph of any possible F_x must pass through the origin in the manner indicated in Fig. 3-41. If we now assume that the graph is smooth at the origin, the curve can be approximated for a short section by a straight line. The equation for this straight line graph is

$$F_x = -kx,$$

where $-k$ is the slope of the straight line. Therefore, the resultant force acting on a particle close to a position of stable equilibrium always obeys Hooke's law. And a small oscillation about an equilibrium position will always be a simple harmonic motion because this is the motion that occurs whenever the resultant force obeys Hooke's law. Since so many objects have an equilibrium position and are usually in the vicinity of this equilibrium position, simple harmonic motion occurs frequently in nature.

Fig. 3-35 shows a mass m fastened to one end of a spring. The force exerted by the spring on this mass obeys Hooke's law even when the stretch or compression (x) is quite large. The *force constant* k gives the force per unit elongation, so the greater the value of k, the stiffer the spring.

But even for springs there are departures from Hooke's law if the stretch is too large. In fact, with any solid object, if the deformation is so large that what is known as the *elastic limit* is exceeded, the deformation ceases to be proportional to the deforming force. It is then observed that when the deforming force is removed the object has acquired a permanent deformation.

Example 21 The force constant k of a spring is measured by observing that the spring is stretched 0.100 m by a 10.0-N force applied at one end. A mass of 0.250 kg is then fastened to one end of the spring, while the other end is fixed with the assembly placed on a smooth table as in

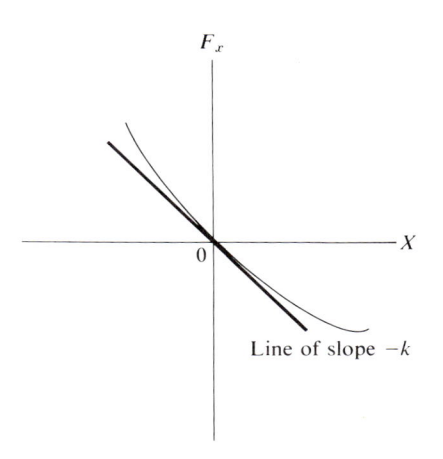

Fig. 3-41 Graph of F_x versus x when O is a position of stable equilibrium. Short segment of any smooth curve can be approximated by a straight line.

Fig. 3-35. The spring is compressed by moving the mass m a distance 0.200 m to the left of its equilibrium position. The mass is then released. Describe the subsequent motion of m.

Solution The force constant k is given by

$$k = F/x,$$

where F is the *magnitude* of the force associated with a stretch x (note that k is a positive number by definition). For this spring we have

$$k = 10.0 \text{ N}/0.100 \text{ m} = 100 \text{ N/m}.$$

The resultant force acting on the mass m is the force exerted by the spring, $F_x = -kx$. Therefore, the mass m executes simple harmonic motion with an angular frequency determined from Eq. (3.39) by m and the spring's force constant k:

$$\omega^2 = k/m = (100 \text{ N/m})/0.250 \text{ kg} = 400 \text{ sec}^{-2}.$$

Therefore, the angular frequency is $\omega = 20.0$ radians/sec, the period is $T = 2\pi/\omega = 0.314$ sec, and the frequency is $f = 1/T = 3.18$ hertz. The amplitude A and the initial phase θ_0 are determined by the initial conditions which, for this motion, are $v_{x0} = 0$ and $x_0 = -0.200$ m. By considering the rotating arrow tip [Fig. 3-36(a)], whose projection on the X-axis would coincide with the motion of m, it is apparent that, in its initial position, the arrow must point horizontally to the left. The initial phase is therefore π radians, and the amplitude is 0.200 m. Since

$$x = A \cos(\omega t + \theta_0), \tag{3.32c}$$

for this motion

$$x = (0.200 \text{ m}) \cos [(20.0 \text{ radians/sec})t + (\pi \text{ radians})].$$

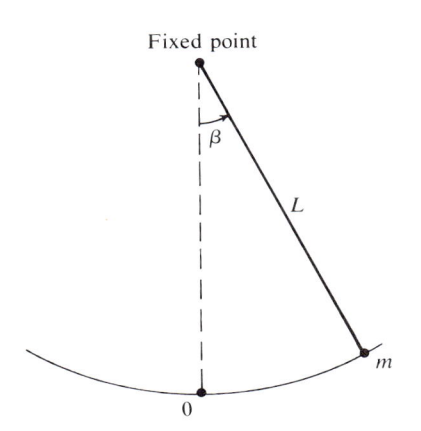

Fig. 3-42(a) Simple pendulum.

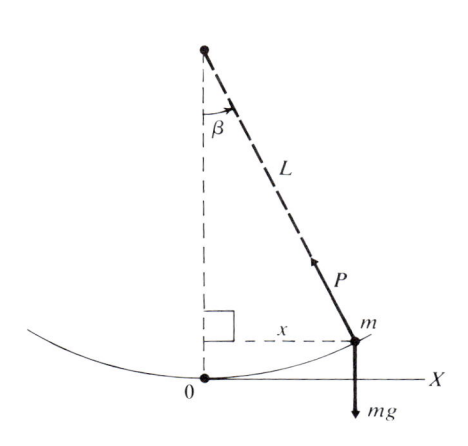

Fig. 3-42(b) Free body diagram for mass m.

The Simple Pendulum

Figure 3-42a shows a particle of mass m suspended from a fixed point by a light inextensible string of length L. Such an idealized system is called a *simple pendulum*. The lowest point accessible to the particle, point O, is obviously the position of stable equilibrium.

It is both interesting and useful to verify that, for a small displacement from the equilibrium, the resultant force acting on the mass m obeys Hooke's law. Figure 3-42b shows the free body diagram for the particle when it is displaced a horizontal distance x from the equilibrium position. Denoting the tension in the string by P we find that the x-component of the resultant force acting on m is given by

$$F_x = -P \sin \beta = -Px/L.$$

136 MECHANICS

As the particle swings back and forth, the tension P will vary, but, provided x/L remains small, the value of P will not depart appreciably from the value (mg) it would have if the particle were at rest at O. Using this approximate value, $P = mg$, we obtain

$$F_x = -\frac{mg}{L} x,$$

a Hooke's law force with a force constant k given by

$$k = \frac{mg}{L}.$$

From our previous studies we now can assert that, provided x/L remains small, the particle will execute simple harmonic motion with a period T given by

$$T = 2\pi \sqrt{m/k}, \qquad (3.40)$$
$$= 2\pi \sqrt{L/g}. \qquad (3.41)$$

Because of its periodic motion, the pendulum is used as the basic time-keeper in pendulum clocks. An exact analysis of pendulum motion shows that the motion is not precisely simple harmonic motion and that the period increases slightly if the amplitude of oscillation is increased. But even for amplitudes so large that the angle β reaches 15°, the error in the expression $T = 2\pi\sqrt{L/g}$ is less than 0.5%.

The period of a pendulum can be measured easily and very accurately and can thereby provide an accurate value of the gravitational field g acting on the pendulum. This is the principle underlying the precise "g-meters" that are so valuable in geophysical detection of possible deposits of ore or oil. Because their density is appreciably different from that of their surroundings, such deposits give rise to local variations in the value of the gravitational field.

QUESTIONS

1. A man weighs 1000 N (225 pounds force). What is the magnitude and direction of the force exerted by the man on the planet earth?

2. A car accelerates down a runway towing a glider. The force exerted by the towline on the glider is 800 N. Find the magnitude and direction of the force exerted by the glider on the towline.

3. The air exerts a horizontal retarding force of 200 N on the glider of Question 2. Draw a diagram showing the forces that determine the glider's acceleration. Find the magnitude and direction of this acceleration if the glider and its contents have a mass of 150 kg.

4. If, in Example 1, the man changes the force he exerts on the stone so that the *velocity of the stone is constant*, then find
   (a) the acceleration of the stone,
   (b) the resultant force acting on the stone,
   (c) the force exerted by the man on the stone,
   (d) the force exerted on the man's hand by the stone.

5. A stone weighing 40 N is thrown vertically upward by a catapult. At an instant when the stone is still in contact with the catapult, the stone's acceleration is 80 m/sec² upward. At this instant find
   (a) the resultant force acting on the stone;
   (b) the force exerted on the stone by the catapult;
   (c) the force exerted on the catapult by the stone.

6. For a particle of mass m, give two different statements of Newton's second law that are equivalent in Newtonian mechanics.

7. A constant resultant force acts on a 2.0-kg mass for 3.0 sec while the velocity changes from 4.0 m/sec to 10.0 m/sec. Find
   (a) the initial and final momentum,
   (b) the rate of change of momentum,
   (c) the resultant force acting on the mass.

8. A system consists of three particles: a 5.0-kg mass going east at 2.0 m/sec, a 3.0-kg mass going east at 10 m/sec, and a 2.0-kg mass going west at 8.0 m/sec. Find the magnitude and direction of the total momentum vector $\vec{P}$ of the system.

9. A 2.0-kg piece of putty travels east at 4.0 m/sec and collides with a 3.0-kg piece of putty traveling west at 5.0 m/sec. They collide and stick together. What is the magnitude and direction of their velocity vector after collision?

10. A 0.01-kg bullet traveling 1000 m/sec east hits a stationary 100-kg block and passes through undeviated but emerges with a velocity reduced to 200 m/sec. What is the block's momentum and velocity as the bullet emerges?

11. If, in the preceding problem, the bullet bounces off the block and retraces its original path with a speed of 200 m/sec, what is the block's momentum and velocity?

12. You are standing at rest at point A on a perfectly smooth sheet of ice armed with a revolver containing two bullets. How can you get from A to a distant point B and stop at B? Explain.

13. A white cue ball strikes a red ball head on and stops dead. Prove that if the two balls have the same mass, the red ball acquires the velocity that the cue ball had just before impact.

14. A 80-kg fullback running north at 10 m/sec is tackled by a 100-kg lineman moving south at 4.0 m/sec. Find the speed and direction of these players after the tackle.

15. A 30-kg boy dives from the stern of a 45-kg boat with a horizontal component of velocity of 3.0 m/sec, north. Initially the boat was at rest. Find the magnitude and direction of the velocity acquired by the boat.

16. A 2.70×10^3-kg Cadillac, traveling east at 3.00 m/sec, collides with a 900-kg Volkswagen traveling north at 30.0 m/sec. They interlock and move off together. Find the magnitude and direction of their common velocity.

17. A 100-kg man and an 80-kg man are at rest at opposite ends of a 30-meter rope on a sheet of ice. They each pull on the rope. How far from the original position of the 100-kg man will they be when they meet?

18. Deduce the expression, thrust $= RV$, for a rocket which ejects fuel at the rate R, at a velocity V relative to the rocket.

19. Explain how a spaceman, drifting along outside a spaceship, can use a tank of compressed air to maneuver. What physical principle is he exploiting?

20. During a burn, a rocket ejects gases at the rate of 150 kg/sec and at a speed of 2.0×10^3 m/sec relative to the rocket.
    (a) What is the thrust of this rocket?
    (b) In a region where the gravitational field is negligible, what is the acceleration of this rocket at the instant when the mass of the rocket and its remaining fuel is 60×10^3 kg?

21. Find the magnitude and direction of the velocity vector of the center of mass of the system of particles described in Question 8.

22. In Fig. 3-7, move the origin of the coordinate system 2.00 m to the left. Determine the new position vector of the 10-kg particle and the 5.0-kg particle and again use Eq. (3.6) to locate the center of mass of the system. Verify that the *same* point C is obtained. (This example illustrates that the location of the center of mass of a system is independent of the choice of the coordinate system.)

23. Verify that if each particle in Fig. 3-7 is moved 2.0 meters to the right, the center of mass of the system moves 2.0 meters to the right. (As illustrated in this question, when all the particles of a system are given the same displacement, the center of mass also experiences this displacement; in other words, the center of mass moves as if rigidly attached to the system.)

138   MECHANICS

24. Show that if you select the origin at the center of mass of a system, so that the position vector of the center of mass is zero, then $m_1 r_1 = m_2 r_2$ for a two-particle system.

25. Prove that the center of mass of two particles of equal mass is midway between them.

26. The mass of the earth is 6.0×10^{24} kg and the mass of the moon is 7.3×10^{22} kg. The separation of the earth and moon is approximately 2.4×10^5 miles. Show that the center of mass of the earth-moon system is 2.9×10^3 miles from the earth's center.

27. (a) What is the law of motion of the center of mass?
    (b) For a system consisting of a single particle of mass m, show that the law of motion of the center of mass is the same as Newton's second law.

28. A 10-kg particle is fastened to a 30-kg particle by a light spring. This system is hurled through the air and at a certain instant its center of mass is moving with a speed of 20 meters/sec. Find
    (a) the total momentum of the system,
    (b) the acceleration of the center of mass.

29. Find the magnitude and direction of the force vectors $\vec{a}$ and $\vec{b}$ if the components in a given coordinate system are:
    $a_x = 4.00$ N, $a_y = -3.00$ N;
    $b_x = -5.00$ N, $b_y = 12.00$ N.

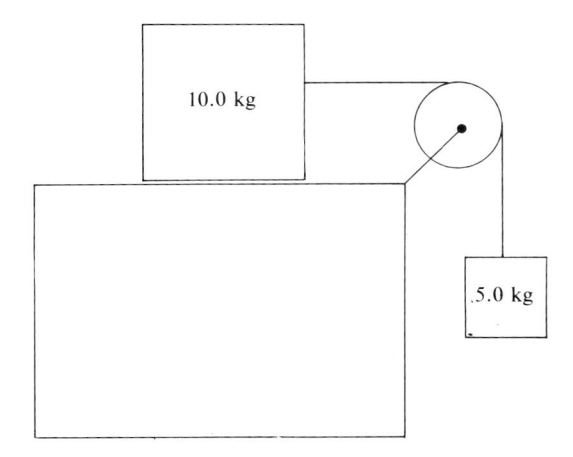

Fig. 3-43 Schematic diagram for Question 38.

30. By adding components, find the vector sum of the following forces: 6.00 N, north; 4.00 N, east; 5.00 N, 30.0° south of east.

31. By adding components, find the magnitude and direction of the resultant of the following forces:
    6.00 N, northeast;
    5.00 N, northwest;
    3.00 N, 15.0° north of west.

32. Find the resultant of the following forces:
    20.0 N, 10° east of north;
    40.0 N, 45° east of north;
    30.0 N, 45° west of north.

33. A trunk, pulled by two different ropes, is acted upon by a force of 200 N, east and a force of 400 N, 53.1° north of east. What is the vector sum of these two forces?

34. What is a free body diagram?

35. (a) In Fig. 3-12, find the force exerted on the first car of the train by the locomotive.
    (b) Find the force exerted on the third car by the second.

36. A 3.0-kg wooden block is connected by a string to a 5.0-kg iron block. This assembly is pulled across a smooth table top by a wire fastened to the iron block which exerts a constant horizontal force of 24 N. Find
    (a) the acceleration of the blocks,
    (b) the force exerted on the wooden block by the string.

37. A locomotive pulling two cars starts from rest and attains a speed of 30.0 m/sec. The tension in the connecting link between the locomotive and the first car is 3.00×10^3 N. Each car has a mass of 1.50×10^4 kg.
    (a) Over what length of track does the train accelerate?
    (b) What is the tension in the connecting link between the cars?

38. A 10.0-kg block is placed on a frictionless horizontal surface. It is connected by a cord passing over a light frictionless pulley to a suspended block whose mass is 5.0 kg (Fig. 3-43).
    (a) Draw the free body diagram for the 10.0-kg block and then for the 5.0-kg block.
    (b) Find the magnitude of the acceleration of the blocks.
    (c) What is the tension in the cord?

39. A block of mass M slides down a smooth plane which is inclined at an angle θ with the horizontal. Find the acceleration of the block.

40. A 20.0-kg trunk is pushed by a *horizontal* force of 200 N as it slides up a smooth plane inclined at an angle of 36.9° with the horizontal. Find the acceleration of the trunk.

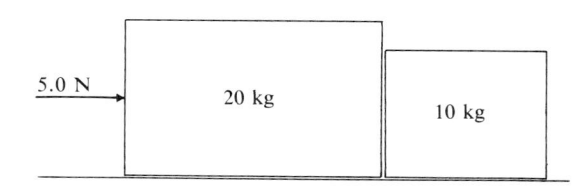

Fig. 3-44 Schematic diagram for Question 41.

41. A 20-kg block and a 10-kg block are placed in contact on a smooth horizontal table as shown in Fig. 3-44.
    (a) A horizontal force of 5.0 N is applied to the 20-kg block. Find the force exerted on the 10-kg block by the 20-kg block.
    (b) The 5.0-N horizontal force is now applied (acting to the left) to the 10-kg block rather than to the 20-kg block. Once again, find the force exerted on the 10-kg block by the 20-kg block.
    (c) Explain why the answers to parts (a) and (b) are different.

42. A heavy rope is given an acceleration a directed vertically upward. Consider a segment of the rope of mass m.
    (a) Show that the tension T_1 at the uppermost cross-section of this segment and the tension T_2 at the lowest portion of this segment are related by
    $$T_1 - T_2 = mg + ma.$$
    (b) If this rope were being accelerated horizontally, what would be the difference in the tensions at the two extremities of the segment of mass m?
    (c) What must be assumed about the rope (as in the solution to Example 10) for the tension to have the same value throughout all segments of the rope?

43. Check the dimensions of the following proposed formula for the tension T of the rope in Atwood's machine:
    $$T = m_1 m_2 g/(m_1 + m_2).$$

44. (a) What are the dimensions of the gravitational constant G?
    (b) Determine which of the following equations are dimensionally correct.
    $$M = g_s r^2/G,$$
    $$v = \sqrt{GM/r}$$
    $$T^2 = 4\pi^2 r^3/GM$$
    $$v = \sqrt{Gr^2/g}.$$

45. An 8.0-kg block slides along a horizontal plane. The coefficient of kinetic friction between the block and the plane is 0.20.
    (a) Find the normal component of the force exerted on the block by the plane.
    (b) Determine the frictional force exerted on the block by the plane.
    (c) Find the acceleration of the block.
    (d) Find the time required for the block to be brought to rest if it is projected along the plane with a speed of 4.0 m/sec.

46. If the plane of the preceding question is at an angle of 36.9° to the horizontal, find the acceleration of the block when
    (a) it is sliding up the plane,
    (b) it is sliding down the plane.

47. In Fig. 3-45, the coefficient of kinetic friction between the 6.0-kg block and the inclined plane is 0.40. The plane forms an angle of 36.9° with the horizontal. Find the tension in the rope connecting the 6.0-kg block to the smooth 2.0-kg block.

48. A 30-kg box is pushed by a horizontal force of 900 N as it slides up a plane inclined at an angle of 30° with the horizontal. The coefficient of kinetic friction between the box and the plane is 0.20. How far will the box travel along the plane in 5.0 sec, if it starts from rest?

49. A boy on a toboggan slides down a hill and attains a speed of 15 m/sec as he reaches the ice at the edge of a lake. The toboggan comes to rest after sliding a distance of 60 m along the ice. What is the coefficient of friction between the ice and the toboggan?

50. The coefficient of static friction between a 60-kg box and the floor is 0.50. A horizontal push is applied to the box and is slowly increased.
    (a) What magnitude of the applied force is required to start the box moving?
    (b) What is the initial acceleration of the box if the coefficient of kinetic friction between the box and the floor is 0.30?
51. In Fig. 3-45, assume the coefficient of static friction μ_s between the 6.0-kg block and the plane is such that the block is at rest but on the point of slipping. The 2.0-kg block is smooth. Find the tension in the rope and the value of μ_s.
52. What is the minimum magnitude of the horizontal force that will hold a 60-kg crate at rest on a plane inclined at an angle of 36.9° to the horizontal, if the coefficient of static friction between the crate and the plane is 0.40?
53. A 100-N weight is suspended from the ceiling by a rope. A horizontal string holds the weight in a position such that the rope makes an angle of 30° with the vertical. Find the tension in the rope and in the string.
54. A 10.0-kg particle is suspended at rest by two ropes. One rope makes an angle of 53.1° with the vertical, and the other makes an angle of 45° with the vertical. Find the tension in each rope.

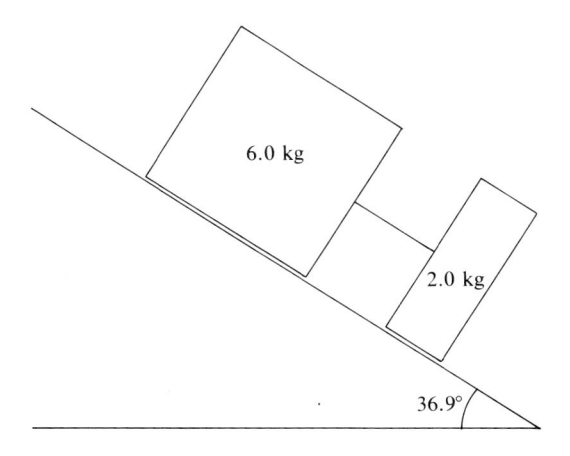

Fig. 3-45 Schematic diagram for Questions 47 and 51.

55. Consider a force $\vec{F}$ lying in a plane and an axis perpendicular to the plane. Define the torque of $\vec{F}$ about the axis.
56. A 1000-N weight is placed on a horizontal seesaw at a point which is 2.0 meters from its supporting axis. What is the torque of the weight about this axis?
57. Give a reason for wishing to know the location of the center of gravity of an object.
58. A uniform door weighs 60 N and is 0.80 meter wide and 2.4 meters high. Find the torque of its weight about a horizontal axis perpendicular to the door and passing through a lower corner.
59. A 2.0×10^3-kg uniform sphere with a radius of 0.50 m touches a vertical wall. What is the torque of the sphere's weight about a horizontal axis along the base of the wall?
60. A 0.20-kg meter stick lies along the positive horizontal X-axis with the zero mark at the origin.
    (a) Find the torque of its weight about a horizontal axis OZ.
    (b) Find the torque of its weight about an axis parallel to OZ and passing through the 70-cm mark on the stick.
    (c) The end marked 100 cm is raised until the meter stick makes an angle of 60° with the horizontal. Find the torque of its weight about the axis OZ.
    (d) What assumption about the meter stick have you made in solving parts (a), (b), and (c)?
61. Give the equations which express the first and second conditions of equilibrium for a body acted upon by forces lying in one vertical plane.
62. A load weighing 160 N is hung 3.0 meters from one end of an 8.0-meter horizontal pole which weighs 80 N. Two men carry the pole, one at each end. Find the forces exerted by the men on the pole.
63. A meter stick which weighs 20 N is suspended in a horizontal position from a vertical string attached at the 100-cm mark and from another vertical string attached at the 40-cm mark. Find the tension in each string.
64. A horizontal plank, 6.00 m long and weighing 200 N, rests on two supports placed 1.2 m from each end. A carpenter who weighs 800 N walks along this plank. How close can he go to the end of the plank without upsetting it?

65. A uniform beam, which weighs 600 N and is 5.00 m long, is hinged to a wall at its lower end by a frictionless hinge (as in Fig. 3-29). A horizontal rope, 4.00 m long, is fastened between the upper end of the beam and the wall. Find the force T exerted by the rope on the beam and also find the horizontal and vertical components, F_h and F_v, of the force exerted by the hinge on the beam.

66. A uniform trapdoor which is 2.00 m long and 2.00 m wide is hinged at one edge and held open at an angle of $60°$ with the horizontal by a rope attached at the opposite edge and pulling perpendicular to the door. The door weighs 180 N. Find the tension in the rope and the horizontal and vertical components of the force exerted on the door at the hinge.

67. The door of Question 58 is supported at one lower corner by a force which has a horizontal component F_h and a vertical component F_v.
    (a) What horizontal force H applied at an upper corner will hold the door in equilibrium?
    (b) Find F_h and F_v.

68. One end of a 600-N beam is hinged at a vertical wall and held horizontal by a rope 5.00 m long fastened to the wall and to the beam at a point 3.00 m from the wall. A 200-N weight is suspended at the free end of the beam, 4.00 m from the wall. Find the tension in the rope and the horizontal and vertical components of the force exerted on the beam by the hinge.

69. A 13.0-m uniform ladder weighing 300 N rests against a smooth wall at a point 12.0 m above the floor. The lower end of the ladder rests on a rough floor. Find the (horizontal) frictional force f and the (vertical) normal force N exerted by the floor on the ladder, as well as the (horizontal) normal force N' exerted on the ladder by the wall.

70. A light uniform ladder is supported on a rough floor and leans against a smooth wall, touching the wall at a point which is at height b above the floor. A man climbs up the ladder until the base of the ladder is at the point of slipping. The coefficient of static friction between the foot of the ladder and the floor is μ. Show that the *horizontal* distance moved by the man is given by $x = \mu b$. [*Hint:* The force exerted on the ladder by the *smooth* wall is normal to the surface of the wall.]

71. A uniform rod 2.00 m long has one end on a rough floor. The rod, resting against the smooth edge of a table which is 1.50 m high, is inclined at a $60°$-angle with the horizontal. If the rod is at the point of slipping, what is the coefficient of static friction between the floor and the rod? [*Hint:* The force exerted on the rod by the smooth edge of the table is perpendicular to the rod.]

72. Characterize stable, unstable, and neutral equilibrium and give an example of each.

73. What is the magnitude of the angular velocity of the second hand of a watch? Of the minute hand?

74. A propeller has a *constant* angular velocity of 1.80×10^3 radians/sec.
    (a) What is the propeller's angular displacement in 0.50×10^{-3} sec?
    (b) How long will it take for the propeller to make 10 revolutions?

75. A grindstone makes 5.0 revolutions in 2.0 sec. Assuming the angular velocity to be constant, what is its value in radians/sec?

76. A circular saw which has a 0.30-m radius is turning at 1600 revolutions per minute. What is the speed of a tooth of the saw?

77. A propeller of 0.80-m radius is rotating counter-clockwise with a constant angular velocity of 2.0×10^3 radians/sec. At the instant when the propeller is vertical, find
    (a) the magnitude and direction of the velocity vector of the propeller tip;
    (b) the magnitude and direction of the acceleration vector of the propeller tip.

78. A particle on the rim of a grindstone is traveling at a speed of 10 m/sec. The grindstone diameter is 0.40 m. What is the angular velocity of the grindstone?

79. (a) In the motion of the mass m in Fig. 3-35, identify the trip that constitutes a cycle starting at the point $x = -A$.
    (b) If the frequency is 4.0 hertz, how long will it take to complete the cycle specified in part (a)?

80. In one step of an experiment designed to measure the gravitational field, a precise determination of the period of a certain pendulum is to be made from the observation that the pendulum completes 200.0 cycles in 440.0 sec. Calculate the frequency and the period.

81. A particle oscillates with a frequency of 0.400 hertz. How long will it take to execute 50 cycles?
82. What is simple harmonic motion?
83. What is the definition of the amplitude of a simple harmonic motion?
84. Show on a diagram representing the simple harmonic motion of a point P, the significance of the following terms: phase angle, initial phase angle, amplitude.
85. An arrow 0.30 m long rotates about its fixed tail with an angular velocity of 0.40 radians/sec.
    (a) Find the period and the amplitude of the simple harmonic motion executed by the projection P of the arrow tip on a diameter of its path.
    (b) The initial phase angle is π radians. How long will it take the point P to move from its initial position to the equilibrium position?
86. The motion of a particle P is described by
$$x = (2.00 \text{ m}) \cos (5.0t + \pi/2).$$
    For this simple harmonic motion find:
    (a) the angular frequency,
    (b) the frequency,
    (c) the period,
    (d) the initial phase angle,
    (e) the amplitude.
87. Draw a diagram showing the length and the initial phase angle of the rotating arrow whose projection on the X-axis will execute the simple harmonic motion described in the preceding question.
88. Give the displacement x as a function of time t for a particle which executes a simple harmonic motion with a period of 0.50 sec. The total range of the particle's motion is 0.16 m. The initial phase angle is zero.
89. (a) What is the speed of the tip of the rotating arrow in Question 85?
    (b) What is the maximum speed attained by the projection P of this arrow tip on a diameter of its circular path?
    (c) Where in its motion does P reach maximum speed?
    (d) What is the minimum speed of P, and where is this attained?
90. A particle executes simple harmonic motion with an *angular* frequency of 5.0 radians/sec. As it passes through the equilibrium position its speed is 2.5 m/sec. What is the amplitude of this simple harmonic motion?

91. (a) Find the magnitude and the direction of the acceleration of the arrow tip of Question 85.
    (b) What is the maximum acceleration of the projection P of this arrow tip on a diameter of its circular path?
    (c) Where in its motion does P have maximum acceleration?
    (d) What is the minimum magnitude of the acceleration of P, and where will this occur?
92. What is the relationship between the acceleration a_x and the displacement x (from a fixed point O) that is characteristic of simple harmonic motion?
93. What is the acceleration of the particle of Example 20 when it is 0.60 m to the left of its equilibrium position?
94. The end of a prong of a tuning fork executes simple harmonic motion with a frequency of 256 hertz and an amplitude of 0.40 mm. Find the maximum acceleration and the maximum speed of the end of the prong.
95. (a) What is the force constant of a spring that is stretched 5.0×10^{-2} m by a force of 10.0 N?
    (b) What is the period of oscillation of a 0.80-kg mass which is fastened to this spring as in Fig. 3-35?
96. (a) When the oscillating mass in Fig. 3-35 is 2.0 kg, the period of oscillation is 1.5 sec. If a 6.0-kg mass is fastened to the 2.0-kg mass and the same spring is used, what will be the period of oscillation?
    (b) What is the force constant of the spring?
    (c) What force is exerted by this spring on the mass fastened to one end when the spring is stretched a distance of 0.10 m?
97. Suppose a mass m is suspended from a vertical spring with a force constant k (Fig. 3-46). When the spring stretches a distance s, the suspended mass m is in equilibrium. We select an axis OX directed downward and the origin O at the equilibrium position. When the mass m has a position coordinate x the total stretch of the spring is $s + x$. Show that the **resultant force acting on the suspended mass is**
$$F_x = -k(s + x) + mg,$$
and that this reduces to
$$F_x = -kx.$$

98. When a 2.0-kg mass is suspended in *equilibrium* from a vertical spring, it is observed that the spring has stretched by 4.0×10^{-2} m. The suspended mass is then pulled down 1.5×10^{-2} m further and released at rest. What is
    (a) the force constant of the spring?
    (b) the period of oscillation of the mass?
    (c) the amplitude of the simple harmonic motion executed by the mass?

99. A spring with a force constant of 39.2 N/m is suspended from the ceiling. A 0.40-kg mass is attached to the lower end of the spring and then released from rest.
    (a) How far will the mass descend before reaching the equilibrium position?
    (b) How far below the equilibrium position will the mass descend?
    (c) Find the frequency and amplitude of the simple harmonic motion executed by the oscillating mass.

100. The 1400-kg body of a car, suspended on springs, executes vertical oscillations with a frequency of 4.0 hertz.
    (a) What is the force constant of this system of springs?
    (b) What will be the frequency of oscillation when there are four passengers in the car, each with a mass of 100 kg?

101. What is the period of small oscillations of a simple pendulum which is 1.00 m long?

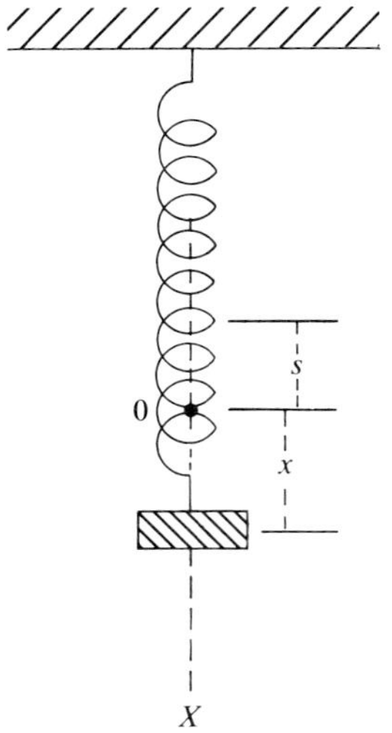

Fig. 3-46 *Resultant* force acting on suspended mass of Question 97 is given by $F_x = -kx$.

102. A clock timekeeper is desired with a period of 2.00 sec. If a simple pendulum is to be used, what length should it have?

103. What is the value of the gravitational field at the location of the pendulum of Question 80? The length of the pendulum is 1.20 m.

SUPPLEMENTARY QUESTIONS

S-1. A neutron of mass M traveling with a velocity of $(1/30)c$ strikes a stationary uranium nucleus of mass $235M$. **The neutron is captured by the uranium** nucleus and they move off together as a compound nucleus.
    (a) What is the velocity of this compound nucleus?
    (b) While moving at this velocity, the compound nucleus decays into two fragments of masses **128M and 108M. The heavier fragment travels with** a velocity of $(1/40)c$ in a direction perpendicular to that of the incident neutron. What is the magnitude and direction of the velocity of the lighter **fission fragment? (c denotes the speed of light** in a vacuum.) [*Hint:* Use the two equations (one for x-components and one for y-components) that are equivalent to the single vector equation, Eq. (3.4).]

S-2. While at rest, a 6.0-kg package explodes into three fragments. A 1.0-kg fragment moves west at 40 m/sec and a 2.0-kg fragment moves south at 15 m/sec. Find the magnitude and direction of the velocity of the remaining fragment.

S-3. A uniform canoe of mass M has seats that are a distance s apart. The canoe is at rest, with a man of mass m seated in the bow. The man then moves to the stern and sits down. Assuming that the horizontal component of the force exerted on the canoe by the water is negligible, find the displacement of the canoe.

S-4. Locate the center of mass of the system composed of three particles placed at the vertices of the right-angled triangle of Fig. 3-47. [*Hint:* Eq. (3.6) implies that the coordinates x_c and y_c of the center of mass are given by $Mx_c = m_1x_1 + m_2x_2 + m_3x_3$, and that $My_c = m_1y_1 + m_2y_2 + m_3y_3$.]

S-5. A dumbbell is constructed from a 5.0-kg sphere and a 3.0-kg sphere joined together by a light rod. The distance between the centers of the spheres is 0.25 m.
 (a) Find the center of mass of this assembly.
 (b) This rigid dumbbell is placed on a sheet of ice and then pulled by a 40-N horizontal force applied to the smaller sphere. Find the acceleration of the center of mass of the dumbbell.

S-6. In calculating the location of the center of mass of a composite system, any portion with mass M and center of mass C can be replaced by a single particle of mass M located at C. [This can be proved from the defining equation of the center of mass, Eq. (3.6).] Use this fact to find the location of the center of mass of a letter T made from two thin uniform bars. The vertical bar has length 0.60 m and mass 3.0 kg. The horizontal bar has length 0.40 m and mass 2.0 kg.

S-7. A 5.0-kg block and a 10-kg block are placed on a smooth table and connected by a spring. The 10-kg block is then pushed east by a horizontal force of 60 N.
 (a) Find the acceleration of the center of mass of the two blocks.
 (b) What is the velocity of the center of mass after 2.0 sec have elapsed? At this time the 10-kg block has a velocity of **6.0 meters/sec east. What** is the velocity of the 5.0-kg block?

S-8. A dog who weighs 100 N has a force of 125 N exerted upwards on his feet by the floor of an elevator. Find
 (a) his mass,
 (b) his acceleration.

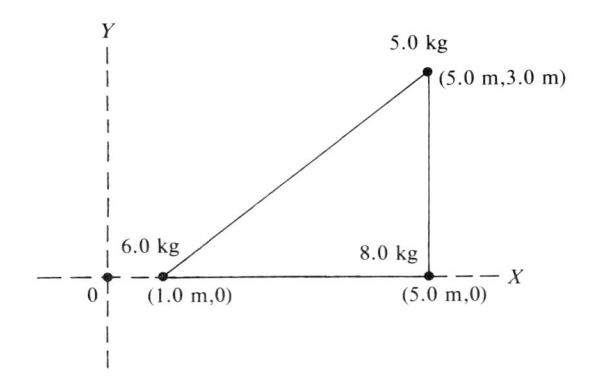

Fig. 3-47 System of particles of Question S-4.

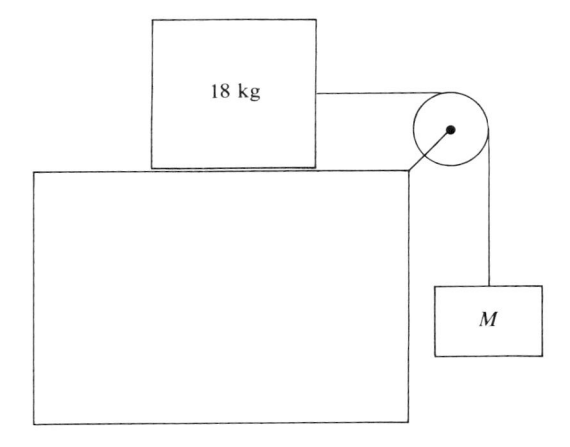

Fig. 3-48 Diagram for Question S-9.

S-9. (a) An 18-kg block is placed on a smooth horizontal table and is subjected to a horizontal constant force of 54 N. Determine the acceleration of the block.
 (b) The motion described in part (a) is achieved in the situation shown in Fig. 3-48 with a light rope and a light frictionless pulley. If the acceleration of the 18-kg block has the value calculated in part (a), find the tension in the rope and the mass M of the object suspended by the rope.

S-10. A railway car rests on a straight and level track. The weight of the car is 9.8×10^4 N. An engine provides a constant horizontal force (on the car) of 1.5×10^4 N.
  (a) Calculate the magnitude of the acceleration of this railway car. (Due to the low coefficient of rolling friction between the steel wheels and the steel rails, a reasonable estimate is achieved even when frictional forces are disregarded.)
  (b) Assume that the acceleration calculated in (a) remains constant. Find the car's speed 3.0 sec after it starts from rest.

S-11. The injuries sustained in most automobile collisions are due to the fact that the persons involved are brought to a stop in a very short distance. This short distance implies that the persons undergo extremely high decelerations and are thus subjected to correspondingly large forces. Consider an example: A car is likely to move from rest through a distance of perhaps 60 m in gaining a speed of 18 m/sec (about 40 mi/hr). In a collision, this speed may be reduced to zero over a distance of, say, 0.60 m. In this case, what is the ratio of the force exerted on a passenger when the car is stopped by collision, to the force exerted as the car originally gained speed?

S-12. What is the minimum acceleration with which a 55-kg boy can safely slide down a rope which will break at a tension of 450 N?

S-13. Consider Fig. 3-49 which shows a 2-kg mass suspended by thread A. Another thread B of the same kind hangs from the bottom. A downward force is applied at C. Explain why, if this force is applied smoothly and steadily, the thread A will break while if a sudden force, a quick jerk, is applied the thread B will break.

S-14. Using Newton's laws of motion, explain why a person changing quickly from a horizontal to a vertical position may temporarily feel dizzy.

S-15. A 10-kg block moves on a horizontal table. The coefficient of kinetic friction between the block and the table is 0.30. A horizontal force of 49.4 N is applied.
  (a) Find the normal component of the force exerted on the block by the table.
  (b) Find the frictional force exerted on the block by the table.
  (c) Compute the acceleration of the block.
  (d) The 49.4-N force is removed when the block is moving with a speed of 4.0 m/sec. How far will the block slide?

S-16. A block of mass 10.0 kg is projected at an initial speed of 5.0 m/sec up a plane inclined at an angle of 36.9° to the horizontal. The coefficients of kinetic and static friction between the block and the plane are 0.20 and 0.30, respectively.
  (a) What is the acceleration of the block as it moves up?
  (b) How far does it move along the plane before coming to rest?
  (c) Show that the block will start down again after coming to rest.
  (d) What is its acceleration down the plane?

S-17. Show that when the brakes are applied in a car moving with a speed v_0, the shortest distance in which the car can be stopped is

$$x = v_0^2/2g\mu_s.$$

(Notice that if there is no sliding between the tires and the road, the frictional force is a force of *static* friction.)

S-18. Consider a man in an elevator which has a vertical acceleration a_y (Y-axis upward). Show that the force

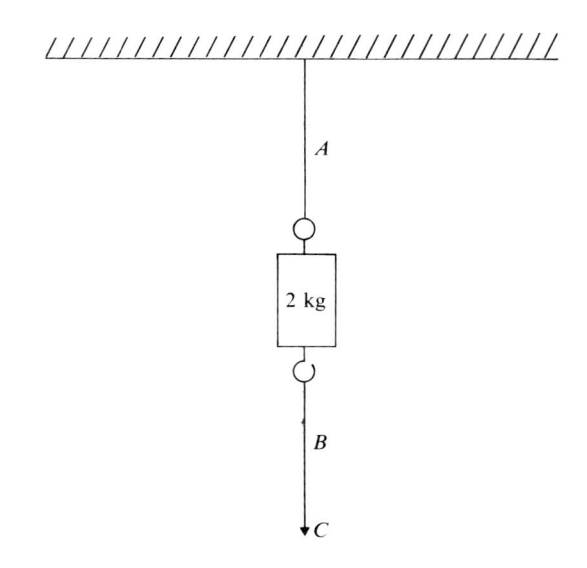

Fig. 3-49 Diagram for Question S-13.

P exerted on the man of mass m by the floor (the magnitude of this force is often called the apparent weight) is given by

$$P = ma_y + mg,$$

whether the acceleration a_y is positive or negative.

S-19. Figure 3-50 shows a conical pendulum—a particle of mass m revolving in a horizontal circle with a constant speed. The particle is attached to the end of a string of length L which makes an angle θ with the vertical. Show that the time required for one complete revolution of the mass is $2\pi\sqrt{L\cos\theta/g}$.

S-20. Consider a car moving at speed v along a road which is curved into a circular arc of radius r. The road is banked at an angle θ to the horizontal. Show that if the car is traveling at a speed such that there is no frictional force exerted on the car by the road, then

$$v^2 = rg\tan\theta.$$

S-21. Show that if a body is in equilibrium when acted upon by three nonparallel forces, the three forces must all pass through a common point.

S-22. A 13.0-m uniform ladder weighing 300 N rests against a smooth wall at a point 12.0 m above the floor. The ladder is at the point of slipping after a man weighing 900-N has climbed a distance of 3.90 m measured along the ladder. Find the coefficient of static friction (μ) between the foot of the ladder and the floor.

S-23. A light stepladder (Fig. 3-51) is constructed from two ladders, each 3.00 m long, which are hinged at the top and joined by a rope 1.00 m long fastened to each ladder at points 1.00 m from the hinge. A man weighing 800 N stands at the midpoint of the ladder on the right. Assuming that the floor is frictionless, find the tension in the rope.

S-24. Analyze the fool's tackle shown in Fig. 3-52. Show that equilibrium cannot be attained because the central pulley must have a downward acceleration of magnitude T/m where T is the tension in the cord and m is the mass of this pulley.

S-25. Two forces have equal magnitude F and opposite directions with a distance D between the lines of action of the forces. Such a pair of forces constitutes a *couple*. Show that the torque exerted by this couple about any axis perpendicular to the plane of the forces has a magnitude FD, and hence is independent of the position of the axis.

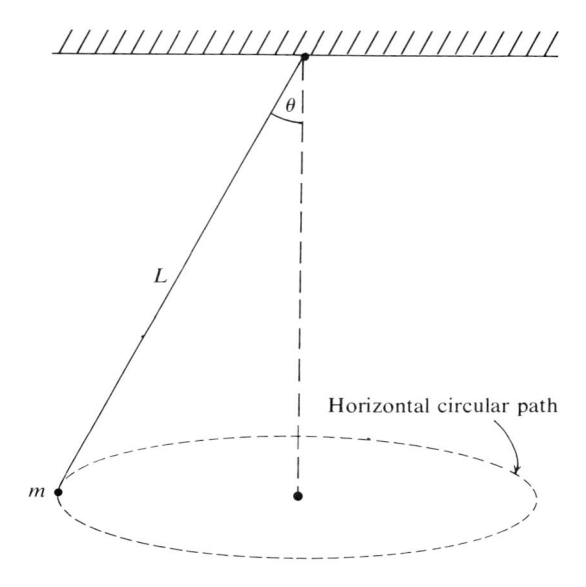

Fig. 3-50 Conical pendulum of Question S-19.

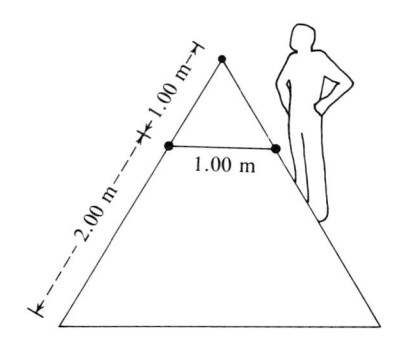

Fig. 3-51 Stepladder of Question S-23.

SUPPLEMENTARY QUESTIONS

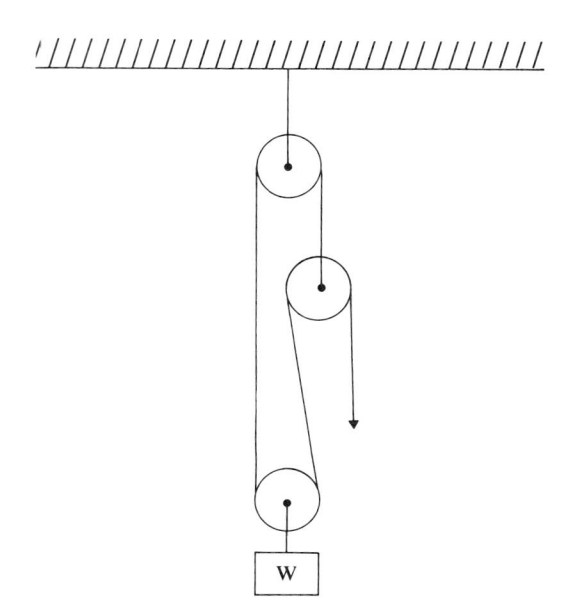

Fig. 3-52 A pulley system called the "fool's tackle."

S-26. Show that the gravitational torque exerted on a uniform rod of mass M and length L (Fig. 3-53) about a horizontal axis through one end is given by

$$\tau = -Mg(L/2)\sin\theta$$

where θ is the angle measured counterclockwise from a direction vertically downward to the rod.

S-27. The orbit of the earth about the sun is approximately circular, with a radius of 1.5×10^{11} m. Consider the earth as a particle and calculate both its angular velocity about the sun and the earth's speed as it moves in its orbit around the sun.

S-28. Assume the earth to be a sphere of radius 6.4×10^6 m. Relative to a frame which does not rotate relative to the distant stars, the earth itself rotates about a polar axis with an angular velocity of 1 revolution per sidereal day (1 year = 366.2 sidereal days).
(a) Evaluate this angular velocity in radians/sec.
(b) Calculate the velocity and the acceleration of a point on the equator.
(c) Calculate the velocity and the acceleration of a point at latitude 45°.

S-29. A high-speed anode of an x-ray tube (of the type shown in Fig. 4-25 which is used for medical diagnostic x-ray work) rotates at 10,000 revolutions per minute.
(a) What is its angular velocity in radians/sec?
(b) If it takes 10 minutes for this rotating anode to come to rest after an x-ray exposure has been terminated, what is the anode's angular acceleration, assuming that it is constant? [*Hint:* $\Delta\omega = \alpha\Delta t$.]

S-30. A merry-go-round, initially at rest, is given a constant angular acceleration of 0.010 radians/sec². Find the angular velocity attained and the number of revolutions that have been made after 30 sec have elapsed.

S-31. The turntable of a record player, initially revolving at 33 revolutions per minute, slows down and stops in 15 sec. Find the angular acceleration (assuming it to be constant) and the number of revolutions of the turntable.

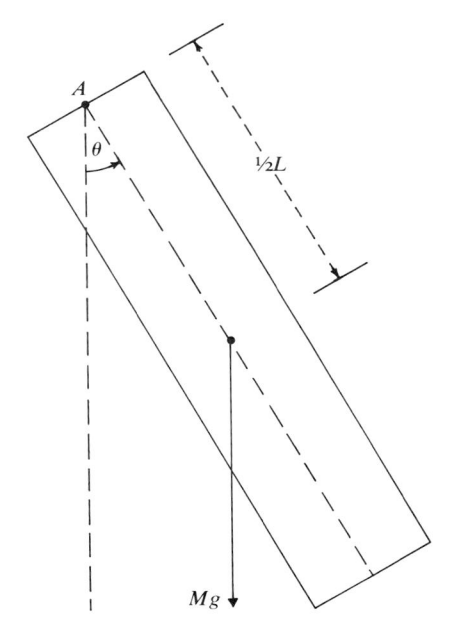

Fig. 3-53 Gravitational force Mg exerts a torque about axis through A.

S-32. (a) Using the equation $\tau = I\alpha$, show that the moment of inertia I has the dimensions $[ML^2]$ and that the mks unit is therefore the kg-m².
(b) A flywheel has a moment of inertia about a fixed axis of 3.0 kg-m². Find the angular acceleration of the flywheel when the resultant external torque about the axis is 90 N-m.
(c) If the flywheel of part (b) is initially at rest, how long will it take for it to reach an angular velocity of 15 radians/sec?

S-33. A resultant external torque of 80 N-m gives a flywheel an angular acceleration of 1.6 radians/sec². What is the moment of inertia of the flywheel about its axis?

S-34. A flywheel 0.80 m in diameter is mounted on a horizontal axis. When a steady pull of 40 N is applied to the free end of a rope which has been wrapped **around the outside of the flywheel, we observe that the flywheel makes 3.0 revolutions in 5.0 sec.**
(a) What is the angular acceleration of the flywheel?
(b) What is the moment of inertia of the flywheel?
(c) What is the final angular velocity?

S-35. A 20-kg pail of water is suspended by a rope wrapped around a windlass which is a solid cylinder 0.40 m in diameter with a mass of 50 kg. The pail is released at rest at the top of a well and descends 20 m before it strikes the water. What is the speed of the pail at this instant? (A uniform cylinder of radius R and mass M has a moment of inertia about its axis given by $I = \frac{1}{2}MR^2$.)

S-36. Assume that the rope in Fig. 3-18 passes over a rough pulley surface. The pulley has a radius R and a moment of inertia I about its axis. At the pulley axis a frictional torque τ_f opposes rotation.
(a) Show that the difference, $T_1 - T_2$, between the tensions in the rope on opposite sides of the pulley is given by
$$(T_1 - T_2) R = \tau_f + I\alpha,$$
when the pulley is rotating with an angular acceleration α.
(b) In mechanics problems with "light frictionless" pulleys we assume that $T_1 = T_2$. In making this assumption about a pulley, what physical quantities are assumed to have negligibly small values?

S-37. The position coordinate of an oscillating particle is given by
$$x = 5.0 \cos(0.50\pi t + \pi/3) \text{ m}.$$

(a) Find the angular frequency, the frequency, and the period.
(b) What are the initial phase angle and the amplitude?
(c) At the instant $t = 2.00$ sec, find the position, velocity and acceleration.

S-38. A block is placed on top of a piston which executes simple harmonic motion along a vertical line with a frequency of 2.0 hertz. The amplitude of oscillation gradually increases. At what amplitude of motion will the block and piston separate?

S-39. A block rests on a rough board which is moved horizontally back and forth in a simple harmonic motion of amplitude 0.15 m. The frequency is slowly increased until, at a frequency of 0.25 hertz, the block just begins to slide. Compute the coefficient of friction between the block and the board.

S-40. A mass is supported by two springs with force constants k_1 and k_2 respectively.
(a) Show that when one end of each spring is connected to the ceiling and the other end is connected to a suspended mass, the pair of springs exert the same force that would be exerted by a single spring with a force constant k given by
$$k = k_1 + k_2.$$
(b) Show that when an end of one spring is joined to an end of the other to form a single long spring, this combination has a force constant k determined by
$$\frac{1}{k} = \frac{1}{k_1} + \frac{1}{k_2},$$
and that this implies that
$$k = \frac{k_1 k_2}{k_1 + k_2}.$$

S-41. (a) At what position will the tension in an oscillating simple pendulum reach its maximum value?
(b) Show that this maximum tension in a pendulum of length L, for small oscillations of amplitude A, has the value $mg[1 + (A/L)^2]$.

S-42. A small block, placed inside a smooth hemispherical bowl of 0.40-m diameter, slides back and forth executing oscillations about its lowest point. What is the period of these oscillations?

ADDITIONAL APPLICATIONS TO MEDICINE AND THE LIFE SCIENCES

Ballistocardiograph

The *ballistocardiograph* is a device used primarily in medical research rather than in regular clinical practice. An instrument which helps to determine the force exerted by the heart in pumping blood, it is a device whose design is very much based on Newton's laws of motion.

In contracting, the ventricles exercise a force to pump blood toward the head. According to Newton's third law there is then a corresponding force of the blood upon the ventricles (and thus the rest of the body) which results in a minute acceleration of the body in the direction of the feet. When the blood reaches the arch of the aorta, the direction changes toward the feet and there is a consequent third-law force on the rest of the body, which recoils toward the head. These minute body motions were detected as early as 1877.

One fairly sensitive device to measure this motion due to blood pumping was made by I. Starr in 1939. The technique was to secure the patient rigidly on a light table suspended by wires from the ceiling. The table thus moves with the body in the body's recoil as blood is pumped. The small recoil motion of the table is then recorded using optical devices. A recording of this motion is called a *ballistogram*; it appears somewhat similar to an ECG.

Question

Upon what factors does this body motion depend (aside from the characteristics of the table and its suspension)?

Answer

Momentum has to be conserved; that is, the momentum of the blood in one direction is equal to the momentum of the rest of the body in the opposite direction. Consequently the recoil velocity of the body depends upon the mass of blood pumped per unit time and the velocity with which it is ejected from the ventricles, as well as on the mass of the rest of the body.

Friction at Body Joints

In the study of degenerative arthritis (osteoarthritis) it is becoming clear that mechanical forces are playing a role in joint degeneration. (It used to be thought that mechanical stress was of no great, if any, significance in joint degeneration because many people did hard physical labor all their lives without exhibiting joint degeneration.) Moreover, these studies of joints have shown that the loads on joints like the hip and the knee are much greater than might be expected from considering body weight as the prime source of load. In fact, the larger part of the compressive forces at a joint is due not to body weight but to the muscles which span the joint and activate the limb. These muscular forces may be of the order of 3 or 4 times the body weight. One might thus expect enormous frictional forces at the joints. However the joints are lubricated exceedingly well by a material called synovial fluid. So well that with a load of some 1800 N (about 400 pounds) the frictional force in a typical joint might be only about 9 N.

Question

What is the coefficient of friction between the bone surfaces of such a lubricated joint?

Answer

$$\text{Coefficient of friction} = \frac{9 \text{ N}}{1800 \text{ N}} = 0.005.$$

SUGGESTED READING

FEYNMAN, R. P., R. B. LEIGHTON, and M. SANDS, *The Feynman Lectures on Physics.* Reading, Mass.: Addison-Wesley, 1963. Chapter 10 on momentum is one of the briefest, most elementary chapters in this book. Highly recommended.

HALLIDAY, D., and R. RESNICK, *Physics Part I.* New York: John Wiley, 1966. Chapters 5, 6, 9, 11, 14, and 15 provide useful collateral reading for students with a knowledge of calculus and vector analysis.

HARRINGTON, E. L., *General College Physics.* New York: D. Van Nostrand, 1952. Students interested in medicine or biology will find particularly valuable the applications of the principles of statics found in this book.

HOLTON, G., and D. H. D. ROLLER, *Foundations of Modern Physical Science.* Reading, Mass.: Addison-Wesley, 1958. Chapter 17 gives an interesting account of momentum at a level appropriate for our readers. Historical background is included.

JENSEN, J. TRYGVE, *Physics for the Health Professions.* Philadelphia: Lippincott, 1975. An up-to-date excellent descriptive text which makes good collateral reading for students interested in medicine and the related health professions.

SEARS, F. W., and M. W. ZEMANSKY, *University Physics.* Reading, Mass.: Addison-Wesley, 1970 (fourth edition). Chapters 1, 2, 5, 8, 9, and 11 are recommended for a more extended and a more advanced treatment of topics in mechanics. Calculus and vector analysis are used.

Enrico Fermi

4 Energy

When treating a truly fundamental concept like energy, it is probably best to talk about it first before attempting to give formal definitions. The consideration of energy by physicists began in Newton's mechanics when it was recognized that in some circumstances the energy of motion (named *kinetic energy*) and energy associated with the configuration or position (named *potential energy*) changed as the motion went on in such a way that their sum (named the *total mechanical energy*) was constant. In short,

Kinetic Energy + Potential Energy = Constant.

This simple equation gave a neat solution to many complicated questions in mechanics and suggested that the total energy, whatever it is, might be important if it were indeed something whose amount is conserved (that is, remains constant) in spite of complex changes in many features of the system.

But the above equation is no longer true; the total mechanical energy decreases *if* there are frictional effects which "dissipate" the energy. However, when this happens it is commonly observed that there is some object that gets warmer.

Following much speculation, thought, and some experimentation, notably by Benjamin Thompson (1753–1814), an American Royalist who later became Count Rumford of Bavaria; a physician, J. R. Mayer (1814–1878); and a physicist, H. von Helmholtz (1821–1894), this whole matter gradually assumed some clarity. And after the series of experiments performed in the 1840s by J. P. Joule (1818–1889), it was recognized that the energy was not lost but simply transformed into an increased energy of random motion of the molecules of the warmer object. Since that time the concept of energy, as a conserved quantity that can be transformed into many guises but never created or destroyed, has emerged as perhaps the most useful idea in all science.

The next major contribution to our understanding of energy came from Einstein's special theory of relativity, which suggested an equivalence of inertia and energy and indicated the possibility of energy transformations that were at once startling and ominous.

"The fundamental point in fabricating a chain reacting machine is of course to see to it that each fission produces a certain number of neutrons and some of these neutrons will again produce fission."

Enrico Fermi
(1901–1954)

At the forefront of the study of the sensational energy transformations that are so characteristic of twentieth-century physics, was Enrico Fermi, whose name joins those of Galileo and Newton as one of the very few who have made outstanding contributions to both theoretical and experimental physics. Any particle, which like the electron has half-integral spin (Section 3.9), is called a *fermion* in recognition of Fermi's fundamental work in 1926 on the behavior of collections of such particles. In 1934, he published a theory of the β-decay of a nucleus, in which theory was included the idea that an electron, and a light neutral particle called the *neutrino* were created at the instant of decay. This detailed and quantitative theory still has a large influence on research in the physics of elementary particles. Fermi's first venture in experimental physics revealed many new radioisotopes produced by neutron bombardment (Section 1.6).

In 1938, the year he was awarded the Nobel prize, Fermi emigrated from Italy to the United States. Here he soon turned his efforts towards ensuring that his adopted country would have mastery of the possibilities suggested by the discovery of uranium fission in 1939. Under Fermi's guidance, the first nuclear reactor was constructed: a lattice of uranium and graphite. On December 2, 1942, a self-sustaining nuclear chain reaction was achieved. The director of this project, Arthur Compton, telephoned to James Conant, president of Harvard University, the famous message, "Jim, you'll be interested to know that the Italian navigator has just landed in the new world."

4.1 ENERGY IN NEWTONIAN MECHANICS

Work

The term *work* is used in physics as a technical term with a precise meaning that must be distinguished from its usage in daily life. The work done by a force $\vec{F}$ when the object on which it acts moves a distance Δs (Fig. 4-1) is defined by

$$\Delta W = F_s \Delta s, \tag{4.1}$$

where ΔW denotes the work, and it is assumed that $\vec{F}$ does not change appreciably during the motion through the distance Δs. F_s denotes the component of $\vec{F}$ in the direction of the motion and can be positive, zero, or negative depending on its direction. Thus

1. If $\vec{F}$ is in the direction of motion,

$$F_s = F, \text{ and } \Delta W = F \Delta s. \tag{4.2}$$

2. If $\vec{F}$ is perpendicular to the direction of motion,

$$F_s = 0, \text{ and } \Delta W = 0. \tag{4.3}$$

3. If $\vec{F}$ is opposite to the direction of motion,

$$F_s = -F, \text{ and } \Delta W = -F \Delta s. \tag{4.4}$$

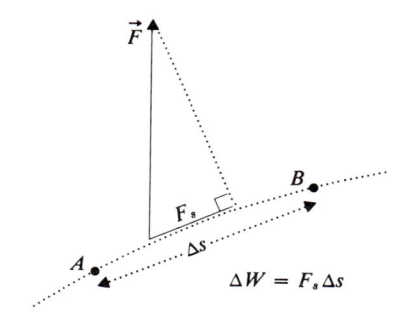

Fig. 4-1 Force $\vec{F}$ does positive work $\Delta W = F_s \Delta s$ on an object which goes from point A to point B.

We see that, as the term "work" is used in physics, it can be positive, zero, or negative. It must be emphasized that if there is no motion, $\Delta s = 0$ and the force does no work in the sense physicists use the term. No work is done by a man who merely supports a hundred-pound weight without moving it up or down.

The mks unit of work, the N-m is named the *joule*. A force of 8.0 newtons moved through 3.0 meters in the direction of the force does a work of $(8.0 \text{ N}) \times (3.0 \text{ m}) = 24$ joules. A much smaller unit of work is used in the physics of molecules, atoms, and elementary particles. This unit is named the *electron volt* (eV), which is related to the joule by the equation,

$$1 \text{ eV} = 1.60 \times 10^{-19} \text{ joule.} \tag{4.5}$$

Commonly used multiples of the electron volt are

$$1 \text{ MeV} = 10^6 \text{ eV},$$
$$1 \text{ BeV} = 1 \text{ GeV} = 10^9 \text{ eV},$$

with M standing for million, and B or G for billion. As we shall show in Section 11.1 the work done by the electric force on an electron accelerated from cathode to anode of a 50,000 volt x-ray tube is 50,000 eV; the work done by the electric force on a proton accelerated in the Berkeley Bevatron is 6.0 BeV.

Example 1 A 50.0-kg stone is dropped from the top of a cliff which has a height h of 10.0 m. After it strikes the ground the stone is dragged along a horizontal road a distance of 80 m. Find the work done by the gravitational force ($m\vec{g}$) on the stone for each portion of this trip.

Solution While the stone is falling the motion is in the direction of $m\vec{g}$. Therefore Eq. (4.2) applies with $F = mg$ and $\Delta s = h$:

$$\Delta W = mgh$$
$$= (50.0 \text{ kg}) \times (9.80 \text{ m/sec}^2) \times (10.0 \text{ m})$$
$$= 4.9 \times 10^3 \text{ joules.}$$

While the stone is being dragged along the road, the direction of motion is perpendicular to the force $m\vec{g}$, and Eq. (4.3) applies:

$$\Delta W = 0.$$

Note that the above statement does not say that no work is being done on the stone. Far from it. Work has to be done by someone or something which drags the stone along the horizontal road. But this work is *not* being done by the gravitational force, which is perpendicular to the direction of motion on this part of the stone's trip.

Kinetic Energy

Picture a Mr. Abel and a Mr. Baker seated at opposite ends of a long smooth table (Fig. 4-2), with Abel pushing a puck of mass m with a constant force F_A through a distance Δx until it attains a speed v. The work done by the force F_A as it moves a distance Δx is

$$\Delta W = F_A \Delta x. \tag{4.6}$$

Newton's second law [Eq. (2.10)] gives $F_A = ma$, so we can substitute for F_A in Eq. (4.6) and obtain

$$\Delta W = ma\Delta x. \tag{4.7}$$

Now we have previously shown that

$$v^2 = 2a\Delta x. \tag{2.7}$$

Substituting $v^2/2$ for the term $a\Delta x$ in Eq. (4.7), we find

$$\Delta W = \tfrac{1}{2} mv^2. \tag{4.8}$$

Having attained this speed v, the puck slides with constant speed across the table until it is caught by Baker who exerts a retarding force $\vec{F}_B$ which brings the puck to rest. By repeating the above analysis, we can show that Baker does work of the amount $-\tfrac{1}{2}mv^2$ on the puck and since, from Newton's third law, the puck exerts a force $-\vec{F}_B$ on Baker, the work done by the puck on Baker is just $\tfrac{1}{2}mv^2$.

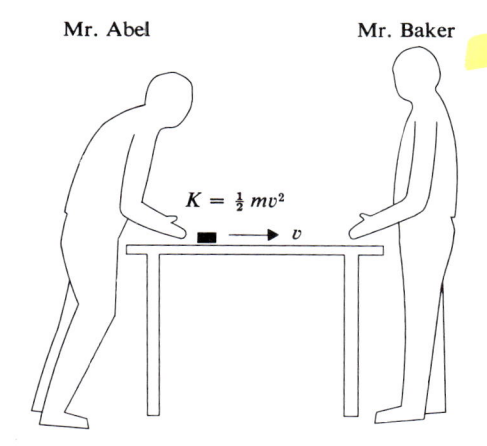

Fig. 4-2 Kinetic energy.

Now let us build up our vocabulary and learn to use it to discuss this situation more fully. We adopt as a definition of energy the statement: *The energy of an object is a measure of its ability to do work.* The puck did work on Baker, and it evidently possessed this ability as it was sliding toward him. So there was energy associated with its motion and the precise amount of energy, the work the moving object could do, was $\frac{1}{2}mv^2$. With this in mind, *we define kinetic energy as energy associated with motion.* We have the result that a mass m moving with a velocity v (without rotating) possesses a *kinetic energy, K*, given by

$$K = \tfrac{1}{2}mv^2. \tag{4.9}$$

Using this language we can say that the force exerted by Abel does work on the puck and supplies to the puck a kinetic energy equal to the work done by this force. Anticipating the idea that energy is conserved, it can be said that the force F_A is the agency by which energy is transferred from Abel to the puck. The puck then *possesses* this kinetic energy as it slides across the table and delivers this energy to Baker, the energy transfer being accomplished by the puck exerting a force and this force doing work on Baker.

Although we have defined energy in terms of work by stating that energy is a measure of the ability to do work, it turns out that in many ways energy is the more fundamental concept. We can characterize work in terms of energy by noting that *work is a measure of the energy transferred by a force.*

Work and Energy are measured in the same units. Thus we speak of an electron having a kinetic energy of 50,000 eV as it strikes the anode of an x-ray tube or of a proton emerging from the Berkeley Bevatron with a kinetic energy of 6.0 BeV. Further examples of energies expressed in electron volts are given in Table 4.1. A mass of 10.0 kg traveling with a velocity of 4.0 m/sec has a kinetic energy of $\frac{1}{2} \times 10.0$ kg $\times$ (4.0 m/sec)^2 = 80 joules. Another energy unit that you often meet in daily life is the *kilowatt hour*. A kilowatt hour is 3.6×10^6 joules (this happens to be the energy supplied to a 100-watt bulb in 10 hours).

Table 4-1. SOME ENERGIES IN ELECTRON VOLTS

Phenomenon	Energy, eV
Average kinetic energy of air molecule at room temperature	4×10^{-2}
Photon of red light	~ 2
Typical x-ray photon	5×10^4
Gamma-ray photon (from Co-60)	1.33×10^6
Energy required to remove a proton or a neutron from the nucleus	$\sim 10^7$

Power

Power is a rate of working or a rate at which energy is supplied. Thus, if a force does work ΔW in a time interval Δt, the average power (P) supplied by this force for this time interval is $\Delta W/\Delta t$, that is,

$$P = \Delta W/\Delta t. \tag{4.10}$$

The mks unit of power, the joule/sec, is named the *watt*.

At this point we summarize some systems of units for the reader's convenience in Table 4-2.

Table 4-2. SOME SYSTEMS OF UNITS

System	Force	Mass	Length (distance)	Work or energy	Power
Engineering	pound (lb)	slug	foot	foot-pound (ft-lb)	ft-lb/sec
cgs	dyne	gram	centimeter	erg	erg/sec
mks	newton (N)	kilogram (kg)	meter (m)	joule (J)	watt (W)

Potential Energy

Let us now consider a system composed of at least two objects with forces exerted by one object on the other. A simple example is the system of the planet earth of mass M and a baseball of mass m. The earth exerts on the baseball a gravitational force $m\vec{g}$. Now if some external agency, say a man, slowly (that is, with negligible acceleration) lifts the baseball a height h by applying an upward force of magnitude mg, he does a positive amount of work equal to mgh. He has then supplied to the earth-baseball system an energy mgh, and the question arises: Where has this energy gone? It has certainly not appeared as kinetic energy because both the baseball and the earth are at rest. It is natural to suggest that this new configuration, with the baseball pushed a distance h further away from the earth, has more energy than the initial configuration. The thought that a definite change in energy is associated with a certain change in configuration is reinforced by the discovery that the same amount of work is done if the man takes the baseball along a different path zigzagging from the same initial point to the same final point. This work can be recovered from the earth-baseball system by allowing the baseball to push on some external object as it is slowly lowered to its initial position. If the baseball is simply dropped, it gets back to its initial position with a very evident form of energy, kinetic energy.

In this way we are led to recognize a new type of energy, *the energy that a system possesses because of its configuration* (the positions of its parts) which is named *potential energy*. The change in the potential energy (U) of a system when its configuration is changed is defined to be the work that would have to be done (by external applied forces) to achieve the change in configuration. Of course, there is no universal formula for the potential energy of every system. Different types of forces will imply different amounts of work to change from one configuration to another.

For a mass m near the earth's surface we have found the formula for the gravitational potential energy U:

$$U = mgh. \qquad (4.11)$$

The height h can be measured from any convenient level; only the *difference* in potential energy between different positions is significant, and this difference is not affected by our choice of the configuration that is assigned zero potential energy. If we measure h from the bottom of a cliff which is 10 m high, the potential energy when a 50-kg stone is on top of the cliff is (50 kg) × (9.8 N/kg) × (10 m) = 4.9×10^3 joules. This tells us that we can extract 4.9×10^3 joules from this stone-earth system by allowing the stone to somehow get to the bottom of the cliff. We would have to supply 4.9×10^3 joules to get the stone from the bottom to the top of the cliff.

The potential energy U for the same system, a particle of mass m and the earth of mass M, is much more difficult to calculate if we no longer confine ourselves to points close to earth where the gravitational force is constant. The general result when m is a distance r from the earth's center (Fig. 4-3) turns out to be

$$U = -\frac{GmM}{r}, \qquad (4.12)$$

which looks something like the formula $F = GmM/r^2$ for the gravitational force involved. But notice that the denominators are different and that the potential energy U is negative. At the earth's surface of radius r_e, U has a minimum value of $-GmM/r_e$, and it increases from this negative value to approach the value 0 as r becomes very large. It is not at all obvious but, if you are patient enough to do the calculation, you can show that the formula $U = mgh$ gives almost the same answers as $U = -GmM/r$ for *changes* in potential energy near the earth's surface.

For a system of two electric charges, q and Q, a distance r apart (exerting Coulomb forces on each other of magnitude $F = kQq/r^2$) the potential energy can be shown to be

$$U = k\frac{qQ}{r}, \qquad (4.13)$$

where U is in joules if q and Q are in coulombs and r in meters with $k = 9.0 \times 10^9$ N m²/coul². This potential energy, U, is positive only if

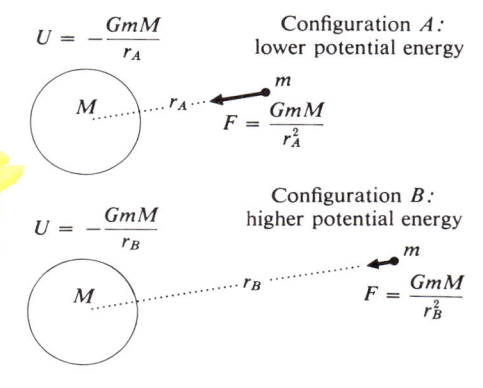

Configuration A: lower potential energy

$U = -\frac{GmM}{r_A}$

$F = \frac{GmM}{r_A^2}$

Configuration B: higher potential energy

$U = -\frac{GmM}{r_B}$

$F = \frac{GmM}{r_B^2}$

Fig. 4-3 Gravitational potential energy.

4.1 ENERGY IN NEWTONIAN MECHANICS

*When the resultant force acting on a particle is a Hooke's law force, $F_x = -kx$, the particle has a potential energy given by

$$U = \tfrac{1}{2}kx^2.$$

This result is obtained by calculating the work that would have to be done by an external applied force, $F = kx$, to bring the particle from the equilibrium position to a point with position coordinate x. The evaluation of this work is complicated by the fact that the force is not constant, but increases during the displacement from the value 0 to the value kx. The correct answer is obtained by multiplying the force occurring at the midpoint of the displacement, the value $k(x/2)$, by the total displacement x to obtain $U = k(x/2)x = \tfrac{1}{2}kx^2$. (This is easily verified using the methods of integral calculus).

*The total mechanical energy of a particle with mass m that is executing simple harmonic motion is

$$E = \tfrac{1}{2}mv_x^2 + \tfrac{1}{2}kx^2.$$

This sum of the particle's kinetic and potential energies is conserved throughout the motion. From this fact we can very easily extract some useful information. Since E is constant it is apparent that maximum speed is reached where x is zero and that x attains its maximum magnitude, A, where the speed is zero. Then by evaluating E at the point $x = A$ where $v_x = 0$ we obtain

$$E = 0 + \tfrac{1}{2}kA^2$$
$$= \tfrac{1}{2}m\omega^2 A^2,$$

having used $k = m\omega^2$ from Eq. (3.39). The total mechanical energy in simple harmonic motion is therefore proportional to the square of the frequency and to the square of the amplitude.

---

the charges have the same sign. In this case the forces are repulsive and you would have to do positive work to push them closer together.

An interesting fact that follows from the definition of potential energy is that *the direction of the force associated with any sort of potential energy is always from positions of higher to lower potential energy*. You can verify that this is consistent with each of the three potential energy formulas that we have encountered.

At this point we should like to offer the reader some relief. These different expressions for the potential energy arising from different forces have been displayed, not with the idea of giving you more formulas to memorize, but rather to emphasize that U is different for different systems.*

Finally, we ask: Can work done by any and all forces produce a change in potential energy? No. There is no potential energy associated with frictional forces. If we drag a rough stone from the bottom to the top of a cliff, the work done by friction will depend on the path we take, so there is no definite amount of frictional work that can be assigned to the change in position.

Total Mechanical Energy

In the mechanics of Newton, the sum of the kinetic energy K and the potential energy U is called the total mechanical energy E.

$$K + U = E.$$

As the motion proceeds, if the only forces doing work are those that gave rise to the potential energy U, then using Newton's second law, it can be shown that the total mechanical energy E is constant. The law of conservation of total mechanical energy can be written

$$K + U = E \text{ (constant)}. \tag{4.14}$$

This tells us that the energy can be transformed back and forth between kinetic energy and potential energy but the total doesn't change. We can equate the values of the total mechanical energy occurring at any two different points A and B:

$$K_A + U_A = K_B + U_B. \tag{4.15}$$

The solution of many mechanics problems is most easily accomplished by exploiting energy conservation.* Consider a stone of mass m on the earth's surface, and suppose it is projected upward, as in Fig. 4-4, with a speed v_0. Let's determine its speed v as it reaches a height h above its initial position. Equating the total mechanical energy at height h to its initial value, we obtain

$$\tfrac{1}{2}mv^2 + mgh = \tfrac{1}{2}mv_0^2 + 0,$$

or

$$v^2 = v_0^2 - 2gh. \tag{4.16}$$

160 ENERGY

For a given projection speed v_0, the speed on different trajectories is the same at points at the same height, as shown in Fig. 4-4.

When the stone is projected vertically upward, its speed at maximum height is zero because it is completely changing its direction. To find this maximum height we equate the total mechanical energy at this point to its initial value:

$$0 + mgh_{max} = \tfrac{1}{2}mv_0^2 + 0,$$

which yields the result

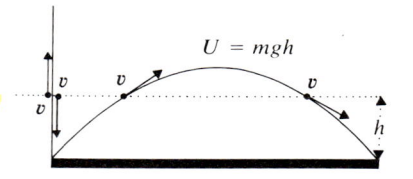

$$h_{max} = \frac{v_0^2}{2g}. \quad (4.17)$$

Fig. 4-4 For a given projection speed v_0, the speed v on different trajectories is the same at points at the same height.

Example 2 What height will be reached by a stone thrown straight up with a velocity of 37 m/sec?

Solution From Eq. (4.17), this stone will reach a height

$$h_{max} = \frac{(37 \text{ m/sec})^2}{2 \times (9.8 \text{ m/sec}^2)} = 70 \text{ m}.$$

Energy transformations occur in all physical processes. The motion we have just studied gives us an opportunity to learn to describe transformations of energy in a very simple situation. Initially the total mechanical energy is entirely the kinetic energy possessed by the stone, but as the stone rises this kinetic energy decreases and the potential energy of the stone-earth system increases. (Incidentally, the potential energy should be associated with the system composed of both the earth and the stone, but many books assign it to the stone alone hoping that the reader will understand that it really belongs to the entire system.) At maximum height, for a vertical trajectory, there is no kinetic energy; the potential energy has reached a maximum value equal to the initial kinetic energy. As the stone falls kinetic energy is acquired at the expense of potential energy. Throughout the entire flight, assuming that negligible work is done by the frictional forces exerted by the air on the stone, the total mechanical energy is constant.

Rocket Escape Velocity

Using energy conservation, it is a simple matter to relate the speed v of an earth satellite at any distance r from the earth's center with the position r_0 and the speed v_0 at any other instant (the rockets being shut off). For this system the potential energy formula has been given in Eq. (4.12). The total mechanical energy at the two different points is the same, so from Eq. (4.15) we get

$$\tfrac{1}{2}mv^2 - G\frac{mM}{r} = \tfrac{1}{2}mv_0^2 - G\frac{mM}{r_0}.$$

4.1 ENERGY IN NEWTONIAN MECHANICS

Particular interest is attached to what is called the *escape speed*, v_e. This is the speed which an object must have at position r_e in order to have energy just sufficient to keep it moving away from the earth forever and reach, with negligible kinetic energy, remote regions where r is so large that GmM/r is negligible. Setting the final kinetic energy and potential energy at the value zero, the energy conservation equation, Eq. (4.15), gives

$$0 + 0 = \tfrac{1}{2}mv_e^2 - G\frac{mM}{r_e},$$

which has the solution

$$v_e = \sqrt{\frac{2GM}{r_e}}.$$

For projection from the earth this gives

$$v_e = 11.2 \times 10^3 \text{ m/sec},$$

which is approximately 25,000 miles per hour.

4.2 THERMAL ENERGY

Dissipation of Mechanical Energy

The law of conservation of total mechanical energy has a serious drawback. This "law" can only be used if the work done by frictional forces is negligible. When frictional forces work on a system, its total mechanical energy decreases. One then speaks of the dissipation of mechanical energy by the work done by dissipative forces. Although mechanical energy is lost, an equivalent amount of energy appears in another form which is called *thermal energy*. The thermal energy of the objects that are rubbed, scraped, or knocked about, increases and they become warmer. This increased thermal energy is just an increase of average kinetic and potential energies of the constituent atoms and molecules engaged in *random* motion.

In the examples using conservation of total mechanical energy we neglected dissipative effects.* In fact, an earth satellite's total mechanical energy is decreased by a frictional drag exerted by the residual atmosphere at orbit altitudes. As the sum of its potential and kinetic energy decreases, the satellite gradually spirals in, and speeds up according to Eq. (2.15), and becomes warmer. Finally, as the more dense atmosphere is encountered, it meets a fiery death unless rockets are used to effect a rapid re-entry. The motion of a stone near the earth's surface is also affected by the frictional drag exerted by the air on the stone. As the stone moves along its trajectory, its total mechanical energy decreases slightly, and it will attain a maximum height a little

---

*The oscillations of a pendulum or of a mass fastened to a spring (Fig. 3-35) actually differ from the idealized motion called simple harmonic motion (Section 3.10). Instead of oscillating forever with a constant amplitude, the amplitude slowly decreases as the mechanical energy of the oscillating mass is dissipated by frictional forces acting on this moving mass. We say that the oscillations are *damped by friction.*

below that predicted by Eq. (4.17). If we replace the stone by a feather, the frictional force is so large even at very low speeds that the dissipation of total mechanical energy cannot be neglected.

In this section we shall discuss briefly processes involving changes in thermal energy and the first law of thermodynamics. Energy conservation is the key idea.

Conversion of Mechanical Energy into Thermal Energy

The thermal energy of a given amount of a pure substance, say a mass m of water, depends upon its temperature and certain other conditions such as the pressure (Section 7.1) and the magnetic field (Section 12.2). The change in thermal energy associated with a temperature increase ΔT (with other conditions, like pressure, unchanged) is easy to measure experimentally. One simply allows a measured amount of mechanical energy to be converted into thermal energy of the substance under investigation. The temperature rise ΔT of the substance is measured. Now the increase in the thermal energy of the substance associated with this temperature increase ΔT can be calculated because, for this process,

> Increase in thermal energy = mechanical energy supplied.

In the years between 1840 and 1868, Joule performed many such experiments, one of which is depicted schematically in Fig. 4-5. While the suspended mass m falls slowly through a distance h, it loses potential energy, $U = mgh$ [Eq. (4.11)]. The falling weight causes a paddle wheel to rotate and churn the water. At the end of the process, when the water has come to rest, its temperature rise ΔT is noted. If corrections for energy going elsewhere are neglected, we can say that this temperature rise of the water is associated with an increase in the water's thermal energy given by mgh. Joule's experiments give the result that, for 1.00 kg of water between 0°C (zero degrees Celsius*) and 100°C, there is an increase in thermal energy of about 4.2×10^3 joules for each 1 C° rise in temperature. For a mass m of water, heated through ΔT in degrees Celsius, the result is

Increase in thermal energy (of water) = $(4.2 \times 10^3 \text{ joule/kg C°}) m \, \Delta T$. (4.18)

Experiments such as the foregoing indicate how changes in the thermal energy of any substance can be measured. But the great historical importance of Joule's experiments lies in the fact that these experiments also provide strong quantitative evidence for our modern notions of energy, namely:

1. Energy is conserved;
2. In heating a substance, we are simply increasing its thermal energy.

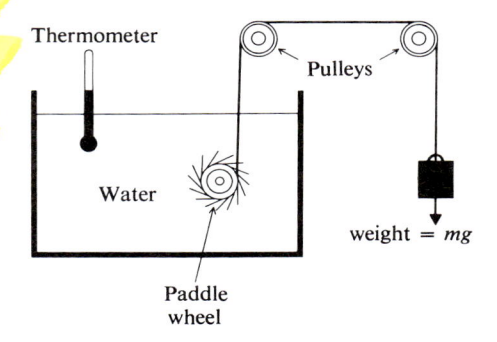

Fig. 4-5 The falling weight causes the paddle wheel to rotate and churn the water. The temperature rise ΔT, associated with the loss of potential energy mgh, is measured by the thermometer.

*The relationship between the Celsius temperature t_C (formerly called centigrade) and the Fahrenheit temperature t_F (which is commonly used in United States and Canada) is $t_C = \frac{5}{9}(t_F - 32°F)$.

4.2 THERMAL ENERGY

These two ideas gained widespread acceptance after Joule's demonstration that, in heating a substance from a certain initial temperature, the disappearance of the same measured amount of *any* sort of energy is always accompanied by the same temperature rise in the substance.

Heat and the First Law of Thermodynamics

Suppose we again consider some definite system, say a mass m of water. But this time we concern ourselves with some process in which not only macroscopic mechanical energy is transferred to the water, but also some energy is transferred to the water because of mere contact with a hotter body. For this system under these circumstances, conservation of energy is expressed by the equation

$$\text{Increase in thermal energy} = (\text{mechanical energy supplied}) + (\text{heat}) \quad (4.19)$$

The statement given by Eq. (4.19) is known as the *first law of thermodynamics.**

The *heat* involved in a process is defined by Eq. (4.19) as the energy which is transferred to the system by nonmechanical means, that is, *heat is energy transferred because of a temperature difference.* The mks unit of heat is, of course, the joule, as is the case for any other energy like potential energy (U), or kinetic energy (K), or work (W).

Example 3 A pot of 5.0 kg of water sitting on a hot stove is stirred violently by a mixer which does 8.0×10^3 joules of mechanical work on the water. The temperature rise of the water is 2.00 C°. Find the heat which flows into the water from the stove during this process.

Solution From Eq. (4.18), for 5.0 kg of water,

Increase in thermal energy =
$$(4.2 \times 10^3 \text{ joule/kg C°}) \times (5.0 \text{ kg}) \times (2.0 \text{ C°})$$
$$= 42 \times 10^3 \text{ joules}.$$

The mechanical energy supplied to the water is 8.0×10^3 joules. The heat is determined by the first law of thermodynamics,

$$\text{Increase in thermal energy} = (\text{mechanical energy supplied}) + (\text{heat}) \quad (4.19)$$

$$42 \times 10^3 \text{ joules} = 8.0 \times 10^3 \text{ joules} + \text{heat}.$$

Therefore,
$$\text{Heat} = 34 \times 10^3 \text{ joules}.$$

Transformations between mechanical energy, heat, and thermal energy are governed by the first law of thermodynamics, no matter what the direction of the energy conversion. In Joule's experiments, mechanical energy is completely converted into thermal energy. On the other

---

Thermodynamics is the name given to the science of heat phenomena, particularly that portion of the subject which deals only with macroscopic quantities without having to make assumptions about the molecular nature of matter. The quantity that we have called thermal energy is usually termed internal energy *in books on thermodynamics.*

hand, in steam engines and automobile engines, heat is extracted from a hot substance, some heat is rejected to cooler surroundings, and the difference is converted into mechanical energy. And this whole process takes place in accordance with the first law of thermodynamics.

Specific Heat

Our knowledge of the thermal behavior of a substance, gained by experiment or by theory, is conveniently expressed by specifying what is called the *specific heat* of the substance. If, in a process performed in a definite way (say with the pressure kept constant), we have heat flowing into a mass m of a substance, thereby causing a temperature rise ΔT, the specific heat s of this substance is defined by

$$\text{Heat} = sm\Delta T. \tag{4.20}$$

For water we find the specific heat is approximately 4.2×10^3 joule/kg C°, the value depending only slightly on the temperature interval for temperatures between 0° C and 100° C.

In a process during which the temperature drops, Eq. (4.20) gives the heat flowing out from the cooling substance.

Example 4 How much heat must flow into 5.0 kg of water to raise its temperature from 20.0° C to 30.0° C?

Solution Here $\Delta T = 10.0$ C° and the specific heat of water is $s = 4.2 \times 10^3$ joule/kg C°. Therefore,

$$\begin{aligned}\text{Heat} &= sm\Delta T \\ &= (4.2 \times 10^3 \text{ joule/kg C°}) \times (5.0 \text{ kg}) \times (10.0 \text{ C°}) \\ &= 21 \times 10^4 \text{ joules.}\end{aligned}$$

Another unit of energy, the *kilocalorie* (sometimes called the *food calorie*), is now defined as follows:

$$1 \text{ kilocalorie} = 4.184 \times 10^3 \text{ joules } \textit{exactly}.$$

In terms of the kilocalorie, the specific heat of water, between 14.5° C and 15.5° C, is 1.000 kilocalorie/kg C°. This apparent simplicity arises from the fact that the kilocalorie originally was defined as that amount of heat which must be supplied to one kilogram of water to raise its temperature through one Celsius degree (i.e., from 14.5° C to 15.5° C precisely). For this reason, heat calculations *involving temperature changes of water* are obviously rather simple in terms of the kilocalorie. However, the reader should not get the impression that the kilocalorie is peculiar to heat. The kilocalorie is just another energy unit. Any energy, say the kinetic energy of a baseball, could be measured in kilocalories. And heat, as we have already illustrated in Examples 3 and 4, can quite properly be measured in joules.

Physics Today

J. P. Joule
(1818–1889)

Table 4-3. SPECIFIC HEATS OF COMMON SUBSTANCES
AT ATMOSPHERIC PRESSURE AND ORDINARY TEMPERATURES

Substance	Specific heat (kilocalories/kg C°)
Water	1.00
Ice	0.49
Hydrogen	3.4
Mercury	0.033
Copper	0.093
Iron	0.12
Window glass	0.20

The specific heats (in kilocalories/kg C°) of various substances are given in Table 4-3. Except for one of the phases of liquid helium (Chapter 9), the substance having the greatest specific heat is hydrogen. Compared to most substances, water also has a very large specific heat, a fact which has an important effect on the earth's climate. Because the specific heat of water is large, the oceans warm very slowly in the spring and cool slowly in the fall. The temperature differences between summer and winter are therefore moderated, particularly in coastal areas.

4.3 SIMPLE MACHINES

Various mechanical devices are used to transform an input force into an output force, often with considerable change in magnitude. The ratio of the magnitudes of the output force F_o to the input force F_i is defined as the *mechanical advantage* of the machine. That is

$$\text{Mechanical advantage} = F_o/F_i.$$

The Lever

An example of a simple machine is the *lever* pictured in Fig. 4-6(a). The lever is supported at a *fulcrum*. The input force F_i is normally supplied by a man, and an upward output force F_o is often applied to lift a stone. The force exerted by the stone on the lever, by Newton's third law, will be downward with a magnitude F_o, as shown in the free body diagram for the lever, Fig. 4-6(b) (we consider only an idealized lever which is rigid and has negligible mass). When the lever is in equilibrium, there is a simple relationship between F_o and F_i. The second condition of equilibrium, Eq. (3.23), for torques about an axis perpendicular

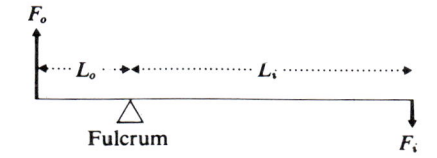

Fig. 4-6(a)   A lever.

166   ENERGY

to the lever and passing through the fulcrum, yields

$$F_o L_o = F_i L_i.$$

The mechanical advantage is therefore given by

$$F_o/F_i = L_i/L_o.$$

By placing the fulcrum close to the output force so that $L_i/L_o = 10$, we can obtain an output force 10 times as large as the input. This law of the lever was discovered by Archimedes, one of the few great thinkers in the Grecian world who took delight in developing practical devices and was not averse to performing experiments.

Levers abound in living plants and animals and in man-made machines and gadgets. In walking or in moving our arms we use bone levers. When we operate a pump, pull a nail with a claw hammer, or cut with scissors, levers come into play.

It is worthwhile examining the operation of a lever from the standpoint of the work done when the end to which the force F_i is applied is depressed a distance s_i and the other end rises a distance s_o (Fig. 4-7). Recall that work is energy transferred by a force. Since there is no energy storage in the lever and very little mechanical energy is converted to thermal energy because of work done by frictional forces, energy conservation enables us to state:

$$\text{Work input} = \text{work output},$$

which, from the definition of work, Eq. (4.2), yields

$$F_i s_i = F_o s_o.$$

Therefore

$$F_o/F_i = s_i/s_o = L_i/L_o,$$

which agrees with our previous result. The equation, $s_i/s_o = L_i/L_o$, follows from the fact that the triangles on either side of the fulcrum in Fig. 4-7 are similar.

Notice that conservation of energy, as expressed by $F_o s_o = F_i s_i$, requires that if the output force is to be greater than the input force, then the output force must move through a smaller distance than the input force. Or if it is desired that the output force move through a greater distance than the input force, then the output force must be less than the input force, in which case the mechanical advantage is less than one.

The Efficiency of Machines

The same principles that apply for the lever can be used to discuss many other machines (which are used to transform force) such as the chain hoist, the block and tackle, the hydraulic press, the wheel and axle, and

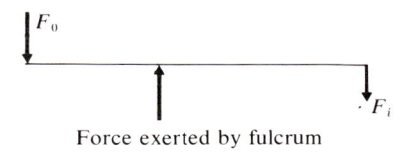

Fig. 4-6(b) Free body diagram for lever of negligible mass.

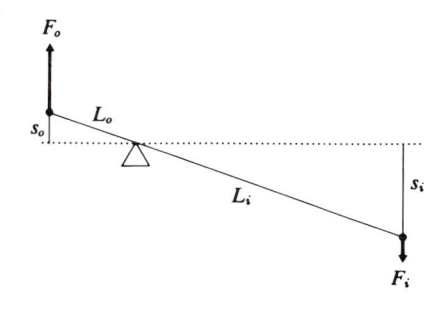

Fig. 4-7 The work input is $F_i s_i$. The lever exerts a force on the load of magnitude F_o directed upward. The work output is $F_o s_o$.

the wedge. In practical machines there is often appreciable energy dissipation because of work done by frictional forces. There may also be a significant change ΔU of energy stored in the machine.

In these circumstances, the law of conservation of energy gives

$$\text{work input} = \text{work output} + W_f + \Delta U,$$

where $-W_f$ is the work done by frictional forces. The positive quantity W_f is often called the *work done against friction*. The change in the stored energy, ΔU, is positive if the machine's stored energy increases. This occurs, for example, if massive pulleys within the machine are raised above their initial positions and thereby have their potential energy increased. If the machine's stored energy decreases, ΔU is negative. This happens when the lever in Fig. 4-6 is moved to the position shown in Fig. 4-7, since the mass center of the lever is lowered. If we exclude from consideration such cases involving the use of energy that had previously been stored in the machine, then ΔU cannot be negative, and the law of conservation of energy implies that the work input $F_i s_i$ is greater than the work output $F_o s_o$. The mechanical efficiency of the machine, defined by

$$\text{efficiency} = \frac{\text{work output}}{\text{work input}},$$

is then less than 100%.

It is convenient to distinguish between the mechanical advantage of the real machine in question, called the *actual mechanical advantage*, and the mechanical advantage of an idealized replica which would operate without friction or energy storage, called the *ideal mechanical advantage*. In such an ideal machine, the input force F'_i and the output force F'_o would be such that the work input equalled the work output; that is,

$$F'_i s_i = F'_o s_o.$$

The mechanical advantage of this ideal machine is the

$$\text{ideal mechanical advantage} = F'_o/F'_i = s_i/s_o.$$

This definition gives two methods for determining the ideal mechanical advantage of a machine. First, from the geometry of the machine or by experimental measurement, we can evaluate the ratio s_i/s_o of the distances moved at the input and at the output. This ratio s_i/s_o equals the ideal mechanical advantage. A second method, which is simpler when the geometry of the machine is complex, is to use the conditions of equilibrium to determine F'_o/F'_i for the idealized machine without friction and with moving parts of negligible mass; that is, for a machine which will neither store nor dissipate mechanical energy.

We will show that the efficiency can be calculated from the ratio of the two easily–measurable quantities, the actual mechanical advantage,

F_o/F_i, and the ideal mechanical advantage, s_i/s_o.

$$\text{Efficiency} = \frac{\text{work output}}{\text{work input}} = \frac{F_o s_o}{F_i s_i} = \frac{F_o/F_i}{s_i/s_o}$$

$$= \frac{\text{actual mechanical advantage}}{\text{ideal mechanical advantage}}$$

Example 5 A set of pulleys shown in Fig. 4-8 is a machine in which the input force pulls down the free end of the cord and the output force raises the suspended mass M of 5.00 kg.
(a) Find the ideal mechanical advantage of this machine.
(b) A force of 26.0 N applied at the free end of the cord raises the 5.00-kg mass, the lower pulley, and the cord with a constant velocity. What is the actual mechanical advantage of this machine?
(c) What is the efficiency of the machine?

Solution

(a) We can find the ideal mechanical advantage by using either of the following methods.

Method 1: We determine the ratio s_i/s_o, where s_i is the distance moved by the free end of the cord upon which the input force F_i is applied, and s_o is the distance moved by the mass M upon which the output force F_o is applied.

From the geometry of Fig. 4-8, we see that if the free end of the cord is pulled down a distance s_i, the portions of the cord on each side of the lower pulley must each shorten by an amount $s_i/2$. The mass M therefore rises a distance $s_i/2$, and we have

$$s_o = s_i/2.$$

Therefore,

$$\text{ideal mechancial advantage} = s_i/s_o = 2.$$

Method 2: We evaluate the ratio F'_o/F'_i of the output to the input force for an idealized replica in which the pulleys are frictionless, and the masses of the pulleys and of the cord are negligible. Then the tension has the same value T in each portion of the cord. In equilibrium, the input force F_i' is then T, and the output force F_o' has a magnitude equal to Mg. Consequently,

$$\text{ideal mechanical advantage} = \frac{F'_o}{F'_i} = \frac{Mg}{T}.$$

We can evaluate the ratio Mg/T by considering the conditions for static equilibrium of a system consisting of the mass M, the lower

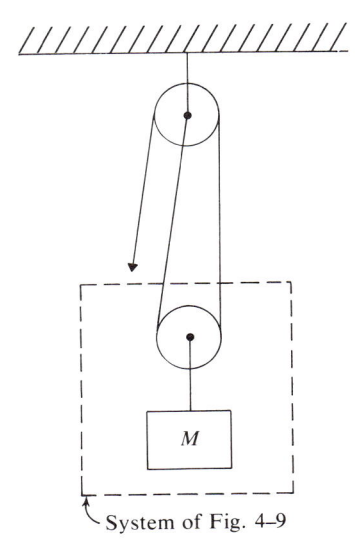

Fig. 4-8 A set of pulleys.

pulley, and the lower portion of the cord. The free body diagram for this system is shown in Fig. 4-9. Equating to zero the algebraic sum of the (vertical) y-components of the forces acting on this system, we obtain

$$T + T - Mg = 0.$$

Therefore, $Mg/T = 2$. Again we find that this machine has an ideal mechanical advantage of 2.

(b) The actual machine has an input force of 26.0 N. The output force remains at a magnitude $Mg = (5.00 \text{ kg})(9.80 \text{ m/sec}^2) = 49.0$ N.

$$\text{Actual mechanical advantage} = \frac{49.0 \text{ N}}{26.0 \text{ N}} = 1.88.$$

(c)

$$\text{Efficiency} = \frac{\text{actual mechanical advantage}}{\text{ideal mechanical advantage}} = \frac{1.88}{2.00} = 0.94.$$

This machine has an efficiency of 94%.

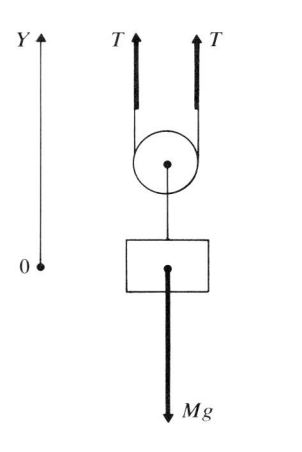

Fig. 4-9 Free body diagram for the system indicated in Fig. 4-8.

4.4 ENERGY AND MASS

Kinetic Energy and Rest Energy

The story of energy and its conservation takes a dramatic turn when we come to consider the contributions of Einstein. Instead of becoming more complicated as you might fear, the task of keeping track of energy becomes extremely simple. Einstein's central idea about energy was suggested to him by his labors in creating, in 1905, the special theory of relativity. However, it is by no means necessary to master this theory before one can appreciate and learn to use the results concerning energy. These results, very strange and speculative when first enunciated by Einstein, can now be regarded as experimental facts, checked and cross-checked by a host of experiments performed in many different physical situations.

The facts are these. For any isolated particle, atom, molecule, or larger structure, the total energy E can be written as a sum of two parts:

$$E = mc^2 + K, \tag{4.21}$$

where c is the speed of light in a vacuum (3.0×10^8 meters/sec) and m is the "everyday" mass of the object. This "everyday" mass m is often called the *rest mass* to distinguish it from the quantity $m/\sqrt{1 - v^2/c^2}$ which is named the *relativistic mass* of an object moving with speed v. Following the present practice of most physicists, we will use the word mass in only one sense, *the rest mass, m*.

The symbol K represents our old friend, the kinetic energy, which is the energy associated with the motion of the object as a whole with a speed v. There is a new feature associated with kinetic energy. An evaluation of this kinetic energy K, by experimental measurement or by Einstein's special theory of relativity, shows that, although $\frac{1}{2}mv^2$ is an excellent approximation for low speeds (say from 0 to $\frac{1}{10}c$), another more complicated formula [see Eq. (5.30)] gives the correct answer at all speeds and must be used for speeds near the speed of light.

The other part of the total energy, the term mc^2, is a completely new feature. If the object is at rest, its kinetic energy K is zero and then the total energy $E = mc^2$. For this reason, mc^2 is named the *rest energy* of the object. One thing to notice immediately is that the rest energy of familiar objects is tremendous. A 0.010-kg bullet has a rest energy of 0.010 kg $\times$ (3.0 $\times$ 10^8 m/sec)^2 = 9 $\times$ 10^{14} joules, which is more than a hundred billion times as much as its kinetic energy at a speed of 1000 m/sec. ($K = \frac{1}{2} \times$ (0.010 kg) $\times$ (1000 m/sec)^2 = 5.0 $\times$ 10^3 joules.) An electron has a rest energy of 512 $\times$ 10^3 eV, which is over ten times the kinetic energy it attains in a 50,000 volt x-ray tube.

What is the significance of the rest energy, mc^2? There are two related points.

1. The very fact of the existence of a mass m implies the presence of rest energy mc^2, so we can regard any particle with rest mass as a localized bundle of energy.
2. If our object of rest energy mc^2 is made up of complicated parts with each part possessing kinetic energy and rest energy and the entire assembly perhaps possessing a potential energy arising from the interaction of its parts, it would seem to be a formidable problem to sum all these contributions and arrive at a formula for the internal energy of the object. But the answer is marvelously simple. This internal or intrinsic energy adds up to just mc^2!

Rest Energy and Energy Conservation

We assume that the total energy of any isolated system is conserved. In the interaction between different objects, we know that the kinetic energy changes (think of a head-on collision of two cars), so there must be compensating changes in the rest mass of the objects involved. Rest energy mc^2 can be changed into kinetic energy and kinetic energy transformed into rest energy! The chemist's famous law of conservation of rest mass in chemical reactions is not true. Instead of having a law of conservation of rest mass and another law of conservation of energy we have only one law, the *law of conservation of energy, in which the energy includes the rest energy.*

To apply the law of conservation of energy to any process we need only to equate the sum of the total energies of all the objects before the process to the sum of the total energies of all the (possibly different) objects after the process. There are some particles, photons and neutrinos, that have no rest mass and therefore no rest energy; such particles always travel at the speed of light and they do possess energy we can regard as kinetic energy. Many of the following examples, besides illustrating the convertibility of rest mass and kinetic energy, are processes of considerable interest.

Example 6: Pair Annihilation (Fig. 4-10) The electron and the positron are a pair of what are called antiparticles. They have exactly equal masses m but opposite electric charges and can annihilate one another when they come close enough ($\sim 10^{-9}$ m) to interact significantly. After annihilation there is nothing left but photons, usually two. Assuming the electron and the positron are at rest before annihilation, the energy conservation equation reads:

$$E_{\text{electron}} + E_{\text{positron}} = E_{\text{photon}} + E_{\text{photon}},$$

which gives

$$(mc^2 + 0) + (mc^2 + 0) = E_{\text{photon}} + E_{\text{photon}}.$$

The electron and the positron each have a rest energy of 0.51 MeV. Inserting these values we get

$$1.02 \text{ MeV} = E_{\text{photon}} + E_{\text{photon}},$$

so there is 1.02 MeV to be shared by the photons. Momentum conservation requires that these photons have momentum vectors in opposite directions and of equal magnitude; this implies that the two photons must have the same energy. The result is then that each photon possesses an energy of 0.51 MeV. This is just what is observed experimentally; the electron and positron disappear and two 0.51 MeV photons are created which go off in opposite directions. The energy mc^2 associated with rest mass has been completely transformed into kinetic energy! The rest mass has disappeared!

Example 7: Pair Production (Fig. 4-11) If a photon has enough energy it can produce a pair of antiparticles, an electron and a positron. (The law of conservation of momentum requires the presence of a third particle, in practice the nucleus of a massive atom, to carry off some momentum.) The energy conservation equation for this pair production process is

$$E_{\text{photon}} = (mc^2 + K_{\text{electron}}) + (mc^2 + K_{\text{positron}})$$

which gives

$$E_{\text{photon}} = 1.02 \text{ MeV} + K_{\text{electron}} + K_{\text{positron}}. \qquad (4.22)$$

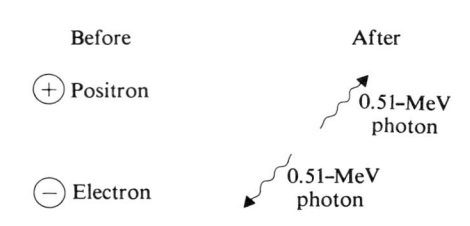

Fig. 4-10 The annihilation of an electron-positron pair.

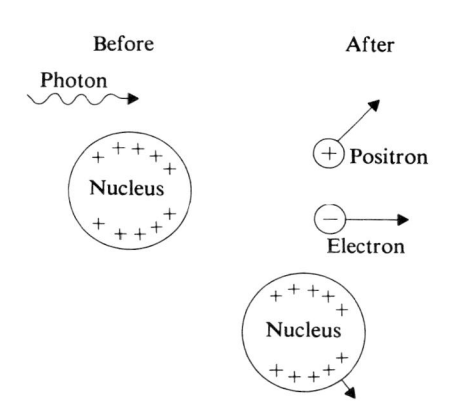

Fig. 4-11 Pair production: the creation of an electron-positron pair.

Evidently 1.02 MeV of the photon's energy goes into the creation of the rest masses of the electron and positron, and the remainder is kinetic energy which is shared by these two particles. Experimentally one does find a threshold photon energy of 1.02 MeV to make this process occur, and many experiments confirm the energy balance of Eq. (4.22). In this pair production process, kinetic energy is transformed into rest energy! Rest mass has been created where none was originally present.

Example 8: Neutron Capture by a Proton to Form a Deuteron A neutron (M_n) and a proton (M_p) which come close enough (within the range of nuclear forces, i.e., approximately 10^{-15} m) will interact strongly and often emit a photon as they become bound together to form a stable system called a deuteron (mass M_d). Assuming that the neutron and proton kinetic energies are negligible when the particles first start to interact, what is the energy of the emitted photon?

Solution Energy conservation applied to this nuclear reaction gives

$$M_n c^2 + M_p c^2 = M_d c^2 + E_{\text{photon}}.$$

The photon energy is therefore determined by

$$E_{\text{photon}} = c^2(M_n + M_p - M_d).$$

It is convenient to do the arithmetic using the *atomic mass unit* (*amu*). By definition one amu equals one-twelfth of the mass of the C-12 atom. The masses are known:

$$M_n = 1.00866 \text{ amu}$$
$$M_p = 1.00728 \text{ amu}$$
$$M_n + M_p = 2.01594 \text{ amu}$$
$$M_d = 2.01355 \text{ amu}$$
$$M_n + M_p - M_d = 0.00239 \text{ amu}$$

Since 1 amu = 1.66×10^{-27} kg, we find that

$$E_{\text{photon}} = \frac{(3.00 \times 10^8 \text{ m/sec})^2 \times (0.00239 \times 1.66 \times 10^{-27} \text{ kg})}{(1.60 \times 10^{-19} \text{ joule/eV})}$$

$$= 2.23 \times 10^6 \text{ eV}$$

$$= 2.23 \text{ MeV}.$$

A 2.23-MeV photon is indeed produced in this neutron-capture reaction. There are many of these "neutron-capture γ-rays" to be found emerging from a nuclear reactor (Section 4.5).

In pair production or annihilation the change in rest mass is 100%. The deuteron reaction of the present example involved a 0.1% change in rest mass. Nuclear fission involves a 0.1% decrease in rest mass and fusion reactions can achieve 0.7% decrease. In every case, the change in rest energy is accompanied by a balancing change in kinetic energy which is quite dramatic, to say the least! Some people have the idea that these changes of rest mass associated with the conversion between rest energy and kinetic energy are peculiar to nuclear physics but, of course, this phenomenon is of perfectly general occurrence. It happens in every chemical reaction, but the changes in rest mass are so small relative to the masses involved (about one part in a billion) that we cannot detect them by direct measurement of mass. We have every reason to believe that these minute rest-mass changes occur in the most mundane situations, but one has only to do the arithmetic to discover why we are not aware of these changes. We shall do this arithmetic next.

Example 9: Collision Between Two Lumps of Putty Assume each lump has a mass $m = 1.0$ kg. The lumps are moving in opposite directions, each with speeds of 3.0 m/sec. They collide and stick together, forming a stationary blob. Their kinetic energy has disappeared, so their rest mass must increase to some value M. Calculate the size of this mass increase.

Solution We equate the sum of total energies before collision to the total energy after

$$(mc^2 + \tfrac{1}{2}mv^2) + (mc^2 + \tfrac{1}{2}mv^2) = Mc^2 + 0.$$

This can be rearranged to give the change in rest mass

$$M - 2m = m\frac{v^2}{c^2}$$
$$= (1.0 \text{ kg}) \times \left(\frac{3.0 \text{ m/sec}}{3.0 \times 10^8 \text{ m/sec}}\right)^2$$
$$= 1.0 \times 10^{-16} \text{ kg}.$$

The change in rest mass is indeed small! No wonder we are not aware of it.

Spontaneous Disintegration

Sometimes a particle of mass M suddenly disintegrates into two (or more) particles of masses M_A and M_B. This happens with radioactive nuclei (Section 1.5) and many elementary particles. The transition of an atom, or a nucleus, from an excited state with rest energy Mc^2 to a lower state with rest energy $M_A c^2$ accompanied by the emission of a photon with zero rest energy can be regarded as just another instance of such a disintegration or decay. Applying the law of conservation of energy to these disintegrations will show us a simple relationship between the mass of the original particle and the sum of the masses of the products of the decay.

Consider the particle of mass M at rest. It decays into two particles which shoot off (in opposite directions because of momentum conservation) with kinetic energies K_A and K_B. The total energy before decay must equal the total energy after the decay:

$$Mc^2 = M_A c^2 + M_B c^2 + K_A + K_B.$$

Since the kinetic energies, K_A and K_B, cannot be negative, this equation tells us that the rest energy Mc^2 must be at least as great as the sum of the rest energies, $M_A c^2$ and $M_B c^2$. Consequently, for the disintegration to be energetically possible, the original mass M must be at least as great as the sum $M_A + M_B$.

But do not get the impression that, just because a process is energetically possible, it must and will occur. In fact, if this were the case, you would disintegrate into a burst of photons! Fortunately, there are physical laws operating, besides energy conservation, which determine the probabilities of energetically possible events. These other laws completely exclude certain processes. As was implied in Section 1.5, for each decay process that *can* occur, there is a definite probability per second (0.69/half-life) that the disintegration will suddenly take place.

We would stress another point. Experiments teach us that particles can be created and destroyed. We do not have to believe that the original particle is actually composed of the particles that appear when it breaks up. In fact, in photon emission or in β-decay, as well as in elementary particle disintegrations, the decay products such as photons, electrons, positrons, and neutrinos are not regarded as previously existing in the original object but instead are considered to be *created* in the decay process.

Example 10: Radioactivity of Radium We consider the possibility of a spontaneous disintegration of the radium nucleus, $_{88}Ra^{226}$, into an α-particle ($_2He^4$) and a daughter radon nucleus, $_{86}Rn^{222}$, according to the equation

$$_{88}Ra^{226} \longrightarrow {_{86}Rn^{222}} + {_2He^4},$$

and ask if this process is possible.

Solution The atomic masses in atomic mass units are

mass of Ra	=	226.02536
mass of Rn	= 222.01753	
mass of He	= 4.00260	
mass of final products	= 226.02013	226.02013
mass difference	=	0.00523

Disintegration is possible since the radium isotope has a greater mass than that of the proposed disintegration products. *We say this decay process is energetically possible.*

The decay process of Example 10 does indeed take place; the half-life of radium is 1620 years. Let us investigate this radium decay a little further. Equating energies before and after decay, we get

$$M_{Ra}c^2 = M_{Rn}c^2 + M_\alpha c^2 + K_{Rn} + K_\alpha.$$

Hence,

$$K_{Rn} + K_\alpha = [M_{Ra} - (M_{Rn} + M_\alpha)] c^2.$$

Now

$$M_{Ra} - (M_{Rn} + M_\alpha) = 0.00523 \text{ amu}.$$

To convert units, we use the fact that 1 amu = 931.5 MeV/c^2 and find that the sum of the kinetic energies is

$$K_{Rn} + K_\alpha = 4.87 \text{ MeV}.$$

This kinetic energy is shared between the α-particle and the radon nucleus which go off in opposite directions, thus conserving momentum. This implies (see Question 63) that the recoil kinetic energy of the massive radon nucleus is small: only about 0.09 MeV. The lighter α-particle, on the other hand, gets almost all the kinetic energy, 4.78 MeV.

Experimentally, determination of α-particle kinetic energy by measurement of the α-particle's momentum (Section 12.3), agrees with this prediction.

We now consider another decay process to indicate that what has been said about energies in decay processes is quite generally applicable. The particle in question this time is the electrically charged π-meson. Such a meson has a mass which is 273.3 times the mass of an electron. The π-mesons are unstable and decay into lighter particles, the μ-meson and the neutrino (Fig. 4-12). The half-life of a π-meson is 1.8×10^{-8} sec. As there seems no good reason for picturing a π-meson to be composed of a μ-meson and a neutrino, we regard these decay products as actually created in the decay process.

Example 11: π-meson Decay Examine the energies involved in this decay process.

Solution Energy conservation requires, for the π-meson at rest decaying into a neutrino with kinetic energy K_ν and a μ-meson with kinetic energy K_μ:

$$M_\pi c^2 = M_\mu c^2 + K_\mu + K_\nu.$$

The rest energy of the π-meson is 139.6 MeV, and that of the μ-meson is 105.7 MeV, while the neutrino having zero rest mass has no rest energy. Energy conservation, therefore, requires

$$139.6 \text{ MeV} = 105.7 \text{ MeV} + K_\mu + K_\nu.$$

The sum of the kinetic energies of the decay products is therefore

$$139.6 \text{ MeV} - 105.7 \text{ MeV} = 33.9 \text{ MeV}.$$

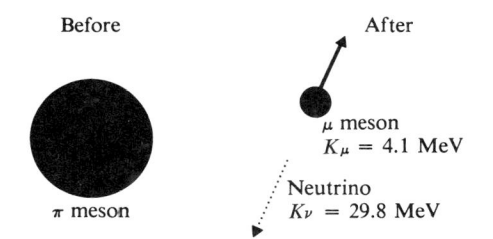

Fig. 4-12 π-meson decay.

The discussion in the solution of Example 11 can be profitably pursued further: the kinetic energy must be shared between the neutrino and the μ-meson in such a way as to give them momentum vectors of equal magnitude pointing in opposite directions. It turns out that this requires $K_\mu = 4.1$ MeV and $K_\nu = 29.8$ MeV. Detection of tracks of 4.1-MeV μ-mesons in photograph emulsions led to the Nobel prize winning discovery of the π-meson in 1947 by C. F. Powell (1903–1969).

The Neutrino and Conservation Laws

In the preceding discussion of the π-meson, it was stated that one of the decay products was a neutrino. In fact, in Powell's experiment, and in similar experiments, the neutrino is *not detected*. Being electrically neutral, it makes no track in a photographic emulsion, bubble, spark, or cloud chamber, or in any of the many other ingenious devices which reveal the trail of charged atoms left in the wake of an energetic particle which carries an electric charge (see Section 1.3). To compound the detection difficulty, the neutrino, unlike the neutron, interacts only very weakly with nuclei.

Although physicists, for very good reasons, were convinced for some 20 years that there must be such things as neutrinos, it was not until during the 1950s that, by Herculean efforts, F. Reines (1918–   ) and C. L. Cowan (1919–   ) actually detected neutrinos initiating a reaction. In fact, the particle detected is named the *antineutrino* (symbol $\bar{\nu}$), and the reaction is

$$\bar{\nu} + p \longrightarrow n + e^+,$$

which symbolizes an interaction between an antineutrino and a proton to produce a neutron and a positron. Using a portion of the approximately 10^{18} neutrinos per second emerging from the Savannah River reactor, they were able to detect about three of the above reactions per hour—the probability of the neutrino interacting is indeed very small! In 1962, using the 33-BeV synchrotron at Brookhaven (Section 4.6), energetic neutrinos were detected producing μ-mesons, and it was shown that these "meson neutrinos" are different from the "electron neutrinos." Now, counting particles and antiparticles, we have four different particles in the neutrino family.

Rather than dwell on these current events, it is interesting to think about the "very good reasons" that sustained physicists' confidence in the existence of neutrinos for two decades. Let us look back at Powell's observations: a π-meson at rest decays as a μ-meson goes tearing off with a kinetic energy of 4.1 MeV. If there is no other particle emitted, then

1. The system has suddenly acquired momentum!
2. 29.8 MeV of energy has been destroyed!

Unless some unseen particle is also emitted, we are faced with a gross violation of the conservation laws of energy and momentum (and angular momentum). The relationship between the missing energy E and the magnitude p of the momentum vector required to cancel out the μ-meson's momentum is found to be

$$E = cp, \qquad (4.23)$$

which is just the relationship for particles with zero rest mass that had been predicted by special relativity and verified for photons. The conservation laws are saved and the energy distribution among the decay particles is understood by postulating that a massless neutral particle is emitted. This ghostly particle, conjured up by Pauli in 1930 and christened the neutrino by Fermi, was first introduced into physics for these reasons, although these physicists were then fretting about the more complicated three-particle β-decay of nuclei rather than the simple two-particle decay of the π-meson that Powell discovered.

Before leaving this matter, we should point out that we are now in a position to draw the reader's attention to a simplification we introduced in Section 1.5 where we first discussed β-decay. At that time we did not wish to confound the reader with additional particles. Suffice it, therefore, now to correct our simplification by stating that β^--decay involves the creation of an electron *and an antineutrino*, thereby not violating any conservation laws.

Binding Energy

Consider a composite particle of mass M which is made up of particles of mass M_A and M_B. In the light of our previous discussion you can see that the particle M cannot spontaneously disintegrate into these constituent particles if M is less than $M_A + M_B$. In this circumstance M is stable, at least as far as this particular way of breaking up (that is, into parts M_A and M_B) is concerned.

A handy measure of the stability of a composite object against separation into specified parts is provided by what is called its *binding energy. The binding energy of an object is the difference between the sum of the rest energies of its parts and the rest energy of the original object.* The binding energy of M relevant to its separation into just two parts, M_A and M_B, is therefore given by

$$\text{Binding energy} = (M_A c^2 + M_B c^2) - Mc^2. \qquad (4.24)$$

Previous conditions for stability or decay can now be rephrased: If the binding energy is positive, the system is stable against a specified disintegration; if the binding energy is negative, decay is energetically possible.

Another interpretation of the binding energy can be given. The defining equation, Eq. (4.24), can be rewritten as

$$\text{Binding energy} + Mc^2 = M_A c^2 + M_B c^2.$$

Now consider supplying from some outside source an energy E to the composite particle sufficient to break it up into parts M_A and M_B which move off with kinetic energies K_A and K_B. Energy conservation requires

$$E + Mc^2 = M_A c^2 + M_B c^2 + K_A + K_B.$$

The minimum value of E corresponds to zero values of the kinetic energies K_A and K_B. Now comparison of this equation with the preceding one shows that the *binding energy is the minimum energy which must be furnished to a composite object in order to separate it into specified parts*. The greater the binding energy, the greater the stability. A deuteron has a binding energy of 2.23 MeV. It is stable; 2.23 MeV of energy must be somehow provided to the deuteron, perhaps by absorption of an energetic photon or by collision with a proton emerging from a cyclotron, in order to separate the deuteron into a proton and a neutron. *Nuclear binding energies are roughly a million times greater than atomic binding energies.* For example, to remove the electron from a hydrogen atom (to ionize the atom), that is, to separate the electron and the proton, an energy of only 13.6 eV is necessary.

From the definition of binding energy, it is clear that this energy can be determined from mass measurements, provided the masses can be measured with sufficient accuracy. For nuclei, this is done rather easily because the binding energies involved are roughly 1% of the rest energies. However, the binding energies of atoms and molecules are roughly one-billionth of the rest energies of these loose structures, and these masses cannot be measured with enough precision to detect the very small difference between the masses of the separated parts and the mass of the composite structure. This is why deviations from the now defunct law of conservation of rest mass were not detected in chemical reactions. (Direct measurement of the minimum energy required to effect separation of an object into parts gives accurate values of binding energies of atoms and molecules as well as nuclei.)

Interaction Energy

Why is it that the mass of a composite structure is different from the sum of the masses of its constituents? To answer this question, we must talk about energies of the constituents inside the structure.

For a start, let us look inside a hydrogen atom at rest in its ground state (that is, its most stable configuration, which corresponds to the lowest possible value of its total energy). Here we find a proton with rest energy $M_p c^2$ and a small kinetic energy that we will neglect. The whirling orbital electron has a rest energy $M_e c^2$ and a kinetic energy

K_e. In accord with our discussions of Section 2.8, the electron and the proton interact by exerting attractive Coulomb forces on each other, thereby giving rise to an *interaction energy*, which in this case is the Coulomb potential energy ($-k\,e^2/r$). These various energies are the ingredients which make up the rest energy Mc^2 of the hydrogen atom:

$$Mc^2 = M_p c^2 + M_e c^2 + K_e + \text{(interaction energy)}. \quad (4.25)$$

If we adopt the Bohr model of the atom (Section 2.8), we can evaluate the last two terms, and find for the ground state,

$$K_e = 13.6 \text{ eV},$$

and the interaction energy (Coulomb potential energy, see Question 18)

$$= -27.2 \text{ eV}.$$

Summing, we obtain

$$K_e + \text{(interaction energy)} = (13.6 - 27.2) \text{ eV}$$
$$= -13.6 \text{ eV}.$$

Thus Eq. (4.25) states that

Rest energy of H atom =
(sum of rest energies of constituents) $-$ 13.6 eV.

The mass of the H atom is therefore less than the sum of the masses of an electron and a proton. A glance at the defining equation for the binding energy, Eq. (4.24), shows that the binding energy of the H atom has the positive value 13.6 eV.

With the above example in mind, we now list some points that happen to be true for all systems:

1. A negative interaction energy arises from an attractive force between the interacting particles. Conversely, a positive interaction energy arises from a repulsive force.
2. The interaction energy becomes negligible when the separation of the interacting particles becomes very large. Moreover, the zero potential energy configuration can no longer be selected at will: infinite separation must correspond to zero interaction energy, that is, zero potential energy due to interaction.
3. Both the interaction energy and the kinetic energies must be considered to determine the binding energy. It is the binding energy itself, rather than any precise division of this energy into kinetic and interaction energy, that is measured experimentally and computed theoretically in quantum mechanics. (In Chapter 16 the negative of this binding energy will be adopted as the most convenient specification of the energy of the composite system.)
4. To achieve a stable or bound system there must be a *negative interaction energy* (attractive forces), and the sum of the kinetic energies of the constituent particles must not be as large as the magnitude of

this interaction energy. In fact, for attractive forces: Binding energy = (magnitude of interaction energy) − (sum of kinetic energies of constituents).

The original question (Why is the mass of a composite structure different from the sum of the masses of its constituent parts?) is now answered. The mass difference in question is just the binding energy, and we see that this binding energy is determined by the interaction energy and internal kinetic energies.

Turning from questions of atomic to nuclear structure, we note that the deuteron is of great interest because it is the simplest nucleus in which we can investigate nuclear forces. In the ground state the deuteron is stable with a binding energy of 2.23 MeV, roughly a million times the binding energy of atoms. Also it is found that the range of these nuclear forces is very short, approximately 10^{-15} m. Outside this range there is practically no interaction at all. These facts tell us that the forces exerted by the proton and neutron on each other are attractive and much stronger than Coulomb forces. (In this connection, we now refer the reader back to Table 2-2.)

4.5 NUCLEAR FISSION AND FUSION

Fission

Fission is a decay process in which a nucleus splits into two fragments of comparable masses (Fig. 4-13). Such occurrences are a mere detail in the panorama of submicroscopic events, but a detail with very valuable as well as devastating consequences. In atomic and hydrogen bombs, as well as in nuclear power reactors which achieve a controlled release of energy, nuclear fission reactions are the key.

The rest energy of a large nucleus like uranium is about 200 MeV greater than the sum of the rest energies of fragments into which uranium might split, say $_{38}Sr^{94}$ and $_{54}Xe^{140}$. According to the discussion of Section 4.4, such a fission reaction is therefore energetically possible, and, if it does occur, roughly 200 MeV of rest energy will be converted into kinetic energy. This energy is about ten million times the rest energy that is converted into kinetic energy when two molecules combine in the course of a chemical explosion of something like TNT. Fission of large nuclei is consequently a formidable source of kinetic energy. In fact, the fissioning of one pound (0.45 kg) of uranium yields about 11×10^6 kilowatt-hours, that is as much energy as is yielded by the burning of three million pounds of coal, or 200,000 gallons of gasoline (Fig. 4-14).

The fact that fission of a certain nucleus is energetically possible does not necessarily mean that this process will occur. A deformation

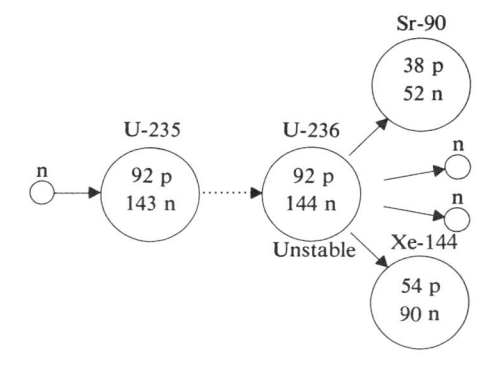

Fig. 4-13 Typical fission of uranium-235. The fission fragments in this case are strontium-90 and xenon-144. Some 90 different fission fragments are possible in the fissioning of U-235. Note that neutrons are released in this process.

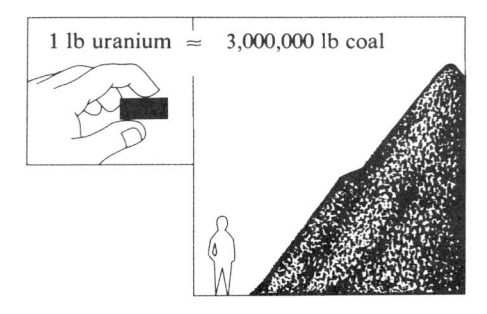

Fig. 4-14 Heat from fission of 1 pound of uranium-235 equals approximately heat from burning 3,000,000 pounds of coal.

of a large nucleus like uranium into two segments which are almost separated involves an energy increase of about 5 MeV above the energy of the nucleus in its ground state configuration. Since such deformation must occur before the nucleus splits, the probability of fission is appreciable only if the nucleus is in an excited state about 5 MeV or more above its ground state. However, the capture of a neutron by a large nucleus often leads to fission, because the new nucleus formed by neutron capture is created in an excited state. The excitation energy is the binding energy of the captured neutron to the new nucleus, augmented by whatever kinetic energy the captured neutron originally possessed.

For example, if the nucleus of the most abundant isotope of uranium, $_{92}U^{238}$, captures a *fast* neutron, say a neutron with a kinetic energy of 1.0 MeV, then the nucleus $_{92}U^{239}$ is formed with an excitation energy of 1.0 MeV plus a 4.9 MeV contribution which is the binding energy of a neutron to $_{92}U^{239}$. This excitation energy is sufficient to permit the nucleus to pass through a deformed configuration 5.5 MeV above the energy of the ground state and proceed to split in two. Fission can, and often does, occur. Fast neutrons are thus efficient agents for causing fission of $_{92}U^{238}$ nuclei. However, if the neutron kinetic energy is less than 0.6 MeV, the excitation following neutron capture by $_{92}U^{238}$ is not enough to allow large nuclear deformations and fission is most improbable. In general, capture of *slow* neutrons by $_{92}U^{238}$ does not lead to fission.

The story for neutron capture by some other nuclei, such as the rare uranium isotope $_{92}U^{235}$ (Fig. 4-13), or the man-made "transuranic" plutonium isotope $_{94}Pu^{239}$, differs in one very significant detail. The binding energy of a captured neutron furnishes, by itself, enough excitation energy to permit fission. It is not necessary that the incoming neutron also possess kinetic energy. Therefore *slow* neutron capture by $_{92}U^{235}$ or $_{94}Pu^{239}$ can produce fission.

Many different fission fragments can result from the fission of a given nucleus. In the case of the fission of $_{92}U^{235}$ some 90 different fission fragments, leading ultimately to about 250 radioactive species, have been found. Division into fragments with mass numbers around 95 and 135 occurs most frequently. As implied above, a fission fragment is generally radioactive and undergoes β- and γ-decay before becoming a stable isotope. It is this radioactivity which makes hazardous the fallout after the explosion of an A-bomb.

Besides the large fragments, there are usually a few neutrons released in fission (the average number per fission is about 2.5 for the fission processes so far mentioned). This is the profound detail that makes fission processes of such practical importance. From one neutron producing one fission, there results, on the average, more than one neutron to go on and produce other fissions. A self-sustaining *chain reaction* is possible (Fig. 4-15).

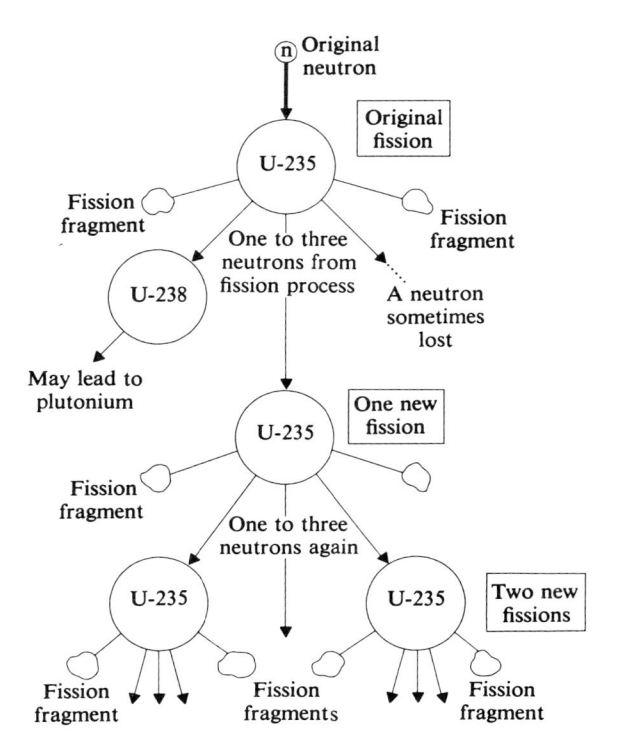

Fig. 4-15 Chain reaction resulting from the fission of a uranium-235 nucleus.

In a bomb, material is chosen and arranged so that each neutron released in a fission has the greatest possible probability of producing another fission. So that neutrons will not be lost by capture in some nonfissionable nucleus, fission bombs have been made of pure $_{92}U^{235}$ or pure $_{94}Pu^{239}$. When, on the average, one fission leads to *more* than one subsequent fission, a rapidly increasing number of fissions occur, each accompanied by the appearance of about 200 MeV kinetic energy which heats the bomb and causes an explosion of infamous violence.

For any particular composition, a bomb must exceed a certain "critical size" and a "critical mass" in order to explode. A chunk which is less than the critical size will have so many neutrons escaping through its surface, compared to those causing fissions within the volume of the material, that each fission gives rise to an average of less than one subsequent fission. A sphere, a few inches in diameter, is the critical size for a bomb made of pure $_{94}Pu^{239}$. A fission bomb is exploded by first driving subcritical parts together to achieve a supercritical mass and size. This supercritical situation is maintained for as long as possible, roughly 10^{-6} sec, so that during the explosion a large fraction of the nuclei will undergo fission and make their contribution to the energy release.

4.5 NUCLEAR FISSION AND FUSION

A nuclear reactor (Fig. 4-16) is designed to achieve a *controlled* release of energy. By the arrangement of the reactor material, a steady chain reaction is maintained, in which each fission produces an average of one new fission.

In a reactor which utilizes natural uranium there are two different uranium isotopes to consider: $_{92}U^{238}$ (99.27%) and $_{92}U^{235}$ (0.72%). A neutron with a kinetic energy less than about 1 MeV, when captured by the abundant isotope $_{92}U^{238}$ usually does *not* cause fission. On the other hand, *slow* neutron capture by $_{92}U^{235}$ does lead to fission. Moreover, the probability of slow neutron capture is much greater for $_{92}U^{235}$ than it is for $_{92}U^{238}$. A design problem is therefore to find a way to rapidly slow down a neutron released in fission and thereby increase the chance it will be captured by $_{92}U^{235}$ and so cause another fission. A solution is to intersperse throughout the reactor a material, such as heavy water (D$_2$O) or graphite, which is called a *moderator*. Within the moderator material, neutrons collide with light atoms and impart a fraction of their kinetic energy to the recoiling atoms. The moderator thus serves to slow down the neutrons without capturing them. (The moderator in the reactor of Fig. 4-16 was graphite.)

Fig. 4-16 Cutaway model of the West Stands of Stagg Field, University of Chicago, showing the world's first nuclear reactor. The reactor consisted of graphite layers in which were embedded small chunks of uranium oxide and uranium metal. The apparatus to the right of the graphite pile was for withdrawing the cadmium control rods. (Courtesy U.S. Energy Research and Development Administration.)

Fine control of the rate at which fissions occur within a reactor, and the consequent power level at which the reactor operates, is achieved by the adjustment of control rods of a material like cadmium whose nuclei readily capture slow neutrons, thereby preventing them from engendering fission of uranium.

Nuclear reactors find extensive use in the world of today. Research reactors are a copious source of neutrons (some 10^{19} neutrons/m² sec) which are useful for production of radioactive isotopes, such as the cobalt-60 referred to in Chapter 1, and for a variety of experiments. Power reactors supply heat to a circulating fluid and this energy is then used to generate electric power. *Breeder* reactors are used to convert material, like $_{92}U^{238}$, which is nonfissionable by *slow* neutrons, into fissionable material like $_{94}Pu^{239}$. One of the possibilities, after neutron capture by $_{92}U^{238}$, is a sequence of γ- and β-decays which lead, after a few days, to the creation of $_{94}Pu^{239}$. This plutonium, which is not found in natural ores since it is radioactive with a half-life of only 24×10^3 years, can thus be "bred" from the abundant uranium isotope.

The foregoing facts of fission have had immediate impact on the sombre history of our times. War was imminent in January 1939 when Otto Hahn (1879–1968) and Fritz Strassman (1902–) in Berlin found chemical evidence for the production of barium by neutron bombardment of uranium. Weeks later Lise Meitner (1878–1968) and Otto Frisch (1904–) in Stockholm, interpreted these experimental results as evidence of fission.* Experiments performed during the next few months in many parts of the world suggested the possibility of the nuclear technology that we have outlined. In the United States, a galaxy of leading physicists, many of them recent refugees from Europe, put aside work directed toward their usual primary objective, an understanding of nature, with the resolve that this new technology not be mastered first in hostile lands.

On December 2, 1942, Enrico Fermi and his colleagues achieved the first self-sustaining chain reaction in a nuclear reactor built in squash courts of the University of Chicago (Fig. 4-16). At Alamogordo, New Mexico, on July 16, 1945, the first A-bomb was exploded. This device was manufactured in the Los Alamos Laboratories under the technical direction of J. Robert Oppenheimer (1904–1967).

The peaceful use of nuclear energy followed shortly. In fact, the first nuclear reactor capable of continual power production (1000 kilowatts) became operational at Oak Ridge, Tennessee, in 1943. This reactor or "pile" known as the "Clinton pile" was the precursor to the research and power reactors which sprang up after the end of the Second World War. Today, three decades later, millions of kilowatts of power are being provided in the United States by several dozen nuclear power stations, and some authorities have estimated that nuclear plants may provide almost 50% of the total U.S. power requirement by the end of this century.

*The word fission was first used for this process by Otto Frisch upon a suggestion by the American biologist, William A. Arnold. In biology the term refers to cell division.

Courtesy of Ulli Steltzer Studio

J. Robert Oppenheimer
(1904–1967)

Nuclear Fusion

In a nuclear fusion reaction, two light nuclei combine to form a single heavier nucleus (Fig. 4-17). Among the many such fusions of light nuclei that have been investigated, the following reactions are of particular interest (the rest energy which is converted into kinetic energy is given in parentheses for each reaction):

$$_1H^2 + {}_1H^2 \longrightarrow {}_2He^3 + {}_0n^1 \quad (3.3 \text{ MeV}) \tag{4.26}$$

$$_1H^2 + {}_1H^2 \longrightarrow {}_1H^3 + {}_1H^1 \quad (4.0 \text{ MeV}) \tag{4.27}$$

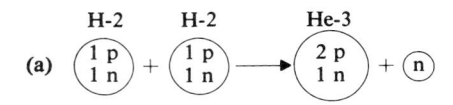

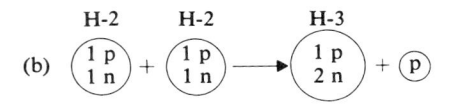

Fig. 4-17 Schematic representation of the fusion reaction. (a) Refers to the reaction of Eq. (4.26). (b) Refers to the reaction of Eq. (4.27).

The nucleus involved in these reactions, the deuteron ($_1H^2$), is very abundant on the surface of our planet. About one in 7000 hydrogen atoms in water is a hydrogen atom with a deuteron as a nucleus (this atom is called *deuterium*) instead of a common hydrogen atom with a proton as nucleus. And deuterium can be separated from water at a low cost (roughly 30¢ per gram). There is therefore the very real hope that, if we can learn to produce controlled fusion reactions, deuterium will serve as the ultimate fuel and fulfill mankind's energy requirements almost indefinitely.

In order for deuterons to react in the manner indicated by either of these two fusion reactions, Eq. (4.26), Eq. (4.27), the two deuterons must come so close together that the short range attractive nuclear forces come into play. This implies the separation of the deuterons be roughly 10^{-14} m. Since each deuteron has a positive electric charge e ($= +1.6 \times 10^{-19}$ coulomb), the deuterons repel each other with a Coulomb force. As two deuterons approach one another their kinetic energy decreases as their Coulomb potential energy increases from the value zero until, at a separation of $R = 10^{-14}$ m, this potential energy is about 0.14 MeV. Evidently the approaching deuterons must start with considerable kinetic energy in order to get close enough for fusion to occur.

At present, the only method envisaged for obtaining self-sustaining fusion reactions is to raise the initial material to a temperature of about 100 million degrees Fahrenheit. At this temperature light elements are stripped down to bare nuclei. The resulting collection of charged particles is called a *plasma*. The particles move in random directions at high speeds with an average kinetic energy of 5×10^3 eV. Then the most energetic nuclei can approach close enough for fusion to occur and the consequent reactions are termed *thermonuclear reactions*.

Such thermonuclear reactions are achieved in a "hydrogen" bomb. Solid lithium deuteride is heated to over 100 million degrees Fahrenheit by the explosion of a fission bomb. At this high temperature, many deuterons furnished by the lithium deuteride undergo the deuteron fusion reactions given by Eq. (4.26) and Eq. (4.27). Some lithium, bombarded by neutrons, splits according to

$$_3Li^6 + {}_0n^1 \longrightarrow {}_1H^3 + {}_2He^4 \quad (4.9 \text{ MeV}),$$

producing a triton, $_1H^3$. These tritons, which are also produced by deuteron-deuteron fusion [Eq. (4.27)], participate in thermonuclear reactions with deuterons and make a large contribution to the energy released:

$$_1H^2 + {}_1H^3 \longrightarrow {}_2He^4 + {}_0n^1 \quad (17.6 \text{ MeV}).$$

The reaction rate and the energy release can be increased by encasing the weapon in the abundant isotope of uranium, $_{92}U^{238}$. Some energetic neutrons, produced in the deuteron-triton fusion, produce fissions of $_{92}U^{238}$ nuclei which release further neutrons. Some of these neutrons split more lithium nuclei and thus produce more tritons. The sequence repeats: fusion, fission, fusion, fission, ... These weapons, apart from the fission bomb required to "ignite" them, use only cheap and readily available deuterium, as well as $_3Li^6$ and $_{92}U^{238}$. There is no theoretical limit to the quantities that can be employed and the consequent energy release. Fission reactions in the uranium casing supply about half the energy released. Uranium fission fragments provide almost all of the radioactive materials that have a half-life long enough to constitute a hazardous fallout. So-called "clean" bombs rely, as much as possible, on fusion rather than fission.

In stars like the sun, thermonuclear reactions within their hot interiors supply most of the energy emitted. The most important reactions involve two distinct chains of events; the end result in each case is that four protons have been fused together to produce a helium nucleus (and one or two positrons and neutrinos), thereby converting 27 MeV of rest energy into kinetic energy. By this thermonuclear burning of protons to form helium, the sun's rest mass decreases by over 4 million tons each second, and the energy output is equal to that of billions of H-bombs exploding each second. Although our planet intercepts less than two billionths of the power radiated by the sun, this fraction is a hundred thousand times greater than the power consumed in all the industries on earth.

4.6 REMARKS ON HIGH-ENERGY PHYSICS

Particle Accelerators

If an individual proton is accelerated to a kinetic energy of several MeV, and then, while traveling through some target material, happens to strike a nucleus, a nuclear reaction will occur. The greater the kinetic energy possessed by the incident proton, the greater the variety of possible outcomes of such a nuclear reaction. Indeed many interesting processes are energetically possible only if the incident projectile furnishes considerable kinetic energy.

In any single interaction between an incident projectile and some target particle, energy and momentum are conserved. Therefore new particles can be created in the interaction only if there is available sufficient energy to furnish the *rest energies* associated with the existence of the new particles. (Not all of the projectile's kinetic energy can be converted into rest energy of new particles because momentum conservation requires that the end products of the interaction move off with appreciable kinetic energy.) The laboratory exploration of this microscopic realm of our universe, the discovery of just which particles can exist and how they interact, is thus seen to require as incident projectiles, a beam of particles which have been accelerated to a high kinetic energy. Consequently, enormous efforts have been directed toward the construction of particle accelerators.

All these accelerators have some common features. The type of particle to be accelerated is chosen to be an electrically *charged* particle such as an electron, a proton, or some heavier nucleus. An *electric field* is always employed to push the charged particle in order to increase its kinetic energy. And the acceleration takes place in a vacuum so that the particle will not fritter away its kinetic energy in collisions with gas molecules.

Acceleration to energies much higher than a few MeV, without having to resort to extremely high voltages, is achieved by the ingenious use of alternating electric fields. In the linear (straight line) accelerator at Stanford University, bunches of electrons are pushed along by the electric field of a confined traveling electromagnetic wave (Chapter 14). After "riding the crest of an electromagnetic wave" for two miles, the electrons acquire a kinetic energy of 20 BeV.

A different family of accelerators, initiated when Lawrence and Livingston devised the cyclotron, employs a magnetic field to exert a deflecting force on a moving charged particle (Section 12.2). This magnetic force curves the particle's path into a circular or spiral trajectory. In this way a charged particle is made to pass many times between the same electrodes, and the (alternating) electric field in the same region of space can be used over and over again to accelerate the particle. Details of the functioning of cyclotrons will be discussed in Section 12.3.

A modern machine of this type is the Fermi National Accelerator Laboratory synchrotron near Batavia, Illinois. After acceleration to an energy of 8 GeV in smaller machines, protons are accelerated to their final energy in the main accelerator (Fig. 4-18). Here the protons travel through an evacuated tube in an orbit four miles in circumference. An accelerating electric field increases a proton's kinetic energy by several MeV in one circuit. After about 10^5 circuits during a 1.6-sec acceleration interval, a proton reaches a final kinetic energy of 200 GeV. The acceleration time is approximately doubled for operation at 500 GeV.

Other major proton accelerators are the 76-GeV synchrotron at Serpukhov in the Soviet Union, the 33-GeV machine at Brookhaven National Laboratory on Long Island, and the 30-GeV machine operated by the European Organization for Nuclear Research (CERN) in Geneva.

Since the cost of accelerators has now reached hundreds of millions of dollars, it is evident that this branch of physics can flourish only in a society with an appreciation of the importance of fundamental research. These costly machines have already partially revealed the existence of a new world of hitherto unknown particles. Experimental study is the essential step in acquiring an understanding of this realm. The experience of the past three hundred years suggests that an understanding of such a fundamental field of physics will prove to be an almost priceless asset.

Fig. 4-18 Aerial view of main accelerator at Fermi National Accelerator Laboratory (Fermilab). The main accelerator is four miles in circumference, 1.25 miles in diameter (FNAL Photo; Tony Frelo, 4/71).

Survey of Elementary Particles

The quest for the ultimate building blocks of nature has led to the discovery of the bewildering profusion of "elementary" particles listed in Table 4-4. Most of these particles are unstable and undergo sudden disintegration into lighter particles.

This domain is a frontier of physics. No comprehensive understanding has yet been achieved. In fact, we call these particles "elementary" only because they are not obviously structures composed of other particles, as is the hydrogen atom, for example.

In spite of our massive ignorance, certain regularities have emerged. For example, we find that for every type of *particle* there exists an *antiparticle* with identical mass and spin but opposite charge. Interaction between a massive particle and its antiparticle leads to annihilation of each and the production of lighter particles, as illustrated by Example 6.

Table 4-4. ELEMENTARY PARTICLES

Name	Symbols Particle	Antiparticle	Mass ($\times M_e$)	Rest energy (MeV)	Half-life (sec)	Spin ($\times \hbar$)	Electric charge ($\times e$)
BARYON FAMILY							
Hyperons:							
omega	Ω^-	Ω^+	3,272	1,672	0.90×10^{-10}	$\frac{3}{2}$	-1
xi	Ξ^-	Ξ^+	2,586	1,321	1.2×10^{-10}	$\frac{1}{2}$	-1
	Ξ^0	$\overline{\Xi}^0$	2,573	1,315	2.0×10^{-10}	$\frac{1}{2}$	0
sigma	Σ^-	$\overline{\Sigma}^+$	2,343	1,197	1.0×10^{-10}	$\frac{1}{2}$	-1
	Σ^0	$\overline{\Sigma}^0$	2,334	1,192	$<10^{-14}$	$\frac{1}{2}$	0
	Σ^+	$\overline{\Sigma}^-$	2,327	1,190	0.56×10^{-10}	$\frac{1}{2}$	$+1$
lambda	Λ^0	$\overline{\Lambda}^0$	2,183	1,115	1.7×10^{-10}	$\frac{1}{2}$	0
Nucleons:							
neutron	n	$\bar{n}$	1,839	939.6	0.65×10^3	$\frac{1}{2}$	0
proton	p^+	p^-	1,836	938.3	stable	$\frac{1}{2}$	$+1$
MESON FAMILY							
kaon	K^0	$\overline{K}^0$	974	498	0.60×10^{-10}	0	0
					3.6×10^{-8}		
	K^+	K^-	966	494	0.86×10^{-8}	0	$+1$
pion	π^+	π^-	273	140	1.8×10^{-8}	0	$+1$
	π^-	π^+	273	140	1.8×10^{-8}	0	-1
	π^0	same	264	135	0.6×10^{-16}	0	0
LEPTON FAMILY							
muon	μ^-	μ^+	207	106	1.5×10^{-6}	$\frac{1}{2}$	-1
electron (β^--particle)	e^-	e^+	1	0.51	stable	$\frac{1}{2}$	-1
μ^-neutrino	ν_μ	$\bar{\nu}_\mu$	0	0	stable	$\frac{1}{2}$	0
e^-neutrino	ν_e	$\bar{\nu}_e$	0	0	stable	$\frac{1}{2}$	0
PHOTON	γ	same	0	0	stable	1	0

The validity of some old and some new conservation laws has been demonstrated. Familiar conservation laws that have been verified in every experimental test involving interactions of these particles are

1. Conservation of energy,
2. Conservation of momentum,
3. Conservation of angular momentum,
4. Conservation of electric charge.

The law of conservation of mass number (Section 1.5), which holds in the radioactive decays and nuclear reactions discussed in Chapter 1, is violated in many processes involving heavy baryons (Table 4-4). A conservation law can be recovered by assigning to each particle a new attribute, a baryonic charge or *baryon number* with the value:

$+1$ for a baryon particle,
-1 for a baryon antiparticle,
0 for all other particles.

We now find that in all processes, the algebraic sum of the baryon numbers before interaction is the same as the algebraic sum of the baryon numbers of the particles which exist after the interaction. In other words, we have a law of *conservation of baryon number*. For example in the reaction

$$\Sigma^+ + \bar{p}^- \longrightarrow \pi^+ + \pi^- + \pi^0$$

the corresponding baryon numbers are

$$(+1) + (-1) \longrightarrow 0 + 0 + 0.$$

We see that this reaction conserves baryon number.

Processes involving leptons (Table 4-4) reveal a completely analogous law of conservation of lepton number. For instance, in the β-decay experienced by a free neutron,

$$n \longrightarrow p^+ + \beta^- + \bar{\nu}_e \quad \text{(recall } \beta^- \text{ means electron; that is, } e^-\text{).}$$

The lepton numbers are

$$0 \longrightarrow 0 + 1 + (-1)$$

and lepton number is conserved.

Quarks

It is tempting to speculate that the profusion of particles listed in Table 4-4 are, in fact, combinations of a few "really elementary" particles. In 1964, at the California Institute of Technology, George Zweig and the 1969 Nobel prize winner, Murray Gell-Mann, hypothesized that

Physics Today

Murray Gell-Mann
(1929–   )

there exist three such truly fundamental particles, dubbed *quarks* (after the phrase in James Joyce's *Finnegans Wake*, "Three quarks for Muster Mark!"). It is assumed that a quark has the peculiar property of having a fractional baryonic number $\frac{1}{3}$ and an electric charge which is a *fraction* of e, either $+\frac{2}{3}e$, $-\frac{1}{3}e$, or $-\frac{1}{3}e$! Anti-quarks are assumed to exist and carry charges of $-\frac{2}{3}e$, $+\frac{1}{3}e$, or $+\frac{1}{3}e$. Then each baryon is a compound of three quarks and each meson is formed from a quark and an anti-quark.

Many consequences of the quark theory are in agreement with experiment, so the quark idea has been taken seriously. Nevertheless, after years of extensive searches for free quarks, with their peculiar electric charge, we still have no decisive experimental evidence for their existence.

QUESTIONS

1. Refer to Fig. 4-19. In which part, (a) or (b), or both, or neither, is work being done (in the sense that work is defined in physics, of course)? Support your answer with an explanation.

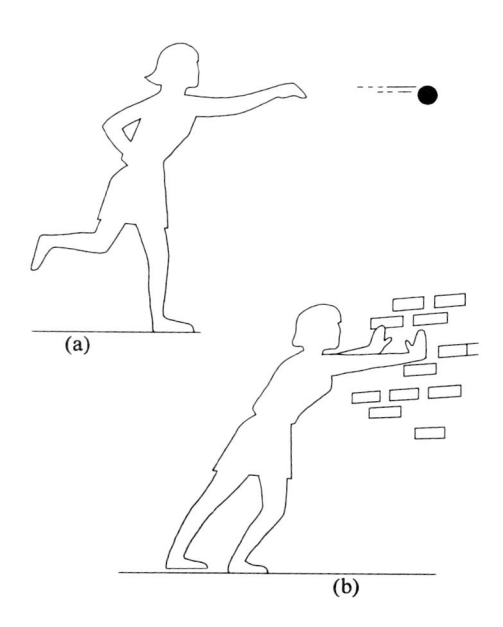

Fig. 4-19 Work or no work?

2. A man exerts a 10-N force horizontally in sliding a box across a horizontal table top through a distance of 0.90 m. Assuming the force is exerted in the direction of motion of the box, how much work is done by this force?

3. A man pushing vertically upward on a stone with a constant force of 200 N moves from point A to B to C to D. Find the work he does for each portion of the trip, given that
   (a) B is 3.0 meters above A,
   (b) C is 13 meters east of B at the same height,
   (c) D is 1.0 meter below C.

4. A constant horizontal force of 900 N pushes on a crate as it slides up a plane inclined at an angle of 60° to the horizontal. What is the work done by this force on the crate during a 3.0 m displacement of the crate along the inclined plane?

5. The crate in the preceding question weighs 400 N. What is the work done by this weight during the 3.0 m displacement along the inclined plane?

6. What is the definition of the kinetic energy of a mass m moving with a speed v?

7. A sailboat (weighing about 900 lb) has a mass of 400 kg. If this boat is moving along at a speed of 2.5 m/sec (which is about 5 knots), what is its kinetic energy in joules?

8. A railroad car is accelerated from rest until its kinetic energy is 5.0×10^5 joules. What is the work done by the resultant force acting on the car? If the force has constant magnitude of 1000 N and is in the direction of the motion, find the distance covered by the car before it attains this kinetic energy.

9. When a system at rest breaks up into two fragments of masses m and M, show that momentum conservation implies that the ratio of kinetic energies is given by

$$\frac{\text{kinetic energy of } M}{\text{kinetic energy of } m} = \frac{m}{M}.$$

10. Use the result of the above problem to show that when you fire a rifle, although the planet earth recoils with a momentum of exactly the same magnitude as the bullet's momentum, the kinetic energy of the recoiling planet is entirely negligible.

11. If 10^5 joules of energy are employed in 10^3 seconds, what is the average power consumption in watts?

12. In a modern industrial nation the average per capita power employed is of the order of 2500 watts. What is the per capita energy consumption, expressed in joules, in one day (24 hr)?

13. The 2.0×10^3-kg hammer of a pile driver is lifted 2.0 m in 3.0 sec. What power does the engine furnish to the hammer?

14. (a) From the definition, 1 watt = 1 joule/sec, show that

$$1 \text{ kilowatt-hour} = 3.6 \times 10^6 \text{ joules}.$$

(b) A commonly used unit of power is the horsepower (hp) which is related to the watt by

$$1 \text{ hp} = 746 \text{ watts}.$$

If 4.0 hp is required to operate a certain electric motor, how much will it cost to run this motor for 2.0 hours at a cost of 3.0 cents per kilowatt-hour?

15. While a force $\vec{F}$ acts on a particle, the particle undergoes a displacement Δs in a small time interval Δt and therefore has a speed $v = \Delta s / \Delta t$.

Show that the power supplied by this force is given by

$$\text{Power} = F_s v,$$

where F_s is the component of $\vec{F}$ in the direction of motion.

16. While a boat is being towed at a speed of 10 m/sec, the tension in the towline is 3000 N. What is the power supplied to the boat by the towline?

17. A jet aircraft engine develops a thrust of 12×10^3 N. What power does it supply when the aircraft is traveling at 300 m/sec?

18. A 40-kg mass hangs from the ceiling by a rope which is 3.0 m long. What is the change in gravitational potential energy when this mass swings out from its lowest point to a position where the rope forms an angle of 60° with the vertical?

19. (a) Is there one formula for the potential energy of a system or are there different expressions for potential energies arising from work done by different types of forces? Comment.

(b) Is there potential energy associated with work done by frictional forces? Comment.

20. A balloon lifts a 200-kg mass with a constant velocity to a height of 1500 meters in 10.0 minutes. Find the work done by the force exerted by the balloon on the mass in this ascent. What is the power supplied by this force to the mass? What is the change in the potential energy during this ascent?

21. (a) What are the dimensions of work?
(b) Show that the expression for kinetic energy, $\frac{1}{2}mv^2$, has the same dimensions as work.
(c) Show that the expressions given in Eqs. (4.11), (4.12) and (4.13) all have the dimensions of work.

22. If a giant were holding at rest a planet (M) and its moon (m), would he have to push to increase their separation from r to an infinite separation? Would he do positive work on the planet-moon system? Would the potential energy of this system increase? If you decided to call the final potential energy zero, would the initial potential energy then be a negative number?

23. In the preceding problem, replace the masses m and M by two positive charges q and Q and then answer the same questions.

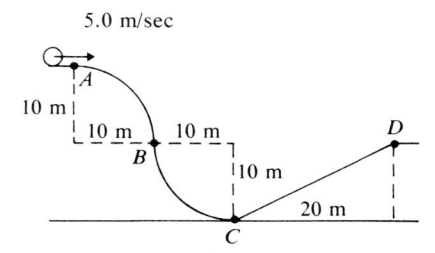

Fig. 4-20 Particle of Question 24 slides along the smooth surface.

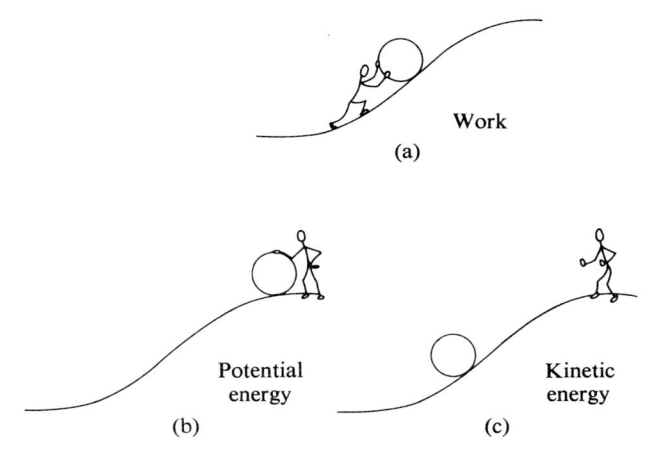

Fig. 4-21 Mechanical energy. Physicist playing with a large snowball.

24. In Fig. 4-20 a 2.0-kg particle slides along a smooth curved surface. The level of the point C is selected as the reference level for the measurement of gravitational potential energy. At each of the points A, B, C, and D find the particle's
    (a) potential energy,
    (b) total mechanical energy,
    (c) kinetic energy,
    (d) speed.

25. Fig. 4-21 portrays the concepts of mechanical energy. Suppose that the vertical difference between the top and the bottom of the incline is 8.0 meters, that the energy loss due to friction is negligible, and that the mass of the snowball is 90 kg.
    (a) How much work (in joules) is done in bringing the snowball from the bottom of the incline to the top?
    (b) What is the increase in potential energy of the snowball at the top of the incline as compared to its potential energy at the bottom of the incline?
    (c) If the snowball were allowed to slide or roll all the way down the incline, what would be its total kinetic energy at the moment of arrival at the bottom of the incline?
    (d) Suppose that frictional effects are not negligible. What effect would this have on your answers to (a), (b), and (c) above?

26. A puck slides down a toboggan slide which dips and flattens out, twists and turns. Relate the puck's speed v_1 at height h_1 to its speed v_0 at height h_0 assuming that the work done by frictional forces is negligible so that mechanical energy is conserved. Show that

$$v_1^2 + 2gh_1 = v_0^2 + 2gh_0.$$

27. A bullet is fired with the rifle held at a position between the horizontal and vertical, and the bullet travels along a parabolic path first rising and then descending. For this general situation show that the relationship between the bullet's speed v_1 at height h_1 and its speed v_0 at height h_0 is the same as for the puck of the preceding problem. (Notice the power of energy conservation arguments. You can ignore all details of the motion if you know the potential energy everywhere and if the total mechanical energy is conserved.)

28. In Bohr's model of the hydrogen atom, in the ground state the electron has a circular orbit of radius $r_0 = 0.53 \times 10^{-10}$ meter. The potential energy of the atom is the Coulomb potential energy given by $U = -ke^2/r_0$ (the proton has a charge $+e$ and the electron has a charge $-e$, $e = 1.60 \times 10^{-19}$ coulomb). Show that this potential energy, U, in electron volts (1 eV = 1.60×10^{-19} joule) is -27.2 eV.

29. The total mechanical energy of the ground state of the hydrogen atom is -13.6 eV. From the preceding problem, Bohr's model predicts the Coulomb potential energy is $U = -27.2$ eV. How much kinetic energy is there in the ground state?

30. If an H atom is forced apart into an infinitely separated electron and proton with both at rest, then the potential energy is zero and the kinetic energy is zero. Find the total mechanical energy and compare it to the total mechanical energy of the H atom in its ground state (-13.6 eV). Now, how much energy must be supplied to an H atom to ionize it (that is to rip off the electron)? The answer is named the *binding energy* of the atom.

31. A space ship returning home out of fuel is traveling slowly when it is a few million miles from the earth so that its kinetic energy is negligible and the potential energy of the ship-earth system is also negligible. The ship is attracted by the earth and comes in on a parabolic orbit which unfortunately intercepts the earth. What is its velocity on impact? Justify your answer.

32. With what speed must an object be projected vertically upward from the earth's surface (radius r_e) in order to reach a maximum altitude of $\frac{1}{2}r_e$? Express this speed in terms of r_e, G, and the mass of the earth, M_e. (Disregard the force exerted on the object by the air.)

33. A block is projected at a speed v_0 along a rough horizontal plane. The coefficient of sliding friction between the block and the plane is μ_k. Using the fact that the decrease in the total mechanical energy of the block must equal the magnitude of the work done by friction, show that the block comes to rest after moving a distance $v^2/2\mu_k g$.

34. A 10-kg block slides down a rough plane inclined at an angle of 53.1° to the horizontal. The block's speed increases from 4.0 m/sec to 8.0 m/sec in a distance of 5.0 m measured along the plane.
    (a) Find the change in the block's total mechanical energy.
    (b) Find the work done by the frictional force acting on the block.
    (c) Compute the coefficient of kinetic friction between the plane and the block.

35. A bullet of mass m traveling at a speed v strikes a block of mass M and is embedded in it.
    (a) Show that the ratio of the total kinetic energy after this collision to the kinetic energy before collision is $m/(m + M)$.
    (b) Find how far the block (with the bullet embedded in it) will slide along a rough horizontal board. The coefficient of kinetic friction between the board and the block is μ_k.

36. Use the law of conservation of total mechanical energy to find the relationship between the maximum velocity and the amplitude in simple harmonic motion.

37. A particle executes simple harmonic motion. Specify, in terms of the amplitude A, the values of the position coordinate at which:
    (a) the kinetic energy is zero;
    (b) the potential energy is zero;
    (c) the potential energy is one-fourth the total mechanical energy;
    (d) the potential energy and the kinetic energy are equal.

38. A 2.0-kg mass is fastened to a spring of force constant 200 N/m and placed on a smooth table as in Fig. 3-35. The mass is pulled 0.100 m to the right of its equilibrium position and then projected to the right at a speed of 0.20 m/sec.
    (a) What is the amplitude of the motion?
    (b) If the mass were projected to the left from the same position at the same speed, what would be the amplitude?

39. A 8.0-kg mass is suspended from one spring and set into oscillation with an amplitude of 0.15 m. A second identical spring suspends a 2.0-kg mass which oscillates with an amplitude of 0.60 m. What is the ratio (E_2/E_1) of the total mechanical energies of these masses?

40. A mass of 50 kg of mercury is poured from a height of 20 m into a vessel containing 10 kg of water. Assuming all the mechanical energy lost by the mercury is converted into thermal energy of the water (i.e., neglect the increase in thermal energy of the mercury), find the temperature rise of the water.

41. A heating coil inside an electric kettle is furnished with 1500 watts. How long will it take to raise the temperature of 0.50 kg of water (about 1 pint) from 20° C to 100° C? (Neglect the increase in thermal energy of the coil and kettle.)

42. (a) What equation expresses the first law of thermodynamics?
    (b) What meaning is ascribed to the term "heat" that appears in this law?

43. (a) How much heat is required to change the temperature of 2.0 kg of mercury from 10° C to 30° C?
    (b) What temperature rise would this amount of heat produce in 2.0 kg of water?

44. Deduce an expression for the mechanical advantage of a lever in terms of the distances from the points of application of the input and output forces to the fulcrum.

45. A man can push down with a force of 800 N on one end of a crowbar which is 2.00 meters long.
    (a) When the fulcrum is placed 0.40 meter from the other end, how great a weight can be supported by the man?
    (b) What is the mechanical advantage of this machine?

46. A complicated machine is known to have an efficiency of 90%. In a certain application an input force of 20 N moves 0.100 meter in the direction of the force. Find
    (a) the work input;
    (b) the work output;
    (c) the magnitude of the output force, given that it moves 0.027 meter.

47. (a) What is meant by the term "ideal mechanical advantage"?
    (b) Find the ideal mechanical advantage and the actual mechanical advantage for the machine in the preceding question.

48. Using pulleys, strings, and the ceiling, show how a suspended 10-kg mass can be used to support a suspended 20-kg mass. Apply the conditions of equilibrium to a clearly indicated system to prove that equilibrium indeed is obtained.

49. A cable is tightened by turning a crank. The cable is wound on the crank spindle of radius 0.020 m, and the crank handle moves in a circle of radius 0.15 m.
    (a) Find the ideal mechanical advantage of this crank.
    (b) If the efficiency is 0.75 (or 75%), calculate the actual mechanical advantage.

50. Show how a pulley system with an ideal mechanical advantage of 6 can be constructed using two blocks of three pulleys each, with the top block attached to the ceiling. Verify that the ideal mechanical advantage is 6 by considering the conditions for equilibrium of an appropriately selected system.

51. Each of the two pulley blocks in the preceding question has a mass of 4.00 kg. This block and tackle is used to raise a 96.0-kg mass.
    (a) What is the input force required if there is negligible friction?
    (b) What is the actual mechanical advantage and the efficiency of this block and tackle?
    (c) Find the efficiency when this block and tackle are used to raise a mass of 6.00 kg.

52. Stevedores slide a 4000-N crate up a rough inclined plane, which is 3.0 m long, to a platform 0.50 m above the ground. The force exerted by the stevedores on the crate is 1000 N, parallel to the inclined plane. For this inclined plane find the ideal mechanical advantage, the actual mechanical advantage, and the efficiency.

53. (a) What is the total energy of an isolated particle with kinetic energy K and rest mass m?
    (b) What is a formula for its kinetic energy which can be used when its speed is much less than c?
    (c) Why is mc^2 called the *rest* energy?
    (d) Show that mc^2 has the dimensions of energy.

54. A convenient unit for measuring masses of atoms and nuclei is the atomic mass unit (amu) which is exactly 1/12 the mass of an atom of carbon whose nucleus has 6 protons and 6 neutrons. The masses of a neutron, proton, and hydrogen atom are only about 1% greater than 1 amu. The rest energy associated with a rest mass of 1 amu is therefore a useful number to know. Calculate this rest energy in MeV using the fact that 1 amu = 1.66×10^{-27} kg; 1 eV = 1.60×10^{-19} joule.

55. The energy transformed into kinetic energy of the exploding fragments in the chemical explosion of one ton of TNT is 4.2×10^9 joules. This amount of energy is now called a "ton," and the kinetic energy produced in fusion and fission bombs is measured in kilotons and megatons (*kilo* implies a thousand, *mega* a million). Show that the rest energy of a 1.0-kilogram mass is 21 megatons.

56. How many cars, each of mass 10^3 kg, traveling at 30 meters/sec are required in order that their kinetic energies add up to the rest energy of a 1.0-kg mass?

57. The neutral pi meson, the π^0, spontaneously emits a pair of photons and disappears (half-life is 10^{-16} sec). The rest energy of the π^0 meson is 135 MeV. Compute the energy (in MeV) of one of the photons observed, assuming the π^0 meson was at rest.

58. Give an example of 100% conversion of rest mass into kinetic energy. Does an H-bomb achieve such a high percentage conversion?

59. Can rest mass be created from objects that have no rest mass? Give an example.

60. Are changes of rest mass confined to nuclear phenomena, or do they occur whenever the kinetic energy of any isolated system changes? Explain the fact that rest mass changes were not observed by such scientists as Newton, Faraday, and Joule.

61. Is it energetically possible for a proton to decay into a positron and a photon? You have evidence that this does not happen. What is it?

62. (a) Is it energetically possible for a neutron (n) to decay into a proton (p), an electron (e), and an antineutrino ($\bar{\nu}$)? Explain.

    $M_n = 1.00866$ amu
    $M_p = 1.00728$ amu
    $M_e = 5.4860 \times 10^{-4}$ amu
    $M_{\bar{\nu}} = 0$

    (b) What is the sum of the kinetic energies (in MeV) of the particles produced if the neutron decays according to the reaction examined in part (a)?

63. Apply the result of Question 9 to justify the stated division of kinetic energy between the α-particle and the recoiling radon nucleus in Example 10.

64. What is the definition of the binding energy of a system of mass M against decomposition into two parts of masses M_A and M_B?

65. Arrange atoms A, B, C in order of stability, given that the binding energies of the outermost electron are

    $(BE)_A = 4.6$ eV,
    $(BE)_B = 13.6$ eV,
    $(BE)_C = 1.5$ eV.

66. It is observed that an H_2 *molecule* can be dissociated into two separated hydrogen atoms if it absorbs a photon of energy at least 4.5 eV. What is the binding energy of the H_2 molecules against this type of breakup? (Chemists call this energy the *dissociation* energy of the molecule.)

67. Nuclear reactions can release about 10 million or more times the energy of a chemical process involving the same quantity of atoms. For example, a single pound of fissionable uranium can produce as much usable energy as 250,000 gallons of diesel oil or three million pounds of coal. Relate these facts to binding energies.

68. If the isotope $_{92}U^{235}$ were more abundant in nature than the isotope $_{92}U^{238}$, would it be easier to manufacture A-bombs and construct nuclear reactors? Explain.

69. Explain what is meant by "critical mass" with respect to nuclear fission.

70. Why do deuteron fusion reactions not occur in a jug of heavy water maintained at room temperature? (A molecule of heavy water is formed from two deuterium atoms and one oxygen atom.)

71. Is the decay symbolized by

    $$\Lambda^0 \to p^+ + \pi^-$$

    possible according to the law of conservation of
    (a) energy?   (b) charge?
    (c) baryon number?   (d) lepton number?

SUPPLEMENTARY QUESTIONS

S-1. An elevator, carrying a man weighing 800 N, is stuck between floors.
  (a) How much work is done by the elevator mechanism during this circumstance?
  (b) If the elevator now rises a distance of 3.00 m to the floor immediately above, how much work is done on the man?
  (c) Comment on what must be known in order to calculate the work done by the elevator mechanism in the situation in part (b).

S-2. If an electric hoist has an efficiency of 72% and the motor operates at a power level of 5.0 kilowatts, how rapidly can the hoist lift a 34-kg load? (Efficiency = power output/power input.)

S-3. A 70.9-kg man runs up a flight of stairs 7.32 m high in 4.00 sec.
  (a) What horsepower does he develop?
  (b) How long should it take a man of his weight to go up these stairs if a heart condition limits him to 0.200 horsepower?

S-4. While a ship is making a constant speed of 16.5 knots, the power delivered by its propellers is 2000 horsepower. Determine the resistive force exerted on the ship by the water. (1 knot = 1830 m/hour, and 1 horsepower = 746 watts.)

S-5. What is the efficiency of a pumping system driven by a 7.5-kilowatt motor which elevates 2.5×10^3 kg of water per minute through an 8.0-meter high vertical pipe and then ejects the water at a speed of 3.0 m/sec? [*Hint:* Consider the total energy gain of the water per second.]

S-6. The gravitational potential energy of a system of particles in a uniform gravitational field can be evaluated as if the entire mass of the system were located at the mass center. Prove that this is so for a system of two particles of masses m_1 and m_2 located at altitudes y_1 and y_2 above the XZ-plane.

S-7. Consider the expression for the gravitational potential energy, $U = -GMm/r$, of a system consisting of the earth of mass M and a particle of mass m.
  (a) Show that at a height h above the earth's surface, where $r = r_e + h$, this potential energy is given by
  $$U = \frac{-GMm}{r_e}(1 + h/r_e)^{-1}$$
  (b) From the binomial theorem it can be shown that $(1 + x)^{-1}$ is approximately equal to $(1 - x)$ when the magnitude of x is much less than one. Using this approximation, show that the result of part (a) implies that, near the surface of the earth,
  $$U = \text{constant} + mgh.$$

S-8. An automobile weighing 14,200 N (about 3200 pounds) is traveling at a speed of 12 m/sec.
  (a) What is the car's kinetic energy?
  (b) How high could the amount of energy calculated in (a) lift the car?

S-9. A long chain of length L rests on a smooth horizontal table. One link is allowed to dangle over the edge. The chain then slides over the edge. Find the speed of the chain as the last link leaves the table.

S-10. A particle of mass m is fastened to one end of a string which is fixed at the other end. The particle moves with varying speed in a vertical circle of radius L (Fig. 4-22). At its uppermost position the particle has a speed v_1.
  (a) What is the tension in the string when the string is horizontal?
  (b) Find the tension in the string when the particle reaches its lowest position.

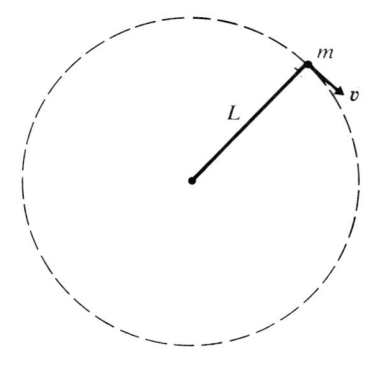

Fig. 4-22 Particle of mass m moves in a vertical circle of radius L (Question S-10).

S-11. Two identical spherical space ships are initially at rest (relative to an inertial frame of reference) in outer space, separated by a distance r. They are each of mass m.
 (a) What is an expression for the initial potential energy of the system of these two space ships in the initial configuration?
 (b) What are the initial kinetic energies of the space ships?
 (c) What is the total mechanical energy of this system?

S-12. Under the influence of the attractive gravitational forces, the space ships of the preceding problem are accelerated toward each other and crash when their separation reaches the value D, the diameter of each spherical ship.
 (a) What is the potential energy of the system at impact?
 (b) Show (by thinking about momentum conservation) that in this motion they each have the same speed at any instant.
 (c) What is the total kinetic energy of this system at impact if each ship has a speed v?
 (d) What is the total mechanical energy at impact?
 (e) Now, using conservation of total mechanical energy, show that the speed of each at impact is given by
$$v = \sqrt{Gm\left(\frac{1}{D} - \frac{1}{r}\right)}.$$

S-13. A 450-metric-ton train traveling at a speed of 108 km/hr is stopped on a horizontal track by application of the brakes to all wheels. The coefficient of sliding friction between iron and iron is 0.200 (1 metric ton = 1000 kg).
 (a) What is the distance traveled by the train from the time the brakes are applied until it comes to a full stop?
 (b) How much energy is dissipated due to friction in the braking process?

S-14. A schematic representation of the human forearm as a lever is shown in Fig. 4-23. The elbow serves as the fulcrum. The "input" force F_i is furnished by the biceps muscles which are attached approximately 5 cm from the elbow.
 (a) If the output force is furnished by the palm of the hand 35 cm from the elbow, what is the mechanical advantage of this forearm?

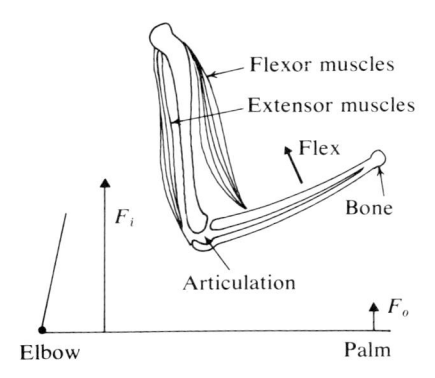

Fig. 4-23 Human forearm as a lever.

 (b) What force must the biceps exert on the forearm to lift a 100-N weight in the palm?
 (c) What movement of the hand will be produced if the biceps contract 1 cm?

S-15. The handle of a vise moves in a circle of 0.25-m radius. While the handle makes one revolution, a jaw of the vise moves 2.0 millimeters toward the stationary jaw.
 (a) What is the ideal mechanical advantage of this vise?
 (b) If the efficiency is 50%, what force will be exerted on a board between the jaws when an 80-N input force is applied at the handle?

S-16. The chain hoist commonly used in machine shops and garages uses a differential pulley, (Fig. 4-24). The upper wheel consists of two connected pulleys of slightly different radii (a and b) which rotate on the same axle. A continuous chain passes over the toothed pulley wheels. The input force F_i is provided by pulling the chain at the location indicated, and the output force F_o raises the suspended weight.
 (a) Show that the ideal mechanical advantage of the differential pulley is $2a/(a - b)$.
 (b) The upper wheel consists of pulleys of diameters 0.300 m and 0.280 m. When lifting a 5.0×10^3-N weight, the efficiency is 40%. What is the input force?

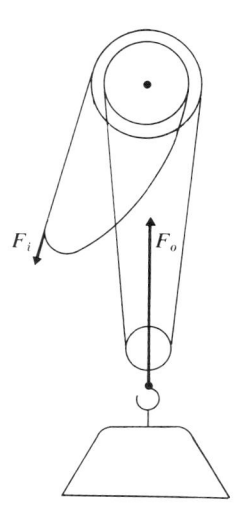

Fig. 4-24 Differential pulley.

S-17. The ideal mechanical advantage of a certain chain hoist is 100, but the machine involves a considerable frictional force of 120 N, which is independent of the size of the load being lifted.
(a) Evaluate the input force (F_i) that would be required to lift a 4000-N weight with this hoist, if there were no friction. Because of friction, the actual input force would be 120 N greater than this figure. Now calculate the actual mechanical advantage and determine the efficiency.
(b) Consider lifting a 400-N weight with this hoist and determine the efficiency in this case. Does the efficiency of the machine decrease as the load gets smaller?

S-18. The mass of Fig. 3-35 executes simple harmonic motion of amplitude A and angular frequency ω. At $t = 0$, the mass has the position coordinate x_0 and velocity v_{x0}. Use the law of conservation of total mechanical energy to show that the amplitude has the following dependence on the initial conditions:
$$A^2 = (v_{x0}^2/\omega^2) + x_0^2$$

S-19. *Kinetic Energy of a Rotating Rigid Body.* Consider a rigid body rotating about a fixed axis with an angular velocity ω. The body is composed of particles of masses $m_1, m_2, \ldots, m_n$ whose respective distances from the axis of rotation are denoted by $R_1, R_2, \ldots, R_n$. Show that the rigid body's kinetic energy K can be expressed as follows:
$$K = \tfrac{1}{2} I \omega^2$$
where $I = m_1 R_1^2 + m_2 R_2^2 + \ldots + m_n R_n^2$. ($I$ is called the moment of inertia of the body about the axis of rotation.)

S-20. A thin hoop has a mass of 2.0 kg and a radius of 0.50 m.
(a) From the expression for the moment of inertia given in the preceding question, find the moment of inertia of this hoop about its axis (a line through the hoop's center and perpendicular to the plane of the hoop).
(b) What is the kinetic energy of this hoop when it is rotating about its axis with an angular velocity of 3.0 radians/sec?

S-21. A 20-kg pail of water is suspended by a rope wrapped around a windlass which is a solid cylinder 0.40 m in diameter with a mass of 50 kg. (A uniform cylinder of radius R and mass M has a moment of inertia about its axis given by $I = \tfrac{1}{2} MR^2$.) The pail is released at rest at the top of a well and descends 20 m before it strikes the water. At the instant it strikes:
(a) How much potential energy has been lost by the pail of water because of its descent?
(b) What is the sum of the kinetic energies of the pail and of the rotating windlass?
(c) What is the ratio of the kinetic energies of these objects? [*Hint:* the speed v of the pail is related to the angular velocity ω of the windlass by $v = \omega R$, where R is the radius of the rotating windlass.]
(d) What is the speed of the pail?

S-22. Consider any reaction in which the particles present before the reaction have kinetic energies whose sum is K_{initial} and the (possibly different) particles which exist after the reaction have kinetic energies whose sum is K_{final}. The Q of the reaction, defined by
$$Q = K_{\text{final}} - K_{\text{initial}},$$
measures the energy released by the reaction.
(a) Show that
$$Q = M_{\text{initial}} c^2 - M_{\text{final}} c^2$$
where M_{initial} is the sum of the rest masses of the particles present before the reaction, and M_{final} is the sum of the rest masses of the particles which exist after the reaction.

(b) When the process under consideration is the breakup of a single particle of mass M into particles of masses M_A and M_B, show that the binding energy is $-Q$.

S-23. Evaluate the Q of the reactions described in Examples 6 to 11.

S-24. When an α-particle interacts with a sodium-23 nucleus, magnesium-26 along with a proton may be formed.
(a) Write this nuclear reaction in symbolic form as shown in Chapter 1.
(b) If the mass of the two original nuclei in this reaction exceeds the mass of the two product nuclei by 0.00194 amu, what is the energy Q, in MeV, released in this reaction? Express this energy also in joules.

S-25. The first study of nuclear reactions produced by energetic particles from a particle accelerator was published by John D. Cockcroft and Ernest T. S. Walton in England in 1932. They had bombarded nuclei of lithium-7 with protons accelerated in their new machine. It was then observed that in some instances a nuclear reaction occurred in which these initial particles (lithium nucleus and proton) were transformed into two α-particles. Moreover, Cockcroft and Walton noted that two α-particles produced in this manner had more kinetic energy than the incident proton.
(a) Using the manner described in Chapter 1, write the nuclear reaction involved.
(b) From the following experimental data evaluate Q, first from the kinetic energies, and then from the rest energies or masses.

   Rest energy of Li^7 ..... 6540.8 MeV
   Rest energy of H^1 ..... 939.6 MeV
   Rest energy of He^4 .... 3731.6 MeV
   Kinetic energy of each
      He^4 nucleus .... 8.6 MeV
   Kinetic energy of proton .. 0.25 MeV

(These data are the rest energies of *neutral atoms*, rather than of *bare nuclei*. However, because atomic numbers balance in the equation for this nuclear reaction, the numbers of orbital electrons also balance, which implies that the error introduced by inclusion of the energies of these electrons cancels out.)
(c) Explain how this experiment can be interpreted to support the relation $E = mc^2 + K$.

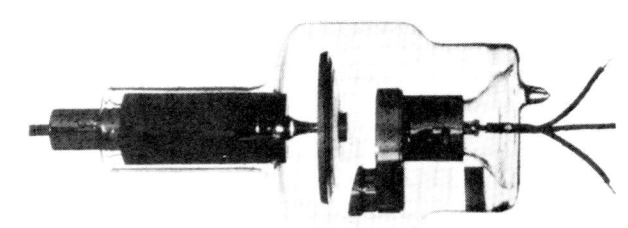

Fig. 4–25 Modern rotating anode x-ray tube used in medical diagnosis. To reduce overheating of the target (or anode, at the left), due to the bombardment by electrons originating from the centrally off-set electrode at the right, the target consists of a rotating disk. (Photograph courtesy Siemens A. G., Bereich Medizinische Technik.)

S-26. (a) Show how to determine the comparative values of the energy yield of fissioning as opposed to the energy yield of gasoline combustion (see p.181). First find the energy yield from the complete fissioning of one pound of U-235 (0.45 kg). The Q for fission of U-235 is about 200 MeV. There are 1.1×10^{24} atoms in 0.45 kg of U-235. Express your answer both in joules and in kilocalories.
(b) Determine the energy yield of the complete combustion of 250,000 gallons of diesel oil which is stated to have a heat of combustion of 1.5×10^6 kilocalories/barrel. (There are 42 gallons in a barrel.) Again express your answer in joules and in kilocalories.

S-27. X-ray production with conventional x-ray tubes is very inefficient. Of the total energy supplied to a medical x-ray tube, usually less than 1 percent is manifested in x-ray energy. The remainder of the energy originally imparted to electrons during their acceleration toward the tube target unfortunately goes into increasing the temperature of the x-ray tube target. (See Fig. 4-25.)

If 700 joules of energy are supplied to a tube during a chest x-ray exposure of 0.10 sec, to what temperature (from a room temperature of 20°C) will a 500-gm tungsten target rise? The average specific heat of tungsten for this process is 0.035 kilocal/kg C°. Assume that all the supplied energy is dissipated as thermal energy in the target and that there is negligible heat loss from the target during the 0.10 sec exposure.

S-28. (a) Calculate the total binding energy (in MeV) of the neutrons and protons in the nucleus $_6C^{12}$. (The necessary data on masses is given in Example 8.)
(b) What is the average binding energy per particle (neutron or proton) in $_6C^{12}$?

S-29. (a) In what way do observations of pion decay provide evidence for the existence of neutrinos?
(b) Why are neutrinos difficult to detect?
(c) What physical properties do neutrinos have?

S-30. A collision is said to be *elastic* when kinetic energy is conserved in the collision. A particle of mass m, traveling along the X-axis with a velocity v_x, strikes a stationary particle of mass M. After an elastic collision, m travels along the X-axis with a velocity v'_x and M with a velocity V'_x.
(a) Show that

$$v'_x = \left(\frac{m-M}{m+M}\right)v_x, \qquad V'_x = \left(\frac{2m}{m+M}\right)v_x.$$

(b) Show in this elastic collision that the fractional decrease in the kinetic energy of the mass m is given by

$$\frac{\tfrac{1}{2}mv_x^2 - \tfrac{1}{2}mv'^2_x}{\tfrac{1}{2}mv_x^2} = \frac{4mM}{(m+M)^2}$$

(c) Evaluate the fractional decrease in the kinetic energy of a neutron in an elastic collision with each of the following nuclei: U^{238}, C^{12}, H^2. (Use the approximate value, 1 atomic mass unit, for the neutron's mass.)

(d) Explain why deuterium is a more effective moderator in a nuclear reactor than graphite (carbon) or uranium. You may assume that neither carbon nor deuterium absorbs an appreciable fraction of the incident neutrons.

S-31. Show that in an elastic collision between two particles of equal mass moving along the X-axis, the particles simply exchange velocities.

S-32. When a proton p interacts with an antiproton $\bar{p}$ the annihilation reaction often results in the creation of one neutral pion and 4 charged pions:

$$p + \bar{p} \to 2\pi^+ + 2\pi^- + \pi^0.$$

(a) Verify that this reaction is in accord with the conservation laws for charge, baryon number, and energy.
(b) Evaluate Q for this reaction.

ADDITIONAL APPLICATIONS TO MEDICINE AND THE LIFE SCIENCES

Hypodermic Needle Injection

To get into body tissue requires work. To avoid pushing after the puncture is made (the patient might move), one jabs hard initially to give the needle adequate *kinetic energy* for sufficiently deep penetration. This kinetic energy of the needle and syringe is converted into thermal energy of the tissue and the needle as work is done in penetrating the tissue.

Effects of Radiation on Forest Ecosystems

It has often been assumed that if the radiation level is such that man will not suffer harmful effects of ionizing radiation, other life would be safe by a wide margin. Unfortunately, this may not be true. Studies show that ionizing radiations of sufficient intensity to harm humans can also have damaging effects on some ecosystems. (An ecosystem is a terrestrial or aquatic unit of the biosphere that comprises one or several plant communities, animal communities, and all the biological and physical factors that influence them.)

Experiments have shown that in forest ecosystems, irradiated for a period of six months, pine trees showed deleterious effects with an exposure of one or two roentgens per day and were killed by exposure of 20 to 30 roentgens per day. It appears that pines are particularly sensitive to radiation damage. Fields of weeds survived exposure of 100 to 200 roentgens per day without major short-term damage. Among major ecosystems forests seem to be among the most sensitive and grasslands among the most resistant.

To place these exposure figures in perspective, one should compare these exposures to those tabulated in Tables

1-2 and 1-3. An exposure of one roentgen per day is substantial, more than 4000 times the radiation exposure endured by New Yorkers during the high nuclear-test fallout in 1958. But the extremely high radiation levels that experiments have shown to be harmful to forests did occur on certain islands 100 miles from the site of the 1954 tests at Bikini.

Further Reading

"The Ecological Effects of Radiation" by George M. Woodwell, in the *Scientific American*, June 1963.

SUGGESTED READING

DUCKWORTH, H. E., *Little Men in the Unseen World*. London: Macmillan, 1963. The last half of this 150-page book deals with matters raised in our Chapter 4, while the first half provides an entertaining review of some of the matters raised in our Chapter 1. The whole book, which is intended as a guide to atomic physics for the general reader, is written in a lighthearted and whimsical manner and is full of interesting anecdotes about scientists. Many instructive and amusing analogies are employed. Illustrations are of the cartoon variety intended to mix levity with instruction. Can be recommended for bedtime reading!

EINSTEIN, A. and L. INFELD, *The Evolution of Physics*. New York: Simon and Schuster, 1950. As the authors say "this is not a textbook of physics.... Our intention was rather to sketch in broad outline the attempts of the human mind to find a connection between the world of ideas and the world of phenomena." This book, which covers the growth of ideas from early concepts to relativity and quanta, is written for the layman. It can be highly recommended. A very elementary introduction to potential and kinetic energy is given on pp. 47-51. Energy and inertia are discussed on pp. 205-210.

FERMI, LAURA, *Atoms in the Family*. Chicago: The University of Chicago Press, 1954. A most delightful book about Enrico Fermi, his life and his role in the development of the controlled nuclear chain reaction, written by his wife. The physics is sound and is woven into the narrative of day-to-day living from Fermi's early days as a physicist through to his part in the University of Chicago cyclotron of 1947. There are many pictures of interest, including a photocopy of the letter to President Roosevelt signed by Einstein, informing the president of the possibility of tremendous energy release through nuclear fission.

HOLTON, G., and D. H. D. ROLLER, *Foundations of Modern Physical Science*. Reading, Mass.: Addison-Wesley, 1958. Chapter 18 gives an interesting account of energy at a level appropriate for our readers. Historical background is included.

INGLIS, D. R., *Nuclear Energy: Its Physics And Its Social Challenge.* Reading, Mass.: Addison-Wesley, 1973. The author "attempts to bring the reader . . . a piercing glimpse into both the scientific and humanistic aspects of nuclear-energy problems, including the problems of nuclear weapons, from the point of view of a scientist who professes enough concern that he may be given heed." The book is organized to provide the basis for understanding both the technical and socio-political sides of the problems without prerequisite knowledge of either.

PRIEST, J., *Energy for a Technological Society.* Reading, Mass.: Addison-Wesley, 1975. In his statement of purpose the author says that the book addresses itself to "(1) examine the consequences of working with finite fuel resources, (2) examine the effects, extent, and control of the by-products that are usually produced in energy conversion processes, (3) examine alternative energy sources and the trade-offs involved and, (4) teaching physical principles as they relate to understanding the energy situation."

TILLEY, D. E., *University Physics for Science and Engineering.* Menlo Park, California: Cummings, 1976. Chapters 10 and 11 give a more complete and a more advanced treatment of energy. Calculus is used.

Albert Einstein

5 Frames of Reference and Relativity

"It strikes me as unfair, and even in bad taste, to select a few . . . for boundless admiration, attributing superhuman powers of mind and character to them. This has been my fate," remarked Einstein in his later years. Physicists have often marveled at the extent to which Einstein's work captures the imagination of the public while the comparable achievements of others are completely ignored. In any case, whatever ranking a historian of physics might propose, it is certain that Einstein would be placed among the foremost of the theoretical physicists of any century.

In 1905 three remarkable papers appeared in *Annalen der Physik*, written, not by a learned university professor, not by a graduate student of recognized brilliance, but instead by a 26-year-old examiner of patents in the patent office in Bern, Switzerland. Albert Einstein, as a student, had revealed great ability in mathematics, but he had a distaste for examinations as well as for the authoritarian atmosphere of most classrooms. Unable to secure an academic appointment, he passed eight happy and fruitful years (1901–1909) earning a living at the patent office and applying his mind to fundamental questions in what are now three large fields of endeavor in theoretical physics: special and general relativity, quantum theory, and the kinetic theory of matter.

The special theory of relativity, published in 1905 in an amazingly complete form, is the major subject of this chapter. In contrast to the development of most physics in this century, the special and general theories of relativity are primarily the work of one man. In quantum theory, Einstein's 1905 paper on the photoelectric effect introduced the idea of quanta of light (photons), and it was this triumphant idea that was cited when he was awarded the Nobel prize in 1921. In a fundamental paper (1917) on the transition probabilities associated with the emission and absorption of photons, the phenomenon of *stimulated emission* was predicted. It is this stimulated emission that is the all important distinctive feature of a laser (Chapter 16). Einstein pioneered the application of quantum theory to the theory of specific heats (1907) while his work on the quantum theory of gases (1924–1925) introduced a new type of statistics now known as Bose-Einstein

". . . *every reference body has its own particular time; unless we are told the reference body to which the statement refers, there is no meaning in a statement of the time of an event.*"

Albert Einstein
(1879–1955)

statistics. In the third field, the kinetic theory of matter, his most celebrated contribution was the 1905 paper on the Brownian motion which is discussed in Chapter 7.

From 1909 until his death in 1955, Einstein held a series of academic positions, mostly in Germany. However in 1933, due to the political climate of Germany at this time, he accepted a professorship at the newly established Institute for Advanced Study at Princeton, N.J., and became an American citizen. It was through Albert Einstein, most prestigious of physicists of the time, that nuclear physicists were able to command the instant attention of President Roosevelt when they felt that the United States must explore the possibility of the construction of an atomic bomb. (The actual letter written to President Roosevelt is reprinted in a number of books, including the one by Laura Fermi cited in the Suggested Reading of Chapter 4.)

5.1 FRAMES OF REFERENCE IN NEWTONIAN MECHANICS

Transformations from One Frame to Another

Certain features of the description of any motion, say that of a baseball thrown around by boys in a bus, will depend on the *frame of reference* employed; a frame fixed in the bus or a frame fixed in the highway. However, in some ways, a whole set of different frames of reference turn out to be equivalent for the description of nature, and one can use this fact, as Einstein did, to discover vital new things about possible laws of nature. First, we must learn how to relate the position, velocity, and acceleration of a particle which is observed from two different frames.

Consider one frame of reference OX, a line painted along the center of a highway, and another frame of reference $O'X'$ painted on the floor from the rear to the front of a bus (Fig. 5-1) which travels down the highway in the direction OX with a constant velocity V. Assume that as the two origins coincide, a stopwatch is started on the bus. This watch ticks away giving readings we shall denote by t'. Another stopwatch reading times t, is started by a stationary policeman at the side of the highway. Now suppose a baseball is rolling forward (Fig. 5-1) in the bus and squishes an ant (event A) and a little later a bug (event B). An *event* in physics doesn't have to be very exciting; it is regarded as completely specified in one frame by answers to two questions: *Where? When?* So event A (see Fig. 5-1) is specified by x'_A and t'_A for those who use the bus as a frame of reference and by x_A and t_A for those who, like the stationary policeman, use the highway as their frame. Looking at Fig. 5-1, our common sense leads us to assert that

$$x_A = x'_A + Vt'_A \tag{5.1}$$

$$t_A = t'_A \tag{5.2}$$

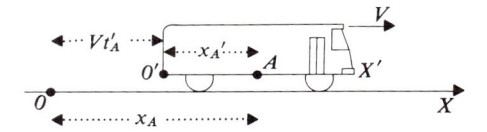

Fig. 5-1 Frame $O'X'$ moves relative to frame OX with a velocity V. $x_A = x'_A + Vt'_A$.

The second assumption, $t_A = t'_A$ [Eq. (5.2)], was tacitly assumed by everyone up to the year 1905. This assumption is built into our vocabulary when we talk of *the* time. For now we will accept it as true, at least for the remainder of this section.

Writing for event B, equations similar to Eq. (5.1) and (5.2), and then subtracting these new equations from Eq. (5.1) and (5.2), we obtain the relationships for the intervals between these events:

$$\Delta x = \Delta x' + V \Delta t' \tag{5.3}$$

$$\Delta t = \Delta t' \tag{5.4}$$

where $\Delta t = t_B - t_A$, $\Delta x = x_B - x_A$, $\Delta x' = x'_B - x'_A$, $\Delta t' = t'_B - t'_A$. To relate velocities we divide Eq. (5.3) by the time interval [i.e., Eq. (5.4)] and obtain

$$\frac{\Delta x}{\Delta t} = \frac{\Delta x'}{\Delta t'} + V \tag{5.5}$$

which states that the baseball's velocity v relative to the highway is related to its velocity v' relative to the bus by

$$v = v' + V. \tag{5.6}$$

Example 1 If a baseball has a forward velocity of 10 meters/sec relative to the bus ($v' = 10$ m/sec) and the bus is traveling at a speed of 30 m/sec ($V = 30$ m/sec) then the velocity of the baseball relative to the highway is the sum:

$$v = (10 + 30) \text{ m/sec} = 40 \text{ m/sec}.$$

It is not hard to show that Eq. (5.6) is valid even if the velocity of the bus V is not constant. Then, by examining the change in velocity between two events, it can be shown that *accelerations* are related by the plausible expression

$$a = a' + a_{\text{frame}} \tag{5.7}$$

where a_{frame} is the acceleration of the bus relative to the highway; that is, a_{frame} is the rate of change of the bus's velocity V. In our example this means that the baseball has an acceleration a relative to the highway and a' relative to the bus.

For a general motion, the accelerations in Eq. (5.7) and the velocities in Eq. (5.6) are to be understood as *components* along the axes OX and $O'X'$. The three-dimensional generalizations of these equations are the vector equations

$$\vec{v} = \vec{v}' + \vec{V} \tag{5.8}$$

$$\vec{a} = \vec{a}' + \vec{a}_{\text{frame}} \tag{5.9}$$

All the equations that we have so far encountered in this chapter allow us to *transform* from a description in one frame of reference to a description in another frame. Such equations are called *transformation equations*.

Inertial Frames of Reference

The *law of inertia* states that when the resultant force acting on a body is zero, the body moves with a constant velocity, i.e., its velocity vector remains unchanged. An *inertial frame of reference* is a frame in which the law of inertia is true. Relative to an inertial frame, force-free objects move with a constant speed in a straight line. Observations of a puck sliding on ice show that any frame fixed on the earth, say our frame OX fixed in the highway, is a good approximation to an inertial frame of reference.

Given one inertial frame of reference, we can find another inertial frame easily. Looking at the equation, $\vec{v} = \vec{v}' + \vec{V}$, it is apparent that any object like the puck, moving with a constant velocity $\vec{v}$ relative to

210 FRAMES OF REFERENCE AND RELATIVITY

the highway, will move relative to the bus with a velocity $\vec{v}'$ which is also constant, provided the velocity of the bus is constant. The conclusion is that the bus constitutes another inertial frame of reference as long as its velocity V is constant. In fact, we now see that given one inertial frame, we can find an infinity of other inertial frames, namely, all frames of reference which have a constant velocity vector relative to the given frame.

When the velocity of the bus is constant, its acceleration $\vec{a}_{\text{frame}}$ is zero and Eq. (5.9), $\vec{a} = \vec{a}' + \vec{a}_{\text{frame}}$, reduces to

$$\vec{a} = \vec{a}'. \tag{5.10}$$

The *acceleration* of the baseball is the same relative to the bus as it is relative to the highway. Thus Newton's second law, $\vec{F} = m\vec{a}$, may also be written $\vec{F} = m\vec{a}'$. We have discovered that Newton's second law still holds true when the bus is used as the frame of reference. This accords with all our experience: we are not conscious of motion with constant speed in a straight line; we become aware that there is something different about a bus as a frame of reference only when it acquires an acceleration by going over a bump, speeding up, slowing down, or rounding a turn.

Principle of Relativity for Newtonian Mechanics

Our discovery is emphasized by formally stating the *principle of relativity for Newtonian mechanics: The laws of mechanics are the same in all inertial frames of reference.* An object has the same acceleration, $\vec{a}$, relative to all inertial frames and Newton's second law holds in all such frames. Any mechanics experiment will have exactly the same outcome when it is done on the bus traveling smoothly at a constant speed of 80 mph in a straight line as it will when the experiment is repeated with the bus parked on the highway.

The relativity principle implies, at least as far as mechanical behavior is concerned, that all inertial frames are on an equal footing. *There is no single preferred frame of reference.* The constant velocity bus provides just as good a frame of reference as does the highway. Nature does *not* seem to single out one particular frame that is "really at rest" and distinguish this frame from other inertial frames because they are moving relative to this particular frame.

However, as will be stressed in the next section, mechanics is very different in a frame which is *accelerated* relative to an inertial frame. In accelerated frames Newton's second law is no longer valid; it must be amended by the addition of new forces called *frame forces* which arise from the acceleration of the frame of reference. Evidently nature does make a distinction between accelerated frames and inertial frames.

The problem remains of finding a frame of reference which is an inertial frame (see Fig. 5-2). A frame fixed in the earth is only approximately inertial. Departures from the predictions of $\vec{F} = m\vec{a}$ are noticed

Acceleration of frame toward earth's axis of rotation
$a = 3.4 \times 10^{-2}$ m/sec^2

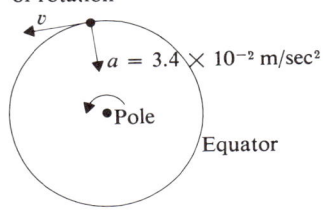

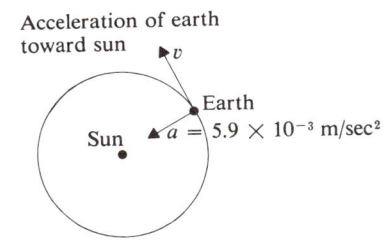

Acceleration of earth toward sun
$a = 5.9 \times 10^{-3}$ m/sec^2

Acceleration of sun toward center of galaxy
$a \approx 10^{-10}$ m/sec^2

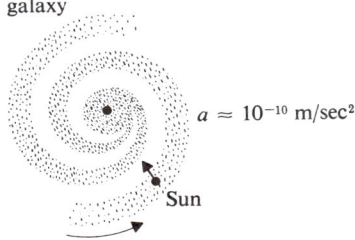

Fig. 5-2 The accelerations of a frame fixed on the earth.

5.1 FRAMES OF REFERENCE IN NEWTONIAN MECHANICS

for motions which involve long time intervals or long trajectories. A better approximation to an inertial frame is one with the origin carried along by the earth in its orbit but with axes not rotating with respect to the distant stars. Relative to such a frame, our highway frame at the equator would have, because of the earth's rotation, an acceleration of 3.4×10^{-2} m/sec² toward the earth's center. Now the earth, moving in its orbit, has an acceleration of 5.9×10^{-3} m/sec² toward the sun. A still better approximation to an inertial frame is provided by what is called the astronomical frame, a frame with its origin near the sun's center (at the mass center of the sun and its planets) and axes not rotating with respect to distant stars. But our sun is way out on an arm of a spiral galaxy which rotates. Participation in this general rotation of the galaxy means that the sun has an acceleration toward the galaxy's axis of rotation of roughly 10^{-10} m/sec². And this galaxy of ours accelerates relative to the mass center of our local group of about ten galaxies. There does not seem to be hardware out in space that provides a frame of reference that we are sure has precisely zero acceleration relative to an inertial frame. Nevertheless, as a practical matter, the astronomical frame can be considered inertial for the mechanics of our solar system and everything in it. Even our highway frame is a useful approximation to an inertial frame, as Galileo discovered, and as present-day civil engineers assume in calculating the banking on highway curves.

5.2 ACCELERATED FRAMES*

Accelerated Frames of Reference and Frame Forces

There is a lot of physics to be shaken out of the simple equation $\vec{a} = \vec{a}' + \vec{a}_{\text{frame}}$. If we look at a baseball of mass m, acted on by a resultant force $\vec{F}$, and write down Newton's second law for the baseball's acceleration, $\vec{a}$, relative to the highway (we assume the highway is an inertial frame of reference), we have

$$\vec{F} = m\vec{a} \qquad (2.10)$$

This equation takes on an interesting form in terms of the baseball's acceleration, $\vec{a}'$, relative to the accelerated bus. Substituting in Eq. (2.10) the result of Eq. (5.9) we get

$$\vec{F} = m\vec{a}' + m\vec{a}_{\text{frame}} \qquad (5.11)$$

which is conveniently written in the form

$$\vec{F} - m\vec{a}_{\text{frame}} = m\vec{a}'. \qquad (5.12)$$

Defining the symbol $\vec{F}_{\text{frame}}$ by

$$\vec{F}_{\text{frame}} = -m\vec{a}_{\text{frame}}, \qquad (5.13)$$

Eq. (5.12) yields

$$\vec{F} + \vec{F}_{\text{frame}} = m\vec{a}'. \qquad (5.14)$$

*This section can be omitted without significant loss of continuity. It treats material which, though interesting and instructive in rounding out one's comprehension of Newtonian mechanics, is often ignored in introductory courses.

This last equation tells us that we can work, and think, and do our physics in the accelerated bus, and the acceleration $\vec{a}'$ relative to the bus can be still determined by the equation

$$Force = mass \times acceleration$$

provided we add to the real resultant force $\vec{F}$, a new quantity, $\vec{F}_{frame}$. The new entity $\vec{F}_{frame}$ evidently plays the role of a force, and from its definition, $\vec{F}_{frame} = -m\vec{a}_{frame}$, we see it arises solely from the fact that the frame of reference we are using, the bus, has an acceleration $\vec{a}_{frame}$ relative to an inertial frame. The quantity $\vec{F}_{frame}$, which must be added to the real force to recover the form of Newton's second law in an accelerated frame, we shall call a *frame force*. Other designations commonly used, synonymous with frame force, are "fictitious," "pseudo," or "inertial" force.

When the brakes of a moving bus are jammed on, it is this frame force that accelerates you forward relative to the bus (Fig. 5-3). While the bus speeds up it is again the frame force that presses you backward into the seat. If the bus is struck violently from behind, this frame force can become large enough to hurl your head backward until your neck breaks: the whiplash effect.

The effects of a frame force are real enough but there is, of course, a distinction between these forces and other types. Frame forces do not arise from other particles but exist only because of the acceleration of the frame of reference. *In an inertial frame of reference there is no frame force.*

You have probably read about pilots blacking out when pulling out of a dive if the centripetal acceleration is maintained at from 5 to 9 g's, or about astronauts being subjected to 5 g's during launching, and so on. In such expressions, g ($= 9.8$ m/sec^2) is being used as a convenient unit in which to give the acceleration of the frame of reference. The number of g's turns out to be the ratio of the frame force acting on a mass m to its weight mg at the earth's surface. Thus, if during launching, a rocket has an acceleration of 5.0 g's, then $a_{frame} = 5.0 \times 9.8$ meters/sec^2, and the frame force acting on an astronaut of mass m has a magnitude $ma_{frame} = 5.0 \times m \times (9.8$ meters/sec$^2) = 5.0$ times his weight at the earth's surface.

Example 2. The bus brakes, slowing down at the rate of 4.0 meters/sec^2. What is the frame force acting on a 100-kg man, and what will be his acceleration relative to the bus if the resultant of the other forces acting upon him is zero?

Solution With the bus velocity vector in the direction OX, its acceleration vector must be in the opposite direction, since the bus is slowing down, that is,

$$\vec{a}_{frame} = 4.0 \text{ m/sec}^2, \text{ backward.}$$

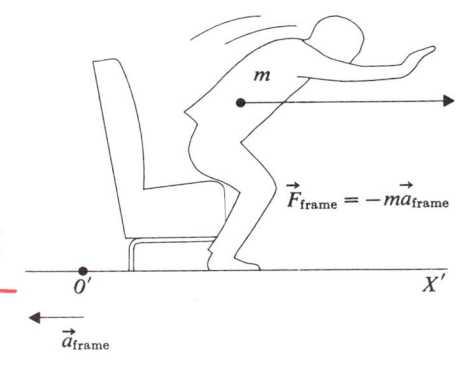

Fig. 5-3 Bus brakes are jammed on. The bus becomes an accelerated frame of reference with an acceleration $\vec{a}_{frame}$ to the left. The passenger experiences a frame force $\vec{F}_{frame} = -m\vec{a}_{frame}$ which hurls him forward.

The frame force acting on the 100-kg man is

$$\vec{F}_{\text{frame}} = -m\vec{a}_{\text{frame}}$$
$$= [100 \text{ kg } (4.0 \text{ m/sec}^2)], \text{ forward}$$
$$= 400 \text{ N, forward.}$$

The frame force points forward. Newton's second law, *amended by the addition of the frame force because our frame of reference is accelerated*, gives

$$\vec{F} + \vec{F}_{\text{frame}} = m\vec{a}', \tag{5.14}$$

or

$$0 + \vec{F}_{\text{frame}} = 100\, \vec{a}';$$

hence

$$\vec{a}' = 4.0 \text{ meters/sec}^2, \text{ forward.}$$

The acceleration vector, $\vec{a}'$, has the same direction as $\vec{F}_{\text{frame}}$, so it too points forward. The man accelerates forward relative to the bus, a fact which seems quite plausible since the only force involved is the frame force, and we have seen earlier that it was directed forward. The advantage of using seat belts in automobiles should be patently obvious.

Example 3 Notice that the preceding problem could have been analyzed from an inertial frame (as was suggested in Chapter 2), the highway, in the following way. The man's acceleration $\vec{a}$ relative to the highway is determined by Newton's second law

$$\vec{F} = m\vec{a}. \tag{2.10}$$

With the resultant of the real forces $\vec{F} = 0$ this gives $\vec{a} = 0$. The man is not accelerated relative to the inertial frame. (Of course this is obvious. He experiences zero resultant force and so moves with constant velocity according to the law of inertia. But since the bus is coming to a halt while the passenger continues to move forward, he will suffer the consequences exactly as implied in Example 2.) Of course, pursuit of the mathematical analysis should show how this approach is equivalent to that of Example 2. Let us see. The accelerations in different frames are related by Eq. (5.9),

$$\vec{a} = \vec{a}' + \vec{a}_{\text{frame}} \tag{5.9}$$

and in the present case $\vec{a} = 0$ while $\vec{a}_{\text{frame}} = 4.0$ meters/sec², backward. Therefore,

$$\vec{a}' = -\vec{a}_{\text{frame}}$$
$$= 4.0 \text{ meters/sec}^2, \text{ forward.}$$

This agrees with the analysis given in Example 2.

Rotating Frames and Centrifugal Force

Employing an inertial frame, say a frame of reference fixed in a biological laboratory, let us describe the motion of a particle of mass m

214 FRAMES OF REFERENCE AND RELATIVITY

in a *centrifuge* [Fig. 5-4 (a)]. The particle moves along with the centrifuge with constant speed v in a circular path of radius r and therefore has an acceleration vector of magnitude v^2/r directed toward the center of the circle: the centripetal acceleration. There are forces exerted on this particle by other particles in the centrifuge. The resultant of these forces, F, is, according to Newton's second law, given by

$$\vec{F} = m\vec{a}. \tag{2.10}$$

Knowing that $a = v^2/r$ (see Section 2.3), we get

$$F = m\frac{v^2}{r} \tag{5.15}$$

and $\vec{F}$ points inward in the direction of the acceleration vector.

Now, it is often natural and convenient to use the centrifuge itself [Fig. 5.4 (b)] as the frame of reference, but this is an accelerated frame and we must learn just what *frame forces* arise in such a *rotating frame of reference*. Writing

$$\vec{F} + \vec{F}_{\text{frame}} = m\vec{a}', \tag{5.14}$$

with $\vec{F}$ the resultant of the real forces acting on the mass m, and $\vec{a}'$ the acceleration of m relative to the rotating frame, we wish to find the expression for $\vec{F}_{\text{frame}}$. To do this we can apply Eq. (5.15) to the situation that we already understand. We had found that a particle of mass m, acted upon by a real inward force

$$F = m\frac{v^2}{r},$$

moves around with the centrifuge and so, being *at rest relative to the centrifuge*, has its acceleration $\vec{a}'$ *relative to the centrifuge* equal to zero.

(a) Motion of m relative to laboratory

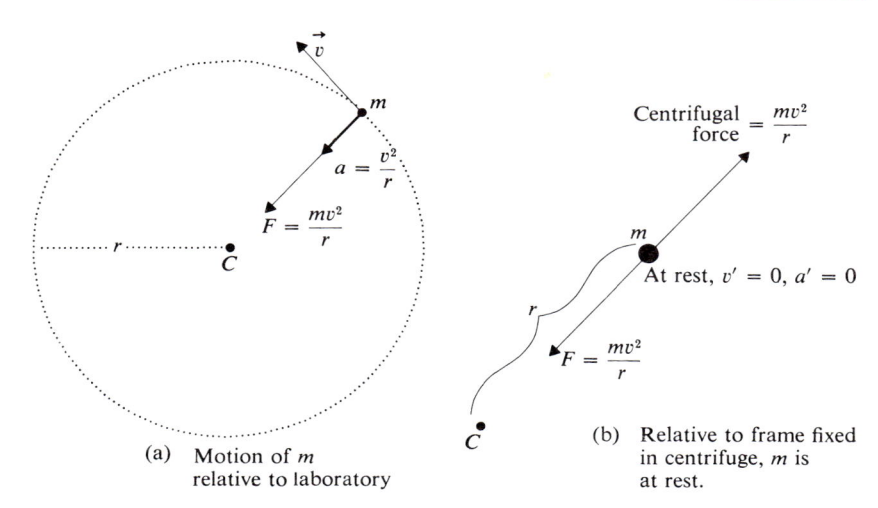

(b) Relative to frame fixed in centrifuge, m is at rest.

Fig. 5-4 The same situation viewed from (a) an inertial frame (the laboratory) and (b) a rotating frame (the centrifuge). Only in the rotating frame is there a centrifugal force.

5.2 ACCELERATED FRAMES

Inserting this data into Eq. (5.14), we obtain

$$\vec{F} + \vec{F}_{\text{frame}} = m \times 0 = 0.$$

Therefore, $\vec{F}_{\text{frame}}$ *points outward and has a magnitude* mv^2/r. This frame force has been given a special name, the *centrifugal force*. To recapitulate, *in a rotating frame of reference there is a frame force, acting on a particle of mass m, called the centrifugal force, of magnitude* mv^2/r, *directed outward* from the axis of rotation (v is the speed of the frame at radius r, relative to an inertial frame). Unfortunately the term centrifugal force is often used incorrectly. We therefore caution the reader to remember that there is no centrifugal force in an inertial frame; centrifugal force is a frame force arising only in a rotating frame of reference.

The once humble centrifuge which we have just discussed has today become a potent research instrument. The Swedish chemist, T. Svedburg (1884–) developed the *ultracentrifuge* in the early 1920s and measured the molecular weights of proteins by observing their radial sedimentation. With the development of a magnetic suspension to replace bearings, J. W. Beams (1898–) has obtained centrifugal forces which are a billion times the force of gravity. These modern centrifuges enable the determination, with a precision of better than 1%, of molecular weights ranging from 50 to 10^8 for giant molecules, such as the tobacco mosaic virus. By comparison, the centrifuges so commonly employed in medical laboratories develop more moderate values of g, generally in the range from a few up to 400,000 g. For example, a bench model clinical centrifuge may operate at 1000 revolutions per minute with the sample being treated experiencing about 120 g's.

Example 4 An ultracentrifuge whirls a particle of mass m around in a circle of radius r with speed v relative to the laboratory. When we take the *centrifuge* to be the frame of reference, the particle experiences balanced forces, a real force acting inward and a centrifugal force, mv^2/r, outward. If this centrifugal force is a billion g's (that is $mv^2/r =$ billion $\times$ mg), compute the speed of the centrifuge at the radius $r = 10^{-4}$ meter.

Solution
$$mv^2/r = 10^9 \, m\text{g},$$
so
$$v^2 = 10^9 \, \text{g}r$$
$$= (10^9 \times 9.8 \times 10^{-4}) \, (\text{m/sec})^2.$$
Therefore,
$$v = 0.99 \times 10^3 \text{ meters/sec}.$$

In this case this speed corresponds to about 1.5 million revolutions per second. It is interesting to note what might happen under such circumstances. As an example, we note that small steel balls of radius 10^{-4}

Courtesy of Eli Aron

J. W. Beams
(1898–)

meter, rotated at such high frequencies, explode when the peripheral speed attains approximately 1000 meters/sec.

Viewed from an "astronomical frame" which is not rotating relative to the distant stars, the earth is seen to rotate making one complete rotation per day. Experiments show that "the astronomical frame" is an inertial frame to an excellent approximation (force-free objects do maintain a nearly constant velocity vector relative to this frame). It follows that any frame, like our highway OX, which is fixed in the earth, is not quite a truly inertial frame and so one should include frame forces for really accurate work.

Thus, to get the acceleration $\vec{a}'$ of a baseball relative to a frame fixed in the ball park, Newton's second law should be written in the amended form of Eq. (5.14) which in this instance leads to

$$m\vec{g} + \vec{F}_{\text{frame}} = m\vec{a}'$$

where $m\vec{g}$ is the gravitational force acting on the ball (its weight). Therefore the acceleration $\vec{a}'$ of the baseball is not quite the same as the gravitational field $\vec{g}$ but the difference is small, the contribution from the centrifugal force ranging from 0.34% of $\vec{g}$ at the equator to zero at the poles.

It should be mentioned at this point that when a particle is moving relative to a rotating frame, another frame force comes into play in addition to the centrifugal force. This additional frame force, called the *Coriolis force*, is a deflecting force at right angles to $\vec{v}'$, the particle's velocity vector relative to the rotating frame. The magnitude of the Coriolis force is proportional to the particle's mass m, speed v', and to the revolutions per second of the frame. Although we are not going to do any calculations with this more complicated frame force, it is interesting to know at least one of its consequences. Air masses moving toward a low-pressure region in the northern hemisphere are deflected by a Coriolis force into a counterclockwise rotation about the "low" in the northern hemisphere and into a clockwise circulation in the southern hemisphere. Look for this prevalent characteristic of cyclones the next time one appears on the weather map.

Weightlessness

An astronaut can "float" along in the middle of his capsule without touching the floor. You could do the same thing for a few seconds in an elevator if the cable broke (Fig. 5-5). This phenomenon, called weightlessness, arises in any frame of reference which has an acceleration, $\vec{a}_{\text{frame}}$, equal to the local value of the gravitational field $\vec{g}$. The weight $m\vec{g}$, our name for the gravitational force exerted on the mass m, still exists but it is canceled out by the frame force $\vec{F}_{\text{frame}} = -m\vec{a}_{\text{frame}} = -m\vec{g}$. The amended form of Newton's second law appropriate for use in a

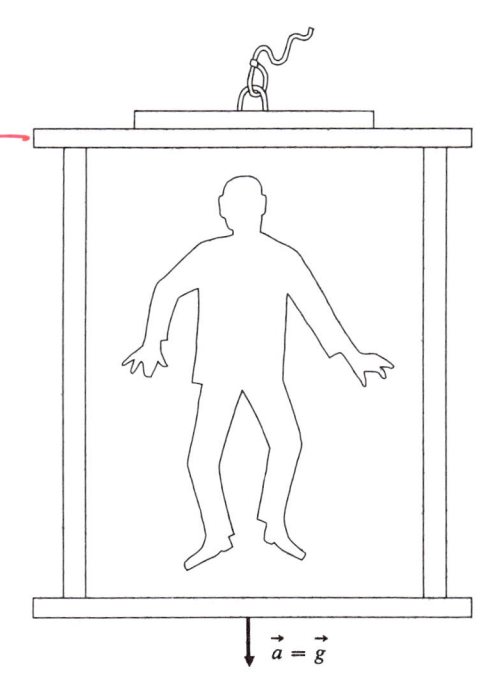

Fig. 5-5 Elevator falling with an acceleration $\vec{g}$. Relative to the elevator, the man's acceleration $\vec{a}'$ is zero.

5.2 ACCELERATED FRAMES 217

frame which has an acceleration $\vec{g}$ relative to an inertial frame is

$$\vec{F} + \vec{F}_{\text{frame}} = m\vec{a}', \qquad (5.14)$$

which gives for the "floating" man

$$m\vec{g} - m\vec{g} = m\vec{a}'.$$

Therefore, since m is obviously not zero, we get

$$\vec{a}' = 0.$$

The man has no acceleration relative to his accelerated frame, be it the capsule or an elevator. Therefore, if his velocity in this frame is zero, he will stay where he is, even if this is three feet above the floor of the falling elevator or of the orbiting space capsule.

5.3 RELATIVITY OF TIME

The Speed of Light: Failure of Velocity Addition

When a bullet is fired forward on our bus of Fig. 5-1 with a speed v' of 500 meters/sec relative to the bus, and the bus's speed V is 30 meters/sec, experimental measurement of the bullet's speed relative to the highway gives the result $v = 530$ meters/sec in accord with Eq. (5.6),

$$v = v' + V. \qquad (5.6)$$

Now when the experiment is repeated with a photon instead of a bullet, the result for the speed relative to the bus is $v' = c = 2.99793 \times 10^8$ meters/sec, and the speed relative to the highway is found to be $v = c = 2.99793 \times 10^8$ meters/sec; *not* $v = c + V$ but just $v = c$! Yes, at this speed, the speed of light, velocities do not add according to $v = v' + V$! *The speed of light is the same in all inertial frames of reference.* A seemingly peculiar effect, indeed!

You will realize, of course, that the effect or lack of effect at ordinary bus velocities V would be very hard to detect, but A. A. Michelson (1852–1931) and E. W. Morley (1838–1923) in 1887 had failed to discover any change in the speed of light in a laboratory because of a change in the earth's orbital velocity vector as the earth moves around the sun. Viewed from the astronomical frame, the earth's velocity vector $\vec{V}$ changes from about 3×10^4 meters/sec in one direction to about 3×10^4 meters/sec in the opposite direction during a six-month interval. These experiments are consistent with the preceding italicized statement, which is the seemingly ridiculous situation Einstein chose to wrestle with in 1905.

How could Einstein explain the fact that light has the same speed relative to different frames in relative motion? The prospect was bleak for an explanation in terms of things we previously assumed to be true

Physics Today

A. A. Michelson
(1852–1931)

because the rule $v = v' + V$ followed directly from our common sense notions of space and time summarized in Eq. (5.3) and Eq. (5.4):

$$\Delta x = \Delta x' + V \Delta t' \qquad (5.3)$$

$$\Delta t = \Delta t'. \qquad (5.4)$$

Instead of trying to explain the fact that light has the same speed in all inertial frames, Einstein decided to simply accept this as a fact or a postulate and then investigate the consequences. Taking a good hard look at the *transformation equations*, Eq. (5.3) and Eq. (5.4), and realizing that there was no direct experimental evidence for their validity at very high speeds, Einstein was struck with the idea that our commonly accepted notion of time is suspect! Perhaps the innocent equation $\Delta t = \Delta t'$ is wrong and time elapses differently in different inertial frames!

Gedanken Experimente and Our Notions of Time

The necessity of revising our notions of time because light has the same speed relative to different inertial frames can be brought out using one of Einstein's favorite devices: the thought experiment (*Gedankenexperiment* in German). Consider a long fast train (Fig. 5-6) moving down the track with a velocity V. Bombs, set to explode when they receive a flash of light, are carried in the caboose and in the locomotive.

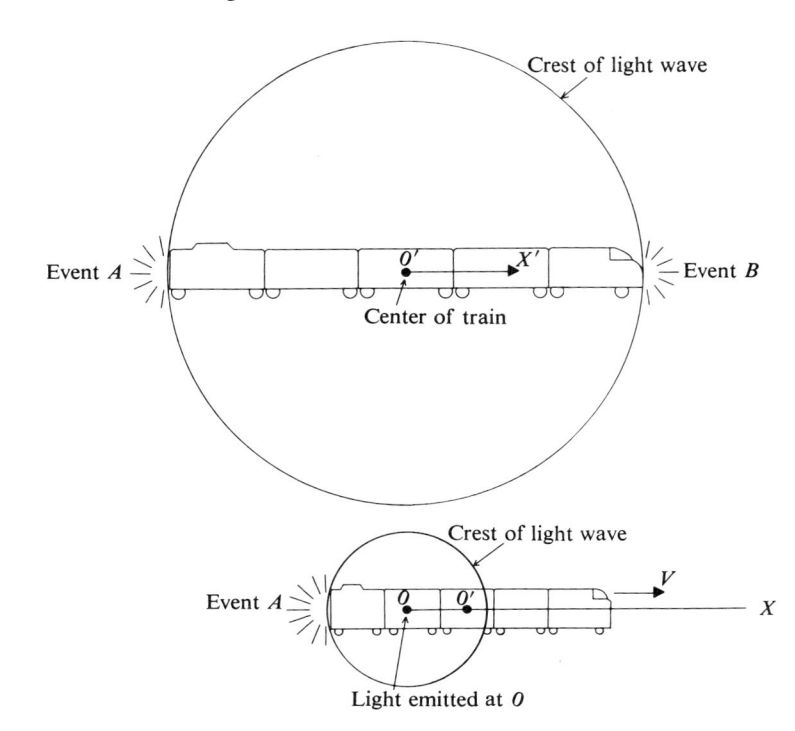

Fig. 5-6 Upper figure shows events A and B which are simultaneous relative to the frame $O'X'$ fixed in the train. Lower figure shows event A occurs before the light reaches the locomotive, relative to the frame OX fixed in the track.

5.3 RELATIVITY OF TIME

A light signal is flashed from the center of the train and eventually the caboose explodes (event *A*) and the locomotive explodes (event *B*). People on the train will agree that these explosions are *simultaneous*, since the light travels with the same speed toward the caboose as toward the locomotive and has the same distance to go. In terms of train time t' this means that

$$t'_A = t'_B$$

or

$$\Delta t' = t'_B - t'_A = 0.$$

Meanwhile, our policeman, parked by the railway track, will have a different record of the proceedings. The caboose, moving toward the light signal will have advanced somewhat before it explodes at a track time t_A. The light traveling at the same speed c forward, must go farther before it reaches the locomotive which therefore explodes at a later time t_B. We have

$$t_B > t_A$$
$$\Delta t = t_B - t_A > 0.$$

Therefore, for these two events,

$$\Delta t \neq \Delta t'$$

and Eq. (5.4), $\Delta t = \Delta t'$, is *not* correct! The conclusion is that *events which are simultaneous in one frame* ($\Delta t' = 0$) *are not simultaneous in another frame* ($\Delta t \neq 0$), *when the frames are in relative motion*. We have encountered what is known as the *relativity of simultaneity*.

With this conclusion we must abandon the idea of absolute time expressed by the equation $\Delta t = \Delta t'$. We cannot simply say that the caboose exploded 2 seconds before the locomotive. This may be true for the time used in a frame moving relative to the train with a certain velocity, but in a different frame the time interval between the same two events might be 1.0 sec, in another frame 0 sec (simultaneous), in yet another frame the locomotive might explode first! We must replace the idea of absolute time, one time t for all observers regardless of the frame, by a different time for each different frame of reference. We must employ a train time t' as well as a track time t.*

In thinking about the physics we are now encountering, it is well to consider the implications of some of the words and phrases we use in daily life. The notion of absolute time lurks behind such phrases as "*the*" time, *when*, *simultaneous*, *before*, *after*. With each one of these words you should, mentally at least, add a specification of the frame of reference.

Since it is no trivial step to give up the simple concept of absolute time, it is worthwhile contemplating from as many viewpoints as possible why this renunciation is necessary. Another instructive situation to think

---

*Fortunately, it is possible to have just one time t' for one entire frame of reference, and there are no peculiarities encountered with time as long as we confine our attention to measurements in this one frame. The time that we assign to an event is always the time on the clock of an observer *located at the event and at rest in the frame in question*. A passenger on the train near the caboose sees *the explosion of the caboose at time t'_A and sees the explosion of the distant locomotive at the later time $t'_B + (\Delta x'/c)$, where $\Delta x' = x'_B - x'_A$*. Nevertheless this passenger, knowing that light from the locomotive explosion takes the time $\Delta x'/c$ to reach him, records the event B as having occurred at time t'_B. He finds that his recorded times t'_B and t'_A are equal and so says the events were simultaneous, even though he saw them at different times. The time we assign to an event is always this recorded time, the time that an observer on the spot would see the event, never the time that the event would be seen by some distant observer.

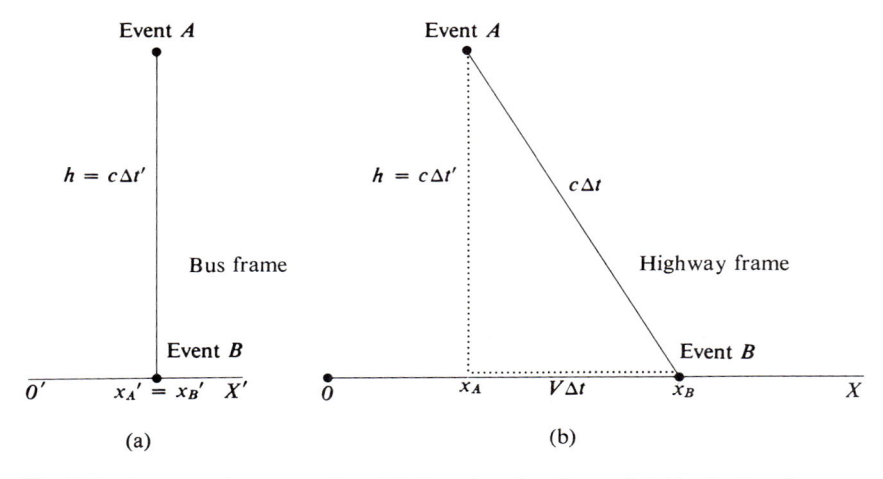

Fig. 5-7 Between the events A and B there is a time interval $\Delta t'$ in the bus frame and Δt in the highway frame. The right-hand figure shows $c\Delta t$ is greater than $c\Delta t'$. The bus is traveling at speed V relative to highway.

through is pictured in Fig. 5-7. A lamp is switched on in the roof of the bus (event A) and light travels a distance h to strike the floor vertically below the lamp (event B). The distance h equals $c\Delta t'$ where $\Delta t'$ is the bus-time interval between these two events, A and B. We can then say $\Delta t' = t'_B - t'_A$, and also $\Delta x' = x'_B - x'_A = 0$ since the events occur at the same x' coordinate in the bus and we choose to use the prime quantities, i.e., x', etc., as those referred to the frame of reference associated with the bus. The unprimed coordinates for the same two events in the highway frame are pictured in Fig. 5-7 (b). We assume that the height h is the same in the two frames. The bus has advanced a distance $V\Delta t$ while the light is in transit from the bus roof to floor and, having in this highway frame a longer distance to go at the same speed c, there will be a longer time interval Δt between events A and B. That is, Δt is greater than $\Delta t'$: time elapses differently in the two frames.

A little geometry and algebra will give us a useful relationship between these time intervals. Using the theorem of Pythagoras and referring to Fig. 5.7 (b) we obtain

$$(c\Delta t)^2 = (c\Delta t')^2 + (V\Delta t)^2$$

which gives

$$(\Delta t)^2 - \frac{V^2}{c^2}(\Delta t)^2 = (\Delta t')^2.$$

Therefore

$$\left(1 - \frac{V^2}{c^2}\right)(\Delta t)^2 = (\Delta t')^2.$$

After taking the square root of each side of this equation and then

5.3 RELATIVITY OF TIME

dividing each side by $\sqrt{1 - V^2/c^2}$, we obtain

$$\Delta t = \frac{\Delta t'}{\sqrt{1 - \frac{V^2}{c^2}}}. \qquad (5.16)$$

Table 5-1

V/c	$\dfrac{1}{\sqrt{1 - V^2/c^2}}$
.01	1.00005
.10	1.005
.50	1.15
.60	1.20
.70	1.40
.80	1.67
.90	2.3
.98	5.0
.99	7.1
.9999	71.

Some values of the factor $1/\sqrt{1 - V^2/c^2}$ are given in Table 5-1 for different values of the frame speed V. Notice that unless V is very large this factor is extremely close to the number one. This means $\Delta t = \Delta t'$ to an excellent approximation for the low velocities that we encounter in our daily life. The factor $1/\sqrt{1 - V^2/c^2}$ gives us a smooth transition to the strange new world of relativity as the velocity V approaches the velocity of light. This factor increases very rapidly when V is close to c but is a well-defined real number only if V is less than c. This suggests that *c is nature's speed limit*. In fact, there is abundant evidence that no particle with rest mass can attain the speed c but with enough pushing one can come arbitrarily close. In the Stanford linear accelerator, electrons reach a speed of $0.9999999997\,c$.

5.4 LORENTZ TRANSFORMATIONS

After pondering our train and bus "experiments" we are driven to consider modifications of the equations

$$\Delta x = \Delta x' + V \Delta t' \qquad (5.3)$$

and

$$\Delta t = \Delta t' \qquad (5.4)$$

that will replace the velocity transformation rule

$$v = v' + V \qquad (5.6)$$

by a rule that transforms $v' = c$ into $v = c$.

Let us therefore find new transformation equations that are in accord with Einstein's two postulates:

1. The principle of relativity: *The laws of nature are the same in all inertial frames of reference.*
2. The invariance of the speed of light: *The speed of light is the same in all inertial frames, regardless of the motion of the source of light.*

In the first postulate, the principle of relativity is asserted to apply to *all natural phenomena*, not to just the laws of mechanics. The essential point for our immediate attention is that all inertial frames are equivalent. There is no preferred frame with special properties that render nature's laws different from those found in any other frame. In particular, we wish to have transformation equations that imply that the frame OX is equivalent to the frame $O'X'$ for the description of nature.

For low speeds Eqs. (5.3), (5.4), and (5.6) are in agreement with experiment, so any new equations must differ from these only for high speeds. As a trial, let us assume modified transformation laws for space intervals and time intervals in the form

$$\Delta x = \gamma(\Delta x' + V \Delta t') \qquad (5.17)$$

and

$$c \Delta t = \gamma(c \Delta t' + \beta \Delta x') \qquad (5.18)$$

where the quantities γ and β are constants whose values we will try to select in such a way that the velocity of light is the same relative to the frame OX as it is relative to $O'X'$. First, putting $\Delta x' = 0$ in Eq. (5.18) gives

$$\Delta t = \gamma \Delta t'. \qquad (5.19)$$

But we know that when $\Delta x' = 0$, Eq. (5.16) holds (because we previously derived this equation under the condition $\Delta x' = 0$):

$$\Delta t = \frac{\Delta t'}{\sqrt{1 - V^2/c^2}}. \qquad (5.16)$$

Comparison of these two equations, Eq. (5.19) and Eq. (5.16), shows that we must select the value

$$\gamma = \frac{1}{\sqrt{1 - V^2/c^2}} \qquad (5.20)$$

It remains to determine β. To do this we examine the velocity transformation law implied by our new equations. Dividing Eq. (5.17) by Eq. (5.18) gives

$$\frac{1}{c} \frac{\Delta x}{\Delta t} = \frac{\Delta x' + V \Delta t'}{c \Delta t' + \beta \Delta x'}.$$

Dividing both numerator and denominator on the right-hand side of this equation by $c \Delta t'$ we get

$$\frac{1}{c} \frac{\Delta x}{\Delta t} = \frac{\frac{1}{c} \frac{\Delta x'}{\Delta t'} + \frac{V}{c}}{1 + \frac{\beta}{c} \frac{\Delta x'}{\Delta t'}}$$

which, using the definition of velocity [Eq. (2.3)], allows us to write

$$\frac{v}{c} = \frac{\frac{v'}{c} + \frac{V}{c}}{1 + \beta \frac{v'}{c}} \qquad (5.21)$$

where v' is the velocity of an object relative to $O'X'$ and v is the velocity of the same object relative to OX. Now let us see if we can select a value of β such that $v' = c$ implies that $v = c$. From Eq. (5.21) this condition requires

$$\frac{c}{c} = \frac{\frac{c}{c} + \frac{V}{c}}{1 + \beta \frac{c}{c}}, \quad \text{or} \quad 1 = \frac{1 + \frac{V}{c}}{1 + \beta}.$$

Therefore,

$$\beta = \frac{V}{c}. \tag{5.22}$$

The job is done. All our immediate objectives have been attained:

1. For V much less than c, the factor

$$\gamma = \frac{1}{\sqrt{1 - V^2/c^2}} \tag{5.20}$$

is almost the number one (see Table 5.1), and β is nearly zero. In these circumstances the new equations, (5.17), (5.18), and (5.21), become the same as the old equations, (5.3), (5.4), and (5.6); the new agrees with the old in the low-velocity region.

2. Eq. (5.21), with $\beta = V/c$, named the *Einstein law for addition of velocities*, does say that for two frames with relative velocity V, a velocity c relative to one frame implies a velocity c relative to the other.

3. It is not obvious, but it can be verified, that these new equations are consistent with the principle of relativity. The frames OX and $O'X'$ are equivalent. The transformation equations, Eq. (5.17) and Eq. (5.18), when solved to express $\Delta x'$ and $\Delta t'$ in terms of Δx and Δt, take exactly the same mathematical form:

$$\Delta x' = \gamma(\Delta x - V\Delta t) \tag{5.23}$$

$$c\Delta t' = \gamma(c\Delta t - \beta \Delta x). \tag{5.24}$$

Notice that these equations can be obtained from Eq. (5.17) and Eq. (5.18) simply by interchanging primed and unprimed quantities and by replacing V by $-V$. This last point is as it should be. Since V is the x-component (the other components are zero) of the velocity of $O'X'$ relative to OX, the x-component of the velocity of the frame OX relative to $O'X'$ should be of the same magnitude but in the opposite direction and therefore equal to $-V$.

The new transformation equations, Eq. (5.17) and Eq. (5.18), with

$$\gamma = \frac{1}{\sqrt{1 - V^2/c^2}}, \tag{5.20}$$

are called the *Lorentz transformation equations*. We summarize our results and nomenclature in Table 5-2.

Table 5-2. SOME IMPORTANT EQUATIONS OF SPECIAL RELATIVITY

Lorentz transformations:

$$\Delta x = \gamma(\Delta x' + V\Delta t') \tag{5.17}$$

$$c\Delta t = \gamma(c\Delta t' + \beta \Delta x') \tag{5.18}$$

where

$$\gamma = \frac{1}{\sqrt{1 - V^2/c^2}} \tag{5.20}$$

and

$$\beta = V/c \tag{5.22}$$

Einstein law for the addition of velocities:

$$\frac{v}{c} = \frac{\frac{v'}{c} + \beta}{1 + \beta \frac{v'}{c}} \tag{5.21}$$

224 FRAMES OF REFERENCE AND RELATIVITY

5.5 CONSEQUENCES OF THE LORENTZ TRANSFORMATIONS

The Relativity of Simultaneity

The most novel feature introduced by the Lorentz transformations is the replacement of the equation $\Delta t = \Delta t'$ by $c\Delta t = \gamma(c\Delta t' + \beta \Delta x')$. Instead of a universally agreed upon time interval between two events, we have space and time intervals scrambled up so that the space and time interval in one frame determine the time interval in another. We can now account for the relativity of simultaneity experienced with the exploding caboose and locomotive. For these two events, although they are simultaneous in the train time so that $\Delta t' = 0$, we see from the Lorentz transformation, Eq. (5.18), that because they occurred with a spatial separation $\Delta x'$, they will be separated by the time interval

$$\Delta t = \gamma \beta \frac{\Delta x'}{c},$$

as measured by track time.

Time Dilation

A most famous effect and one for which we have abundant experimental confirmation is what is called *time dilation*. A clock at rest in the bus of Fig. 5-8 measures some time interval $\Delta t'$ between the event "tick" and the event "tock" and $\Delta x' = 0$ for these two events which are both located at the clock. The Lorentz transformation, Eq. (5.18), now gives

$$\Delta t = \frac{\Delta t'}{\sqrt{1 - V^2/c^2}}. \tag{5.16}$$

For any V this implies Δt (on highway) $> \Delta t'$ (on the bus) so, according to highway time, the bus clock is running slow. Do not think that this implies that bus passengers would find that highway clocks run fast.

Fig. 5-8 Relative to the highway frame, the clock on the moving bus runs slow, that is $\Delta t'$ is less than Δt.

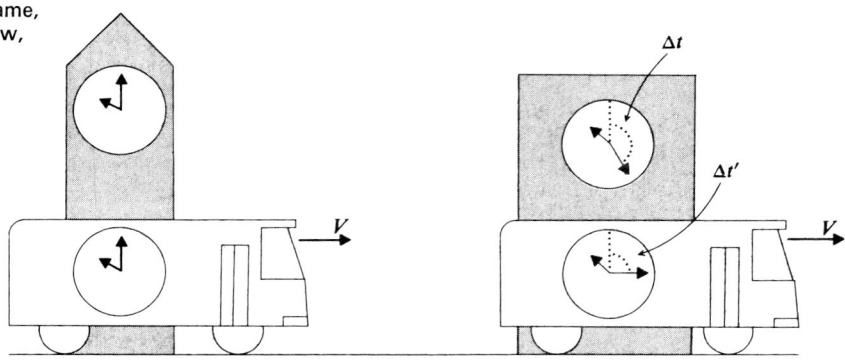

Highway clocks move relative to the bus, and if our frames are really equivalent for the description of nature, bus passengers must find that the highway clocks run slow. You are asked in Question 37 to verify that this happens. The situation is summarized by the statement "*moving clocks run slow.*"

The fact that the time really can be dilated or stretched out by moving briskly along is demonstrated for us by some short-lived unstable particles like Powell's π-mesons. At rest these π-mesons disintegrate after an average life of about 10^{-8} sec. Now return to Eq. (5.16) and Table 5-1. We see there that when traveling through a laboratory at 99% of the speed of light (which corresponds to $1/\sqrt{1 - V^2/c^2} = 7.1$), the average life of these mesons is expected to be increased by a factor of 7.1. This amazing fact is indeed found to be so by the court of last appeal, the experiment!

Length Contraction of a Moving Object

Measurement of the dimensions of a moving object reveals a *contraction in the dimension parallel to its velocity*. Traveling along in our bus we pass a parked car of manufacturer's length Δx. In our bus we call the car's length the difference $\Delta x' = x'_B - x'_A$ of the coordinates of its extremities measured at the same time relative to the bus, so $\Delta t' = t'_B - t'_A = 0$. The first of the Lorentz transformation equations, Eq. (5.17), gives

$$\Delta x = \frac{1}{\sqrt{1 - V^2/c^2}} \Delta x',$$

or

$$\Delta x' = (\sqrt{1 - V^2/c^2}) \Delta x. \tag{5.25}$$

Therefore, $\Delta x'$ is less than Δx. Bus passengers' measurements show that the car is shorter than the manufacturer's specifications! Of course, the effect is reciprocal since the frames are equivalent, and the car's owner will claim the bus length is less than advertised.

Transverse Dimensions of Moving Objects

In our earlier Gedanken Experiment dealing with the light signal emitted from the ceiling of the bus, we assumed that the bus height h would be the same in both frames of reference. We can show that this assumption is required if the frames are to be equivalent. Suppose the height of the bus were contracted for measurements made in the highway frame. Then the equivalence of the frames of reference demands that the height of an object like a tunnel would be contracted for measurements made in the bus frame. The highway policeman would find the low bus roof did not touch the top of the tunnel but the bus driver would discover the tunnel ceiling was so low that it knocked off his roof! This is indeed a genuine contradiction, so we conclude that transverse dimensions like the bus height are the same in both frames.

The Invariant Interval

So far we have seen that measurements of the car's length give different answers in different frames of reference. Time intervals between two given events are different in different frames of reference. In this treacherous state of affairs it is useful to find entities that are *invariant*. A quantity is called *invariant* if it has the same value in all inertial frames. The electric charge and the rest mass of a particle are two important invariant quantities. Although lengths and time intervals are *not invariant* it can be shown by some relatively involved algebra, employing the Lorentz transformation equations, that for any two events A and B,

$$(\Delta x)^2 - c^2(\Delta t)^2 = (\Delta x')^2 - c^2(\Delta t')^2 \tag{5.26}$$

where $\Delta x = x_B - x_A$, $\Delta t = t_B - t_A$, $\Delta x' = x'_B - x'_A$, $\Delta t' = t'_B - t'_A$.

This result shows that the combination of a distance and a time interval, $(\Delta x)^2 - c^2(\Delta t)^2$ is invariant. This important invariant, $(\Delta x)^2 - c^2(\Delta t)^2$, called the *relativistic interval* between two events, emphasizes that space intervals and time intervals should be considered together to get quantities that retain their significance in different inertial frames. This idea was pursued by H. Minkowski (1864–1909), who demonstrated the convenience of thinking in terms of a four-dimensional *space-time* (three spatial dimensions plus one more for time), rather than in terms of *space* and of *time* separately.

The Einstein Law of Addition of Velocities

The Einstein law of addition of velocities,

$$\frac{v}{c} = \frac{\frac{v'}{c} + \beta}{1 + \beta \frac{v'}{c}} \tag{5.21}$$

does the job of transforming a speed c relative to one frame into a speed c relative to another frame. Let's be sure that nature's speed limit is observed and that one doesn't get relative speeds greater than c. Refer to the following situation.

Example 5 A proton moves to the right with a speed relative to the laboratory of 3/4 the speed of light, and an electron moves to the right with a speed *relative to the proton* of 4/5 the speed of light. What is the speed of the electron relative to the laboratory?

Solution It is not just $3/4 + 4/5 = 31/20$ of the speed of light. The rule $v = v' + V$ has been replaced by Einstein's

$$\frac{v}{c} = \frac{\frac{v'}{c} + \beta}{1 + \beta \frac{v'}{c}} \tag{5.21}$$

First we let the frame OX be attached to the laboratory. In this frame the proton has the speed V/c. We now take the proton as being at the origin of the frame $O'X'$. Thus relative to OX the frame $O'X'$ has the speed V/c ($= 3/4$). But the electron's speed relative to $O'X'$ is $v'/c = 4/5$. Thus Einstein's velocity addition law gives

$$\frac{v}{c} = \frac{4/5 + 3/4}{1 + (4/5 \times 3/4)}$$
$$= \frac{31}{32}.$$

Evidently v is less than c in this case. And it can be shown that Einstein's law always gives this result: that $v/c < 1$ or, stated differently, $v < c$.

5.6 RELATIVISTIC MECHANICS

Einstein, in proposing the special theory of relativity, adopted two basic postulates. We restate these postulates here:

1. The principle of relativity: *The laws of nature are the same in all inertial frames of reference.*
2. The invariance of the speed of light: *The speed of light is the same in all inertial frames of reference.*

This second postulate is, as we have seen, satisfied by the Lorentz transformations. With transformations from one inertial frame to another to be accomplished by these transformations, it becomes apparent that Newton's mechanics, if true in one inertial frame, will be false in another frame because, among other things,

$$\Delta t' \neq \Delta t.$$

Einstein therefore set about modifying mechanics in such a way that it could be correct in all inertial frames. Instead of patiently working out the dictates of the principle of relativity, we shall content ourselves with a mere listing of some of the most important consequences.

In order to have observers in different frames agree that momentum is conserved in a collision or any other process, the momentum of a particle of rest mass m moving with velocity $\vec{v}$ must be redefined by

$$\vec{p} = \frac{m\vec{v}}{\sqrt{1 - v^2/c^2}}. \tag{5.27}$$

Recall that the factor $1/\sqrt{1 - v^2/c^2}$, evaluated in Table 5-1, is imperceptibly different from the number one for the speeds we encounter in daily life. For low speeds, therefore, $\vec{p}$ has just about Newton's value, $m\vec{v}$. But the closer we come to the speed of light, the larger this factor becomes. Very near the speed of light, say for v greater than 99% of c, huge increases in momentum occur when the velocity increases a little. The role of the speed c as *nature's speed limit* is most apparent. The value $v = c$ would correspond to infinite momentum.

With this definition of momentum, the form of Newton's second law which is correct at all speeds is, as we pointed out in Section 3.2: *the rate of change of the momentum vector $\vec{p}$ is equal to the resultant force vector $\vec{F}$.*

The total energy of a particle is related to its momentum by

$$E^2 = (mc^2)^2 + (cp)^2. \tag{5.28}$$

For particles with no rest mass, like the photon and the neutrino, this simplifies to

$$E = cp. \tag{4.23}$$

For these "massless" particles the expression $p = mv/\sqrt{1 - v^2/c^2}$ gives $p = 0$ unless $v = c$, but then $mv/\sqrt{1 - v^2/c^2}$ cannot be used since we have a zero in the denominator. The theory is consistent only if zero rest mass particles do carry momentum $p = E/c$ and always travel at nature's speed limit c in any intertial frame.

In Chapter 4 we examined the vital role of the rest energy mc^2 expressing the total energy by

$$E = mc^2 + K. \tag{4.21}$$

Another useful expression for this total energy E of a particle with rest mass m traveling at a speed v is

$$E = \frac{mc^2}{\sqrt{1 - v^2/c^2}}. \tag{5.29}$$

Notice again that a particle with rest mass cannot travel at the speed c. This would, according to Eq. (5.29), imply an infinite energy.

Using Eq. (5.29), the kinetic energy defined by Eq. (4.21) can now be expressed in terms of the speed. First we rearrange Eq. (4.21)

$$K = E - mc^2;$$

then,

$$K = \frac{mc^2}{\sqrt{1 - v^2/c^2}} - mc^2. \tag{5.30}$$

This expression for K can be shown to be very well approximated by $\frac{1}{2}(mv^2)$ for speeds much less than c. On the other hand, for high speeds, K given by Eq. (5.30) is much greater than $\frac{1}{2}(mv^2)$ and it is this new formula, Eq. (5.30), that fits the experimental facts. Let us now use some of these newly acquired equations in an example.

Example 6 For an electron (rest energy 0.51 MeV) with speed of $0.98c$, find the total energy (E), the kinetic energy (K), and the momentum (p).

Solution From Table 5-1 we find that for $v = 0.98c$, $1/\sqrt{1 - v^2/c^2} = 5.0$. Then Eq. (5.29) gives

$$E = \frac{mc^2}{\sqrt{1 - v^2/c^2}} = (5.0 \times 0.51) \text{ MeV} = 2.55 \text{ MeV}.$$

Since
$$E = mc^2 + K \tag{4.21}$$

we have
$$K = E - mc^2 = (2.55 - 0.51) \text{ MeV} = 2.04 \text{ MeV}.$$

The momentum p is found from
$$E^2 = (mc^2)^2 + (cp)^2, \tag{5.28}$$

which yields
$$cp = \sqrt{E^2 - (mc^2)^2} = \left(\sqrt{(2.55)^2 - (0.51)^2}\right) \text{ MeV}$$
$$= 2.50 \text{ MeV}.$$

We say $p = 2.50$ MeV/c.

5.7 REMARKS ON SPECIAL AND GENERAL RELATIVITY

An Assessment of Special Relativity

The relativistic expressions for momentum, energy, and time dilation, are among our best tested formulas and can now be presented as engineering facts instead of interesting speculation. In fact, hundred million dollar particle accelerators are confidently designed on these assumptions and, fortunately, these accelerators work.

   Aside from its practical value for nuclear and particle physics, special relativity teaches us all some important lessons about science. A theory like Newtonian mechanics can work wonderfully in a certain domain of experience (low speeds) and yet fail utterly in other circumstances (speeds near c). When forced to revise our theories, experimental results are the guide and also the touchstone of truth. However, one can sometimes very fruitfully generalize and postulate on the basis of flimsy experimental evidence, and so construct tentative theories to be tested.

   In this process, and generally in our thinking about nature, our common sense cannot always be trusted! Our common sense itself derives from a very limited range of experience. When we explore new realms there may be real surprises. It has turned out that at high speeds things are *not* more or less as you might expect, that is, just faster. And in studying quantum physics you will discover that very small objects display a most astonishing behavior; an electron is not merely a small version of a charged billiard ball.

General Relativity

The word "special" in Einstein's special theory of relativity refers to the fact that the theory is limited to consideration of *inertial frames of reference*. Einstein spent most of the last 45 years of his life working on generalizations considering transformations between any kinds of frames of reference. Some of the fruits of these efforts comprise what he called the general theory of relativity.

General relativity is a theory of gravitation. Einstein's starting point was a fact that we noticed in discussing motion in gravitational fields: Different freely falling bodies experience exactly the same acceleration at a given point in space. We obtained this result because the gravitational force $m\vec{g}$ is strictly proportional to the mass, so that when we write down Newton's second law, $m\vec{g} = m\vec{a}$, the mass m cancels out, leaving $\vec{a} = \vec{g}$, regardless of the value of m. One often describes this situation by saying that the "gravitational mass," the quantity m that determines the force $\vec{F} = m\vec{g}$ in a gravitational field, is the same as the "inertial mass" m that appears in Newton's second law, $\vec{F} = m\vec{a}$.

Einstein compared gravitational forces with the frame forces, $\vec{F}_{\text{frame}} = -m\vec{a}_{\text{frame}}$, that arise simply because the frame of reference is accelerated relative to an inertial frame. These frame forces are also strictly proportional to m, and all free bodies in such a frame possess the same acceleration, $-\vec{a}_{\text{frame}}$. Einstein was thus led to interpret gravitational forces as frame forces and assert that the existence of a gravitational field at a point in space is equivalent to an acceleration of the coordinate frame at that point. This led him eventually to a theory in which gravitational effects were interpreted in terms of the geometry of the universe, a universe with a geometry more complicated than ordinary Euclidean geometry.

Einstein has by this "geometrical" means sought to explain gravity, to understand what lies behind the very successful description of motion in a gravitational field that Newton achieved merely by postulating a gravitational force

$$F = G\frac{Mm}{r^2}. \tag{2.14}$$

In situations we can presently observe and analyze with confidence, the predictions of Einstein's theory of gravitation almost coincide with those of Newton's, and this makes it very difficult to find compelling experimental evidence in favor of general relativity. Three of Einstein's famous predictions, a correction to the orbital motion of the planet Mercury, the bending of a light beam near the sun, and a shift to the red in the spectra of light coming from atoms in a strong gravitational field, have indeed been verified. But these effects are minute and difficult to measure with a precision sufficient to decide definitely between general relativity and competing explanations. At the moment, special relativity is a necessary part of a physicist's workaday toolkit, but the same cannot yet be said of general relativity.

QUESTIONS

1. (a) A river flows east at a speed of 2.0 m/sec. A man rows a boat east at a speed of 3.0 m/sec relative to the water. Find the velocity vector of the boat relative to the land. (Consider a frame of reference $O'X'$ moving with the water in the river at a speed $V = 2.0$ m/sec relative to a frame OX fixed on the land.)
   (b) In the above situation, the man turns the boat around and rows west at the same speed relative to the water. Find the velocity vector of the boat relative to the land.

2. The stairs of an escalator extend 15.0 m up an incline. The escalator moves upward at a speed of 1.5 m/sec. A man walks with a speed relative to the escalator of 2.0 m/sec. Find how long the man will be on the escalator when
   (a) he walks up the escalator,
   (b) he walks down the escalator.

3. An airplane flies east with an "airspeed" of 100 m/sec (airspeed is speed relative to a frame $O'X'$ floating along with the air), and the "wind velocity" is 20 m/sec east (the wind velocity is the velocity of the frame $O'X'$ relative to earth). Find the "ground speed" of the aircraft (ground speed is the velocity relative to earth).

4. What is the "point of no return" for an aircraft with a flying time of 3.0 hours, when the aircraft flies east with an airspeed of 100 m/sec and the wind velocity is 20 m/sec, east?

5. A boat moves at a constant speed relative to the water in a river which flows at a constant velocity relative to the land (a constant current). On a trip between two towns which are 60 miles apart, the boat takes 4.0 hours to travel upstream and 3.0 hours to travel downstream. Find the current and the speed of the boat relative to the water.

6. A bus travels north at a speed of 20 m/sec. Inside the bus a ball is rolled straight across the bus from the west side to the east side, traveling 3.0 m in 0.30 sec. Find the magnitude and direction of the ball's velocity vector relative to the bus and also relative to the ground. [*Hint:* In this question, as well as in the following three questions, perform the addition of velocity vectors indicated in Eq. (5.8) by adding corresponding *components*, as outlined in Section 3.5.]

7. An aircraft heading 30° north of east has an airspeed of 90 m/sec. The wind velocity is 30 m/sec, east. Find the magnitude and direction of the aircraft velocity relative to the ground.

8. A river which is 60 m wide flows east at 2.0 m/sec. A motor boat starts from the south bank traveling at a speed of 4.0 m/sec relative to the river.
   (a) If the boat is heading 30° north of east, find the magnitude and direction of its velocity relative to the land. Where and when will the boat reach the opposite bank?
   (b) In what direction should the boat be headed in order to cross in the smallest possible time? What is this time?
   (c) If the boat reaches a point on the opposite bank due north of its starting point, in what direction was the boat heading as it crossed the river?

9. A car travels at 30 m/sec in a direction 36.9° north of east. Find the magnitude and direction of this car's velocity vector relative to a bus traveling north at 20 m/sec.

10. A bus, heading east, speeds up with an acceleration of 2.0 m/sec². A baseball is pushed forward in the bus with an acceleration of 40 m/sec² relative to the bus. What is the acceleration of the baseball relative to the ground?

11. The bus driver of Question 10 stops accelerating and travels with a constant velocity of 30 m/sec. The baseball is again given an acceleration relative to the bus of 40 m/sec². What is the acceleration of the baseball relative to the ground?

12. While the bus is traveling at a constant velocity of 30 m/sec, an ice cube slides along the bus floor

in a straight line with constant speed relative to the bus. Will the motion of the ice cube, as viewed from the highway, be motion in a straight line with constant speed? Will this speed be different from the speed relative to the bus? What is the acceleration of the ice cube relative to the bus? The highway?

13. What is the law of inertia?

14. What is an inertial frame of reference?

15. On a certain trip, a car
    (a) speeds up,
    (b) travels with constant speed in a straight line,
    (c) goes around a turn,
    (d) jams on the brakes.
    For each portion of the trip, decide whether or not the car is an inertial frame of reference, assuming that the highway is an inertial frame.

16. State the principle of relativity for Newtonian mechanics.

17. What is a frame force?

18. Refer to Fig. 2-30 and describe what occurs in the accelerated frame of reference provided by the car, making reference to a frame force.

19. During a car crash, the car slows down at the rate of 490 m/sec^2. Using the car as a frame of reference, what is the magnitude and direction of the frame force acting on a 100-kg driver? What is the ratio of this force to his weight?

20. In the preceding question, if you use the highway as a frame of reference, what is the frame force acting on the driver? From the viewpoint of this highway frame, account for the fact that the driver smashes into the steering wheel.

21. A 70.0-kg man is standing on scales in an elevator that has an upward acceleration of 0.80 m/sec^2. Using the elevator as the frame of reference, determine the reading of the scales (in newtons).

22. A block rests on a horizontal board. The coefficient of static friction between the block and the board is 0.40. The board is given a horizontal acceleration which is increased until the block starts to slip along the board. What is the value of the board's acceleration when the block starts to slip?

23. A small test tube is whirled in a centrifuge in a circle of radius r with a constant speed v.
    (a) Taking a frame of reference fixed in the laboratory, draw a force diagram showing the forces acting on the test tube. What is its acceleration relative to this frame?
    (b) Taking a rotating frame of reference fixed in the centrifuge, draw the force diagram showing the forces acting on the test tube. What is the acceleration of the test tube in this frame?

24. What is the speed of an object in an ultracentrifuge which is spun in a circle of radius 2.0×10^{-2} m, with an acceleration of a billion g's?

25. If the earth were not rotating, would a freely falling object have a greater acceleration relative to the earth at the equator? Explain.

26. A mass is suspended by a string from the roof of an automobile which is traveling north at 20 m/sec. The mass remains at rest when the string makes an angle of 15° with the vertical.
    (a) What is the magnitude of the acceleration of the automobile?
    (b) If the mass is deflected toward the north, what is the direction of the automobile's acceleration?
    (c) If the mass is deflected toward the west, what is the direction of the acceleration? From the magnitude of this acceleration what can be said about the path of the automobile?

27. Determine the correct angle θ of banking of a highway curve of radius r for a specified speed v. Analyze the problem in a frame of reference with origin at the center of the curve and rotating so that the car is at rest relative to this frame. If the angle of banking is correct, no frictional force is necessary to obtain equilibrium.

28. When an orbiting astronaut drifts about in the capsule, is his weight $m\vec{g}$ zero? Explain the fact that his acceleration relative to the capsule is zero.

29. An atom, moving east at a speed relative to the laboratory of 2.0×10^4 m/sec, emits a photon. The photon moves east with a speed c relative to the atom. The nucleus of the atom undergoes β-decay and emits an electron which goes east with a speed of 4.0×10^5 m/sec relative to the atom. From the facts presented at the beginning of Section 5.3, compute the speed of the photon and of the electron relative to the laboratory. Assume the atom is so massive that it maintains a practically constant velocity during the emission of the photon and the electron.

30. Give an example of two events which are simultaneous in one frame of reference but not simultaneous in another frame. What formerly trusted equation is contradicted by this example?

31. Show that if a pulse of light bouncing back and forth from the roof to the floor of a bus is used as the timing element of a clock, with the clock ticking each time the light is at the roof, the time interval $\Delta t'$ between two ticks in the bus is related to the time interval Δt between these same two events in the highway frame by

$$\Delta t = \frac{\Delta t'}{\sqrt{1 - V^2/c^2}}$$

where V is the speed of the bus relative to the highway. Does this clock run slow as far as highway time is concerned?

32. Two rocket ships are approaching head on. The speed of one relative to the other is $0.98c$. A pulse of light travels from the roof to the floor of one ship, a vertical distance of 3.0 meters as determined in this ship.
    (a) What is the time interval between emission of the light from the roof and its arrival at the floor, as determined in the ship containing the light pulse?
    (b) What is the time interval between the same two events, determined in the other ship? [*Hint:* Use Eq. (5.16) and the data of Table 5-1.]

33. Two rocket ships, traveling in the same direction, have a relative speed of $0.98c$. A light pulse travels **from tail to nose of the leading ship, a distance ($\Delta x'$) of 20 m as measured in this ship (frame $O'X'$).**
    (a) What is the time interval ($\Delta t'$) between the emission and arrival of this light pulse, as determined in the ship containing the pulse (frame $O'X'$)?
    (b) Use the Lorentz transformations, Eqs. (5.17) and (5.18), to determine the distance Δx and time interval Δt between these same two events, as determined by observers in the other rocket ship (frame OX).
    (c) Verify from your calculation of Δx and Δt in part (b) that $\Delta x/\Delta t = 3.0 \times 10^8$ m/sec; that is, that light has the same speed relative to each rocket.

34. Imagine a train of rest length 6.0×10^3 m traveling relative to the ground at a speed of $0.98c$. As the train speeds through a station, the stationmaster's clock and the train conductor's watch both read zero. The conductor is at the midpoint of the train. The engineer observes lightning striking the locomotive and the adjacent ground (event A) when his watch reads 4.0μ sec (1μ sec $= 10^{-6}$ sec). A second bolt of lightning is observed by the brakeman in the caboose to strike the caboose and the adjacent ground when his watch reads 4.0μ sec (event B).
    (a) The conductor should *report* these events as having occurred at what times on his watch?
    (b) When the conductor sees these events, what time interval (on his watch) will have elapsed since the occurrence of these events?
    (c) Find the time interval between these events as recorded in the system of the stationmaster.
    (d) Find the distance the stationmaster will measure along the ground between the two marks made by the lightning strokes.

35. Particle counters are placed at the top and bottom of a tower 27 m high. A particle moving with a velocity of $0.9c$ relative to the counters triggers the counter at the top (event A) and later the counter at the bottom (event B).
    (a) What is the distance and the time interval between these two events in the laboratory frame of reference?
    (b) What is the distance and the time interval between these two events in a frame of reference moving upward at a speed of $0.98c$?
    (c) What is the velocity of the particle relative to the second frame?
    (d) What is the distance between these two events in a frame of reference moving with the particle?

36. Pions are produced when a synchrotron beam of high energy protons strikes a target. Consider a pion which decays after a lifetime of 3.0×10^{-8} sec in a frame in which the pion is at rest. If this pion emerged from the target at a speed of $0.98c$ how far would it travel in the laboratory before decaying? What distance would be expected by someone who believes that the lifetime must be the same in all frames of reference?

37. Show that bus passengers will find that highway clocks run slow by considering two events at the location of a highway clock ($\Delta x = 0$ for these events) and using the appropriate Lorentz transformation equation.

38. A laboratory measurement of the coordinates of the ends of a moving meter stick, taken at the same time in the laboratory, yields the result that the "meter stick" is 0.20 m long. Evaluate the factor γ for this motion and use Table 5-1 to find the corresponding velocity of the stick.

39. Show that measurements of the length L of a moving bus made in a highway frame (OX) will reveal that that length L is less than the length L_0 of the bus as measured in the bus frame $(O'X')$.

40. Verify that the relativistic interval between the two events A and B of Fig. 5-7 is invariant. That is, show that

$$(\Delta x)^2 - c^2(\Delta t)^2 = (\Delta x')^2 - c^2(\Delta t')^2$$

is satisfied for these events.

41. Evaluate the relativistic interval between the events A and B of Question 34 for both a frame fixed in the train and a frame fixed in the ground.

42. Measurements in one frame of reference show that two events occur simultaneously, 4.0 m apart. What is the time interval between these same two events in another frame of reference in which these events are 5.0 m apart?

43. Use the fact that the relativistic interval between two events is invariant to deduce the time dilation formula

$$\Delta t = \frac{\Delta t'}{\sqrt{1 - \frac{V^2}{c^2}}}$$

where $\Delta t'$ is a time interval on a clock moving at a speed V.

44. For the electron in Question 29, calculate its velocity relative to the laboratory using Einstein's law for the addition of velocities. Compare the answer with that obtained using the non-relativistic expression, $v_x = v_x' + V$.

45. Repeat the calculation of the preceding question for the case when the atom's velocity relative to the laboratory is $0.50c$, east, and the electron's velocity relative to the atom is $0.95c$, east.

46. An atomic nucleus, while traveling east at $0.60c$, emits an electron which has a speed of $0.80c$ relative to the nucleus and which travels west. Find the magnitude and direction of the electron's velocity vector relative to the laboratory.

47. A proton travels west at speed $0.80c$ and an electron travels east at speed $0.90c$, both speeds being measured relative to the laboratory. Find the speed of the electron relative to the proton.

48. Using the data in Table 5-1, show that a particle's momentum is ten times as large at a speed of $0.9999c$ as it is at a speed of $0.99c$.

49. An electron (rest energy 0.51 MeV) has a momentum of 1.0 MeV/c (that is $cp = 1.0$ MeV). Find
    (a) the total energy of the electron,
    (b) the kinetic energy of the electron.

50. What is the total energy of an electron whose speed is $0.9999c$?

51. The rest mass of a proton is 938 MeV. If a proton has a kinetic energy of 500 MeV, find
    (a) its total energy in MeV,
    (b) the magnitude of its momentum in MeV/c,
    (c) its speed in terms of c.

52. At what speed will a particle's kinetic energy be four times its rest energy?

53. A μ meson (muon) is created in the atmosphere and travels vertically downward and decays as it reaches the earth's surface. An observer on the earth measures the muon's velocity as $0.98c$, and the thickness of the atmosphere traversed as 3000 m [muon's rest mass = 207 × (electron's rest mass)]. For this observer, find
    (a) the total energy,
    (b) the kinetic energy,
    (c) the momentum,
    (d) the lifetime of this muon.

54. For an observer traveling with the muon of the preceding question, find
    (a) the total energy,
    (b) the kinetic energy,
    (c) the momentum,
    (d) the lifetime,
    (e) the thickness of the atmosphere traversed.

55. A pulse of light from a large laser has an energy of 2.0×10^3 joules.
    (a) What is the momentum of this pulse?
    (b) If this light were absorbed in a 4.0×10^{-3}-kg marble, what speed would the marble acquire if it were initially at rest?

56. A photon with an energy of 0.485 MeV collides with a stationary nucleus of rest mass 11.65×10^{-27} kg and is absorbed by the nucleus.
    (a) What is the momentum (in MeV/c) of the recoiling nucleus?
    (b) What is the kinetic energy of the recoiling nucleus? (Since the recoil speed is small compared to c, the expressions for momentum and kinetic energy from Newtonian mechanics can be used for this nucleus.)

57. In Example 6 of Chapter 4 we discussed the annihilation of an electron-positron pair. Considering such an event from a frame in which the total momentum vector of the system is zero, show that it is impossible for the process to result in the creation of just *one* photon.

58. (a) What does it mean to say that a quantity is invariant?
    (b) What does it mean to say that a quantity is conserved?
    (c) Give an example of a conserved quantity that is not invariant, and another example of an invariant quantity that is not conserved.

59. An atom of mass M emits a photon of energy E_{photon}. The atom recoils with a momentum MV and a kinetic energy $\frac{1}{2}MV^2$ (expressions of Newtonian mechanics can be used because the atom's speed is small compared to c). Show that the recoil kinetic energy of the atom is given by

$$K = E^2_{photon}/2Mc^2.$$

SUPPLEMENTARY QUESTIONS

S-1. (a) A train 200 m long takes 100 sec to pass another train 800 m long which is traveling in the same direction. What is the velocity of the shorter train relative to the longer?
    (b) When these two trains are traveling with the same speeds as in part (a) but in opposite directions, it takes 10.0 sec for them to pass. Find the speed of each train relative to the ground.

S-2. A man rowing upstream passes a floating beer bottle, and keeps on rowing upstream for 15 min. He then turns around and rows downstream, always maintaining the same speed relative to the water. When he overtakes the beer bottle he finds it has drifted 2 mi downstream. What is the current in the river? (Don't get involved in heavy algebra; this is a good cocktail party question.)

S-3. Rain is falling vertically. When a car is traveling at 10 m/sec, the tracks made by raindrops on the vertical side windows are inclined at an angle of 60° with the vertical. Find the speed of a raindrop relative to the car and also relative to the ground.

S-4. Two ships move toward each other on intersecting lines at speeds which will cause them to collide. Examine this situation from a frame of reference fixed on one of these ships. Explain how observers on this ship can tell that they are on a collision course by successive observations of the bearing of the other ship.

S-5. A ship travels north at 4.0 m/sec. The wind velocity is 3.0 m/sec, east. A pennant flies from the masthead. In what direction does this pennant point?

S-6. While a bus is traveling along a highway, a ball is thrown from one passenger to another. Find the ball's in-flight acceleration relative to the bus and also relative to the ground when:
    (a) the bus has a constant velocity;
    (b) the bus has an acceleration of 3.0 m/sec², east.

S-7. A block slides to the right, down a smooth plane inclined at an angle of 30° with the horizontal. Without changing the angle of inclination, this plane is accelerated horizontally to the right at 7.0 m/sec². Find the acceleration of the block relative to the plane and relative to the ground.

S-8. A marble is placed in a hemispherical bowl of radius **0.200 m. The bowl is placed upon a turntable and rotated at 75 revolutions per minute about its vertical axis. At what height above the bottom of the bowl can the marble remain stationary relative to the bowl?**

S-9. A simple pendulum has a period of 1.20 sec when swung on the ground. What would its period be when measured in an elevator with an upward acceleration of 2.00 m/sec²?

S-10. The hypothetical train in Fig. 5-6 has a rest length of 600 m and travels in the positive x-direction with a speed of $0.800c$ relative to the track. The light signal originated at O and O' when these points coincided. Verify that the lower portion of Fig. 5-6 correctly represents values of the position coordinates (of O, O' and the extremities of the train) for measurements made at the same instant in the frame OX.

S-11. In the preceding question assume that the origins coincide at $t = 0$ and at $t' = 0$. Then calculate the position coordinates and times of events A and B in $O'X'$ and also in OX.

S-12. Suppose we assume transformations of the form

$$\Delta x = \gamma_1 (\Delta x' + \alpha \Delta t'),$$
$$c \Delta t = \gamma_2 (c \Delta t' + \beta \Delta x'),$$

which are more general than Eqs. (5.17) and (5.18). Then show that the requirement that the point O move with a velocity $-V$ *relative to* $O'X'$ (and therefore that the two events located at O have $\Delta x = 0$ and $\Delta x' = -V \Delta t'$) leads to the conclusion that $\alpha = V$, in agreement with Eq. (5.17). Then show that the requirement that the point O' move with a velocity V relative to OX implies that $\gamma_1 = \gamma_2$.

S-13. An interesting alternative approach to the Lorentz transformations is to assume, instead of our Eqs. (5.17) and (5.18), that

$$\Delta x = \gamma (\Delta x' + V \Delta t'),$$

and also that

$$\Delta x' = \gamma (\Delta x - V \Delta t),$$

where γ is a coefficient independent of the values of the space and time intervals in these equations. Show that these assumptions, together with our analysis of Fig. 5-7, imply that

$$\gamma = \frac{1}{\sqrt{1 - \dfrac{V^2}{c^2}}}$$

and that

$$\Delta t = \gamma \left(\Delta t' + \frac{V \Delta x'}{c^2} \right).$$

S-14. (a) From Eqs. (5.17) and (5.18), which express Δx and Δt in terms of $\Delta x'$ and $\Delta t'$, deduce Eq. (5.23) and Eq. (5.24), which express $\Delta x'$ and $\Delta t'$ in terms of Δx and Δt.
(b) Show that

$$v_x' = \frac{v_x - V}{1 - v_x V/c^2}.$$

S-15. Prove that if v_x' and V are both less than c, then the Einstein law for addition of velocities yields v_x, which is also less than c. For simplicity, assume that all velocities are in the positive direction.

S-16. In a hypothetical train moving with a constant velocity of $0.80c$ relative to the straight tracks, a light is flashed in one of the cars just as the car passes an observer in the station. The light is flashed again after a time interval of 1.60 sec, as measured in a frame fixed in the train. The observer in the station starts his stopwatch when he sees the first flash and stops the watch when he *sees* the second flash. What is the final reading of his watch?

S-17. Show that if two events A and B occur at the same place in $O'X'$ (so $\Delta x' = x'_B - x'_A = 0$) with event A before event B (so $\Delta t' = t'_B - t'_A$ is positive), then in any other frame OX these events occur in the same temporal order, that is $\Delta t = t_B - t_A$ is positive, so event A occurs before B. (In this circumstance the statement that event A occurs before event B has absolute significance in the sense that it is true in every inertial frame of reference.)

S-18. Show that if two events A and B occur at the same time in $O'X'$ (So $\Delta t' = t'_B - t'_A = 0$) with event A to the left of event B (so $\Delta x' = x'_B - x'_A$ is positive), then
(a) in frame OX, event A occurs before event B if V is positive (i.e., if $O'X'$ moves to the right relative to OX).
(b) In frame OX, event A occurs after event B if V is negative (i.e., if $O'X'$ moves to the left relative to OX).
(For this pair of events one cannot make any statement about the temporal order of events that is true in all inertial frames of reference.)

S-19. From the Lorentz transformations, prove Eq. (5.26):

$$(\Delta x)^2 - c^2 (\Delta t)^2 = (\Delta x')^2 - c^2 (\Delta t')^2.$$

S-20. At what speed will a particle's momentum equal mc?

S-21. Show that for speeds near the speed of light the factor $\sqrt{1 - v^2/c^2}$ is approximately $\sqrt{2}\sqrt{1 - v/c}$. Use this approximation to evaluate the total energy of an electron at a speed which is $10^{-6} c$ less than the speed of light.

S-22. In what is called the *extreme* relativistic range, a particle's total energy E is much greater than its rest energy. Show that this implies that E is approximately equal to cp, and that the particle's speed is close to c.

S-23. At low speeds, the factor $1/\sqrt{1 - v^2/c^2}$ is approximately $1 + \frac{1}{2}(v^2/c^2)$.
(a) Justify this assertion using the binomial theorem.
(b) Using this approximation, evaluate the difference in readings between an earth-bound clock and a clock carried by an astronaut in an earth satellite which travels at 7.0×10^3 m/sec for 1.0×10^5 sec, as measured by earth-bound clocks. (Assume that a frame fixed in the earth is inertial, to a sufficiently good approximation for this calculation.)

S-24. What is the fractional error made when momentum is calculated using mv rather than $mv/\sqrt{1 - v^2/c^2}$ for a satellite at a speed of 7.0×10^3 m/sec?

S-25. While at rest, a particle of mass M decays into one particle of mass m and one particle of zero rest mass which has energy E'. Show that the laws of conservation of energy and momentum applied to this process (using the relativistically correct expressions) imply that
$$E' = \Delta Mc^2 \left(1 - \frac{1}{2}\frac{\Delta M}{M}\right),$$
where $\Delta M = M - m$.

S-26. Apply the result of the preceding question to the decay of a pion into a muon and a neutrino (Example 11 of Chapter 4) and evaluate the energy of the neutrino and the kinetic energy of the muon.

S-27. From the Lorentz transformation equation for a transverse coordinate, $\Delta y = \Delta y'$, deduce the transformation equation for a transverse component of velocity, v_y.

S-28. Show that a velocity vector with components $v_y' = c$, $v_x' = 0$ is transformed by the velocity transformation equations into a velocity vector $\vec{v}$ with a magnitude $\sqrt{v_x^2 + v_y^2}$ which is equal to c.

S-29. When an incident photon interacts with a single free electron, show that the electron cannot simply absorb all the energy of the photon.

AN EXAMPLE FROM THE MEDICAL LABORATORY

Centrifuge in the Medical Laboratory: Rotating Frames and Centrifugal Force

Common rotational speeds for centrifuges in medical laboratory work are from about 1000 to 3000 r.p.m. A common use is the determination of the *hematocrit*, that is, the volume of erythrocytes packed by centrifugation in a given volume (also known as a packed cell volume) of blood (45-48% by volume of erythrocytes is considered normal).

Question

Suppose a hematocrit is carried out at 3000 r.p.m., how many "g's" are developed at this speed? Assume the effective radius to be 0.20 m.

Answer

A particle traveling in a circle of 0.20 m radius travels a distance of $2\pi \times 0.20$ m each revolution. It makes 3000 revolutions in 60 seconds so its speed is given by.
$$v = (3000 \times 2\pi \times 0.20/60) \text{ m/sec}$$
$$= 62.8 \text{ m/sec}.$$
The particle's acceleration is
$$v^2/r = [(62.8)^2/0.20] \text{ m/sec}^2$$
$$= 19.7 \times 10^3 \text{ m/sec}^2.$$
The ratio of this acceleration to g is
$$(v^2/r)/g = (19.7 \times 10^3 \text{ m/sec}^2)/(9.8 \text{ m/sec}^2)$$
$$= 2.0 \times 10^3.$$
The acceleration of a particle is about two thousand g's.

SUGGESTED READING

BORN, MAX, *Einstein's Theory of Relativity*. New York: Dover, 1965. For the more ambitious student whose interest in relativity may have been aroused by our Chapter 5. This Dover book is a translation of the one Born published in 1920 and which was essentially based on a series of popular lectures he gave in Frankfurt am Main in the wake of the spread of Einstein's fame. Although the treatment is very thorough, the author has restricted himself to a minimum of mathematical tools, no more than we have used in our book. Born says in his preface that he feels his presentation should have an appeal "to those who, without knowing higher mathematics and modern physics, remember something of what they learned at school and are willing to do a little thinking." Max Born himself was awarded the Nobel prize in physics in 1954.

DURELL, CLEMENT V., *Readable Relativity*. New York: Harper & Brothers, 1960 reprinting. The quality of this little book is aptly described in the preface written in 1960 by Freeman J. Dyson: ". . . it is certainly the best layman's introduction to relativity that has ever been written by anybody. . . . Durell . . . writes in a simple and humorous style, with a lightness of touch which can come only to one who is absolute master of his subject."

EINSTEIN, ALBERT, *Relativity, the Special and the General Theory*. London: Methuen, 1957 (fifteenth edition). This translation from the German by R. W. Lawson is a revised, enlarged, and updated version of a small book Einstein first wrote in 1916. The purpose of the book was to provide, as he himself stated it, "an exact insight into the theory of Relativity for those readers who, from a general scientific and philosophical point of view, are interested in the theory, but who are not conversant with the mathematical apparatus of theoretical physics." The 58 pages of Part I of this book deal with the special theory of relativity, using mathematics no more complicated than that used in our book. Part II deals with the general theory of relativity and employs some calculus in the discussion. Part III, the last part of the book, deals with certain cosmological aspects. This book by the master himself is well worth consulting, even for Part I alone, since after all there is no substitute for "original" originality.

FRENCH, A. P., *Special Relativity*. New York: W. W. Norton & Company Inc., 1968. This text is addressed to advanced freshman science students. It is particularly valuable for its many discussions of modern experiments.

FRENCH, A. P., *Newtonian Mechanics*. New York: W. W. Norton & Company Inc., 1971. Chapter 12, entitled "Inertial forces and non-inertial frames" is probably the most suitable supplementary reading for our readers on the subject of frames of reference in Newtonian mechanics.

GAMOW, G., *Mr. Tompkins in Wonderland*. Cambridge, England: Cambridge University Press, 1964. An entertaining and stimulating account of Mr. Tompkin's adventures in a wonderland where relativistic and quantum effects are apparent in the everyday life of its citizens.

KITTEL, C., W. D. KNIGHT, and M. A. RUDERMAN, *Mechanics—Berkeley Physics Course*, Vol. 1, New York: McGraw-Hill, 1965. Inertial frames of reference and fictitious or pseudo forces (called frame forces in our book) are discussed in Chapter 3. Chapters 10 to 14 are concerned with relativity.

Archimedes

6 Fluid Mechanics

We now turn abruptly from the realm of extremely high speeds to the investigation of fluids at rest or in steady flow. Our study of these topics will introduce principles and concepts necessary for the comprehension of many practical devices. In fluid statics pressure is defined and then used to describe a variety of phenomena. Pressure will be found to be of considerable importance in the chapters on heat and thermal processes.

Pioneering contributions in fluid mechanics were made by Archimedes of Syracuse, who is renowned as perhaps the greatest of the Greek mathematicians and scientists. He hit upon what we call *Archimedes' principle,* a neat prescription for the buoyant force exerted by a fluid on any immersed object. In this connection he introduced the notion of density and developed procedures for its measurement. After demonstrating the utility of the idea of a *center of gravity*, he originated mathematical methods to locate such a point for objects of various shapes. One of his most famous discoveries is the *law of the lever.* This achievement is widely reported to have brought forth the exclamation quoted on the title page of this chapter. He made many first-rate mathematical contributions, introducing ideas and methods that are at the heart of the integral and differential calculus that was to be developed some nineteen hundred years later by Newton and Leibnitz. Archimedes had the rare gift of being able to create both mathematics and science, using each of these as a source of discovery in the other.

During the Second Punic War when Syracuse was besieged by a Roman army under Marcellus, Archimedes' super-catapults and other ingenious devices were instrumental in repelling enemy attacks. Eventually the city fell. Archimedes was slain, reportedly while he was engaged in the study of the mathematical figures that he habitually drew in the dust. The Greek world is weighed against the Roman Empire in Whitehead's observation, "No Roman lost his life because he was absorbed in the contemplation of a mathematical diagram." [A. N. Whitehead (1861–1947), English mathematician and philospher.]

In fluid mechanics, in addition to the names Archimedes and Newton, one encounters others such as Pascal and Bernoulli. Blaise Pascal (1623–1662) might be described as an intellectual whose particular interests were mathematics, physics and religion. Before the age

"Give me a fulcrum on which to rest, and I will move the earth."

Archimedes of Syracuse
(287–212 B.C.)

of 16 Pascal had managed to prove one of the most important theorems in a branch of mathematics known as *projective geometry*. Of his important contributions to physics probably the most significant is what is known as *Pascal's principle* which is involved in the operation of devices, such as the hydraulic press, that use fluids to link the motions of different rigid bodies.

Daniel Bernoulli (1700–1782) was a member of the renowned Bernoulli family which numbered many outstanding mathematicians. After obtaining his doctorate in medicine (with a thesis on the action of the lungs) this Swiss genius started his academic career as Professor of Mathematics at the University in St. Petersburg. A few years later he returned to Switzerland to the university in Basel. In 1733 he published his major work, *Hydrodynamica*, in which he used Newton's laws of motion to develop the subjects of hydrostatics and hydrodynamics.

6.1 FLUID STATICS

Pressure

A fluid is a substance that can flow. The term therefore includes both *liquids* and *gases*. In fluid statics we assume that the fluid and any other relevant objects, such as containers, are at rest. Then the fluid in contact with any surface exerts a force which is a *push perpendicular to the surface*. If we measure the force exerted by the fluid on a small plane surface, we find that this force has the same magnitude no matter what the orientation of the surface, provided the small surface is kept at the same height in the fluid. The ratio of the magnitude of the force exerted by the fluid on a small surface to the area of the surface is called the *pressure* in the fluid. We thus associate a number called the pressure with each point in the fluid. This number, as all divers know, increases with increasing depth.

If the pressure P has the same value at all points on a surface of area A, then the fluid exerts a force F (see Fig. 6-1), which is a push perpendicular to the surface with a magnitude

$$F = PA. \qquad (6.1)$$

The surface A may be the surface of a solid within the fluid or it may be any mathematical surface we care to imagine within the fluid. When the pressure varies over a surface, as it does on the vertical surface of a dam, one should apply Eq. (6.1) only to an area A small enough so that the pressure changes are relatively small over the surface.

The unit in which pressure is expressed is evidently a force unit divided by an area unit. In the mks system, this unit is the N/m^2.*

The weather bureau reports atmospheric pressure in millibars. The pressure of the atmosphere at the earth's surface is approximately 1000 millibars or 1 bar. This pressure unit is defined in terms of the N/m^2 by 1 bar = 10^5 N/m^2. Another common unit of pressure is the *standard atmosphere* (1 atm), which is almost the same size as a bar:

$$1 \text{ atm} = 1.01 \times 10^5 \text{ N/m}^2 = 1.01 \text{ bar}.$$

In everyday life one also encounters pressure expressed in the *Engineering System of Units*, that is, in pounds-force per square inch. It can be shown that

$$1000 \text{ millibars} = 10^5 \text{ N/m}^2 = 14.5 \text{ lb-force/in.}^2,$$

so that

$$1 \text{ atm} = 14.7 \text{ lb-force/in.}^2$$

When people speak of tires with 30 lb pressure, what is really meant is that the pressure is 30 lb-force/in². Moreover, this pressure is the so-

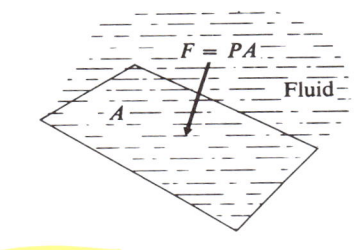

Fig. 6-1 The force exerted by a fluid on a surface is a push of magnitude $F = PA$, where P is the pressure in the fluid at the surface of area A.

*In the SI system this unit of pressure is named the pascal: 1 pascal = 1 N/m^2.

called gauge pressure, and thus is really 30 lb-force/in.² above the prevailing atmospheric pressure.

Example 1 A horizontal glass pane with an area of 0.20 m² is exposed to air pressure of 1.00×10^5 N/m² on its lower surface. The upper surface is exposed to a vacuum. What is the force exerted by the air on the glass?

Solution The air exerts an upward force on the lower surface given by

$$F = PA \qquad (6.1)$$
$$= (1.00 \times 10^5 \text{ N/m}^2)(0.20 \text{ m}^2)$$
$$= 2.0 \times 10^4 \text{ N}.$$

This is a force of about 4.5×10^3 lb! Fortunately, we usually have about the same air pressure on *both sides* of a pane of glass and therefore the *resultant* force exerted by the air (when no wind is blowing) on a window pane is more or less zero.

Density

The *density* (ρ) of any homogeneous substance, be it a solid or a fluid, is defined as its mass (M) divided by its volume (V); that is,

$$\rho = M/V. \qquad (6.2)$$

The mks unit of density is the kg/(meter)³. Water has a density of 1.00×10^3 kg/m³; the density of gold is almost twenty times greater; while air at 32°F at a pressure of 1.00 atm has a density of only 1.29 kg/m³, which is slightly more than 1/1000 the density of water. Mercury, with a density of 13.6×10^3 kg/m³, has the greatest density of any substance that is a *liquid* at room temperature and normal pressures.

The density of gases changes markedly when the temperature and pressure are changed, as will be discussed in Chapter 7. The densities of solids and liquids change relatively little for modest temperature and pressure changes, so it is usually an adequate approximation in fluid statics to consider such densities as constant. Typical densities of gases, liquids, and solids are given in Table 6-1.

Table 6-1 DENSITIES (KG/METER³) AT 0°C AND 1 ATM PRESSURE

Air	1.29	Wood, white pine	0.42×10^3
Hydrogen	0.0899	Aluminum	2.70×10^3
Helium	0.179	Iron	7.8×10^3
Oxygen	1.43	Silver	10.5×10^3
Water	1.00×10^3	Lead	11.3×10^3
Olive oil	0.92×10^3	Gold	19.3×10^3
Mercury	13.6×10^3	Platinum	21.4×10^3

Usually we have a situation in which we know the density ρ and we wish to compute the mass M of a volume V of the substance. This is accomplished by solving $\rho = M/V$ to give

$$M = \rho V.$$

Thus the mass of 5.0×10^{-2} m³ of water is

$$(1.00 \times 10^3 \text{ kg/m}^3)(5.0 \times 10^{-2} \text{ m}^3) = 50 \text{ kg}.$$

Change of Pressure with Depth

The first condition of equilibrium enables us to understand the pressures encountered at different points in a fluid at rest. In fluid statics every portion of the fluid is in equilibrium. We select as the system to which we apply the equilibrium condition, the portion of the fluid within an imaginary cylindrical surface of height Δy and cross-sectional area A (Fig. 6-2). The free body diagram is shown in Fig. 6-3. The system has a volume equal to $A\Delta y$. From $M = \rho V$ its mass is given by $\rho A \Delta y$. Therefore its weight, Mg, is equal to $(\rho A \Delta y)g$. The fluid outside this cylindrical region exerts

1. Horizontal forces on the vertical sides of the cylinder;
2. A downward force, $F = PA$, on the upper surface. (P denotes the pressure on the upper surface);
3. An upward force $(P + \Delta P)A$ on the lower surface, where the pressure on the lower surface is denoted by $P + \Delta P$.

For equilibrium we require

Sum of forces acting upward = sum of forces acting downward. For this system, this means

$$(P + \Delta P)A = PA + (\rho A \Delta y)g.$$

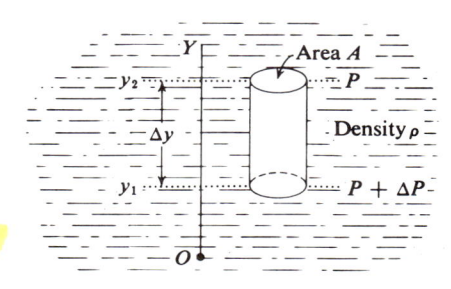

Fig. 6-2 A system in equilibrium is the fluid within a cylinder of cross-sectional area A and height Δy.

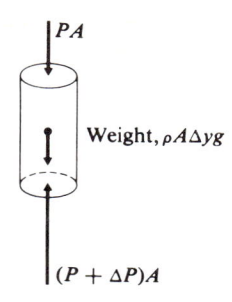

Fig. 6-3 Free body diagram for system of Fig. 6-2. Only the vertical forces are shown. Equilibrium requires $\Delta P = \rho g \Delta y$.

6.1 FLUID STATICS

Dividing both sides of this equation by A, and subtracting the term P from each side, we obtain

$$\Delta P = \rho g \Delta y.$$

This gives the pressure increase ΔP associated with an increase in depth Δy in a fluid of density ρ.

Thus in descending a distance of 10.0 meters through water of density 1.00×10^3 kg/m^3, the increase in pressure is

$$(1.00 \times 10^3 \text{ kg/m}^3)(9.8 \text{ N/kg})(10.0 \text{ m}) = 0.98 \times 10^5 \text{ N/m}^2,$$

which is about 1 atmosphere. If the pressure at the surface of a lake is 1 atmosphere, the pressure at a depth of 10 meters is thus approximately 1 atmosphere greater, namely 2 atmospheres.

By way of contrast, the pressure change associated with a 10-meter change in altitude in *air* of density 1.29 kg/m^3 is merely

$$(1.29 \text{ kg/m}^3)(9.8 \text{ N/kg})(10.0 \text{ m}) = 126 \text{ N/m}^2,$$

which is only about 0.1% of typical atmospheric pressure. This result illustrates the fact that for *gases* in laboratory containers the difference in pressure at two points is usually negligible and the pressure can be taken to be the same throughout the container. However, if we are concerned with an increase in altitude of a few kilometers or more, there is an appreciable drop in the pressure of the air. (But because the air density changes when the pressure changes, the equation $\Delta P = \rho g \Delta y$ can be applied only to small changes of altitude.)

We can extend the proof of the equation $\Delta P = \rho g \Delta y$, which was concerned only with the pressures at points like A and B on the same vertical line (Fig. 6-4), to apply to a much more general case. First, one can show that equilibrium of horizontal forces acting on a horizontal cylindrical portion of the fluid requires that the pressures at points like B and C on the same horizontal line are the same. It then follows that the above equation applies to points A and C since it applies to A and B, and $P_B = P_C$.

In this way we obtain the desired generalization, *the fundamental law of hydrostatics*. The equation

$$\Delta P = \rho g \Delta y \qquad (6.3)$$

gives the pressure difference between two points (like A and D in Fig. 6-4) with a vertical separation Δy, regardless of their horizontal separation, provided that the two points can be connected by a path lying in the fluid of constant density. The student should carefully note that this law applies for static fluids. In general, it is not true if the fluid inside the container is moving.

According to this law, the shape of the containing vessel does not affect the pressure and the pressure is the same at all points at the same

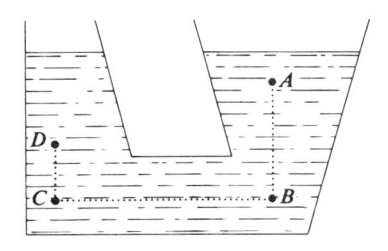

Fig. 6-4 The fundamental law of hydrostatics is the statement that the pressure change between two points like A and D is equal to $\rho g \times$ (vertical distance between A and D).

horizontal level in a connected static fluid. This is in accord with the readily observable fact that the levels of the liquid in the left- and right-hand portions of a container such as that of Fig. 6-4 are the same.*

The blood in the human circulatory system is a connected fluid of considerable interest. Blood pressure in parts of the body at different heights will be different because of the hydrostatic* change of pressure with height implied by

$$\Delta P = \rho g \Delta y. \quad (6.3)$$

Physicians measuring blood pressure usually select a point on the arm at the level of the heart which serves as a standard reference level. A proper distribution of blood throughout the body is achieved by vascular control. When this control is lost by shock or other causes, the body must be placed in a horizontal position to avoid large imbalances.

*We are assuming that the adhesive forces exerted by the container on the fluid are negligible.

*The pumping action of the heart and the motion of the blood also influence the blood pressure.

Pressure Gauges

The *open-tube manometer* shown in Fig. 6-5 is used to determine the pressure P of a gas in the tank. The liquid in the manometer has a pressure P at the liquid surface exposed to the gas in the tank, while at the other surface the pressure is the atmospheric pressure P_a. With the vertical separation of these two liquid surfaces denoted by Δy, the fundamental law of hydrostatics, Eq. (6.3), with $\Delta P = P - P_a$ gives

$$P - P_a = \rho g \Delta y. \quad (6.4)$$

As mentioned earlier, the pressure P is called the *absolute pressure* to distinguish it from the difference $P - P_a$ which is named the *gauge pressure*. Knowledge of ρ and a measurement of the difference in height Δy of the liquid columns evidently allows calculation of the gauge pressure.

Blood pressure is often measured by means of a mercury *sphygmomanometer* (see Fig. 6-6). This is a modification of the open-tube manometer of Fig. 6-5. A cistern has been substituted for one half of the U-tube. The advantage of this device is that a pressure reading is given by noting the level of only the upper mercury surface. The cistern has a cross section so large that the cistern mercury level changes can generally be neglected.

The *mercury barometer* (Fig. 6-7) is a glass tube, some 80 centimeters long, which has been filled with mercury and then inverted in a dish of mercury. The space above the mercury column contains only mercury vapor whose pressure at room temperature is so small that it can be neglected. This instrument was invented in 1643 by Galileo's pupil, Evangelista Torricelli (1608–1647). His barometer showed that the air of our atmosphere exerts a pressure and also provided a means for measuring changes in this pressure.

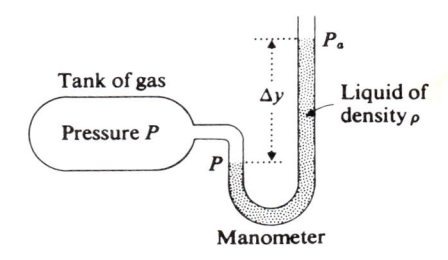

Fig. 6-5 The manometer. The difference between the tank pressure P and the atmospheric pressure P_a is $\rho g \Delta y$.

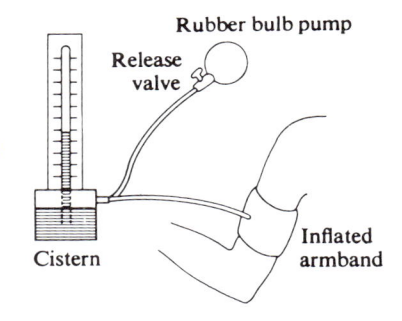

Fig. 6-6 Mercury sphygmomanometer used to measure blood pressure.

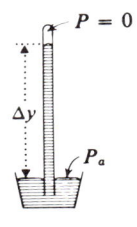

Fig. 6-7 The mercury barometer.

6.1 FLUID STATICS 249

The "Torricellian vacuum" above the mercury column of this barometer created a sensation because it violated the Aristotelian tenet that "nature abhors a vacuum." With this sterile idea out of the way, the development of vacuum pumps proceeded and knowledge of gases consequently increased rapidly, as we shall see in Chapter 7. The first practical vacuum pump, invented by Otto von Guericke in Germany around 1650, was used to stage the Magdeburg hemisphere demonstration (Fig. 6-8), a masterpiece of scientific showmanship.

The theory underlying Torricelli's barometer is provided by the fundamental law of hydrostatics. If we denote by Δy the height of the mercury column above the mercury level in the dish, Eq. (6.3) (with $\Delta P = P_a - 0$) gives the pressure of the atmosphere at the mercury surface in the dish:

$$P_a = \rho g \Delta y. \tag{6.5}$$

Laboratory workers using mercury barometers and manometers find it convenient to specify pressures directly in terms of the height Δy of the column of mercury. The pressure difference in a vertical mercury column which is exactly 76 centimeters long under specified conditions (density of mercury being 13.595×10^3 kg/m³, and the gravitational field g being 9.80665 N/kg) is called *one standard atmosphere* (1 atm). With the above density of mercury Eq. (6.5) gives

$$1 \text{ atm} = (13.595 \times 10^3 \text{ kg/m}^3)(9.80665 \text{ N/kg})(0.7600 \text{ m})$$
$$= 1.013 \times 10^5 \text{ N/m}^2.$$

The various equivalent expressions for normal atmospheric pressure are listed below:

1 atm; 76 cm of mercury*; 14.7 lb-force/in.²; 1.01×10^5 N/m².

Example 2 On a day when the barometer reads 77 cm of Hg, a mercury manometer (Fig. 6-5) connected to a tank of compressed air has the top of the mercury column which is exposed to the tank at a level 33 centimeters below the top of the column which is open to the atmosphere. What is the absolute pressure in the tank in N/m²?

Solution The gauge pressure $(P - P_a)$ of the air in the tank is given by

$$P - P_a = 33 \text{ cm of Hg.}$$

The barometer gives $P_a = 77$ cm of Hg, so $P = 110$ cm of Hg. This is the absolute pressure in the tank. We can express this pressure in N/m² by using the previous result that 76 cm of Hg represent a pressure of 1.01×10^5 N/m².

$$P = (110 \text{ cm of Hg}) \left\{ \frac{1.01 \times 10^5 \text{ N/m}^2}{76 \text{ cm of Hg}} \right\}$$
$$= 1.46 \times 10^5 \text{ N/m}^2.$$

Fig. 6-8 The Magdeburg hemispheres (bronze, approximately 2 ft in diameter) and pump devised by Otto von Guericke (1602–1686). Using these hemispheres, von Guericke demonstrated in a most striking manner the reality of atmospheric pressure. *Once the air had been evacuated from the bronze sphere by an air pump, apparently two teams of eight horses each were unable to pull the two halves of the sphere apart.* This public scientific drama was staged before the Reichstag at Regensburg on May 8, 1654, during von Guericke's tenure as burgomaster of Magdeburg in Saxony. (Photo courtesy of Deutsches Museum, Munich.)

*This is frequently expressed as 760 torr in modern vacuum technology, 1 torr being equal to 1 mm of Hg.

6.2 PASCAL'S PRINCIPLE

An immediate consequence of the fundamental law of hydrostatics is a fact discovered in the seventeenth century by the French philosopher, mathematician, and physicist, Blaise Pascal (1623–1662). This fact is now known as *Pascal's principle*: *If added pressure is applied anywhere to a confined fluid, this added pressure is transmitted undiminished to every portion of the fluid and to the walls of the confining vessel.*

To understand why this is so, we need only to notice that any increase in pressure at one point in the fluid must result in the same increase at all points because, according to the fundamental law of hydrostatics ($\Delta P = \rho g \Delta y$), pressure *differences* do not change. The pressure *difference* between two points in a connected fluid is determined only by the difference in vertical height Δy and the fluid density ρ.

Pascal's principle is exploited in the functioning of a *hydraulic press*, a machine using a confined liquid. This machine is used in car lifts, hospital operating tables, barber and dentist chairs, as well as in hydraulic automobile brakes, to obtain in a convenient way a very large ratio of output to input force (mechanical advantage). A small input force f applied to a piston of small area a results in a large output force F exerted by a piston of large area A (Fig 6-9). The pressure P is the same under both pistons since these are points at the same horizontal level in the same liquid. According to Eq. (6.1) the liquid exerts a small force

$$f = Pa$$

on the piston with a small area, while the liquid exerts on the piston with the large area A, the force

$$F = PA.$$

Dividing the second of these equations by the first, we obtain

$$F/f = A/a, \qquad (6.6)$$

which shows that the mechanical advantage F/f is given by the ratio of the areas of the pistons, A/a.*

In hydraulic brakes, when the brake pedal pushes on a piston in the master cylinder, the pressure increase is transmitted through the brake fluid to the four wheel cylinders. At each wheel a small piston then moves to activate a brake shoe or a disk.

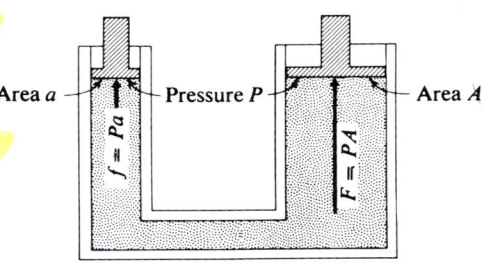

Fig. 6-9 Principle of the hydraulic press.

*Since the effects of friction have been ignored, this result is the ideal mechanical advantage. The actual mechanical advantage will be slightly smaller.

6.3 ARCHIMEDES' PRINCIPLE

The Buoyant Force

A fluid exerts a buoyant force on a body which is totally or partially immersed in it. Why this is so is not hard to discover. The fluid exerts pressure on the body's surface, causing an upward force on the bottom and a downward force on the top of an immersed object. Since the pressure is greater at greater depths, the upward force will be the greater and the resultant force will be an upward or *buoyant* force.

A very simple expression for the precise magnitude of this buoyant force was discovered over 2000 years ago by Archimedes. His result can be deduced from the laws of fluid statics as follows.

First consider a fluid in equilibrium without any immersed object. Focus attention on a certain portion of this fluid occupying a volume V with a surface S (Fig. 6-10). This volume V is held in equilibrium by the forces exerted by the remainder of the fluid on the surface S. Therefore these forces must have a resultant force which is directed vertically upward through the center of gravity of the fluid in V and has a magnitude equal to the weight ($Mg = \rho V g$) of the fluid in V. If we now replace the fluid inside V by some other object, since the pressures in the outside fluid are unchanged at each point on the surface S, the forces exerted by the outside fluid are unchanged. That is, *the fluid of density ρ exerts a buoyant force B given by*

$$B = \rho V g \qquad (6.7)$$

on an object which displaces a volume V of the fluid. The force $\vec{B}$ passes through the center of gravity (which is named the center of buoyancy) of this volume V of the displaced fluid. This law for ascertaining the buoyant force is known as *Archimedes' principle*.

A balloon of volume V which has an average density ρ_b less than the density ρ of the surrounding air will experience a resultant upward force

$$B - Mg = \rho V g - \rho_b V g = (\rho - \rho_b) V g$$

and therefore will be accelerated upward. The hydrogen or helium inside the balloon does not lift the balloon; it merely serves to keep the balloon distended.

A body whose average density is less than that of a liquid's density can float partially submerged in the liquid. Such a body sinks until its weight is balanced by the buoyant force. According to Archimedes' principle, this buoyant force is equal to the weight of the displaced liquid. Hence the weight of the displaced liquid is equal to the weight of a body floating in it. Saying that a ship displaces 10,000 tons of water is just another way of saying that the loaded ship weighs 10,000 tons.

The buoyant force exerted by water on a floating human is exploited in a type of *hydrotherapy* involving underwater gymnastics. In air, the lifting of an arm can bring into play forces in shoulder joints in excess of 450 N (about 100 lb). Such motion can cause considerable pain for an arthritis patient. For muscles weakened by poliomyelitis, normal exercise is often impossible. However, when the limbs of a person are immersed in water, the buoyant force "cancels out" all but

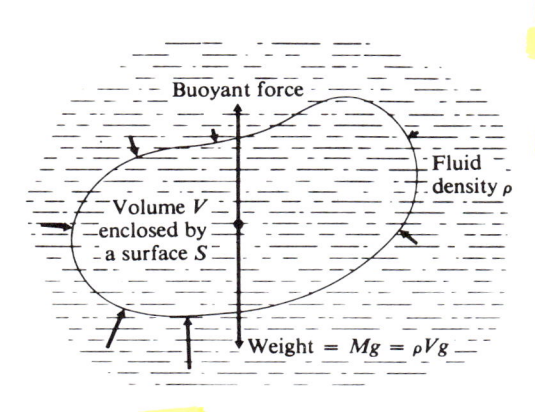

Fig. 6-10 Archimedes' principle. The fluid outside the surface S exerts a buoyant force $\rho V g$ on this surface.

about 4% of the weight of the limbs and beneficial exercise can then be undertaken.

Specific Gravity

The density of a liquid is easily measured by floating a *hydrometer* [Fig. 6-11 (a)] in it and noting the depth to which the hydrometer sinks. The hydrometer is usually a glass cylinder with a weight at the bottom and a thin calibrated tube at the top. The denser the liquid the less the depth to which the hydrometer must sink to achieve a buoyant force equal to its weight.

The glass stem is usually calibrated in terms of the *specific gravity* rather than the density of the liquid. The specific gravity of a substance is defined as the ratio of the density of the substance to the density of water. Pure water, therefore, has a specific gravity of exactly 1. From the data of Table 6-1, we see that mercury has a specific gravity of 13.6.

A very common diagnostic procedure in medicine, urinalysis, involves among other things a determination of the specific gravity of the urine with a hydrometer (called a urinometer) especially calibrated for this purpose. The specific gravity of urine is usually between 1.015 and 1.030. Some illnesses are accompanied by an increase of salts in the urine and a correspondingly higher specific gravity.

Blood tests also often include a measurement of the specific gravity, which normally ranges from 1.040 to 1.065. The higher the proportion of red blood corpuscles, the greater the specific gravity. Anemia is signaled by a decrease in the blood's specific gravity.

The most familiar hydrometer is the battery tester [Fig. 6-11 (b)] used in every garage to ascertain the state of charge of a car's storage battery. When the lead plates of the battery have reacted with much of the battery acid, the specific gravity of the liquid drops to around 1.14, the "no charge" condition. After a charging current has reversed these chemical reactions to produce again sufficient acid to raise the liquid's specific gravity to 1.28, "full charge" is indicated.

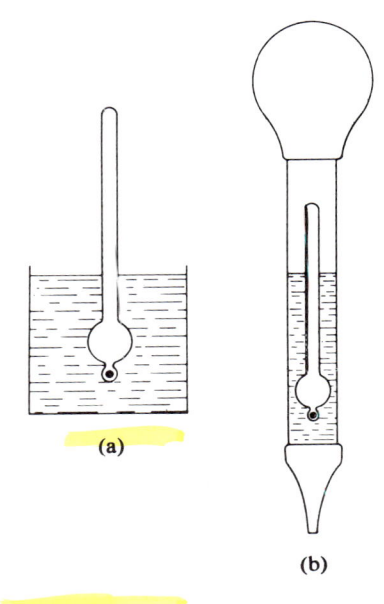

Fig. 6-11 (a) Hydrometer.
(b) Hydrometer used as battery tester.

6.4 FLUID DYNAMICS

Fluid dynamics, the study of the motion of a fluid and the related forces, forms one of the foundations of aeronautics, mechanical engineering, meteorology, marine engineering, civil engineering and bio-engineering. In fluid dynamics, the research scientist can still find fascinating experimental phenomena that are far from being understood. Although the underlying physical principles are the well-known laws of Newtonian mechanics, the application of these laws to describe a general motion of a real fluid often leads to problems of such complexity that attempts to gain a theoretical understanding are defeated. Difficulties in extracting general answers from the Newtonian mechanics of fluid motion arise from the fact that, except for the gravitational force, the forces acting on different portions of the fluid are not known in advance but instead are determined by the fluid motion. And the motions of all portions of the fluid have to be determined in a single calculation. Nevertheless, we shall see that with several simplifying assumptions it is quite easy to obtain a result known as *Bernoulli's equation*, which proves to be very useful in analyzing many situations.

Streamlines, Equation of Continuity

To describe fluid motion, we consider first a small volume ΔV of the fluid, called a volume element. The velocity $\vec{v}$ of the center of mass of this volume element is termed the *fluid velocity* at the location of the volume element. In this way, at any instant, a velocity vector is associated with each point in a region occupied by a fluid. A *streamline* is a line drawn so that at each point its direction is the direction of the fluid velocity at that point [Fig. 6-12(a)].

A fluid flow is said to be *steady* if the fluid velocity at every point does not change as time goes on. Steady flow occurs at low speeds in pipes and streams. A given volume element may have different velocities at different points, but in a steady flow all volume elements have the same velocity as they pass a given point. Streamlines then are stationary, and the path of a given volume element is a streamline.

A tubular region of a fluid, with sidewalls that are streamlines, is called a *tube of flow* [Fig. 6-12(b)]. Since the fluid velocity is parallel to these sidewalls, no fluid flows through the sidewalls. A tube of flow is therefore like a pipe in that the fluid that enters at one end must leave at the other.

This simple observation leads to a relationship between the cross-sectional area A_1, the speed v_1, and the density ρ_1, at one end of a thin tube of flow and the corresponding quantities A_2, v_2, and ρ_2 at the other end of the tube. In a small time interval Δt, the fluid at the input moves a distance $v_1 \Delta t$. The volume of fluid entering through A_1 is therefore $A_1 v_1 \Delta t$, and the mass of this fluid is

$$\Delta M = \rho_1 A_1 v_1 \Delta t. \tag{6.8}$$

In the same time interval, the fluid at the output moves a distance $v_2 \Delta t$. The volume of fluid leaving through A_2 is therefore $A_2 v_2 \Delta t$, and the output mass is

$$\Delta M = \rho_2 A_2 v_2 \Delta t. \tag{6.9}$$

Since mass is conserved (in Newtonian mechanics), the mass which

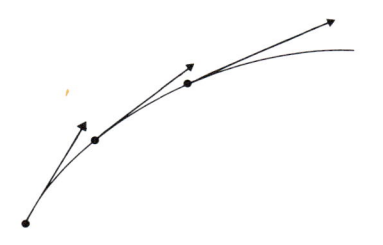

Fig. 6-12(a) A streamline. The arrows indicate fluid velocity vectors at different points along the streamline.

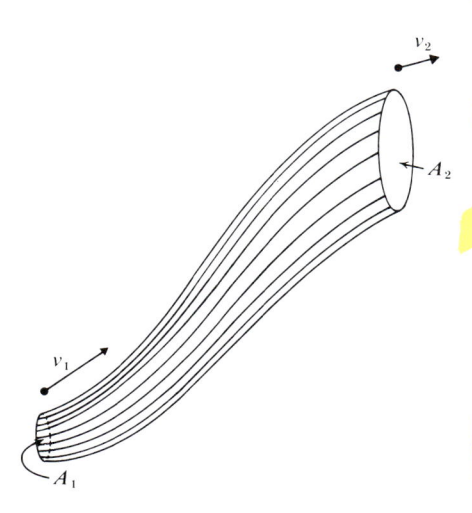

Fig. 6-12(b) The tube of flow has sidewalls that are streamlines.

enters through A_1 in the interval Δt must equal the mass which leaves through A_2 in the same interval. Therefore,

$$\rho_1 A_1 v_1 = \rho_2 A_2 v_2, \quad (6.10)$$

a result which is known as the *equation of continuity*.

The variations in density that occur in the flowing fluid are often relatively small, and it is a good approximation to assume that $\rho =$ constant; that is, to assume that the fluid is *incompressible*. This is usually an excellent approximation for flowing liquids. And even though gases are highly compressible, it often happens that the pressure changes encountered by a flowing gas are rather small and the consequent changes in gas density are unimportant. For instance, at velocities well below the speed of sound in air, the flow of air past an airfoil does not involve large changes in air density. Consequently, a useful theory of subsonic aerodynamics can be developed by assuming that the air is incompressible.

For an incompressible fluid, $\rho_1 = \rho_2$, and the equation of continuity simplifies to

$$v_1 A_1 = v_2 A_2. \quad (6.11)$$

This shows that where the tube area is large, the fluid speed is small, and vice versa. And in a stream of constant width, the water flows rapidly where the stream is shallow, but we observe that "still waters run deep."

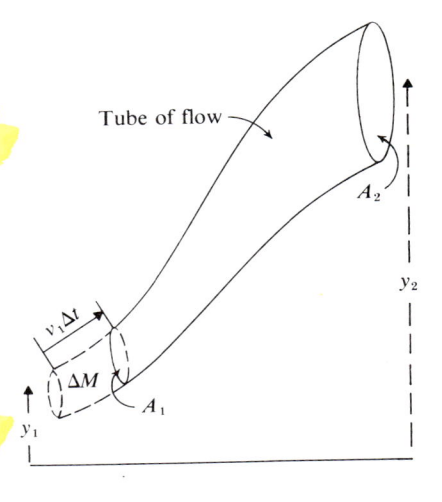

Fig. 6-13(a) Position at the beginning of the time interval Δt of the system comprising the fluid within the tube and the mass ΔM.

Bernoulli's Equation

A particularly useful theorem of fluid mechanics was discovered in 1738 by Daniel Bernoulli. This theorem is now recognized as a statement of the law of conservation of energy. It can be derived by examining the energy transformations that occur as a portion of a fluid moves through a tube of flow.

It will be assumed that the fluid is *nonviscous*. When one layer of a real fluid slides over another layer, this slipping is opposed by a frictional force called a *viscous* force which results in the dissipation of mechanical energy. Although viscosity is of great importance in some circumstances, there are many situations in which viscous effects are small enough so that a good approximation is obtained by assuming that the flow is *nonviscous*.

The situation is further simplified by assuming that the fluid is incompressible ($\rho =$ constant), so that there is no need to take account of the volume changes of a portion of the fluid or of the associated work done on this portion as it compresses or expands.

We focus now on the system shown in Fig. 6-13(a) which comprises, not only the mass of the fluid within the flow tube between the ends A_1 and A_2, but also the mass ΔM that is about to enter the tube through A_1. After a time interval Δt, the fluid of this system has moved to the position shown in Fig. 6-13(b).

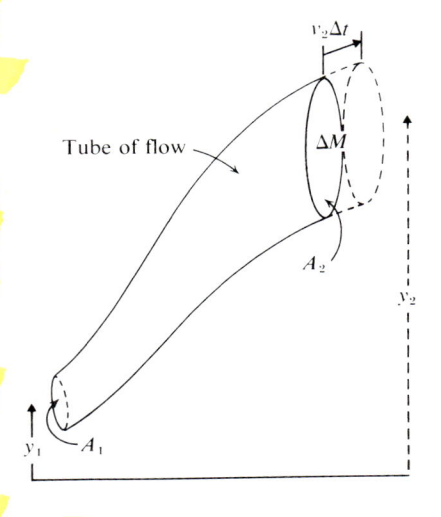

Fig. 6-13(b) Position at the end of the time interval Δt of the system in Fig. 6-13(a).

6.4 FLUID DYNAMICS

In the configuration shown in Fig. 6-13(a), this system has a total mechanical energy

$$E_1 = \tfrac{1}{2}(\Delta M)v_1^2 + (\Delta M)gy_1 + \text{(mechanical energy of fluid within tube)}.$$

The ends of the tube have vertical coordinates y_1 and y_2 measured from some convenient reference level. After the time interval Δt, the total mechanical energy of this system is

$$E_2 = \tfrac{1}{2}(\Delta M)v_2^2 + (\Delta M)gy_2 + \text{(mechanical energy of fluid within tube)}.$$

We assume the flow is *steady*. Then the fluid velocity at a given location remains unchanged, and the tube is stationary. Consequently, the mechanical energy of the fluid within the tube is constant; and the change in the system's mechanical energy, from the preceding equations, is therefore

$$E_2 - E_1 = \tfrac{1}{2}(\Delta M)v_2^2 + (\Delta M)gy_2 - \tfrac{1}{2}(\Delta M)v_1^2 - (\Delta M)gy_1.$$

Energy is transferred to the system by forces exerted on the system by the surrounding fluid. The work done by these forces measures the energy transferred to the system. The surrounding fluid exerts a force P_1A_1 on the mass ΔM of Fig. 6-13(a) in the direction of motion of this mass. During the time interval Δt, this mass moves a distance $v_1\Delta t$, and the work done on the system by the force P_1A_1 is

$$\begin{aligned} W_1 &= P_1 A_1 v_1 \Delta t \\ &= P_1 \Delta M/\rho, \end{aligned}$$

where we have used $A_1 v_1 \Delta t = \Delta M/\rho$ from Eq. (6.8). At the other end of the tube, the surrounding fluid exerts a force $P_2 A_2$ in the direction opposite to that of the motion of the system, which moves a distance $v_2 \Delta t$ in the time interval Δt. The corresponding work done by this force on the system is therefore the negative quantity

$$\begin{aligned} W_2 &= -P_2 A_2 v_2 \Delta t \\ &= -P_2 \Delta M/\rho, \end{aligned}$$

since $A_2 v_2 \Delta t = \Delta M/\rho$ from Eq. (6.9). No work is done by the forces exerted on the sidewalls by the surrounding fluid because these forces are perpendicular to the sidewalls (assuming no viscosity) and are therefore perpendicular to the direction of motion of the fluid on which they act.

Conservation of energy now implies that the energy $(W_1 + W_2)$ transferred to the system by the forces exerted by the surrounding fluid is equal to the change $(E_2 - E_1)$ in the system's total mechanical energy; that is,

$$W_1 + W_2 = E_2 - E_1, \tag{6.12}$$

which gives

$$P_1 \Delta M/\rho - P_2 \Delta M/\rho = \tfrac{1}{2}(\Delta M)v_2^2 + (\Delta M)gy_2 - \tfrac{1}{2}(\Delta M)v_1^2 - (\Delta M)gy_1.$$

Canceling ΔM from each term, multiplying by ρ, and then rearranging, we obtain the famous *Bernoulli equation*,

$$P_1 + \tfrac{1}{2}\rho v_1^2 + \rho g y_1 = P_2 + \tfrac{1}{2}\rho v_2^2 + \rho g y_2, \qquad (6.13)$$

for any two points (1 and 2) on the same streamline in a steady flow of a nonviscous incompressible fluid.

Notice that this fundamental result of fluid dynamics includes the fundamental law of hydrostatics as the special case that obtains when v_1 and v_2 are equal:

$$P_1 + \rho g y_1 = P_2 + \rho g y_2,$$

which yields

$$P_1 - P_2 = \rho g(y_2 - y_1),$$

in agreement with Eq. (6.3).

A fluid flow is called *irrotational* if there would be no rotation of a small paddle wheel immersed anywhere in the fluid. This excludes not only whirlpools but also a flow which involves a variation of the velocity vector in a transverse direction. For irrotational flow, it can be shown that Bernoulli's equation applies to any two points within the fluid, whether or not they lie on the same streamline.

Applications of Bernoulli's Equation

Fig. 6-14 represents a steady flow of water through a horizontal pipe which has a varying cross-section. The equation of continuity, $v_1 A_1 = v_2 A_2$, requires that the speed of the water be greatest at the constriction. The pipe very nearly constitutes a tube of flow, so Bernoulli's equation may be applied to the pipe, if we disregard viscous effects. For points 1 and 2 at the same horizontal level, $y_1 = y_2$, and Bernoulli's equation gives

$$P_1 + \tfrac{1}{2}\rho v_1^2 = P_2 + \tfrac{1}{2}\rho v_2^2.$$

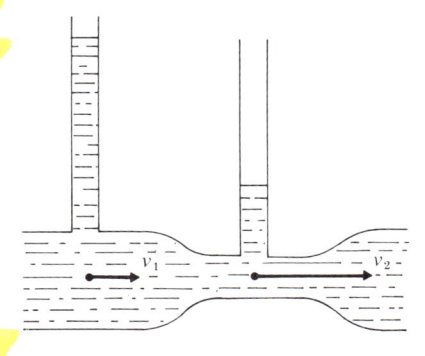

Fig. 6-14 Venturi effect. Where the fluid speed is greatest, the pressure is least.

This implies that, at the constriction where the speed is the greatest, the pressure is a minimum, in accord with the pressure measurements indicated by the heights of the vertical columns of water in Fig. 6-14.

This effect is exploited in a *Venturi meter* which is used to measure the speed of flow of a fluid. The meter introduces a constriction in the flow tube. A manometer measures the amount, $P_1 - P_2$, by which the pressure P_2 at the constriction drops below the pressure P_1 where the pipe has its normal cross-section A_1. Manometer readings can be calibrated to give the corresponding speed of flow v_1 within the pipe. The calculation of v_1 from knowledge of $P_1 - P_2$ and A_1/A_2 can be effected using both the equation of continuity and Bernoulli's equation (see Question S-10).

In the *aspirator pump* shown in Fig. 6-15, the pressure within the flowing stream of water drops at the throat (the constriction) to a value below atmospheric pressure, and suction occurs through a side tube. This simple pump can reduce the pressure to values as low as a few centimeters of mercury.

The speed with which a liquid flows out of an orifice in a large container was observed by Torricelli to be such that the emerging stream, when it is directed vertically upward, can rise almost to the level of the liquid in the container. Bernoulli's equation permits a simple analysis of this phenomenon. Fig. 6-16 shows a streamline which connects the upper surface at point 1 to a point 2 just outside the container. If the container cross-section is large compared to that of the outlet, the speed v_1 of water at the upper surface is approximately zero. The pressure at the upper surface and on the sides of the emerging jet is the atmospheric pressure P_0. Bernoulli's equation therefore yields

$$P_0 + 0 + \rho g y_1 = P_0 + \tfrac{1}{2}\rho v_2^2 + 0,$$

where y_1 is the height of the surface of water above the level of the output. Solving for the flow speed just outside the outlet, we obtain Torricelli's law,

$$v_2 = \sqrt{2gy_1}, \tag{6.14}$$

which is exactly the speed of vertical projection necessary for a projectile to rise to a height y_1.

Example 3 Water flows in glass tubing (with a 2.0 mm inner diameter) located in the basement of a laboratory. The tubing rises a height of 7.0 m to the second floor, with the diameter of the tubing tapering to 1.0 mm. What is the water pressure P_2 and velocity v_2 in the pipe at the second floor when within the pipe in the basement the water pressure is $P_1 = 2.00 \times 10^5$ N/m² and the water velocity is $v_1 = 1.0$ m/sec? Assume that Bernoulli's equation applies in this example.

Solution The equation of continuity, $v_2 A_2 = v_1 A_1$, allows us to determine v_2 from v_1 and the ratio of the pipe cross-section, A_1/A_2.

$$A_1/A_2 = (2.0 \text{ mm})^2/(1.0 \text{ mm})^2 = 4.0,$$

so

$$\begin{aligned} v_2 &= (A_1/A_2)v_1 \\ &= 4.0 \times (1.0 \text{ m/sec}) \\ &= 4.0 \text{ m/sec.} \end{aligned}$$

Disregarding the effects of viscosity, and assuming that the glass

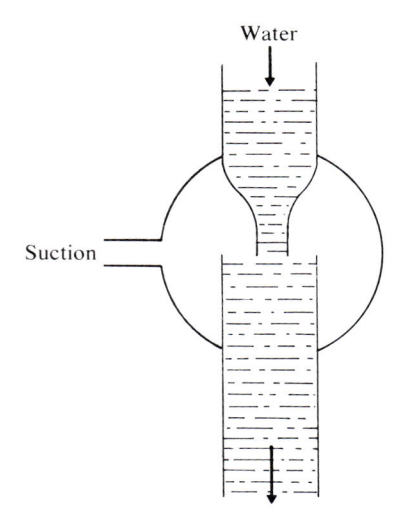

Fig. 6-15 Aspirator pump.

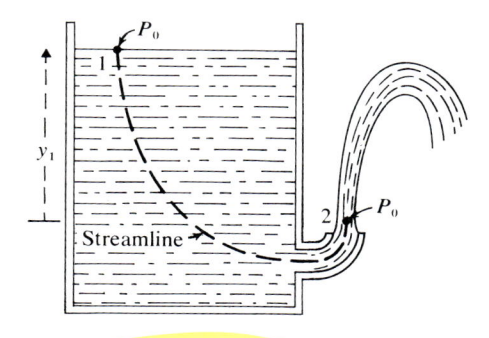

Fig. 6-16 Flow from an orifice in a large container.

tubing constitutes a tube of flow, we may apply Bernoulli's equation which gives

$$P_2 + \tfrac{1}{2}\rho v_2^2 + \rho g y_2 = P_1 + \tfrac{1}{2}\rho v_1^2$$

where y_2 is the height of the second floor above the basement. Therefore,

$$\begin{aligned}
P_2 &= P_1 + \tfrac{1}{2}\rho v_1^2 - \tfrac{1}{2}\rho v_2^2 - \rho g y_2 \\
&= 2.00 \times 10^5 \text{ N/m}^2 + \tfrac{1}{2}(1.0 \times 10^3 \text{ kg/m}^3)(1.0 \text{ m/sec})^2 \\
&\quad - \tfrac{1}{2}(1.0 \times 10^3 \text{ kg/m}^3)(4.0 \text{ m/sec})^2 \\
&\quad - (1.00 \times 10^3 \text{ kg/m}^3)(9.8 \text{ m/sec}^2)(7.0 \text{ m}) \\
&= 2.00 \times 10^5 \text{ N/m}^2 + 0.005 \times 10^5 \text{ N/m}^2 \\
&\quad - 0.08 \times 10^5 \text{ N/m}^2 - 0.69 \times 10^5 \text{ N/m}^2 \\
&= 1.24 \times 10^5 \text{ N/m}^2.
\end{aligned}$$

Viscosity and Bernoulli's Equation

All the effects that have been discussed in this section are modified by the presence of viscous forces. Because of work done by viscous forces in a flowing fluid, mechanical energy is dissipated in heating the fluid. Then Eq. (6.12) is replaced by

$$W_1 + W_2 = E_2 - E_1 + \text{(energy dissipated)},$$

which leads to the result that, for a point 2 downstream from the point 1 and on the same streamline,

$$P_1 + \tfrac{1}{2}\rho v_1^2 + \rho g y_1 > P_2 + \tfrac{1}{2}\rho v_2^2 + \rho g y_2, \qquad (6.15)$$

an expression that we will call Bernoulli's *inequality*.

In viscous flow through a horizontal pipe of constant cross-section, the pressure is observed to drop as one moves downstream (Fig. 6-17). This accords with the Bernoulli inequality, which for this case, with $y_1 = y_2$ and $v_1 = v_2$, gives

$$P_1 > P_2.$$

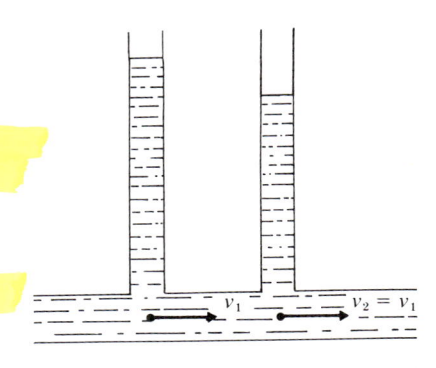

Fig. 6-17 Viscosity causes reduction of the downstream pressure.

And because of viscosity, the downstream pressure P_2 in Example 3 will actually be less than the value we computed using Bernoulli's equation.

The reason that the emerging jet of water in Fig. 6-16 does not quite attain the level in the container is that there is a dissipation of mechanical energy associated with viscosity. And when the analysis that led to the Torricelli law is repeated, using the Bernoulli inequality instead of the Bernoulli equation, we find that v_2 is less than $\sqrt{2gy_2}$.

These examples indicate that in many real situations the results obtained using Bernoulli's equation, when uncorrected for energy dissipation, should not be regarded as anything more than estimates. In this respect, one of the most important phenomena of fluid mechanics

6.4 FLUID DYNAMICS

must be mentioned, and that is *turbulence*. When a fluid moves sufficiently rapidly down a pipe or past an obstacle, a churning of the fluid occurs, eddies are formed, and the detailed motion of a volume element of the fluid generally becomes chaotic. This is *turbulent flow*. Considerable mechanical energy is dissipated within a fluid where the flow is turbulent. In such turbulent regions there can be pronounced departures from the predictions of Bernoulli's equation.

QUESTIONS

1. The pressure in the water next to a 9.0×10^{-3} m² flat tile is 1.2×10^5 N/m². Find the magnitude of the force exerted by the water on the tile. Draw a diagram showing the tile, the water, and the direction of the force on the tile.

2. While skating, a man applies a force of 1000 N distributed over the 1.00×10^{-3} m² area of the bottom of a skate.
   (a) What is the pressure on the surface of the ice underneath this skate?
   (b) If the man were to replace the skate by a shoe, would this pressure decrease? Explain.

3. Assume the diameter of the Magdeburg hemispheres (Fig. 6-8) to be 0.60 meter. At normal atmospheric pressure what would be the force pressing the one hemisphere against the other if, inside the sphere, there were a perfect vacuum? [*Hint:* By first considering the equilibrium of a single closed hemispherical surface in air, you can deduce that the force exerted by the air on the curved surface has the same magnitude as the force exerted on the flat surface of area $\pi(0.30 \text{ m})^2$.]

4. The piston of a hydraulic lift for automobiles has a radius of 0.15 meter. What pressure is required to lift a car weighing 12×10^3 N?

5. What is the mass in kilograms, and the weight in newtons, of a volume of 1.0 cubic centimeter (10^{-6} m³) of water?

6. On a certain day the pressure at the surface of a lake is 1.00×10^5 N/m². How far must one descend to encounter a pressure of 3.00×10^5 N/m²?

7. A tube contains a 2.00-m layer of olive oil floating above a 1.50-m layer of water which in turn floats above a 0.500-m layer of mercury. The upper surface of the olive oil is exposed to the pressure of the atmosphere which is 1.00×10^5 N/m². What is the absolute pressure at the bottom of the mercury layer?

8. When a man who is 2.00 m tall is standing, what is the hydrostatic difference between the blood pressure in his feet and the blood pressure in his brain? Assume that his blood has a density of 1.06×10^3 kg/m³.

9. A long thin vertical tube passes through a rubber stopper into a flask. When 100 cubic centimeters of water are poured into this assembly, the flask and tube are filled so that the height from the base of the flask to the free water surface is 1.50 meters. The area of the flask base is 2.00×10^{-2} m². Find the force exerted by the water on the flask base. (Ignore the effect of the air pressure which contributes equal forces to the top and bottom surfaces of the base of the flask.) Show that the force exerted by the water on the base is much greater than the weight of 100 cm³ of water. Explain this "hydrostatic paradox."

10. The manometer in Fig. 6-5 contains mercury. The level difference is as shown in the figure with Δy = 19 cm. The barometric pressure is 76 cm of Hg.
    (a) Give the gauge pressure and the absolute pressure of the gas in the tank in centimeters of Hg.
    (b) What is the absolute pressure in N/m²?

11. After hearing of Torricelli's mercury barometer, Blaise Pascal made a water barometer. What is the height of the water column in such a barometer on a day when the atmospheric pressure is 1.00×10^5 pascal?

12. A mercury barometer at the top of a building reads 1.00 cm less than a barometer on the ground floor. How tall is the building?

13. Deduce an expression for the ideal mechanical advantage of a hydraulic press by finding the ratio of distances moved by the two pistons.

14. (a) If the hydraulic press of Fig. 6-9 were constructed with pistons of areas of 0.10 m² and 2.0×10^{-4} m², what would be the mechanical advantage?
    (b) What force would have to be applied to the small piston to lift an automobile weighing 15×10^3 N?
    (c) What would be the pressure underneath the pistons?

15. A nurse applies a push of 44 N (about 10 lb) to the piston of a syringe. The piston has a radius of 1.0 cm. What is the pressure increase (in N/m²) in the fluid within the syringe?

16. State Archimedes' principle.

17. Will a solid iron ball float in mercury? Will a solid gold ball float in mercury? Explain.

18. A block of aluminum with a volume of 0.100 m³ is completely immersed in water. The block is suspended by a rope. Find
    (a) the mass and weight of the aluminum block;
    (b) the buoyant force exerted on the block by the water;
    (c) the force exerted on the block by the rope. (This force is equal in magnitude to what is called the *apparent weight* of the submerged block.)

19. A barge, loaded with coal, has a total mass of 2.40×10^5 kg. The barge is 15.0 m long and 8.0 m wide.
    (a) Using Archimedes' principle, calculate the depth D of the barge below the water line.
    (b) Using Eq. (6.3) calculate the pressure increase at this depth.
    (c) From $F = PA$, Eq. (6.1), calculate the force exerted on the barge by the water, and compare this with the weight of the barge.

20. (a) The density of ice is 0.92×10^3 kg/m³. What fraction of the volume of a floating ice cube is above water?
    (b) What volume of an ice floe in a river is just large enough to carry a boy who weighs 400 N?

21. (a) What is the volume of a helium-filled balloon that will support a weight of 1000 N?
    (b) How great a weight would the same balloon support if the helium were replaced by hydrogen?

22. (a) What is the definition of the specific gravity of a substance?
    (b) From the data of Table 6-1, compute the specific gravity of gold.
    (c) Successive readings of a battery tester are taken as a storage battery is being recharged. Will the hydrometer float higher or lower? Explain.

23. The apparent weight of an ore sample is 206 N when the ore is completely immersed in water, and 230 N when the ore is in air. What is the specific gravity of the ore?

24. A stone has an apparent weight of 5.40 N when it is completely immersed in water, and 6.00 N when it is completely immersed in oil of density 0.80×10^3 kg/m³. What is the stone's specific gravity?

25. Define the terms *streamline* and *tube of flow*.

26. What is *steady* flow? Is a gently flowing stream an example of steady flow? Is a babbling brook?

27. Show that the fluid volume per unit time passing through a cross-section A of a pipe is given by

$$V = Av$$

where v is the fluid speed in the pipe.

28. The water entering a house flows with a speed of 0.10 m/sec through a pipe of 21 mm inside diameter.
    (a) What is the speed of the water at a point where the pipe tapers to a diameter of 7 mm?
    (b) What mass and what volume of water enter the house in 60 sec?

29. (a) Show that the work done during the time interval Δt by the external fluid forces on the fluid of the system shown in Fig. 6-13 is given by

$$W = (P_1 - P_2) \Delta V,$$

where ΔV is the volume of the mass ΔM (the volume that flows through the tube in the time interval Δt).
    (b) How much work is done by the pressure in forcing 5.0 m³ of water through a pipe when the pressure difference between the ends of the pipe is 1.2×10^5 N/m²?

30. Give the dimensions of each quantity appearing in Bernoulli's equation, and verify that the equation is dimensionally correct.

31. Water flows upward through a tapered vertical tube. Where the cross-section is 4.0×10^{-5} m², the pressure is 2.0×10^5 N/m², and the fluid speed is 1.5 m/sec. At a point which is 3.0 meters higher, the cross-section is 12.0×10^{-5} m².
    (a) Find the fluid speed at the higher point.
    (b) Use Bernoulli's equation to estimate the pressure at the higher point.

32. The constriction in Fig. 6-14 has an inner diameter of 0.5 cm, and the normal diameter of the pipe is 1.0 cm. When the volume of water per second flowing through the pipe is 0.20×10^{-4} m³/sec find
    (a) the fluid speed at the constriction and also at a point far from the constriction;
    (b) the pressure difference between the normal fluid pressure P_1 and the pressure at the constriction, P_2.

33. (a) What is the speed at which water flows from a small hole 3.0 m below the water level in a full tank which is 5.0 m high?
    (b) How far will the jet of water be projected horizontally before striking the ground at the level of the bottom of the tank?

34. In a closed pressure tank, the air pressure above the water surface is 2.0×10^5 N/m². A jet of water is squirted vertically upward from an aperture 4.0 m below the water surface. How high will this jet rise?

SUPPLEMENTARY QUESTIONS

S-1. An ocean-going ship is loaded at a port in Lake Ontario. Will the ship sink lower or rise higher when it reaches salt water? Explain. (The density of salt water is greater than that of fresh water.)

S-2. The gauge pressure of the air within the tires of a 2000-kg automobile is 2.0×10^5 N/m². How much total surface area of the tires is in contact with the road?

S-3. Fig. 6-18 shows the setup commonly used for intravenous infusions. (In this aspect of medical practice, pressures are usually expressed in mm H$_2$O rather than in mm Hg. For example, a pressure of 50 mm H$_2$O means that the pressure is that exerted by a column of water 50 mm high.)
   (a) How high above the needle must the intravenous fluid level be in order that, at the needle, the net pressure (the intravenous fluid pressure minus venous pressure, that is, minus the opposing pressure of 80 mm H$_2$O of the blood in the veins) is 700 mm H$_2$O? Assume the fluid being used for "intravenous feeding" is simply water.
   (b) Show whether the height calculated in (a) is significantly different from the height that would be determined if, instead of water, the fluid were the commonly used 5% Dextrose solution. (This means that each 100 milliliters of water contains 5 gm of dextrose; the combined volume of the water and the dextrose is itself 100 milliliters to a good approximation.)

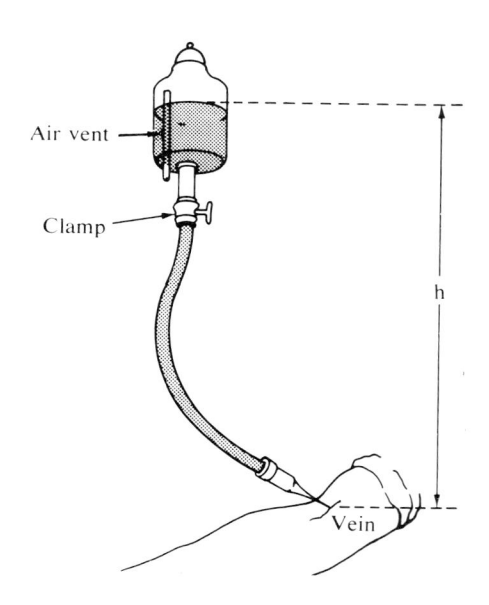

Fig. 6-18 An IV (intravenous) infusion setup.

S-4. The cerebrospinal fluid in the cranial and spinal cavities exerts a pressure of 100 to 200 mm H$_2$O. Suppose the patient in Fig. 6-19 has a cerebrospinal fluid pressure of 150 mm H$_2$O above the prevailing atmospheric pressure.
   (a) How high in the tube in Fig. 6-19 will the fluid rise? (The tube is inserted into the spinal cavity by means of a hollow needle.)
   (b) What assumption have you made in answering part (a)?
   (c) In the Queckensted test, the veins in the neck are compressed and the pressure transmitted to the cerebrospinal fluid. What would one expect to happen to the height of the fluid column in the tube? Why? What conjecture might one make if the fluid level remained constant?

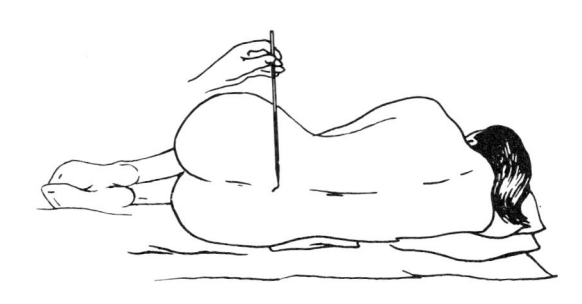

Fig. 6-19 Cerebrospinal puncture.

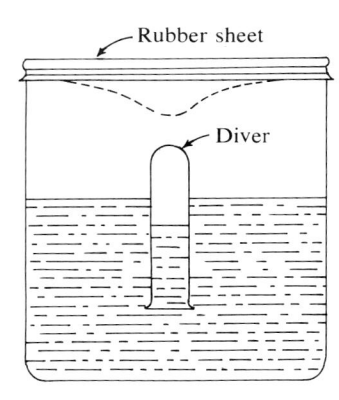

Fig. 6-20 Cartesian diver.

S-5. Archimedes devised a nondestructive test which would determine whether his king's crown was pure gold. A balance was used to select a quantity of pure gold which had a mass equal to that of the crown. He then immersed the crown in a container which had been completely filled with water and measured the amount that overflowed. After refilling the container, the pure gold was immersed in the water and again the overflow was measured. From these measurements he determined that the crown was not pure gold. Explain how this was possible. If the ratio of the mass of water which overflowed when the pure gold was immersed to the mass which overflowed when the crown was immersed was 0.900, what was the density of the crown?

S-6. An ingenious toy called a cartesian diver is represented in Fig. 6-20. The diver is an inverted test tube which contains just enough air to float in a large tank of water. The tank is closed by a rubber sheet stretched over the top. When this sheet is pushed down slightly, the diver sinks to the bottom. When the rubber sheet returns to its original position, the diver rises to the surface. Explain this behavior and account for the fact that the diver cannot find a stable position between the surface and the bottom of the tank.

S-7. A beaker of water on one pan of a balance and weights on the other pan are in equilibrium. Without touching the beaker, a man places his finger tip in the water. Explain what happens.

S-8. An ice cube floats in a glass filled to the brim with ice water. When the ice melts will the water overflow? Do the experiment and *explain* your observation.

S-9. The drop in pressure encountered in a one-kilometer length of a certain oil pipeline is 1.0×10^5 N/m^2.
   (a) How much work is done by the pressure in forcing 1.0 m^3 of oil through one kilometer of this pipeline?
   (b) The pipeline is horizontal and has a constant cross-section. What drop in pressure is predicted by Bernoulli's equation? Comment.
   (c) What is the mechanical energy dissipated per cubic meter of oil transported per kilometer of pipeline length?

SUPPLEMENTARY QUESTIONS

S-10. (a) A Venturi flowmeter introduces a constriction of cross-section A_2 in a pipe of normal cross-section A_1. The meter determines the pressure difference $P_1 - P_2$ between the normal fluid pressure P_1 and the pressure at the constriction, P_2. Using Bernoulli's equation and the equation of continuity, show that where the cross-section has the normal value A_1, the fluid speed v_1 is given by

$$v_1 = A_2 \sqrt{\frac{2(P_1 - P_2)}{\rho(A_1^2 - A_2^2)}}$$

where ρ is the fluid density.

(b) Show that the fluid volume per unit time passing through any cross-section of the pipe is given by

$$V = A_1 A_2 \sqrt{\frac{2(P_1 - P_2)}{\rho(A_1^2 - A_2^2)}}$$

S-11. Water enters the aspirator pump in Fig. 6-15 at a rate of 0.16 m³/minute at a pressure of 2.0×10^5 N/m² through a pipe with a cross-section of 4.0×10^{-4} m². In order to use this pump to achieve pressures as low as 3.0×10^4 N/m², what should be the cross-sectional area of the constriction?

S-12. The siphon shown in Fig. 6-21 transfers water from one container to a lower container but first raises the liquid to a height h_2 above the level in the first container. Consider a siphon of cross-section 4.0×10^{-4} m² with $h_2 = 3.0$ m and the lower end of the siphon (point 3) such that $h_3 = 5.0$ m. At point 3 the pressure is the atmospheric pressure, $P_0 = 1.0 \times 10^5$ N/m². Find
(a) the pressure P_1 at point 1 within the siphon;
(b) the pressure P_2;
(c) the speed v of the water within the siphon;
(d) the volume per second of water flowing through the siphon.

S-13. Show that, if the flow of water in a siphon is assumed to be governed by Bernoulli's equation, one can make the following predictions about siphon operation.
(a) The flow rate is proportional to $\sqrt{h_3}$. (The distance h_3 is shown in Fig. 6-21.)
(b) The siphon can operate only if the sum $h_2 + h_3$ is less than the height h of a column of water in a water barometer; that is, we must have

$$h_2 + h_3 < h$$

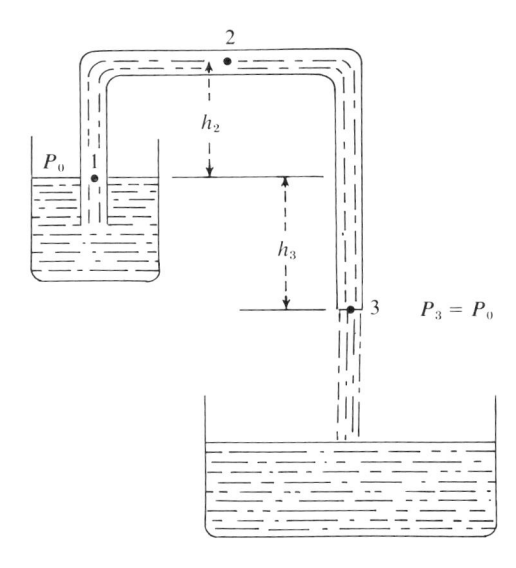

Fig. 6-21 Siphon.

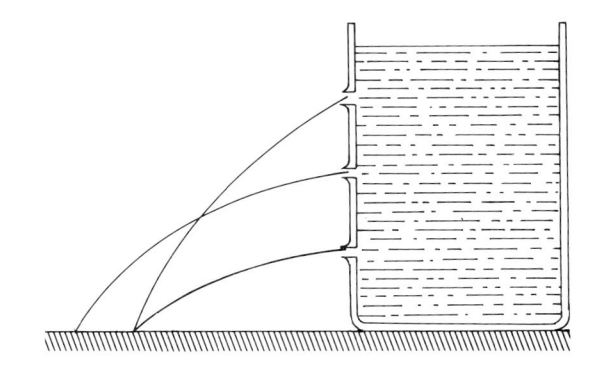

Fig. 6-22 Diagram for Question S-14.

where
$$\rho g h = P_0.$$

[*Hint:* Minimum fluid pressure, which occurs at point 2, must always remain a positive quantity.]

S-14. During a discussion of how the water spurting from a hole in a can is dependent on the water pressure at the hole, a physics student sketches a picture such as shown in Fig. 6-22. The holes are assumed respectively to be ¼, ½, and ¾ of the way up to the water level (which is held constant by allowing a gentle inflow at the top of the can). Is the student more or less correct in his sketch? Support your answer with the appropriate analysis.

S-15. Here are two problems in fluid statics that are easily analyzed from an accelerated frame of reference (Section 5.2).
(a) A bottle of water containing air bubbles is falling freely with an acceleration g. Do the bubbles rise through the water?
(b) A balloon filled with helium is fastened to the floor of a car by a length of string. While the car is going around a bend, which way will the balloon move relative to the car?

ADDITIONAL APPLICATIONS TO MEDICINE AND THE LIFE SCIENCES

Pumps

You are undoubtedly acquainted with the basic fundamentals of pumps of the piston type. The piston can be moved in a cylinder to increase the pressure on the fluid in the cylinder, thereby forcing the fluid out of the cylinder. In the case of the vacuum pump, the fluid that is forced out is gaseous, usually air. In any event, the key aspect of pump operation is the creation of a pressure gradient. The fluid in question then moves from the region of high to the region of low pressure.

Although Otto von Guericke is generally credited with the discovery of the air pump of the piston type, circa 1650, some record of a certain type of pump in Greece appears as early as the third century B.C.

For most laboratory vacuum work the piston-type pump has been replaced by the *rotary oil pump*. However, to obtain the extremely low vacuum required by modern x-ray and electronic tubes these pumps are used in conjunction with a so-called *diffusion pump* (for details see the book by Harrington cited in the *Suggested Reading*).

In the body, the pumping system which propels the blood does not involve the production of a vacuum, but rather a sequential periodic increase of the pressure exerted first by the right part of the heart (which operates the pulmonary circulation through the lungs) and then by the left part (which receives the re-oxygenated blood from the lungs and recirculates it throughout the body). This is to some extent what is called *peristaltic pressure*. A more literal example of *peristalsis* is the wave-like progression of alternate contraction and relaxation of the muscle fibers of the esophagus and intestines by which contents are propelled along the alimentary tract.

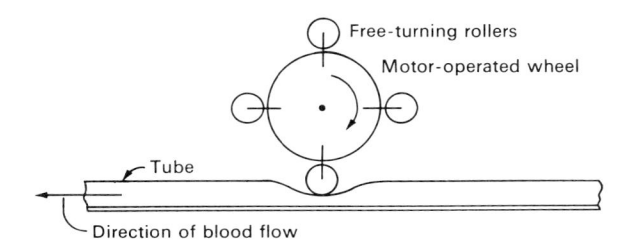

Fig. 6-23 The essence of a roller-type pump.

Some pumps designed for pumping blood into heart-lung machines simulate heart action somewhat: rubber or plastic tubing is compressed by one or more rollers (see Fig. 6-23). Pumps working on this principle are known as *peristaltic pumps*. Aside from medical applications, such pumps have also found general laboratory and industrial use. Peristaltic pumps have been found ideal for transferring corrosive liquids, for proportional mixing, for on-stream sampling and for long-term infusions.

Dizziness Associated with Upward Acceleration: Hydrostatic Change of Blood Pressure

Earlier we have discussed a connected fluid which is at rest relative to an inertial frame of reference. It is interesting to consider how the situation changes when the fluid is accelerated upward.

Question

Why is a patient who is suffering from *postural hypotension* (myocardial insufficiency, that is, the pumping action of the left part of the heart is weak) likely to become dizzy when he is in an upward accelerating elevator?

Answer

To give each portion of the patient's blood an upward acceleration a, the difference in the blood pressure ΔP between the pressure at the portion's lower surface and at its upper surface must be greater than the pressure difference which would merely produce static equilibrium. The modification of the fundamental law of hydrostatics, appropriate for an accelerated fluid, is

$$\Delta P = (g + a)\rho \Delta y.$$

This follows from the application of Newton's second law to the system shown in Fig. 6-3.

Consequently the "hydrostatic" increase of blood pressure from the heart up to the brain in a vertical passenger is greater when the elevator is accelerating upward than when it is at rest.

The patient with a myocardial insufficiency may have difficulty in maintaining a proper supply of blood to the head under normal circumstances. The upward elevator acceleration accentuates this difficulty, thus possibly causing dizziness and, in extreme cases, fainting.

Blood Pressure: Pressure Gauges

When determining the blood pressure, two distinct values of pressure are generally considered: the maximum pressure *(systolic pressure)* occurring while the heart is actually pumping, and the pressure when the heart is at rest *(diastolic pressure)*. The diastolic pressure is largely due to the contraction of blood vessels which dilate during the systolic process.

Systolic pressure is considered normal near 120 mm Hg and diastolic pressure near 80 mm Hg. The 40 mm Hg pressure difference is called the *pulse pressure*.

The heart is a quite active pump, coping with about 7000 liters of blood in 24 hours.

The most sophisticated and precise measurement of **blood pressure involves inserting a transducer (Chapter 10) into an artery and obtaining a readout via relatively complex electronic circuitry.**

A more convenient and common way to measure blood pressure is by the use of a sphygmomanometer. This may be either an aneroid type or a mercury type. In either case a rubber cuff, or armlet, is wrapped around the bare upper arm of the patient. This rubber cuff is inflated with air and thereby compresses the brachial artery. When the pressure exerted by the cuff exceeds the blood pressure, circulation is stopped. The air is now slowly released and the pressure is read on the sphygmomanometer just at the moment blood resumes flowing. This procedure yields the systolic pressure.

Question

(a) Is the 80 mm Hg diastolic pressure absolute or gauge pressure?
(b) What would you expect to find with respect to systolic pressure in persons suffering from arteriosclerosis (hardening of the arteries)?

Answer

(a) Any pressure measured with the sphygmomanometer is a gauge pressure.
(b) Increased systolic pressure while the heart pumps, because of the lack of dilation of the blood vessels during this stage.

Bed Sores and Water Beds: Pascal's Principle

Patients confined to bed for long periods often develop bed sores on parts of the body which support most of the body's weight. It has been found that this problem can be alleviated by the use of a so-called *water bed*, a flexible plastic bag containing water.

Question

Why are bed sores less likely to develop when a water bed is used?

Answer

The pliability of this mattress ensures a more or less uniform contact over a relatively large area, and since the pressure throughout the water is essentially the same (Pascal's Principle), the forces exerted by the mattress on the patient are uniformly distributed over the contact area. Large supporting forces confined to a small area, the type which are most likely to lead to bed sores, are therefore avoided.

It should be noted, however, that such water beds are exceedingly heavy and thus are not likely to be used as extensively as perhaps desirable. An alternate approach to dealing with bedridden patients is the use of beds in which flexible tubes (about 4 cm in diameter) are "snaked" back and forth across a special layer on top of an ordinary mattress. Through these tubes either water or air is pulsed. This device does not provide as uniform a force distribution as the water bed, but it has a massaging effect on the patient.

SUGGESTED READING

HALLIDAY, D., and R. RESNICK, *Physics Part I.* New York: John Wiley, 1966. Chapters 17 and 18 give clear and concise treatments of the statics and dynamics of fluids. Calculus and vector analysis are used.

HARRINGTON, E. L., *General College Physics.* New York: D. Van Nostrand, 1952. Students interested in medicine or biology will find particularly valuable the applications of the principles of hydrostatics found in this book.

JENSEN, J. TRYGVE, *Physics for the Health Professions.* Philadelphia: Lippincott, 1975. An up-to-date excellent descriptive text which makes good collateral reading for students interested in medicine and the related health professions.

SEARS, F. W., and M. W. ZEMANSKY, *University Physics.* Reading, Mass.: Addison-Wesley, 1970 (fourth edition). A particularly clear treatment of hydrostatics is given in Chapter 12. The very few equations involving calculus can be skipped without missing the essential points.

Ludwig Eduard Boltzmann

7 Disordered Energy

"For the molecules of the body are indeed so numerous, and their motion so rapid, that we can perceive nothing more than average values."

Ludwig Eduard Boltzmann
(1844–1906)

The next three chapters are devoted to the physics of *mascroscopic* systems. Such systems are large compared to a single molecule—large enough to be directly perceived by our senses. An ice cube containing 10^{24} molecules is a typical macroscopic system. All our experiments and all our applications of science ultimately rely on understanding the behavior of such comparatively large objects. We shall be particularly concerned with molecular motions inside objects which themselves reveal little or no bulk motion.

Physicists have learned how to apply the physical laws governing the behavior of individual molecules to such aggregates of particles. And so man has discovered an interpretation for such quantities as *temperature* and *pressure* in terms of the mechanics of such aggregates of molecules. Instead of letting the analysis become lost in a fog of some 10^{24} equations, the appropriate thing to do is to use statistical methods and compute averages.

Statistical physics is a vast field. It deals with aggregates of particles which constitute a gas, a liquid, a solid, or perhaps a plasma. Applications of recently acquired knowledge abound: for example, the transistor, which has revolutionized the electronics industry, is an invention arising from an understanding of the behavior of an aggregate of electrons in a solid. But in this chapter we concentrate attention on a simple and most instructive collection of molecules, a gas.

This "kinetic theory of gases," which views a gas as a collection of many molecules in incessant random motion, not only gives a detailed understanding of gaseous behavior but also brings out the remarkable fact that regularity in physical phenomena on the macroscopic scale can result from molecular chaos.

The Swiss mathematician and physician, Daniel Bernoulli (1700–1782), was the first to develop the consequences of the idea that a gas consists of molecules in random motion. But his brilliant work, presented in 1738, came more than a hundred years too soon to interest other scientists. Essentially the same theory was rediscovered independently in the middle of the nineteenth century by James Prescott Joule, A. Krönig and Rudolf Clausius. During the latter half of the nineteenth century the kinetic theory of gases and the interpretation of thermodynamics in terms of the statistics of a multitude of molecules was developed in depth, most notably by James Clerk Maxwell, Ludwig Boltzmann, and the first American to achieve eminence in theoretical physics, J. Willard Gibbs (1839–1903) of Yale University.

Boltzmann's life work was centered on statistical physics. He contributed such fundamental insight into the molecular explanation for macroscopic behavior that today his name looms large in the annals of science. But during the later years of his life his work was exposed to crushing attacks by such reputable scientists as Ernst Mach (1838–1916) and Wilhelm Ostwald (1853–1932) on the grounds that the proper concern of physical theory was with macroscopically observable physical quantities. Boltzmann, they felt, was contaminating the subject with the unnecessary and unverified hypothesis of the atomic nature of matter. The direct experimental evidence vindicating Boltzmann's point of view came tumbling in only after his death by suicide in 1906.

The objective in this chapter is to give the reader some appreciation of the behavior of large collections of atoms while acquiring a useful knowledge of gases and to use this knowledge as a stepping stone to some fundamental ideas often encompassed in the simple word, *heat*.

7.1 GASES

Pressure

Consider a gas confined within a cylinder fitted with a piston (Fig. 7-1). The gas exerts a force, a push, on the cylinder walls and on the inner surface of the piston. This force F exerted by the gas on a flat surface is perpendicular to the surface and proportional to the area A of the surface in question. The ratio F/A, which is characteristic of the condition of the gas, is defined to be the pressure P within the gas: that is,

$$P = F/A. \tag{7.1}$$

The mks unit for pressure is the N/m^2.

The pressure in the air of our atmosphere near the earth's surface varies by a few percent from day to day, a typical value being 10^5 N/m^2.

Example 1 A gas exerts an outward force of 3.6×10^3 N on a piston whose cross-sectional area is 1.2×10^{-2} m^2. What is the gas pressure P?

Solution
$$P = F/A \tag{7.1}$$
$$= \frac{3.6 \times 10^3 \text{ N}}{1.2 \times 10^{-2} \text{ m}^2}$$
$$= 3.0 \times 10^5 \text{ N/m}^2.$$

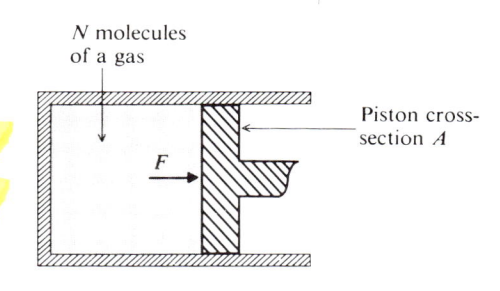

Fig. 7-1 The gas pressure P equals F/A.

Ideal Gas Law

The number N of molecules* in a mass M of a sample of gas containing identical molecules can be calculated if the mass m of an individual molecule has been determined. Molecular masses can be measured using a mass spectrometer (Section 12.3). Evidently

$$M = Nm.$$

Therefore,
$$N = M/m. \tag{7.2}$$

Now if we measure the volume occupied by different amounts of a given type of gas at constant pressure and temperature, we discover that, as we would expect, this volume is proportional to the number N of gas molecules present.

Suppose that with a definite amount of a gas contained in the cylinder of Fig. 7-1 we measure the gas pressure P corresponding to various values of the volume V occupied by the gas. We find that if the volume be decreased, the pressure increases in such a way that the product PV is a constant if the temperature remains constant. This result, discovered in 1660 by Robert Boyle (1627–1691), is known as Boyle's law.

Continuing our experimental investigation, we observe that when the gas is heated, the product PV increases. A most convenient temperature scale, called the Kelvin temperature scale, can be defined in such

---

*Instead of referring directly to the number N of the molecules, it is common practice, particularly in chemistry, to specify the quantity of a substance in terms of the mole. This is a measure of quantity of matter that is convenient for macroscopic amounts. The mole is defined as the amount of substance of a system containing as many elementary entities as there are atoms in exactly 0.012 kg of carbon-12. The elementary entities must be specified, and may be molecules, atoms, ions, electrons or other particles. The number of atoms in exactly 0.012 kg of carbon-12 is called Avogadro's number and is denoted by N_0. A mole of any specified kind of molecule (or atom) consists of N_0 molecules (or atoms) of this kind. The experimental value of Avogadro's number N_0 is 6.022×10^{23} $(mole)^{-1}$. This truly enormous number is the order of magnitude of the number of atoms in a handful of carbon or of the number of molecules in a test tube full of water. Avogadro's number is therefore typical of the number of molecules involved in chemical laboratory experiments. The number of molecules in n moles of a substance is given by $N = nN_0$. To determine the number of moles of a measured mass of a pure substance we must know the molecular weight of the molecules of the substance (or the atomic weight of the atoms). This is the ratio of the mass m of a molecule (or atom) of the substance to one-twelfth the mass of an atom of carbon-12. In other words, a molecular or atomic weight is a dimensionless ratio of masses given by $\mu = m/(\frac{1}{12}$ mass of carbon-12 atom). From this

definition, the atomic weight of carbon-12 is exactly 12. By experimental measurement of mass ratios we obtain results such as $\mu = 1.007825$ for hydrogen-1 and $\mu = 15.99491$ for oxygen-16, which illustrates that the atomic weight of an atom is approximately the same as the mass number of the atom's nucleus. These definitions imply that one mole of carbon-12 has a mass of exactly 12 grams and in general that one mole of a substance of molecular weight μ has a mass of μ grams. Consequently, the number of moles n of a measured mass of a substance composed of molecules of molecular weight μ (or of atoms of atomic weight μ) is given by

$$n = \frac{\text{mass in grams}}{\mu}.$$

a way that the Kelvin temperature T of the gas (at low pressures and sufficiently high temperature) is proportional to this product PV.

All these statements are neatly summarized in the one equation

$$PV = NkT, \tag{7.3}$$

where k, known as Boltzmann's constant, will shortly be discussed in detail. This equation is known as the *ideal gas law* and a fictitious substance that would behave this way for all temperatures and pressures is called an *ideal gas*. For real gases, the ideal gas law is an excellent approximation when the pressure is not too high and the temperature is well above the condensation temperature, but deviations from this behavior do occur under other conditions. Certainly when a gas is liquefied the ideal gas law is inapplicable.

Various Temperature Scales

Before further discussion of the important ideal gas law, we will digress briefly to give the relationship between the Kelvin temperature T which appears in the ideal gas law and other commonly used temperature scales. Household and medical thermometers in the United States still use the Fahrenheit scale. On this scale, when the pressure is one atmosphere (1.01×10^5 N/m²), one finds that the temperature of melting ice is 32.0° F (32.0 degrees Fahrenheit) and the temperature of boiling water is 212.0° F. The corresponding temperatures on the Celsius temperature scale, commonly used in scientific work, are 0.00° C (0.00 degrees Celsius) and 100.00° C. The relationship between the Celsius temperature t_C and the Fahrenheit temperature t_F is

$$t_C = \tfrac{5}{9}(t_F - 32°\text{ F}).$$

For example, what is often referred to as normal room temperature, 68° F, corresponds to $t_C = \tfrac{5}{9}(68 - 32)° \text{C} = 20°$ C.

The Celsius temperature scale, and, indirectly, the Fahrenheit temperature scale, are defined in terms of the Kelvin temperature scale by the equation

$$t_C = T - 273.15°\text{ K}$$

where T is the Kelvin temperature corresponding to the Celsius temperature t_C. Thus 0° C corresponds to 273.15° K (273.15 degrees Kelvin), and normal room temperature is 293.15° K. The unit of temperature, the degree, has the same magnitude on the Kelvin and Celsius scales. The Fahrenheit, Celsius, and Kelvin temperature scales are compared in Fig. 7-2.

A precise definition of the Kelvin temperature scale will require an analysis of thermal phenomena which can only be given later in Section 9.5. We explore in the following sections the interesting relationship between the Kelvin temperature and molecular motion.

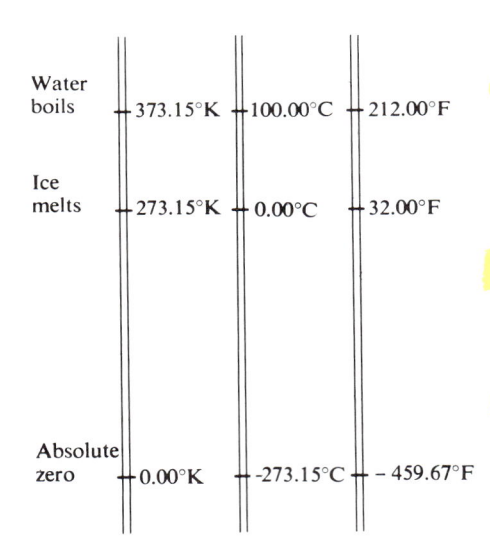

	Kelvin	Celsius	Fahrenheit
Water boils	373.15°K	100.00°C	212.00°F
Ice melts	273.15°K	0.00°C	32.00°F
Absolute zero	0.00°K	-273.15°C	-459.67°F

Fig. 7-2 Comparison of the Kelvin, Celsius, and Fahrenheit scales of temperature.

Boltzmann's Constant k

We return to the ideal gas law, $PV = NkT$, and consider different types of gases. In 1811, some hundred years before the number of molecules in a sample of gas could be determined, A. Avogadro (1776–1856), after considering evidence accumulated from study of chemical reactions between gases, advanced the following hypothesis: equal volumes of different gases at the same temperature and pressure contain the same number of molecules. Experiments confirm Avogadro's hypothesis; at a pressure of 1.01×10^5 N/m² and a temperature of 273° K, a cubic meter of any gas contains 2.68×10^{25} molecules. From this experimental result Boltzmann's constant can be evaluated. Solving Eq. (7.3) for k we obtain

$$k = \frac{PV}{NT}.$$

Therefore,

$$k = \frac{(1.01 \times 10^5 \text{ N/m}^2) \times (1.00 \text{ m}^3)}{(2.68 \times 10^{25} \text{ molecules}) \times (273° \text{ K})}$$

$$= 1.38 \times 10^{-23} \frac{\text{joule}}{\text{molecule-° K}}.$$

Boltzmann's constant, according to Avogadro's hypothesis, has this same value for all gases. We say that k is a universal constant.

The use of the ideal gas law is illustrated in the following examples.*

Example 2 A mass of 32.0×10^{-3} kg of oxygen is contained in a cylinder at a pressure of 1.01×10^5 N/m² and a temperature of 273° K. What volume does the gas occupy? (The mass of an oxygen molecule is 5.32×10^{-26} kg.)

Solution The number of oxygen molecules in the cylinder is

$$N = \frac{M}{m} \quad (7.2)$$

$$= \frac{32.0 \times 10^{-3} \text{ kg}}{5.32 \times 10^{-26} \text{ kg}}$$

$$= 6.02 \times 10^{23} \text{ molecules}.$$

The ideal gas law, $PV = NkT$, gives

$$V = \frac{NkT}{P}$$

$$= \frac{(6.02 \times 10^{23} \text{ molecules}) \left(1.38 \times 10^{-23} \frac{\text{joule}}{\text{molecule-° K}}\right)(273° \text{ K})}{(1.01 \times 10^5 \text{ N/m}^2)}$$

$$= 22.4 \times 10^{-3} \text{ m}^3.$$

---

*Making the substitution $N = nN_0$ in Eq. (7.3), we can express the ideal gas law in terms of the number of moles n of the gas:

$$PV = nRT,$$

where R, called the gas constant, is given by

$$R = N_0 k = 8.314 \text{ joules/mole °K}$$

and N_0 is Avogadro's number. To illustrate the use of the ideal gas law in the alternative form, $PV = nRT$, we shall reconsider Example 2. Suppose that, instead of knowing the mass of the oxygen molecules, we are given that their molecular weight is $\mu = 32.0$. [The molecules of gaseous oxygen (O_2) are formed from two oxygen atoms, and the molecular weight is approximately $2 \times 16 = 32$.] A mass of 32.0×10^{-3} kg of oxygen therefore comprises

$$n = \frac{32.0 \text{ grams}}{32.0} = 1.00 \text{ mole}.$$

The volume occupied by this gas at a pressure of 1.01×10^5 N/m² and a temperature of 273°K is therefore given by

$$V = \frac{nRT}{P}$$

$$= \frac{(1.00 \text{ mole})(8.31 \text{ joules/mole °K})(273°\text{K})}{1.01 \times 10^5 \text{ N/m}^2}$$

$$= 22.4 \times 10^{-3} \text{ m}^3.$$

It is worth noting that, as this calculation illustrates, a mole of any ideal gas occupies 22.4 liters under standard conditions of pressure and temperature (that is, one atmosphere at 0°C).

7.1 GASES

Example 3 A gas which occupies a volume of 2.24 m³ at a pressure of 1.01 × 10⁵ N/m² and a temperature of 273° K is compressed to a volume of 1.12 m³ and heated to a temperature of 819° K. Find the pressure.

Solution Denoting the pressure, volume, and temperature in the first state by P_1, V_1, and T_1 and in the second state by P_2, V_2, and T_2, we shall first deduce a simple relationship between these quantities which is applicable whenever we are concerned with the same number (N) of molecules in two different states. The ideal gas law, Eq. (7.3), gives

$$kN = \frac{P_1 V_1}{T_1} = \frac{P_2 V_2}{T_2}.$$

The desired relationship is

$$\frac{P_1 V_1}{T_1} = \frac{P_2 V_2}{T_2}. \tag{7.4}$$

To find the pressure P_2, we use Eq. (7.4) and obtain

$$P_2 = \left(\frac{V_1}{V_2}\right)\left(\frac{T_2}{T_1}\right) P_1$$

$$= \left(\frac{2.24 \text{ m}^3}{1.12 \text{ m}^3}\right) \times \left(\frac{819° \text{ K}}{273° \text{ K}}\right) \times (1.01 \times 10^5 \text{ N/m}^2)$$

$$= 2.00 \times 3.00 \times (1.01 \times 10^5 \text{ N/m}^2)$$

$$= 6.06 \times 10^5 \text{ N/m}^2.$$

7.2 KINETIC THEORY INTERPRETATION OF TEMPERATURE

We now attempt to understand the behavior of gases (as described by the ideal gas law, $PV = NkT$) in terms of molecular motion. A gas is a collection of billions of molecules engaged in random motion. The impact of these molecules with the container walls gives rise to the force exerted by the gas on the walls. This force and the associated gas pressure can be calculated from the average of the translational kinetic energy ($\frac{1}{2}mv^2$) possessed by the gas molecules. Then comparison with the ideal gas law reveals that temperature is a measure of the average translational kinetic energy of the molecules.

Temperature and Molecular Kinetic Energy

Let us first examine a very simple situation—one molecule which travels with a speed v back and forth in an empty box of length L with ends of cross-sectional area A (Fig. 7-3). The molecule exerts a force on the end of the box as it bangs into it and rebounds. The average force arising from such molecular impacts, divided by the area of the box surface, is

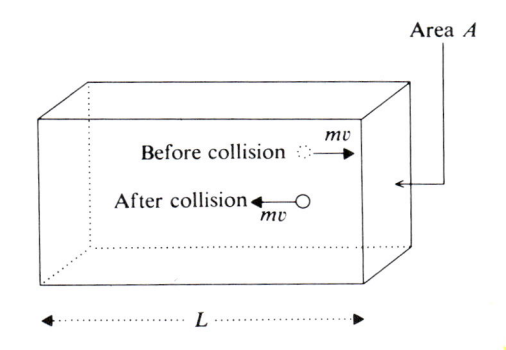

Fig. 7-3 Molecule exerts a force on right-hand wall as it collides and rebounds.

the pressure P contributed by this molecule. This pressure is clearly related to the molecule's speed v. In a collision with the right-hand wall we assume that the molecule's momentum vector changes from a vector of magnitude mv directed to the right to a vector of magnitude mv directed to the left, which implies that the *change* of momentum Δp is $2mv$. The time interval Δt between collisions is $2L/v$. The magnitude of the average force F exerted by the wall on the molecule is, according to Eq. (3.2),

$$F = \frac{\Delta p}{\Delta t} = \frac{2mv}{2L/v} = \frac{mv^2}{L}.$$

Now, by Newton's third law, F is also the magnitude of the average force exerted by the molecule on the right-hand wall. The average pressure on this wall is therefore

$$P = F/A = mv^2/LA.$$

Since LA is volume V of the box, our last equation can be rewritten to yield

$$PV = mv^2. \tag{7.5}$$

This result shows that, at least in the case of a gas consisting of this one molecule moving with constant speed, the product PV is determined by the molecular kinetic energy, $\frac{1}{2}mv^2$.

In a more general situation we find a similar result. It can be shown that with N molecules in the box with different molecules moving with different speeds in various directions, colliding with one another and with the walls, the product PV is given by

$$PV = \tfrac{2}{3}N \times (\text{average of } \tfrac{1}{2}mv^2), \tag{7.6}$$

provided that the range of the intermolecular forces is small compared with the average distance between molecules. A comparison of this kinetic theory equation with the ideal gas law,

$$PV = NkT, \tag{7.3}$$

shows that

$$(\text{average of } \tfrac{1}{2}mv^2) = \tfrac{3}{2}kT. \tag{7.7}$$

Here is the kinetic theory interpretation of temperature. *The Kelvin temperature T is proportional to the average translational kinetic energy of the molecule's random thermal motion.*

The molecular kinetic energy that determines the temperature is the so-called *translational* kinetic energy, $\frac{1}{2}mv^2$, associated with the motions of the centers of mass of the molecules. Kinetic energy associated with vibrations or rotations of a molecule's atoms relative to its center of mass does not appear in Eq. (7.7). Such kinetic energy does not contribute to the temperature of a substance.

Example 4 Evaluate in joules and also in electron volts the average translational kinetic energy of a molecule in a gas at a temperature of 273° K (0° C or 32° F).

Solution From Eq. (7.7),

(average of $\tfrac{1}{2}mv^2$) $= \tfrac{3}{2}kT$

$$= \tfrac{3}{2} \times \left(1.38 \times 10^{-23} \frac{\text{joule}}{\text{molecule-°K}}\right) \times (273° \text{ K})$$

$$= 5.65 \times 10^{-21} \text{ joule}$$

$$= \frac{(5.65 \times 10^{-21} \text{ joule})}{(1.60 \times 10^{-19} \text{ joule/eV})}$$

$$= 0.0353 \text{ eV}.$$

Thermal Speed

The molecules of a substance undergo incessant thermal motion. At any instant, different molecules are generally moving with different speeds. A typical molecular speed encountered at temperature T is given by the thermal speed v_T defined by

$$\tfrac{1}{2}mv_T^2 = \text{average of } \tfrac{1}{2}mv^2 = \tfrac{3}{2}kT.$$

Therefore,

$$v_T = \sqrt{\frac{3kT}{m}}. \tag{7.8}$$

For hydrogen molecules, $m = 3.35 \times 10^{-27}$ kg, so at room temperature (293° K) their thermal speed is

$$v_T = \sqrt{\frac{3 \times \left(1.38 \times 10^{-23} \frac{\text{joule}}{\text{molecule-°K}}\right) \times (293° \text{ K})}{3.35 \times 10^{-27} \text{ kg}}}$$

$$= 1.9 \times 10^3 \text{ m/sec}.$$

This is a speed of over 4000 miles per hour.

Eq. (7.8) shows that the more massive the particle, the smaller its thermal speed. Thermal motion is very violent for molecules and is still appreciable in the Brownian motion of visible particles discussed below, but is entirely negligible for massive bodies.

Brownian Motion

In 1827, an English botanist, Robert Brown, peering through a microscope, noticed that tiny grains of plant pollen suspended in a liquid were

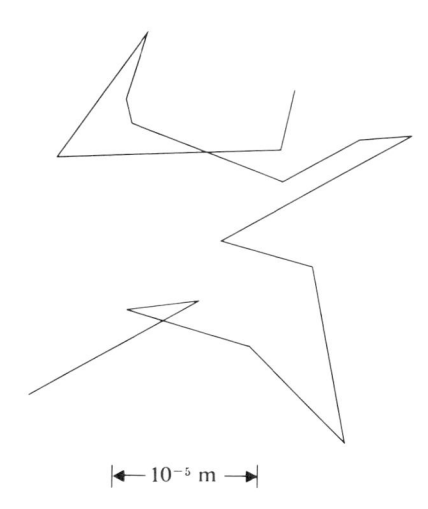

Fig. 7-4 Brownian motion. The line joins positions of a small particle (mass 10^{-16} kg) observed at 2-minute intervals, as it is buffeted about by water molecules.

$\leftarrow 10^{-5}$ m $\rightarrow$

in continual motion, zigzagging about in a random fashion (Fig. 7-4). Brown, suspecting that this motion was connected with the fact that the pollen was living matter, was surprised to find that a similar random motion is executed by a small speck of dust or a smoke particle suspended in still air or in a liquid.

Kinetic theory gives the explanation for this "Brownian motion" of visible specks. Such particles (each composed of some million molecules) are buffeted about by molecular impacts, and just like individual molecules, acquire an average kinetic energy of $\frac{3}{2}kT$. This was the central idea in the quantitative treatment of Brownian motion advanced by Einstein in 1905. He showed how to calculate the average kinetic energy, $\frac{3}{2}kT$, of a visible particle executing Brownian motion from measurement of the distance it has wandered in a measured time. Once $\frac{3}{2}kT$ and hence k was known, it was possible to make a reliable calculation of the number N of molecules in a sample of gas, using [from Eq. (7.3)]

$$N = \frac{PV}{kT},$$

and then determine the mass m of the individual molecules in a sample of mass M from Eq. (7.2),

$$m = M/N.$$

Consequently, before the invention of mass spectrometers (Section 12.3) in the second decade of this century, Brownian motion served as an important source of molecular data. And it was the recognition of Brownian motion as a visible manifestation of molecular impacts that convinced even Ostwald and Mach of the existence of molecules and of the value of Boltzmann's life work.

7.3 THERMAL ENERGY AND TEMPERATURE

The thermal energy of many substances at room temperature comprises important contributions from

1. The translational kinetic energy of molecules associated with their random thermal motion,
2. The rotational and vibrational kinetic energy arising from the motion of the atoms of a molecule relative to the mass center of the molecule,
3. The potential energy associated with the interaction of the atoms within a molecule, and
4. The potential energy arising from the interaction of one molecule with another. When molecules are packed tightly together, as in a liquid or a solid, they continually exert appreciable forces on each other. The associated potential energy is then by no means the negligible quantity that it is assumed to be for an ideal gas.

There is an important distinction between the temperature of an object and its thermal energy. The temperature, according to the result,

$$\text{average of } \tfrac{1}{2}mv^2 = \tfrac{3}{2}kT \tag{7.7}$$

is a measure of the average translational kinetic energy *only*. This is true for any substance, not just an ideal gas. The thermal energy generally includes not only these translational kinetic energies but also the mechanical energy of motions inside the molecules and intermolecular potential energies.

The zero of temperature on the Kelvin scale, called *absolute zero*, corresponds to zero translational kinetic energy of the molecules. Thermal motion ceases entirely at absolute zero, at least according to Eq. (7.7). This is the prediction of classical mechanics. However, the Heisenberg uncertainty relation of quantum mechanics* requires that, even at absolute zero, there must remain some vibrational motion of atoms within the molecules, or vibrations of atoms about crystal lattice points. These vibrations are called *zero-point vibrations,* and the energy associated with these vibrations is called the *zero-point energy.*

Energy levels are important in quantum phenomena. Any bound system, be it composed of atoms rotating or vibrating within a molecule or of electrons bound to an atom has as possible values of its total energy only certain discrete values which are named the *energy levels* of the system. As the molecules of a gas are banged about by the thermal agitation at a Kelvin temperature T, some molecular collisions occur where the molecules approach at high speed but limp away from each other. During the collision, translational kinetic energy is converted into excitation energy of one of the molecules. In absorbing an energy ΔE, this molecule makes a transition from a lower energy level E_ℓ to an upper

---

*At sufficiently low temperatures and correspondingly low energies, the conditions for the validity of Newtonian mechanics are no longer satisfied and quantum phenomena (Chapter 16) become evident.

energy level E_u, where
$$\Delta E = E_u - E_\ell.$$

In subsequent collisions, this molecule may lose this energy passing back from level E_u to E_ℓ; or alternatively it may be promoted to a still higher energy level.

Eventually, after many collisions an equilibrium distribution is obtained, with the number of molecules in level E_ℓ fluctuating about an equilibrium population N_ℓ with a lesser equilibrium population N_u in the upper level E_u. The ratio of equilibrium populations N_u/N_ℓ will be small unless the temperature is high enough so that the average translational kinetic energy, $\frac{3}{2}kT$, is large enough to be comparable to the energy difference, $\Delta E = E_u - E_\ell$. For instance, if we let E_ℓ represent the *ground* state energy and E_u the first *excited* state, and if the temperature is so low that $\frac{3}{2}kT$ is much less than ΔE, then the population of molecules in the ground state is very much greater than the population in the excited state. At a temperature this low, molecules hardly ever collide with enough kinetic energy to make excitation possible. We say the possibility of excitation to the energy level E_u is *frozen out*.

The way that the thermal energy of any substance is partitioned among the various possibilities therefore is dependent on the size of $\frac{3}{2}kT$ relative to the energy difference ΔE between different energy levels of the particles. As the temperature is increased, an increasing variety of excitations becomes possible. The situation at various temperatures is indicated in Table 7-1.

Table 7-1 TYPICAL PHENOMENA AT VARIOUS TEMPERATURES

Temperature (°K)	$\frac{3}{2}kT$	Phenomena
0	0 eV	Zero-point vibration
273	0.035 eV	Ice melts
2000	0.26 eV	Most substances gaseous
6000	0.78 eV	Temperature at sun's surface
10^6	130 eV	Light elements largely ionized
10^7	1.3 KeV	Temperature at sun's center
10^{10}	1.3 MeV	Nuclei undergo thermal excitation and disintegration

In terms of the known temperatures, man lives in a relatively restricted temperature region. Most people have not experienced the $-35°C$ temperatures that occur in some cities in Northern Canada or the unbearable 40°C or more* of Death Valley. Yet this is a range of only 75 Celsius (or Kelvin) degrees. Compare this limited range with that shown for various conditions in Fig. 7-5 on the next page.

*Hottest temperature recorded in Death Valley was 134°F (i.e., about 57°C), on July 10, 1913.

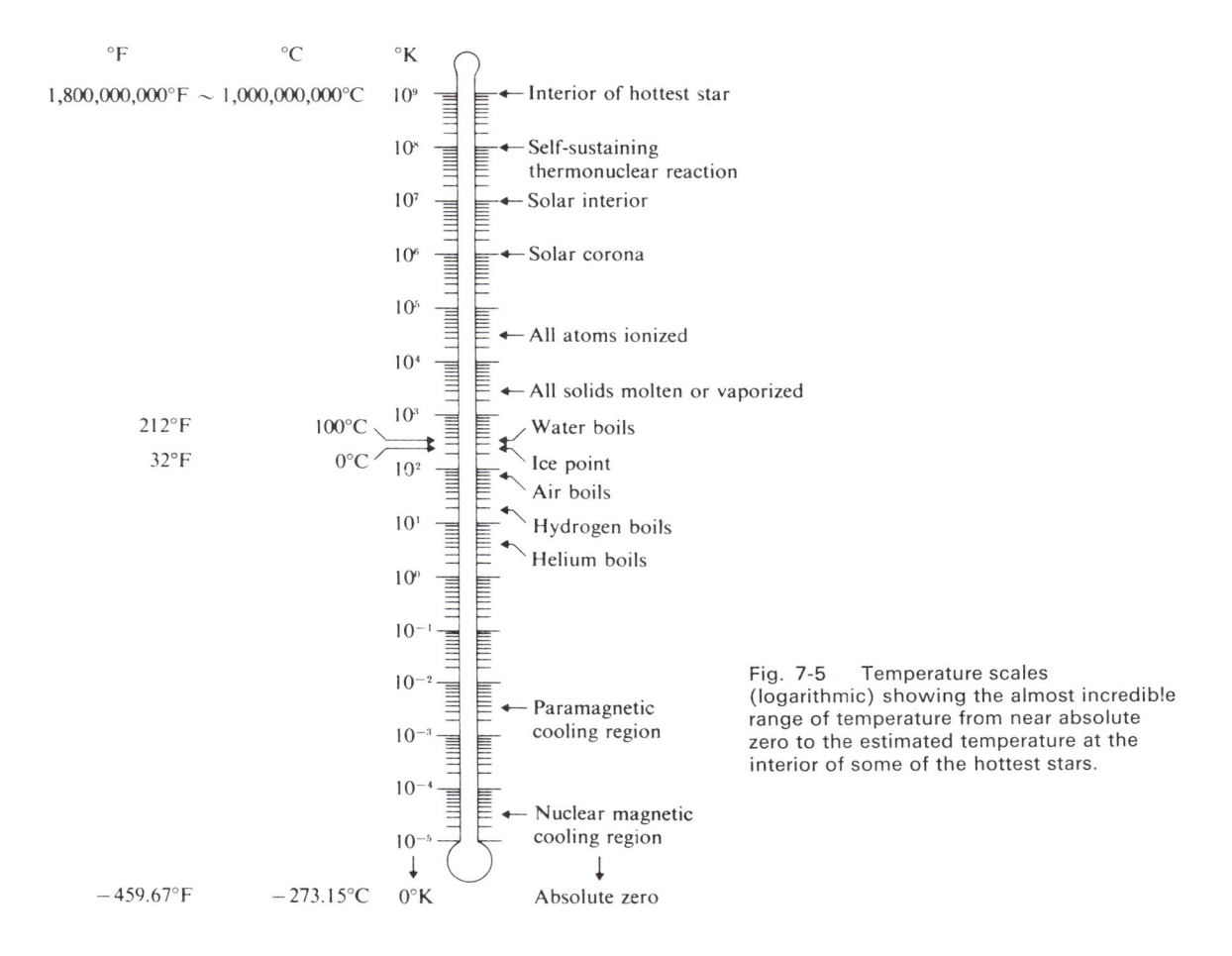

°F	°C	°K	
1,800,000,000°F ~ 1,000,000,000°C		10^9	← Interior of hottest star
		10^8	← Self-sustaining thermonuclear reaction
		10^7	← Solar interior
		10^6	← Solar corona
		10^5	
		10^4	← All atoms ionized
			← All solids molten or vaporized
212°F	100°C	10^3	Water boils
32°F	0°C	10^2	Ice point
			Air boils
		10^1	Hydrogen boils
			Helium boils
		10^0	
		10^{-1}	
		10^{-2}	← Paramagnetic cooling region
		10^{-3}	
		10^{-4}	
		10^{-5}	← Nuclear magnetic cooling region
−459.67°F	−273.15°C	0°K	Absolute zero

Fig. 7-5 Temperature scales (logarithmic) showing the almost incredible range of temperature from near absolute zero to the estimated temperature at the interior of some of the hottest stars.

7.4 CALCULATION AND MEASUREMENT OF THERMAL ENERGY $\bar{E}$

The energy transfers between *macroscopic systems* can be understood only if we know how to keep track of changes in the thermal energies of these systems. While the thermal energy of a very few macroscopic systems can be calculated rather simply in terms of their temperature and volume or pressure, more generally such a theoretical calculation is a formidable problem. Fortunately, as Joule demonstrated (Section 4.2), the experimental measurement of changes in the thermal energy of a system is a straightforward matter. It requires measuring the mechanical work done on the system in a process in which this system neither gives off nor absorbs heat.

The symbol $\bar{E}$ will be used to denote the *thermal energy* of a macroscopic system. Consider a macroscopic system such as a bouncing ball:

280 DISORDERED ENERGY

the system has a kinetic energy due to its bulk motion and also a potential energy because of the altitude of the ball. This energy associated with bulk motion or position of the ball as a whole is to be sharply distinguished from the thermal energy $\bar{E}$ of the ball. The bulk motion of the ball involves an *ordered* motion of the molecules of the system. But $\bar{E}$, the thermal energy, is the internal energy of the ball arising from randomly directed molecular motions and randomly distributed internal excitations. *Thermal energy $\bar{E}$ is disordered energy.* To evaluate $\bar{E}$, we use a reference frame in which the macroscopic system is stationary.

Thermal Energy of Gases

It is possible to calculate the thermal energy of some ideal gases using the ideas discussed in Section 7.3. If the only contributions to the thermal energy are due to a molecule's translational kinetic energy, then a system of N such molecules will have a thermal energy which is just N times the average translational kinetic energy of each molecule. Using Eq. (7.7) we therefore obtain

$$\bar{E} = \tfrac{3}{2}NkT \tag{7.9}$$

This is all there is to the thermal energy of the monatomic gases: helium, neon, argon, krypton, xenon, and radon. At usual terrestrial temperatures, kT is much less than the approximately 10-eV energy required to cause one of these atoms to make a transition from its ground state to its first excited state. Therefore, each of the atoms of such a gas acquires an average translational kinetic energy $\tfrac{3}{2}kT$, but no internal excitation of an atom occurs.

The thermal energy of molecules formed from more than one atom can include contributions from excitation of *atomic motions* within a molecule. Such contributions will be significant only at temperatures high enough that kT is comparable to the separation of the energy levels involved. Hydrogen is an interesting example. Below about 70° K, its thermal energy is given by $\bar{E} = \tfrac{3}{2}NkT$. As the temperature increases, a molecule spends more and more time rotating like a dumbbell (Fig. 7-6). These dumbbell rotations increase the value of $\bar{E}$ until, above room temperature, one finds $\bar{E} = \tfrac{5}{2}NkT$. At temperatures over 700° K the excitation of still another atomic motion acquires a significant probability. The hydrogen atoms within a molecule can *vibrate* towards and away from each other. The excitation of these vibrations raises the thermal energy of hydrogen gas to nearly $\tfrac{7}{2}NkT$ at temperatures over 3000° K. At still higher temperatures the molecule is knocked apart into "dissociated" hydrogen atoms.

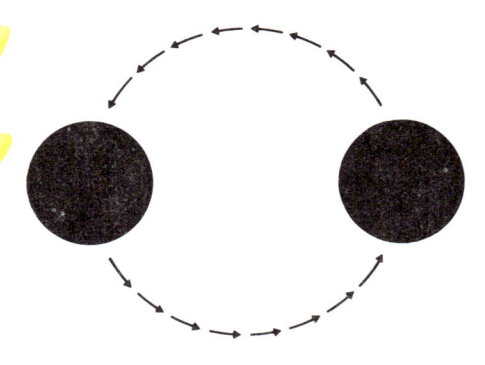

Fig. 7-6 A representation of the "dumbbell rotations" of a molecule composed of two atoms.

Thermal Energy of Crystals

In a crystalline solid each atom vibrates about an *equilibrium position*. At low temperatures the possibility of excitation of these vibrations is "frozen out," but at high temperatures each atom acquires an average kinetic energy of $\frac{3}{2}kT$ and an average potential energy due to its interaction with other atoms of $\frac{3}{2}kT$. The result is that at sufficiently high temperatures, the thermal energy of any crystalline solid with N atoms is given by the simple expression

$$\bar{E} = 3NkT.$$

Tables of Thermal Energy

For a given amount of most substances, the thermal energy depends not only on temperature but also on pressure. When the molecules are squeezed closer together, there is a change in the potential energy due to the interaction between molecules. We can rarely find a simple formula for the thermal energy of a substance at specified values of temperature and pressure. However, we can always build up a table of values (Table 7-2) which summarizes experimental measurements or theoretical calculations. Only *changes* in thermal energy can be measured. Consequently we record the changes from any convenient reference condition (97° C and 1 atmosphere in Table 7-2). Although not at all necessary, it is sometimes convenient to follow the common practice in this branch of science of using the kilocalorie as the unit of energy. By international agreement, the thermochemical *kilocalorie* is *defined* by

$$\text{one kilocalorie} = 4.184 \times 10^3 \text{ joules, } exactly.$$

Table 7-2. THERMAL ENERGY PER KILOGRAM OF H_2O AT 1.00 ATMOSPHERE WHEN WATER AT 97°C IS ASSIGNED ZERO THERMAL ENERGY

	Temperature	Thermal energy per kilogram (kilocalorie/kg)
Water	97° C	0.0
Water	98° C	1.0
Water	99° C	2.0
Water	100° C	3.0
Steam	100° C	501.7

Data for the thermal energy tables for any substance can be acquired, as described in Section 4.2, by doing a measured amount of macroscopic work, W, on the substance and measuring the initial and final temperatures and pressures. Precautions must be taken to avoid any loss of thermal energy during the process.

Work Done during a Compression or Expansion

When the piston of Fig. 7-1 is pushed in by an external force, the force does work on the gas within the cylinder. Similarly whenever a

282 DISORDERED ENERGY

substance is compressed, work is done on the substance.

When a substance expands against external forces, work is done by the substance on whatever exerts the external force. Equivalently we can say that there is a negative amount of work done *on* the substance during such an expansion.

Example 5 One kilogram of H_2O is contained in a large cylinder fitted with a piston. The water temperature is 100°C and the pressure is 1.013×10^5 N/m². Heat is applied until the water is entirely converted to steam at the same temperature and pressure. Because of this process, the change in the volume occupied by the H_2O molecules is 1.672 m³. Evaluate the work done by the steam on the piston.

Solution The steam exerts a force

$$F = PA \qquad (7.1)$$

on the piston of cross-section A. When the piston moves out a distance Δs, the change in cylinder volume (Fig. 7-7) is given by

$$\text{Volume change} = A\,\Delta s.$$

The work ΔW done by the force F when the piston moves a distance Δs is

$$\Delta W = F\Delta s. \qquad (4.2)$$

Substituting PA for F this yields

$$\begin{aligned}\Delta W &= PA\,\Delta s \\ &= P \times \text{(volume change)} \qquad (7.10)\\ &= (1.013 \times 10^5 \text{ N/m}^2)(1.672 \text{ m}^3) \\ &= 1.694 \times 10^5 \text{ joules} \\ &= 40.5 \text{ kilocalories.}\end{aligned}$$

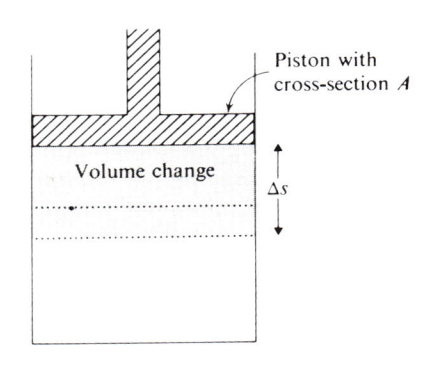

Fig. 7-7 When the piston moves out a distance Δs, the cylinder volume changes by an amount equal to $A\Delta s$.

Thus the steam does 40.5 kilocalories of work on the piston during the expansion. We could also say that the work done on the steam is -40.5 kilocalories.

In the following discussions of the first and second laws of thermodynamics we shall assume that the only way work is done is by expansion or compression of the system. *Then during any process in which the system's volume is maintained constant, no work is done.*

7.5 HEAT

The term *heat* which was introduced in Section 4.2 is the name given to the *energy transferred* from one system to another because of a *temperature difference* between the systems.

It is interesting to digress for a moment and realize that in the eighteenth century there was a very understandable confusion about the nature of heat. Calorimeter experiments, to be described in the next section, were in accord with the theory that heat was a light invisible fluid called "*caloric*" which flowed from a hotter to a colder body. Since a fluid has volume, this caloric theory also explained quite nicely why an object expanded when heated. But after experiments by Rumford near the end of the eighteenth century, particularly his demonstration that an apparently unlimited amount of "caloric" could be produced by continual boring of a cannon with a dull drill, the caloric theory became implausible.* Following Joule's experiments some fifty years later, heat was recognized as a form of energy and the notion of caloric faded into history.

*Readers having a bent for historical pursuits will enjoy the book by Sanborn Brown listed in the Suggested Reading.

In a general interaction the thermal energy of any system, say a gas in a cylinder, can be increased in two ways:

1. Macroscopic mechanical energy can be supplied to the gas by external forces which do work in compressing the gas;
2. Thermal energy can be transferred from the cylinder walls to the gas if the wall temperature is above the gas temperature. The energy transferred in this second way is what is called the *heat absorbed* by the gas or the *heat given off* by the cylinder walls.

For any such process, the law of conservation of energy requires that the change in thermal energy, $\Delta \bar{E}$, of the system be given by

$$\Delta \bar{E} = (\text{work for process}) + (\text{heat for process}). \quad (7.11)$$

This equation is a statement of the *first law of thermodynamics* and provides us with a quantitative definition of heat in terms of a thermal energy change and a macroscopic mechanical work, both of which can be determined by direct laboratory measurement.

When Eq. (7.11) is used, the work and the heat for a process are counted positive for energy supplied to the system and negative for energy taken away from the system. That is, the heat for a process is positive when the system absorbs heat and negative when the system gives off heat. Work for a process is positive when the system is compressed and negative when the system expands against external forces.* No work is done during a process in which the volume is constant.

*A different sign convention and a different notation are used in some of the books in the Suggested Reading. Many authors write the first law of thermodynamics in the form $Q = \Delta U + W$ where Q denotes the heat absorbed by the system, W denotes the work done by the system on the surroundings, and ΔU is the change in the thermal energy of the system. The quantity that we have called thermal energy is often termed *internal energy* in other books.

Example 6 Find the heat absorbed by the H_2O molecules in the process described in Example 5.

Solution In thermodynamics, as in mechanics, we must specify the system to which we apply physical laws such as Eq. (7.11). In this case we select as the system the contents of the cylinder, that is, the 1.00 kg

284 DISORDERED ENERGY

of H_2O molecules. Then from Example 5, we have

$$\text{Work for process} = -40.5 \text{ kilocalories.}$$

Notice the minus sign. The expansion of the system against external forces transfers 40.5 kilocalories of macroscopic mechanical energy to the matter outside our system. From Table 7-2, this system's final thermal energy is

$$(1.00 \text{ kg})(501.7 \text{ kilocalories/kg}) = 501.7 \text{ kilocalories,}$$

and the initial thermal energy is

$$(1.00 \text{ kg})(3.0 \text{ kilocalories/kg}) = 3.0 \text{ kilocalories.}$$

The change in thermal energy is therefore

$$\Delta \bar{E} = (501.7 - 3.0) \text{ kilocalories}$$
$$= 498.7 \text{ kilocalories.}$$

The heat for this process, according to the definition of heat, given by Eq. (7.11), is

$$\text{Heat for process} = \Delta \bar{E} - (\text{work for process})$$
$$= 498.7 \text{ kilocalories} - (-40.5 \text{ kilocalories})$$
$$= 539.2 \text{ kilocalories.}$$

The sign of this term is positive so the system *absorbs* heat from its surroundings. The quantity that we have evaluated is called the *latent heat of vaporization* of water. It will be discussed in Section 8.4.

Example 7 The piston of Example 5 is clamped so that the volume of the cylinder's contents remains constant. Now heat is applied and the pressure of the steam and water within the cylinder rises to 10.0 atmospheres. The piston is now unclamped and allowed to move out slowly with the pressure maintained at 10.0 atmospheres as heat continues to be applied. When the steam has expanded to the same volume attained in Example 5, the cylinder is again clamped. Now heat is given off by the steam until its temperature falls to 100°C and the pressure is again 1 atmosphere. Find the work and the heat for this entire process.

Solution In the initial state there is 100% water at 100°C and at a pressure of 1 atmosphere. In the final state there is 100% steam at 100°C and at a pressure of 1 atmosphere. The system therefore has the same initial and final states as in Example 6, so the initial and final thermal energies are the same as in Example 6. Consequently, for the entire process,

$$\Delta \bar{E} = 498.7 \text{ kilocalories}$$

as in Example 6.

No work is done during the processes in which the volume is constant. During the expansion at a pressure of 10.0 atmospheres we obtain, from Eq. (7.10), a work which has a magnitude ten times that of Example 5. Thus for the entire three-stage process the work done on the system of H_2O molecules is

$$\text{Work for process} = 0 + (-405 \text{ kilocalories}) + 0$$
$$= -405 \text{ kilocalories}.$$

The heat for the entire process, from Eq. (7.11), is

$$\text{Heat for process} = \Delta \bar{E} - (\text{work for process})$$
$$= (498.7 \text{ kilocalories}) - (-405 \text{ kilocalories})$$
$$= 904 \text{ kilocalories}.$$

The system absorbs almost twice as much heat in this process as it did in the process of Example 6.

Examples 6 and 7 illustrate an important point. When the transformation of a system from a given initial state (such as water, 100°C, 1 atmosphere) to a given final state (steam, 100°C, 1 atmosphere) is accomplished by *different* processes, *different* amounts of heat and work are involved. One kilogram of steam at 100°C and at a pressure of 1 atmosphere contains a definite amount of thermal energy, but this collection of H_2O molecules can have acquired this thermal energy in processes requiring varying amounts of work and heat. Consequently no meaning can be attached to phrases like "the heat contained by the system" or the "work contained by the system." The first phrase is an unfortunate remnant of the old caloric theory of heat. Although this phrase still crops up frequently, particularly in popularized literature, the word "heat" is then being used colloquially in place of the correct scientific term, which is "the thermal energy of the system."

7.6 CALORIMETERS AND SPECIFIC HEATS ATS

Calorimetry

Laboratory measurements of macroscopic work and thermal energy changes, which serve to determine the heat involved in a process, constitute the science of *calorimetry*. Besides being of historical importance and a source of necessary information for engineers, calorimeters are still a vital research instrument for investigations in modern solid-state physics and the studies of such phenomena as superconductivity and superfluidity (Section 9.6).

Calorimeters can be used to determine an important thermal property of a substance, its *specific heat*. Recall from Section 4.2 that for a substance of mass m the specific heat s for a process is defined in terms of the heat for the process and the associated temperature change ΔT:

$$\text{Heat for process} = ms\Delta T. \qquad (7.12)$$

The mks unit for this specific heat is the joule/kg C°, but it is common practice to measure heat in kilocalories and the specific heat per unit mass in kilocalories/kg C° (Table 4-3).

A water calorimeter (Fig. 7-8) allows measurement of average specific heats in the temperature range from 0°C to 100°C using a procedure called the *method of mixtures*, which is illustrated in the following example.

Example 8 The inner can in the calorimeter (Fig. 7-8) is made of aluminum ($s = 0.217$ kilocalorie/kg C°) and has a mass of 0.400 kg. The can contains 0.140 kg of water. This water is thoroughly stirred and its temperature is measured to be 10.0°C. A 0.200-kg block of a new alloy, the substance under investigation, is heated to 100°C in a steam bath and then quickly transferred to the calorimeter. The water is again stirred and the common temperature of the alloy and water is found to be 30.0°C.* Find the specific heat (s) of this alloy.

Solution We assume that the heat-insulating jacket surrounding the inner calorimeter container reduces the heat loss to the surroundings to a negligible amount. Thus

(Heat given off by alloy) = (heat absorbed by water)
    + (heat absorbed by calorimeter).

We evaluate each term of this equation using Eq. (7.12), which gives the heat absorbed in a temperature rise or the heat given off in a temperature drop:

s (0.200 kg)(70.0 C°) = (0.140 kg)(1.00 kilocalorie/kg C°)(20 C°)
    + (0.400 kg)(0.217 kilocalorie/kg C°)(20 C°).

This yields

$$s = 0.324 \text{ kilocalorie/kg C°}$$

as the average value of the alloy's specific heat in the temperature range from 30°C to 100°C.

Modern calorimetry employs the Joule heating (Section 11.5) by an electric current to measure specific heats at temperatures ranging from 0.1°K to the melting point. A heater coil of resistance wire is wound around the sample under investigation. A current I is established in the heater coil of resistance R for a short time t and thereby supplies a thermal energy I^2Rt to the sample. The temperature rise ΔT of the

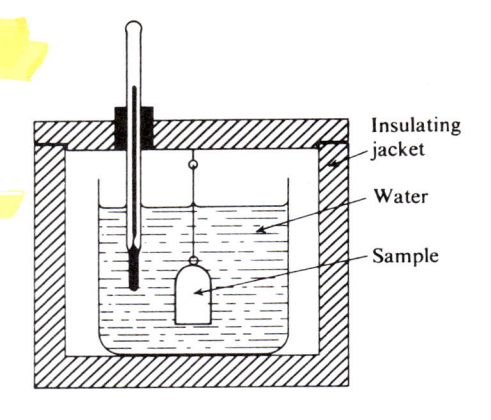

Fig. 7-8 A water calorimeter. The sample whose specific heat is under investigation is immersed in the water.

*In the following chapter we stress that different objects in contact in a thermally isolated system eventually reach a common temperature.

sample is measured. The specific heat per unit mass is then calculated from

$$I^2Rt = ms\Delta T.$$

In the realm of biology and medicine the method of calorimetry finds use as the most direct method of measuring the energy absorbed (absorbed dose) by the object of interest from a radiation beam, such as an x-ray beam. In the past, the most limiting factor in such calorimetric dosimetry has been the inadequate thermometry systems. However, since the middle of the present century a very sensitive temperature measuring device known as the thermistor (to be described in Section 8.3) has been increasingly improved. Correspondingly, so has the field of calorimetric dosimetry. Calorimetry also finds use in medical and physiological studies involving metabolism. Usually in this area one resorts to a more or less indirect calorimetry in that one measures oxygen consumption rather than attempting a direct measurement of the heat involved in the process.

Specific Heat per Molecule

The heat that must be absorbed by a pure substance to produce a given temperature rise ΔT is expected to be proportional to the number N of particles of the substance. It is convenient, therefore, to define a specific heat per molecule C by an equation similar to Eq. (7.12) but with m replaced by the number of molecules N:

$$\text{Heat for process} = NC\Delta T. \quad (7.13)$$

The specific heat per unit mass s for a pure substance is related (see Question 48) to the specific heat per molecule C by

$$C = (\text{mass of single molecule}) \times s$$

For a given temperature increase, the heat absorbed by a substance is less when the substance is confined to a constant volume and the pressure allowed to rise than when the pressure is maintained constant and the substance expands doing macroscopic mechanical work on the surroundings. *Evidently a given substance has different specific heats for processes performed in different ways.* Of particular interest is the specific heat for a process in which the substance is confined to a constant volume. This specific heat is called the *specific heat at constant volume* and is denoted by C_V.*

The specific heat per molecule C_V of a substance is simply related to the thermal energy, as we shall now show. For processes that involve no change in volume there will be no macroscopic mechanical work done on the substance, so the first law of thermodynamics [Eq. (7.11)] yields

$$\Delta \bar{E} = 0 + \text{heat for process}. \quad (7.14)$$

---

*Experiments usually determine a specific heat for a process in which the pressure has been maintained constant. The corresponding value of the specific heat at constant volume C_V can then be calculated (see the book by Zemansky in the Suggested Reading).

288 DISORDERED ENERGY

Equations (7.13) and (7.14) imply

$$\Delta \bar{E} = N C_V \Delta T. \qquad (7.15)$$

This important relationship gives us a way of calculating changes in thermal energy from calorimeter measurements of specific heats. Theoretical expressions for thermal energy can be checked by calculating C_V and comparing the predicted value with experiment.

Example 9 Calculate the specific heat at constant volume C_V for a monatomic ideal gas. (A molecule of a monatomic gas consists of a single atom.)

Solution We assume that the only contribution to the thermal energy of the gas is the translational kinetic energy of its molecules so that

$$\bar{E} = \tfrac{3}{2} NkT. \qquad (7.9)$$

For a process in which the temperature increases an amount ΔT,

$$\Delta \bar{E} = \tfrac{3}{2} Nk \Delta T.$$

Substitution of this result in Eq. (7.15) gives

$$\tfrac{3}{2} Nk \Delta T = N C_V \Delta T.$$

Therefore $C_V = \tfrac{3}{2} k$, a beautifully simple result, confirmed by experimental measurement with helium, neon, and other monatomic gases.

A calculation similar to that of Example 9 gives the specific heats corresponding to the expressions for the thermal energy of polyatomic gases and of crystalline solids that were discussed in Section 7.4. The result (see Question 50) for a crystalline solid at high temperatures,

$$C_V = 3k,$$

is again remarkably simple and independent of the nature of the atoms and the geometry of the crystal. In 1819 Dulong and Petit had noticed this regularity in the specific heats of crystalline solids at high temperatures. But at low temperatures C_V is less than $3k$, and near absolute zero C_V is essentially zero. The "freezing out" of various excitations as the temperature is lowered was a great mystery from the point of view of classical mechanics. Einstein in 1907 recognized that this could be understood if the possible changes in the energy of an oscillating atom could not be arbitrarily small. Specific heats thus provided early evidence of the quantum mechanical behavior of matter. In quantum mechanics any bound system such as an oscillating particle has certain discrete energy levels as the only possible values of its total mechanical energy. At low temperatures, when kT is much less than the energy required for transitions from one energy level to another, excitation of oscillations is a rare event.

QUESTIONS

1. What is the force exerted by the gas on the piston of Fig. 7-1 if the gas pressure is 5.0×10^5 N/m² and the piston cross-sectional area is 1.5×10^{-2} m²?

2. The air of the atmosphere exerts a force of 20.4 N on a 2.0×10^{-4} m² metal surface of a pressure gauge. What is the air pressure?

3. The mass of a hydrogen molecule determined by mass spectrometer measurement is 3.35×10^{-27} kg. How many hydrogen molecules are there in 2.0 kg of hydrogen?

4. What is the ideal gas law and what is an ideal gas?

5. The temperature of the human body is normally 98.6° F. Find the corresponding temperatures on the Celsius and the Kelvin scales.

6. Low pressures achieved in a vacuum system are often given in terms of a unit called the *torr* (in honor of Torricelli), which is related to the N/m² by

   $$1 \text{ torr} = 1.33 \times 10^2 \text{ N/m}^2.$$

   At an "ultra-high vacuum" of 10^{-10} torr, at a temperature of 300°K, how many molecules still remain in a volume of 1.00 cubic centimeter (1.00×10^{-6} m³)?

7. By pushing in the plunger, the volume of air within a syringe is compressed to one-fourth of its original volume. Initially the pressure was 1.01×10^5 N/m². Find the final pressure for the case when
   (a) the compression is performed so slowly that the temperature of the air within the syringe remains about the same as that of the surroundings,
   (b) the temperature of the air within the syringe increases from an original value of 300°K to a final value of 330°K.

8. Nitrogen is contained in a steel cylinder at a pressure of 2.0×10^5 N/m² when the temperature is 300°K. What is the pressure when the temperature is raised to 500°K?

9. By what percentage will the pressure in a car's tire increase if the temperature of the air in the tire is raised from 300°K to 400°K? Assume that the change in the volume of the tire is negligible.

10. A hydrogen balloon at the earth's surface has a volume of 5.0 m³ on a day when the temperature is 300°K and the pressure is 1.00×10^5 N/m². The balloon rises and expands as the pressure drops. What is the volume of the balloon (at an altitude of about 40 kilometers) when the pressure is merely 0.33×10^3 N/m² and the temperature is 260°K?

11. During the compression stroke of the piston, air within a cylinder of a diesel engine is compressed to 1/16 of its initial volume, and the pressure of this air increases from 1.0 atmosphere to 42 atmospheres. The initial temperature of the air was 300°K. Find its temperature at the end of the compression stroke.

12. Gas in a cylinder is heated until both the pressure and the volume have tripled. If the initial temperature was 300°K, what is the final temperature?

13. How many molecules of H_2O are there in a test tube containing 3.0 moles of water?

14. A test tube contains 48×10^{-3} kg of water. The molecular weight of water (H_2O) is approximately $2 + 16 = 18$. How many moles of water are there in the test tube?

15. What is the pressure in a 50-liter tank containing 0.44 kg of carbon dioxide (CO_2) at a temperature of 20°C? (1 liter = 10^{-3} m³)

16. What is the mass of the carbon dioxide contained in a 50-liter tank at a pressure of 3.0 atmospheres and a temperature of 30°C?

17. What is the mass of helium in a helium-filled balloon which has a volume of 6.00 m³ at a pressure of 0.50×10^5 N/m² and a temperature of 250°K?

18. What pressure is exerted by 0.600 kg of oxygen at 300°K within a 50-liter tank?

19. What interpretation of temperature is given by the kinetic theory?

20. What is the temperature of a substance whose molecules in random motion have an average translational kinetic energy of 1.0 eV?

21. (a) What is the thermal speed of oxygen molecules (mass 5.32×10^{-26} kg) at 300°K?
    (b) At what temperature will this thermal speed be doubled?

22. What is the relationship between the thermal speed v_T and $\overline{v^2}$? The thermal speed is often called the root mean square speed. Explain.

23. Explain how the mass of an individual molecule in a gas can be calculated, once the value of Boltzmann's constant has been determined.

24. Calculate the thermal speed of a smoke particle, of mass 1.0×10^{-16} kg, which executes Brownian motion in air at a temperature of 300°K.

25. Explain how Avogadro's number can be calculated once the value of Boltzmann's constant has been determined.

26. What is the distinction, as far as kinetic theory is concerned, between the temperature and the thermal energy of a collection of molecules.

27. (a) Calculate the thermal speed of argon molecules ($\mu = 40$) at 0°C.
    (b) What is the kinetic energy of a mole of argon gas at this temperature?
    (c) What is the total momentum relative to the laboratory of this amount of argon in a container which is at rest in the laboratory?

28. (a) Deduce an expression for a temperature at which the population of molecules in an excited state E_u will be comparable to the population in the ground state E_l.
    (b) If $E_u - E_l = 1.0$ eV, what is the value of the temperature found in part (a)?
    (c) At room temperature will there be an appreciable fraction of these molecules excited to level E_u? ($E_u - E_l = 1.0$ eV.)

29. At 300°K, what is the thermal energy in joules within a balloon containing 6.0×10^{23} molecules of neon (0.020 kg)?

30. (a) What is the thermal energy of 0.207 kg of lead (6.0×10^{23} atoms) at 300°K?
    (b) If this could be converted into gravitational potential energy, how high would the lead rise?

31. Gas at a pressure of 8.0×10^5 N/m² is confined within a cylinder fitted with a piston of radius 3.0×10^{-2} m. As the gas is warmed, the piston moves out in such a way that the pressure within the cylinder is kept constant. How much work is done on the piston by the gas during an expansion in which the piston moves 0.12 m?

32. One kilogram of ice at 0°C is placed in a cylinder. A piston maintains a constant pressure of 1.00 atmosphere (1.01×10^5 N/m²) on the cylinder contents. Heat is applied and the ice melts, turning into water at 0°C. As the ice melts, the piston moves inward. When all the ice has melted, the volume of the cylinder contents has decreased by 8.6×10^{-5}.m³. Find the work done by the piston on the contents of the cylinder.

33. (a) How is the heat for a process defined in thermodynamics?
    (b) What equation is a statement of the first law of thermodynamics?

34. What is the increase in the thermal energy of a system in a process in which 8.0×10^3 joules of work are done on the system and 12.0×10^3 joules of heat are absorbed by the system?

35. The thermal energy of a system increases by 3.0 kilocalories in a process in which 5.0 kilocalories of heat are absorbed by the system. What is the work done on the system in this process?

36. For the process described in Question 32, the internal energy increase is given by
    $$\Delta \bar{E} = 79.7 \text{ kilocalories/kg}.$$
    Find the heat for the process.

37. For the system of Question 32, assume the initial and final temperature and pressure are still 0°C and 1 atmosphere, but it is arranged to have the compression take place with a high constant pressure of 1.0×10^4 atmospheres.
    (a) Find the change in thermal energy of the cylinder contents.
    (b) Find the work done on the cylinder contents in this process.
    (c) Find the heat for the entire process.

38. Use the words *work*, *heat*, or *thermal energy* to correctly fill in the blanks in the following sentences.
    (a) Water contains more _____ at 20°C than at 0°C at the same pressure.
    (b) The _____ in two stones is increased by rubbing them together.
    (c) The _____ in an insulated copper wire increases at the rate I^2R when there is an electric current I in the wire.

39. A lead bullet traveling at 500 m/sec strikes a board and is brought to rest. Calculate the temperature rise of the bullet, assuming that the heat loss to the board is negligible. (Specific heat of lead is 0.031 kilocalorie/kg C°.)

40. What is the temperature rise in water that passes over Niagara Falls, tumbles 50 m down, and splashes into the river below?

41. A system is compressed by a constant pressure of 5.0×10^5 N/m² in a process in which the system gives off 2.0 kilocalories of heat and the thermal energy of the system remains unchanged. Find the decrease in the volume of the system.

42. (a) How much heat is required to raise the temperature of 3.0 kg of water from 15°C to 20°C in a process for which the specific heat of water is 1.00 kilocalorie/kg C°?
    (b) What temperature rise would the same amount of heat produce in mercury? (The specific heat of mercury is 0.033 kilocalorie/kg C°.)

43. A 10.0-kg copper block at 100°C is placed in 10.0 kg of water at 20°C. What is the final temperature of the copper and water? (The specific heat of copper is 0.093 kilocalorie/kg C°.)

44. The calorimeter of Example 8, with the calorimeter container containing 0.120 kg of water at 18.0°C, is used to determine the specific heat of a metal whose mass is 0.300 kg. The metal is heated to 98.0°C and placed in the calorimeter container. The final temperature of the water and metal is 28.0°C. Find the specific heat of the metal.

45. Explain how the specific heat of water can be measured using an electric heater coil immersed in the water. What quantities must be measured?

46. A heater coil of resistance wire is immersed in 0.600 kg of a liquid, and electrical energy is dissipated for 60 sec at a constant rate of 80 watts. The temperature of the liquid increases from 18.12°C to 21.62°C. Find the average specific heat s of the liquid in this temperature range.

47. A 0.600-kg sample of an alloy at a temperature of 100°C is dropped into 0.500 kg of water at 15.1°C which is contained in a 0.200-kg calorimeter made of the same alloy. The water, the calorimeter, and the sample reach a temperature of 20.1°C. Find the specific heat of the alloy.

48. Show that the defining equations, Eqs. (7.12), and (7.13), imply that the specific heats s and C are related by
$$C = (\text{mass of single molecule}) \times s.$$

49. Show that the first law of thermodynamics and the definition of C_V imply that
$$\Delta \bar{E} = NC_V \Delta T.$$

50. Show that the expression
$$\bar{E} = 3NkT$$
for the thermal energy of a crystalline solid implies that
$$C_V = 3k.$$

51. From the discussion of the thermal energy of hydrogen in Section 7.4, what values of C_V would you expect hydrogen to have in various temperature ranges?

52. A chunk of lead containing 6.03×10^{23} atoms has a mass of 0.207 kilogram. The specific heat of lead at constant volume is 0.029 kilocalorie/kg C°.
    (a) Compute the specific heat per lead atom, C_V.
    (b) Show that C_V is approximately equal to $3k$.

53. The *specific heat per mole C'* for any substance is defined by
$$\text{heat for process} = nC'\Delta T,$$
where n is the number of moles of the substance in question.
    (a) Show that the specific heat per mole C' is related to the specific heat per molecule C by
$$C' = N_0 C,$$
where N_0 is Avogadro's number.
    (b) From the equation $C_V = 3k$ for a crystalline solid at high temperatures, show that
$$C'_V = 3R.$$
This result is known as the law of Dulong and Petit.
    (c) From the equation $C_P - C_V = k$ for an ideal gas (derived in Question S-16) show that
$$C'_P - C'_V = R.$$

SUPPLEMENTARY QUESTIONS

S-1. Birds constitute a non-trivial hazard to present-day air traffic. The speed and the nature of the jet aircraft have resulted in the problem being more serious now than with propeller-driven aircraft. Consider a 5.0-kg goose flying at 20 m/sec head-on into the windshield of an aircraft traveling at 200 m/sec.

(a) Suppose the bird's speed relative to the aircraft is brought to zero in an impact time of 1.0×10^{-3} sec. What is the average force acting on the bird during this interval?

(b) If the impact area on the windshield is 2.00×10^{-2} m², what is the average pressure on this part of the windshield during the impact?

S-2. Calculate the ratio of the final to the initial volume of an air bubble that rises from the bottom of a lake which is 30 m deep. Assume that the temperature change of the bubble is negligible and that the atmospheric pressure is 1.00×10^5 N/m².

S-3. Air is trapped in a barometer tube above the mercury column. On a day when the pressure of the atmosphere is 76.0 cm of Hg, the height of the mercury column is merely 72.0 cm and the air space above the column is 12.0 cm long. On a day when this mercury column is 70 cm long, what is the pressure of the atmosphere?

S-4. For one state of a given type of ideal gas, denote the density, pressure and temperature respectively by ρ_1, P_1 and T_1. In a second state denote them by ρ_2, P_2 and T_2. Show that

$$\rho_1 T_1 / P_1 = \rho_2 T_2 / P_2.$$

S-5. What is the density of air at the top of a mountain on a day when the temperature is 17°C and the pressure is 0.90 atmosphere? (At 0°C and 1 atmosphere, air has a density of 1.29 kg/m³.)

S-6. As oxygen is withdrawn from a 50-liter tank, the pressure within the tank drops from 21 atmospheres to 7 atmospheres, and the temperature drops from 300°K to 280°K. How many kilograms of oxygen were withdrawn from the tank?

S-7. (a) Assume that the temperature in the upper portions of our atmosphere is 0°C, and calculate the thermal speed of hydrogen molecules (H$_2$) in this region.

(b) Compare this thermal speed to the speed required to escape from the earth (the escape speed of Section 4.1). Can the absence of hydrogen in our atmosphere be attributed to the ease with which hydrogen molecules can escape from the earth? Comment.

(c) Will the rate of escape be greater for hydrogen than for nitrogen and oxygen? Explain.

S-8. (a) Calculate the thermal speed of hydrogen *atoms* at the sun's surface where the temperature is 6000°K.

(b) Compute the escape speed for projectiles at the sun's surface and, by comparing this with the answer to part (a), show that these facts are consistent with the observation that there is an abundance of hydrogen in the sun's atmosphere.

S-9. After calculating relevant thermal speeds and escape speeds, explain why the moon cannot have an atmosphere.

S-10. (a) In a gas of uranium hexafluoride, calculate the ratio of the thermal speeds of the two different isotopes, $U^{235}F_6$ and $U^{238}F_6$, which have molecular weights 349 and 352, respectively.

(b) When uranium hexafluoride is allowed to diffuse through a porous barrier into an evacuated space, which constituent will diffuse through the barrier at the greater rate? Explain.

S-11. **In Section 7.2 we showed (and in succeeding sections used) the relation that the Kelvin temperature T is proportional to the average translational kinetic energy of random molecular motion, i.e., (*average* of $\frac{1}{2}mv^2$) $= \frac{3}{2}kT$. In the special theory of relativity (Chapter 5) the upper limit of v is the speed of light c. Does this fact force us to predict that there must be some theoretical upper limit on the value of the Kelvin temperature T? Comment.**

S-12. A cylinder fitted with a piston contains 1.00×10^{22} molecules of argon. This gas is heated from 27°C to 77°C with the pressure maintained constant at the value of 2.0×10^5 N/m².

(a) Calculate the work done by the gas in this expansion.

(b) Find the change in the thermal energy of the argon.

(c) Find the heat for this process.

(d) Verify that the result of part (c) implies that $C_p = \frac{5}{2}k$

S-13. Einstein's expression for the rest energy of an object,

$$\text{Rest energy} = mc^2,$$

gives us a universally valid expression for changes $\Delta \bar{E}$ in the thermal energy of any macroscopic system:

$$\Delta \bar{E} = \Delta mc^2,$$

where Δm is the change in rest mass. From the

data of Table 7-2, compute the percentage change in rest mass ($100 \times \Delta m/m$) when water at 100°C is converted to steam at 100°C. (Notice that the percentage change is so small that we cannot measure Δm directly.)

S-14. The Saint Lawrence River drops 75 m as it flows from Lake Ontario to the sea. A letter published in a Montreal newspaper suggested that freezing of the river be prevented by establishing a chain of hydroelectric plants along the river and by using their output to power electric heaters immersed in the river. Evaluate the temperature rise that could be produced in this way and comment on the merits of the proposal.

S-15. The heat capacity of an object is the heat per unit temperature change of the object.
(a) Show that the heat capacity of an object of mass m and specific heat s is given by

$$\text{Heat capacity} = ms.$$

(b) Compare the heat capacities of equal masses of water and lead. (Specific heat of lead is 0.031 kilocalorie/kg C°.)
(c) Compare the heat capacities of equal volumes of water and lead. (Specific gravity of lead is 11.3.)

S-16. The thermal energy of an ideal gas is independent of the volume occupied by the gas and is a function of the temperature only [for a monatomic gas this is illustrated by the function given in Eq. (7.9), $\bar{E} = \frac{3}{2}NkT$]. Consequently for an ideal gas, the validity of the result

$$\Delta \bar{E} = NC_V \Delta T$$

is not restricted to processes taking place at constant volume but may be applied to *any* process.
(a) Use this fact to show that, for a process in which an ideal gas expands at a constant pressure P, the first law of thermodynamics gives

$$NC_V \Delta T = NC_P \Delta T - P \Delta V,$$

where ΔV is the volume change of the gas.
(b) From the ideal gas law, express $P \Delta V$ in terms of ΔT and use the result of (a) to show that the specific heats of an ideal gas satisfy

$$C_P - C_V = k.$$

S-17. The question is sometimes raised as to whether tissue damage due to radiation, such as x-rays, can be attributed to tissue temperature changes resulting from the energy absorption. Show that these temperature changes appear inconsequential by calculating the increase in temperature of 1.0 gm of body tissue caused by an absorbed dose of 400 rads. (This is the whole-body dose which is lethal to 50% of a large group of mammalians so irradiated.) By definition, the rad = 10^{-5} joule/gm. Assume that body tissue, on the average, has the specific heat of water.

ADDITIONAL APPLICATIONS TO MEDICINE AND THE LIFE SCIENCES

Temperatures of Warm-Blooded Organisms: Thermal Energy and Temperature

In terms of temperature ranges, man is a fairly delicate creature. Normal body temperature is about 37°C, but this can only be maintained without external aid such as clothing, central heating, air conditioning, etc., within a relatively limited range of air temperatures. In air beyond the so-called critical air temperatures man fails to maintain his normal constant body temperature. Man's range of critical air temperatures is from about 17°C to 32°C. As these critical temperatures are approached, the body responds with certain compensatory chemical and physical reactions. For example, shivering occurs at the low temperature and perspiring at the higher temperature.

Failure to maintain the normal internal body temperature does not, of course, necessarily spell death, which will occur only if the internal temperature change exceeds certain limits. In man, heat strokes and brain lesions become evident as the body temperature rises through say 41°C to 44°C, the latter figure marking roughly the upper survival limit on a body temperature spectrum.

Other warm-blooded organisms are much hardier. The cow, for instance, with a normal body temperature of about 38°C, can maintain this temperature without external aid in air temperatures from about −40°C to 27°C. And the chicken, with a normal body temperature of 41°C, can do almost as well. Of the common animals about us, it seems that the mouse is the one with almost as limited a critical air temperature range as man.

Basal Metabolic Rate: Calorimetry

In this chapter reference has been made to direct and indirect methods for determining a person's metabolic rate. The direct method involves a calorimeter the size of a small room. One version of this large scale calorimeter involves a well insulated room along one or more walls of which runs a series of pipes through which water circulates. By noting the difference in the temperature between the inflowing water and the outflowing water as well as the amount of water, it is possible to calculate the thermal energy generated by the person.

Since this giant calorimeter is not highly practical for large-scale use, the metabolic rate is usually estimated by measuring the amount of oxygen a person consumes in a given time interval. The apparatus employed for this procedure is called a *spirometer* This apparatus dispenses pure oxygen to the person through a mask. The expired CO_2 returns through some CO_2 absorber. Experience has shown that the consumption of one liter of O_2 (with foods in which carbohydrates, fats, and proteins are present in normal amounts) indicates metabolic processes developing about 4.8 kilocalories.

In medicine the spirometer is used for measuring the *basal metabolic rate* (BMR), which is the thermal energy per hour per square meter of body surface produced by a fasting individual at rest 12-15 hours after his last meal. Since on the average a healthy person develops about 40 kilocalories/hr/m^2, the BMR can serve as a diagnostic aid. There is a great increase in the BMR in leukemia and fevers, for example, while a pronounced reduction occurs in, among other conditions, cretinism and malnutrition.

SUGGESTED READING

BROWN, SANBORN C., *Count Rumford, Physicist Extraordinary*. Garden City, New York: Doubleday, 1962. This 170-page paperback, one of the Science Study Series, is a lively presentation of the life of a most amazing adventurer, inventor, and scientist: the American, Benjamin Thompson, who later became Count Rumford of Bavaria. Professor Brown wrote this little book while engaged in a full-scale comprehensive biography of Rumford.

NEWMAN, J. R., *The World of Mathematics*, Vol. 2. New York: Simon and Schuster, 1956. The kinetic theory of gases, as developed by Daniel Bernoulli more than a century before the scientific world was ready to appreciate it, is presented on pp. 774-777.

Scientific American, September 1954. New York: Scientific American Inc. The nine articles of this issue are devoted to different aspects of thermal energy. Particularly valuable is the essay by Professor Freeman J. Dyson titled "What Is Heat?"

ZEMANSKY, M. W., *Heat and Thermodynamics*. New York: McGraw-Hill, 1968 (fifth edition). This is a most famous textbook in thermodynamics for advanced undergraduate physics students. Although the mathematical level is too advanced for our readers, there are many descriptive sections which can be consulted for an authoritative and an instructive presentation.

Joseph Black

8 Thermal Processes

Several thermal processes that have an immediate relevance in our daily lives have been selected for discussion in this chapter. Applications in thermometry are given. We emphasize that a macroscopic system, left undisturbed, will eventually reach an *equilibrium state*. This important fact is introduced by considering thermal equilibrium and phase equilibrium which are attained, for instance, when ice and water coexist indefinitely.

A landmark in the early history of these subjects was the discovery by Dr. Joseph Black, a Scottish physician, of the role temperature plays in the distribution of thermal energy. Although Black first studied languages and natural philosophy and completed a degree in medicine, his main interest seems to have been in the physical sciences. His comprehension of the role of temperature is best left in his own words: "By the use of thermometers, we have learned that, if we take a thousand, or more, different kinds of matter—such as metals, stones, salts, woods, cork, feathers, wool, water and a variety of other fluids—although they be all at first of different temperatures, and if we put them together in a room without a fire, and into which the sun does not shine, the heat will be communicated from the hotter of these bodies to the colder, during some hours perhaps, or the course of a day, at the end of which time, if we apply a thermometer to them all in succession, it will give precisely the same reading." In modern terms we would say the objects had reached a state of *thermal equilibrium,* a condition characterized by each having the *same temperature.*

Dr. Black emphasized a significant fact about the processes of melting and freezing when he wrote, "When ice or any other solid substance is melted, I am of the opinion that it receives a much larger quantity of heat than what is perceptible in it immediately afterwards by the thermometer. A large quantity of heat enters into it, on this occasion, without making it apparently warmer, when tried by that instrument . . . . On the other hand, when we freeze a liquid, a very large quantity of heat comes out of it, while it is assuming the solid form, the loss of which heat is not to be perceived by the common manner of using the thermometer . . . ." He had apparently used a crude

"All bodies communicating freely with one another, and exposed to no inequality or external action, acquire the same temperature, as indicated by a thermometer."

Joseph Black
(1728–1799)

calorimeter to measure the large "latent heat" involved in melting ice or freezing water. However, this discovery was not formally published. In fact Black published very little, and hence much of his work survived only through his students' notes. (A complete set of such notes was published in 1803 by John Robinson.) Fortunately such notes were plentiful because Black's lectures were greatly esteemed and well attended.

8.1 THERMAL EXPANSION

Solids

The atoms of a crystalline solid are arranged in a regular array (Fig. 14-11) whose geometry is determined by the interatomic forces. Each atom oscillates back and forth with an amplitude of about 0.1 Å and a frequency of about 10^{13} hertz. When the temperature of the crystal is raised, there is an increase in the average translational kinetic energy of the atoms and in the amplitude of their oscillations. The motion with larger amplitude brings about a change in the interaction between adjacent molecules which usually leads to a slightly greater average distance between atoms. Therefore the entire solid expands as its temperature is raised.

Experiments show that the increase ΔL in the length L of a solid when its temperature is raised by an amount ΔT is given by

$$\Delta L = \alpha L \Delta T \tag{8.1}$$

where α, named the *coefficient of linear expansion*, has a value characteristic of the type of solid (Fig. 8-1). To an approximation adequate for most applications, α can be assumed to be a constant independent of the temperature interval or the reference length appearing in Eq. (8.1). Values of α for several solids are shown in Table 8-1. The order of magnitude of the expansion of most solids is easy to remember, being about 1 millimeter per meter length per 100 Celsius degrees.

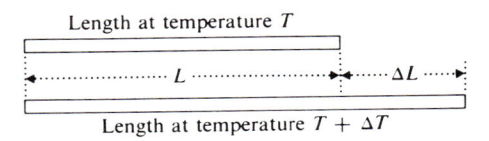

Fig. 8-1 Thermal expansion. When the temperature is raised by an amount ΔT, the length increases by an amount $\Delta L = \alpha L \Delta T$. In this figure the expansion has been exaggerated for clarity.

Table 8-1. APPROXIMATE COEFFICIENTS OF LINEAR EXPANSION (α) NEAR 20°C

Substance	α (per C°)	Substance	α (per C°)
Steel	12×10^{-6}	Granite	8×10^{-6}
Brass	19×10^{-6}	Glass (pyrex)	3×10^{-6}
Aluminum	24×10^{-6}	Glass (ordinary)	9×10^{-6}
Lead	29×10^{-6}	Brick	9×10^{-6}
Invar	0.9×10^{-6}	Hard rubber	80×10^{-6}

Example 1 A steel railway track is 20.000 m long on a winter day when the temperature is $-15°C$. What is its length on a summer day when the temperature is 35°C?

Solution The temperature increase is

$$\Delta T = 35°C - (-15°C) = 50 \text{ C°}.$$

Therefore,

$$\Delta L = \alpha L \Delta T \tag{8.1}$$
$$= (12 \times 10^{-6}/\text{C°})(20.000 \text{ m})(50 \text{ C°})$$
$$= 0.012 \text{ m}.$$

8.1 THERMAL EXPANSION 299

The new length is
$$L + \Delta L = 20.000 \text{ m} + 0.012 \text{ m}$$
$$= 20.012 \text{ m}.$$

This example illustrates that engineers must take thermal expansion into consideration. Railway tracks would buckle if enough space were not left between sections to permit their length to increase by an amount $\alpha L \Delta T$ where ΔT is the largest temperature increase that is anticipated. For the same reason, spaces must be left between the concrete slabs used in highway construction and even between adjacent sections of large buildings.

Thermostats are made from thin strips of different metals which are welded or riveted together. When the temperature changes, since one metal expands more than the other, the combination must bend in an arc (Fig. 8-2).

Uneven heating and the consequent uneven expansion gives rise to large forces within a solid. Ordinary glass shatters when heated unless it is warmed so slowly that appreciable temperature differences between different regions of the glass do not occur. Pyrex has the merit of having a lower coefficient of linear expansion (Table 8-1), and thus is not subjected to such markedly different expansions at different temperatures.

Materials which are composed of millions of minute crystals orientated at random will expand equally in all directions. For such materials, a temperature increase ΔT will produce increases of any area A and any volume V given by

$$\Delta A = 2\alpha A \Delta T \qquad (8.2)$$

and

$$\Delta V = 3\alpha V \Delta T. \qquad (8.3)$$

These equations can be deduced from $\Delta L = \alpha L \Delta T$ and are good approximations only when the fractional expansion ($\Delta A/A$ or $\Delta V/V$) is small. Area expansion is exploited when one loosens the metal lid of a glass jam jar by heating it under hot water.

Liquids

Liquids generally expand when heated. The approximate increase in a volume V associated with a temperature rise ΔT is given by

$$\Delta V = \beta V \Delta T, \qquad (8.4)$$

where β is called the *coefficient of volume expansion* of the liquid in question.

Example 2 On a summer day when the temperature reaches 31°C, a steel gasoline tank at 20°C is filled with cold gasoline pumped from underground storage tanks where the temperature is also 20°C. What frac-

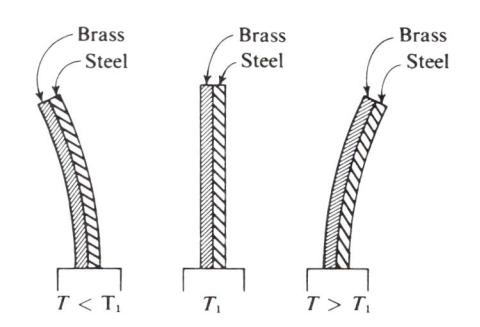

Fig. 8-2 Principle of a thermostat element. Each metal strip expands a different amount when the temperature changes.

Table 8-2. COEFFICIENTS OF VOLUME EXPANSION (β) AT 20°C

Liquid	β (per C°)
Mercury	18×10^{-5}
Gasoline	95×10^{-5}
Benzene	124×10^{-5}
Ether, ethyl	166×10^{-5}

tion of the gasoline will have overflowed when the gasoline and the tank temperatures have attained 31°C?

Solution From Eq. (8.3), the expansion of the volume V of the steel tank is

$$\Delta V_{\text{tank}} = (3 \times 12 \times 10^{-6}/\text{C}°)(11\ \text{C}°)\ V$$
$$= 4.0 \times 10^{-4}\ V.$$

The expansion of the volume V of the gasoline, from Eq. (8.4), is

$$\Delta V_{\text{gasoline}} = (95 \times 10^{-5}/\text{C}°)(11\ \text{C}°)\ V$$
$$= 105 \times 10^{-4}\ V.$$

Therefore,

$$\text{Overflow volume} = \Delta V_{\text{gasoline}} - \Delta V_{\text{tank}}$$
$$= (105 - 4) \times 10^{-4}\ V$$
$$= 0.0101\ V.$$

About 0.01 or 1% of the gasoline overflows.

Anomalous Expansion

After having been exposed to numerous examples of thermal expansion, one may be led to believe that materials always expand when heated. But this is not true. At some temperatures water contracts when heated, and so do certain other substances. Water is particularly interesting in this regard since its coefficient of volume expansion is negative at very low temperatures (about 0°K to 70°K), as well as in the region from 0°C to 4°C.

As for other substances, a certain type of iron expands as its temperature increases from 0°K up to about 1100°K, at which point it contracts slightly. (Subsequently it continues expansion with further temperature increase.) At 1100°K certain molecular rearrangements take place in the iron crystal, resulting in contraction. Examples of other materials which at certain temperatures have negative expansion coefficients are silicon, selenium, tellurium, and a certain cobalt-iron-chromium alloy. Evidently, though relatively rare, there *is* such a phenomenon as contraction with increased temperature.

An interesting and simple demonstration of the negative thermal expansion coefficient of ordinary elastic bands can be done this way: Suspend a small mass by means of a rubber band; then hold a flaming match near the rubber. The small mass will be seen to rise noticeably as the rubber contracts on being heated.

That water displays this anomalous thermal expansion has profound consequences for life on our planet. As implied earlier, the volume of a given mass of water does not increase regularly as its temperature is increased. Instead the volume *decreases* as the water is warmed from 0°C to 4°C. Above 4°C the water expands when heated. The volume occupied by a given mass of water is therefore a minimum at 4°C.

The reason for the contraction of water as it is heated from 0°C to 4°C involves the structure of ice. Ice crystals are a rather open structure with empty spaces. When a chunk of ice melts, ice crystals collapse and the volume is reduced. But many ice crystals remain in water at 0°C. If the temperature is raised, more and more ice crystals collapse. This is the dominant effect up to 4°C.

These properties determine the sequence of events that occur in the cooling and freezing of lakes. The implications of Archimedes' Principle (Section 6.3) are most easily understood in terms of water and ice *densities* rather than volumes occupied by a given mass. Two important facts are

1. Ice is less dense than water. The buoyant force exerted by water on ice is therefore sufficient to cause the ice to float.
2. Water has a maximum density at 4°C. Below 4°C, cooler portions of water will be forced upward by the buoyant forces exerted by warmer portions. Therefore when the surface of a lake is cooled below 4°C, mechanical equilibrium will be attained with the coldest water at the surface and the temperature rising as the depth increases.

Consequently ice forms first at the surface and, because ice floats, it remains there. With continued cooling the ice thickens but, unless all the water freezes, the water temperature near the bottom will remain at 4°C. Most solids expand when melted and liquids expand when heated so that the solid and the coldest liquid sink. If water and ice behaved in the usual manner lakes would freeze from the bottom up and easily become a solid mass of ice in colder climates. The effects on marine life and on our weather would be drastic.

8.2 HEAT TRANSFER

Thermal energy can be transferred from one place to another by three different mechanisms: conduction, convection, and radiation. *Conduction* is a relatively slow process by which thermal energy is transferred by electron motion or molecular interaction. *Convection* is a more rapid process which occurs when a portion of a heated fluid moves from one place to another. *Radiation* involves the emission and absorption of electromagnetic waves which travel at the speed of light.

Conduction

Thermal energy is transferred by conduction from the hot to the cold end of a metal rod. Conduction involves a transport of thermal energy without the conducting material itself being moved. Metals, good conductors of electricity, are also good thermal conductors. The electrons in the conduction band, which account for the metals' conduction of electricity as described in Section 16.7, effect a rapid transfer of thermal energy from a hot spot to a cooler region. In solid insulators and semiconductors a transfer of thermal energy is accomplished by the interactions of neighboring atoms. Relatively slow thermal conduction, due to molecular collisions, occurs in nonmetallic liquids and gases.

Convection

Convection consists of the actual motion of a portion of a heated fluid. Natural convection currents are set up if a heated region of fluid expands, becoming less dense than the surrounding fluid. Then, according to Archimedes' principle (Section 6.3), the buoyant force exerted by the surrounding cooler fluid is greater than the weight of the heated portion. The heated fluid thus rises. Similarly, a cooler portion sinks. The thermal energy from a hot-water radiator is carried to other parts of a room by such natural convection currents. In forced convection, the fluid motion is either caused or increased by a blower or pump.

Thermal Radiation

All objects like a glowing tungsten wire in a light bulb, the surface of the sun, or even human skin, continuously radiate energy as electromagnetic waves.* The rate at which a body emits this thermal radiation increases rapidly as its temperature increases, being approximately proportional to T^4, where T is the Kelvin temperature of the body. At high temperatures, radiation is frequently the principal mechanism of heat loss.

A detailed discussion of electromagnetic waves and of the electromagnetic spectrum is given in Chapter 14.

The distribution of energy among the different wavelengths is characteristic of the temperature of the emitting surface. The predominant wavelengths of thermal radiation decrease when the temperature of the radiating object is increased. Below about 900°K, these wavelengths are too long to be detected by the eye but such *infrared* electromagnetic waves do cause a sensation of heat in the skin. At 1000°K, an object appears dull red; evidently some radiation is visible. At still higher temperatures, the radiating body emits relatively more of the shorter wavelength blue light and at sufficiently high temperatures appears bluish-white. White-hot is hotter than red-hot. [The study of thermal radiation had a profound impact on the history of physics. Planck's constant h and the first evidence of the quantum mechanical nature of the world made their appearance in 1900 in Planck's effort to understand the way that electromagnetic energy was distributed among the different wavelengths in thermal radiation (Section 16.2).]

Heat Transfer in Humans

The human body provides interesting examples of heat transfer. Conduction, convection, and radiation maintain a remarkably constant body temperature in spite of extreme environmental temperature changes. A fairly uniform temperature of 37°C throughout the body's interior is achieved by forced convection, with the heart serving as the pump and the blood as the heater fluid. Heat is brought to the skin by convection and conduction within the body. The skin then loses heat to the surroundings. Some 2000 kilocalories in a day is a typical figure for the heat loss of a sedentary man.

The relative importance of the different mechanisms of heat loss varies greatly according to circumstances. Experiments show that over 50% of the heat loss from a dry nude in still air is in the form of infrared radiation; this radiation loss, however, amounts to only about 5% of the total heat loss by a person properly attired for outdoor activity in the winter. (The presence of perspiration and the temperature and dryness of the surrounding air greatly affect the rate at which heat is lost, because for each kilogram of perspiration evaporated nearly 600 kilocalories of heat are absorbed from the body. Evaporation and the "latent heat" required to change water to vapor will be discussed in Section 8.4.)

Dewar Flask

An instructive practical example of efforts to circumvent convection, conduction, and radiation is provided by the Dewar flask, or thermos bottle (Fig. 8-3). The flask is a double-walled glass vessel which is silvered on the inside. The space between the walls is evacuated to prevent convection and conduction. Because glass is a poor conductor, little heat is conducted through the glass walls over the neck. The silvered surfaces reflect most of the thermal radiation that would leave from the inside or enter from the outside.

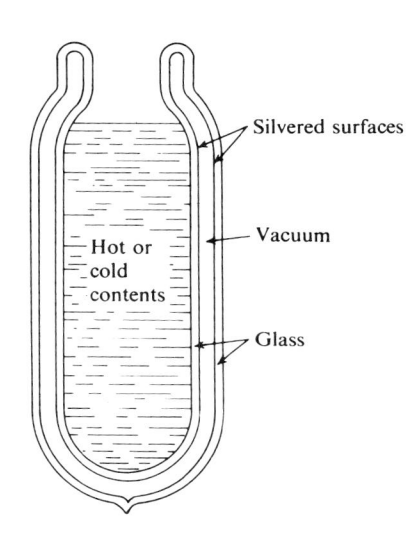

Fig. 8-3 Dewar flask.

8.3 THERMOMETERS

Most of the phenomena involved in the operation of different practical thermometers have now been discussed. Before examining the details of these useful devices we pause to consider the familiar but fundamental concept of temperature.

We have already discussed a microscopic interpretation of temperature as a measure of random translational kinetic energy of molecular motion. In Chapter 9 we will give a very general, though somewhat abstract, definition of temperature in terms of quantities related to the microscopic description of matter. But it is interesting to put aside briefly all microscopic notions and consider temperature as a purely macroscopic quantity, a fundamental concept in the macroscopic science of thermal phenomena, thermodynamics.

The Concept of Temperature

Our sense of touch involves a temperature sense which enables us to distinguish hot bodies from cold bodies and even to decide on the order of hotness of several different objects. This life-saving temperature sense, although it is too subjective and unreliable to provide a basis for scientific work, does give us the useful intuitive notion that a temperature, which is a measure of the hotness of an object, can somehow be assigned.

The macroscopic temperature concept can be made precise by considering Dr. Black's discovery, *thermal equilibrium*. Any object such as a copper bar which is insulated from its surroundings by an asbestos sheet soon reaches a state of thermal equilibrium in which its macroscopically observable properties such as its length or volume remain unchanged as time elapses. If an iron plate, also in equilibrium, is placed in contact with the copper bar, it is usually observed that equilibrium is disturbed. Lengths, volumes, and other properties alter as time goes on. In this case we say the two objects are not in thermal equilibrium with each other. After a sufficiently long time the copper bar and the iron plate will reach a state of thermal equilibrium in which the observable properties of the objects are constant. We then say that these objects [A and B in Fig. 8-4 (a)] are in thermal equilibrium with each other.

We now consider several different macroscopic objects and examine thermal equilibrium between different pairs of these objects [Figs. 8-4 (b) and 8-4 (c)]. All our experience with such experiments is expressed concisely in the *zeroth law of thermodynamics*: *Two systems in thermal equilibrium with a third are in thermal equilibrium with each other.*

The *temperature* of an object is identified in thermodynamics *as the property which determines whether or not an object will be in thermal equilibrium with other objects.* Objects in thermal equilibrium have the same temperature. When objects with different temperatures are placed in an enclosure and allowed to exchange thermal energy by any means, conduction, convection, or radiation, thermal equilibrium eventually will be reached with all objects at the same temperature. This is the essence of Dr. Black's simple but significant observation. Moreover he realized that thermal radiation from the exterior could disturb thermal equilibrium, for he added the proviso "a room without a fire, and into which the sun does not shine. . . ."

Thermometers and the Kelvin Temperature Scale

A *thermometer* is simply a relatively small macroscopic system which is arranged so that when the system absorbs or gives off heat, one property of the system changes in an evident and readily measurable manner. The reading of a thermometer is the temperature of all systems in thermal equilibrium with it.

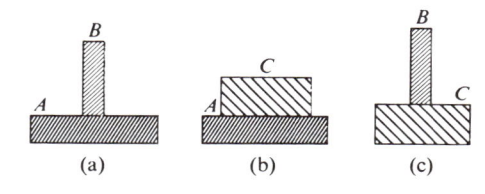

Fig. 8-4 Thermal equilibrium, the zeroth law of thermodynamics, and temperature.
(a) Objects A and B are in thermal equilibrium.
(b) Objects A and C are in thermal equilibrium with the state of A the same in both cases.
(c) Experiment shows that objects B and C then will be found to be in thermal equilibrium with each other. Objects A, B, and C have the same *temperature*.

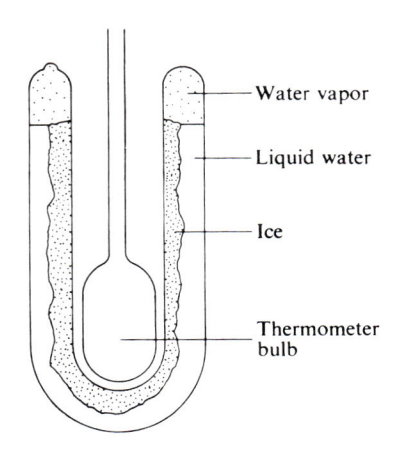

Fig. 8-5 Triple-point cell containing ice, liquid water, and water vapor.

*This value, adopted in 1954 by international convention, has been selected so that the modern scale will agree as closely as possible with the less accurate values obtained using other procedures.

In different types of thermometers, different measurable properties are selected to indicate the temperature. Each of the thermometers to be described has features that make it particularly useful in certain applications.

To assign numbers to the various temperatures indicated by a thermometer, we must establish a temperature scale. In setting up the most important scale, the Kelvin temperature scale, the temperature at which pure water, ice, and water vapor coexist in equilibrium (Fig. 8-5) is arbitrarily assigned the value 273.16 degrees Kelvin. Water in this state is said to be at its *triple point*. The reason for choosing the triple point of water as a standard state is that there is only one definite value of temperature and pressure (4.58 mm Hg) at which water will remain in equilibrium with ice and water vapor. Changes in the relative amounts of the solid, liquid, and gaseous forms do not affect the temperature or pressure. This state of water therefore furnishes a reproducible standard of temperature which, by definition, is

$$T_{\text{triple point}} = 273.16°K \; exactly.*$$

Numbers are now assigned to other temperatures on the Kelvin temperature scale in accordance with the definition of temperature to be given in Section 9.5. Over a wide range of temperatures, these Kelvin temperatures can be obtained directly from readings of the pressure of helium contained in the constant volume gas thermometer that we shall now describe. The other thermometers can then be calibrated.

Constant-Volume Gas Thermometer

A constant-volume gas thermometer can be used to determine temperatures as low as 1°K. A small amount of helium is put into a glass bulb and the pressure of the gas is measured at various temperatures. Provision is made to keep the gas volume constant as both its temperature

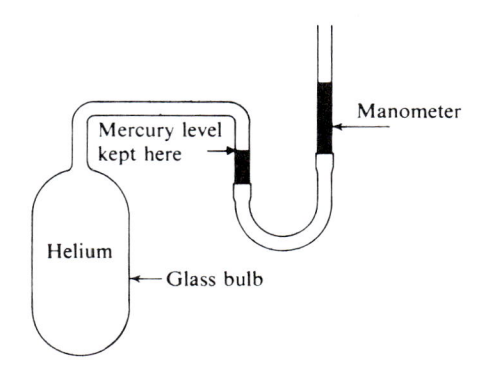

Fig. 8-6 Constant-volume gas thermometer. The manometer is adjusted so that the mercury level in the left tube is maintained at a fixed mark. In this way the volume of the helium is kept constant. The manometer pressure reading determines the temperature of the helium in the bulb.

306   THERMAL PROCESSES

and pressure change (Fig. 8-6). If the helium gas obeys the ideal gas law,* then

$$PV = NkT \qquad (7.3)$$

where P is the pressure reading of the thermometer at the temperature T. The pressure reading P_t of the thermometer at the triple-point temperature of water (273.16°K) therefore satisfies

$$P_t/273.16°K = Nk/V = P/T,$$

where we have used the fact that N and V are maintained constant in this thermometer. Therefore

$$T/273.16°K = P/P_t.$$

In this way the Kelvin temperature T can be calculated after measurement of the ratio of the pressures P and P_t. We can check that the helium density is low enough to ensure that the ideal gas law is obeyed by verifying that the same pressure ratios are obtained with a similar thermometer containing less helium.

Gas thermometers have the disadvantage of being cumbersome and slow in reaching thermal equilibrium. Their use is therefore confined to research and standards laboratories.

Liquid-in-Glass Thermometers

The thermal expansion of a liquid, usually mercury or alcohol, is exploited in the most common type of thermometer. The liquid fills a thin-walled bulb at the end of a capillary tube. A very small expansion of the liquid produces an appreciable increase in the height of the liquid column in the capillary tube.

The *clinical thermometer* is a mercury-in-glass thermometer with a constriction (Fig. 8-7) between the bulb and the capillary tube. This constriction inhibits the return flow of mercury from the capillary to the bulb, so the thermometer indicates the highest temperature that the bulb has attained.

Liquid-in-glass thermometers are portable, relatively inexpensive, and easy to read. For accurate work, however, many corrections must be applied, and for this reason this type of thermometer has been almost entirely replaced in research laboratories by resistance thermometers or thermocouples.

Bimetallic Thermometer

A rugged portable thermometer (Fig. 8-8) can be made using a strip formed of two metals which expand by different amounts when their temperature is raised. As the strip bends, it moves a pointer over a scale.

*With temperature defined by Eq. (9.14), it is possible to deduce that gases of sufficiently low density must obey this law.

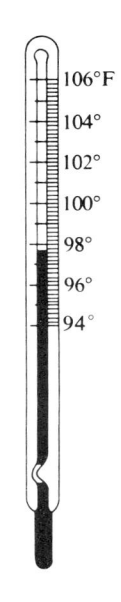

Fig. 8-7 Clinical thermometer.

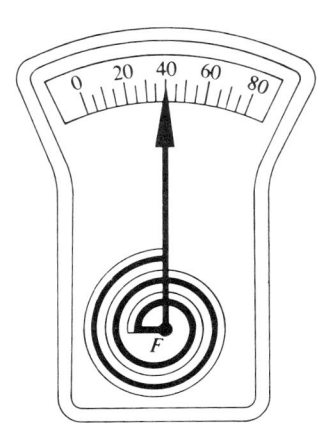

Fig. 8-8 A bimetallic thermometer. The bimetallic strip, which employs the principle of differential expansion to activate the pointer, is in the form of a spring centered at point F.

8.3 THERMOMETERS 307

*The electrical terms current, resistance, EMF, and voltage that are used in this section are discussed in detail in Chapter 11.

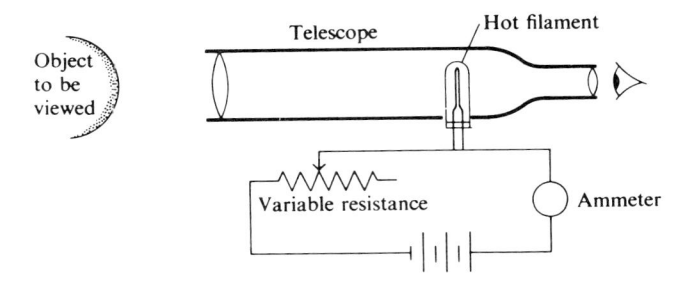

Fig. 8-9 Optical pyrometer. A glowing object is viewed through the telescope. The filament current is adjusted until the filament brightness matches that of the glowing object.

Optical Pyrometer

When an object is so hot that it emits visible thermal radiation, its temperature can be measured using the *Disappearing Filament Optical Pyrometer* (Fig. 8-9). The glowing object is viewed through the pyrometer telescope and the observer compares the brightness of the object with the brightness of an electric lamp filament. The filament current* is increased until the filament brightness matches that of the glowing object. The unknown temperature then can be determined from the ammeter reading of the filament current if the instrument has been previously calibrated at known temperatures.

Thermography

Modern infrared scanners can detect small variations in the temperature of radiating surfaces and convert this information into a picture called a thermograph in which each region is displayed with a shading characteristic of its temperature. An obvious military application is the nighttime detection of hostile troops. In industry, one of many applications of thermography involves the testing of electronic apparatus. Any flaw which leads to a change in an electric current distribution will also alter the thermal pattern.

Thermography also finds use in medicine. The human body, with a skin temperature of about 31°C, radiates approximately 0.05 watt of power per cm^2 of skin surface. This energy is largely in the range of 60,000 to 150,000 Å, that is, the wavelengths are in the infrared region. An instrument known as a thermograph can scan small segments of the body, record the emitted energy, and thus pinpoint local temperature differences. Such localized temperature differences may be exploited in the diagnosis and monitoring of breast cancer, peripheral vascular disorders, wound healing, arthritis, burns, frostbite, plastic surgery, and even in obstetrics and gynecology for both pregnancy determination and placental localization. Since present thermography instrumentation is capable of a thermal sensitivity of less than 0.07C°, the slight tempera-

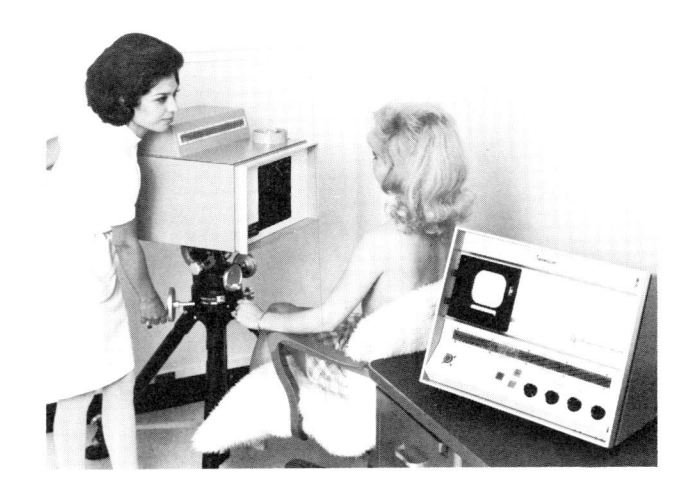

Fig. 8-10 A very promising application of thermography is the routine scanning of women to detect breast cancers while they are still in what is called the subclinical stage—a time when the growth is of the order of a centimeter or less in diameter and when therapeutic measures are most effective. Since the scan simply involves reading the infrared radiation emitted by the body this method of diagnosis avoids hazards such as those involved in x-ray examinations.

ture increase generally associated with breast cancers (a few tenths of a Celsius degree) can be readily detected (Fig. 8-10).

Resistance Thermometers

The electrical resistance of a conductor changes when the temperature changes. Extremely precise measurements of resistance can be made by standard electrical methods, so resistance thermometers furnish some of the most precise temperature measurements. A thermometer to measure high temperatures employs a coil of platinum wire (melting point 1770°C) wound on a silica spool (melting point 1420°C). At extremely low temperatures the platinum wire is replaced by a germanium crystal or a carbon cylinder.

Thermistor *

A resistance thermometer which can measure temperature changes as small as $\frac{1}{1000}$ C° is obtained by measuring the resistance of a *thermistor* —a small bead of semiconductor material placed between two wire leads. A small temperature increase leads to a very large increase in the number of electrons in the conduction band of a semiconductor (Section 16.8). This greatly increases the electrical conductivity of the semiconductor. The electrical resistance of a semiconductor is therefore a very sensitive indicator of temperature.

The physical mechanisms responsible for the functioning of a thermistor are not discussed until Section 16.8. This useful device is mentioned here to complete the survey of thermometers commonly used in modern science and engineering.

Thermocouple

Physicists and engineers generally rate the *thermocouple* as the most useful thermometer. This device is based on an effect that we have not yet described. When wires of two different metals are joined together at

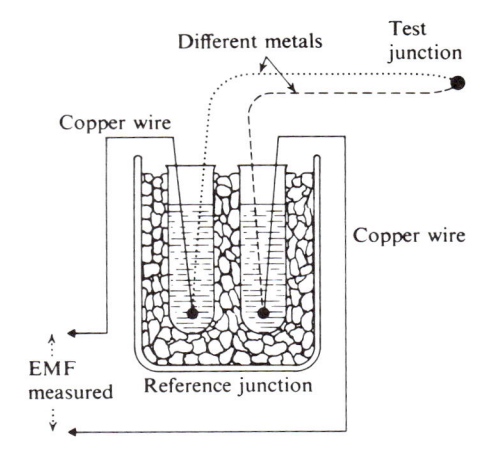

Fig. 8-11 Thermocouple.

*Some authors make a distinction between water vapor and steam by defining steam as vapor condensed on dust particles in the air (like fog). However, we use the terms steam and water vapor synonymously.

both ends and the two junctions are maintained at different temperatures, an EMF is set up in the circuit. This EMF depends on the temperature difference between the junctions. Thus the EMF can be used to measure this temperature difference.

In operation as a thermometer, one thermocouple junction called the "test junction" is embedded in the material whose temperature is to be determined. At the other end of each thermocouple wire, copper wires are joined and these two junctions (Fig. 8-11), maintained at any desired reference temperature, constitute the reference junction. The temperature difference between the test junction and the reference junction is determined from measurement of the EMF between the two copper wires. Previous calibration with known temperatures is obviously necessary.

The thermocouple has many merits. It can follow temperature changes rapidly because the test junction has such a small mass that it comes quickly into thermal equilibrium with the material under investigation. The reading of the instrument, the EMF measurement, can be performed at a location remote from the test junction.

Thermocouples with one test junction wire of copper and the other wire of an alloy called constantan are used to measure temperatures ranging from $-190°C$ to $300°C$. This *copper-constantan* thermocouple gives the comparatively large EMF of 40 microvolts per degree temperature difference between the junctions. Since this EMF can be measured with an accuracy of about 1 microvolt, using an instrument called a *potentiometer*, temperature differences can be determined with an accuracy of about $\frac{1}{40}$ degree. Thermocouples of platinum and a platinum-rhodium alloy are used for temperatures up to $1600°C$.

8.4. CHANGES OF PHASE AND PHASE EQUILIBRIUM

Latent Heats

Substances can exist in several different *phases* or forms. Thus a collection of H_2O molecules can exist in the *solid phase* as ice, the *liquid phase* as water, or the *gaseous phase* as steam.* For a given substance a change from one phase to another occurs at a well-defined temperature and involves a volume change as well as the absorption or liberation of heat. As Dr. Black observed, the magnitude of the heat absorbed in melting ice is impressive and has pronounced effects on our climate.

The main features of phase changes are brought out by considering the sequence of events that occurs when heat is continuously supplied to a mass m of ice contained in a cylinder fitted with a piston, with the pressure on the piston maintained at one atmosphere. We start with ice, fresh from the refrigerator, at a temperature of $-20°C$. The H_2O molecules of the ice vibrate about definite equilibrium positions with an

310 THERMAL PROCESSES

average kinetic energy of $\frac{3}{2}kT$. This energy is insufficient to allow a molecule to break away from its equilibrium position which is determined by the strong intermolecular forces. We now apply heat. The ice does not immediately begin to melt but rather warms up in accordance with

$$\text{Heat for process} = ms\Delta T \qquad (7.12)$$

where s (approximately 0.5 kilocalorie/kg C°) is the specific heat of ice in this range of temperatures. As the ice warms, its molecules jiggle more violently.

At 0°C, the melting point for ice (at a pressure of one atmosphere), the temperature ceases to rise, even though heat continues to be supplied to the ice. Instead, the most energetic molecules enter the liquid phase which is characterized by the molecules still being closely packed together but no longer with fixed relative positions. We emphasize that as heat is applied, the temperature remains at 0°C until all the ice has melted.

The heat absorbed by the mass m which changes from the solid to the liquid phase without undergoing a change of temperature is given by

$$\text{Heat for process} = mL_f \qquad (8.5)$$

where L_f is called the *latent heat of fusion*. For an ice-water transition at 0°C,

$$L_f = 80 \text{ kilocalories/kg.}$$

In other words, 80 kilocalories of heat are required to change 1 kg of ice at 0°C into water at 0°C.

When this process is reversed so that water freezes, each kilogram of water at 0°C *gives off* 80 kilocalories of heat in changing to ice at 0°C. Eq. (8.5) yields the heat given off in freezing *or* the heat absorbed in melting.

Now with all the ice melted so that the cylinder contains only water at 0°C, we continue to apply heat. The water molecules execute a still more violent random thermal motion and the temperature increases in accordance with

$$\text{Heat for process} = ms\Delta T, \qquad (7.12)$$

where s is now the specific heat of water (1.00 kilocalorie/kg C°).

At 100°C and at a pressure of one atmosphere, the most energetic molecules near the surface of the water have enough energy to escape and form water vapor or steam. Evaporation takes place. Again, since only the most energetic molecules escape from the liquid phase, energy, named a *latent heat*, must be supplied to accomplish the phase transition.

The heat absorbed by the mass m which changes from the liquid phase to the gaseous phase without change in temperature is given by

$$\text{Heat for process} = mL_v \qquad (8.6)$$

where L_v is called the *latent heat of vaporization* of the substance. For water L_v is large, nearly 539 kilocalories/kg when the pressure is one atmosphere. For the reverse process, condensation, Eq. (8.6) still applies, but now the heat is given off by the condensing vapor.

Example 3 How much heat is required to change 5.0 kg of ice at $-8.0°C$ into water at 20°C in a process which takes place at a pressure of one atmosphere? Assume $s_{ice} = 0.50$ kilocalorie/kg C°.

Solution To warm the ice from $-8.0°C$ to $0.0°C$ requires heat:

$$\text{Heat for process} = sm\Delta T \qquad (7.12)$$
$$= (0.50 \text{ kilocalorie/kg C°})(5.0 \text{ kg})(8.0 \text{ C°})$$
$$= 20 \text{ kilocalories.}$$

The phase transition from ice at 0°C into water at 0°C again requires heat:

$$\text{Heat for process} = mL_f \qquad (8.5)$$
$$= (5.0 \text{ kg})(80 \text{ kilocalories/kg})$$
$$= 400 \text{ kilocalories.}$$

To heat the water from 0°C to 20°C requires still more heat:

$$\text{Heat for process} = (1.00 \text{ kilocalorie/kg C°})(5.0 \text{ kg})(20 \text{ C°})$$
$$= 100 \text{ kilocalories.}$$

The *total heat* required for the process is therefore

$$(20 + 400 + 100) \text{ kilocalories} = 520 \text{ kilocalories.}$$

The fact that the temperature of an ice-water system does not drop below 32°F (0°C) until all the water has frozen can be used to provide a handy method of temperature control. An old trick, known to farmers (see Question 23), has been revived by the United States Air Force in sending fresh fruits and vegetables to Arctic outposts. Freezing and spoilage of most of these foods starts at 30.4°F. By using a water-soaked layer of material sealed between inner and outer insulating layers, it is found, even when the outside temperature is as low as $-65°F$, that the interior is maintained at a safe 32°F for about six hours, until the water has completely frozen.

Phase Equilibrium

Consider again the ice and water in the cylinder at a temperature of 0°C and a pressure of 1 atmosphere. If the heat transfer to or from the cylinder is stopped, the ice and water can coexist indefinitely. The escape of energetic molecules from the solid phase is offset by the arrival, from the liquid, of molecules which become bound to the solid. At the melting or fusion temperature (0°C for ice at a pressure of 1 atmosphere), these two competing processes proceed at equal rates. We say *phase equilibrium* is obtained and that the solid phase is in equilibrium with the liquid phase.

Similarly, we can obtain equilibrium between liquid and vapor phases. At a pressure of 1 atmosphere, water and steam confined to a cylinder and isolated will coexist indefinitely at a temperature of 100°C. Of course a liquid tends to evaporate at any temperature, and in most circumstances the vapor is not in equilibrium with the liquid. However, if the liquid and vapor are confined, the vapor pressure alters until the evaporation of the liquid is balanced by the condensation of the vapor, and equilibrium is attained. The pressure of the vapor at which the vapor and the liquid coexist in equilibrium is called the *saturation vapor pressure*. For reasons that we shall now investigate, *the saturation vapor pressure depends only on the temperature*, not on the volume of vapor or liquid.

Evaporation and Condensation

The approach to equilibrium between water and its vapor is illustrated in Fig. 8-12. Some water is placed in a cylinder. A vacuum pump is used to remove the air and the vapor from the region above the water [Fig. 8-12 (a)].

The valve is then closed so the pumping stops [Fig. 8-12 (b)]. The collection of H_2O molecules within the cylinder is not in an equilibrium state and will not remain entirely in the liquid phase. Within the liquid, the most energetic of the molecules which approach the surface will have sufficient energy to cross the surface and break away from the attractive forces exerted by the molecules remaining in the liquid phase. Some of the most energetic molecules thus enter the vapor phase [Fig. 8-12(b)]. They have been *evaporated* or *vaporized*.

The rate of evaporation per unit liquid surface area will be determined entirely by the translational kinetic energies (average value $\frac{3}{2}kT$)

Fig. 8-12 (a) Molecules of water vapor are pumped away as fast as they are evaporated from the liquid. (b) The valve is closed. The number of molecules of water vapor increases. (c) Phase equilibrium is attained. The pressure now exerted by the vapor is called the *saturation vapor pressure*.

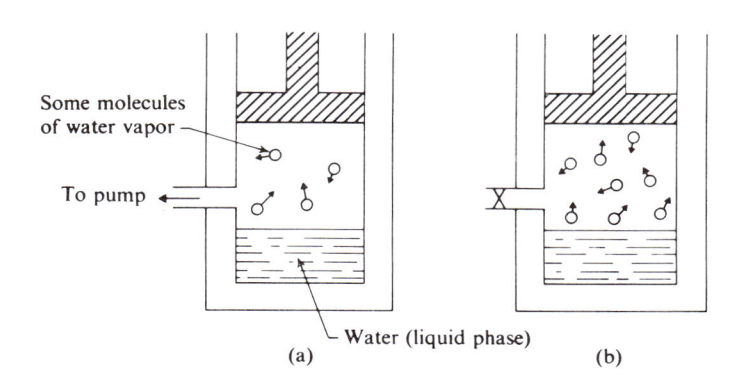

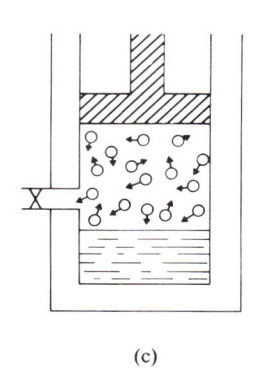

8.4 CHANGES OF PHASE AND PHASE EQUILIBRIUM

of the molecules within the liquid, and these energies are determined by the temperature. Consequently the *rate of evaporation* per unit liquid surface area is determined by the liquid *temperature* only.

As the molecules in the vapor phase bounce around, some will strike the liquid surface and re-enter the liquid phase. We say such molecules undergo *condensation* from the vapor phase to the liquid phase. The number of vapor molecules striking the surface is proportional to the pressure of the vapor. Consequently the *rate of condensation* per unit area of liquid surface is proportional to the *pressure of the vapor*.

Evaporation and condensation compete. Evaporation proceeds at a rate determined by the liquid temperature. In Fig. 8-12(b) evaporation is the dominant effect because the pressure of the vapor is so low that the condensation rate is low. The pressure of the vapor thus increases until, at a value called the *saturation vapor pressure* [Fig. 8-12 (c)], *phase equilibrium* is attained with

Evaporation rate = condensation rate.

Since the evaporation rate depends only on the temperature, and the condensation rate depends on the pressure of the vapor, this phase equilibrium equation implies that the *saturation vapor pressure depends only on the temperature*. Experiments confirm this expectation; results are given in Table 8-3 and the vaporization curve is shown in Fig. 8-13.

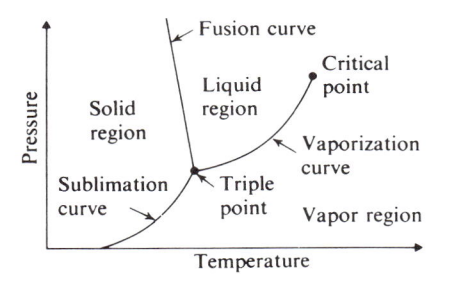

Fig. 8-13 Phase diagram for pure water.

Table 8-3. SATURATION VAPOR PRESSURE OF WATER

Temperature °C	Pressure mm Hg	Temperature °C	Pressure atmospheres
0.0	4.58	100	1.00
5.0	6.54	120	1.96
10.0	9.21	150	4.70
15.0	12.8	200	15.4
20.0	17.5	250	39.3
40.0	55.3	300	84.8
60.0	149	350	163.2
80.0	355	374.15*	218.4*
100.0	760	*critical point	

If the piston is suddenly pushed down a certain distance the pressure of the vapor will momentarily rise, but condensation now occurs at a greater rate. Equilibrium will soon be reached with less vapor and more liquid but with the *same saturation vapor pressure*, provided the temperature is maintained constant.

If the vapor is removed as fast as it evaporates, phase equilibrium is never attained. Continued evaporation will take place. This happens when pumping as in Fig. 8-12(a) or when dry air blows over a lake or past wet laundry. Evaporation is a very efficient cooling process, as we all discover when we stand soaking wet in a breeze. A large *latent heat of vaporization* is absorbed from a wet object as water is vaporized. The energetic molecules that escape into the vapor phase deplete the liquid's thermal energy.

Sublimation

Solids also evaporate. A direct transition between the solid and vapor phase, without passing through the liquid phase, is called *sublimation*. The evaporation of ice, which accounts for the drying of wet clothes hung outside in below freezing temperatures, is a familiar example. As with other phase transitions, a latent heat is involved in sublimation. There is also a very definite temperature at which equilibrium is obtained between the solid and the vapor at a specified pressure. For instance, solid carbon dioxide (*dry ice*) reaches equilibrium with its vapor at a pressure of 1 atmosphere, when the temperature falls to $-79°C$.

Sublimation is exploited in the process of *freeze-drying* foods. A solidly frozen food is exposed to a low pressure (approximately 0.1 mm Hg), and heat is supplied. The ice in the food continues to sublime as the water vapor is pumped away. By this process the moisture in the food is removed without appreciably altering the shape, color, or taste of the product. Vacuum packed, the food will keep for long periods at room temperature. Reconstitution is accomplished simply by adding water.

Amorphous Solids

The phase changes that we have been discussing for ice are typical of the behavior of many solids which have a definite crystal structure. However, there are certain materials, called *amorphous solids*, like glass and various resins which lack a crystal structure. These amorphous substances do not possess a definite melting point. As glass is heated it gradually softens. For this process there is no phase transition and no latent heat of fusion. An amorphous solid is more closely related to a very viscous liquid than to a crystalline solid.

Phase Diagram

Many of the foregoing facts about the behavior of a substance that relate to fusion, vaporization, and sublimation can be depicted on a phase diagram (Fig. 8-13). This diagram shows the phase or phases of a *pure* substance that are present in equilibrium at every possible combination of pressure and temperature.

Each point on a phase diagram refers to the substance in what is called an *equilibrium state*, the condition reached by the substance when the external conditions have been left unchanged for a sufficiently long time. An *equilibrium state* of a typical system, such as a collection of 10^{24} molecules of H_2O confined within a cylinder, is characterized by

1. A uniform temperature T throughout the cylinder,
2. A uniform pressure P throughout the cylinder, and
3. A definite mass of H_2O in each possible phase.

The interpretation of the phase diagram of Fig. 8-13 is clarified by singling out certain particular states, as shown in Fig. 8-14. Thus the equilibrium state of H_2O at a temperature of 110°C and a pressure of 1.00 atmosphere is represented by the point A. In this state there is only water vapor or steam. The point B corresponds to an equilibrium state at a temperature of 100°C and a pressure of 1.00 atmosphere. Under these conditions, water and its vapor can coexist in equilibrium, and the 1.00 atmosphere pressure is called the *saturation vapor pressure*. At the point C (50°C, 1.00 atmosphere) there is only water. At the point D (0.0°C, 1.00 atmosphere) ice and water can coexist in equilibrium, while at E (−20°C, 1.00 atmosphere) there is only ice. The line $EDCB$ represents the process described at the beginning of this section: ice is warmed and then melted to water, which is warmed and then vaporized.

Evidently there is a region of the phase diagram corresponding to equilibrium states in which the substance is entirely solid, a second region corresponding to the liquid phase, and a third region corresponding to the vapor phase. The liquid and vapor regions are separated by the *vaporization curve* which gives all the values of temperature and pressure for which the liquid and vapor can coexist in equilibrium. In other words, the vaporization curve gives the saturation vapor pressure at different temperatures. The data from which this curve is plotted is given in Table 8-3. The *sublimation curve* separates the solid and vapor regions and gives the values of temperature and pressure at which solid and vapor can coexist in equilibrium. States with solid and liquid in equilibrium are given by the *fusion curve* which separates the solid and liquid regions. A phase change involves passing from one region to another by crossing one of these three curves.

The vaporization, sublimation, and fusion curves intersect in a point called the *triple point*. At the triple-point temperature and pressure the three phases can coexist in equilibrium (Fig. 8-5). This state is easy to achieve and gives an accurately reproducible temperature which is now the basis for modern thermometry (Section 8.3).

The vaporization curve of Fig. 8-13 does not extend indefinitely. With a liquid and its vapor in equilibrium, it is found that as the temperature increases, the distinction between the two phases decreases and

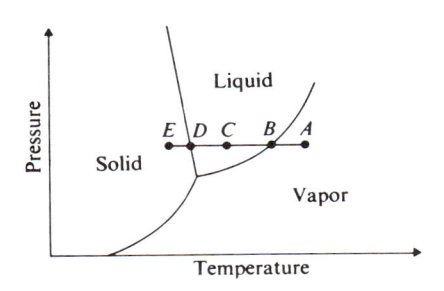

Fig. 8-14 Various equilibrium states of H_2O.

finally disappears at a certain temperature and pressure (and volume per unit mass) that we call the *critical point*.

Boiling

Boiling occurs when bubbles of vapor form throughout the volume of a liquid. Since incipient bubbles do not collapse, the vapor is in phase equilibrium with the liquid. Therefore, at the location of the bubble the pressure must be the saturation vapor pressure. That is, *the pressure in a boiling liquid is the saturation vapor pressure at the temperature of the liquid.* The vaporization curve thus acquires an added significance. The vaporization curve gives the corresponding temperatures and pressures of a boiling liquid. From the vaporization curve (Fig. 8-13 or Table 8-3) we can conclude that at room temperature water will boil at a pressure of 17.5 mm Hg, while at 100°C boiling occurs at 1.00 atmosphere pressure. One can control the temperature of boiling by adjusting the pressure. The valve in a pressure cooker is usually set to maintain the cooker pressure at about 1.0 atmosphere (14.7 pounds/in.^{2}) above the kitchen pressure of 1.0 atmosphere. The pressure inside the cooker will then be about 2.0 atmospheres and the water inside the cooker will boil at the corresponding temperature on the vaporization curve. This is about 120°C. The cooking time is thereby reduced by as much as a factor of ten.

8.5 WATER VAPOR IN AIR

In our atmosphere, water vapor is mixed with air. The presence of the air has practically no effect on the behavior of the water vapor. The total pressure exerted by the mixture is the sum of the pressures that each gas would exert if it alone were present.

    The presence of air above the surface of water does not change the equilibrium conditions shown by Table 8-3 and the corresponding phase diagram (Fig. 8-13) for pure water. The pressure of water vapor *in equilibrium with water* is still the *saturation vapor pressure* given by Table 8-3, the same as if the air were not there at all.

    But in our atmosphere, the water vapor is usually *not* in equilibrium with water. We are confronted with a variety of different situations. Water vapor at a given temperature is said to be *unsaturated, saturated*, or *supersaturated* according to whether the pressure of the water vapor is less than, equal to, or greater than the saturation vapor pressure at the temperature in question.

Example 4 The pressure of the water vapor in the atmosphere on a certain day is 17.5 mm Hg. Classify this vapor as unsaturated, saturated, or supersaturated when the temperature is (a) 10°C, (b) 20°C, (c) 40°C.

Solution

Temperature	Saturation vapor pressure from Table 8-3	Condition of vapor with pressure of 17.5 mm Hg
10°C	9.21 mm Hg	Supersaturated
20°C	17.5 mm Hg	Saturated
40°C	55.3 mm Hg	Unsaturated

The supersaturated state of a vapor is not an equilibrium state. The excess water vapor eventually condenses into water, usually forming tiny drops on minute particles in the air, until the pressure of the water vapor has been reduced to the saturation vapor pressure.

Relative Humidity

Unsaturated vapor is the most common situation in our homes and in the atmosphere. The drier the air, the more rapid will be the evaporation from any moist objects. A useful characterization of the degree of saturation is furnished by the *relative humidity* defined by the equation

$$\text{Percentage relative humidity} = 100 \times \frac{\{\text{pressure of vapor at temperature in question}\}}{\{\text{saturation vapor pressure at same temperature}\}}.$$

The relative humidity is 100% when the vapor is saturated and zero if no water vapor is present.

Example 5 Find the relative humidity within a house when the temperature is 20°C and the pressure of the water vapor is 5.0 mm Hg.

Solution From Table 8-3 the saturation vapor pressure at 20°C is 17.5 mm Hg. Therefore,

$$\text{Relative humidity} = 100 \times \frac{5.0 \text{ mm Hg}}{17.5 \text{ mm Hg}}$$
$$= 29\%.$$

This is an uncomfortably dry environment. Both for comfort and health a humidifier should be used to introduce more water vapor and maintain the relative humidity around 40–50%.

When air with unsaturated vapor is cooled sufficiently, a temperature called the *dew point* is reached. At this temperature the vapor is saturated. Further cooling will cause condensation. Clouds, fog, and rain result from the cooling of a portion of the atmosphere below its dew point. On a clear night, the earth's surface is cooled by emitting thermal radiation to space without receiving a balancing radiation from clouds. If this cooling brings the water vapor near the earth below its dew point, grass is covered with dew.

The most accurate method of measuring relative humidity is simply to determine the dew point by observing the temperature at which dew forms on the polished surface of a cool metal container.

Example 6 The air temperature inside a house is 20°C. There is dew on the surface of a metal can containing beer at 10°C. No dew forms on warmer surfaces. Find the relative humidity within the house.

Solution The dew point is 10°C. A thin film of air next to the metal surface is at 10°C and the water vapor in that layer is saturated. Therefore the vapor pressure in this film is the saturation vapor pressure at 10°C, 9.21 mm Hg, from Table 8-3. The vapor pressure of the warm air elsewhere in the house, being the same as the vapor pressure in the cool film, is also 9.21 mm Hg. But at 20°C, the vapor pressure necessary *for saturation,* again from Table 8-3, is 17.5 mm Hg. Therefore

$$\text{Relative humidity} = 100 \times \frac{9.21 \text{ mm Hg}}{17.5 \text{ mm Hg}}$$
$$= 53\%.$$

Cloud Chamber

If saturated vapor is rapidly cooled, it becomes supersaturated. This is a nonequilibrium state, and during the next few seconds a gradual condensation into a uniform fog occurs. The formation and growth of minute water drops is enhanced if the drops carry an electric charge. This fact is exploited in that important instrument of particle physics, the *Wilson cloud chamber* (Section 1.3). The chamber is filled with saturated vapor and then suddenly expanded. The expansion is accompanied by a rapid cooling, so the vapor becomes supersaturated. For the next few seconds growing water drops will form on the ions left in the wake of any high-energy charged particle that passes through the chamber. Such cloud-chamber tracks (Fig. 1-4) are a prime source of information about the subatomic world.

8.6 EQUILIBRIUM STATES: MICROSCOPIC AND MACROSCOPIC VIEWPOINTS

A macroscopic system in mechanical equilibrium, say some apparently motionless water and water vapor in a test tube, represents an enormously complicated system from a microscopic point of view. There are some 10^{24} interacting molecules in motion. A description which involves some 10^{24} details is far beyond the capacity of the largest computers and would hardly be very illuminating, even if it were available.

Fortunately, as we have seen, the gross characteristics of a macroscopic system often can be described in terms of a few *macroscopic quantities* such as *temperature*, *pressure*, and *volume* that can be determined by direct experimental measurement. It is found that if a system is left undisturbed by interaction with anything else, that is, if the system is *isolated*, it will eventually reach a situation in which no more changes are apparent. This special situation, obtained simply by waiting long enough, is called an *equilibrium state*. The universal tendency of macroscopic systems to approach an equilibrium state has been illustrated by the behavior of a confined liquid and its vapor. Except for minute fluctuations, the pressure, volume, temperature, and all other macroscopic quantities have constant values in an equilibrium state.

When a system attains an equilibrium state, three types of equilibrium are obtained:

1. *Mechanical equilibrium.* There is no acceleration of any macroscopic portion of the system. Mechanical wave motion, turbulence, and eddies do not occur.
2. *Thermal equilibrium.* All portions of the system are at the same temperature, which is constant.
3. *Chemical equilibrium.* The chemical composition of a macroscopic portion of the system at any location does not change as time elapses.

From the microscopic viewpoint, equilibrium requires that the change of a macroscopic property caused by the motion of some molecules must be counterbalanced by the change caused by other molecules. Equilibrium evidently is a very special situation as far as molecular motions are concerned. From the condition of exact balancing it is reasonable to hope that microscopic theory could deduce the special macroscopic properties of substances in equilibrium states. Using what is called *statistical mechanics*, this has been done with considerable success.

Thermodynamics is the name of the science that is concerned with a macroscopic description only. It deals with systems in equilibrium states and processes in which a system changes from one equilibrium state to another. Thermodynamics is a practical subject, to a large extent developed and first used with eyes riveted upon applications in engineering and chemistry. But the laws of thermodynamics are of such generality that the subject merits the attention of every natural philosopher.

In the next chapter, our emphasis first will continue to be placed on the macroscopic description provided by thermodynamics. Then the corresponding microscopic point of view will be presented as we seek an understanding of a unique and much discussed law of nature.

QUESTIONS

1. (a) What is the increase in length in a steel "meter" stick when its temperature is raised 10C°?
   (b) In what way is an invar measuring tape superior to a steel measuring tape?

2. A steel cable supporting a suspension bridge has a length of 500 m at 30°C. What is the change in its length when the temperature is decreased to −10°C?

3. A locomotive wheel is 1.30 m in diameter at 20°C. A steel tire with a diameter 0.50 millimeter undersize is to be shrunk on. To what temperature must the tire be heated to make its diameter 0.50 millimeter oversize?

4. The control element in a thermostat is a bimetallic strip. Explain how it works.

5. A steel steam pipe has a cross-sectional area of 8.00×10^{-3} m^2 at 20°C. What will be its area when filled with superheated steam at a temperature of 170°C?

6. Explain why the column of mercury first descends and then rises when a mercury-in-glass thermometer is thrust into boiling water.

7. A 0.100 m^3 bottle made of ordinary glass is completely filled with benzene at a temperature of 15.0°C. What volume of benzene overflows when the temperature is raised to 25.0°C?

8. What is unusual about the thermal expansion of water? Comment on some of the practical implications of this peculiarity.

9. What water temperature would you expect to find near the bottom of a lake which has ice on its surface? Why?

10. Why is a thick-walled bottle made of ordinary glass prone to shatter when heated?

11. What is the principal means of heat transfer from a hot object to a cold object when they are separated by
    (a) a vacuum;
    (b) a solid metal;
    (c) a gas when the warmer object is beneath the cooler object?

12. (a) What method of heat transfer is employed to heat the upstairs of a house with a furnace in the basement?
    (b) How is heat transfer from the sun to the earth accomplished?
    (c) Name at least one household situation which relies on conduction to transfer heat.

13. A certain man loses 1.2×10^3 kilocalories in a day by evaporation of his perspiration.
    (a) Approximately how much water does his body lose in this process?
    (b) If this energy could have been retained and used with 100% efficiency to climb a mountain, how high would this energy permit a 70 kg man to climb?

14. Sketch a Dewar flask and explain the purpose of each feature in its design.

15. Give an example of objects which are in thermal equilibrium, and another example showing objects which are not in thermal equilibrium.

16. What is the zeroth law of thermodynamics?

17. A constant-volume gas thermometer registers a pressure of 0.200 atmosphere when its bulb is in thermal equilibrium with water at the triple point. What is the Kelvin temperature of a bath if this thermometer registers a pressure of 0.225 atmosphere when its bulb is in thermal equilibrium with the bath water?

18. Assume the gas in the bulb of a constant-volume gas thermometer obeys the ideal gas law. If T_1 and T_2 are the Kelvin temperatures of two objects for which the thermometer registered pressures P_1 and P_2 respectively, show that

    $$T_1/T_2 = P_1/P_2.$$

19. You wish to monitor the temperature within a factory chimney. Which type of thermometer would you select and why?

20. Describe two different types of thermometers and discuss their merits and their drawbacks.

21. A 12.0-kg block of ice at 0°C absorbs 400 kilocalories of heat in a certain process which takes place at a pressure of 1.00 atmosphere. What is the final temperature? Explain.

22. How much heat must be supplied to 3.0 kg of water at 20°C in order to convert it to steam at 100°C? Assume the process takes place at a constant pressure of 1.00 atmosphere.

23. A farmer places a barrel containing 200 kg of water in a storage room to prevent his produce from freezing.
    (a) How many kilocalories of heat are given out by the water in cooling from 20°C down to 0°C?
    (b) How many kilocalories of heat are given off while half the water in the barrel freezes, changing from water at 0°C to ice at 0°C?
    (c) If the barrel of water were to be replaced by a 1000 watt electric heater, for how long would the heater have to operate to furnish the same energy as that given off in part (b) by the freezing of 100 kg of water?

24. How much heat is given off in a process in which 5.0 kg of steam at 100°C are converted into ice at −10°C? (The pressure is maintained constant at one atmosphere.)

25. A mass of 0.100 kg of ice at 0.0°C is introduced into a calorimeter can containing 0.800 kg of water at 30.0 °C. At what temperature will thermal equilibrium be attained? (Neglect the heat given off by the calorimeter can.)

26. Give an example of phase equilibrium for a solid and a liquid, a liquid and a vapor, and a solid and a vapor.

27. What is the definition of the term "saturation vapor pressure"?

28. Pure water is introduced into a closed container from which all the air has been removed. The temperature is maintained at 20°C. The pressure of the vapor is measured as time goes on. Describe what will be observed.

29. (a) Give an example of sublimation.
    (b) What is "freeze-drying"?

30. Sketch a phase diagram for *pure* water and label the sublimation, fusion, and vaporization curves as well as the triple point and the critical point. What is the significance of these curves and points?

31. Draw a line on a phase diagram to show a process, taking place at constant pressure, in which a solid is warmed and then undergoes sublimation, whereupon the vapor is warmed.

32. Draw a line on a phase diagram to show a process in which a vapor is liquefied by increasing the pressure while the temperature is maintained constant.

33. The phase change from water to ice can lead to the cracking of rocks and the destruction of highways. Explain.

34. The temperature of phase equilibrium between ice and water decreases 0.0075°C for each atmosphere increase in pressure. A large sample of ice at 0.0°C is insulated and the pressure is increased from 1.00 atmosphere to 1.00×10^3 atmospheres (typical of the pressure underneath the blade of a skate). Some of the ice melts. At what temperature will phase equilibrium be attained between the ice and water?

35. At the top of a mountain which is 20,000 ft (about 9000 ft less than Mount Everest) above sea level, it is observed that water boils at 80°C. What is the pressure of the atmosphere at this location?

36. What information about boiling is provided by the vaporization curve? Explain.

37. What purpose does the water fulfill in the cooking of a soft-boiled egg?

38. Heat is applied to a pressure cooker which is set to allow the pressure within the cooker to rise to 2 atmospheres. When the pressure reaches 2 atmospheres what will be the temperature if
    (a) there is water in the cooker?
    (b) there is air but no water in the cooker? (Before heat was applied this air was at a pressure of 1 atmosphere and a temperature of 27°C.)

39. Define the terms "relative humidity" and "dew point."

40. On a certain winter day, the outside temperature is 0°C, and the water vapor is saturated at this temperature. Some of this air is taken inside and warmed to 20°C without changing the pressure of the water vapor. What is the relative humidity? (This value will be typical of the environment inside buildings that do not use humidifiers.)

41. On a day when the temperature is 15°C, the dew point is found to be 10°C. Calculate the relative humidity. Explain your reasoning at each step of the calculation.

42. Some air is confined to the left half of a rigid box by a thin partition. Within the right half there is a vacuum. The partition is suddenly ruptured and the air undergoes what is called a "free expansion."
    (a) While it is expanding, is the air in an equilibrium state? Explain.
    (b) Describe the approach to equilibrium and the equilibrium state attained.
    (c) In this "free expansion" do the contents of the box (the air) perform any macroscopic work on the objects outside the *rigid* box?

SUPPLEMENTARY QUESTIONS

S-1. Consider the thermal expansion of a rectangular area A with sides of lengths L_1 and L_2. When the fractional expansions $\Delta L_1/L_1$ and $\Delta L_2/L_2$ are small, the approximate increase in area is given by

$$\Delta A = 2\alpha A \Delta T.$$

Prove this algebraically from Eq. (8.1) and draw a figure showing the relevant quantities.

S-2. Consider the thermal expansion of a cube with sides of length L and show, from Eq. (8.1), that the approximate increase in volume is given by

$$\Delta V = 3\alpha V \Delta T.$$

S-3. A thermostat, which is based on the fact that mercury is a good electrical conductor, is constructed as illustrated in Fig. 8-15. Two wires from an electrical circuit are inserted in the tube, one at the junction with the bulb and the other 15 cm away. The volume inside the bulb is 1.00 cm³ and at 0.00°C the mercury just fills it. The area of the cross-section of the glass tube is 3.00×10^{-4} cm². The mean value (for the temperature in question here) of the thermal coefficient of volume expansion of mercury is 18×10^{-5} per C°. Disregarding the expansion of the glass, calculate the temperature at which the circuit will be completed and the current will commence.

S-4. Burns caused by steam are often severe. To understand why this is so, consider the heat that would be given off to a person's skin during the transformation of 0.010 kg of steam at 100°C to water at 37°C, and compare this heat to that which would be given off in cooling 0.010 kg of hot water from 100°C to 37°C.

S-5. The specific heat of a metal is to be determined in an ice calorimeter experiment. A 0.600-kg block of the metal at a temperature of 100°C is introduced into a calorimeter which contains a large quantity of ice at 0°C. When contents of the calorimeter have reached thermal equilibrium, 0.082 kg of ice has melted and the final temperature is 0°C. Find the specific heat of the metal and state the assumptions that are made in this calculation.

S-6. In an experiment to measure the latent heat of vaporization (L_v) of water, the water is boiled at atmospheric pressure, and steam is led into a calorimeter and allowed to condense. The calorimeter is made of aluminum ($s = 0.217$ kilocalorie/kg C°) and has a mass of 0.400 kg. Initially the calorimeter contains 0.300 kg of water at 15°C. When the temperature within the calorimeter has risen to 65°C, the total mass of the calorimeter and its contents is found to be 0.733 kg. Calculate L_v from these data.

S-7. One type of solar heating system exploits the fact that sodium sulphate has a convenient melting point (31°C) and a sizable latent heat of fusion (51.3 kilocalories/kg).
(a) Explain how sodium sulphate could be used to store thermal energy received during periods of sunshine and to give off heat to a house at other times.
(b) If it takes about 3×10^5 kilocalories per day to heat a house, what mass of sodium sulphate would be required to supply heat for one day?

S-8. Explain why it is that the saturation vapor pressure depends only on the temperature and is independent of the volume of the vapor or the liquid.

S-9. It is possible to start with water in the liquid phase and to transform it into a vapor without ever reaching a state where the liquid and the vapor phases are present simultaneously. Indicate such a transformation by drawing on a phase diagram a path from an initial state in the liquid region, through the intermediate states, to a final state in the vapor region. Then describe in words how such a transformation could be accomplished.

S-10. A closed bathroom with a volume of 30 m³ is maintained at a temperature of 20.0°C. The relative humidity in the room is 30%. A basin is then filled with water at 20.0°C. What mass of this water will evaporate?

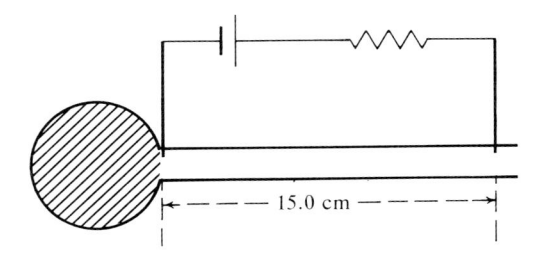

Fig. 8-15 Thermostat of Question S-3.

ADDITIONAL APPLICATIONS TO MEDICINE AND THE LIFE SCIENCES

The Human Thermostat

Experiments performed during the past few decades have finally located the sensory organ in the brain that functions as the human thermostat. The *hypothalamus*, centrally located under the great hemispheres of the brain, contains this sensory organ. Here the temperature of blood which bathes the brain cells is measured and the heat-dissipating mechanisms which regulate this temperature are tripped.

Some recent evidence suggests that aging of the human body may be retarded if the body temperature is depressed by one to two C°.

Further Reading
"The Human Thermostat," by T.H. Benzinger, in the *Scientific American*, Jan., 1961.

Some Patient Care Procedures: Heat Transfer

Knowledge of heat transfer processes is applied in numerous mundane procedures involved in the care of patients. For example, flannel has a relatively low *coefficient of thermal conductivity* (0.000023 kcal/m-sec C°; as compared with copper, 0.092; glass, 0.00025; water, 0.00014). It therefore makes good sense to wrap hot water bottles in flannel before applying them to patients. On the other hand, bedpans being usually metallic are good conductors, so it helps to warm them slightly before use.

Incidentally, human skin is not a particularly good thermal conductor. It has a coefficient near that of wood: about 0.00004 kcal/m-sec C°.

Convection is also employed in numerous ways. One example is the *sedative bath*. In this bath heat is transferred to the patient by conduction, but the water temperature is kept relatively uniformly warm by continuous currents due to warm water inflow.

Question
Give an example of some hospital procedure depending primarily, not upon conduction or convection, but rather on radiation as the heat-transfer mechanism.

Answer
The use of heat lamps (e.g., to increase blood supply to an area and thus promote healing).

Factors Affecting Skin Temperature: Heat Transfer in Humans

The maintenance of a relatively constant internal body temperature in the face of significant changes in heat production (e.g., exercise, eating) and heat loss (e.g., cold environment, wind) involves multitudinous interdependent processes such as shivering, thermal conduction through tissue, sweating, vasoconstriction, and vasodilation.

As indicated earlier heat is transported within the body partly by thermal conduction and partly by a controlled convection, controlled by the mechanism of vasoconstriction and vasodilation.

Question

Does a warm environment promote vasoconstriction or vasodilation?

Answer

Vasodilation. This is the controlling mechanism in convection, which results in high blood flow to the skin and thus maximizes the cooling effect.

Mercury Disposal: Liquid-in-Glass Thermometer

The widespread use of mercury in hospitals from manometers and thermometers to electric switches, not to mention mercurous and mercuric compounds used for medicinal purposes, has led to a new awareness of dangers inherent in mercury poisoning. For example, in July 1971, the American Hospital Association issued a bulletin devoted exclusively to the "Control of Mercury Pollution."

One of the problems with Hg is that it is relatively volatile. At an international *Special Symposium on Mercury in Man's Environment*, held in Ottawa, Ontario, in Feb. 1971, W.B. Lewis, Senior Vice-president of Atomic Energy of Canada, related this story: "In 1941 in Calgary, Miss A and Miss B, stenographers working in the same office, were taken ill and died. Miss A became ill on August 16 and died on September 24. Miss B sickened on September 5 and died on October 17. The cause adduced was that in the office where they worked for 3 to 5 months the air contained about 1 milligram of mercury per m^3 (it may have been rather more in the heat of July and August). To appreciate the scale, 1 mg/m^3 is about 1/10 of the room temperature vapor concentration in equilibrium with liquid mercury and is about 10 times the recommended industrial maximum tolerable concentration. (But it was not liquid mercury that was exposed; it was an organic mercury compound improperly packaged and stored nearby. Moreover, it was one of the more highly toxic mercury compounds then coming into wide use in agriculture to protect seed from attack by fungi.)"

Question

(a) Assume a clinical mercury thermometer contains 2 gm of Hg (this is probably on the high side because while some thermometers contain more, many contain as little as 1 gm). Investigations show that in Canadian hospitals about 8.5 thermometers are broken per hospital bed per year. It is likely that U.S. figures would be the same. Finally, assume some 2×10^6 hospital beds in the United States and calculate the mass of mercury "lost" through broken clinical thermometers per year in the U.S.

(b) Drug stores apparently sell for home use about 0.75 as many clinical thermometers per year as are used by hospitals. On this basis what is the total metallic mercury spillage into the environment per year due to clinical thermometer breakage?

Answer

(a) Mass of Hg = $(2 \times 8.5 \times 2 \times 10^6)$ gm
$= 34 \times 10^6$ gm
$= 34 \times 10^3$ kg.

(b) Hospital breakage: 34×10^3 kg.
Home breakage: $(34 \times 10^3 \times 0.75)$ kg = 25×10^3 kg.
Total is therefore 59×10^3 kg per year.

Dermal and Internal Temperature Measurements in Medicine: Thermocouple and Thermistor

Thermocouples have found use in medical practice and research. In fact, it was using a relatively primitive thermocouple that Dr. Ray Lawson, a Montreal surgeon, first found in 1955 that the skin over a malignant tumor in the breast is generally warmer (from a few tenths of C° to about 3C°) than the surrounding tissue. So in a sense the thermocouple played a role in the development of the medical thermography described earlier in this chapter.

In medicine the *thermistor thermometer* has some advantages over the thermocouple: Various probes are availabe for both dermal and internal temperature measurements (and as in the case of the thermocouple, the probe may be located some distance—as much as 100 m—from the main unit); the probe can be very fine and can have various shapes; the readings can be made in a matter of seconds; and the precision is greater than that of the thermocouple, and much greater than that of clinical mercury thermometers.

As an adjunct to thermography, fine thermistor probes can be used to verify hot spots as indicated in a thermogram.

Alcohol Rubs: Latent Heats

Question

In patient care it is not uncommon to sponge ethyl alcohol on patients for a soothing cooling effect. Since the cooling effect is due to evaporation and the latent heat of vaporization of water at the temperature of the human body is about 580 kilocalories/kg, while that of alcohol is only about 300 kilocalories/kg, why does one not use water rather than alcohol?

Answer

The cooling effect depends upon both the latent heat as well as the amount of liquid evaporating per unit time; in other words, on the product mL_V. Because alcohol evaporates at a much greater rate than water, the difference in latent heats is more than overcome.

SUGGESTED READING

MARSHALL, J. S., E. R. POUNDER, and R. W. STEWART, *Physics*. Toronto: Macmillan, 1967 (second edition). Of particular interest is the chapter "Water Substance and the Atmosphere."

SEARS, F. W., and M. W. ZEMANSKY, *University Physics*. Reading, Mass.: Addison-Wesley, 1963 (third edition). A very clear and complete treatment of thermal phenomena. The descriptive portions of Chapters 15 and 18 will be most useful to our readers.

STANLEY, H. E., *Biomedical Physics and Biomaterials Science*. Cambridge, Mass.: M.I.T. Press, 1972. An interesting up-to-date selection of specific applications of physics and bio-engineering to medicine. This reference is relevant to almost all of our chapters.

Sadi Carnot

9 Disorder and Its Increase

Thermal energy is disordered energy. The amount of energy is measured by a quantity $\bar{E}$ and the amount of disorder by a new quantity S called the *entropy*. In any process which occurs in nature, the total energy is conserved and the *total entropy increases*. The quantitative formulation of these ideas and the study of their implications is the topic of this chapter.

The *principle of entropy increase* is one of the most profound laws of physics. Its ramifications extend throughout chemistry and biology as well as physics and engineering.

The historical roots of this subtle subject lie in a study of the all-important device of the Industrial Revolution, the steam engine. In 1824 a young French engineer, Nicolas Leonard Sadi Carnot, published a brochure titled *Reflections on the Motive-Power of Heat, and on Machines Fitted to Develop That Power.* Near the beginning of his paper he questioned ". . . whether the motive power of heat is unbounded, whether the possible improvements in steam engines have an assignable limit—a limit which the nature of things will not allow to be passed by any means whatever." He had asked the right question. Penetrating analysis and plausible assumptions led Carnot to the result that all "heat engines" have an efficiency limit determined entirely by the temperature at which they take in heat and the temperature at which they exhaust or reject heat.

Carnot's single paper attracted little attention; no contemporary reference to his work can be found in any published paper. Quite unknown in his day, he died of cholera following an attack of scarlet fever.*

The significance of Carnot's discovery was not recognized until the middle of the nineteenth century. It was then that the English physicist, William Thomson (1824–1907), later to become Lord Kelvin, and the German physicist, R. J. E. Clausius (1822–1888), realized that Carnot had pointed the way to a new law whose generality ranged far beyond the confines of steam engines. The law they formulated, called the *second law of thermodynamics,* has been stated in many different but logically equivalent ways. Two phrasings of the second law are

---

"The phenomenon of the production of motion by heat has not been considered from a sufficiently general point of view . . . . It is necessary to establish principles applicable not only to steam engines but to all imaginable heat engines, whatever the working substance and whatever the method by which it is operated."

Sadi Carnot
(1796–1832)

*Coincidentally, another exceptionally talented young innovator, Evariste Galois, died in Paris only a few days before Carnot. He too was unrecognized and rejected by the scientific authorities of the time, yet was probably the most brilliant French mathematician of his day.

R. J. E. Clausius
(1822–1888)

William Thomson (Lord Kelvin)
(1824–1907)

Kelvin statement. It is impossible by means of inanimate material agency to derive mechanical effect from any portion of matter by cooling it below the temperature of the coldest of the surrounding objects.

Clausius statement. No process is possible whose *sole* result is the transfer of heat from a cooler to a hotter body.

Clausius also stressed the utility of entropy and showed that the second law was the same as saying that the total entropy increases in any process. The microscopic interpretation of entropy as a measure of molecular disorder was discovered by Boltzmann.

Our approach to the concept of entropy follows a route in keeping with William Thomson's frequently quoted general view: "I often say that when you can measure what you are speaking about, and express it in numbers, you know something about it; but when you cannot express it in numbers, your knowledge is of a meagre and unsatisfactory kind; it may be the beginning of knowledge, but you have scarcely, in your thoughts, advanced to the stage of Science, whatever the matter may be."

9.1 CALCULATION OF ENTROPY CHANGES

Entropy and Disorder

Consider a typical macroscopic system, such as one kilogram of H_2O molecules confined within a cylinder. When the external conditions are unchanged for a sufficiently long time, the system will eventually reach an equilibrium state which can be characterized by a temperature, a pressure, and the mass of H_2O in each possible phase. In this state, the system occupies a definite volume and possesses a definite value of thermal energy. The system also has a definite value of a new quantity S called *entropy*. The entropy of 1.00 kg of H_2O in several different states is given in Table 9-1.

Table 9-1. ENTROPY PER KILOGRAM OF H_2O AT 1.00 ATMOSPHERE WHEN ICE AT $-10°C$ IS ASSIGNED ZERO ENTROPY

Phase	Temperature	Entropy in kilocalories/kg°K
Ice	−10.0°C	0
Ice	0.0°C	0.021
Water	0.0°C	0.313
Water	20.0°C	0.383
Water	40.0°C	0.449
Water	60.0°C	0.511
Water	80.0°C	0.569
Water	100.0°C	0.624
Steam	100.0°C	2.069
Steam	250.0°C	2.23

Entropy is a measure of molecular disorder. Thus a system consisting of one kilogram of H_2O molecules has little entropy when the molecules are arranged in an orderly fashion as in a crystal at low temperature. As the crystal is warmed, there is increased disorder associated with the random thermal motion of its molecules, so the entropy increases. In a liquid the molecules are free to roam about. The liquid phase is considerably more disordered than the solid phase. Consequently, melting involves an appreciable increase in entropy. In the gaseous phase there is practically no correlation between the motions of different molecules. This is an even more disordered situation than when the molecules are in the liquid phase. A large entropy increase therefore is associated with vaporization.

Clausius Equation

By considering a system of a large number of molecules from the point of view of quantum mechanics, we can establish the numerical measure of the system's disorder that we call the *entropy of the system*. **This we will do later.** But decades before entropy was understood from a molecular viewpoint, Clausius had learned that substances could be assigned a rather mysterious something called entropy S. Its *changes* ΔS in a process could be determined by the heat the substance absorbed or gave off, divided by the Kelvin temperature of the substance. The Clausius equation for the change in the entropy of a system during a given process is

$$\Delta S = (\text{heat for process})/T, \qquad (9.1)$$

where T is the Kelvin temperature of the system during the process. The heat for the process is positive when the system absorbs heat and negative when the system gives off heat. (Although the validity of the Clausius equation is restricted to certain idealized "reversible processes" discussed in Section 9.2, it can always be used to construct entropy tables like Table 9-1.)

Notice that the Clausius equation is at least roughly in accord with our interpretation of entropy as a measure of molecular disorder. Heating a substance increases the random molecular motion and therefore increases the disorder; the Clausius equation states that heating a substance increases its entropy.

The importance of entropy lies in the fact that a grand new physical law which determines the direction in which energy transformations occur can be neatly formulated in terms of entropy changes. But before we can use this new law we must learn how to use the Clausius equation to calculate entropy changes.

Example 1 A mass of 10.0 kg of water at 100°C is transformed into steam at 100°C. Find the entropy change ΔS of these 10.0 kg in this process.

Solution The Kelvin temperature is 373°K throughout the process.

$$\text{Heat absorbed} = mL_v \qquad (8.6)$$
$$= (10.0 \text{ kg})(540 \text{ kilocalories/kg})$$
$$= 5.40 \times 10^3 \text{ kilocalories}.$$

The Clausius equation gives the entropy increase:

$$\Delta S = (\text{heat for process})/T \qquad (9.1)$$
$$= 5.40 \times 10^3 \text{ kilocalories}/373°\text{K}$$
$$= 14.5 \text{ kilocalories}/°\text{K}.$$

Example 2 A mass of 10 kg of water at 98.0°C is warmed to 100.0°C.
(a) What is the entropy change of the water during this process?
(b) What is the entropy change when the water is cooled from 100.0°C to 98.0°C?

Solution Because the temperature changes, the question arises as to what temperature should be used in the Clausius equation. To solve the problem exactly, we must use integral calculus. But when the temperature change is small compared to the Kelvin temperatures T occurring, the Clausius equation gives a useful approximation using either the initial or final temperature. It is better still to use an intermediate temperature such as $99°C = (273 + 99)°K$.

(a) From Eq. (7.12):

Heat absorbed by water = (10 kg)(1.00 kilocalorie/kg C°)(2.0°C)
= 20 kilocalories.

The Clausius equation gives

$$\Delta S = \text{(heat for process)}/T \quad (9.1)$$
$$= 20 \text{ kilocalories}/(273 + 99)°K$$
$$= 0.054 \text{ kilocalories}/°K.$$

This is the entropy *increase* of the water.

(b) This cooling process brings the water back to the *same* condition or thermodynamic state that it was in at the beginning. Therefore the entropy must return to the same value it had initially. Consequently

Entropy *decrease* of water = 0.054 kilocalories/°K.

Of course this could be computed from Eq. (9.1).

These examples illustrate how entropy changes can be calculated from measurements of specific heats and latent heats. Such calculations provide the data from which entropy tables like Table 9-1 are constructed. Notice that since we are concerned only with *changes* of entropy, we can arbitrarily select some state as a reference state and tabulate only the differences in entropy between other states and this reference state.

The entropy of a composite system is the sum of the entropies of its parts. This is implied by the Clausius equation. Consequently the entropy of a mass m of a substance is simply m times the entropy per unit mass of the substance.

Heat Reservoirs

A heat reservoir is a system so large compared to the systems with which it interacts that its temperature remains essentially unchanged when it absorbs or gives off heat. The Clausius equation allows the calculation of the entropy change of the reservoir to be performed very simply. For a reservoir at *Kelvin* temperature T we obtain

Entropy increase of reservoir = (heat absorbed by reservoir)/T, (9.2a)
or
Entropy decrease of reservoir = (heat given off by reservoir)/T. (9.2b)

These equations give the entropy changes of a heat reservoir correctly for *any* process without restriction. We will use these equations frequently in the following sections.

Example 3 A hot stone is placed in a lake which has a temperature of 290°K. As the stone cools, it gives off 58 kilocalories. What is the change in entropy of the water in the lake?

Solution To an excellent approximation the lake can be considered to be a heat reservoir. The initial and final water temperature is 290°K. Therefore the entropy increase of the lake water is given by

$$\Delta S = \text{(heat absorbed by reservoir)}/T \quad (9.2a)$$
$$= 58 \text{ kilocalories}/290°K$$
$$= 0.20 \text{ kilocalorie}/°K.$$

9.2 THE SECOND LAW OF THERMODYNAMICS

Principle of Entropy Increase

Many processes that would seem possible according to the first law of thermodynamics, because they do conserve energy, in fact never occur. For example, a brick resting on a horizontal road never spontaneously accelerates along the road by gathering up thermal energy from the road. (See Fig. 9-1.) If this process occurred, the energy which is randomly distributed over many molecules of the road would be converted into an ordered motion of the brick as a whole. This transformation from disorder to order is quite compatible with the first law, but for some reason it does not happen.

The reverse process (Fig. 9-2) does occur and illustrates a feature present in all our experience. When a brick is projected along the road, it slows down as the kinetic energy due to the *ordered* motion of the brick as a whole is transformed into thermal energy associated with *disordered* molecular motion within the road and the brick.

Examination of this and many other processes reveals that there is a general rule which determines the direction in which energy transformations occur. Ordered energy is transformed into disordered energy; that is, using an arrow to denote the direction of the process,

$$\text{order} \longrightarrow \text{disorder}.$$

But the reverse transformation is not observed. The disorder in one object can be reduced in a process only if there is a more than compensating increase in the disorder of other objects that participate in the process.

This law of nature which determines the direction in which processes occur can be given a quantitative formulation in terms of *entropy*.

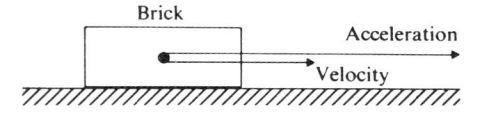

Fig. 9-1 The spontaneous acceleration of the brick along a horizontal road does *not* occur. This would involve a transformation of *disordered* molecular motion into *ordered* motion of the brick as a whole.

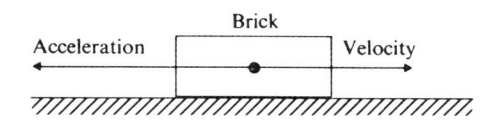

Fig. 9-2 The reverse of the process shown in Fig. 9-1. The brick slows down as the kinetic energy associated with the *ordered* motion of the brick as a whole is transformed into *disordered* molecular motion (thermal energy) of the brick and the road. This process occurs.

The entropy of an object is a measure of its molecular disorder. Increasing disorder implies increasing entropy. The rule is that processes take place only in the direction that increases the *total entropy* of the participating objects. Stated formally, this is the *principle of entropy increase*: *In any process, the total entropy of the participating objects is increased or* (in an idealized "reversible process") *unchanged. The total entropy never decreases.*

The Arrow of Time, Irreversible Processes

Life is a one way street. We are born, we grow old, and die. The physical law that distinguishes the past from the future, that puts the arrow on the time in Fig. 9-3 is the *principle of entropy increase*. The situation with lower entropy (less disorder) must come before that with higher entropy (more disorder). This is how we can tell that if a movie shows a broken egg reassemble into a perfect egg, the movie must be running backwards.

Fig. 9-3 The situation with less disorder comes before the siutation with greater disorder.

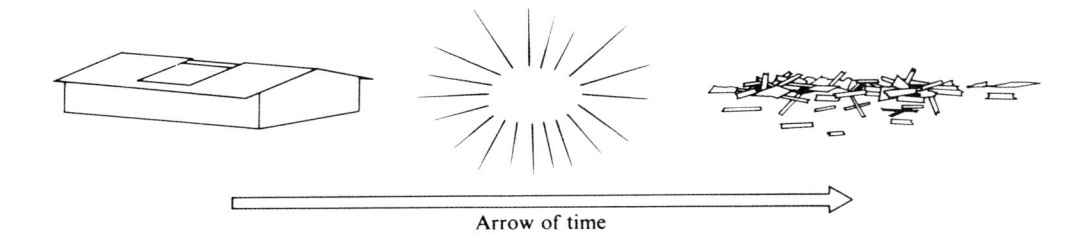

Arrow of time

Any real process involves an increase in the total entropy. A reversal of such a process, leading to the restoration of the initial entropies of all the participating objects, would therefore involve a decrease in entropy. The entropy principle thus implies that entropy-increasing processes are *irreversible* processes.

A detailed examination shows that *irreversible* entropy increases occur when

1. Macroscopic mechanical energy is dissipated into thermal energy. This happens when work is done by a frictional force or when Joule heating occurs in a resistor.
2. Heat is transferred from an object at a higher temperature to an object at a lower temperature.
3. Mechanical equilibrium of all portions of a system is not maintained; therefore, accelerations, eddies, or turbulence develop.
4. Phase equilibrium (Section 8.4) is not maintained. (We will not consider chemical reactions and the irreversibility arising from failure to maintain chemical equilibrium.)

Reversible Processes

The principle of entropy increase leads to the classification of processes displayed in Table 9-2. Processes for which there is no change in the total entropy can occur in *either* direction. They are called *reversible processes*.

Table 9-2. CLASSIFICATION OF PROCESSES ACCORDING TO PRINCIPLE OF ENTROPY INCREASE

ΔS_{total} increases	Process is possible; it will be irreversible.
ΔS_{total} decreases	Process is impossible (statistical mechanics says improbable).
$\Delta S_{total} = 0$	Process is reversible; it may occur in either direction.

A reversible process is an idealization never observed in macroscopic experiments. However, it can be closely approximated by (1) minimizing dissipative effects resulting from frictional forces or electrical resistance, (2) transferring heat only through very small temperature differences, (3) avoiding accelerations and turbulence, and (4) avoiding appreciable departures from phase or chemical equilibrium.

The Clausius equation,

$$\Delta S = (\text{heat for process})/T, \qquad (9.1)$$

is valid only for reversible processes. This is why reversible processes are so important in thermodynamic theory. To calculate the entropy of a substance in different states and thus build up the entropy table for the substance, we always consider reversible processes and employ the Clausius equation. Once we know the entropy of a substance in different states, we can use this data to calculate the substance's change in entropy, no matter what process brings it from one state to another.

Example 4 A kettle contains 1.00 kg of hot water at 40.0°C. This is poured into a pot containing 1.00 kg of cold water at 0.0°C. After thorough mixing, the water has a uniform temperature of 20.0°C throughout. Find the change in the total entropy which occurred in this process.

Solution We use the entropies of water given in Table 9-1:

Final total entropy = (2.00 kg)(0.383 kilocalorie/kg°K)
 = 0.766 kilocalorie/°K.

Initial entropy of hot water = (1.00 kg)(0.449 kilocalorie/kg°K);

Initial entropy of cold water = (1.00 kg)(0.313 kilocalorie/kg°K);

Initial total entropy = 0.762 kilocalorie/°K.

$$\Delta S_{\text{total}} = 0.766 \text{ kilocalorie}/°K - 0.762 \text{ kilocalorie}/°K$$
$$= 0.004 \text{ kilocalorie}/°K.$$

The total entropy increased so the process is irreversible. This is no surprise; we know we cannot "unmix" the hot and cold water.

Notice that the Clausius equation was used only in the construction of the entropy table, as outlined in Section 9.1. Having once done this job, we can use this data to compute entropy changes in any process, no matter how irreversible the process is.

Entropy and the Second Law of Thermodynamics

The most important properties of entropy for macroscopic theory are gathered together in a summary that constitutes a statement of the *second law of thermodynamics*:

1. We can assign a definite value of a quantity S called *entropy* to each equilibrium state of a system.
2. *Entropy changes* ΔS of a system during a reversible process are determined by the Clausius equation,

$$\Delta S = (\text{heat for process})/T, \tag{9.1}$$

where T is the Kelvin temperature of the system.

3. *Principle of entropy increase*: In any process, the total entropy of the participating objects is increased or (in an idealized reversible process) unchanged. The total entropy never decreases.

The second law is relevant not only in engineering and physics, but also in chemistry and biology. The direction in which chemical reactions proceed is governed by the principle of entropy increase. Consequently the entropy of different chemicals is of vital interest to chemists.

In biological processes, as random collections of molecules are exquisitely organized into living matter, reduction of disorder is often obvious. Do such processes violate the second law? Apparently not. If we look at *all* the objects participating in the process, we find that the obvious entropy decreases have been more than matched by entropy increases elsewhere.

Entropy Increase Due to Heat Conduction

The quantitative implications of the second law are most clearly brought out by studying certain simple physical processes. Heat conduction provides a very instructive example.

We consider a process in which a heat Q flows spontaneously from a reservoir at temperature T_1 to a reservoir at temperature T_2 (Fig. 9-4). The process takes place by heat conduction through a copper wire joining the reservoirs. We examine all the entropy changes associated

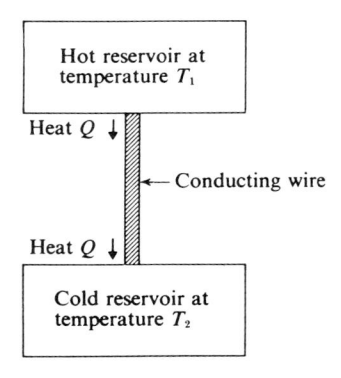

Fig. 9-4 Conduction. A heat Q is given off by the hot reservoir and is absorbed by the cold reservoir. The resulting entropy change is $(Q/T_2) - (Q/T_1)$.

with this process, computing the entropy changes of reservoirs from Eq. (9.2):

a. Entropy *increase* of cold reservoir = Q/T_2;
b. Entropy *decrease* of hot reservoir = Q/T_1;
c. We neglect the entropy change of the wire. This entropy change is proportional to the mass of the wire and will be small if this mass is small. In fact, if each portion of the wire is maintained in the same thermodynamic state (constant temperature and pressure) its entropy will not change at all.

The change in the total entropy because of this process is therefore given by

ΔS_{total} = (entropy increase of cold reservoir) −
(entropy decrease of hot reservoir)
$= (Q/T_2) - (Q/T_1).$ (9.3)

Therefore

$$\Delta S_{\text{total}} = Q\left(\frac{T_1 - T_2}{T_2 T_1}\right).$$

This result shows that if T_1 is greater than T_2, that is, if the heat flow is from higher to lower temperature, ΔS_{total} is positive. Then the total of the entropies of the participating objects increases. This verifies that our everyday experience with the direction of heat flow agrees with the principle of entropy increase. It is worth emphasizing that when heat flows from a higher to a lower temperature, the total entropy always is increased; it never merely remains unchanged.

The process is irreversible. A heat flow from the lower temperature T_2 to the higher temperature T_1 would result in a decrease in the total entropy in violation of the second law of thermodynamics. The *Clausius statement* of the second law, "No process is possible whose *sole* result is the transfer of heat from a cooler to a hotter body," has emerged as a consequence of our formulation of the second law in terms of the principle of entropy increase. But some interesting speculations on ways and means to circumvent this stubborn one-way process have been made from time to time. Consider the following.

A well-informed little athlete with lightning reflexes, called a Maxwell demon, is pictured in Fig. 9-5. This figment of Maxwell's imagination devotes his life to confounding Clausius by opening the door at the correct times to permit the fastest molecules to move to the right half, and the slowest molecules to move to the left half of the box. The right half gets hotter and the left half cooler.

Maxwell pointed out that if this demon or his mechanized equivalent existed and could continue indefinitely without intervention from outside the box, then the second law of thermodynamics could be violated. But so far no demons have been detected and none are expected!

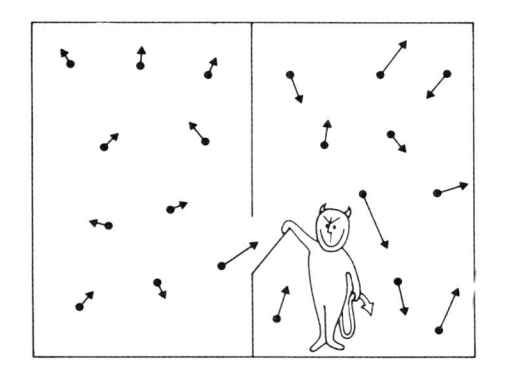

Fig. 9-5 Maxwell demon.

Perpetual Motion Engines of the Second Kind

Another immediate and important consequence of the second law is that *no process is possible whose sole effect is to extract heat from one reservoir and convert this heat entirely into work.** The key word in this statement is *sole*. If, at the end of a process, all objects except the heat reservoir (Fig. 9-6) are returned to the equilibrium states in which they started, the entropy of each object will return to its initial value. Therefore the entropy *change* for all these objects for the process will be zero. The only entropy change will be the *entropy decrease* of the heat reservoir. This process would therefore cause a decrease in the total entropy of the participating objects. According to the second law, this is impossible.

Inventors who have given up trying to beat the first law by creating energy with a "perpetual-motion engine of the first kind" will be equally frustrated in their efforts to devise a "perpetual-motion engine of the second kind" which conserves energy but violates the second law. An engine which can run forever solely by extracting heat from the ocean unfortunately is impossible. In the following section we will find that to make a "heat engine" that works, a temperature difference is required and the engine which would use heat from the ocean must give off some heat at a lower temperature. To persistent seekers of perpetual motion engines, the first law says that you can't win, then the second law adds that you can't even break even!

*The impossibility of such a process is often taken as the (Kelvin) statement of the second law of thermodynamics. It is then possible, by following a long and interesting chain of reasoning, to deduce the statements about entropy that we have adopted as the second law.

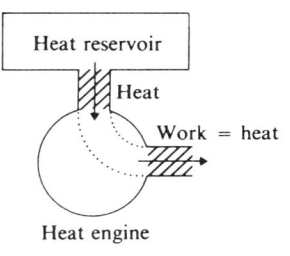

Fig. 9-6 Schematic diagram of a perpetual motion engine of the second kind which extracts heat from one reservoir and converts this heat entirely into work. Such a device is an impossibility according to the second law of thermodynamics.

9.3 HEAT ENGINES AND REFRIGERATORS

Heat Engines

A heat engine is a device that absorbs a heat Q and delivers useful macroscopic mechanical energy by performing a work W. Practical examples are the steam, gasoline, and diesel engines. Let us consider an engine that runs for an indefinite period and therefore operates in what are called *cycles*, returning after each cycle to exactly the same condition as that in which it started. The entropy and the thermal energy of the engine and its contents therefore return to their initial values after each cycle. In other words, the change of the entropy and the change of the thermal energy of the engine and its contents are zero after one cycle. That is, for one cycle

$$\Delta S_{\text{engine}} = 0, \qquad (9.4)$$

and

$$\Delta \bar{E}_{\text{engine}} = 0. \qquad (9.5)$$

The *thermal efficiency* of a heat engine is defined in terms of the heat Q furnished to the engine in one cycle and the work output W in one

cycle:
$$\text{Thermal efficiency} = W/Q.$$
This is the ratio of the output desired to the input furnished.

The second law of thermodynamics is such a sweeping generalization that it is possible to discover interesting limitations on the efficiency of a heat engine without any consideration of detailed mechanisms. An examination of entropy changes is necessary. The extraction, during each cycle, of a heat Q_{hot} from a hot reservoir at Kelvin temperature T_{hot} decreases the entropy of this reservoir by an amount $Q_{\text{hot}}/T_{\text{hot}}$ [from Eq. 9.2 (b)]. Therefore, a compensating entropy increase must be provided. This is not to be found in the engine itself nor in its contents, because after one cycle these return to the equilibrium state in which they started and their entropies consequently return to their initial values ($\Delta S_{\text{engine}} = 0$). Therefore some heat Q_{cold} must be given off by the engine in each cycle to increase the entropy of some other object (Fig. 9-7). If this heat Q_{cold} is absorbed by a second heat reservoir at Kelvin temperature T_{cold}, the entropy of the reservoir will increase by an amount $Q_{\text{cold}}/T_{\text{cold}}$ [from Eq. 9.2(a)]. To achieve the largest possible compensating entropy increase ($Q_{\text{cold}}/T_{\text{cold}}$) for a given quantity of heat Q_{cold}, the temperature T_{cold} should be as low as possible. Entropy considerations thus show that the reservoir at temperature T_{cold} should be the coldest available. As far as the reservoir at temperature T_{hot} is concerned, the hotter the better.

The principle of entropy increase requires that the entropy decrease of the hot reservoir be not greater than the entropy increase of the cold reservoir. That is

$$Q_{\text{hot}}/T_{\text{hot}} \leq Q_{\text{cold}}/T_{\text{cold}},$$

which implies

$$T_{\text{cold}}/T_{\text{hot}} \leq Q_{\text{cold}}/Q_{\text{hot}}. \tag{9.6}$$

It is important to examine the thermal efficiency of this engine. Let us apply the first law of thermodynamics to the system consisting of the engine and its contents. For one cycle we know that

$$\Delta \bar{E} = 0. \tag{9.5}$$

So from

$$\Delta \bar{E} = (\text{work for process}) + (\text{heat for process}) \tag{7.11}$$

we get

$$0 = -W + (Q_{\text{hot}} - Q_{\text{cold}}),$$

where W is the work output of the engine in one cycle. (The work done *on* the engine and its contents is therefore $-W$.) Solving this equation for W/Q_{hot}, we obtain

$$\text{Thermal efficiency} = W/Q_{\text{hot}}$$
$$= 1 - (Q_{\text{cold}}/Q_{\text{hot}}).$$

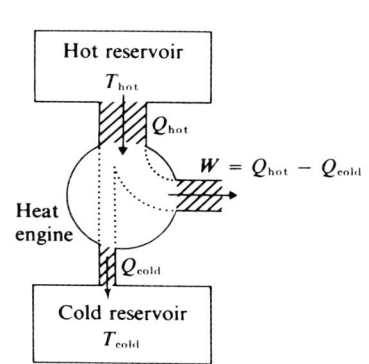

Fig. 9-7 Schematic diagram of a heat engine which, in one cycle, extracts heat Q_{hot} and rejects heat Q_{cold}, performing a work W equal to $Q_{\text{hot}} - Q_{\text{cold}}$. The entropy increase of the cold reservoir ($Q_{\text{cold}}/T_{\text{cold}}$) more than compensates for the entropy decrease of the hot reservoir ($Q_{\text{hot}}/T_{\text{hot}}$).

Together with Eq. (9.6), this implies

$$\text{Thermal efficiency} \leq 1 - (T_{\text{cold}}/T_{\text{hot}}). \qquad (9.7)$$

The maximum possible thermal efficiency of an engine operating between two given heat reservoirs, called the *Carnot efficiency*, is then seen to be given by

$$\text{Carnot efficiency} = 1 - (T_{\text{cold}}/T_{\text{hot}}). \qquad (9.8)$$

Sadi Carnot obtained this important result in the early 1820s before the nature of heat was understood. His ideas were taken up decades later by Kelvin and Clausius and thus contributed to the modern formulation of the second law of thermodynamics.

The foregoing analysis provides guidelines in the design of practical heat engines. At some portion in a cycle, an engine must exhaust some heat to a cool body, usually to the atmosphere or to a nearby river. With the "cold reservoir" temperature determined by nature, the only way an engine designer can increase the Carnot efficiency is to have the heat intake occur at as high a temperature as possible. For this reason steam engines utilize high-pressure boilers to obtain steam superheated to temperatures well above 100°C.

Example 5 What is the Carnot efficiency of a steam engine with a boiler temperature of 177°C which exhausts heat to a condenser at a temperature of 27°C?

Solution
$$\text{Carnot efficiency} = 1 - (300°\text{K}/450°\text{K})$$
$$= 1 - 0.666$$
$$= 0.333 \text{ or } 33.3\%.$$

The overall efficiency achieved in practice will be considerably less than this, being typically about 20%.

Only when there is no increase in the total entropy does the engine's thermal efficiency attain its maximum possible value—the Carnot efficiency. [The equality signs then hold in Eqs. (9.6) and (9.7).] When $\Delta S_{\text{total}} = 0$, the process is *reversible*. For this to be true, each process in the engine's operation must be reversible and therefore subject to all the limitations mentioned in Section 9.2. An engine capable of operating in this idealized way is called a *Carnot engine*. The efficiency of any real engine is reduced below the Carnot efficiency because of work done by frictional forces, heat transfer through rather large temperature differences, turbulence and acceleration of gases in the cylinders, and failure to maintain phase equilibrium or chemical equilibrium.

Thermal Pollution

When a river, lake, or ocean is used as a cold reservoir, the heat (Q_{cold}) that it absorbs causes what has been named "*thermal pollution*." A typical case has been investigated in Florida where two oil-fueled installations for generating electric power discharge about 10,000 gallons of cooling water a second at about 100°F. This causes the ocean temperature in the vicinity to rise from about 85°F to nearly 100°F. Elsewhere, recent studies of the Great Lakes suggest that a continuation of the present trend will increase the heat discharged into the lakes elevenfold by the year 2000.

The increased temperature reduces the amount of oxygen in the water. Fish are adversely affected in many ways: metabolism is speeded up so they must breathe more rapidly, motion in pursuit of food is slowed, reproduction is inhibited.

At elevated temperatures normal vegetation may be killed while the growth of algae is stimulated. There is danger of a general disruption of the plant-fish-insect-bird "food chain" underlying the ecology of a region.

Our study of heat engines has shown that although the rejection of the polluting heat (Q_{cold}) seems a senseless waste of energy as far as the first law of thermodynamics is concerned, this rejection is absolutely necessary according to the second law. The compensating entropy increase of the cold reservoir (Q_{cold}/T_{cold}) must be provided for the engine to operate. Thermal pollution is an unpleasant consequence of the second law of thermodynamics.

Refrigerators

Significant general features of the operation of a refrigerator can be obtained from an analysis similar to that given for heat engines. In fact a refrigerator can be regarded as a heat engine run in reverse (Fig. 9-8).

Consider a refrigerator which has a *work input W* in one cycle during which it extracts a heat Q_{cold} from a cold reservoir at Kelvin temperature T_{cold} and gives off a heat Q_{hot} to a hot reservoir at Kelvin temperature T_{hot}. The purpose of the refrigerator is to extract as much heat Q_{cold} as is possible for a given work input W. In a home refrigerator the work is done by an electrically driven compressor. The ice and food inside the refrigerator constitute the cold reservoir; the kitchen air is the hot reservoir.

The lower limit to the work required to operate the refrigerator is easy to determine. The first law of thermodynamics is applied to one cycle of operation of the device represented by the circle of Fig. 9-8. For one cycle

$$\Delta \bar{E} = 0. \qquad (9.5)$$

So the first law,

$$\Delta \bar{E} = \text{(work for process)} + \text{(heat for process)}, \qquad (7.11)$$

gives

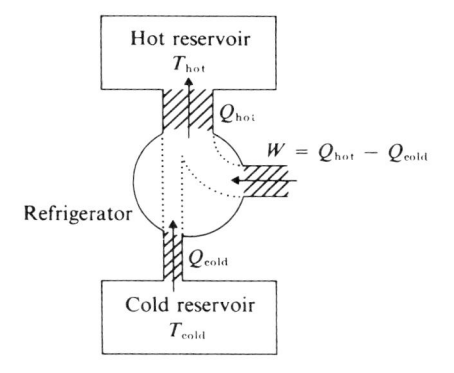

$$0 = W + (Q_{cold} - Q_{hot}).$$

Fig. 9-8 Schematic diagram of a refrigerator (a heat engine run in reverse). In one cycle there is a work input W equal to the difference between the heat rejected to the hot reservoir and the heat extracted from the cold reservoir.

Therefore
$$W = Q_{hot} - Q_{cold}$$
$$= Q_{cold}\,[(Q_{hot}/Q_{cold}) - 1]. \qquad (9.9)$$

The entropy of the hot reservoir increases and that of the cold reservoir decreases. For the compressor and contents after one cycle,
$$\Delta S_{engine} = 0. \qquad (9.4)$$

The entropy changes of the reservoirs are calculated from Eq. (9.2). From the second law of thermodynamics we conclude that the entropy increase Q_{hot}/T_{hot} of the hot reservoir must be greater than the entropy decrease Q_{cold}/T_{cold} of the cold reservoir. That is
$$Q_{hot}/T_{hot} \geq Q_{cold}/T_{cold},$$
which implies
$$Q_{hot}/Q_{cold} \geq T_{hot}/T_{cold}.$$

This result, together with Eq. (9.9), gives
$$W \geq Q_{cold}\,[(T_{hot}/T_{cold}) - 1]. \qquad (9.10)$$

Example 6 What is the minimum amount of power that must be supplied to a refrigerator that freezes 2.00 kg of water at 0°C into ice at 0°C in a time interval of 5.00 minutes? The room temperature is 20°C.

Solution The heat Q_{cold} which is extracted from the water is given by
$$Q_{cold} = mL_f. \qquad (8.5)$$
$$= (2.00 \text{ kg})(80 \text{ kilocalories/kg})$$
$$= 160 \text{ kilocalories}.$$

From Eq. (9.10), the minimum work which must be supplied is given by
$$W = Q_{cold}\,[(T_{hot}/T_{cold}) - 1]$$
$$= (160 \text{ kilocalories})[(293°K/273°K) - 1]$$
$$= (160 \text{ kilocalories})(20/273)$$
$$= 11.7 \text{ kilocalories}$$
$$= (11.7 \text{ kilocalories})(4.18 \times 10^3 \text{ joules/kilocalorie})$$
$$= 48.9 \times 10^3 \text{ joules}.$$

$$\text{Power} = 48.9 \times 10^3 \text{ joules}/5.00 \times 60.0 \text{ sec}$$
$$= 163 \text{ watts}.$$

Those who are interested in the ingenious methods used to make practical engines and refrigerators may be disappointed to find that we have not examined how particular devices work. But the beauty of thermodynamics is that one can make useful statements about *all* such devices no matter what the details.

9.4 DEGRADATION OF ENERGY

The entropy increases that occur in natural processes measure the increased disorder. For engineers and philosophers, another interpretation of the significance of entropy increases is perhaps more useful. A total entropy increase is associated with a lost opportunity to obtain useful mechanical work. When a process leads to an increase in the total entropy, some energy which could have been used to perform macroscopic mechanical work has been *degraded to a form in which it is unavailable for work*. Natural processes occur in a direction such that on a scale of availability for mankind's use, energy runs downhill.

Heat conduction (Fig. 9-4) provides a clear example. We found in Section 9.2 that when heat Q passes from a hot reservoir at temperature T_1 to a cold reservoir at temperature T_2 there is an increase of total entropy given by

$$\Delta S_{\text{total}} = Q/T_2 - Q/T_1. \tag{9.3}$$

If, instead, this heat Q had been used as the input to a heat engine operating with the Carnot efficiency $(1 - T_2/T_1)$, the work output that would have been obtained is

$$\begin{aligned} W &= Q[1 - (T_2/T_1)] \\ &= Q[(T_2/T_2) - (T_2/T_1)] \\ &= T_2[(Q/T_2) - (Q/T_1)]. \end{aligned}$$

Comparison with Eq. (9.3) shows that

$$W = T_2 \Delta S_{\text{total}}. \tag{9.11}$$

We see that when the heat conduction process transferred all the heat Q into the cold reservoir, the opportunity to achieve a work $T_2 \Delta S_{\text{total}}$ was lost. An amount of energy $(T_2 \Delta S_{\text{total}})$ proportional to the entropy increase, ΔS_{total}, was made unavailable for work by the entropy-increasing process.

The mixing of hot and cold water which results in lukewarm water is an everyday example of the lost opportunity associated with a total entropy increase (see Example 4). Before mixing, a heat engine could have performed useful work by taking in heat from the hot water and rejecting heat to the cold water. This cannot be done with the lukewarm water at a uniform temperature.

The entropy-increasing direction of all natural processes tends to bring about uniformity of temperature, pressure, and composition. A gloomy topic for natural philosophers is the possibility of an ultimate "heat death" of the universe when a situation of complete uniformity is reached. Since no process could then increase the total entropy, physical, chemical, and biological change would cease.

Instead of pursuing this perhaps unjustified extrapolation of our knowledge, we shall survey our present situation. Our modern society has an ever-increasing requirement for macroscopic mechanical energy. To fulfill this requirement, we take energy that has been stored in an available form and ultimately degrade it into thermal energy of our environment. Apart from nuclear fuel, our planet's main source of available energy is solar radiation. Most of the solar energy arriving at our planet is immediately degraded into thermal energy. Fortunately, there are natural processes which store a significant amount of the incident solar energy in very available forms. Short-term storage as potential energy of water is found in elevated lakes and rivers. Long-term storage of available energy occurs as complex molecules are formed and become constituents of trees and eventually of fossil fuels.

9.5 ENTROPY IN STATISTICAL MECHANICS

Microscopic Definition of Entropy

Preceding sections made evident the utility of entropy calculations in situations of practical interest. We now consider the significance of entropy from a molecular point of view and show how the entropy of a collection of molecules is defined.

Consider a typical macroscopic system such as 0.0180-kg of water in a test tube. A specification of the temperature and volume determines the macroscopic equilibrium state of this water. When this same system, a collection of 6.02×10^{23} H_2O molecules, is described using Newtonian mechanics or quantum mechanics, we find that there are many different microscopic states* which correspond to essentially the same macroscopic equilibrium state. Roughly speaking, there are many different molecular motions that produce the same behavior on the macroscopic scale. When we know only the macroscopic equilibrium state, the system can be in any one of many microscopic states. These are called the *accessible* states for this particular macroscopic state.

In Newtonian mechanics, a microscopic state of a system of N particles is specified by giving the position and velocity of each particle. In quantum mechanics a microscopic state is specified by giving the wave function (Chapter 16) of the system.

The disorder of a macroscopic system is measured by the number n of accessible states for the macroscopic state in question. This number n is just the number of *microscopically different situations* that correspond to the given macroscopic equilibrium state. As Feynman puts it, "we measure 'disorder' by the number of ways that the inside can be arranged, so that from the outside it looks the same."*

The Feynman Lectures on Physics, Vol. 1, p. 46-7, Addison-Wesley, 1963.

The entropy S of a macroscopic system in a macroscopic equilibrium state is defined in terms of the number n of its accessible states by the following equation:

$$n = 10^{S/ka} \quad (9.12)$$

where k is Boltzmann's constant and $a = 2.30$.* According to this definition, the system's entropy S in a given macroscopic state is simply an exponent which gives the corresponding number of accessible

The number a in this expression has been introduced merely to make Eq. (9.12) agree with the conventional definition of entropy, in which S is an exponent of "the base of the natural logarithms" rather than of the familiar number 10.

a = natural logarithm of 10 = 2.30

states. Evidently the entropy of a system can be calculated from knowledge of the microscopic states of the system.

The definition of entropy implies that a system composed of parts has an entropy which is the sum of the entropies of each part. This can be demonstrated using two mathematical rules which we illustrate by numerical examples:

1. The law of exponents gives
$$10^3 \, 10^5 = 10^{3+5}.$$

2. The number n of different ways one couple (a boy and a girl) can be selected from a group of 2 boys and 3 girls is given by
$$n = 2 \times 3.$$

Now consider a macroscopic state of a composite system with two parts. One part of the system can be found in n_1 microscopically different situations corresponding to an entropy S_1. The other part can be found in n_2 microscopically different situations corresponding to an entropy S_2. The number of microscopically different situations for the composite system is given by
$$n = n_1 n_2,$$
since each of the n_1 situations of one part can occur in conjunction with the n_2 different situations of the other part. Therefore the entropy S of the composite system must satisfy
$$10^{S/ka} = n = n_1 n_2 = 10^{S_1/ka} \, 10^{S_2/ka} = 10^{(S_1+S_2)/ka},$$
where the last equation follows from the law of exponents. We have shown
$$10^{S/ka} = 10^{(S_1+S_2)/ka},$$
so the exponents on each side of this equation must be equal. Consequently
$$S = S_1 + S_2.$$

This equation states that the entropy of the composite system is the sum of the entropies of its parts. This is the advantage of working with the entropy S instead of with the related number n of microscopically different states.

Entropy and Probability

The development of the microscopic theory of the behavior of large collections of molecules constitutes the subject of *statistical mechanics*. In this subject, the second law of thermodynamics acquires a new and interesting interpretation.

Statistical mechanics focuses attention on the number n of accessible states of a given system. As time goes on, a macroscopic system makes transitions from one of its microscopic states to another. Ultimately the system will have spent some time in each of its accessible states. The situation achieved after many transitions is analogous to that produced by repeated shuffling of a pack of cards. No matter how the deck is stacked initially, after sufficient shuffling no one arrangement of cards is particularly favored. All possible arrangements become equally likely. We assume that, similarly, after a sufficient time has elapsed, a collection of many molecules reaches a situation where the system is found with *equal probability* in each one of its n accessible states. This situation characterizes an *equilibrium* state of the macroscopic system.

Suppose a system which had been confined to only a portion n_1 of n microscopic states has new possibilities opened up so that all of these n states become accessible. Transitions occur and after sufficient time the system is equally likely to be in each of the n accessible states. This is a spontaneous change from an entropy S_1 (determined by $n_1 = 10^{S_1/ka}$) to a larger entropy S (determined by $n = 10^{S/ka}$). The direction of spontaneous change in an isolated system is seen to be the direction which increases the entropy.

After an isolated system has reached equilibrium with n accessible states, the probability of its being found in some situation corresponding to a restriction to only n_1 of these n states, is the fraction n_1/n. From this fraction one can compute the probability of fluctuations of a given size in macroscopic quantities such as the pressure and density. Fluctuations occur continually, but if the system contains a large number of molecules, the fluctuations will be small.

In general, we find that in statistical mechanics the second law of thermodynamics and all its consequences are statements about probabilities, not certainties. Entropy-decreasing processes are not impossible, merely improbable. The principle of entropy increase is toned down to say "it is highly improbable that entropy-decreasing processes will be observed." But computations of the probabilities involved show that the measurable entropy decreases of isolated macroscopic systems considered in Section 9.2 are indeed so highly improbable that the second law can be regarded as stating engineering certainties. Eddington* knew how to give us a feeling for these small probabilities. Contemplating a group of wild monkeys punching randomly on a set of typewriters, he deduced that they could be expected to reel off without error Shakespeare's complete works many quadrillion times in succession with about the same frequency as an observer would see a stone leap into the air by converting one calorie into macroscopic mechanical energy.

*A. S. Eddington (1882–1944). English physicist. A pioneer in studies of stellar structure and the general theory of relativity. He belongs to that small group of geniuses with the ability and inclination to convey complex mathematical ideas to the layman.

Temperature Defined as $\Delta \bar{E}/\Delta S$

When describing thermal phenomena using statistical mechanics, there is no need to introduce temperature by means of a separate postulate.

Temperature can be *defined* in terms of the entropy change associated with a change of thermal energy.

The number n of microscopically different situations corresponding to a given macroscopic state of a system is related to the system's entropy S by the equation

$$n = 10^{S/ka}. \tag{9.12}$$

This definition determines the value of S/k. Notice that S can be calculated only when the value of Boltzmann's constant k has been fixed.

We now consider a process in which the thermal energy of the system is changed a small amount $\Delta \bar{E}$ by a small flow of heat while the volume of the system is maintained constant. This process changes the system's entropy by an amount ΔS. For this system the "absolute temperature," which we denote by the product, kT, is *defined* by

$$kT = \frac{\Delta \bar{E}}{\Delta S/k}. \tag{9.13}$$

This is equivalent to defining T by

$$T = \frac{\Delta \bar{E}}{\Delta S}. \tag{9.14}$$

This definition implies that the "hotness" of an object is determined not by its thermal energy nor by its entropy but rather by the amount that its entropy *changes* when its thermal energy is changed. For a given change $\Delta \bar{E}$ in thermal energy, the entropy change of a hot object is smaller than the entropy change of a cold object. In other words, the hot object has a larger value of $\Delta \bar{E}/\Delta S$ than the cold object. We burn our fingers if we touch something whose entropy changes too little when its thermal energy changes.

The familiar properties of temperature that were outlined in Section 8.3 can now be deduced from the definition $T = \Delta \bar{E}/\Delta S$. Consider the transfer of thermal energy $\Delta \bar{E}$ from an object at temperature $T_1 = \Delta \bar{E}/\Delta S_1$ to another object at temperature $T_2 = \Delta \bar{E}/\Delta S_2$. If the first object is warmer ($T_1 > T_2$), the entropy decrease of the first object has a smaller magnitude than the entropy increase of the second. The transfer of thermal energy then increases the total entropy. Thus our new definition of temperature, together with the rule that entropy-increasing processes will occur spontaneously, implies this: When two objects are placed in thermal contact, thermal energy will be transferred from the warmer object to the cooler until both reach equilibrium states at the same temperature, that is, the same value of $\Delta \bar{E}/\Delta S$.

In the processes that we have been considering, the changes in thermal energy were entirely due to a thermal energy transfer (heat).

That is,
$$\Delta \bar{E} = \text{(heat for process)}.$$

For such a process the definition, $T = \Delta \bar{E}/\Delta S$, implies that
$$\Delta S = \text{(heat for process)}/T.$$

This is the familiar Clausius equation. With some labor it can be deduced that the Clausius equation remains valid for a reversible process involving both heat and work so that the thermal energy change is
$$\Delta \bar{E} = \text{(work for process)} + \text{(heat for process)}.$$

It turns out that we still have
$$\Delta S = \text{(heat for process)}/T.$$

Both the work and the heat increase a system's thermal energy, but in a reversible process only the heat increases its entropy.

Kelvin Temperature Scale and Boltzmann's Constant

The *Kelvin* temperature T is defined by arbitrarily assigning the value 273.16 degrees Kelvin as the Kelvin temperature of water at its triple point.* This particular choice, 273.16°K, fixes the value of Boltzmann's constant k. We now take any substance whose thermal behavior is theoretically understood (so that $\Delta \bar{E}/(\Delta S/k)$ can be expressed in terms of macroscopically measurable properties) and make experimental measurements when the substance is in thermal equilibrium with water at its triple point. Then k can be calculated from

$$kT = \Delta \bar{E}/(\Delta S/k). \tag{9.13}$$

One simple substance is any collection of N weakly interacting particles. Theory shows that for this substance the right-hand side of Eq. (9.13) is PV/N. This equation then becomes the ideal gas law:

$$kT = PV/N.$$

Therefore Boltzmann's constant can be determined by measurement of N, the pressure P, and the volume V, for any sufficiently rarefied gas in equilibrium with water at its triple point. (The experimental data was given and calculation of k performed in Section 7.1.) Furthermore, the measurements of pressures, P_1 and P_2, of a sufficiently rarefied gas in a constant-volume gas thermometer at different Kelvin temperatures, T_1 and T_2, can be used to determine the ratio of these temperatures. This follows from the ideal gas law, which implies that

$$T_1/T_2 = P_1/P_2$$

when N and V are constant.

---

*By international agreement in 1967 the name of the unit of temperature was changed from degree Kelvin (symbol °K) to kelvin (symbol K). The older name and symbol are still used so frequently that they have been retained in this book.

9.6 LOW-TEMPERATURE PHENOMENA

At very low temperatures the disorder associated with thermal agitation is very much reduced. Some substances then display remarkable properties such as *superconductivity* or *superfluidity*.

Superconductivity was discovered by H. Kamerlingh Onnes in Leiden in 1911, three years after he had achieved the first liquefaction of helium. The electrical resistance of many metals was found to drop abruptly to an unmeasurably small value when the metal was cooled below a sharply defined temperature (7.2° K in the case of lead). An electric current induced in a superconducting lead ring persists (without any battery) for years—in fact the current continues as long as the lead is kept in liquid helium (Fig. 9-9). [This astonishing phenomenon now finds practical application in the huge electromagnets used in many branches of physics (Fig. 9-10). The magnetic field of the magnets is created by electric currents established in superconducting coils. The necessary refrigerating equipment increases the capital cost, but the operating cost arising from I^2R losses (see Joule heating in Section 11.5) is practically zero.]

Superfluidity is another amazing low-temperature phenomenon. Below 2.18° K, liquid helium exhibits a completely frictionless flow which permits it to pass easily through tiny holes (less than 10^{-8} m in diameter) and to show a most unusual behavior in many situations. For instance, this superfluid will not stay in an open container! It manages to climb up the inside walls, pass over the top edge, and travel down the outside.

These strange effects at low temperature have posed some most

Fig. 9-9 Superconductivity. Persistent currents induced in the superconducting lead dish create a magnetic field which supports the bar magnet. (Courtesy National Aeronautics and Space Administration.)

Fig 9-10 Superconducting solenoid which produces a magnetic field of 7.5 weber/m². The solenoid, about 40 cm in diameter and about 60 cm long, is located between the two protruding circular flanges with strengthening webs. The equipment to the left of the left support flange is a lathe for winding the superconducting filaments upon the solenoid. These superconducting filaments are composed of niobium-titanium kept at about 4°K by means of liquid helium circulating through channels in the solenoid structure. (Photo courtesy Ferranti Packard Ltd.)

instructive puzzles for physicists. It seems that with thermal disorder markedly reduced, we are observing on a *macroscopic* scale the peculiar effects associated with the quantum mechanical behavior of matter.

The laboratory achievement of low temperatures is accomplished in stages. A temperature of 1° K is attained in a bath of liquid helium, boiling under reduced pressure. By using only the rare isotope, helium-3, temperatures of 0.3° K can be reached. Removal of a strong magnetic field from a thermally insulated system is used to create still lower temperatures, all the way down to 10^{-6} °K.

QUESTIONS

1. (a) What does entropy measure?
   (b) Would you expect a kilogram of solid lead to have more or less entropy than a kilogram of molten lead? Explain on the basis of your answer to part (a).

2. Using the data of Table 9-1, calculate the change in entropy of 10.0 kg of H_2O molecules when they are transformed from water at 20.0°C to steam at 250°C. (Pressure = 1 atmosphere.)

3. Using the data of Table 9-1, calculate the change in entropy of 5.00 kg of H_2O molecules when they are transformed from water at 20.0°C to ice at -10.0°C. (Pressure = 1 atmosphere.)

4. Find the change in entropy of 2.00 kg of CO_2 during sublimation from "dry ice" to gas at a pressure of 1.00 atmosphere and a temperature of -78.5 °C. The latent heat of sublimation under these conditions is 138 kilocalories/kg.

5. Find the change in entropy of 3.00 kg of H_2O at a pressure of 1.00 atmosphere when its temperature is raised from 26.0°C to 28.0°C. Assume the specific heat of water in this temperature interval is 1.00 kilocalorie/kg C°.

6. What is the entropy change of 50.0 kg of copper sulphate when it is melted at 31°C? (Latent heat of fusion is 51.3 kilocalories/kg.)

7. A mass of 5.0 kg of water at 37°C is vaporized (latent heat of vaporization at this temperature is 577 kilocalories/kg). Find the entropy change of the water in this process.

8. What is a heat reservoir?

9. A heat reservoir at 100°C gives off 2.00 kilocalories of heat. What is the entropy change of this reservoir?

10. (a) What entropy change will be produced in a heat reservoir at 0°C if it absorbs the 2.00 kilocalories of heat given off by the reservoir of the preceding question?
    (b) What is the total of the entropy changes in the two reservoirs because of this process?

11. Consider the possibility of 3.00 kg of water in a pail suddenly spontaneously cooling itself from 28.0°C to 26.0°C and springing up into the air.
    (a) If in this way thermal energy could be converted into kinetic energy of ordered motion upward, how high would the water rise?
    (b) What is the entropy change of the water in this process? (See Question 5.)
    (c) Are there other entropy changes due to this process?
    (d) Is the process possible according to the first law of thermodynamics?
    (e) Is the process possible according to the principle of entropy increase?

12. Describe three different processes that you have observed which are irreversible, as far as your experience would indicate.

13. During a certain process a system absorbs 600 kilocalories of heat from a heat reservoir at 400°K. The consequent increase in the entropy of the system is 1.80 kilocalories/°K. Is this process reversible? Explain.

14. During a certain process the entropy of a system **decreases by 1.20 kilocalories/°K and the system** gives off 360 kilocalories of heat to a reservoir at a temperature of 300°K. Is this process reversible? Explain.

15. A mass of 2.00 kg of hot water at a temperature of 100.0°C is mixed with a mass of 3.00 kg of cold water at a temperature of 0.0°C. After stirring, the temperature of the mixture is 40.0°C.
    (a) Find the change in the total entropy for this process.
    (b) Is the process irreversible?

16. Specify a system which interacts with other systems in such a way that its entropy decreases. Then show that a more than compensating increase of entropy occurs in another system.

17. State the second law of thermodynamics.

18. Show that if a Maxwell demon could operate without increasing his own entropy, the demon would produce a decrease in the total entropy.

19. Calculate the changes in the total entropy that occur when 2.0 kilocalories of heat are conducted from a hot reservoir at 100.0°C to a cold reservoir at 0.0°C. The conduction takes place through a copper wire of negligible mass. Is the process reversible?

20. Prove that the second law of thermodynamics, as **stated in this text on page 337, implies the Kelvin** statement: "No process is possible whose sole effect is to extract heat from one reservoir and convert this heat entirely into work."

21. What is a perpetual motion engine of the second kind?

22. In one cycle, a heat engine extracts 1.20 kilocalories from a hot reservoir and rejects 0.90 kilocalories to a cold reservoir.
    (a) What is the engine's work output in one cycle?
    (b) What is the efficiency of this engine?

23. If a heat engine gives off heat to the atmosphere which is at 27°C, how high must be the temperature of the hot reservoir in order to achieve a Carnot efficiency of 50%?

24. Prove that any heat engine, operating between a hot reservoir at temperature T_{hot} and a cold reservoir at temperature T_{cold}, has a thermal efficiency less than or equal to the Carnot efficiency $[1 - (T_{cold}/T_{hot})]$.

25. A power plant uses heat engines which exhaust heat to the environment. If thermal efficiency were the only consideration, when should the heat engines exhaust to the air and when to the nearby ocean? Assume that average temperatures in the winter are 0°C for the air and 6°C for the ocean, while in the summer these are 20°C for the air and 15°C for the ocean.

26. Why are elevated boiler temperatures desirable in steam engines? Contrast the Carnot efficiency of a steam engine having a boiler temperature of 100°C with that of the engine in Example 5. Assume the condenser temperature is 27°C.

27. Compare the Carnot efficiency of a steam engine which has a boiler temperature of 170°C and a condenser temperature of 40°C to the Carnot efficiency of a gasoline engine which has a combustion temperature of 1500°C and an exhaust temperature of 400°C.

28. Identify several processes which occur in practical heat engines that result in the engine's efficiency being less than the Carnot efficiency.

29. Is it theoretically possible to devise a heat engine which will create no "thermal pollution"? Explain.

30. The refrigerator of Example 6 has a work input of 11.7 kilocalories. Calculate the heat given off by this refrigerator to the room which is at 20°C. Compare this heat to the work input.

31. On a hot day a housewife leaves the refrigerator door open in an effort to cool the kitchen. Will this work? Explain.

32. A refrigerator absorbs heat from water at 0°C and gives off heat to the room at a temperature of 27°C. When 20 kg of water at 0°C have been converted to ice at 0°C, what is the minimum amount of energy that has been supplied to operate the refrigerator? What is the minimum amount of heat that has been given off to the room?

33. Consider the entropy changes of the cold and the hot reservoirs and show that the principle of entropy increase requires that a refrigerator give off more heat to the hot reservoir than it takes in from the cold reservoir.

34. How much energy becomes unavailable for work because of the heat conduction process described in Question 19?

35. Energy which is carried to the earth by electromagnetic radiation from the sun is ultimately used to heat an element in a kitchen stove. Examine the energy transformations involved and point out the degradations of energy that occur.

36. The entropy of a system composed of two parts is the sum of the entropies of each part. Show that this statement follows from the definition
$$n = 10^{S/ka}$$

37. A system, initially in a macroscopic state which can occur in n_1 microscopically different ways, undergoes a process which brings it into a different macroscopic state which can occur in n_2 microscopically different ways. If the initial entropy is S_1 and the final entropy is S_2, show that
$$n_2/n_1 = 10^{(S_2 - S_1)/ka}$$

38. How can the temperature of a system be defined when the entropy changes corresponding to changes in the system's thermal energy are known?

39. Contrast the entropy changes that occur in hot objects with those that occur in cold objects when they experience a given small change of thermal energy.

40. If water at the triple point were assigned the value of exactly one degree "Jones," what would be the value of Boltzmann's constant in joules per kilogram Jones degree?

SUPPLEMENTARY QUESTIONS

S-1. A 4.0-kg block of ice at 0°C is placed in 8.0 kg of water at 20°C. This system is insulated and maintained at a pressure of 1.00 atmosphere. Find the final temperature and the change in entropy for this process.

S-2. If the heat engine of Fig. 9-6 could be built, in violation of the Kelvin statement of the second law of thermodynamics, then show how this could be used in conjunction with a refrigerator to produce a device which violates the Clausius statement of the second law of thermodynamics.

S-3. An engine runs at 600 revolutions per minute and in each revolution takes in 10.0 kilocalories of heat from a boiler which is at a temperature of 167°C. The exhaust temperature is 107°C. The efficiency of the engine is 50% of the Carnot efficiency. What is the power output of this engine?

S-4. The electric power input to a certain refrigerator is 1.5 kilowatt. Heat is taken from the cold reservoir at a rate of 0.60 kilowatt. At what rate is heat given to the hot reservoir?

S-5. The coefficient of performance K of a refrigerator (Fig. 9-8) is defined by
$$K = Q_{cold}/W.$$
Show that the maximum possible value of K is
$$K_{Carnot} = \frac{T_{cold}}{T_{hot} - T_{cold}}$$

S-6. What is the relationship between the coefficient of performance K_{Carnot} of an ideal refrigerator and the Carnot efficiency of the same device run backwards as a heat engine?

S-7. The heat leakage in a certain refrigerator is 150 kilocalories/hour. The cooling element operates at $-13°C$ and the condenser at 37°C. The coefficient of performance of the refrigerator is 40% of the Carnot value K_{Carnot} discussed in Question S-5. What is the average power consumption?

S-8. The best way of heating any building, from the point of view of operating cost, is provided by a "heat pump." A heat pump works in cycles and absorbs a heat Q_o from a cool outside environment at a temperature T_o and gives off a heat Q_r to a warm room at temperature T_r.
   (a) Show that the heat pump requires a work input
$$W = Q_r - Q_o.$$
   (b) Assume that all processes are reversible, that is $\Delta S_{total} = 0$. Then show that
$$Q_r = W \left(\frac{T_r}{T_r - T_o} \right)$$
   (c) Assume that ocean water is available at a temperature of 2°C in the winter and that the room temperature is 22°C. Calculate how much heat the reversible heat pump will furnish if $W = 2.0$ kilocalories? How much heat would an electric heater furnish when supplied with 2.0 kilocalories?

S-9. Explain how a refrigerator could be employed as a heat pump to heat a house in the winter.

S-10. In a house heated by hot water radiators, the radiator temperature is maintained at 70°C by supplying 5.0 kilowatts to electrical heaters immersed in the water within the radiators. If instead, an ideal (reversible) heat pump were used with the heat intake at 0°C, what power would be needed to operate the heat pump?

S-11. The energy of a single green photon (2.3 eV) is absorbed as heat by a system at room temperature such that $kT = 1/40$ eV. The entropy of the system thus increases by an amount $S_2 - S_1 = \Delta S = 2.3$ eV/T. Use the result of Question 37 to determine that the number of microscopically different states increases by a factor (n_2/n_1) of 10^{40}.

S-12. (a) Show that when N atoms of a monatomic ideal gas are confined to a constant volume and heated ($C_V = \frac{3}{2}Nk$) so that the temperature rises a small amount ΔT, the factor n_2/n_1 defined in Question 37 is given by

$$n_2/n_1 = 10^{0.65 N \Delta T/T}.$$

(b) Evaluate this ratio when $T = 325°$K, $\Delta T = 1$ K°, and the gas contains a mere 500 atoms.

ADDITIONAL APPLICATIONS TO MEDICINE AND THE LIFE SCIENCES

Thermodynamics in Biology: Disordered Energy

Thermodynamics is widely applied to the study of living biological systems. It is probably safe to say there is virtually no field in the realm of the life sciences which does not lend itself in some manner to analysis depending on the basic physics known as thermodynamics. Some examples of basic studies in biology involving thermodynamics are (1) behavior of enzyme systems, (2) transport of molecules against chemical gradients, (3) molecular transport against electric fields, (4) molecular basis of nerve conduction, and (5) information theory in such seemingly widely disparate areas as sensory perception and protein structure. [Information theory is closely related to the ideas of thermodynamics, particularly to the concept of entropy.]

Thermal Energy and the Ecosystem of the Earth*

Matter goes around in circles through the ecosystem. Some of the air molecules exhaled by a football player in Green Bay, Wisconsin, may well be inhaled by a football fan in Ottawa, Ontario. There are no international boundaries in an ecosystem. The calcium of the soil becomes the calcium of milk. The water in one person's urine may well appear, downstream, in someone else's coffee. It is possible to describe a *cycle* for every chemical in the ecosystem.

---

* Adapted from R.H. Horwood, *Inquiry into Environmental Pollution*, Macmillan of Canada, Toronto, 1973, with permission of the author and the publishers.

But one very important part of the ecosystem does not cycle: *Energy*. Energy enters the earth's ecosystem primarily as radiation energy, light. Much of this radiation energy is ultimately transformed into complex chemical bond energy by organisms the ecologist calls "primary producers" (e.g., trees in the process of photosynthesis). This energy is used later by "consumers," "decomposers," or even by the "primary producers" themselves. Such use may result in the conversion of the chemical bond energy into other forms such as kinetic energy, or electrical energy, or it may even be reradiated as radiation energy. Ultimately, however, most of the conversion and use of chemical bond energy results mainly in the production of thermal energy. Once energy has been degraded to thermal energy, it has spelled out its eventual loss to the ecosystem. Thermal energy cannot recycle because the "primary producers" have no way to convert thermal energy into energy useful to the ecosystem. So the ecosystem submits to the second law of thermodynamics. And, indeed, in looking at thermodynamics and the ecosystem, one can see the sense in the frequently found reference to the second law of thermodynamics in connection with *The Arrow of Time*.

Cryosurgery: Low-Temperature Phenomena

Strictly speaking, *cryosurgery* would not be classed in the realm of "low-temperature phenomena." However, *cryogenics* is the study of the various aspects of low-temperature production and so it seems suitable to mention a few words about cryosurgery here.

It has been found that localized extremely low temperatures ($-20°C$ or lower) can be a valuable physical agent in surgery. In order to cool a given body tissue volume sufficiently, it is necessary to create a negative thermal gradient from the preselected volume to the instrument employed. The instrument itself has at its "working end" a 2-3 mm diameter cannula (a hollow probe—see Fig. 9-11) through which flows the cooling agent, say liquid nitrogen.* (It boils at $-196°C$ at atmospheric pressure.) The liquid is allowed to evaporate through a tiny hole at the tip of the cannula, thereby lowering the temperature of the cannula tip to a temperature which can be regulated by controlling the flow rate of liquid nitrogen. The vaporized nitrogen is generally conducted back up the cannula in a space between the thin capillary which permits the flow of liquid nitrogen and the insulation of the cannula (insulation protects those tissues not being surgically attacked). The use of a compressed liquid gas cooling cannula to cause freezing and tissue destruction is called *cryosurgery*.

The cryosurgical cannula described above may also have a thermocouple inside the insulation layer and thus the actual temperature employed may be monitored.

According to some medical authors, cryosurgery fulfills the following criteria for an ideal method of destroying biologic tissue: reversibility (it is possible to cool tissue sufficiently to precipitate physiologic inhibition without cooling sufficiently to produce pathologic changes), consistent reproducibility (the creation of either physiologic inhibition or a therapeutic lesion is dependent upon known physical laws), sharp definition of the treatment volume, flexibility, safety, and rapidity of application.

The actual pathologic effect of cryosurgery presumes cell destruction by the following mechanisms: dehydration and the resulting concentration of electrolytes, crystallization and the consequent rupture of cellular membranes, denaturation of the lipidprotein molecules within the cell membrane, thermal shock, and vascular stasis (stoppage of fluid flow in the various vessels). The latter mechanism is possibly the most significant of those listed.

Examples of very successful applications of cryosurgery are found in both neurosurgery and ophthalmology. In the former there has been much success with Parkinson's disease (a paralytic condition involving tremors, known commonly as "shaking palsy").

In the treatment of this ailment, the object is to destroy a small volume of brain cells in the thalamus, at the base of the brain. The "reversibility" aspect of cryosurgery comes very much to the fore here as the patient can be kept awake (after the opening in the skull has been made). At first only a small liquid nitrogen flow is permitted into the cannula—enough to cause physiological inhibition but not enough to kill the cells. With the patient's assistance (he can be asked to hold up his hand and perform certain actions), the surgeon can locate the particular volume which needs to be destroyed. Then an increased nitrogen flow can lower the temperature sufficiently to produce the required pathological effect. Temperatures in the order of $-70°C$ may be used for this purpose.

In ophthalmology, cryosurgery can be employed to treat detached retinas by producing controlled scarring to seal the tear. The cryosurgical system can also be applied to the problem of lens extraction in the case of cataract surgery (Fig. 9-12 shows a simple device for this purpose). The effect is essentially that experienced by anyone who

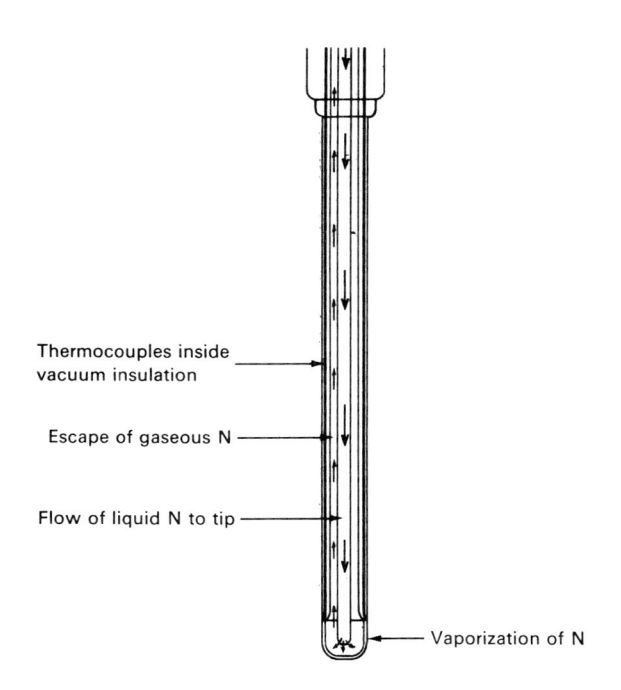

Fig. 9-11 Cross-section of vacuum-insulated cryosurgical cannula.

(Labels: Thermocouples inside vacuum insulation; Escape of gaseous N; Flow of liquid N to tip; Vaporization of N)

* Present trends are to replace liquid nitrogen with carbon dioxide and with freon (largely due to cost factors). However, the principle with these liquefied gases is the same as that with nitrogen which was used in much of the pioneering in this field.

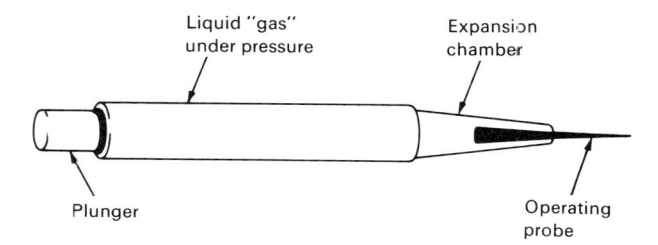

has touched a very cold metal object with damp fingers. The tissue freezes to the cannula tip. Cryosurgery seems to be effective also in the treatment of certain types of glaucoma. Temperatures used in ophthalmological work are of the order of $-20°$ to $-70°C$, depending upon the particular treatment involved.

Fig. 9-12   Cataract cyroextractor. Pressure on the plunger permits the flow of the liquefied gas into the expansion chamber. Cooling takes place here. The tip of the metallic operating probe is cooled by conduction. Temperatures of the order of $-20°C$ can be achieved with this relatively simple device.

SUGGESTED READING

BENT, H. A., *The Second Law, An Introduction to Classical and Statistical Thermodynamics.* New York: Oxford University Press, 1965. A novel and illuminating introduction to thermodynamics. Mathematical difficulties are avoided, at least for the first ten chapters.

CARNOT, S., *Reflections on the Motive Power of Fire.* New York: Dover, 1960. This paperback provides the reader with an English translation of Carnot's 1824 paper, "Reflexions sur la Puissance Motrice du Feu . . . ." Most of the paper is in a descriptive vein, and only toward the end is mathematics involving logarithms and some calculus employed. The frontispiece is a reprint of the actual title page of this famous 1824 paper. Also included in the book are two shorter papers on thermodynamics, one by E. Clapeyron and one by R. Clausius.

KAEMPFFER, F. A., *The Elements of Physics, A New Approach.* Waltham, Mass.: Blaisdell, 1967. Chapters 12, 13, and 22 discuss and extend topics we have introduced.

KAVALER, L., *Freezing Point.* London: David and Charles, Limited, 1973. The science of cryogenics has within a generation brought about some revolutionary changes in technology. Though low temperatures can cause death, cryogenics now holds an honorable place in the battle against disease. In this book the main emphasis is on how this physical phenomenon (called "cold" in the popular sense) affects life and particularly human life.

MENDELSSOHN, K., *The Quest for Absolute Zero.* New York: McGraw-Hill, 1966. The meaning and developments of low-temperature physics explained by a foremost authority in the field. This book is part of the World University Library, an international series of books, each of which has been commissioned with a view to providing an authoritative introductory treatment for university students. The historical approach employed makes much of the physics spring to life—photographs of the leading physicists involved in the relevant developments are included.

REIF, F., *Statistical Physics, Berkeley Physics Course,* Vol. 5. New York: McGraw-Hill, 1967. The entirely descriptive first chapter is excellent collateral reading. Section 7.4 gives a valuable summary of the laws of thermodynamics.

WILSON, M., and the editors of *Life, Energy.* New York: Life Science Library. Pages 57 to 69 describe and illustrate degradation of energy, "heat death" of the universe, transitions to greater disorder, and manifestly irreversible processes.

John William Strutt, Lord Rayleigh

10 Mechanical Waves: Sound

The mechanics of Newton provides the underlying theory of *sound*. This is not to say that nothing was understood of the nature of sound prior to Newton. Far from it. In fact, essentially correct ideas on this subject were held by the ancient Greeks.

Natural philosophers in those ancient times were quite aware of the fact that sound arises from the motion of parts of bodies, that it is transmitted through air by some sort of air motion and ultimately impinges on the ear, giving rise to the sensation of hearing.

In many ways the development of *sound*, in a sense a practical aspect of the study of mechanics, resembles the development of that practical aspect of light, *optics* (which the reader will encounter in a later chapter). Both of these fields, sound and optics, rather than being readily highlighted by a few or even a single major breakthrough, resemble a historical quilt, a patchwork of a multiplicity of contributions, assembled and stitched together by some great scientist: Kepler in the case of optics and Lord Rayleigh in the case of sound.

And sound, as we indicated earlier, quite like optics, has its roots in antiquity. It is speculated that Pythagoras discovered that two similar strings under the same tension when plucked simultaneously provide the ear with a pleasant harmonious sound if the lengths of the strings are in the ratios of small integers. But perhaps he knew this from the Egyptians. In any event, Pythagoras made a deep impression by citing this relationship in support of a mystic belief in the great significance of numbers. Here was perhaps mankind's first example (outside the field of plane geometry) of nature submitting to a numerical relationship.

Galileo also provided an impetus to the study of sound when he declared that the frequency of vibration is the physical aspect which determines the physiological sensation of pitch. Then in 1636 Galileo's former pupil, Marin Mersenne (1588–1648), made the first measurement of the speed of sound in air. And Otto von Guericke demonstrated that, in contrast to light, sound is not transmitted through a vacuum.

In the experimental field an eminent position must be accorded Ernst Friedrich Chladni (1756–1824), who is often referred to as "the

"When once we have discovered the physical phenomena which constitute the foundation of sound, our explorations are in a great measure transferred to another field lying within the dominion of the principles of Mechanics."

Lord Rayleigh
(1842–1919)

father of modern acoustics." Among his many investigations were such matters as the comparison of the transverse vibration of strings with longitudinal and torsional vibrations. He also measured the speed of sound in gases other than air.

Incontrovertible evidence that, counter to popular belief, sound was transmitted through liquids was provided by Jean Daniel Colladon (1802–1893) who, in collaboration with J. F. Sturm (1803–1855), determined the speed of sound in Lake Geneva to be 1.435×10^3 meters per second. (The belief that sound was not transmitted through liquids was held for a considerable time because it was assumed that liquids were incompressible.)

The physiological aspects of sound (and musical tones) were brought to light by one of the most versatile of the nineteenth-century scientists, H. L. F. von Helmholtz (1821–1894), a German medical doctor and physiologist who in his later years became a renowned theoretical physicist.

Most of these various contributions to the study of sound, at the time only to be found in scattered periodicals and transactions of scientific societies, all in various languages all over the world, were gathered and presented for the first time in a comprehensive, connected and mathematically rigorous exposition in 1877. The physicist who accomplished this impressive feat of intellectual organization, and in the process developed a solid theoretical basis for the subject, was John William Strutt, Baron Rayleigh. Lord Rayleigh was born in England in November 1842 into well-to-do circumstances and was offered excellent schooling of which he made the most. He succeeded James Clerk Maxwell as Cavendish Professor at Cambridge and in due course displayed conspicuous success in the establishing of courses in experimental physics at Cambridge University. In 1904 he was awarded the Nobel prize for physics, principally for his work leading to the discovery of argon and other inert gases.

Of course the story of sound moved on past Rayleigh's *Theory of Sound*, published in its second edition in 1894. This was particularly so in the realm of technology, with the development of the microphone, the telephone, electro-acoustics (i.e., the transmission of sound by means of electrical waves), the phonograph, underwater depth-sounding, and ultrasound. But most of the basic physics on the subject of sound as published by Rayleigh is as relevant today as it was at the time of publication.

Since sound is a form of wave motion, we consider first the familiar motion of waves on the surface of water to develop some useful concepts applicable to the study of wave motion in general.

10.1 WAVE PHENOMENA

Everyone is acquainted with the kind of disturbance that is created when a stone is thrown into a pond. A wave is propagated outward from the point at which the stone strikes the water, and when this disturbance reaches some object floating on the water, say a duck, the duck bobs up and down (Fig. 10-1). The duck receives energy that has been propagated by the wave. A mechanical wave motion can *transfer energy* over great distances *without the transport of matter*. The water near the initial disturbance caused by the stone's impact remains in that vicinity rather than being transported to the location of the duck.

Figure 10-2 illustrates another familiar type of mechanical wave—a transverse wave on a stretched rope. If one end of the rope is disturbed briefly, then a wave form of limited extent called a *pulse* is propagated along the rope. The *wave speed* is the speed at which this wave form travels. The motion of a particle of the rope is quite different from the motion of the wave form. The wave travels along the length of the rope at a constant speed, if the rope is homogeneous and has a constant tension. The particle P of the rope in Fig. 10-2 is motionless until the pulse reaches it, whereupon the particle accelerates in a direction perpendicular to the rope, the y-direction of Fig. 10-2. The particle thus acquires a velocity v_y and a displacement y from its equilibrium position. As the pulse moves on, this particle reverses the direction of its motion. After the pulse has passed by, the particle has returned to its equilibrium position and is at rest. Such waves, which involve a disturbance at right angles to the direction of propagation, are called *transverse* waves.

When the free end of the rope is moved up and down with simple harmonic motion of amplitude A and period T, we obtain the *train* of waves illustrated in Fig. 10-3(a). (To avoid complications due to a wave train reflected from the other end, we will assume the rope to be infinitely long.) The graph of Fig. 10-3(a) is a snapshot of the rope showing the displacements of different particles of the rope at one instant of time. Such a wave is called a sinusoidal wave because y is a sine or a cosine function of x and t. The highest points are called *crests* and the lowest points *troughs*. The wave amplitude A is the height of a crest or the depth of a trough, as measured from the equilibrium position of the rope.

In any periodic wave, the disturbance is repeated in equal spatial intervals of length λ, which is called the *wavelength*. The wavelength of a sinusoidal wave is the distance between adjacent crests. More generally, the wavelength is the distance between any two successive points that are in precisely the same state of disturbance at the same instant.

Each particle of the rope mimics the motion that is imposed on the free end, but only after a time delay which is just long enough for the wave to travel from the free end to the particle in question. Fig. 10-3(b) shows the graph of the displacement y as a function of t for one particle

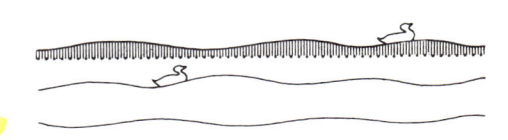

Fig. 10-1 Waves in the duck pond.

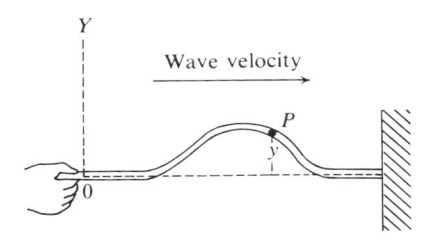

Fig. 10-2 A transverse wave of limited extent (a pulse) propagating along a stretched rope.

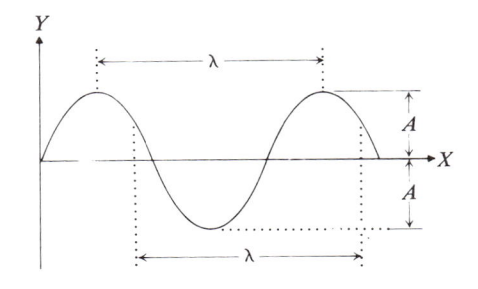

Fig. 10-3(a) A sinusoidal wave.

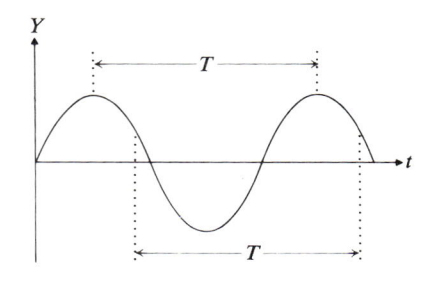

Fig. 10-3(b) The graph showing the displacement of one particle of the rope at different times.

of a rope that is transmitting a sinusoidal wave. Each particle of the rope executes simple harmonic motion with the same amplitude A and the same period T. But because a disturbance propagating along the rope reaches different particles at different times, we find that the simple harmonic motions of particles at different locations on the rope have different phase angles. Two simple harmonic motions are said to be *in phase* if the difference of their phase angles is zero or an integral multiple of 360°; otherwise the motions are *out of phase*. In the traveling wave depicted in Fig. 10-3(a), particles one wavelength apart oscillate in phase while particles separated by one-half of a wavelength are 180° out of phase.

The wave speed v is related to the wavelength λ and the period T. To determine the wave speed we focus on a given point on the *wave form*, say the crest which is observed first at the point P in Fig. 10-4 and travels to the right occupying the successive positions P_1, P_2, P_3, where the distance PP_3 is one wavelength λ. The time required for this crest to move from P to P_3 is exactly one period T because during this time the particle of the rope at P_3 has completed exactly one cycle of its oscillation. Since the crest moves a distance λ in a time interval T, the speed of this crest, the *wave speed* v, is given by

$$v = \lambda/T. \tag{10.1}$$

Expressed in terms of the frequency, $f = 1/T$, this becomes

$$v = \lambda f. \tag{10.2}$$

This important result, which is simply a consequence of our definitions of v, λ and T, is valid for any type of sinusoidal wave.

The wave forms encountered in practice are usually much more complex than the simple sinusoidal wave in Fig. 10-3(a). Fortunately, by using a mathematical procedure known as *Fourier analysis*, any wave can be expressed as a sum of sinusoidal waves of appropriately selected amplitudes, frequencies, and phases. Since the study of sinusoidal waves thus provides the basis for an understanding of complex wave motion, we can restrict our attention to sinusoidal waves.

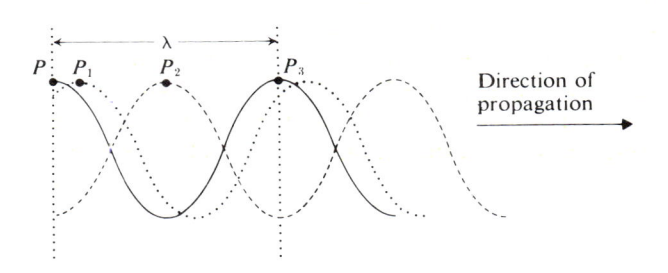

Fig. 10-4 A sinusoidal wave propagating to the right at a speed $v = \lambda/T$.

360 MECHANICAL WAVES: SOUND

10.2 SOUND

The Nature of Sound

It is a familiar fact that vibrating bodies are sources of what we call sound. This suggests that sound is a wave motion.

A simple experiment (Fig. 10-5) demonstrates that a material medium is required to transmit sound. An electric bell, suspended by fine wires inside a jar, is audible when there is air inside the jar. As the air is removed by a vacuum pump the sound fades. Nothing can be heard when a good vacuum is achieved. On the other hand, if the bell touches the glass walls, the ringing is again quite audible. Evidently sound is transmitted by gases and solids. One can easily devise a variation of this experiment to demonstrate that liquids also transmit sound.

Many experiments have made it clear that sound is a *mechanical wave motion in a medium*. Small displacements of the molecules of the medium occur as a sound wave travels. This displacement refers to an ordered collective motion in which all the molecules in a small volume move together. (This ordered motion is superimposed on the random thermal agitation of the molecules.) In thinking about a sound wave we therefore visualize molecules of the medium jiggling back and forth in a regular fashion.

A theoretical understanding of the vibrations of solids and the propagation of sound waves is obtained from Newtonian mechanics. We select a small chunk of the medium as the system to which we apply Newton's second law. The motion of this chunk is determined by the forces exerted by neighboring portions of the medium. It turns out that each chunk vibrates back and forth in such a way that a disturbance, a wave, propagates with a speed which is characteristic of the type of wave and of the medium under specified conditions. As is the case with any wave motion, mechanical waves are described using terms such as wavelength, period, frequency, amplitude and speed of propagation.

In a gas, the only possible type of wave is a longitudinal wave. That is, the molecules of the gas vibrate back and forth along the direction in which the waves are moving.

The description that was given of the *transverse* waves traveling along a stretched rope can be used to help us visualize sound waves which are *longitudinal*. If we denote by y the displacement of the matter in a small volume from its undisturbed position and draw a graph showing this displacement of this particular matter at different times, we get a graph just like that of Fig. 10-3(b). The vibrations back and forth of this one small chunk of matter correspond to the particle motion shown in Fig. 10-3(b).

In a sound wave, the amplitude of oscillation of the air molecules at any one point is very small—only $\frac{1}{100}$ of a millimeter for a sound wave that is so loud that it hurts the ear (when the frequency is 1000

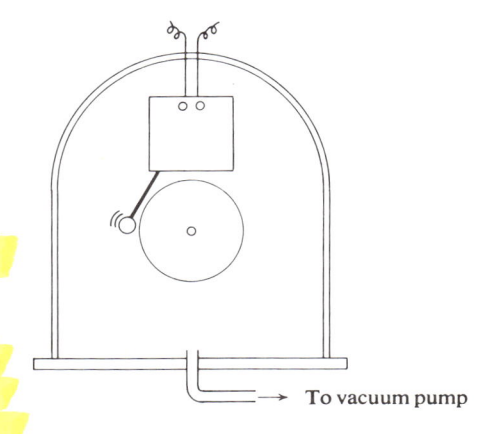

Fig. 10-5 Sound is not transmitted through a vacuum.

hertz). In such a wave, the maximum velocity attained by an oscillating chunk of matter is merely 0.06 m/sec.

If we now consider *one instant of time* and draw a graph showing the displacement *y* of *different small chunks of matter*, labeled by the different *x* values of their equilibrium positions, we get a graph like that of Fig. 10-3(a) for a wave propagating in the direction of the *X*-axis. The fact that matter at different locations has different displacements leads to regularly spaced regions where the molecules are crowded together and air is compressed, separated by regions where the molecules are abnormally far apart and the air pressure is below its normal value [Fig. 10-6(a)]. These pressure variations in space associated with a sound wave, called *compressions* and *rarefactions,* are shown in a graph in Fig. 10-6(b). At any one point in space, the air pressure varies periodically with time. At a frequency of 1000 hertz the maximum pressure variation is about 28 N/m² for an extremely loud sound with a displacement amplitude of $\frac{1}{100}$ of a millimeter. Although this pressure change is only 0.03% of atmospheric pressure, it is sufficient to cause pain to the ear.

Fig. 10-6 Pressure variations (compressions and rarefactions) associated with a sound wave in air.

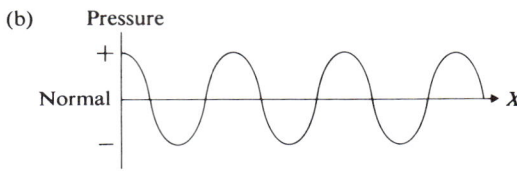

The wave form at one instant of time is shown in Fig. 10-3(a). For a traveling wave, at a later instant, one finds a similar wave form but in a different position in the medium, as shown in Fig. 10-4. The *speed of propagation of this wave form* is what is called the *speed of sound in the medium*.

In a solid, the vibrations can be transverse as well as longitudinal. It turns out that the speed of the longitudinal wave is always greater than that of the transverse wave. This fact is exploited in seismology. An estimate of the distance of an earthquake from an observation station can be obtained by noting the difference in arrival time of these two types of waves.

The Speed of Sound

In air at 0°C the speed of sound is 331 meters/sec. This is about 1 mile in 5 seconds (a handy fact to remember if you wish to compute your distance from a lightning stroke).

Table 10-1 shows the speed of longitudinal sound waves in some common substances. The speed is greater the greater the stiffness of the medium and the smaller its density. In general it is found that sound travels faster in solids and liquids than it does in gases.

Table 10-1

Medium	Speed of sound (kilometers/second)
Air (0°C)	0.331
Water (15°C)	1.45
Lead (20°C)	1.23
Iron (20°C)	5.13
Granite (20°C)	6.0

In the table above a temperature is specified in each case because the speed of sound in a medium depends upon the temperature of the medium. If we assume the stiffness of the material is not changed appreciably by a given temperature change, then this temperature dependence of sound can be implied from our earlier statement that the speed is greater as the density becomes smaller because, as a rule, a temperature increase results in an expansion of a material with the consequent decrease in the material's density. In air at normal temperatures for each degree Celsius increase in temperature, the speed of sound increases by about 0.6 m/sec.

The speed of sound in a given medium may be determined by direct measurement or by indirect experimental methods, or it may be calculated on purely theoretical grounds. As usual, the experiment is the court of last appeal with respect to a theoretical calculation. In the case of sound, theory is convincingly supported by experimental measurements.

Early experimenters resorted to direct measurement procedures which were, of course, attended by many errors that were reduced as experimental techniques became more sophisticated. One of the usual procedures in early measurements was simply to have one experimenter fire a gun while others acted as observers stationed at various distances. These observers would measure the time interval between the flash of the gun and the hearing of the sound. Needless to say, matters like subjective judgments as to *when* one *hears* and *sees* something led to inaccuracies, superimposed upon those introduced by such physical variables as wind velocity, air temperature, and humidity.

Another interesting direct measurement technique was that which Colladon and Sturm used to measure the speed of sound in Lake Geneva. A bell was struck under water by a hammer and at the same instant a flash of gunpowder was ignited. The time interval between gunpowder flash and sound arrival was noted by an observer nine miles away, using an underwater ear trumpet known as a *hydrophone*.

Direct measurements of the speed of sound are based on time-interval measurements. The development of modern electronics (both for the precise measurement of small intervals and also for sound

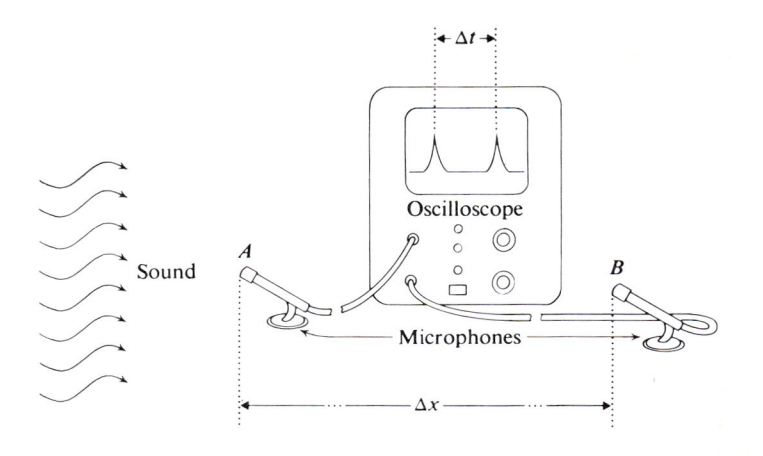

Fig. 10-7 An example of a technique for the direct measurement of the speed of sound. The time interval between the reception of the sound signals from microphones A and B is measured by means of an oscilloscope, and then the speed of sound v is determined by using Eq. (2.3): $\Delta x = v\Delta t$.

amplification) has led to precise measurements involving experimental setups along the lines suggested in Fig. 10-7.

In some indirect measurements, the speed of sound is calculated from the equation

$$v = \lambda f \qquad (10.2)$$

after the wavelength λ and the frequency f have been measured.

Frequency of Sound Waves

We encounter mechanical vibrations of a tremendously wide range of frequencies (see Fig. 10-8). However, the human ear generally detects only those frequencies lying between 20 hertz and 20×10^3 hertz. Frequencies of this order are called sound, but the term *sound* is actually used by physicists to mean also mechanical waves with frequencies well outside this audible spectrum. Above the audible range, we call sound waves *ultrasonic*. Frequencies up to 600×10^6 hertz are conveniently

Fig. 10-8 Spectrum of mechanical vibrations indicating the approximate frequency regions of various applications.

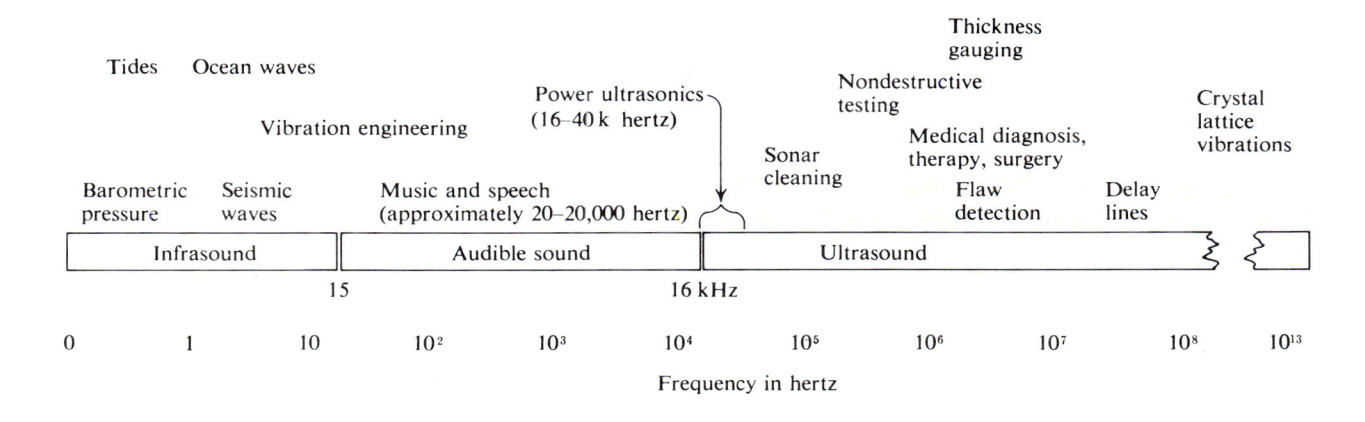

364 MECHANICAL WAVES: SOUND

generated in the laboratory. Waves with frequencies below the audible spectrum are called *infrasonic*. Such waves are generated by earthquakes.

The wavelength of a sound wave in air at 0°C can be calculated from Eq. (10.2): $v = \lambda f$. With $v = 331$ m/sec and $f = 20$ hertz, this gives a wavelength of 17 meters, about the length of a house. When the frequency is 20,000 hertz, the wavelength is merely 1.7 cm. In fact, ultrasonic wavelengths may be as short as light waves. (At $f = 5 \times 10^8$ hertz, $\lambda = 6 \times 10^{-7}$ m, which is the same wavelength as that of orange light.)

Superposition of Sound Waves

When different sound waves of small amplitude travel through a medium, a *superposition principle* holds. That is, the displacement, at any instant, of a small chunk of matter, when two waves are present, is the sum of the displacements that would occur if each wave alone were present. The wave obtained by this addition is called the *superposition* of the two original waves.

The superposition of different waves leads to what are called *interference* and *diffraction* phenomena. The detailed treatment of these subjects we postpone to Section 14.2. For the present we shall merely mention the interesting fact that a wave which passes through a hole of diameter D undergoes a spreading and fans out into a cone with a vertex angle having an order of magnitude given by

$$\theta = (10^2 \text{ degrees}) \lambda/D, \qquad (10.3)$$

as depicted in Fig. 10-9. This effect, termed "diffraction spreading," is illustrated by the water waves shown in Fig. 10-10. For an aperture of 1 meter and wavelength of 0.3 meter, this diffraction spreading has a cone vertex angle of about 30 degrees. Consequently, for sound waves with wavelengths longer than about 0.3 meter (frequencies lower than about 1000 hertz), there is a considerable diffraction spreading when a wave passes through a door or a window. This is why a speaker doesn't have to be in sight to be heard.

In certain applications, a narrow beam of sound waves is desired. The beam will have a small diffraction spreading only if the wavelength is much less than the diameter of the source. Therefore short wavelength ultrasonic waves must be used.

The superposition of two sound waves which have slightly *different frequencies* produces throbbing sounds, called *beats*. This effect is produced by two aircraft propellers which have not been precisely synchronized. At the ear of the listener, for a fraction of a second, crests from one source arrive at the same time as crests from another source. The superposition of the two waves then has an amplitude which is the sum of the amplitudes of the individual waves. The sound is loud. But since the two waves have different frequencies, this situation changes.

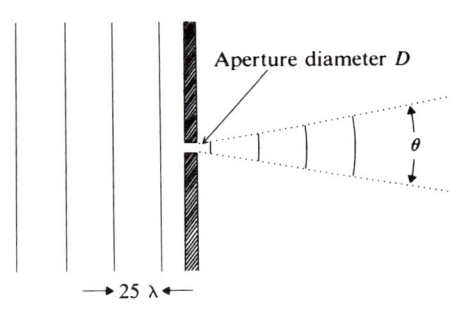

Fig. 10-9 Diffraction spreading behind an aperture when plane waves are incident. In order to have as little spreading as this, the wavelength would have to be about 1/25 of the distance between the indicated wavefronts.

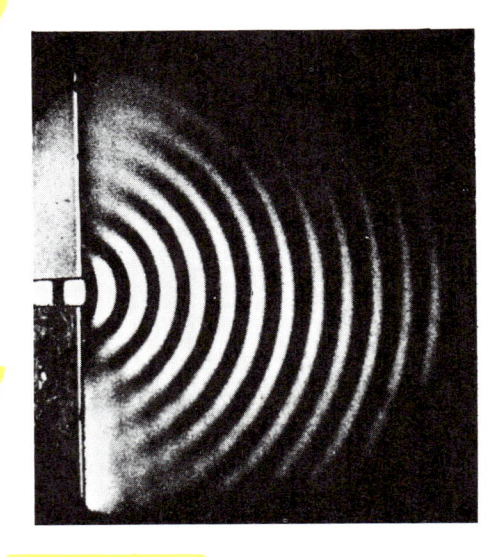

Fig. 10-10 Diffraction of a water wave which passes through a small aperture. (From Alonso and Finn, *Physics*, Addison-Wesley, 1970.)

After many vibrations there will come a time when crests from one source arrive together with troughs from the other, giving a superposition with an amplitude which is the *difference* of the amplitudes of the two waves. Little or no sound will be heard. The loudness of the sound therefore rises and falls periodically. Analysis shows that the number of *beats per second* is just the *difference of the frequencies* of the two waves.

Comparison of two almost equal frequencies is conveniently accomplished by detecting beats. A stringed instrument can be tuned in this way. The closer two frequencies become, the fewer the number of beats per second. The beats disappear when the frequencies are the same.

10.3 VIBRATING SYSTEMS AND RESONANCE

Natural Frequencies

When Newtonian mechanics is applied to the analysis of the motion of any system executing small vibrations, it is found that any possible motion can be regarded as a superposition of certain relatively simple wave motions called *normal modes of vibration*. (See Figs. 10-11 and 10-12.) These normal modes are therefore particularly important. The frequency of oscillation in a normal mode is named a *natural frequency* of the system. The lowest natural frequency is termed the *fundamental frequency* and the higher natural frequencies are called *overtones*. In a normal mode of vibration, there are certain fixed locations called *nodes* where there is no vibration at all and other places called *loops* (or anti-nodes) where the vibration has a maximum amplitude. The type of wave occurring in a normal mode is called a *standing wave* because, in contrast to a traveling wave like that shown in Fig. 10-4 a crest does not move from one location to another; instead there are fixed locations of loops and nodes.

An interesting way to produce a standing wave is to have present simultaneously two traveling waves of the same wavelength λ and the same amplitude but traveling in opposite directions. As shown in Fig. 10-13, the superposition is then a standing wave and the distance between adjacent nodes is $\lambda/2$.

Fundamental f_1
$\frac{\lambda_1}{2} = L$

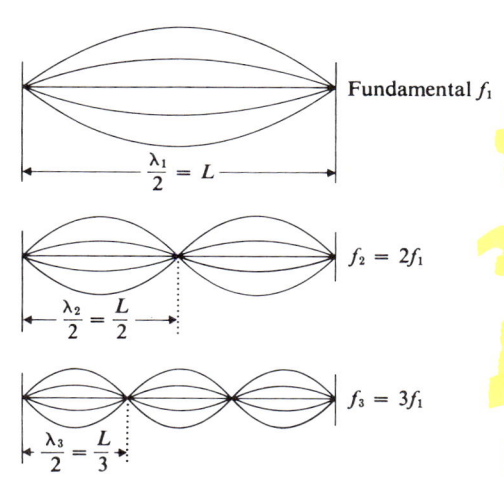

$f_2 = 2f_1$
$\frac{\lambda_2}{2} = \frac{L}{2}$

$f_3 = 3f_1$
$\frac{\lambda_3}{2} = \frac{L}{3}$

Fig. 10–11 Normal modes of vibration of a string of length L, fixed at both ends. These are standing waves. The string position is shown at four different instants.

Fig. 10–12 Normal modes of vibration of a drumhead corresponding to the three lowest natural frequencies. Shaded portions are moving down while the unshaded portions are moving up.

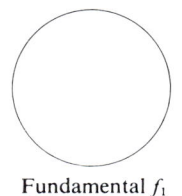

Fundamental f_1

First overtone
$f_2 = 1.59 f_1$

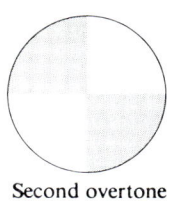

Second overtone
$f_3 = 2.30 f_1$

Normal modes for a system consisting of a string of length L fixed at both ends (the prototype of any stringed instrument) are simple. In any motion of the string, the ends must be nodes. Now if one regards the standing wave as a superposition of two traveling waves of wavelength λ which travel in opposite directions, then the distance between adjacent nodes is $\lambda/2$. In the fundamental mode (Fig. 10-11) this implies a wavelength, λ_1, such that

$$\lambda_1/2 = L.$$

This gives

$$\lambda_1 = 2L. \tag{10.4}$$

In the next mode of vibration, called the *first overtone*, the wavelength λ_2 must (see Fig. 10-11) satisfy

$$\lambda_2/2 = L/2$$

or

$$\lambda_2 = 2L/2.$$

Proceeding in this way, we find that wavelengths in the different normal modes of vibration form a sequence:

$$2L/1,\ 2L/2,\ 2L/3,\ \ldots.$$

The frequency f of oscillation in any normal mode is determined from Eq. (10.2), $v = \lambda f$, where v is the speed of traveling waves on the string.

The frequency f_1 of oscillation in the fundamental mode, the fundamental frequency, is therefore given by

$$f_1 = v/\lambda_1.$$

Since $\lambda_1 = 2L$, this implies

$$f_1 = v/2L. \tag{10.5}$$

Using the wavelength appropriate for each normal mode we obtain in this way the sequence of possible frequencies: $v/2L$, $2(v/2L)$, $3(v/2L)$, ..., which can be written as f_1, $2f_1$, $3f_1$, ...

Overtones that are *integral* multiples of the fundamental frequency are called *harmonics*. Our result shows that for a string fixed at both ends all the overtones are harmonics.* Note that the overtones for the vibrating drum head shown in Fig. 10-12 are not harmonics.

The details of the way a system is set into oscillation, for example, the position at which a stretched string is plucked, will determine which of its natural frequencies will be present in the oscillation. But no matter how complex, the resultant motion is always just a superposition of the various normal modes of vibration. A vibrating body, exposed to air, is a source of longitudinal traveling waves in the air which carry away energy. These radiated sound waves will have the frequencies present in the source, i.e., the natural frequencies.

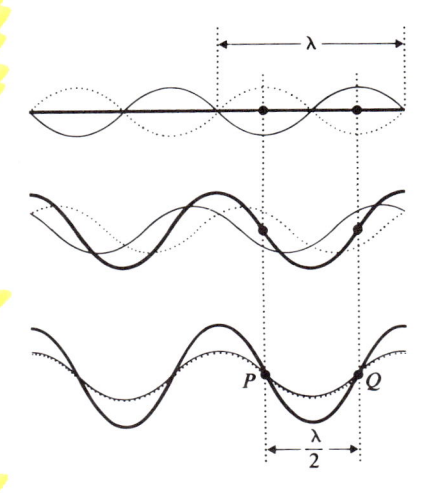

Fig. 10-13 The standing wave (heavy line) formed by the superposition of two waves of the same wavelength λ and the same amplitude, traveling in opposite directions. The wave indicated by the dotted line travels to the left while that indicated by the light solid line travels to the right. At the instant depicted in the uppermost diagram, cancellation occurs. The situation at later instants is shown in the lower diagrams. Points like P and Q, which never move, are called *nodes*.

*When the natural frequencies are f_1, $2f_1$, $3f_1$, ..., it is customary to speak of the fundamental frequency as the first harmonic, the first overtone as the second harmonic, and so on.

Example 1 A steel piano wire is 1.00 m long. Its tension is adjusted until the velocity v of transverse traveling waves (as shown in Fig. 10-13) is 500 m/sec.

(a) Find the fundamental frequency of this wire.
(b) What is the frequency of the third harmonic?
(c) What is the frequency of the second overtone?

Solution
(a) The fundamental frequency is given by

$$f_1 = v/2L \qquad (10.5)$$
$$= (500 \text{ m/sec})/2.00 \text{ m}$$
$$= 250 \text{ hertz.}$$

(b) The third harmonic has a frequency $3f_1 = 750$ hertz.
(c) The second overtone is the third harmonic.

Resonance

Suppose that instead of considering a body as a source of waves, we study its *absorption* of energy when sound waves from elsewhere are incident upon it. The body soon settles into *forced oscillations* at the frequency of the incident wave, the so-called *driving frequency*. The amplitude of these forced oscillations becomes exceptionally large when the driving frequency is near a *natural frequency* of the vibrating body. This large response at certain driving frequencies is termed *resonance* and the corresponding frequencies are called *resonant frequencies*.

Resonance is a widespread and important phenomenon. An everyday example is the occurrence of a shimmy or rattle in a car at a certain speed. An irregularity in a tire can result in a periodic driving force which, at a certain speed, matches a natural frequency of part of the car. Resonant interaction of light with atoms and molecules will be described in Section 16.10. Resonances of great variety occur in the interactions of atoms, nuclei, and elementary particles.

10.4 CHARACTERISTICS OF SOUND

Particularly from the standpoint of sound as a stimulus detected by the ear, certain characteristics such as *loudness*, *pitch*, and *quality* are of prime interest.

Loudness, Intensity, and the Decibel

A traveling sound wave transports energy. The intensity, I, of a sound wave at a given point in space is defined to be the average power crossing a small area A, divided by A:

$$I = \text{(average power)}/A,$$

where the area A is the area of a small surface through the point in question and oriented so that it is perpendicular to the direction in which the waves are traveling. The mks unit of intensity is the watt/meter^2. It can be shown that the intensity of a sound wave is proportional to the square of the amplitude of the wave.

The physiological sensation of *loudness* in the human ear is intimately related to the *intensity* of the incident sound wave. At a frequency of about 1000 hertz, the ear detects sounds varying from the barely audible intensity of 10^{-12} watt/m^2 to the pain threshold of 1 watt/m^2, a truly huge range. Although different people will have a somewhat different judgment as to whether one sound is twice as loud as another, it has been found that a scale proportional to the logarithm of the intensity corresponds roughly with the response of the ear.

Typical values of intensities on such a logarithmic scale, called the decibel* intensity level (abbreviation db), are shown in Table 10-2. A clearly noticeable change in loudness corresponds to about 3 db.

*The term bel is derived from the name of Alexander Graham Bell, inventor of the telephone.

Table 10-2 SOUND LEVELS

Situation	Decibel intensity level	Intensity (watts/m^2)	Sensation
Threshold of feeling	120	1	Painful
Thunder	110	10^{-1}	
Artillery			Deafening
Riveting	100	10^{-2}	
Elevated train	90	10^{-3}	
Noisy factory			
Unmuffled truck			Very loud
Pneumatic drill	80	10^{-4}	
Busy street traffic	70	10^{-5}	
Noisy office			Quite
Average street noise			pronounced
Conversation in home	60	10^{-6}	
Background noise in "active" home	50	10^{-7}	
Average office			Moderate
Quiet radio in home	40	10^{-8}	
Quiet home	30	10^{-9}	
Private office			Faint
Desirable level for patients in a hospital			
Whisper	20	10^{-10}	
Rustle of leaves	10	10^{-11}	
Soundproof room			Scarcely
Threshold of audibility	0	10^{-12}	audible

Pitch and Quality

Especially in music one is concerned with the several auditory sensations mentioned earlier: *loudness*, *pitch*, and tone *quality*. Each of these sensations corresponds to a physical quantity associated with a sound wave. We have already discussed the relationship between loudness and intensity. We turn next to pitch.

To each distinct *pitch* there corresponds a distinct frequency (except for very loud sounds). Higher frequency implies higher pitch. In doubling the frequency, the pitch is increased by one *octave*.

By the tone quality we refer to the property by which we can distinguish a given note, say middle C on the piano, from middle C on a violin, or a flute. Tone quality is determined by the overtones present and by their relative intensities. The fundamental frequency alone is not a very interesting sound. It is the overtones which provide "character."

10.5 THE EAR AND HEARING

Although the ear is most comfortable with sound intensities in the range of 40-60 decibels and is most sensitive to frequencies of the order of 2000-3000 hertz, it can handle intensities and frequencies which vary by factors of a thousand billion and a thousand respectively: a remarkable feat. Let us look briefly at some of the structures and mechanisms involved.

Several anatomical details are illustrated in Fig. 10-14. Sound waves travel down the ear canal, which is terminated by a thin membrane called the *eardrum*. Vibrations of the eardrum are transmitted through the air-filled middle ear by a complicated linkage of bones: the hammer, the anvil, and finally the stirrup, which acts on the membrane of the oval window. This set of bones has a mechanical advantage of 3. The ratio

Fig. 10-14 The ear.

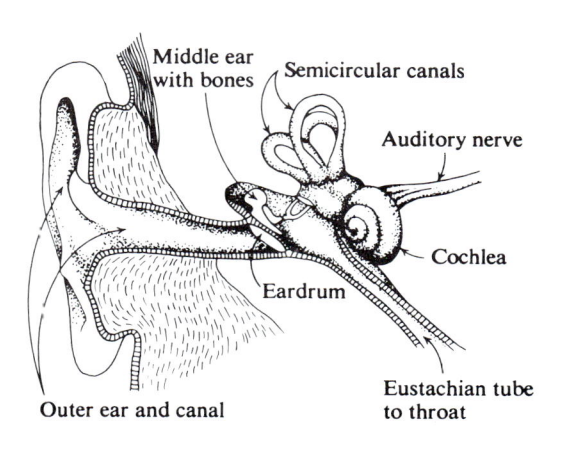

370 MECHANICAL WAVES: SOUND

of the area of the eardrum to the area of the oval window is typically 15 to 20. The result is that pressure changes in the liquid of the inner ear behind the oval window are some 45 to 60 times the pressure changes of the incident wave at the eardrum.

The eardrum vibrations are converted into nerve impulses in the cochlea of the inner ear. This liquid-filled cochlea is shaped like a snail shell of $2\frac{3}{4}$ turns. Uncoiled it is about 3.5 cm long. The cochlea is divided into two canals by a structure that includes the flexible basilar membrane, along which are some 30,000 basilar fibers or "Corti rods." Vibration of the basilar membrane moves these rods. Each rod is terminated by a hair cell which stimulates the nerve ending at its base. From there electrochemical pulses are believed to be transmitted along the auditory nerve to the brain.

The *position* along the basilar membrane at which the vibration causes resonance, that is, produces the maximum amplitude, determines the sensation of *pitch*.* If this resonance occurs at the inner tip of the cochlea, the hair cells and their nerve endings in that region receive more stimulation than those in other positions along the basilar membrane. Maximum stimulation in this region is interpreted by the brain as a sound of low pitch. The closer to the other end of the cochlea that the maximum stimulation occurs, the higher the pitch sensed.

While the location of the stimulated basilar fibers determines pitch, the extent of their stimulation determines *loudness*. The greater the amplitude of motion of these fibers (or rods) the greater the stimulation effected on the nerve endings by their respective hair cells. The greater this stimulation the more electrical pulses per second are transmitted to the brain and the louder is the sensation.

A very useful aspect of hearing is that it is binaural. Ordinarily one hears most effectively when facing the source of sound. This is so because when the face is turned to one side the ears are no longer equidistant from the source and the sound waves received by the two ears differ in intensity. Also a given crest of a sound wave does not reach both ears at the same time. For these reasons one can form judgments as to the direction of the source of a sound.

*G. von Békésy received the 1961 Nobel prize in physiology and medicine for work on this subject carried out at Harvard University.

10.6 DOPPLER EFFECT

One hears an abrupt drop in the pitch of the sound from the horn of a car as it whizzes by. Any such change in the observed frequency arising from motion of the source or the observer is called the *Doppler effect*.

The effect was first analyzed by an Austrian physicist, J. C. Doppler (1803–1853), who pointed out that similar frequency shifts should occur for light waves. Experiments later proved he was right. And as we shall describe in Section 14.4, "Doppler shifts" in the frequency of light coming from distant stars can be used to provide us with a vital item of astronomical information: the velocity of the star relative to the earth. Sound waves show a Doppler effect for two distinct reasons:

1. The wavelength in the air is changed by motion of the source. This is illustrated in Fig. 10-15 which shows crests of the successive waves emitted by the moving source. After emission, the wave moves with a speed v determined entirely by the mechanical characteristics of the transmitting medium, usually air. The sound wave thus forgets the source is moving, and a spherical crest expands with its center located at the position that the source occupied at the instant when the crest was emitted. The result, as shown in Fig. 10-15, is that the wavelength λ_{air} in front of the source is reduced by the source motion while the wavelength behind the source is increased. A stationary listener in front of the source will measure a frequency v/λ_{air} corresponding to this reduced wavelength λ_{air}. Therefore he will measure a frequency *higher* than the frequency of vibration of the approaching source of the sound.

2. If the observer is moving through the air toward the source he will encounter more crests per second than if he had remained stationary relative to the air. He will measure a frequency given by

$$f_o = \frac{\text{speed of waves relative to observer}}{\text{wavelength}}.$$

By moving with a speed v_o toward the source, the observer increases the speed of the sound waves relative to him to the value $v + v_o$ and consequently hears a higher pitch.

In a following example it will be shown that it is a straightforward matter to deduce an exact expression for the frequency f_o which is measured when both the source and the observer are moving through the air. Although the exact formula for f_o in terms of the source frequency f is not too complicated, the approximate relationship

$$\frac{\Delta f}{f} = \frac{\text{velocity of source relative to observer}}{v} \qquad (10.6)$$

is worthwhile exhibiting because it is so simple. The symbol Δf is the frequency shift, $f_o - f$. The relative velocity in the numerator is counted positive when the source is approaching the observer and negative when it is receding. Eq. (10.6) is a good approximation only when the source and observer velocities are much less than the speed of sound v relative to the transmitting medium. In this approximation, the frequency shift depends only on the velocity of the source relative to the observer.

Example 2 A train approaches a station platform at a speed of 10 m/sec, blowing its whistle at a frequency of 100 hertz, as measured by the engineer. The speed of sound in the air is 330 m/sec.

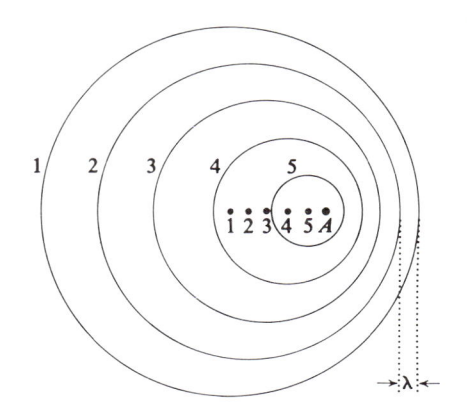

Fig. 10-15 Crests of waves emitted by a moving source A. Crest number 1 was emitted when A was at point number 1. Notice that in front of the source the wavelength λ is reduced, and behind the source it is lengthened.

372 MECHANICAL WAVES: SOUND

(a) What is the wavelength λ_{air} in front of the train?
(b) What is the frequency that will be measured by a man standing on the platform?
(c) What frequency will be measured by the driver of a car proceeding at 20 m/sec relative to the highway, approaching the station from a direction opposite to that of the train?

Solution

(a) The time interval between the emission of successive crests from the whistle is the period T (Section 3.10) of its vibration given by

$$T = 1/f = 1/(100 \text{ hertz}) = 1.00 \times 10^{-2} \text{ sec}.$$

In this time the first crest advances a distance

$$(330 \text{ m/sec})(1.00 \times 10^{-2} \text{ sec}) = 3.30 \text{ m},$$

while the whistle advances a distance

$$(10 \text{ m/sec})(1.00 \times 10^{-2} \text{ sec}) = 0.10 \text{ m}.$$

The distance between successive crests (the wavelength λ_{air}) in front of the train is therefore

$$\lambda_{air} = 3.30 \text{ m} - 0.10 \text{ m} = 3.20 \text{ m}.$$

(b) A man on the platform measures a frequency

$$\begin{aligned} f_o &= v/\lambda_{air} \\ &= (330 \text{ m/sec})/(3.20 \text{ m}) \\ &= 103 \text{ hertz}. \end{aligned}$$

Another approach is to use the approximate relationship:

$$\Delta f/f = (\text{velocity of source relative to observer})/v \quad (10.6)$$
$$= (10 \text{ m/sec})/(330 \text{ m/sec})$$
$$= 0.030.$$

This gives

$$\begin{aligned} \Delta f &= 0.030 \, f \\ &= (0.030)(100 \text{ hertz}) \\ &= 3.0 \text{ hertz}, \end{aligned}$$

and therefore

$$f_o = 103 \text{ hertz},$$

in agreement with the previous calculation.

(c) The speed of the sound waves relative to the driver of the car is

$$(330 \text{ m/sec}) + (20 \text{ m/sec}) = 350 \text{ m/sec}.$$

The wavelength from part (a) is 3.20 m. Consequently this driver measures a frequency

$$\begin{aligned} f_o &= (350 \text{ m/sec})/(3.20 \text{ m}) \\ &= 109 \text{ hertz}. \end{aligned}$$

In order to check this calculation using Eq. (10.6), we first compute the velocity of the whistle relative to the driver. This is

$$10 \text{ m/sec} + 20 \text{ m/sec} = 30 \text{ m/sec}.$$

Then Eq. (10.6) gives

$$\Delta f/f = (30 \text{ m/sec})/(330 \text{ m/sec})$$
$$= 0.091.$$

Consequently

$$\Delta f = (0.091)(100 \text{ hertz})$$
$$= 9.1 \text{ hertz},$$

in agreement with the alternate calculation.

10.7 SUPERSONIC SPEEDS AND SHOCK WAVES

The crests of sound waves emitted from an object which moves faster than the speed of sound are shown in Fig. 10-16. As in Fig. 10-15, the center of each crest is located at the position that the source occupied at the instant when the crest was emitted. Along the cone which touches these wave fronts, these waves reinforce each other leading to a concentration of energy that constitutes a *shock wave*. The same sort of thing happens in the production of *bow waves* by a boat traveling at a speed greater than that of waves on the water's surface.

The moving object does not have to be a vibrating source of sound to create shock waves. As soon as any object's speed passes that of sound, a shock wave is formed. Common examples are the shock waves that we hear as the crack of a passing bullet or the *sonic boom* from a distant supersonic aircraft. The energy carried by the conical shock wave that accompanies a high speed aircraft is sometimes sufficient to smash windows and shake buildings.

Each observer underneath the route of a supersonic aircraft hears a loud bang as a shock wave passes him.* No sound from vibrating sources in the aircraft can be heard by an observer outside the cone of Fig. 10-16. An observer at P will hear the sound emitted by the aircraft when it was at S.

The development of shock waves is determined by the ratio of the source speed to the speed of sound in the air at the existing temperature. This ratio is called the *Mach* number [after Ernst Mach (1838–1916)]. A missile at Mach 5 is traveling 5 times as fast as sound travels through the air in its locality. Shock wave formation commences at Mach 1.

A Russian physicist, P. A. Cerenkov (1904–   ), discovered that a fast charged particle would radiate light waves whenever the particle passed through a material at a speed greater than the speed of light in the material. This phenomenon is analogous to the production of shock

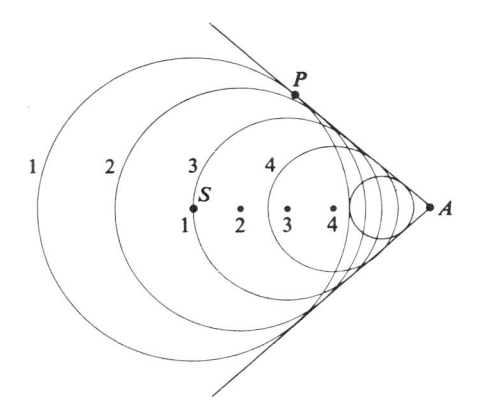

Fig. 10-16 Shock wave produced by object A moving to the right with supersonic speed. The shock wave is the cone with vertex at A. This cone touches the spherical crests emitted when A was at previous positions such as the points labeled 1, 2, 3, 4. The angle SAP must satisfy sin $\angle SAP = SP/SA = v/v_A$ where v is the wave speed and v_A is the speed of the object A.

*Two principal shock fronts, one at the nose and one at the tail, accompany a supersonic aircraft. These cause a double boom.

waves. Indeed the wave fronts of *Cerenkov radiation* are exactly as depicted in Fig. 10-16. A practical method of determining the speed of fast charged particles is based on measurement of the angle at which they emit this bluish Cerenkov radiation as they pass through transparent material.

10.8 SONIC SPECTRUM

Earlier, in reference to Fig. 10-8, we have indicated the wide range of the spectrum of mechanical vibrations. Most of our discussion to this point has, however, been applied to sound in the sense that audible frequencies were considered. We now look at the wider range of the spectrum and also consider some of the hazardous implications under the heading "Sound Pollution."

Ultrasonics

Sounds with frequencies above 20×10^3 hertz are called *ultrasonic*. This is the high-frequency limit of human audibility. Dogs can detect higher frequencies, and so can be quietly summoned with an ultrasonic whistle. Bats find their way about by emitting ultrasonic pulses and detecting the echoes.

In our century, men also have begun to exploit these high-frequency sound waves. Ultrasonic waves are used by ship's *sonar* to locate submarines in much the same way that a bat navigates. A pulse of ultrasonic waves is emitted. The transmitter then functions as a detector of reflected waves. The distance from the reflected object is determined from the time interval between transmission and detection. The same principle enables a ship to determine the depth of the sea. Ship-to-ship underwater communication is conveniently accomplished with ultrasonic waves. Frequencies from 20×10^3 hertz to 100×10^3 hertz are commonly employed. For all these applications ultrasonic waves have the advantage of being silent. Moreover, with shorter wavelengths a well-defined narrow beam can be obtained. The relevant phenomenon is *diffraction*. As we have previously stressed, the diffraction spreading [Eq. (10.3)] and bending of beams is small for wavelengths small compared to the obstacles and apertures involved.

Industrial and medical applications of ultrasound are now plentiful. First we consider examples of the use of mechanical vibration to perform work.

In plastic welding high-speed vibrations are applied, by means of a horn or probe, to two similar surfaces in contact with each other. The combination of regularly applied pressure and the actual vibratory action results in the generation of sufficient thermal energy (through the process of friction) to bring the two surfaces to be bonded up to a

Physics Today

P. A. Cerenkov
(1904–   )

temperature at which these surfaces become molten. Upon cessation of the application of the ultrasonic vibratory action, the welds become strong and stable almost instantaneously.

Another of this type of application (i.e., the direct use of ultrasonic energy) is involved in ultrasonic cleaning. Peculiar as it may seem on first realization, ultrasonic cleaning is the only known method suitable for thoroughly cleaning the minute and complex nooks and crannies found in laboratory and medical apparatus. This cleaning process is achieved by a mechanism known as *cavitation*. At the risk of a little oversimplification, here is how cavitation is effected: If the ultrasonic vibrations are produced in a liquid, tiny bubbles (hence the word *cavitation*) will form as the rarefaction of the ultrasound wave appears. Now when the compression follows, these bubbles are compressed until they implode (i.e., explode inward). Within the regions of these small bubbles there are tremendous internal pressures and temperatures. What we have then is many localized violent shock waves. And it is this process which is called cavitation or "cold boiling." This cavitation process literally blasts contaminants from the object which has been placed in the liquid in order to be cleaned. To achieve cavitation in a liquid, the transmission of ultrasonic energy of sufficient magnitude at a suitably high frequency is all that is required. Since the cleaning effect is more pronounced at lower frequencies than at extremely high ultrasonic frequencies, equipment for ultrasonic cleaning is usually designed to operate in the range of 20,000 to 30,000 hertz. This cleaning process is so effective in killing many kinds of bacteria that it qualifies in a sense as a sterilization process.

Ultrasonography

The second category of the application of ultrasonics involves the sending of a signal of ultrasonic frequency and the observation of the transmission, or refraction, or reflection of this signal. In this connection we have already mentioned sonar used by ships. A similar use of ultrasound is finding increasing application in medical diagnostic procedures. It is interesting to examine this facet of ultrasonics in some detail.

Since ultrasound is used to probe the internal structures of the body, the reader may wonder what the advantage of this procedure is over the better-known procedures of x-ray examinations. "Probably the primary advantage of ultrasound is that its capability for distinguishing interfaces between objects of different density is far more sensitive than x-ray. There are a large number of substances that are not radiopaque, but which can be detected easily with ultrasound. For example, it is very simple for ultrasound to distinguish between liquid and human soft tissue. This differentiation is virtually impossible with x-ray."[*]

[*] *Harvey Feigenbaum, M.D., of the Indiana University Medical Center, in Annals of Internal Medicine, Vol. 65, No. 1, July 1966.*

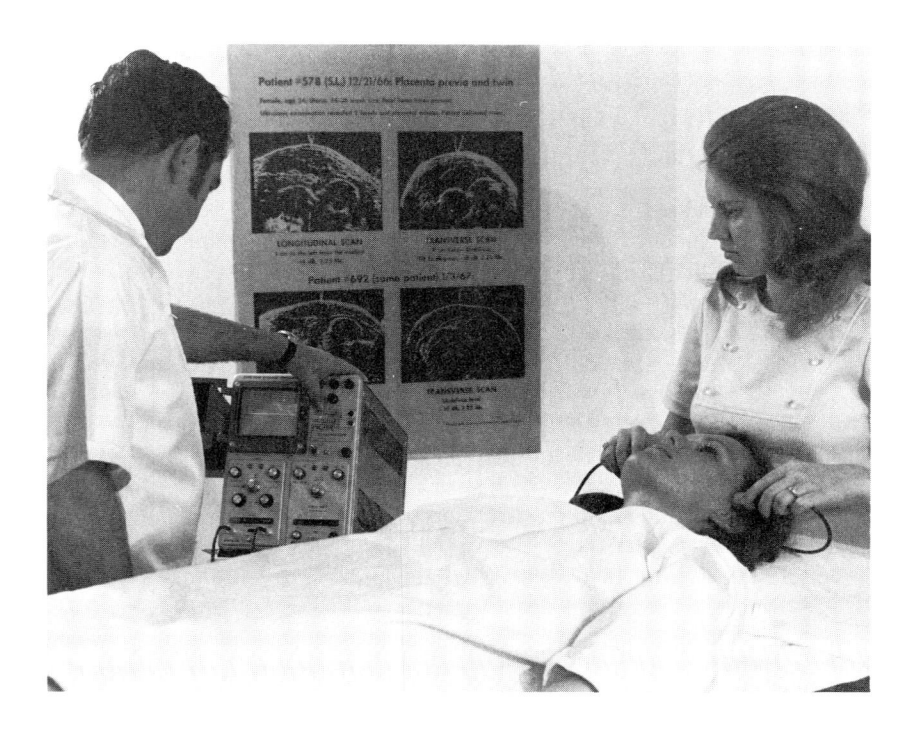

Fig. 10-17 Ultrasonography used in medical diagnosis. The technician on the right is applying the probe which is transmitting ultrasonic pulses through the patient's head. The information gained is displayed on the oscilloscope.
(Photo courtesy Picker Electronics, Inc.)

In particular, ultrasound can serve in diagnosis (see Fig. 10-17) in a number of ways such as revealing mitral valve action (the mitral valve is between the left atrium and ventricle of the heart), revealing fetal development, differentiating between cysts and tumors, permitting certain examinations of the liver, spleen, and kidneys; and locating the midline structures of the brain, a procedure known as *echoencephalography* (see Fig. 10-18).

Fig. 10-18 Ultrasonography used to locate the midline structures of the brain. Pathological abnormalities can cause a detectable shift in the midline structure. This type of diagnosis is very helpful in looking for brain lesions in psychiatric patients.

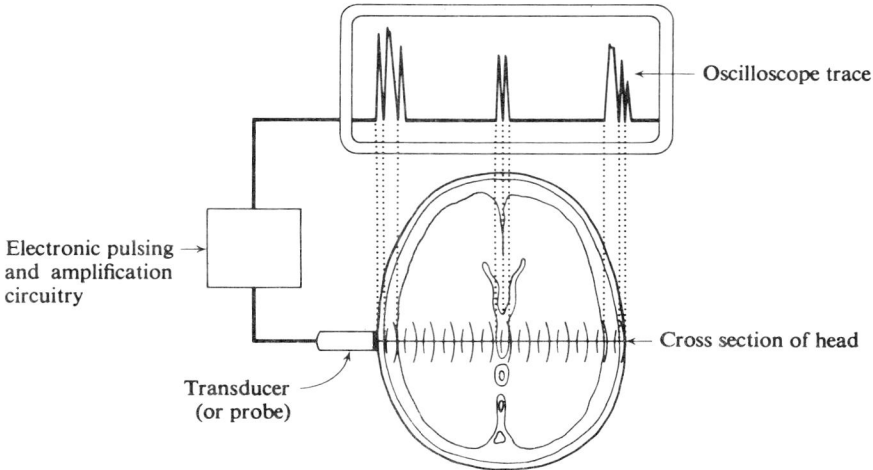

10.8 SONIC SPECTRUM

*A transducer is simply a device by means of which energy can flow from one system to another (in this case from the electrical circuit generating electrical pulses to the mechanical system of the human body). Often, as in this situation, the transducer is used not only as a sort of systems coupling device but also as an energy converter.

*The piezoelectric effect refers to the phenomenon of the production of a potential difference across opposite surfaces of certain crystals when subjected to mechanical pressure.

The essentials of the way ultrasound is used in such procedures, known as *ultrasonography*, are as follows. The ultrasonic energy is converted from electrical energy by a device called a *transducer*.* Pulses of ultrasonic energy are passed from the transducer through the skin into the internal anatomy. When these ultrasonic vibrations are reflected at interfaces between materials of different densities the reflections are picked up by the transducer which converts the returning ultrasonic (mechanical) energy into electrical energy. The electrical voltage so "created" is then amplified and displayed on an oscilloscope.

The actual transducer (or probe) may contain a crystal of lead zirconate titanate, about a centimeter in diameter and 2–3 mm thick. This crystal is connected to an electronic circuit that applies a voltage pulse across the crystal at the rate of about 500 times a second. Each voltage pulse causes the crystal to vibrate at a frequency which depends, among other things, on the dimensions of the crystal. For medical diagnostic purposes the sought-for frequencies are in the range of 1–10 megahertz. If such a vibrating crystal, in the form of a probe, is placed in contact with the body (see Figs. 10-17 and 10-18), 500 energy pulses per second will travel through the body and be scattered and reflected by the structures within the body.

Now each voltage pulse and its resultant pulse of ultrasonic waves lasts in the order of a few microseconds. In the interval between pulsations the transducer acts as a listener (about 99% of the time the transducer is in this passive phase). The compressions and rarefactions of the reflected waves cause the transducer to be alternately compressed and expanded and in this process generate a voltage (piezoelectric effect*) which is amplified and displayed somehow, usually on an oscilloscope (Fig. 10-18).

While it is always dangerous to speculate regarding the future, it is tempting to suggest that the versatility of ultrasonics will lend itself to increasing medical applications as both equipment and techniques continue to be refined. Undoubtedly also, on the basic research front, study of the interaction of ultrasonic radiation with biological tissue will yield new knowledge.

On the other hand, this very knowledge may lead to a more restricted use of sound in medicine because possibly as yet unidentified hazards may come to light.

The studies to date with ultrasound have indicated that sound levels of a few watts/cm^2 can cause cavitation in a tissue and thereby harm it. But ultrasonic diagnostic work is done at much lower power levels, in the milliwatt/cm^2 region. All of which begs the question, "Is there a *threshold* level for damage?" Or, as seems perhaps more likely, are we faced here with somewhat the same dilemma as was discussed under the heading *Ionizing Radiation and Biological Tissue* in Section 1.7?

Infrasound

Infrasound, that is mechanical vibrations at the long-wavelength (low-frequency) end of the spectrum, is the latest part of the sonic spectrum to be studied. This is not to say that certain phenomena in this region have not been investigated for some time (such as seismic waves and the pressure waves of low frequency at the mouth of artillery pieces) but rather that concerted laboratory investigations in the infrasonic region are relatively in their infancy.

For example, only in recent years have the dangers associated with infrasound been revealed. Accidental observation of vibration-induced sickness led to studies which have revealed that at very low frequencies (in the region of 5–10 hertz) the organs of the body tend to resonate. And this rubbing of one organ upon another has caused noticeable illness at even relatively low power levels.

Sound Pollution

The possibly significant detrimental effects due to excessive exposure to sound, ultrasound or infrasound, excessive in both the sense of duration and of intensity, seem not to have been fully explored. Yet it has been established that there is cause for concern. The risks may be serious for people working at high noise levels for extensive periods and for those subjected frequently to the sonic booms from aircraft.

While there is no doubt that prolonged exposure to loud sounds can precipitate deafness* and nervous disorders, relatively recent experiments on animals have shown that deformed offspring may be born as the result of the mother being exposed to sonic booms during pregnancy.

*Some medical studies have revealed that certain teen-age "rock musicians" developed boilermaker's ear. *This means the hearing of these young musicians is so impaired that when only in their teens their hearing is on a par with that of the average 65-year-old.*

QUESTIONS

1. Distinguish between the speed of a transverse pulse on a stretched rope and the speed of a particle of this rope.
2. The ducks of Fig. 10-1 move up and down through a total vertical distance of 0.6 m.
   (a) What is the amplitude of the water waves in question?
   (b) If the amplitude of the water waves were changed to 0.4 m without changing the frequency, would more energy or less energy be imparted to the ducks than in case (a)?
3. (a) Explain briefly what is meant by the wavelength λ and the frequency of a wave.
   (b) What is the relationship between the period T and the frequency?
   (c) Show with the aid of a diagram that the wave speed is given by $v = \lambda/T$.
4. The shortest wavelength (in air at 0°C) of the ultrasonic waves that a bat emits is approximately 3.3 millimeters. What is the frequency of these waves? Is this the highest or the lowest frequency emitted by the bat?

5. A duck bobs up and down completing 7.0 cycles in 10 sec. One crest moves from the duck to the shore, 9.0 meters away, in 5.0 sec.
   (a) What is the frequency of the duck's oscillation?
   (b) What is the wave speed?
   (c) What is the wavelength?

6. There is a 1.2 m distance between a crest and an adjacent trough in a series of waves on the surface of a lake. In 30 sec, 35 crests pass a buoy anchored in the lake. What is the speed of these waves?

7. Describe an experiment which demonstrates that sound is not transmitted through a vacuum.

8. Sound waves are said to be longitudinal. Describe how these waves differ from transverse waves.

9. Describe the molecular motion that occurs when a sound wave travels through air, pointing out the relevance of Figs. 10-3(a) and 10-3(b).

10. The time interval between seeing a flash of lightning and hearing the associated thunder is 10 sec. Estimate the distance to the lightning stroke.

11. Calculate the distance that a sound wave actually travels in 5.00 sec in air at 0°C. What percentage error is made in assuming that this distance is one mile?

12. When a stone is dropped into a well, the sound of the splash is heard 3.00 sec later. How deep is the well? Assume that the speed of sound in the air is 340 m/sec.

13. To determine the depth of water below a ship's keel, a short pulse of ultrasonic waves is sent out and the time to the first echo is measured and found to be 0.150 sec. The speed of the sound waves in the water is 1.50×10^3 m/sec. What is the depth below the keel?

14. Why is it that men marching in step to the music of a band are seen to be progressively more and more out of step with the band as their distance from the band increases?

15. The "international pitch" used in music is based on a frequency of 440 hertz for the middle A on the piano.
    (a) What is the period of the vibration of the piano string which produces this tone?
    (b) At room temperature the speed of sound in air is about 340 m/sec. Calculate the wavelength of the vibration which yields middle A.

16. Ultrasonic waves with a wavelength of 5×10^{-7} meter in air (as short as that of visible light) can be generated by the vibrations of a thin quartz crystal plate which is driven by an alternating electric field. What is the frequency of such sound waves?

17. A beam of sound waves of wavelength λ passes through an aperture of 10 millimeters diameter. As the beam emerges and travels through air, most of the beam intensity remains confined within a cone of 10° vertex angle. Estimate the wavelength of the beam. What would happen if the wavelength were made 5 times as long?

18. When and why do beats occur?

19. A tuning fork of unknown frequency makes 2.0 beats per second with a standard tuning fork of frequency 256 hertz.
    (a) What are the two possible values of the unknown frequency?
    (b) A small piece of wax is placed on the prong of the first fork. Then it is observed that the number of beats per second is reduced. Which value calculated in part (a) is correct?

20. A piano tuner using a 256-hertz tuning fork hears 6 beats per second when a certain note on the piano is sounded. When he uses a 260-hertz tuning fork he hears 2 beats per second when the same note is sounded. What is the frequency of that note?

21. Define the following terms: natural frequency, fundamental frequency, overtones, harmonics.

22. What is the difference between a traveling wave and a standing wave? (Draw sketches of each type of wave at different instants.)

23. The fundamental frequency of a string which is 0.50 m long and fixed at both ends is 400 hertz. Find the frequency of vibration in the third normal mode of vibration (that is, the third harmonic or second overtone). Sketch the typical appearance of the string at different instants, pointing out the locations of the nodes and loops.

24. A stretched string, 0.80 m long, has a fundamental frequency of 100 hertz.
    (a) What is the velocity of the traveling waves whose superposition produces the standing wave?
    (b) What is the wavelength of the traveling waves in (a)?
    (c) What is the frequency of the first overtone?
    (d) What is the frequency of the fourth harmonic?

25. Find the acoustic power output of a loudspeaker which radiates a sound wave such that there is a uniform intensity of 0.80 watt/m² on the surface of a hemisphere located 20 m from the loudspeaker.

26. While a man is shouting, the average power of his voice passing through a 2.0-m² window is 1.6×10^{-4} watt. If the power is uniformly distributed over this 2.0-m² area, what is the intensity at the window?

27. A 33-rpm record is accidentally played at 45 rpm. Will the pitch of the recorded human voices be altered? Explain.

28. What physical properties of a sound wave correspond to the sensations of loudness, pitch, and tone quality?

29. A policeman, blowing a whistle which vibrates with a frequency of 512 hertz, runs at a speed of 6.0 m/sec in the direction of a group of people standing at an intersection. The speed of sound in the air is 340 m/sec.
    (a) What is the wavelength in front of the policeman?
    (b) What frequency is heard by the people on the corner?

30. The policeman in the preceding question stops, but one of the people on the corner runs away at a speed of 6.0 m/sec. Find the frequency heard by this person as he departs.

31. A car approaching a cliff at a speed of 40 m/sec sounds its horn, which vibrates at a frequency of 320 hertz. The speed of sound in air is 340 m/sec.
    (a) Find the wavelength of the sound waves which travel from the horn to the cliff. Will the waves reflected from the cliff have the same wavelength?
    (b) Relative to the driver, what is the speed of the waves reflected from the cliff?
    (c) What frequency of these reflected waves will be detected by the driver?

32. The train of Example 2 passes through the station and proceeds down the track, still blowing its whistle.
    (a) What is the wavelength behind the train?
    (b) What is the frequency that will be measured by the man standing on the platform?
    (c) What frequency will be measured by the driver of the car receding from the train at 20 m/sec relative to the highway?

33. Consider a source of sound that vibrates with a frequency f and period T and moves relative to the air with a speed v_s. Denote the speed of sound relative to the air by v. Following the method of Example 2, derive the general theory of the Doppler effect for sound waves by showing that:
    (a) the wavelength in front of the source is given by
    $$\lambda_{air} = vT - v_s T = (v - v_s)/f_s\,;$$
    (b) an observer, approaching the souce with a speed v_0 relative to the air, detects a frequency given by
    $$f_0 = \frac{v + v_0}{\lambda_{air}} = \left(\frac{v + v_0}{v - v_s}\right) f_s\,.$$

34. Show that when v_0 and v_s of the preceding question are speeds of recession instead of approach, the observed frequency is given by
    $$f_0 = \left(\frac{v - v_0}{v + v_s}\right) f_s\,.$$

35. Deduce the relationship expressed in Eq. (10.6) for the special case when the source is stationary and the entire frequency shift arises from the motion of the observer through the still air.

36. With what speed is a plane traveling toward a stationary whistle of frequency 500 hertz when the pilot measures a frequency of 600 hertz? Assume the speed of sound is 330 m/sec.

37. Sketch a scale drawing of the shock wave produced by an aircraft traveling at Mach 2. Show the position of the aircraft and the position of one crest that was emitted at a previous instant. What is the vertex angle of the cone formed by this shock wave?

38. A bullet travels through air at a speed of 660 m/sec in a region where the speed of sound is 330 m/sec.
    (a) What is the Mach number for this bullet?
    (b) Find the angle made by the shock wave with the velocity vector of the bullet.

39. The wave front of the bow wave of a ship makes an angle of 10° with the ship's velocity vector. The ship's speed is 7.0 m/sec. Calculate the speed of the water waves.

40. The ratio of the speed of light in a vacuum to the speed of light in water is about 1.33. What is the cutoff speed below which a charged particle traveling through water will not emit Cerenkov radiation?

41. At what speed through water will electrons emit Cerenkov radiation having a conical wavefront which makes an angle of 60° with the electron's velocity vector?

42. Briefly describe the essentials of ultrasonic cleaning.

43. A common power rating for an ultrasonic washing machine is 5.0 watts/in², that is, about 7.6×10^3 watts/m² of liquid surface. What is the energy supplied to the cleaning liquid if the tank is 0.50 m long and 0.30 m wide and the cleaning operation takes 20.0 sec?

44. Consider the ultrasonic washing machine of Question 43. What is the minimum power requirement for the generator which is required to drive the vibrating mechanism if the entire process is 45% efficient?

45. Among the medical applications of intense ultrasound is the modification and/or destruction of soft-tissue tumors. Suppose that a frequency of 10 megahertz is employed.
    (a) What would be the wavelength of such a wave in air? Assume that the speed of sound in air is 340 m/sec.
    (b) In body tissue would the wavelength be greater or less than in (a)?

46. It has not been clearly established what, if any, harmful effects are associated with the use of ultrasound in medical diagnosis. Speculate upon why one might expect there could possibly be some deleterious effect associated with the use of ultrasound.

SUPPLEMENTARY QUESTIONS

S-1. A transverse wave of 2.00-m wavelength travels along a stretched rope.
    (a) Locate two particles of the rope that are in phase with the particle at the origin.
    (b) Find a particle that is 180° out of phase with the particle at the origin.

S-2. A workman strikes an iron rail with a hammer, and the sound travels through both the air and the rail to reach an observer 1000 m away.
    (a) If the wavelength of the sound in air is 0.500 m, what is the frequency of the sound?
    (b) How much time separates the arrival of the sound through the air and that of the sound through the rail?

S-3. A convenient measure of loudness is provided by the *decibel* (dB) *intensity level* defined by the following equation:

Intensity level = $10 \log_{10} [I/(10^{-12} \text{watt/m}^2)]$ dB.

Use this definition of intensity level to verify the decibel levels corresponding to each intensity listed in Table 10-2.

S-4. Show that the difference in intensity levels of two sounds of intensities I_1 and I_2 is $10 \log_{10} I_1/I_2$ dB.

S-5. If the intensity level in a room is 55 dB when one person is talking, what will the intensity level be when ten people are talking simultaneously? (The intensity due to a number of such independent sound sources is the sum of the individual intensities.)

S-6. The intensity of sound radiated from a small source is measured at a point 50 m from the source and found to be 2.0×10^{-6} watt/m².
    (a) What will be the intensity in the same direction from the source at a point which is 150 m further from the source?
    (b) Does the answer to part (a) depend on the following assumptions?
        (i) The source is isotropic.
        (ii) The power absorbed in the intervening air is negligible.
        (iii) All the power comes directly from the source rather than from reflecting surfaces.

S-7. An observer stands 60 m from a loudspeaker. How far must he walk toward the speaker in order to encounter an intensity level greater by 10 dB?

S-8. The intensity level of sound is 25 dB lower at a point 300 m from a foghorn than it is at a point 30 m from the foghorn.
    (a) What decrease in intensity level would occur if the intensity were proportional to the inverse square of the distance from the foghorn?
    (b) What part of the actual decrease in intensity level can be attributed to the absorption of sound by the fog?

S-9. A train passes a stationary observer at a speed of 30 m/sec on a day when the speed of sound in air is 340 m/sec. What percentage drop in the frequency of the train whistle is detected by the observer as the train passes?

S-10. A student holding a 256-hertz tuning fork approaches a laboratory wall at a speed of 5.0 m/sec. The speed of sound in the air is 340 m/sec.
   (a) What frequency will the student detect from the waves emitted from the fork and reflected from the wall?
   (b) How many beats per second will he hear between the reflected waves and the waves coming directly from the fork?

S-11. Show that the approximation given by

$$\frac{\Delta f}{f} = \frac{\text{velocity of source relative to observer}}{v}$$

follows from the general result given in Question 33(b) for the case when v_0 and v_s are small compared to v.

S-12. If an observer and a source of sound are both stationary relative to the earth, will there be a change in the observed frequency when there is a wind? Explain.

S-13. An object moving at supersonic speeds sets up a conical shock wave. Show that the acute angle θ between the shock wave and the velocity vector of the moving object must satisfy the equation

$$\sin \theta = \frac{v}{v_A}$$

where v is the wave speed and v_A is the speed of the moving object.

S-14. A supersonic aircraft, flying at an altitude of 4000 m, has a speed such that its Mach number is 1.25.
   (a) Find the angle made by the shock wave with the velocity vector of the aircraft.
   (b) How long after the aircraft has passed overhead will the boom be heard by an observer on the ground? (Assume that the shock wave travels at a speed of 330 m/sec.)

S-15. Consider the medical application of ultrasound and the phenomenon of resonance. What significance might a knowledge of the phenomenon of resonance have in the use of ultrasound in medicine? [*Hint:* What, for example, is the implication of the fact that biological cells, like other physical structures, have some natural frequency of vibration?]

ADDITIONAL APPLICATIONS TO MEDICINE AND THE LIFE SCIENCES

Mechanical Resonances of Biological Cells

Most mechanical structures have resonances. Such resonances may be employed to study crystal structures, properties of liquids, internal friction in metals and polymers, etc. Similar resonant phenomena seem to be applicable to the study of the physical properties of biological cells. In fact, it has been noted that at a characteristic frequency, the relative rate at which a given cell is destroyed is greatly increased. Optical studies have shown that the shapes of cells exposed to mechanical vibrations, in the ultrasonic frequency range, are distorted most at specific frequencies related, in part, to the geometry of the cell in question. Such resonance studies are valuable not only in the pursuit of basic biology, but also have a medical significance in the light of the increasing use of ultrasound in medical diagnosis and therapy.

Question

After reading about the medical use of ultrasound, the reader might consider what the implication of the above information is in the field of medicine.

Answer

To minimize body damage in diagnostic use of ultrasound, the resonant frequencies of the most critical cells need to be determined. And for the most efficacious therapeutic use, the resonant frequencies of such objects as gallstones are useful to know. The destruction of gallstones by the application of ultrasonic mechanical vibrations is more effective if the vibration frequency is a resonant frequency of the stones.

Ultrasonic Doppler Effect in Medicine

In addition to the comments made earlier in this chapter it is interesting to note that ultrasonographic investigations depending upon the *Doppler effect* are also possible.

In ophthalmology, Doppler shifts due to the pulsation of the ophthalmic artery may yield information about blood flow to the eye (see Fig.10-19).

In the study of fetal physiology, particularly the fetal cardiovascular system, the Doppler effect is also useful in that it yields information about fetal activity at early stages in gestation (at a time of 10-11 weeks when the fetal heart has a mass no greater than 0.5-1.0 gm). In the *ultrasonic Doppler inspection* of the fetal heart a continuous ultrasonic beam at 5.0 MHz (megahertz) is applied at the maternal abdominal skin surface. When this beam is reflected from a moving interface such as the fetal heart, the Doppler shift in frequency of the received echo creates a beat with the input signal at the maternal skin surface. This beat frequency is monitored to give information about fetal heart action.

Question

(a) In the case of ultrasonic inspection of the intrauterine fetal heart, ultrasound at 5 MHz is often used. If one assumes the velocity of sound in tissue to be about that in water, that is about 1500 m/sec, what is the velocity of motion of the fetal heart if a beat frequency of 500 Hz (hertz) is observed?

(b) If such beat frequencies were observed periodically at a rate of 160/min, what conclusion might one draw?

Answer

(a) The number of beats per second (Hz) is the difference (Δf) between the 5.0×10^6 Hz frequency of the ultrasound source and the frequency of the reflected signal from the heart. This Δf is the sum of two Doppler shifts, Δf_1 and Δf_2.

The fact that there are two Doppler shifts may be a bit puzzling.* Let us see how the two shifts come about. First, suppose we imagine that the 5.0×10^6 Hz source were affixed to the heart as the heart moves toward the observer at the surface of the body. Then at the surface of the body we would note a Doppler shift which we could calculate by using Eq. (10.6). This way, with the heart moving at a speed v_h (v_h being the velocity of the source relative to the observer in this case), we get

$$\Delta f_1 = \frac{v_h \times (5.0 \times 10^6 \text{ Hz})}{(1.5 \times 10^3 \text{ m/sec})}.$$

Now the plot thickens. The frequency sent out by the heart as it is moving toward the observer is not 5.0×10^6 Hz, but is slightly larger than this by the amount Δf_2. This is so because the heart is moving toward the signal which has originated not at the heart but at the surface of the body. In that sense the heart is the observer who sees a shift, and then sends out a signal incorporating this shift. To get the approximate shift of the signal we again use Eq. (10.6), and get

$$\Delta f_2 = \frac{v_h \times (5.0 \times 10^6 \text{ Hz})}{(1.5 \times 10^3 \text{ m/sec})}.$$

So the signal as received at the body surface has undergone a frequency shift of

$$\Delta f = \Delta f_1 + \Delta f_2 = \frac{2v_h \times (5.0 \times 10^6 \text{ Hz})}{(1.5 \times 10^3 \text{ m/sec})}.$$

* While one can derive an equation to yield the total Doppler shift, in the case of an observer and a moving reflector the approximation we get by using Eq. (10.6) is good—and hopefully more revealing of what is actually going on.

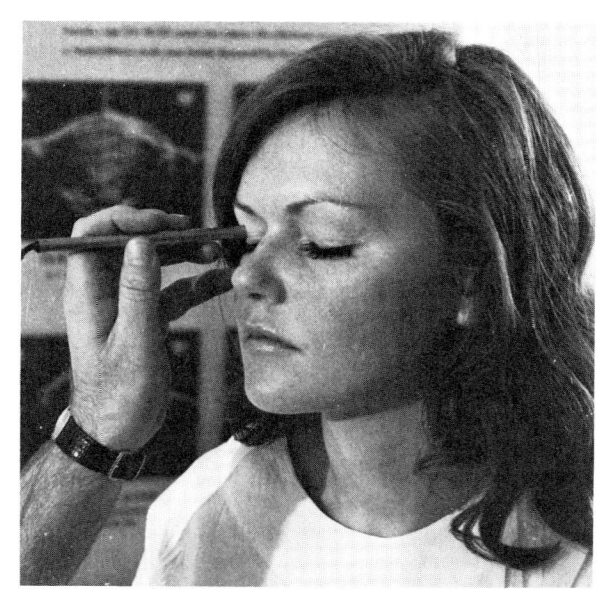

Fig. 10-19 Ultrasonic probe. (Courtesy Picker Electronics, Inc.)

We now solve for v_h with $\Delta f = 500$ Hz:

$$v_h = \left(\frac{500 \text{ Hz}}{5.0 \times 10^6 \text{ Hz}}\right)\left(\frac{1.5 \times 10^3 \text{ m/sec}}{2}\right)$$

$$= 0.075 \text{ m/sec}.$$

(b) The reflected wave will have a frequency *increased* by 500 Hz while the heart is expanding so that its surface approaches the observer at a speed of 0.075 m/sec. The frequency of the reflected wave will be *decreased* by 500 Hz while the heart is contracting and the heart surface receding at 0.075 m/sec from the observer. Therefore a 500 Hz-beat will occur twice during each cycle of the heart's motion. Consequently,

$$2 \times \text{heart rate} = 160/\text{min, or}$$

$$\text{heart rate} = 80/\text{min}.$$

Cerenkov Radiation in Eyes

Astronauts on lunar missions have reported observing flashes of light in their eyes during their flights. While this phenomenon is obviously of interest to physicists, physiologists, and psychologists, it also possibly poses a potential health hazard of interest in the realm of space medicine.

One possible mechanism which could explain these flashes of light is *Cerenkov radiation* (this hypothesis has been proposed by a number of physicists, although very tentatively). In translunar flight the astronauts are subjected to the primary cosmic rays (principally nuclei of $Z = 1$, up to such heavy nuclei as those of iron, $Z = 26$). These cosmic ray particles will traverse the astronaut's eyes at possibly such velocities that they will generate Cerenkov radiation in the vitreous humor, the fluid which fills the vitreous cavity (see Chapter 15). Since this Cerenkov radiation is highly concentrated near the particle track, the retinal region nearest to the point at which the cosmic ray particle passes through the retina will receive the most light. The effect would be that the observer will report only a very small point, streak, or flash of light, rather than a general bright illumination—and this is what has been reported.

Bats and Ultrasonics

While night birds, such as owls, cannot fly in complete darkness without risking crashing into objects, bats are expert fliers in pitch-black darkness. They hunt prey and avoid obstacles with the greatest of skill. To accomplish this navigational feat, bats send out sound waves, rather ultrasonic waves, and detect the echoes. With most species the basic transmission frequency lies somewhere between 20-100 kHz. These ultrasonic frequencies are emitted intermittently. When the bats are hunting prey, the transmissions are repeated irregularly from 10 to 30 times a second. These ultrasound pulses are emitted through the mouth (except in the case of the horseshoe bats—with them the nose acts as a directional transmitter).

Question

(a) How small an insect can a bat locate?
(b) Is there any defence possible against the bat's locating system?

Answer

(a) To answer this question we need to seize upon some of the implications of diffraction. Since this aspect of diffraction has not been explicitly pursued in the relevant text discussion we shall employ a water-wave analogy. It is possible to get some information about a big ship anchored, say in a bay, by observing the waves. On the leeward side of the ship (the opposite side from which the wind and the waves are coming) there would be a relatively calm region, a sort of "shadow" of the ship. Around the ends of the ship the waves would be diffracted in a way characteristic of their wavelength—but this diffraction would not obscure the "shadow," so we could determine something about the size and shape of the ship from the "shadow." On the other hand, if these same waves rolled by a buoy in the bay, the behavior of the waves would barely be disturbed, there would be no "shadow," at least not enough to yield much information about the buoy. Perhaps we might be able to infer that some relatively small object was there, but no more. If instead of a buoy we now consider a fishing line hanging straight down into the water, we would likely not even be able to detect anything at all simply by studying the waves after they rolled by the line. The crucial point is that the wavelength of the detecting signal (water waves in this case) is too large for the small object (the fishing line) to be detected. Of course, we could see the fishing line by means of light waves, but these waves have a wavelength small compared to the dimensions of the fishing line. For purposes of detecting any object with any wave motion, the wavelength has to be less than or of the order of the size of the object to be detected. It is this very issue which places a limit on the "smallness" of something which is to be seen with a microscope. If what we want to distinguish is much smaller than the wavelength of light we are out of luck.

This same situation pertains to sound waves. Assuming a transmission frequency of 100 kHz and the speed of sound in air at 331 m/sec, we get

$$\lambda = \frac{v}{f} = \frac{331}{100 \times 10^3} \text{ m}$$
$$= 3.31 \times 10^{-3} \text{ m}$$
$$= 3.31 \text{ mm}.$$

So insects in the millimeter region can be detected. In fact, experiments have shown that bats can locate minute fruit flies (Drosophila) which are only about 2.5 mm long.

(b) Yes. Avoid surfaces which produce regular reflection. Moths, a favorite prey of the bat, in fact, have this valuable survival mechanism in the form of a furry covering on their bodies which results in irregular reflection and consequently much sound absorption in the covering hairs. Moths are most vulnerable when they have their wings spread, for from these the bats get a useful echo. It is interesting at this point to note the mysterious and complex interactions in nature: Moths are apparently equipped with an organ of hearing sensitive to ultrasonic energy. If the moths sense this ultrasonic energy they fold their wings instantly and drop, thereby, perhaps, eluding the bat.

Further Reading

Griffin, D.R., *Listening in the Dark*, Yale University Press, New Haven, Conn., 1958, deals with many aspects of the natural sonar of bats, birds, and porpoises.

SUGGESTED READING

FEYNMAN, R. P., R. B. LEIGHTON, and M. SANDS, *The Feynman Lectures on Physics*, Vol. 1. Reading, Mass.: Addison-Wesley, 1963. An entirely descriptive Chapter 51 discusses such things as shock waves, bow waves, and a tidal bore.

GRIFFIN, D. R., *Echoes of Bats and Men*. Garden City, New York: Doubleday, 1959. An excellent descriptive introduction to sound waves and their application.

HARRINGTON, E. L., *General College Physics*. New York: Van Nostrand, 1952. Although this book is over 20 years old, few books can match it in details of physics as applied to physiology and medicine. Chapters 25–28 are well worth consulting.

STEVENS, S. S., FRED WARSHOFSKY, and editors of *Life, Sound and Hearing*, New York: Life Science Library, Time, Inc., revised edition, 1967. Excellent illustrations. Particularly suited for those readers interested in such topics as musical instruments and the physical and physiological aspects of hearing.

Georg Simon Ohm

11 Electric Potential and Circuits

We now pause in the introduction of new fundamental principles of physics to examine in this chapter the behavior of electric circuits. These circuits, which so conveniently effect a rapid transfer of energy, are of enormous practical importance in the technology that we encounter in our daily lives. We shall therefore study electric charges in some sort of continuous motion or flow. The simplest of such considerations involves a steady drift of charge, always in the same direction, so-called direct current (DC) or, more precisely, steady direct current.

This chapter consequently finds its historical root in some experiments performed in England around 1730 by Stephen Gray (c 1667–1736). His discovery opened the door to the electrical technology upon which we are now dependent; he found that electricity, and consequently electrical energy, could be transferred from one body to another over fairly long distances by some materials, but not by others. Thus, out of experimentation was born the distinction between conductors and nonconductors.

A real impetus to the era of current electricity may be ascribed to the subsequent invention of the voltaic pile. This discovery of Alessandro Volta (Italy, 1745–1827) permitted the production of fairly steady continuous currents.

The important mathematical relationship which underlies electrical technology as far as conductors are concerned came a century after Gray's discovery of conduction when a German physicist, professor, and high school teacher, Georg Simon Ohm, formulated the law now known by his name. Ohm's work on this topic of the flow of electric charge in a conductor was apparently prompted by an analogy with heat flow, a topic which had been worked out a few years earlier by the French mathematician, Jean-Baptiste Fourier (1768–1830). In retrospect this analogy upon which Ohm leaned was indeed well conceived, for, as we understand it now, both heat conduction and electrical conduction involve the so-called "free" electrons in a conductor. And in fact, generally good electrical conductors are inclined to be good heat conductors, while poor electrical conductors tend also to be poor heat conductors.

The equations of electricity "are so similar to those given for the propagation of heat . . . that even if there existed no other reasons, we might with perfect justice draw the conclusions that there exists an intimate connection between these natural phenomena; . . . ."

Georg Simon Ohm
(1789–1854)

Ohm's paper, which laid the foundation for what is now known as Ohm's law, was completed during a year's leave of absence from his teaching position at a Jesuit college in Cologne. This paper, entitled *The Galvanic Circuit Investigated Mathematically*, was published in May 1827. Ohm's success in solving the then baffling problem of the electric circuit depended on his realization that the circuit must be considered in its entirety including the battery—that dealing only with the wires external to the battery was not adequate.

While Ohm's contribution to physics is far from being on the same scale as that of many of the other scientists featured in these chapter introductions, one must nevertheless recognize that modern electrical technology owes much to Ohm. Whereas Ohm's law is not what one would call a fundamental general physical law, this law does suffice, and is indeed much used, to solve many mundane electrical problems of great practical value.

11.1 ELECTRIC POTENTIAL

The concept of electric potential is probably the most commonly employed concept in electrical technology. We must understand this idea because it contains the key to much of what is to be discussed in the realm of electric currents. The student would therefore do well to acquire a very firm understanding of such terms as *potential*, *potential difference*, and *potential energy*.

Potential Difference and Potential

We shall address ourselves first to the difference of electric potential, commonly called voltage, measured with an instrument called a *voltmeter* and expressed numerically in terms of a unit known as the *volt* (after Alessandro Volta).

Voltages encountered in everyday life may range from the 1.5 volts potential difference between the terminals of an ordinary flashlight cell to the kilovoltages (1 kilovolt = 10^3 volts) required in television sets. Those readers who have experienced an electroencephalograph may be interested to know that the potential difference between the electrodes placed on the scalp are of the order of microvolts (1 microvolt = 10^{-6} volt). Small as they are, such voltages are measurable and of considerable significance in medical diagnosis. Ordinary household voltages, such as encountered when plugging in a vacuum cleaner are, in North America, usually from 110 to 120 volts.

We shall use the concept of electric field to establish the notion of potential difference. Any electric charge placed in a region in which there is an electric field experiences a force. If the charge moves under the influence of this force, the electric field does work on it.

We have seen that an electric field is described by associating a vector, $\vec{E}$, with each point in space, as in Fig. 2-25. It is also very useful to associate a *number* called the *electric potential* (e.g., V_A) with each point in space in such a way that if a charge q were transported from any point A to any other point B, and the electric force did a work W on the charge during this trip, then the *difference of potential* $V_A - V_B$, between these two points is

$$V_A - V_B = \frac{W}{q}. \qquad (11.1)$$

Eq. (11.1) defines* the *difference of potential* between any two points of space (see Fig. 11-1). This potential difference ($V_A - V_B$) is frequently written as V_{AB} or even simply as V, the given two points A and B being understood. Thus one sees this equation sometimes written simply as

$$V = \frac{W}{q}. \qquad (11.2)$$

Courtesy of Burndy Library

Alessandro Volta
(1745–1827)

*It is assumed that W is independent of the path taken from A to B. This is always true when the charges producing the field are at rest.

As for the actual potential at a point in space, say, for example, V_A, its value depends on what point is *arbitrarily* assigned the value of zero potential. In the case of electrical circuits, the point in the circuit which is grounded (i.e., connected to the earth; see Section 11.6) is assigned zero potential.

The unit for potential at a point in space or for the potential difference between two points is, according to Eq. (11.1), the joule/coulomb, which is named the *volt*.

As an example of these ideas, a tube in a TV set might have at one point A a potential of 110 volts and at another point B a potential of -10 volts; that is, $V_A = 110$ volts and $V_B = -10$ volts. The difference of potential between A and B is

$$V_A - V_B = 110 \text{ volts} - (-10 \text{ volts}) = 120 \text{ volts}.$$

When we know the difference of potential between two points in space (such as points A and B in Fig. 11-1) we can calculate the work done by the electric forces on a charge q that moves from one point to the other, by solving Eq. (11.2) explicitly for W:

$$W = q(V_A - V_B). \tag{11.3}$$

Thus, in a radio or TV tube such as mentioned in the preceding paragraph, if a proton (charge $= +1.60 \times 10^{-19}$ coul) were transported from point A to B, the electric forces would do the work

$$W = (1.60 \times 10^{-19} \text{ coul})(120 \text{ volts})$$
$$= 192 \times 10^{-19} \text{ joule}.$$

The definition of potential difference given by Eq. (11.1) implies that *a positive charge is pushed by the electric force from a region of higher to lower potential*. A negative charge is, of course, pushed in the opposite direction.

Electrical Potential Energy

The reader should now note that we associate an electric potential with a point in space, whether or not there is any electric charge at that point. However, if a charge q is placed at a certain point A where the potential is V_A, then this charge q possesses an electric *potential energy*, U_A, which can be taken to be qV_A. To justify this we note that if the charge moves from A to B the change in potential energy, $qV_A - qV_B$, is, according to

$$W = q(V_A - V_B), \tag{11.3}$$

equal to the work W done by the electric forces. Thus we see that the work done in moving a charge q in an electric field can be associated with changes in its potential energy. We say that the charge's *potential energy*, U, (when this charge is located at a point where the potential is V) is given by

$$U = qV. \tag{11.4}$$

Example 1 In Fig. 11-1, a proton is accelerated by the electric field from the lower plate to the upper plate. Point A is at a potential of 110 volts and point B is at -10.0 volts.
(a) Evaluate the proton's potential energy at point A and then at point B.
(b) By how much does the proton's kinetic energy (K) increase as it moves from A to B?

Solution
(a) From Eq. (11.4), since the proton's charge is 1.60×10^{-19} coul

$$U_A = (1.60 \times 10^{-19} \text{ coul}) \times (110 \text{ volts})$$
$$= 17.6 \times 10^{-18} \text{ joule,}$$
$$U_B = (1.60 \times 10^{-19} \text{ coul}) \times (-10.0 \text{ volts})$$
$$= -1.60 \times 10^{-18} \text{ joule.}$$

(b) Because the proton is accelerated through a *vacuum*, it does not lose energy in collisions with gas molecules and so its total mechanical energy is conserved:

$$K_A + U_A = K_B + U_B. \qquad (4.15)$$

Therefore, the kinetic energy increase is given by

$$K_B - K_A = U_A - U_B$$
$$= (17.6 \times 10^{-18} \text{ joule}) - (-1.60 \times 10^{-18} \text{ joule})$$
$$= 19.2 \times 10^{-18} \text{ joule.}$$

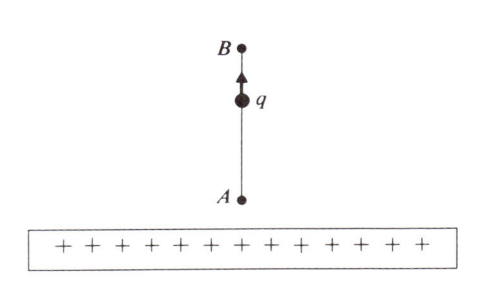

Fig. 11-1 The difference of potential, $V_A - V_B$, is related to work, W, done by the electric field when a charge q is moved from A to B, by the expression $W = q(V_A - V_B)$.

The Electron Volt

We have often used the electron volt ($1 \text{ eV} = 1.60 \times 10^{-19}$ joule) in earlier chapters. The utility of the electron volt as an energy unit is now readily made apparent in situations like that of Example 1.

Example 2 Repeat the calculations of Example 1, evaluating all energies in electron volts.

Solution

(a) $U_A = \dfrac{(1.60 \times 10^{-19} \times 110 \text{ joule})}{(1.60 \times 10^{-19} \text{ joule/eV})}$
$= 110 \text{ eV.}$

Similarly,

$U_B = -10 \text{ eV.}$

(b) $K_B - K_A = U_A - U_B$
$= (110 \text{ eV}) - (-10 \text{ eV})$
$= 120 \text{ eV.}$

We have found that a particle of charge e ($=1.60 \times 10^{-19}$ coul), when accelerated through a difference of potential of 120 volts, has its kinetic energy increased by 120 electron volts.

11.1 ELECTRIC POTENTIAL

The results of Example 2 clearly can be generalized to the statement that for a particle of charge $q = ne$, the *potential energy* in electron volts is just n times the potential in volts. Furthermore, when total mechanical energy is conserved, the increase in kinetic energy evaluated in electron volts, of a particle with charge $q = ne$, is just n times the potential difference (in volts) through which the particle is accelerated.

Example 3 In Fig. 11-1, the potential of the lower plate is set at 3.0 volts and that of the upper plate is 0.0 volt. An electron is ejected with a kinetic energy K from the lower plate and travels straight up. It is slowed down by the electric field and stops just short of the upper plate. Find the electron's initial kinetic energy, K.

Solution At the lower plate the electron has a potential energy

$$U = \frac{(-1.60 \times 10^{-19} \text{ coul}) \times (3.0 \text{ volts})}{(1.60 \times 10^{-19} \text{ joule/eV})}$$
$$= -3.0 \text{ eV}.$$

At the upper plate the electron's potential energy is zero and, since the electron has stopped, its kinetic energy is also zero. Because the total mechanical energy of the electron is conserved for this motion, we can equate the sum of the electron's kinetic and potential energies near the lower plate to this sum near the upper plate:

$$K + (-3.0 \text{ eV}) = 0 + 0.$$

Therefore, $K = 3.0$ eV.

Potential Difference and Electric Field

In a region where there is a *uniform* electric field $\vec{E}$ directed from a point A at potential V_A toward a point B at potential V_B, as in Fig. 11-1, the electric field equals the potential difference between the points divided by the distance between them.* To prove this we note that the work W done by the electric force $\vec{F} = q\vec{E}$ [Eq. (2.18)] on a charge q which is transported a distance (L) from A to B is the product of the force (qE) and the distance (L):

$$W = qEL.$$

Also, from Eq. (11.3) with $V = V_A - V_B$,

$$W = qV.$$

Therefore, equating the above two expressions for W, we get

$$qV = qEL$$

which implies

$$V = EL, \qquad (11.5)$$

and

$$E = V/L. \qquad (11.6)$$

---

*In the most general case, the electric field at a point A has a component E_s in the direction from point A to a nearby point B which is given by

$$E_s = -\Delta V/\Delta s,$$

where $\Delta V = V_B - V_A$ and Δs is the distance from A to B. It is assumed that Δs is small enough so that E_s does not change appreciably along the line AB.

Equation (11.6) brings out the fact that the mks unit for the electric field can be chosen to be the volt/meter. This unit, in fact, is much more commonly employed than the unit, N/coul, introduced in Chapter 2. In any case,

$$1 \text{ volt/meter} = 1 \text{ N/coul}.$$

Example 4 There is a potential difference of 120 volts between the points A and B (Fig. 11-1) which are 5.0×10^{-2} m apart. Point A is at a higher potential than B. Find the magnitude and direction of the electric field ($\vec{E}$) in the region between these points.

Solution The field is uniform so Eq. (11.6) can be applied.

$$E = V/L$$
$$E = \frac{(120 \text{ volts})}{(5.0 \times 10^{-2} \text{ m})}$$
$$= 24 \times 10^2 \text{ volt/m}.$$

The field is directed from higher to lower potential; that is from the point A toward B.

11.2 CURRENT ELECTRICITY

The term *current electricity* refers to a continuous transfer or drift of electric charge through a conductor. By *direct current*, we mean that the average motion of the electric charges is always in the same direction.

Mechanism of Charge Transfer

What is the mechanism involved in an electric current? What, if anything, undergoes motion?

Contrary to the arbitrary assumptions made in Benjamin Franklin's day, it is not the positive electrical charges which move in a solid conductor (such as metal), but rather the negative charges, that is, the electrons. An ingenious experiment known as the Hall effect (E. H. Hall, 1879) conclusively demonstrates that in metallic conductors only the electrons are free to move. Positive charge in a piece of copper, for example, is as immobile as it is in glass or any other insulating material.

Now it should not be construed that in a metal wire all the electrons pass along in a nice orderly fashion from one end to the other, depending on the direction of the electric field established in the wire. Rather, in the absence of an electric field, some electrons, the so-called free electrons, undergo a random motion through the metal. However, as soon as an *electric field is applied*, say from one end of the wire to the other, although the electrons do not give up their random motion, there is nevertheless a drift, or average motion, through the wire in a direction specified by the *applied electric field*.

The speed of this drift is surprisingly low. For example, even in a wire carrying a quite reasonable current, the drift speed may be so low that it takes some 30 sec for the conduction electrons to drift a distance of 1 cm. It is important at this point to emphasize that the drift speed of the electrons is not the speed with which electric field disturbances pass along the wire. This latter speed is close to the speed of light in a vacuum. And it is for this reason that one obtains almost instantaneous response when one switches on and off electric appliances.

Although in metallic conductors the mechanism of charge transfer involves only electrons, it is quite true that in certain cases positive charges may also move. Such is the case in electric currents through electrolytes (such as a salt-water solution) and through ionized gases such as in mercury vapor lamps, for example. A difference between conducting solutions and gases is that in conducting solutions there are no free electrons present, as there are in ionized gases. In solutions, negative charge carriers, like the positive charge carriers, are all in the form of ions.

Further, when we consider materials called *semiconductors*, which are in the limbo between conductors and nonconductors, we find that to describe the charge transfer it is often convenient to refer to the motion of quasi-positive charges, that is, holes (places in the material where electrons are lacking).

Conventional Current Direction

Since a positive charge moving in one direction is, in almost all situations, equivalent to a negative charge moving in the opposite direction, one may continue with the Franklin convention of assuming all charge carriers to be positive, the Hall effect notwithstanding. Consequently, we shall always draw current arrows in the direction a positive charge would move were it the charge carrier. (See Fig. 11-2 and Fig. 11-3.) This direction is commonly called the *conventional current direction* and is

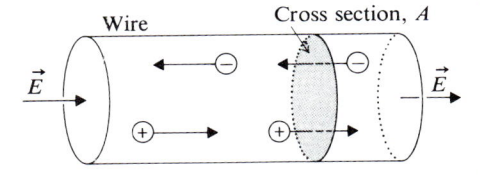

Fig. 11-2 Negative charges drifting to the left are equivalent to positive charges drifting to the right, in the direction of the electric field $\vec{E}$. Conventional current direction is that of the drift of positive charges, that is, in the direction of $\vec{E}$.

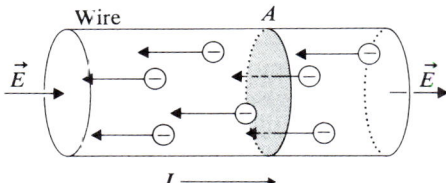

Fig. 11-3 Electric current. The charge carried by the electrons drifting through the cross section, A, of the wire per unit time is called the *current*. Note that I, indicating the current, is in the conventional direction, as though positive charges were moving.

used almost exclusively in the world of physics and electrical engineering, although for a time there was a move afoot to introduce electron flow as a standard convention.

Current and Charge

We now state in mathematical terms what is meant by electric current:

$$I = \frac{q}{t}, \qquad (11.7)$$

where I is the current, q is the electric charge, and t is the time involved in the drifting of the charge, q, through any given cross section of the conductor (see Fig. 11-3).

Eq. (11.7) may also serve as a definition of the unit of charge, the coulomb, for we shall later (Chapter 12) independently define the unit of current, the ampere:

$$1 \text{ coulomb} = (1 \text{ ampere}) \times (1 \text{ second}).$$

By a current of one ampere we mean that a charge amounting to one coulomb drifts past some point in an electric circuit (that is, through a given cross section of a conductor) in one second. For very small currents one uses the milliampere (mA) or microampere (μA) as current units (1 mA = 10^{-3} amp; 1 μA = 10^{-6} amp). In keeping with these terms for current, the instrument for measuring current is called an *ammeter;* for small currents one uses a *milliammeter* (mA meter), and for still smaller currents a *microammeter*. And, as we shall indicate in Chapter 12, a more sensitive instrument of this type, the instrument used to detect very small currents, is called a *galvanometer*.

Example 5 Suppose in a case such as shown in Fig. 11.3 a charge of 20 coulombs drifts through the cross section, A, in 2.0 sec. What is the current?

Solution

$$I = \frac{q}{t} = \frac{20 \text{ coul}}{2.0 \text{ sec}} = 10 \text{ amperes.}$$

Example 6 How many electrons would pass through this cross section in 1.0 sec to produce this current of 10 amp?

Solution

$$q = (10 \text{ amp}) \times (1.0 \text{ sec})$$
$$= 10 \text{ coul,}$$

but the charge on one electron is about 1.6×10^{-19} coul, hence:

$$\text{number of electrons} = \frac{10 \text{ coul}}{1.6 \times 10^{-19} \text{ coul/electron}}$$

$$= 6.3 \times 10^{19} \text{ electrons.}$$

Conductors, Semiconductors, and Insulators

Our previous discussion of the mechanism of charge transfer has dealt with the concept of electrical conduction. The best and most important electrical conductors are metals. It is characteristic of metals to contain free electrons, free to move through the crystal, not bound to a particular nucleus. For example, each neutral copper atom has 29 electrons and within a copper wire two of these electrons are free to move.

In the case of insulators (or dielectrics), on the other hand, there are no such free electrons. This is not to say that all of the electrons in an insulator are entirely bound to their own atoms, but rather that the motion they may perform is one of simply exchanging places. Such motion, of course, is of no merit in the conduction of a current.

None of the above is to be construed as implying that there is some magic cut-off point which separates conductors from insulators. Rather, materials fall in line in a sort of conduction spectrum from very good conductors to very good insulators; but there is no such thing as a perfect conductor* (one which offers no opposition to current) nor a perfect insulator. Moreover, if the voltage across any dielectric material (i.e., insulator) is made large enough, the dielectric will break down, and this material will now conduct. The situation is like that of a spark in the air; the air is a fairly good insulator but at high enough voltages we get a spark. It is this fact which has posed many serious insulating problems in the development of the high-voltage equipment associated with the transmission of electric power.

The difference in electrical conducting properties of materials, that is, the extent of the spectrum mentioned above, is tremendous. A look ahead at Table 11-1 will reveal that the electrical conductive property of a good conductor such as copper is 10^{23} times that of a good insulator such as hard rubber.

Present-day electrical technology is becoming increasingly more dependent on a group of materials in the middle of the conduction (or resistance) spectrum. These materials are the semiconductors which are used in the manufacture of transistors which have, in a large measure, replaced the vacuum radio tubes in a multitude of electronic devices.

Some semiconductors are elements, such as germanium and selenium; others are the oxides and sulfides of certain metals, for instance Cu_2O and PbS.

*Superconductivity, achieved only at low temperatures, is mentioned in Chapter 9.

That certain materials should behave like conductors, others like semiconductors, and still others like insulators, may be explained in quantum mechanical terms by what is called the band theory of conduction.

11.3 THE COMPLETE CIRCUIT

To obtain some understanding of the multitudinous electrical applications which have had unquestionable and unprecedented impact on the whole of society, one needs a very basic concept, that of the electric circuit. Implicit in the term *circuit* is the notion that there is a complete, closed path for the electric charges in order that they may drift through the circuit continuously, like a runner around a looped racetrack.

Circuit Symbolism

In order to facilitate an understanding of electrical circuits, certain conventional symbols are used. Reference to Fig. 11-4 reveals those symbols that are pertinent at the moment.

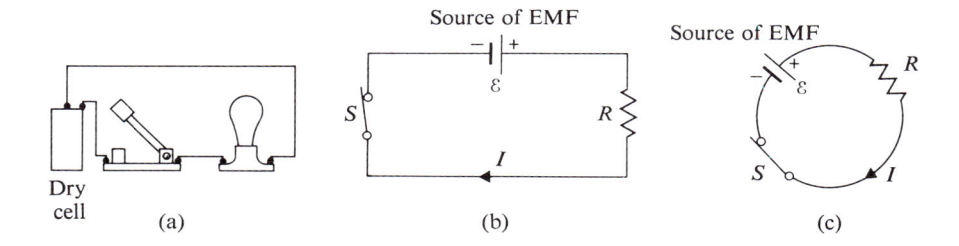

Fig. 11-4 Complete circuit showing a resistor which has resistance R; a source of electrical energy, that is the source of EMF, which has an EMF ε; and a switch S. The direction of the current, I, is indicated by the arrow on the connecting wire. Diagram (b) is electrically equivalent to diagram (c); both are representative of the pictorial arrangement (a).

Note that the conventional way is to draw circuits in more or less rectangular configurations, as in Fig. 11-4(b). But there is no electrical significance to this. The circuit of Fig. 11-4(c) also has a switch, a resistor, and a source of EMF connected in series; it is entirely equivalent to the circuit of Fig. 11-4(b).

By a *resistor* one actually means a specific circuit component designed to offer a certain resistance or opposition to charge flow, that is, current. However, the symbol employed for resistors as shown in Fig. 11-4 is also employed in a more general sense to denote any circuit resistance. For example, the same symbol might be used to indicate the presence of a light bulb, which offers a resistance to current. Sometimes one may also see this symbol next to a source of EMF to indicate the internal resistance of the source. Such a symbol may even be employed to represent the resistance of connecting wires if this resistance is not negligible. In short, the symbol for a resistor is employed any time one wants to denote resistance to current. However, unless otherwise stated, the connecting wires represented by the single connecting lines in a circuit diagram are assumed to offer no resistance to the flow of electric charge.

It must be emphasized that the current I is the same throughout any "one-loop" circuit, like that displayed in Fig. 11-4. This is so because we are considering a *steady state* with no electric charge piling up in any region, and charge is conserved. Therefore the current leaving the resistor is precisely the same as the current entering the resistor, and this is also the current through the dry cell. Hence it suffices to speak of *the* current, I, in the circuit.

Source of EMF

We now concern ourselves with the electrical component called the *source of EMF* in Fig. 11-4. We attack this concept by returning to Fig. 11-1 for the moment.

If the two plates in Fig. 11-1 were joined by a piece of copper wire, electrons would migrate from the upper plate to the lower one. One would say that in this wire there is an electric current (conventional) from the lower plate to the upper one. The reason this current is possible is that work was first done in charging the plates. As soon as the charge from one plate has neutralized the charge on the other plate, the current stops. To keep the current going, we could remove electrons from the lower plate as fast as they arrive and carry them to the upper plate. To do this job we would have to exert some nonelectrical force to move these charges against the forces exerted by the electric field between the plates. This would require work. We would act like a source of EMF, a converter of some other form of energy into the electrical energy which is required to keep a steady flow of charge going, that is, produce a continuous current. In a way, a source of EMF is a sort of "charge pump" which can move charges against the force exerted by the electric field within the source.

In keeping with the above descriptive aspects, we define *a source of EMF* as a device which can change other forms of energy (such as mechanical energy or chemical energy) into electrical energy.

The source of EMF in a flashlight is a dry cell in which chemical energy is converted into electrical energy. A similar energy conversion occurs in a car's storage battery while it is being "discharged." On the other hand, the source of EMF providing the electrical energy for your home lighting may convert mechanical energy into electrical energy.

This energy conversion can be achieved by water falling and thus turning electric generators.

We now indulge in a short aside on the term EMF. As many students will probably be aware, EMF is an abbreviation for "electromotive force," a most misleading term, since EMF refers to a quantity related to energy and *not* to force. It is for this reason we shall, as is common practice, refer only to "EMF," represented by the symbol $\mathcal{E}$, and not to the misnomer "electromotive force."

Now consider a source of EMF like a storage battery. There is a low-potential terminal indicated by a minus sign and a high-potential terminal marked with a plus sign. The difference of potential between these terminals, the so-called *terminal voltage*, we indicate by V_t. When the battery is used to establish a current I in a circuit like that of Fig. 11-4, positive charges are forced by the electric field within the wire from the high-potential terminal through the resistor to the low-potential terminal.* Within the source of EMF, positive charges are pushed by nonelectrostatic forces against the electrostatic forces* from the low-potential terminal to the high-potential terminal. Suppose a work W is done by these nonelectrical forces within the source of EMF on a positive charge q which is transported through the source from the low-potential terminal to the high. The quotient W/q is a constant, characteristic of the source of EMF, and is named *the* EMF, $\mathcal{E}$, of the source. That is,

$$\mathcal{E} = W/q. \tag{11.8}$$

*If the conduction is by electrons, as it is in metallic conductors, then one needs to remember that the electrons are pushed in the opposite direction, that is, from low potential to high potential.

*Electrostatic forces are the (Coulomb) forces exerted by charges at rest.

The mks unit of $\mathcal{E}$ is seen to be the joule per coulomb which we have already named the *volt*. The EMF of your flashlight dry cell is probably 1.5 volts, and of your car's storage battery 6 or perhaps 12 volts.

The terminal voltage, V_t, of a source of EMF is closely related to the EMF, $\mathcal{E}$, of the source. If all the work done by the nonelectrical forces is converted into electrical potential energy, then $V_t = \mathcal{E}$. But when the source of EMF is used to establish a large current, an appreciable amount of energy may be dissipated as heat within the source. In this case, the terminal voltage, V_t, drops below the value, $\mathcal{E}$, as we shall show in Section 11.6.

An Analogy

We now look at an analogy with the intent of clarifying some of the key electrical concepts associated with an electric circuit such as that shown in Fig. 11-4. The analogy, like any analogy, is bound to be somewhat imperfect. However, shortcomings notwithstanding, some insight into electrical affairs may be gained by a perusal of this mud and marble analogy.

Schematic representation

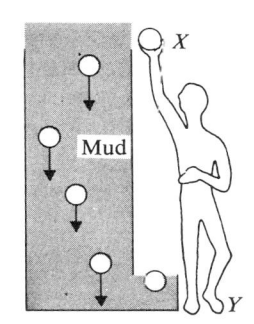

Fig. 11-5 Marble-lifter system: mud, marbles, and marble lifter.

The situation
1. The marble lifter lifts the marbles up and drops them into the tube containing thin mud, at point X (see Fig. 11-5).
2. The marbles ooze down through the mud from point X to point Y, giving up their energy to the mud (by heating it).
3. The marbles are lifted up again.

Particle involved
Marble

Quantitative unit of particles
Mass: A number of marbles have a total mass equal to some mass unit; for example, one kilogram.

Current
Current refers to the number of mass units passing through a given cross section of the circuit per unit time.

Unit of current
kilogram/second

Energy supply
The marble lifter supplies the energy. He does a certain amount of work per unit mass in lifting the marbles from Y to X (see Fig. 11-5).

We concern ourselves not with the total work done by the marble lifter, but rather with the work per unit mass. The unit would thus be joule/kilogram.

Potential difference
The work per unit mass the marbles can do at X minus that they can do at Y may be called the difference in gravitational potential between points X and Y.

Resistance
If the mud is thin, the resistance is small, the marbles ooze down rapidly, and consequently the rate of motion of the marbles through the cycle is fast. Many marbles—that is, many mass units—pass through the mud per unit time.

Conversely, few mass units per unit time will go through the cycle if the mud is thick, that is, if the resistance is large.

(To keep this analogy more or less correct, we must assume that the number of marbles in the circuit is fixed *and* that the marble lifter lifts each marble immediately upon its arrival at the bottom.)

Schematic representation

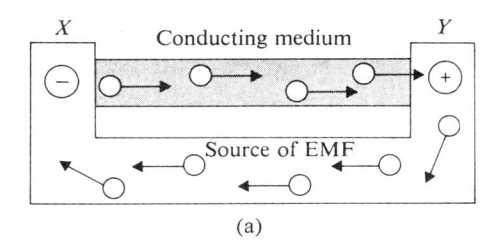

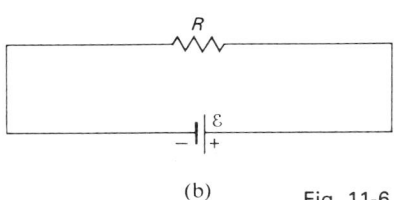

Fig. 11-6 Conducting medium, electrons, and source of EMF. (a) Electrical model corresponding to marble-lifter system. (b) Electrical circuit symbolism for (a), showing R, the resistance of the conducting medium, and ε, the EMF.

1. The source of EMF forces electrons from positive terminal in source to negative terminal at X (see Fig. 11-6).
2. The electrons drift through the wire from the negative terminal X to the positive terminal Y, gaining energy from the electric field set up by the source and losing it to the conducting medium, that is, the wire (by heating it).
3. The electrons are again moved from the positive terminal to the negative terminal inside the source of EMF.

The situation

Electron

Particle involved

Charge: A number of electrons have a total charge equal to some charge unit; for example, one coulomb.

Quantitative unit of particles

Current is defined as the number of charge units passing through a given cross section of the circuit per unit time. If I is the symbol for current, q for charge, and t for time, then $I = q/t$.

Current

(coulomb/sec) = ampere

Unit of current

The electrical energy is supplied by the source of EMF, such as a dry cell or a generator, for example. The source of EMF does a characteristic work per unit charge in transporting the electrons, against the electrostatic forces, from the positive terminal Y to the negative terminal X inside the source of EMF (see Fig. 11-6).

Energy supply

The work per unit charge done by the source is called the EMF, which is symbolized by ε. The unit is (joule/coulomb) = volt.

The work per unit charge the electrons can do when at X minus that they can do when at Y defines the electrical potential difference between points X and Y. The symbol for potential difference is V; and since it represents work per unit charge, V is expressed in volts.

Potential difference

If the electrical resistance is low, that is, if the loss of energy to the conducting medium is low, the electrons will pass through it readily. Consequently with a given energy source many charge units pass through the circuit per unit time; that is, the current is large. Conversely, the electrical current will be small if the electrical resistance is large.

Resistance

Resistance

The last electrical term appearing in the preceding analogy warrants further consideration. We have already discussed conductors and insulators and inferred that conductors have a low electrical resistance and insulators a very high one.

We define resistance as follows:

$$R = \frac{V}{I}, \tag{11.9}$$

where R is the resistance of some circuit element to the flow of charge, V is the potential difference across the circuit element, and I is the current through it (see Fig. 11-7). The name *resistance* for the quantity we have just defined seems appropriate because with a given potential difference V across a circuit element, a small resistance R implies that the current I is large.

Eq. (11.9) also serves to define the unit of resistance, the ohm, in this way:

$$1 \text{ ohm} = \frac{1 \text{ volt}}{1 \text{ ampere}}.$$

Thus, if a potential difference of 1 volt across the resistor in Fig. 11-7 results in a current of 1 ampere, the resistor is said to have a resistance of 1 ohm. (The Greek letter omega, Ω, is generally used as the symbol for ohm.)

The potential difference V between the ends of a resistor R carrying current I, according to Eq. (11.9), is given by

$$V = IR.$$

The current is directed from the end at the higher potential toward the end at the lower potential. In other words, as we traverse a resistor in the *direction of the current*, the potential always drops, and the amount of the potential drop is given by $V = IR$. This change in potential therefore is called an *IR-drop*.

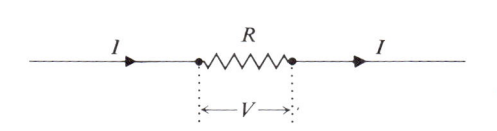

Fig. 11-7 The resistance, R, of a circuit element is defined in terms of the potential difference, V, across it and the current, I, through it.

Example 7 Suppose the current through the resistor of Fig. 11-7 is 4.0 amp and the resistor has a resistance of 2.0 Ω. What is the *IR*-drop across the resistor?

Solution Using the expression for the *IR*-drop, we have

$$\begin{aligned}V &= IR \\ &= (4.0 \text{ amp}) \times (2.0 \text{ ohms}) \\ &= 8.0 \text{ volts.}\end{aligned}$$

This is the voltage across the resistor. The potential at the left end of the resistor is 8.0 volts above that at the right end.

Notice that the IR-drop equation employed above derives simply from the definition of resistance. Nothing in this equation at this stage, for example, tells us that if we increase the voltage across the resistor, the current will increase correspondingly. It may well be that, depending on the nature of the resistor, increasing the voltage across it will increase the value of the resistance and thus perhaps result in a negligible change in current. This fact, that R may vary as either V or I are changed, must be contrasted with the "restricted" version of this expression we shall identify as Ohm's law in Section 11.4.

Resistivity

The resistance of a wire depends on the nature of the material as well as on the length L and the cross-sectional area A (see Fig. 11-8). To single out the dependence on the nature of the material, we define the resistivity (ρ) of the material by

$$\rho = \frac{E}{(I/A)},$$

where E is the electric field within the wire at any point and I/A is the current at that point divided by the cross section through which the current passes.

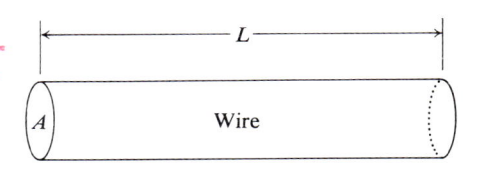

Fig. 11-8 The resistance of a wire is given by $R = \rho L/A$.

The resistance R of a wire can be related to its resistivity, length, and cross section. First we shall recall that the voltage V between the ends of the wire is related to the uniform field E within the wire by

$$E = V/L. \qquad (11.6)$$

Therefore

$$\rho = \frac{E}{(I/A)} = \frac{V/L}{I/A}.$$

Next, since (V/I) is the resistance R, the above equation leads to

$$R = \rho \frac{L}{A}. \qquad (11.10)$$

For some materials, the value of the resistance R of a specimen depends on the value of the current I through it. The value of the resistivity ρ of such a material depends upon the value of I/A. However, metals and some other materials have the same value of the resistivity ρ (and hence the same value of R) no matter what the value of I/A. Such materials are said to obey Ohm's law, the topic of the next section.

Perusal of Eq. (11.10) shows that if L is in meters, A is in m², and R is in ohms, then the unit for ρ must be the ohm-meter. Table 11-1 on the following page gives various values of ρ.

Table 11-1. APPROXIMATE RESISTIVITIES OF CERTAIN MATERIALS AT ROOM TEMPERATURE (20°C)

Material	Resistivity, ohm-meter
Silver	1.6×10^{-8}
Copper	1.7×10^{-8}
Aluminum	2.7×10^{-8}
Tungsten	5.6×10^{-8}
Lead	2.1×10^{-7}
Constantan (Ni and Cu alloy)	4.5×10^{-7}
Carbon	3.5×10^{-5}
Salt water (sea water)	$2. \times 10^{-1}$
Germanium	$5. \times 10^{-1}$
Copper oxide (CuO)	$1. \times 10^{3}$
Distilled water	$5. \times 10^{3}$
Glass (varies widely with type of glass)	$1. \times 10^{12}$
Transformer oil	$2. \times 10^{14}$
Rubber (varies widely with type of rubber)	$1. \times 10^{15}$

11.4 OHM'S LAW

Ohm's Discovery

Ohm discovered that the resistance $R(=V/I)$ of a metallic conductor at a given temperature is *the same* no matter what the value of the difference of potential, V, between the terminals of the conductor (or the current through the conductor). This constancy of the resistance R is known as *Ohm's law*.

Hence, with the strict proviso that R be a constant, not determined by either V or I, we can state *Ohm's law* in the form

$$V = IR. \tag{11.11}$$

The reader is now asked to recall our earlier comments regarding this equation because as presently cited it is a vastly more powerful tool. For if R is constant, then, using this equation, one can immediately calculate what happens to the current through a resistor as the voltage across it is changed or vice versa.

Ohmic Circuit Elements

If we were to make voltage and current measurements involving a metallic resistor, what would the results yield? Consider a circuit as in Fig. 11-9(a). We note that as the voltage across the resistor is increased, the current through this resistor increases proportionally. Ideally, when the voltage is doubled the current will be doubled. Plotting the readings of the voltmeter on the abscissa and the corresponding ammeter readings on the ordinate results in a straight line, as in Fig. 11-9(b). The fact that

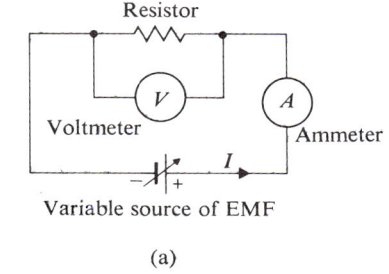

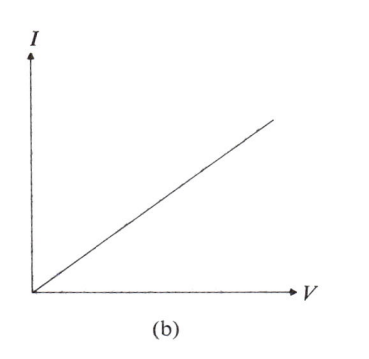

Fig. 11-9 Verification of Ohm's law. (a) Circuit for measuring voltage-current relationship. (b) Graphical representation of the results of voltage-current measurements made in a circuit such as (a).

this line is straight and passes through the origin shows at a glance that the ratio V/I is constant, that is, R is constant. The resistor in question is therefore said to be an *ohmic* circuit element; it obeys Ohm's law. Since this information is implied by the straight line on the graph, ohmic circuit elements are commonly also called *linear* circuit elements. In our present "electrical culture," probably the most common ohmic circuit element is the copper wire used in all home, office, and industrial electrical installations.

Finally we note that circuit elements which are less than perfectly ohmic are frequently treated as if they were, because the results obtained by using Ohm's law in such cases are close enough to the truth for many practical applications.

Nonlinear Circuit Elements

Many conductors do not follow Ohm's law. This means that their resistances are not constant, but vary noticeably with the applied voltage. A voltage-current graph of such devices does not yield a straight line as that of Fig. 11-9(b). Rather, the graph is a curve of some kind depending on the particular circuit element concerned. For this reason, such conductors are called nonlinear circuit elements.

Fig. 11-10 shows examples of the voltage-current curves of two nonlinear or nonohmic devices. Fig. 11-10(a) refers to a vacuum tube such as a radio tube and (b) is the case of a rectifier, a device which readily passes current in one direction but has a high resistance to current in the opposite direction. (For metallic or ohmic conductors, changing the polarity of the applied voltage simply changes the current's direction but not its magnitude.)

Fig. 11-10 Nonlinear circuit elements. (a) Vacuum tube. (b) Selenium rectifier.

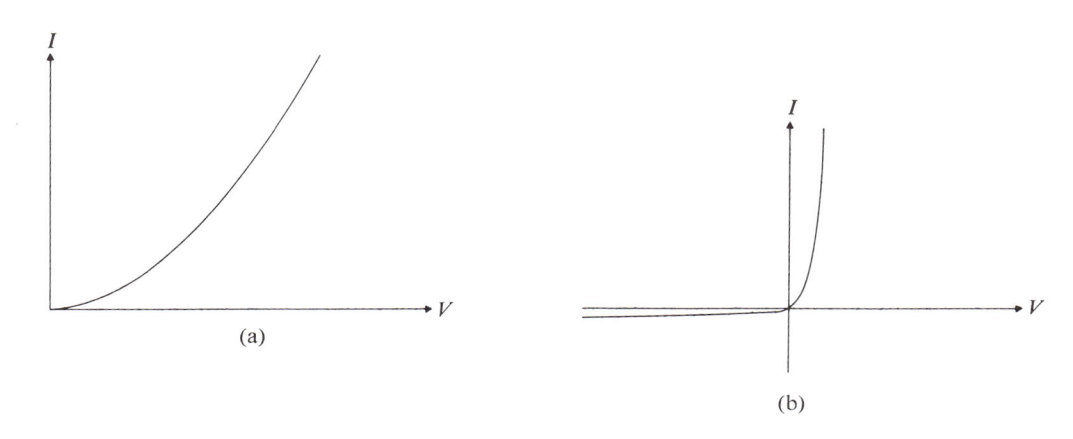

11.4 OHM'S LAW

11.5 POWER AND ENERGY IN ELECTRIC CIRCUITS

Energy

Consider any portion AB of a circuit across which there is a potential difference $V_A - V_B = V_{AB}$, and through which there is a current I directed from A to B. In a time interval t, a quantity of charge $q = It$ enters terminal A at potential V_A while the same amount of charge leaves terminal B at potential V_B. The charge passing through this portion of the circuit therefore experiences a change of potential energy given by

$$W = qV_A - qV_B = (V_A - V_B)It = V_{AB}\,It. \quad (11.12)$$

If V_A is greater than V_B, the charge loses potential energy in passing through AB. Therefore, there is an energy *input* to this portion of the circuit. If, on the other hand, V_A is less than V_B, then the circulating charge gains potential energy in AB. This implies that there is an energy *output* from this portion of the circuit.

Power

The corresponding *power* P is the *rate*, W/t, at which the energy W is transferred. Eq. (11.12) therefore yields

$$P = V_{AB}\,I \quad (11.13)$$

as the general expression for the electrical power input or output to the portion of the circuit between A and B.

In Eq. (11.13) if V is in volts and I in amperes, then P is in watts:

$$(\text{volt}) \times (\text{amp}) = \frac{\text{joule}}{\text{coul}} \times \frac{\text{coul}}{\text{sec}} = \frac{\text{joul}}{\text{sec}} = \text{watt}.$$

Example 8 How much power is supplied to an electric heater across which there is a potential difference of 100 volts, and through which the current is 15 amp?

Solution

*kW stands for kilowatt.

$$P = VI = (100 \text{ volts})(15 \text{ amp}) = 1.5 \times 10^3 \text{ watts} = 1.5 \text{ kW}^*.$$

Example 9 What is the energy W supplied to the heater in Example 8 during 1.0 hour of operation?

Solution

$$W = Pt = (1.5 \text{ kW})(1.0 \text{ hr}) = 1.5 \text{ kW-hr},$$

or in mks units,

$$W = (1.50 \times 10^3 \text{ watts})(36 \times 10^2 \text{ sec})$$
$$= 5.4 \times 10^6 \text{ joules}.$$

Joule's Law

We now consider a circuit element that is purely resistive. In this case $V_A - V_B$ is always positive (an IR-drop from A to B), so there is an energy input to the resistor. The potential energy lost by the circulating charges appears as thermal energy within the resistor, and its temperature rises until an outward flow of heat matches the energy input. In this process, called *Joule heating*, we say the electrical potential energy is *dissipated* in the resistor.

From a microscopic point of view, electrons lose potential energy as they drift through the crystals that comprise a metallic resistor. Instead of appearing as increased kinetic energy of the electrons, this energy is transferred to the atoms of the crystal and leads to an increased amplitude of thermal vibrations of the crystal lattice.

For an ohmic circuit element, $V_{AB} = IR$. Then, from Eq. (11.13), the power *dissipated* in a resistor is

$$P = I^2 R, \qquad (11.14)$$

a result known as *Joule's law*.

While in the case of an electrical heater it is desired to change electrical energy into thermal energy, in many other cases this thermal dissipation of electrical energy is most undesirable. For example, power is lost because of the heating of the conducting wires which carry current from the generating station to the home. Since this power loss is given by Eq. (11.14) one refers to this power loss as the *I^2R loss* (or the Joule heating loss). To reduce the I^2R loss, one provides conducting wires with a low R (e.g., copper) and, even more significant, transmits the power at the lowest feasible current.

A glance back at Eq. (11.13), $P = VI$, shows that if the power transmitted to the user is to be kept constant and yet the current is to be low, then the voltage of the transmission system must be high: many thousands of volts. Therein lies the reason for high-voltage transmission lines. And yet in our homes, schools, and offices, etc., largely for reasons of safety, we use power at a relatively low voltage. This implies that there must be a system for stepping up and stepping down voltages. (Further discussion of this matter will be found in following chapters.)

EMF and Power Conversion

When a positive charge q is transported within a source of EMF from the low to the high-potential terminal, the energy that is converted from some other form into electrical potential energy is, from Eq. (11.8),

$$W = \mathcal{E}q.$$

If this charge transport requires a time t, the current I is q/t, and the power (P_c) converted is given by:

$$P_c = W/t = \mathcal{E}q/t = \mathcal{E}I.$$

This important expression for the power converted,

$$P_c = \mathcal{E}I, \tag{11.15}$$

is valid no matter what the direction of the current through the source of EMF.

In normal usage, when the source is used to supply electrical power, the conventional current is directed from low to high potential *within* the source. And energy is converted into electrical potential energy from *chemical energy* if the source is a storage battery or dry cell, or *mechanical energy* if the source is a dynamo or generator (see Chapter 12).

However, if another source of EMF is used to drive a current "backwards" through the source from its high- to low-potential terminal, then energy is converted from electrical potential energy into chemical energy of a storage battery or into mechanical energy of a dynamo. In the latter case we would call the dynamo an electric motor.

Example 10 An automobile storage battery which has an EMF of 6.0 volts establishes a current of 500 amperes in an automobile starter.
(a) What is the electrical power converted within the battery?
(b) What energy is converted from chemical energy into electrical energy within the battery in 100 seconds?

Solution
(a) The power converted into electrical power within the battery is, from Eq. (11.15),
$$P_c = \mathcal{E}I = (6.0 \text{ volts}) \times (500 \text{ amps})$$
$$= 3.0 \times 10^3 \text{ watts}.$$

(b) The energy W converted within the source of EMF in 100 seconds is
$$W = P_c t$$
$$= (3.0 \times 10^3 \text{ watts}) \times (100 \text{ sec})$$
$$= 3.0 \times 10^5 \text{ joules}.$$

11.6 SERIES AND PARALLEL CIRCUITS

We now look at different circuit arrangements and consider different combinations of EMFs and resistances.

A Single-Loop Circuit

The circuit of Fig. 11-4, which is a single loop, is simple to analyze. Since energy is conserved, we can write

$$\begin{bmatrix} \text{Electrical power supplied} \\ \text{by source of EMF} \end{bmatrix} = \begin{bmatrix} \text{power dissipated as} \\ \text{thermal energy in resistor} \end{bmatrix}$$

Using $P_c = \mathcal{E}I$, Eq. (11.15), and $P = I^2R$, Eq. (11.14), the above power equation gives

$$\mathcal{E}I = I^2R.$$

Dividing each side by I, we obtain

$$\mathcal{E} = IR. \tag{11.16}$$

This important result we shall call the *single-loop equation*.

In words: The *EMF* of the loop *equals* the *IR-drop* encountered along the loop. (Recall that the current is directed from higher to lower potential through the resistor. In other words, the current is directed from the positive (+) terminal to the negative (−) terminal in the circuit *external* to the source of EMF, while the current direction is from the negative terminal to the positive terminal *within* the source of EMF.)

Example 11. Suppose a 12-volt battery, having negligible internal resistance (internal resistance is the resistance to current within the source of EMF itself), is connected to a 4.0-ohm resistor by wires of negligible resistance (see Fig. 11-11). Find the current, I, in the circuit.

Solution For this circuit, $\mathcal{E} = 12$ volts and $R = 4.0\ \Omega$. The single-loop equation, $\mathcal{E} = IR$, gives

$$12 \text{ volts} = I \times (4.0 \text{ ohms})$$

Therefore

$$I = \frac{12 \text{ volts}}{4.0 \text{ ohms}}$$

$$= 3.0 \text{ amp.}$$

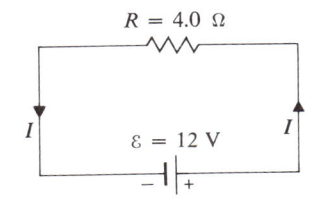

Fig. 11-11 Single-loop circuit. The current, I, is 3 amp.

Certain very common types of circuits can be completely analyzed merely using the single-loop equation, $\mathcal{E} = IR$ [Eq. (11.16)], where $\mathcal{E}$ is the single EMF equivalent to the combination of EMFs in the circuit, and R is the single resistance equivalent to that of the network of resistors of the circuit. In studying such circuits we will encounter important facts about the addition of voltages across different portions of a circuit and the addition of currents at a junction in a circuit.

11.6 SERIES AND PARALLEL CIRCUITS

Series Circuits

Circuit elements are said to be connected in series when they are connected end to end as shown in Fig. 11-12. Each resistor carries the same current, I. The single resistance R, equivalent to this series combination as far as external connections are concerned, is given by

$$R = \frac{V}{I},$$

where V is the voltage across the combination. From the definition of potential difference (see Question 4), it follows that

$$V = V_1 + V_2 + V_3$$

where V_1 is the voltage across R_1, V_2 across R_2, and V_3 across R_3. Dividing both sides of this equation by I (I is the same through each resistor) gives

$$\frac{V}{I} = \frac{V_1}{I} + \frac{V_2}{I} + \frac{V_3}{I}.$$

But $R = V/I$, so

$$R = R_1 + R_2 + R_3. \quad (11.17)$$

Evidently one simply adds resistances in series to find the single resistance equivalent to the series combination.

Application of the law of conservation of energy leads immediately to the fact that the single EMF equivalent to several sources of EMF connected in series [see Fig. 11-12(b)] is the sum of the individual EMFs:

$$\mathcal{E} = \mathcal{E}_1 + \mathcal{E}_2 + \mathcal{E}_3 + \ldots. \quad (11.18)$$

It should be noted that EMFs connected in series normally implies that the positive terminal of one source is connected to the negative terminal of the next one. If two equal sources are connected in series with their like terminals joined, then, of course, one would oppose the other with the net effect that $\mathcal{E} = 0$.

Series circuits are commonplace. In a flashlight the switch is connected in series with the bulb and the source of EMF, which in turn usually consists of a battery of two or more dry cells connected in series.

Another example of a series connection is one involving the current-measuring device, the ammeter. This instrument gives a reading of the current through it and therefore (as shown in Fig. 11-22 in connection with Question 60) must always be connected in series with the circuit element carrying the current one wishes to measure. The ammeter resistance is very low, so the IR-drop across the ammeter is usually negligible compared to other voltages in the circuit. (The ammeter will be discussed when we deal with electrical measuring devices in Chapter 12.)

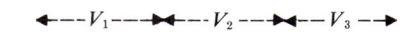

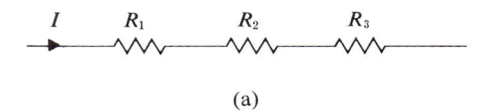

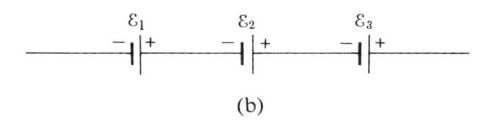

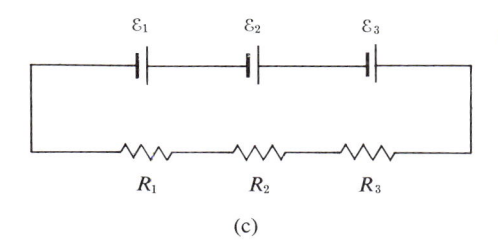

Fig. 11-12 Series connections. (a) Resistors in series: R (total) $= R_1 + R_2 + R_3$. (b) Sources of EMF in series: $\mathcal{E}$ (total) $= \mathcal{E}_1 + \mathcal{E}_2 + \mathcal{E}_3$. (c) Series circuit.

412 ELECTRIC POTENTIAL AND CIRCUITS

Example 12 In Fig. 11-12, the series combination of three sources of EMF is connected across the combination of the three resistors in series [see Fig. 11-12(c)]. Each EMF is 2.0 volts. Each resistor has $R = 4.0\ \Omega$. Find the current.

Solution The single EMF equivalent to the three EMFs in series is given by $\mathcal{E} = \mathcal{E}_1 + \mathcal{E}_2 + \mathcal{E}_3 = 6.0$ volts. The single resistance equivalent to the resistance of the three resistors in series is given by $R = R_1 + R_2 + R_3 = 12.0\ \Omega$. The circuit is therefore equivalent to a single loop with $\mathcal{E} = 6.0$ volts and $R = 12.0\ \Omega$. The single-loop equation, $\mathcal{E} = IR$ [Eq. (11.16)] gives

$$6.0\ \text{volts} = I \times (12.0\ \text{ohms}).$$

Therefore

$$I = 0.50\ \text{amp.}$$

Parallel Circuits

Parallel connections are "side-by-side" connections, as shown in Fig. 11-13. There is the same potential difference V across each of the resistors in parallel. Since charge is conserved and does not pile up at a junction, we must have

$$I = I_1 + I_2,$$

where I_1 is the current through R_1, and I_2 is the current through R_2. Dividing this equation by the common potential difference V gives

$$\frac{I}{V} = \frac{I_1}{V} + \frac{I_2}{V}.$$

Therefore, because $R = V/I$ and hence $1/R = I/V$, we get

$$\frac{1}{R} = \frac{1}{R_1} + \frac{1}{R_2}. \qquad (11.19)$$

The *reciprocal* of the single equivalent resistance R is the sum of the reciprocals of the resistances R_1 and R_2, which are connected in parallel. A simple numerical example will illustrate that the above addition theorem, Eq. (11.19), gives an equivalent resistance R which is less than the resistance of the smallest of the individual resistances.

Fig. 11-13 Parallel connections. (a) Resistors in parallel. (b) Sources of EMF in parallel.

11.6 SERIES AND PARALLEL CIRCUITS

Example 13 Consider just two resistances, 5.0 and 10 ohms respectively, connected in parallel. Find the single resistance equivalent to the parallel combination.

Solution

$$\frac{1}{R} = \frac{1}{R_1} + \frac{1}{R_2}$$

$$= \frac{1}{5.0\,\Omega} + \frac{1}{10\,\Omega} = \frac{3.0}{10\,\Omega}.$$

Therefore,

$$R = \frac{10\,\Omega}{3.0} = 3.3\,\Omega,$$

which is less than 5.0 Ω.

Parallel connected EMFs (positive terminal to positive terminal, and negative to negative) are a different matter: the situation is simple only if the sources have the same EMF and the same internal resistance. The total EMF is then simply the EMF of one source. The total current is the sum of the currents through each source.

Parallel circuitry appears in many places in the home. For example, the various lights in the house must obviously be connected in parallel rather than in series, because in the series connection all the lights would have to be on or off together. One faulty bulb would darken the whole household. No doubt the reader has encountered this problem with certain old-fashioned circuits of Christmas tree lights.

A voltmeter is designed to give a reading of the potential difference between its terminals. Therefore, to measure the voltage across a given circuit element, one terminal of the voltmeter is connected to one end of the circuit element, and the other voltmeter terminal is connected to the other end of the circuit element. That is, the voltmeter is always connected in parallel, across the circuit element across which the voltage is to be measured. (See Fig. 11-22 pertaining to Question 60.) The resistance of a voltmeter is high, so the current through the voltmeter is usually negligible compared to other currents in the circuit.

Short Circuit

With the single-loop equation, $\mathcal{E} = IR$, at our disposal, it is easy to clarify a term quite frequently encountered in everyday life: *short circuit*. A short circuit is a current path which bypasses the load (i.e., the principal resistance) in the circuit. In other words, an electrical circuit is shorted when the terminals of the source of EMF are connected directly, the resistance in the external part of the circuit now being essentially the very small resistance of the connecting wires. From a consideration of the implications of the single-loop equation, $\mathcal{E} = IR$, it now becomes obvious that if the main circuit resistance is bypassed by a very low resistance wire so that the circuit resistance R is drastically reduced while ε, the EMF of the source, remains constant, then the current must increase in a similarly drastic manner. This large current gives rise to considerable heat and hence a pronounced danger of fire. To obviate this fire danger, *fuses* or *circuit breakers* are installed in domestic and industrial electrical circuits. These safety devices open the circuit when the current gets excessively large.

Electric Shock

Another practical application of analysis involving the single-loop equation, $\varepsilon = IR$, which, like the topic of "short circuit," relates to safety, is the matter of electric shock. We wish to capitalize on our familiarity with Eq. (11.16) to clarify what is involved purely from the standpoint of circuit analysis. This bit of analysis will also stand us in good stead when we consider the matter of "grounding" in the next section.

We refer the reader to Fig. 11-14, where in part (a) the actual situation is sketched, and in part (b) the corresponding circuit diagram is outlined.

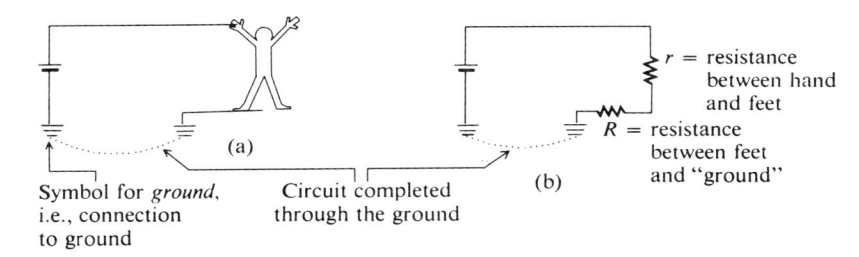

Fig. 11-14 Electric shock situation. (a) The physical situation. (b) The associated circuit diagram.

Without delving into the matter of "grounding" at this point, we simply ask the reader to accept the fact that if the high-potential side as well as the low-potential side of a source of EMF are connected to the ground, then the ground can provide a current path to complete a circuit of the single-loop type. Now suppose the person in Fig. 11-14(a) is touching one terminal of an EMF source, then the single-loop equation and our knowledge of the addition of resistances in series allows us to write

$$I = \frac{\varepsilon}{r + R},$$

where r is the body resistance between the hand touching the voltage terminal and the feet, and R is the resistance between the feet and ground.

If the person is standing on a good insulating surface, such as a dry wooden floor, then R is very large and I is consequently very small. In other words, the person is well protected and will experience the passage of only a very small current, i.e., suffer a very mild electric shock.

If, on the other hand, the person's feet are on a wet floor which is in contact with metal water pipes (which ultimately pass through the ground), as is often the case in a damp basement, then there exists grave danger, for $R \approx 0$, and hence the current through the body is large. One might say the person is providing a short circuit in this case.

Grounding

As implied in the previous paragraphs, in electrical terminology a *ground* is a conductor connected to the earth, the absolute potential of which is arbitrarily taken as zero. A common electrical symbol for a ground is shown in both Fig. 11-14 and Fig. 11-15.

An important aspect of grounding relates to safety. Take the case of an electric washing machine (Fig. 11-16). Should some fault develop in the electric motor wiring (e.g., breakdown of insulation), it is possible that the metal washing machine case could become connected to the electrical supply. And if it becomes connected to the high-potential side of the circuit, then the unsuspecting housewife could, by touching the machine inadvertently, have a high-potential difference V established between different parts of her body. An electric current, $I = V/r$, will then pass through this part of her body which has an electrical resistance r. If the current I is large enough, she will suffer a possibly dangerous electrical shock. On the other hand, if the washing machine is properly grounded, then, when the housewife touches the machine case with her hand, she suffers no passage of electric current, since there exists no potential difference between her hand and other parts of her body which are at ground potential.

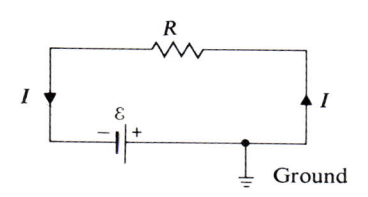

Fig. 11-15 Grounding one side of a circuit.

Fig. 11-16 The use of an independent third wire to ground the frame or casing of an electrical appliance: a safety measure. (a) Normal operation without the safety ground wire. (b) Electric shock: insulation breakdown and casing not properly grounded. Current I through the lady is $I = V/r$, where r is the resistance between her hand and foot. (c) Protection against electrical fault: casing is grounded. Potential difference, V, across the lady is zero, hence current through lady is zero. Note that (as explained with reference to Fig. 11-14) in both cases (b) and (c) the current flows through the ground which provides the completion of the circuit.

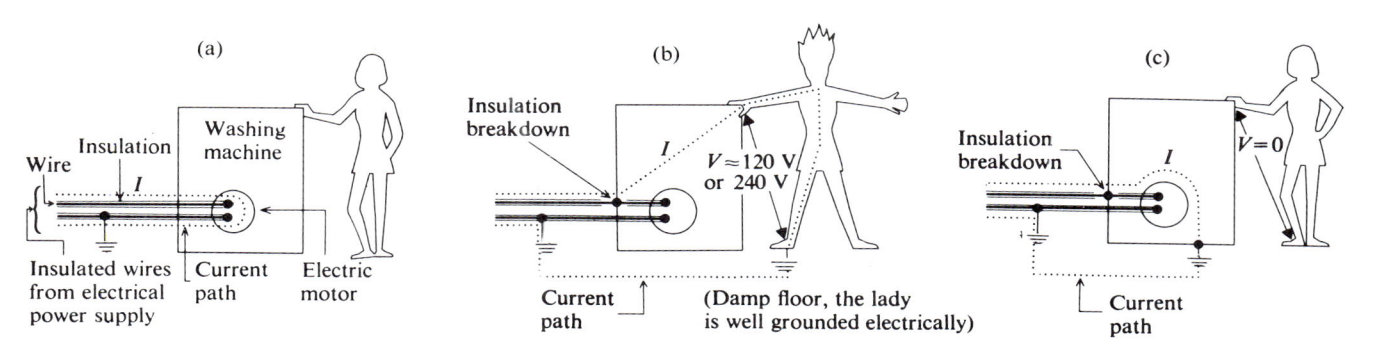

416   ELECTRIC POTENTIAL AND CIRCUITS

It is now common practice to provide a separate ground wire even in ordinary housewiring. This is why one now finds the three-pronged electric plug in common usage. Fig. 11-17 shows how this ground wire is employed to ground the case and plate of an ordinary household electric outlet. Somewhere this ground wire is ultimately connected to either a metal rod driven deep into the ground or simply to a water pipe (providing the pipe is metal and not plastic!), since the pipe usually travels well below ground level from the house to the main water supply.

Why does no current flow into the ground in a connection such as is shown in Fig. 11-15? Recall that for a continuous charge flow (current), one needs a complete, continuous path or loop with a source of EMF included in the loop. In contrast to the situation shown in Fig. 11-14, such a complete loop including the source of EMF is not provided if only one side of the circuit is grounded, as shown in Fig. 11-15.

A Summarizing Example

We now summarize the main concepts of this chapter by considering many questions regarding the circuit of Fig. 11-18. (We assume all circuit elements to be ohmic.)

Example 14
(a) What is the equivalent resistance of the parallel resistors?

$$\frac{1}{R} = \frac{1}{12.0 \, \Omega} + \frac{1}{6.0 \, \Omega},$$

hence

$$R = 4.0 \, \Omega.$$

(b) What is the total circuit resistance including the internal resistance of the two batteries?

$$R = 4.0 \, \Omega + 6.0 \, \Omega + 2.0 \, \Omega = 12.0 \, \Omega.$$

(c) What is the total EMF?

$$\mathcal{E} = 12 \text{ volts} + 12 \text{ volts} = 24 \text{ volts}.$$

(d) What is the current in the circuit? To answer this we apply the single-loop equation, $\mathcal{E} = IR$, with $\mathcal{E} = 24$ volts and $R = 12.0 \, \Omega$. We obtain

$$24 \text{ volts} = I \times (12.0 \, \Omega).$$

Therefore,

$$I = 2.0 \text{ amp}.$$

(e) What is the voltage (or voltage drop, or potential drop, or IR-drop, or potential difference) across the parallel resistors? From the definition of resistance (or Ohm's law in this case, since all the Rs are presumed constant) we get

$$V = IR = (2.0 \text{ amp})(4.0 \, \Omega) = 8.0 \text{ volts}.$$

Insulated (black) wire, "live" or high-potential side of electrical circuit proper

Insulated (white) wire, part of electrical circuit proper, usually at ground potential

Safety ground wire, either bare or insulated (often green)

Containing box for receptacle

Double outlet or receptacle

Fig. 11-17 Double electrical outlet showing how both the actual receptacle (i.e., those parts of it which do not form part of the normal electric circuit) and its containing box are connected directly to ground by the third wire which is not part of the regular circuit.

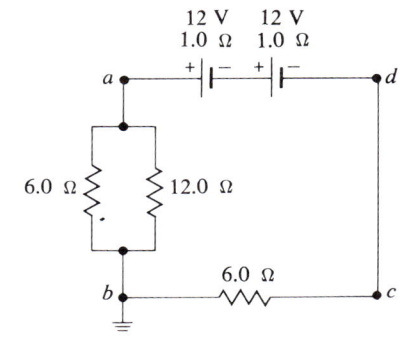

Fig. 11-18 A circuit with series and parallel connections.

11.6 SERIES AND PARALLEL CIRCUITS 417

(f) What is the current in each of the parallel resistors? Consider the 6.0 Ω resistor,

$$I = \frac{V}{R} = \frac{8.0 \text{ volts}}{6.0 \text{ Ω}} = 1.33 \text{ amp.}$$

Since 1.33 amp is the current in the 6.0-Ω resistor of the parallel group, and 2.0 amp go through the two parallel resistors combined, then the current through the 12.0-Ω resistor must be 0.67 amp. Let us check this result using Ohm's law again:

$$I = \frac{V}{R} = \frac{8.0 \text{ volts}}{12.0 \text{ Ω}} = 0.67 \text{ amp.}$$

(g) What is the *IR*-drop across the resistor to the right of point *b*?

$$V = IR = (2.0 \text{ amp})(6.0 \text{ Ω}) = 12.0 \text{ volts.}$$

(h) What is the internal *IR*-drop across the battery combination, that is, the voltage drop due to the internal resistance of the batteries?

$$V = IR = (2.0 \text{ amp})(2.0 \text{ Ω}) = 4.0 \text{ volts.}$$

(i) What is the terminal voltage of the battery combination? Recall that the terminal voltage is the voltage between the terminals of the source of EMF. When current is being drawn, the terminal voltage is less than ε because of the *IR*-drop [see Question (h)] associated with the internal resistance of the source of EMF.

Terminal voltage = ε − (internal *IR*-drop)
$$V = 24 \text{ volts} - 4.0 \text{ volts} = 20 \text{ volts.}$$

(j) What is the current through the battery, through the resistor between *b* and *c*, and through the parallel resistor group? In all these cases, it is simply the current in the circuit; that is, 2.0 amp.

(k) What is the current from *b* to ground? Zero.

(l) What are the absolute potentials at points *a*, *b*, *c*, and *d*? Graph these voltages as a function of position in the circuit.

By definition the absolute potential at point *b* is zero. Hence, employing the results of Questions (e) and (g), we get

$$V_a = +8.0 \text{ volts,}$$
$$V_b = 0,$$
$$V_c = -12.0 \text{ volts,}$$
$$V_d = -12.0 \text{ volts.}$$

For the graphical presentation, see Fig. 11-19. Note that the potential at the positive side of the battery of cells has the highest positive value. We therefore speak of the positive terminal of a source of EMF as being on the high-potential side. Current in the external part of the circuit is said to be from positive to negative, or from high potential to low potential.

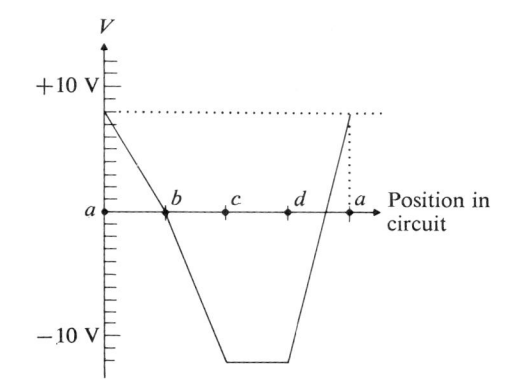

Fig. 11-19 Voltages at different points in the circuit of Fig. 11-18.

418 ELECTRIC POTENTIAL AND CIRCUITS

(m) What is the power dissipated in the 12.0-Ω resistor?
$$P = VI = (8.0 \text{ volts})(0.67 \text{ amp}) = 5.4 \text{ watts.}$$

(n) How much energy is converted into thermal energy in the 6.0-Ω resistor between b and c in 5.0 sec?

$P = VI$, and $W = Pt$, hence
$$W = VIt = (12.0 \text{ volts})(2.0 \text{ amp})(5.0 \text{ sec}) = 120 \text{ joules,}$$
or
$$W = I^2Rt = (2.0 \text{ amp})^2 (6.0 \text{ Ω})(5.0 \text{ sec}) = 120 \text{ joules.}$$

11.7 MULTILOOP CIRCUITS

Many circuits are not merely series and parallel combinations. One systematic procedure for analyzing more complex circuits is based on the following two statements, known as *Kirchhoff's rules:*

Junction Rule. The sum of the currents entering any junction equals the sum of the currents leaving that junction. This rule follows from the law of conservation of electric charge and from the fact that electric charge does not accumulate at the junction.

Loop Rule. The algebraic sum of the EMFs in any loop equals the algebraic sum of the IR products in the same loop. A loop is any closed conducting path that we care to consider in the circuit. This rule is a generalization of Eq. (11.16) and is a consequence of energy conservation.

To apply the loop rule we arbitrarily choose one direction around the loop (clockwise or counterclockwise) as the positive direction. Then, in the equation for this loop,

Algebraic sum of EMFs = algebraic sum of IR products,

the EMFs and the currents in this positive direction are considered positive; EMFs and currents in the opposite direction are negative.

A general procedure which applies to the most complex circuits is outlined in the following example.

Example 15 Find the current in every branch of the circuit of Fig. 11-20.

Solution Unknown branch currents are assigned symbols and *directions*, the junction rule being used as these assignments are made. For example, having made the assignments of I_1 and I_2, indicated in Fig. 11-20, application of the junction rule to junction A tells us that the current in the remaining branch is $I_1 + I_2$ directed away from the junction.

We now apply the loop rule first to loop 1 and then to loop 2, having

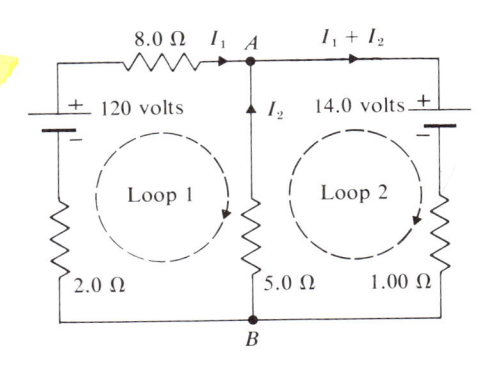

Fig. 11-20 The 2-loop circuit with the branch currents I_1, I_2 and $I_1 + I_2$ that have been assumed for the application of Kirchhoff's loop rule. Solution gives $I_2 = -4.00$ amp, which implies that the actual current in the 5-Ω resistor is opposite to the indicated direction.

indicated that the clockwise direction has been selected as the positive direction for each loop:

$$+120\ V = (8.0\ \Omega + 2.0\ \Omega)I_1 - (5.0\ \Omega)I_2$$
$$-14.0\ V = (1.00\ \Omega)(I_1 + I_2) + (5.0\ \Omega)I_2,$$

which can be rewritten as

$$120\ V = 10.0\ \Omega I_1 - 5.0\ \Omega I_2$$
$$-14.0\ V = 1.00\ \Omega I_1 + 6.0\ \Omega I_2.$$

The solution of this pair of equations is $I_1 = 10.0$ amp and $I_2 = -4.0$ amp. The significance of the minus sign in the answer for I_2 is that the actual current in the 5.0 Ω resistor is in the direction opposite to that assumed for I_2. Therefore, the current in this branch is 4.0 amp directed from junction A toward junction B. The current in the 1.00 Ω resistor is $I_1 + I_2 = (10.0 - 4.00)$ amp = 6.00 amp, directed toward junction B.

QUESTIONS

1. What is the definition of the difference of potential between two points in space?

2. In a region in which there is an electric field the electric forces do 3.00 joules of work on a positive charge of 1.50 coulombs in moving the charge from point A to point B.
   (a) What is the potential difference between points A and B in volts?
   (b) Which point, A or B, is said to be at the higher potential?

3. A negative charge of 0.50 coul is accelerated from point B to point C by an electric field and thereby acquires 2.0 joules of energy.
   (a) What did the work on the charge that it might gain this energy?
   (b) What is the potential difference between points B and C?
   (c) Which point is said to be of the higher potential, B or C?
   (d) If point C is assigned zero potential (in the case of an electric circuit this means this point is grounded), what is the actual potential (absolute potential) of point B?

4. Assigning potentials V_A, V_B, and V_C to three points in space A, B, and C, show that, if V_1 is the potential difference between A and B, and V_2 is the potential difference between B and C, then

$$V = V_1 + V_2$$

where V is the potential difference between A and C.

5. Point A is at a potential of 300 volts, and point B is at a potential of 100 volts.
   (a) Evaluate, in joules and in electron volts, the kinetic energy of an α-particle which starts at rest at point A and is accelerated to B.
   (b) Calculate the kinetic energy at B, in joules and in electron volts, of an electron which has a kinetic energy of 600 eV at A.

6. All points of the lower plate in Fig. 11-1 are at a potential of 500 volts and those of the upper plate at a potential of -100 volts. The separation of the plates is 3.0×10^{-2} m.
   (a) What is the magnitude and direction of the electric field between the plates?
   (b) What is the potential at a point between the plates which is 2.0×10^{-2} m above the lower plate?

7. If the potential is constant throughout a given region of space, what can you say about $\vec{E}$ in that region?

8. In the Millikan oildrop experiment (Section 2.8), an oil drop is suspended in equilibrium when 2.00×10^3 volts are applied across the horizontal plates which are separated by 2.00×10^{-2} m. If the drop has a charge of 1.60×10^{-19} coul, what is the mass of the oil drop?

9. The electric field in a certain region which includes the point P is 5.0×10^3 volts/m directed vertically upward. The potential at P is 200 volts. Estimate the potential at points with the following locations:
    (a) 3 mm above P.
    (b) 4 mm below P.
    (c) 2 mm north of P in the same horizontal plane.
    [*Hint:* Use $E_s = -\Delta V/\Delta s$.]

10. The potential at a point P is 500 volts. The potential is also 500 volts at points near P on a vertical line through P and on a north-south line through P. The potential at a point 3.0×10^{-4} m west of P is 490 volts; and at a point 3.0×10^{-4} m east of P, the potential is 510 volts. Estimate the magnitude and direction of the electric field at P. [*Hint:* Use $E_s = -\Delta V/\Delta s$.]

11. Starting from a given point P in space it is found that, in a displacement of 2.0×10^{-3} m north, the potential decreases by 80 volts; in a displacement of 1.0×10^{-3} m west, the potential increases by 30 volts; and in any small vertical displacement there is no change in potential. Estimate the magnitude and direction of the electric field at P.

12. In a region where there is a uniform electric field, an electron starts from rest and moves northeast, acquiring a kinetic energy of 60 eV after traveling 2.0 mm. Find the magnitude and direction of the electric field in this region.

13. (a) Describe a situation in which the mechanism of charge transfer is not restricted to electrons.
    (b) In what types of materials do electrons alone provide the mechanism of charge transfer?

14. Is the direction of electron drift (not conventional current direction) from high to low potential, or from low to high potential?

15. If the electric current in a wire is 7.00 amp, what is the charge in coul which drifts past a given point in the wire in 1.00 sec?

16. In a two-cell flashlight about 1.08 coul of charge pass by any given point in the flashlight circuit in 2.00 sec. What is the circuit current in amp?

17. Calculate the charge (in coulombs) passing through an x-ray tube during a one-fifth-second exposure at 100 mA.

18. Determine the number of electrons required to yield one coulomb of electric charge.

19. The total charge that an automobile storage battery should circulate in normal operation is 120 ampere-hours (that is 1 amp for 120 hours, 12 amp for 10 hours, etc).
    (a) How many coulombs is a charge of 120 ampere-hours?
    (b) How long could a starter be driven, if the starter current is 500 amp?

20. The central terminal of a flashlight dry cell is positive with respect to the outer electrode. When the dry cell is connected across a flashlight bulb, what is the direction of the (convential) current within the bulb and within the dry cell?

21. When a charge of 3.0 coul passes through a certain source of EMF, the work done on this charge by non-electrical forces is 24 joules. What is the EMF of the source?

22. (a) Under what circumstances is the terminal voltage of a battery equal to the EMF?
    (b) When is the terminal voltage less than the EMF?

23. (a) State the definition of *resistance*.
    (b) Employ this definition to determine the current through a 100-ohm resistor if the voltage across it is 50 volts.

24. A certain toaster operates with a current of 6.0 amp when the voltage across it is 120 volts. What is the resistance of this toaster? (This resistance is essentially the resistance of the heating elements of the toaster.)

25. Arrange the following materials in order, starting with the best conductor and ending with the best insulator: copper, silver, salt water, oil, aluminum, tungsten, glass.

26. What is the resistance in ohms of a piece of wire 30.5 meters long (about 100 ft), such as is used in ordinary electrical wiring in the home? Such wire is copper and has a cross-sectional area of about 20×10^{-7} m². (Assume room temperature of 20°C.)

27. Use the definition, $\rho = E/(I/A)$, to determine the dimensions of ρ.

28. Verify that the equation $R = \rho L/A$ is dimensionally correct.

29. If there is a 0.50 volt IR-drop along a 10.0 m length of a copper wire, find
    (a) the electric field within the wire,
    (b) the current density within the wire,
    (c) the current in the wire, if the cross-section is 20×10^{-7} m².

30. An aluminum wire and a copper wire of the same length have the same resistance.
    (a) What is the ratio of their radii?
    (b) What is the ratio of their masses? (The density of copper is 8.9×10^3 kg/m³; the density of aluminum is 2.7×10^3 kg/m³.)

31. If copper and aluminum wires of the same length and resistance were to cost the same amount, what would be the ratio of their costs per kilogram?

32. The dimensions of a carbon bar with a square cross-section are $(2.0 \times 10^{-2}$ m$) \times (2.0 \times 10^{-2}$ m$) \times (30 \times 10^{-2}$ m$)$.
    (a) The bar is placed in a circuit in such a way that the current is directed parallel to the long dimension from one square face to the other. What is the resistance of the bar?
    (b) When the current is directed from one rectangular face to the opposing rectangular face, what is the bar's resistance?

33. (a) State Ohm's law and use it to solve the following problem.
    (b) The potential difference across a certain resistor is 12 volts while the current through it is 2.0 amp. What is the current through this resistor if the voltage across it is changed to 18 volts?

34. The current in an ordinary household hot-water tank is 25 amp when the voltage across the heating coils is 240 volts.
    (a) What is the resistance of the heating coils?
    (b) What would be the current in the coils if the voltage across them were reduced to 120 volts?

35. (a) What is meant by a "nonlinear" circuit element?
    (b) Suggest a use for such a circuit element.

36. Show that the product VI has the dimensions of power and verify that 1 volt-amp = 1 watt.

37. At 3 cents per kilowatt-hour, what is the cost of forgetting to switch off a 100-watt lamp before an 8.0-hour sleep?

38. Calculate the power dissipated thermally for part (a) and for part (b) of Question 34.

39. (a) Write an expression for the power dissipated in heating an electric circuit as a function of the current and the circuit's resistance. Use this expression to find the power, in watts, if the current is 5.0 amp and the circuit resistance is 10 ohms.
    (b) How much energy (in joules) is received by the circuit resistor if the current is on for 5.0 minutes?

40. (a) What is meant by joule heating?
    (b) What is the object of the transmission of electric power at very high voltages?

41. The voltage across a resistor is V, and the current in the resistor is I. Express the power dissipated in terms of (a) V and I, (b) R and I, (c) V and R.

42. If the voltage across a resistor is doubled, and the resistance does not change, by what factor does the power dissipated in the resistor increase?

43. What is the resistance of a 100-watt bulb connected across a 120-volt DC power line?

44. How much energy will a 120-ampere-hour, 12-volt storage battery deliver without recharging?

45. A car battery which has an EMF of 12.0 volts provides a temporary 600-amp current to the engine starter. How much energy is converted from chemical energy into electrical energy within the battery if the driver presses on the starter for 10 sec?

46. A car has a storage battery with an EMF of 12 volts. If the battery furnishes 5.0×10^3 joules to the starter and an additional 1.0×10^3 joules are dissipated within the battery, what electric charge has passed through the battery?

47. What current is furnished by a 12-volt storage battery to a starter which develops 3.0 horsepower? Recall that 1 hp = 746 watts.

48. (a) State the single-loop equation.
    (b) Use this equation to find the current through a single-loop circuit, such as that of Fig. 11-11, if the EMF is 500 volts and the resistance is 10,000 ohms.

49. Would it be possible to light a Christmas tree using 6-volt bulbs if the only electric power available were the usual household 110-120 volts? Explain.

50. Calculate the single resistance equivalent to three resistances in series if they are 8.0 Ω, 4.0 Ω, and 24 Ω, respectively.

51. Calculate the equivalent resistance of the resistances listed in Question 50, if the resistances are connected in parallel.

52. Employ the concept of potential difference to explain why two dry cells of equal EMF, when connected in parallel, provide simply an EMF equal to that of each dry cell alone, rather than extra voltage, that is, a higher EMF.

53. Refer to Fig. 11-21. What is the resistance, as measured from point a to point b, of this combination of resistances? [*Hint:* Find points which are at the same potential and then redraw the diagram.]

54. Consider a 1200-watt heating element operated on a 120-volt line.
    (a) What current is drawn by the heater?
    (b) What would the current be in the case that the insulation of the cord to the heater had become worn away and developed a short? (Assume the connecting cord wires to have a resistance of 0.012 Ω.)

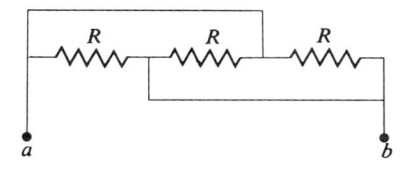

Fig. 11-21 A combination of resistances, each one ohm.

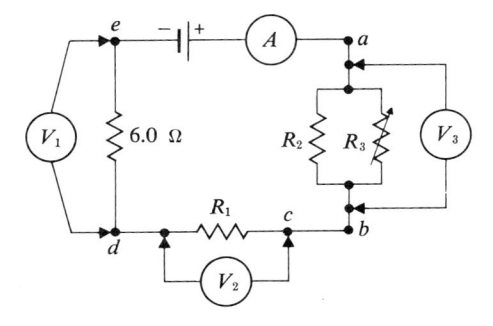

Fig. 11-22 A circuit with resistors in both series and parallel combination. Three voltmeters and one ammeter are shown.

(c) Would the current ever reach the "astronomical" figure determined in part (b)? Explain.

55. A 12-volt source of EMF having an internal resistance of 0.0093 ohm is shortcircuited with a wire of resistance 6.7×10^{-4} ohm.
    (a) What is the magnitude of the current through this source?
    (b) How much energy is dissipated in joule heating in the circuit if the short circuit is maintained for 15 sec?

56. Refer to the diagram presented on p. 415 in connection with electric shock, Fig. 11-14. How large a current would pass through a person who touched a "live" wire at 120 volts (above-ground potential) if the body resistance were 10,000 ohms and the resistance between feet and "ground" were 100 ohms?

57. Consider a circuit in which the following components are connected in series: a 6.00-volt battery of negligible internal resistance, a resistor of 10.0×10^3 ohms, another resistor of 20.0×10^3 ohms and an ammeter of negligible resistance.
    (a) Determine the current in the circuit.
    (b) Determine the IR-drop across the 20.0×10^3-ohm resistor.
    (c) What is the current in the circuit when a voltmeter with a resistance of 30.0×10^3 ohms is connected across the 10.0×10^3-ohm resistor?
    (d) What is the current through the voltmeter itself?

58. A battery of EMF 45 volts, a resistor of 100 ohms, and an ammeter of zero internal resistance are placed in series. If a voltmeter is placed across the resistor, the current in the ammeter is 0.50 amp. What is the resistance of the voltmeter?

59. Why is it unsafe to turn a light switch on or off while reclining in a bath?

60. The following questions refer to Fig. 11-22. The voltmeter V_1 reads 12 volts, V_2 reads 4.0 volts, and the total resistance of the two parallel resistors is 4.0 ohms. (Assume the meters do not affect the circuit behavior.)
    (a) What assumption are you making about the internal resistance of the source of EMF?
    (b) What assumption are you making about the connecting wires?
    (c) What is the current through the 6.0-Ω resistor?
    (d) What is the reading of the ammeter A?
    (e) What is the reading of the voltmeter V_3?
    (f) How large is the EMF?

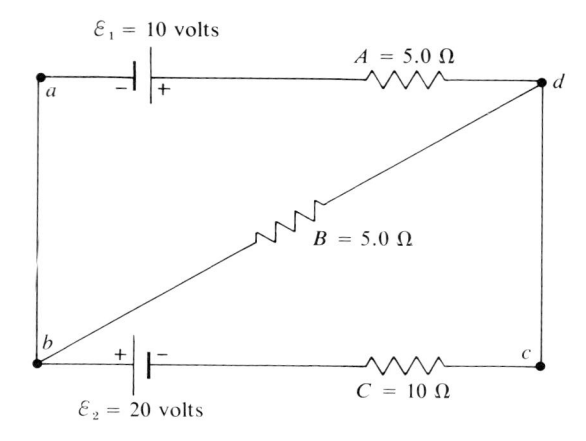

$\mathcal{E}_1 = 10$ volts
$A = 5.0\ \Omega$
$B = 5.0\ \Omega$
$C = 10\ \Omega$
$\mathcal{E}_2 = 20$ volts

Fig. 11-23 Two-loop circuit of Question 61.

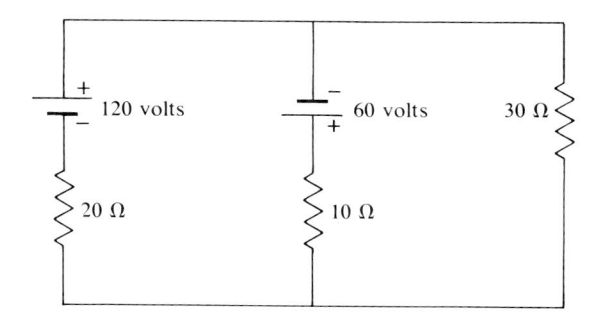

120 volts
60 volts
30 Ω
20 Ω
10 Ω

Fig. 11-24 Circuit of Question 62.

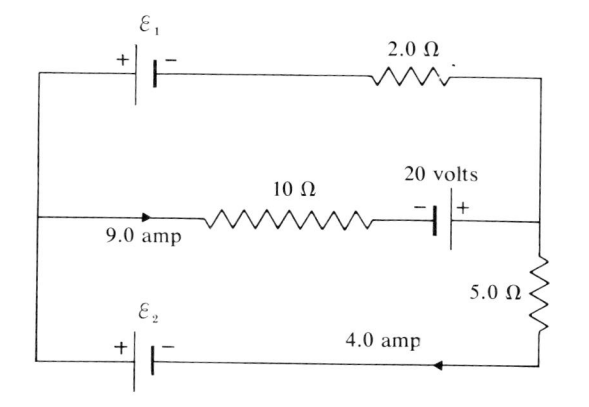

$\mathcal{E}_1$
2.0 Ω
10 Ω
9.0 amp
20 volts
5.0 Ω
$\mathcal{E}_2$
4.0 amp

Fig. 11-25 Unknown EMFs of Question 65.

(g) How large is the resistance of the resistor R_1?
(h) If the resistor R_2 has a resistance of 8.0 ohms, what is the resistance set on the rheostat R_3?
(i) What is the current through R_3?
(j) What is the potential difference between points a and b?
(k) Which point, a or b, is at the higher potential?
(l) What is the potential (absolute potential) of point d if point e is grounded?
(m) What is the potential difference between points c and b?
(n) What would happen to the total current in the circuit if the rheostat R_3 were turned to a lower resistance?
(o) What could be said about the respective currents through R_2 and R_3 in the case of (n) above?
(p) What would you do to "short circuit" this circuit?
(q) What danger is inherent in a "short circuit"?
(r) On the basis of the assumptions you made, what would happen to the total current in the circuit if voltmeter V_2 were removed?
(s) What would happen to the total current if the rheostat were disconnected?
(t) What would happen to the total circuit current if a second source of EMF, identical to the one in Fig. 11-22, were connected in series to the one pictured, with the positive terminal of the second one connected to the negative terminal of the first one?
(u) What would happen to the total circuit current if a source of EMF such as in (t) above were in series as in (t), but the positive terminal of one were connected to the positive terminal of the other?
(v) What would happen to the total circuit current if a source of EMF identical to the one in Fig. 11-22 were added in a standard parallel connection to the original one (that is, positive terminal of one to positive terminal of the other, etc.)?
(w) Does the answer to part (v) above depend on the assumption stated in the answer to part (a) above? Explain.
(x) Is there any significance in the fact that the voltmeter V_1 is connected with wires at an angle and V_2 is connected with wires in a rectilinear configuration?

61. Consider Fig. 11-23. Neglecting the internal resistance of the sources, give the direction and magnitude of the current in each resistor, A, B, and C when $\mathcal{E}_1 = 10$ volts, $A = 5.0$ ohms, $\mathcal{E}_2 = 20$ volts, $C = 10$ ohms, and $B = 5.0$ ohms.

62. Find the current in each source of EMF and in the 30-Ω resistor in Fig. 11-24.

63. Find the current in each branch of Fig. 11-24 for the case in which the polarity of the 60-volt source is reversed from that indicated in the figure.

64. A 120-volt source with an internal resistance of 2.0 Ω is connected in parallel with a source which has an internal resistance of 3.0 Ω but whose EMF is unknown. When this combination is connected across a 10-Ω load, the current in the 120-volt source is 40 amp. Find the size and polarity of the unknown EMF.

65. In the circuit of Fig. 11-25, find $\mathcal{E}_1$ and $\mathcal{E}_2$.

SUPPLEMENTARY QUESTIONS

S-1. In a particle accelerator, an α-particle is accelerated through a difference of potential of 3.0×10^6 volts. Find its kinetic energy and its speed.

S-2. A surface in space whose points are all at the same potential is called an *equipotential surface*.
(a) From the relationship
$$E_s = -\Delta V/\Delta s,$$
show that the component of $\vec{E}$ along an equipotential surface is zero.
(b) From the result of (a) what is the relationship between the direction of the vector $\vec{E}$ at any point in space and the equipotential surface passing through that point?

S-3. In the study of *electrostatics*, we consider the situation when all electric charges are assumed to be at *rest*, so that there is no electric current in any of the conductors under consideration. This implies that the electric field is zero within any conductor (in electrostatics). From this observation show that in electrostatics:
(a) A conductor is an equipotential region. That is, any two points of the same conductor are at the same potential.
(b) At any point just outside the conductor, the electric field is perpendicular to the surface of the conductor.

S-4. Assume that the plates in Fig. 11-1 are copper and that the electric charges are at rest. With potentials assigned as in Question 6, in the region between the plates, sketch and label equipotentials corresponding to 500 volts, 350 volts, 200 volts, 50 volts and -100 volts.

S-5. How is it that one obtains an almost instantaneous response when an electrical appliance is switched on, even though the drift speed of the electrons in the wires is very small (say 1 mm/sec)?

S-6. The resistance of a given conductor depends upon the conductor's temperature. Over a limited range of temperatures, ΔT, the change in resistance, ΔR, may be expressed by the relation $\Delta R = \alpha R_0 \Delta T$, where R_0 is the resistance at the reference temperature, T_0, and α is the thermal coefficient of resistance (or thermal coefficient of resistivity).
(a) Express in words the definition of α implied by this relation.
(b) A copper wire has a resistance of 1.50 ohms at 20°C. If α for copper in this temperature region is 0.004 per C°, what will be this wire's resistance at 100°C?

S-7. Consider a source of constant EMF $\mathcal{E}$ and constant internal resistance r connected to an external circuit of resistance R which can be varied.
(a) Deduce an expression for the power P supplied to R in terms of $\mathcal{E}$, r, and R.
(b) Sketch a graph to show P as a function of R, including the values of P for $R = \tfrac{1}{2}r$, $R = r$, and $R = \tfrac{3}{2}r$.
(c) From the graph, what value of R is suggested as the value for which the power supplied is a maximum? Express the maximum power in terms of $\mathcal{E}$ and r.

S-8. A storage battery with an EMF of 12.0 volts has an internal resistance of $6 \times 10^{-2} \Omega$ (commercial batteries in good condition actually have an internal resistance of only a few thousandths of an ohm).

The battery is to be recharged by connecting 120 volt-DC across a series combination of the battery and a resistor whose resistance R is adjusted to limit the charging current to 10.0 amp.
(a) Which terminal (+ or −) of the battery should be connected to the high-potential side of the power lines?
(b) What is the value of R?
(c) Find the terminal voltage of the battery while it is being charged.
(d) What is the power input from the lines?
(e) What is the power dissipated in R and also within the battery?
(f) What is the power converted within the storage battery?
(g) If a charge of 40 ampere-hours circulates through the battery, what will be the cost at 3 cents per kilowatt-hour?

S-9. A DC motor, with an internal resistance of 3.0 Ω, is connected across 120-volt power lines. When its speed has built up to normal operating speed, the current is 10.0 amp.
(a) What EMF is generated by the motor and what is its polarity relative to the direction of the motor current? (Such an EMF is called a "back EMF.")
(b) What electrical power is converted to mechanical power by the motor?
(c) What is the power dissipated within the motor?
(d) What is the ratio of the mechanical power developed to the total electrical power input?

S-10. During a one-month period a certain electric refrigerator, connected to a 120-volt supply, runs for a total of 150 hours.
(a) If the current required by the refrigerator were 3.2 amp, how much energy would be supplied by the electric power company?
(b) If the electrical energy were supplied at 1.50 cents per kilowatt-hour, what would be the monthly cost of operating this refrigerator?
(c) Compare this cost with that of a television set which requires 2.4 amp at 120 volts and which runs for 90 hours during the month.

S-11. In Fig. 11-26, what is the resistance between the terminals A and B?

S-12. Prove that the single resistor equivalent to a parallel combination of resistors is less than the smallest resistance in the parallel combination.

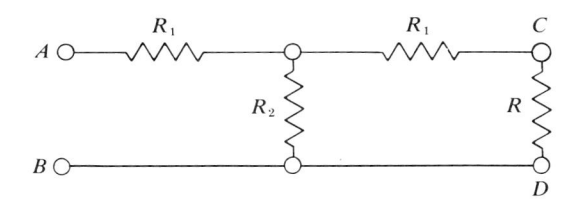

Fig. 11-26 Network of resistors.

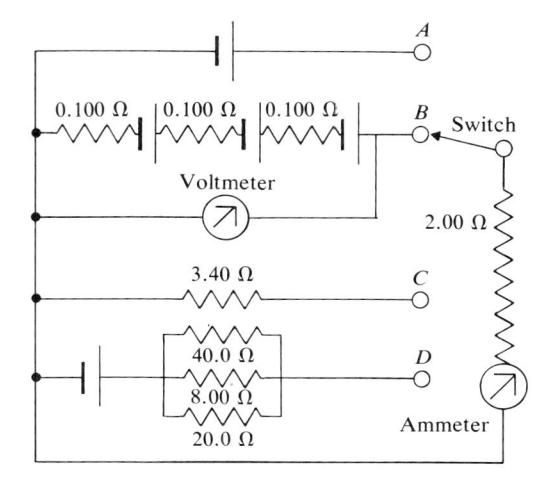

Fig. 11-27 Circuit for Question S-14.

S-13. A man who had been having difficulties with a 15-amp fuse "blowing" frequently in a certain circuit, noticed that the store from which he had been buying the 15-amp fuses also carried 30-amp fuses. He reasoned that his troubles would be solved if he used a 30-amp fuse, since a 30-amp fuse should last twice as long as a 15-amp fuse. Indicate a fallacy in this reasoning and outline in a few sentences how you would explain this person's mistake to him.

S-14. Consider the circuit arrangement in Fig. 11-27 where each of the sources of EMF has an EMF of 1.50 volts.
(a) What should the resistances of the voltmeter and the ammeter be so that these instruments would not disturb the behavior of the circuit at all?

(b) For each of the possible positions (A,B,C,D) of the switch, give the readings on these two "ideal" meters (which are fixed in the positions shown throughout this entire hypothetical experiment).

S-15. Redraw the circuit in Example 14 (p. 417), Fig. 11-18, and add a resistanceless connection from point b to point d. Now, with reference to this modified circuit, answer the same questions asked in Example 14.

S-16. A source with an EMF of 60 volts and an internal resistance of 0.20 Ω is connected in series with a second source which has an EMF of 40 volts and an internal resistance of 0.10 Ω. If this series combination is connected across a 0.70-Ω resistor, what is the current when:
(a) the negative terminal of one source is connected to the positive terminal of the other?
(b) the positive terminals of the sources are connected?

S-17. A voltmeter which has a resistance of 960 ohms is connected across a 120-ohm resistor which is in turn connected in series with a 100-ohm resistor and an ammeter of negligible resistance. The source of EMF is a 3.00-volt battery of negligible internal resistance.
(a) What is the current in the circuit?
(b) What is the current through the voltmeter?
(c) What is the IR-drop indicated by the voltmeter?
(d) What is the (absolute) error in the IR-drop measured? That is, what is the difference between the IR-drop across the 120-ohm resistor with the voltmeter connected and the IR-drop across the same resistor with the voltmeter disconnected?
(e) What is the relative error (error/reading) in the IR-drop introduced by the measuring device, the voltmeter?

S-18. The nine cells of the source of EMF (Fig. 11-28) are identical. Each cell has an EMF of 2.00 volts and an internal resistance of 0.100 ohm. The current I is 5.00 amp. The light bulb filament is made with a tungsten wire 4.8 cm long and 0.15 mm thick. Its resistance at operating temperature (2000°C) is 1.5 ohms. The mean value (between 0 and 2000°C) of the resistance coefficient α is 0.0045 per C° (see Question S-6). When the lamp is operating normally (2000°C), give the value of

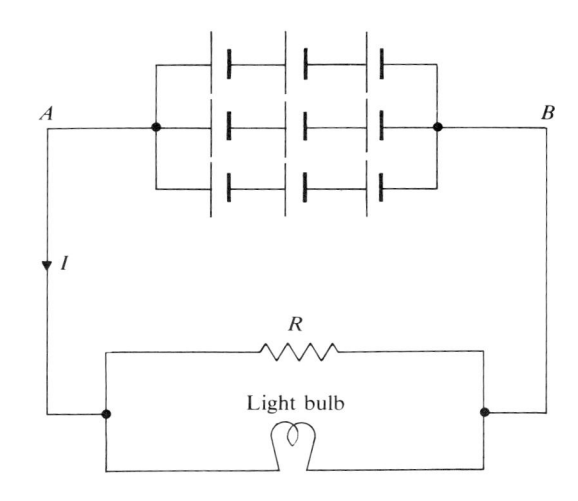

Fig. 11-28 Circuit of Question S-18.

(a) the total EMF of the source;
(b) the equivalent resistance of the source;
(c) the voltage across the source (V_{AB});
(d) the voltage across the light;
(e) the current through the light;
(f) the current through the resistor R;
(g) the resistance of the resistor R;
(h) the power dissipated in the form of thermal energy in the light.
(i) Find the resistance of the tungsten filament at 0°C.
(j) Find the resistivity of tungsten at 0°C.

S-19. Fig. 11-29 shows a circuit, called a *Wheatstone bridge*, which is used to measure an unknown resistance. Suppose, for example, that R_1 is an unknown resistance and that R_2, R_3 and R_4 are known variable resistors. Their variable resistances are adjusted until the galvanometer indicates zero current (the bridge is then said to be "balanced"). Show that when the bridge is balanced

$$R_1/R_2 = R_3/R_4.$$

S-20. When the circuit elements in Fig. 11-29 have the values $R_1 = 10.0\Omega$, $R_2 = 20.0\Omega$, $R_3 = 3.0\Omega$, $R_4 = 12.0\Omega$, $R_G = 2.0\Omega$, $r = 0$, and $\mathcal{E} = 10.0$ volts, find the current through the galvanometer.

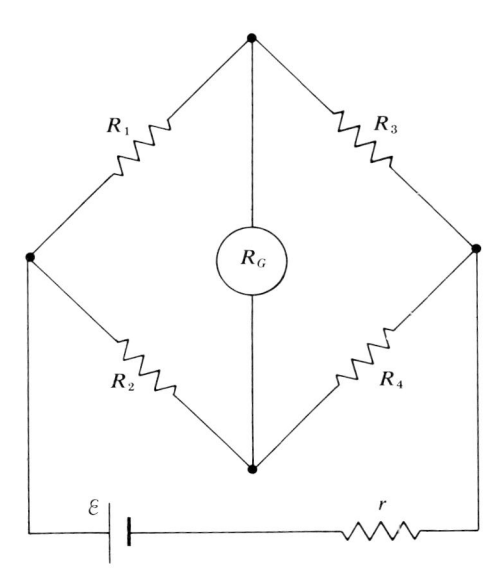

Fig. 11-29 Wheatstone bridge.

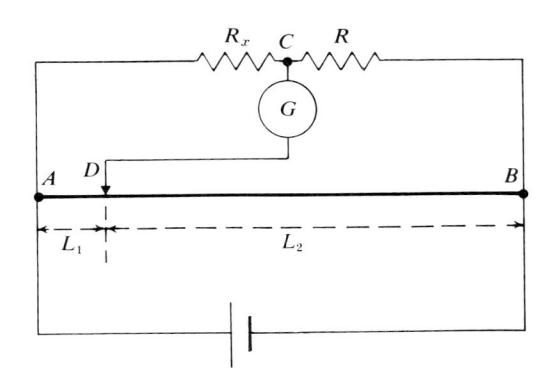

Fig. 11-30 Slide-wire Wheatstone bridge.

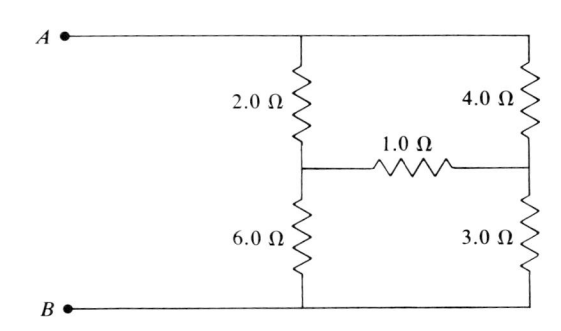

Fig. 11-31 Network of resistors for Question S-22.

S-21. In the slide-wire bridge (Wheatstone) in Fig. 11-30 the resistance R_x is given by the formula

$$R_x = R \frac{L_1}{L_2}.$$

This relationship holds when the bridge is balanced: that is, when the sliding connection D is in a position such that the galvanometer G gives a null reading. The fact that lengths L_1 and L_2 can be used in place of resistance values, say R_1 and R_2, is due to the fact that in this instrument the resistance wire AB is made sufficiently uniform.
(a) Derive the relation expressed above.
(b) Give the value of R_x if the bridge is balanced for $AB = 30.0$ cm; $L_1 = 2.0$ cm; $R = 1.00 \times 10^4$ ohms.

S-22. Find the resistance between the terminals A and B of the combination of resistors shown in Fig. 11-31. [*Hint:* Place a 1-volt source of EMF between A and B and calculate the current through the source using Kirchhoff's rules.]

ADDITIONAL APPLICATIONS TO MEDICINE AND THE LIFE SCIENCES

Electrocardiogram: Potential Difference and Potential

The medical applications cited in Chapter 2 made reference to a physiological monitor which could trace out an ECG pattern. The electrocardiogram is simply a tracing of the alternating potential differences between the two body points to which the instrument's electrodes are attached. The ECG potential differences (at the surface of the body) are of the order of 1 mV. These alternating potential differences **appear every time the heart beats (see Fig. 11-32).**

Since these potential differences associated with the heart action spread to the surface of the body, electrodes affixed to almost any pair of points on the surface of the body will show these periodic alternating potential differences. (In medical language the term "lead" refers to the potential differences. A "lead" requires two electrodes, that is, two wires. Sometimes three wires are used in a "lead".)

Electrocardiograms are usually made with the patient lying down, partly because muscular activity also gives rise to body potential differences.

Question

From Fig. 11-32, estimate the voltage of the pulse which precedes ventricular ejection.

Answer

About 0.5 mV.

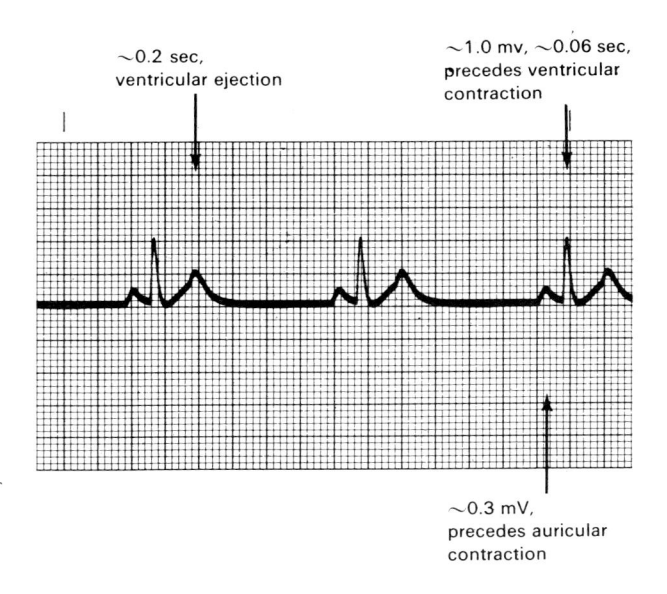

Fig. 11-32 ECG trace. This trace was made with an electronic simulator. The voltage preceding auricular contraction is normally closer to 0.1 mV than to 0.3 mV.

Electroencephalograms and Brain Waves: Potential Difference and Potential

An electroencephalogram (EEG), a record of electrical wave patterns of brain activity measured from the surface of the human head, can be obtained by exploiting essentially the same physical principles and instruments as those used in electrocardiography. There are three different frequency ranges of normal brain waves: α-waves of about 10 Hz, β-waves from about 15 to 60 Hz, and δ-waves below about 8 Hz. Still lower frequencies, typically from 1/2 to 2 Hz, are associated with brain lesions. The EEG finds clinical use for the diagnosis of epilepsy and other brain damage.

Scientists seeking a fundamental understanding of brain activity find these brain waves a complex puzzle. Nevertheless, with the aid of modern computers, progress has been made in decoding the EEG and showing how it is related to the functioning of the nervous system. Particularly interesting are the changes in the brain's natural rhythm which follow some outside stimulus such as a light flash. These changes are called the *evoked response*. The frequency, amplitude, and other parameters of these evoked waves are correlated with different psychological states.

A brain-wave monitor to test human intelligence has been developed by the Canadian psychologist, John Ertl. The test, based on evoked response, has the advantage that results cannot be distorted by ethnic bias because the subject is not required to read, write, or even speak. To take the Ertl test, the subject puts on a helmet containing electrodes and then simply watches a light flash 200 times. The brain wave changes caused by the light are picked up by the electrodes and analyzed by a computer. The whole process takes but five minutes and is claimed to be more accurate than conventional intelligence tests in a significant number of cases.

Further Reading

"The Electrical Activity of the Brain" by W. Grey Walter, in the *Scientific American*, June 1954, and "The Analysis of Brain Waves" by Mary A.B. Brazier, in the *Scientific American*, June 1962.

Bioelectricity: Mechanism of Charge Transfer

Bioelectricity is usually pursued from two different viewpoints: (a) the source of the electrical energy within the cell; and (b) the *electrolytic current* (charge flow) due to electric fields outside the cell. In *electrophysiology* one sometimes penetrates the cell to investigate its internal potential. More frequently, however, measurements are made external to a group of cells which are causing electrolytic current flow. The electrolytic current, or ionic current, appears essentially to involve sodium and potassium ions in solutions.

The interior of a cell contains concentrations of sodium and potassium ions. And the cell is the basic source of all bioelectric potentials. (A bioelectric potential is, to be precise, actually a potential difference: the potential difference between the inside and the outside of the cell.) A difference of potential exists across the semipermeable membrane called the cell wall. When a cell receives a stimulus from an outside source, the characteristics of the cell membrane at the point of stimulation are sufficiently altered to in turn effect a change in the ionic current.

Checking Surgical Gloves: The Complete Circuit

An interesting application of "the complete circuit" is found in a *surgical glove tester*, developed by A.R. Morse and P.J. Serada of the National Research Council of Canada. This glove tester was devised to fill the need for an adequate sterile method of testing surgical gloves for pinholes in order that bacteriological control might be as complete as possible during an operation.

The tester (see Fig. 11-33) includes the wearer as part of the testing system. For example, a surgeon puts on the sterile glove and then places his gloved hand into a container of sterile saline solution. To the surgeon's wrist a body

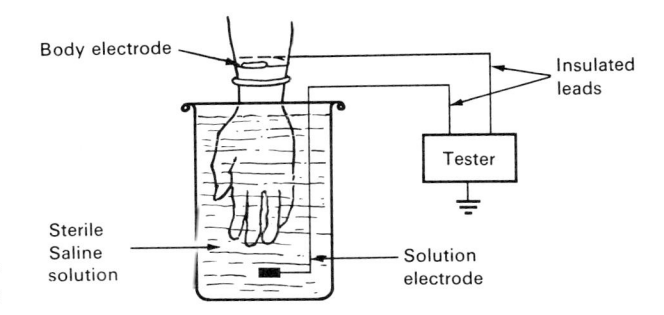

Fig. 11-33 Surgical glove tester.

electrode, a sort of bracelet, is clamped; the second electrode is in the sterile solution. The two electrodes are part of a DC circuit including a microammeter and some high-resistance resistors. If there is no hole in the glove, the resistance of the glove is essentially "infinite"; the circuit is open; there is no reading on the microammeter. If there is even a minute (nonvisible) hole, then some of the saline solution comes into contact with the surgeon's hand; there is now a complete circuit; the microammeter indicates a reading, which is, in part, dependent on the size of the hole. The power source is designed such that the short circuit current (that is, current when glove resistance is zero) is no more than about 20 μamp.

Question

(a) Sketch the circuit labeling the essential components.
(b) What is the purpose of the high-resistance resistors in the circuit?

Answer

(a) This is a series circuit involving components in order: one terminal of EMF source, resistors, body electrode, hand, glove, saline solution, solution electrode, meter, other terminal of EMF source.
(b) To afford protection; that is, to limit the current to a safe value even when the glove resistance is zero.

SUGGESTED READING

EFRON, ALEXANDER, *Magnetic and Electrical Fundamentals.* New York: John F. Rider Publisher, Inc., 1959. For beginning physics students. Chapter 5 provides useful additional examples relating to DC circuits.

HALLIDAY, DAVID, and ROBERT RESNICK, *Physics.* New York: John Wiley, 1966. Chapters 31 and 32 are for the more advanced student seeking examples in DC circuits.

STEWART, A. T., *Perpetual Motion, Electrons and Atoms in Crystals.* Garden City, N.Y.: Doubleday, 1965. This is one of the Science Study Series paperbacks designed to offer to students and to the general public the writing of distinguished authors on fundamental topics of science. Chapters 4 and 5 in which the author describes electrons in metals and electric conductivity in terms of an "electron soup" make delightful and instructive reading.

THWAITS, RUSSEL R., *Fundamentals of Electricity.* Toronto: McGraw-Hill, 1964. This book is an introduction to applied electricity rather than a physics text. It is for the student interested in practical electricity. For example, if you want to know what's inside a circuit breaker and how to make minor electrical repairs, this is the book for you. Such topics as practices in domestic wiring, testing simple domestic appliances, the telephone circuit, and an introduction to electronics are presented.

Michael Faraday

12 Electromagnetism

"... I have always endeavored to make experiment the test and controller of theory and opinion...."

Michael Faraday
(1791–1867)

At the time when a poor and poorly educated youth, Michael Faraday, began probing nature's secrets, knowledge of electricity and magnetism was so fragmentary that, except for the mariner's compass, applications were confined to parlor tricks. However, within his lifetime electric dynamos and motors were invented, the possibility of radio waves was revealed, and the importance of electric forces and electric charge in chemistry became evident. These achievements are but a sample of the consequences of Faraday's experimental research.

Young Faraday, working at the Royal Institution in London, had already made great discoveries in chemistry when, in 1820, the attention of the scientific world was captured by the phenomenon of electromagnetism. In Denmark a physics professor, Hans Christian Oersted (1777–1851), while performing a demonstration experiment before a small group of advanced students, noticed that a compass needle was deflected when it was in the vicinity of a wire carrying an electric current (see Fig. 12-1 on the following page). The compass needle set itself perpendicular to the wire. The first connection between magnetism and electricity had been discovered: magnetic effects are produced by charges in motion. (In 1823 Oersted and Faraday met in London and thereafter remained in correspondence for nearly 30 years.)

This magnetic effect produced by electric current was promptly investigated by the French physicists, J. B. Biot (1774–1862) and F. Savart (1791–1841), as well as by André-Marie Ampère who showed how the magnetic effect of any distribution of currents could be described in quantitative detail.

Faraday took up the study of electromagnetism. His way of thinking in terms of magnetic fields, magnetic flux lines, and magnetic flux proved extremely fruitful and still permeates the most modern treatments of this subject. In this description of electromagnetic phenomena, what Oersted had discovered was that an electric current sets up a magnetic field in the surrounding space. The force on a compass needle is determined by the magnetic field thus set up in its vicinity; the compass needle is forced into alignment with this magnetic field.

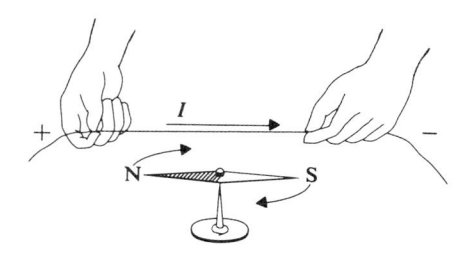

Fig. 12-1 The essence of Oersted's discovery.

Courtesy of the Smithsonian Institution

Joseph Henry
(1797–1878)

On Christmas Day, 1821, Faraday found that a magnetic field exerts a force on a wire carrying a current. This magnetic force has the peculiarity of being a side thrust, always perpendicular to the wire and also perpendicular to the magnetic field. Using this force, Faraday showed that a conducting wire could be made to rotate about a bar magnet, that is, electric motors were possible and it is precisely this magnetic force, Faraday's Christmas present, that drives every electric motor.

Oersted had shown that an electric current gives rise to magnetic effects. Maybe there was an inverse effect. Perhaps a magnet could be made to produce an electric current. Faraday's early researches in this direction, performed in 1824, detected no such effect. No current was induced in a stationary wire near a stationary magnet.

Returning to this matter in 1831, Faraday observed a transitory flick of the needle of his current detector (a galvanometer) while *change* was occurring. *Change* was the key to induced EMFs and the associated induced currents. While a magnet was moving near a stationary coil, electric current was induced in the coil. Again, a current was induced while the coil moved near a stationary magnet. Furthermore, if the current in one coil were changing, a current was induced in another coil nearby but electrically insulated from the first. The results of many such experiments were summarized in Faraday's law of electromagnetic induction: the EMF induced in a coil is proportional to the *rate of change* of the magnetic flux enclosed by the coil. (Faraday visualized magnetic fields by thinking of space as threaded with magnetic flux lines with their density in each region proportional to the local strength of the magnetic field. The magnetic flux through a coil is proportional to the number of flux lines enclosed by the coil.) It is this EMF which is induced in the whirring dynamos that generate almost all of the electric power consumed in industry and in our homes.

Any history which emphasizes the contributions of one man runs a great risk of being unfair to others. Credit for the elucidation of the laws of electromagnetic induction also goes to H. F. E. Lenz (1804–1865), a German scientist working in Russia, and to an American teacher at the Albany Academy, Joseph Henry (1797–1878). Unfortunately, Henry's onerous teaching and administrative duties delayed his publication of the discoveries made during a one-month summer "holiday" in 1830.

12.1 SOURCES OF A MAGNETIC FIELD $\vec{B}$

Moving Charges

With two charges in *motion*, in addition to the electric forces, there are new forces which come into play that are termed magnetic forces. To describe magnetic interaction, physicists again employ the notion of a *field*.

As usual, to describe the interaction of several objects we separate the account into two distinct steps. The force exerted on one particular object is entirely determined by the *field* at its location which is produced by the other (remote) objects. These other objects are regarded as *sources* of a field at all points of space. This description of an interaction requires two steps:

1. A specification of the field produced at each point in space by a given distribution of sources,
2. A rule determining the force exerted by the field at a certain point in space on some object—an object other than the sources considered in Step (1)—which happens to be at this point.

In this section we consider only Step 1, and give only a qualitative description of the fields produced by various sources.

With each point in space, we associate a vector $\vec{B}$, called the *magnetic field** at the point in question. A moving charge is a *source* of a magnetic field at all points of surrounding space. Fig. 12-2 shows the

**Other names for this vector used in the literature of electromagnetism are the* magnetic induction, *the* magnetic field strength *and the* magnetic flux density.

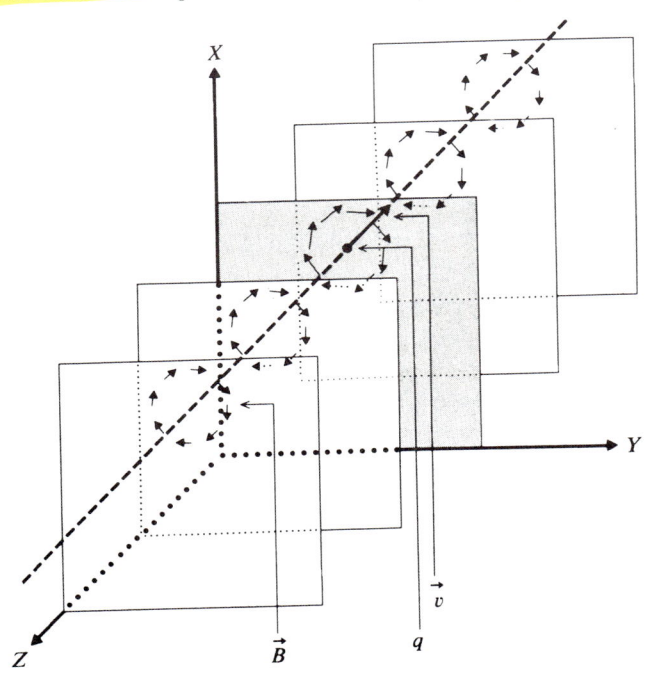

Fig. 12-2 The magnetic field produced by a positive charge which is moving with a constant velocity along the dashed line. The field, $\vec{B}$, is shown at several points located in planes perpendicular to the charge's velocity. Note that $\vec{B}$ is less in magnitude the further the plane in question is from the position of the moving charge.

12.1 SOURCES OF A MAGNETIC FIELD $\vec{B}$

magnetic field produced by a positive charge q which is moving with a constant velocity $\vec{v}$. The field $\vec{B}$ is shown at several points located in planes perpendicular to the charge's velocity vector. At any given point, the magnetic field $\vec{B}$ is proportional to the amount of charge q and to its speed v.

Like other fields, magnetic fields obey a *superposition principle:* the field produced at a certain point in space when there are many different sources is the *vector sum* of the fields that each source would individually contribute, if it alone were present.

A wire carrying an electric current contains a stream of moving charges, each of which contributes to the magnetic field in surrounding space. The superposition of such contributions, in the space near a long, straight section of a wire carrying a current I is displayed in Fig. 12-3.

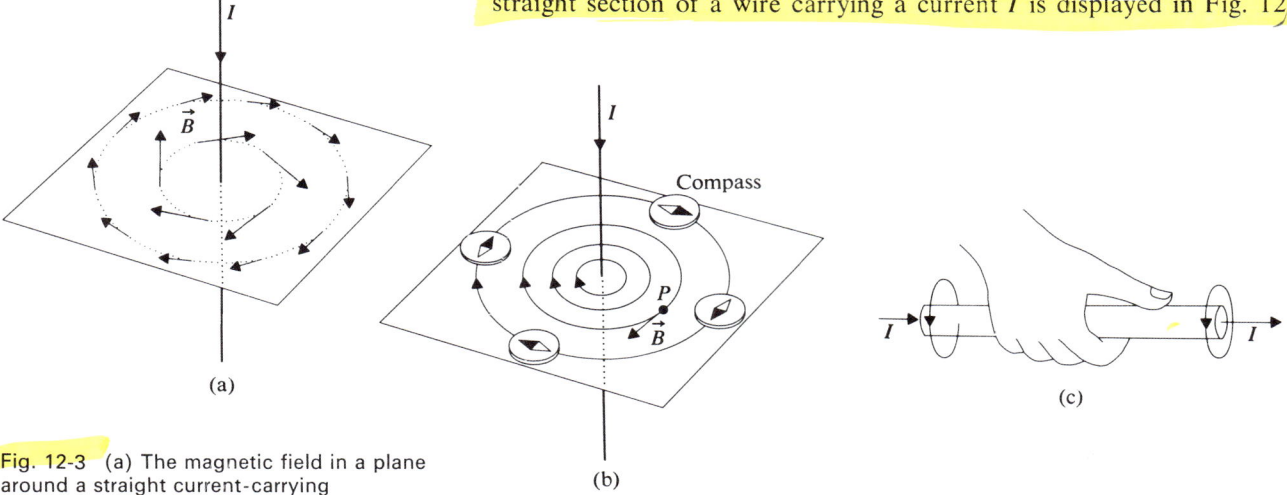

Fig. 12-3 (a) The magnetic field in a plane around a straight current-carrying conductor as represented by the magnetic field vector $\vec{B}$. (b) The same magnetic field indicated in terms of flux lines. A compass needle is forced into alignment with $\vec{B}$. The end of the needle which points in the direction of $\vec{B}$ is called a north pole and the other end a south pole. In the earth's magnetic field the direction in which the north pole of a compass needle points is called magnetic north. (c) A handy *right*-hand rule. With the thumb in the direction of the current, the fingers point in the direction of $\vec{B}$.

Flux Lines

A convenient alternative to the representation of magnetic fields by drawing field vectors $\vec{B}$ at several points in space is given by drawing instead what are called *magnetic flux lines*. A flux line is a line whose tangent at any point is in the direction of $\vec{B}$ at that point. The flux lines representing the field produced by a long, straight wire carrying a current I are shown in Fig. 12-3(b). This figure illustrates certain things that apply to the magnetic field set up by *any* distribution of currents:

1. Each magnetic flux line is a closed curve in space. This closed curve always encircles some electric current.
2. Each flux line is provided with an arrow. From knowledge of the flux lines, the direction of the magnetic field vector $\vec{B}$ at any given point in space [see point P in Fig. 12-3(b)] is determined as the tangent to the flux line through the point in question.

436 ELECTROMAGNETISM

3. The magnetic field is strong ($\vec{B}$ has a large magnitude) where the flux lines are dense or crowded together. Where the field is weaker the flux lines are further apart.

The direction of the arrow associated with the flux lines of the magnetic field produced by a current in a long straight wire can be determined from the right-hand rule illustrated in Fig. 12-3(c). If the wire is grasped in the *right* hand with the thumb pointing in the direction of the current, the *fingers* will curl about the wire in the direction of the flux lines.

Ferromagnetism

The magnetic field produced by a current flowing in a small circular loop [Fig. 12-4(a)] is of particular interest. Ampère recognized that the permanent magnetism displayed by lodestone and magnetized iron bars, with no electric current evident, might arise from tiny current loops in atoms [Fig. 12-4(b)]. Modern theory supports this viewpoint. Some types of atoms do set up external magnetic fields because of circulating electron currents within the atom. And, unless all the electron spin angular momenta cancel out, there is a magnetic field contributed by spinning electrons within an atom.

In what are called *ferromagnetic* materials, notably iron, cobalt, nickel and a few alloys, atoms interact in such a way that tiny but microscopically visible *magnetic domains* are formed, each domain consisting of some 10^{15} aligned atoms. At external points, a magnetic domain produces a strong magnetic field, since each of its aligned atoms furnishes aligned contributions. A chunk of ferromagnetic material normally contains billions of these magnetic domains. When the chunk is unmagnetized, although the alignment of atoms within each domain is nearly perfect, different domains are aligned in different directions at random [Fig. 12-5(a)]. When different domains are given more or less the same alignment by the application of some external magnetic field, the material is said to be magnetized [Fig. 12-5(b)].

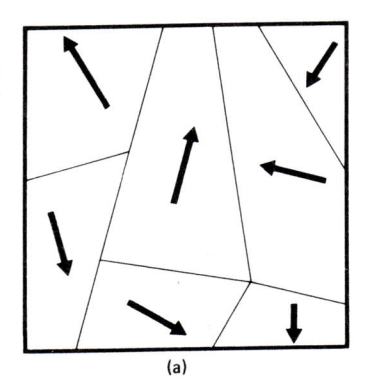

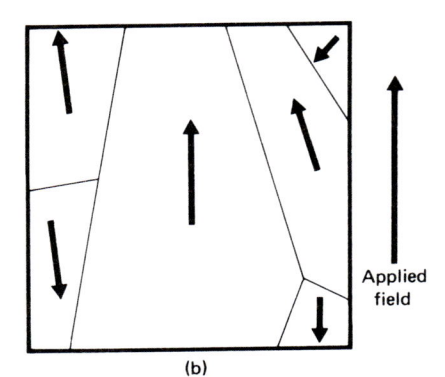

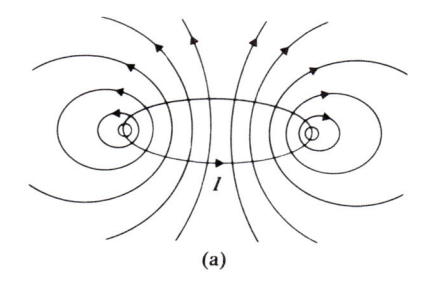

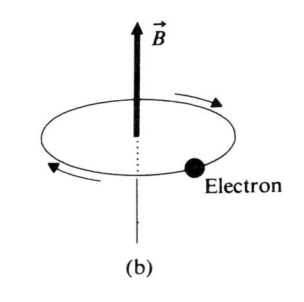

Fig. 12-4 (a) Some magnetic flux lines of the magnetic field produced by a current *I* flowing in a circular loop.
(b) A "tiny current loop" in an atom.

Fig. 12-5 Magnetization of a ferromagnetic specimen. Domain walls shift and domains rotate to produce alignment with the applied field. If the applied field is sufficiently strong, these processes continue until complete alignment is achieved. The material is then magnetically saturated. (a) Unmagnetized. (b) Magnetized, but not to saturation.

12.1 SOURCES OF A MAGNETIC FIELD $\vec{B}$

An unmagnetized ferromagnetic material, placed in a magnetic field, becomes magnetized and thereby makes a substantial alteration in the magnetic field that would otherwise be present, typically increasing the field by a factor of a thousand at points in or near the material. Permanent magnets retain the alignment of their different domains. Other "softer" ferromagnetic materials tend to revert to a random domain alignment when a magnetizing field is removed.

Nonferromagnetic materials such as copper, aluminum, wood, and air do not possess magnetic domains. The alteration in a magnetic field produced by the presence of these nonferromagnetic materials is less than 0.01%. Consequently such materials are hardly different from a vacuum, as far as magnetic effects are concerned. These materials are therefore commonly called "nonmagnetic" materials.

A typical *electromagnet* and its magnetic field are shown in Fig. 12-6. The electric current in the coils produces a magnetic field which aligns domains in the iron. The magnetic field at any point is the superposition of the contributions from the moving charges in the coils and the much greater contributions from the aligned domains of iron. Between the polepieces, not too close to the edges, $\vec{B}$ has the same value at all points. Such a region, called a region of uniform magnetic field, is often required in physics experiments.

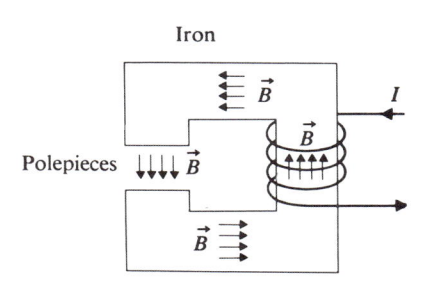

Fig. 12-6 Electromagnet. Within the gap there is a region of uniform magnetic field.

12.2 THE FORCE EXERTED BY $\vec{B}$ ON A MOVING CHARGE

So far we have discussed only half the story of magnetic fields, that part describing the field $\vec{B}$ produced by given sources. To complete the field description of magnetic interaction we now consider what $\vec{B}$ does.

If there happens to be a moving charged particle at a certain point where there is a field $\vec{B}$ produced by sources located elsewhere, the particle will experience a magnetic force. This magnetic force exerted by the magnetic field $\vec{B}$ at a certain point in space on a charge q at that point depends on the velocity vector $\vec{v}$ of the charge. This force is *zero* if the charge is at rest or if the charge is moving along a flux line ($\vec{v}$ parallel to $\vec{B}$). As illustrated in Fig. 12-7 (b), when the velocity vector $\vec{v}$ makes an angle θ with $\vec{B}$, the magnetic field $\vec{B}$ exerts on the moving charge q a "magnetic force" $\vec{F}$ of magnitude

$$F = qvB \sin \theta. \tag{12.1}$$

The direction of $\vec{F}$ is interesting (see Fig. 12-7). This magnetic force is a *deflecting force*; $\vec{F}$ is always *perpendicular* to $\vec{v}$. And the magnetic force, in contrast to electric and gravitational forces, is *not* in the direction of the field. Instead $\vec{F}$ is perpendicular to the magnetic field $\vec{B}$. The further specification of the direction of this magnetic force $\vec{F}$ is given by

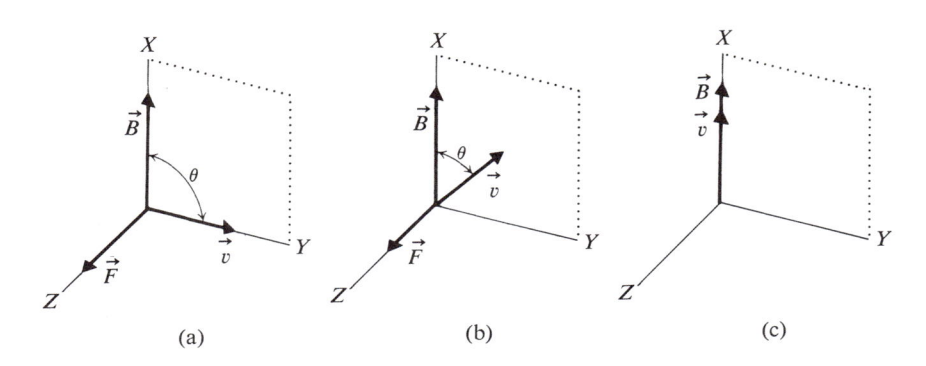

Fig. 12-7 The relationship of $\vec{F}$ to $\vec{v}$ and $\vec{B}$, where $\vec{F}$ is the force experienced by a positive charge moving at velocity $\vec{v}$ in a magnetic field $\vec{B}$. (X, Y, Z are mutually perpendicular axes.) (a) Here $\theta = 90°$, $\sin \theta = 1$, and F therefore has maximum value. (b) $\theta < 90°$, thus $\sin \theta < 1$, and F is therefore smaller than in part (a). (c) $\theta = 0°$ ($\vec{B}$ and $\vec{v}$ are parallel), $\sin \theta = 0$, and hence $F = 0$.

the *right*-hand rule displayed in Fig. 12-8. With the thumb in the direction of the velocity vector and the outstretched fingers pointing in the direction of the magnetic field $\vec{B}$, the magnetic force $\vec{F}$ exerted on a *positive* charge is perpendicular to the palm in the direction that this palm would push. The magnetic force on a negative charge is in the opposite direction.

We summarize this rather complicated rule for the direction of the magnetic force by

$$\vec{F} \perp \vec{v}, \tag{12.2a}$$

$$\vec{F} \perp \vec{B}, \tag{12.2b}$$

and the further specification of the direction of $\vec{F}$ is given by the above right-hand rule.

Before considering the marvelous things that can be done with this magnetic force we must pause to clarify the logic of this description of magnetic interaction. We have yet to define the magnetic field $\vec{B}$. To accomplish this we need merely to point out that the prescription given by Eqs. (12.1), (12.2a), (12.2b) and the right-hand rule of Fig. 12-8 can be taken as a *definition* of the field $\vec{B}$ at a given point, since this prescription shows how, by laboratory measurement of force, velocity, and charge, one could determine $\vec{B}$. Solving Eq. (12.1), we obtain

$$B = F/qv \sin \theta,$$

which shows that the mks unit of B is the (N-sec)/(coul-m), which is named the weber/m².*

$$1 \text{ weber/m}^2 = 1 \text{ (N-sec)/(coul-m)}.$$

A field of 1 weber/m² is a very strong magnetic field. The field between the polepieces of common permanent magnets is often about 0.1 weber/m². A typical value of the earth's magnetic field encountered on the earth's surface is 5×10^{-5} weber/m². Magnetic fields are often given in terms of a much smaller unit, the gauss: 1 gauss = 10^{-4} weber/m².

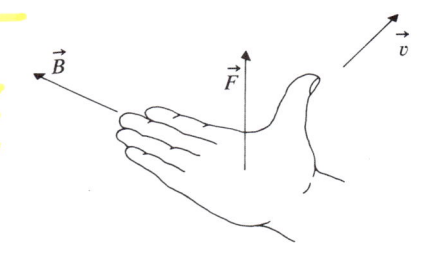

Fig. 12-8 Right-hand rule to determine the direction of the magnetic force $\vec{F}$ exerted by a magnetic field $\vec{B}$ on a *positive* charge moving with a velocity $\vec{v}$.

*The motivation for the name weber/m² will be apparent in Section 12.5. Then the reader will see that

flux in webers = (B in webers/m²) × (area in m²)

In the SI system of units the weber/m² is called the tesla.

12.2 THE FORCE EXERTED BY $\vec{B}$ ON A MOVING CHARGE

Example 1 A proton is moving horizontally north at a speed of 4.0×10^7 m/sec between the polepieces of a magnet where there is a magnetic field of 1.5 weber/m² pointing vertically downward. Find the magnitude and direction of the magnetic force exerted on the proton.

Solution Here $q = 1.6 \times 10^{-19}$ coul. The angle θ between $\vec{v}$ and $\vec{B}$ is $90°$ so $\sin \theta = 1$. Then the magnetic force exerted by $\vec{B}$ on the proton has a magnitude

$$F = qvB \qquad (12.1)$$
$$= (1.6 \times 10^{-19} \text{ coul}) \times (4.0 \times 10^7 \text{ m/sec}) \times (1.5 \text{ weber/m}^2)$$
$$= 9.6 \times 10^{-12} \text{ N}.$$

Since $\vec{F}$ is perpendicular to $\vec{B}$, this force must lie in a horizontal plane. To be perpendicular to $\vec{v}$, the force must then lie along a horizontal line running east-west. Now a right hand with the thumb pointing north (direction of $\vec{v}$) and the fingers downward (direction of $\vec{B}$) will have the palm facing west (direction of force is "out of palm," see Fig. 12-8). The magnetic force on the *positively* charged proton is therefore given by

$$\vec{F} = 9.6 \times 10^{-12} \text{ N, } west.$$

12.3 CIRCULAR MOTION OF A CHARGE IN A UNIFORM MAGNETIC FIELD

Circular Orbits

From the study of the trajectories of charged particles in magnetic fields has come a rich harvest of knowledge about atoms and elementary particles. Fig. 12-9 shows a particle with velocity $\vec{v}$ moving perpendicular to a uniform magnetic field $\vec{B}$. During any very short time interval, the particle moves in the direction of $\vec{v}$. Since this movement is perpendicular to the magnetic force, *no work is done by the magnetic force* [see Eq. (4.3)]. Consequently the magnetic force does *not* change the kinetic energy of the particle: the particle moves with constant speed. The magnetic force, $F = qvB$, therefore has a constant magnitude if the particle continues to encounter the same value of $\vec{B}$ (i.e., if $\vec{B}$ is constant both in direction and in magnitude). While this magnetic force does not alter the magnitude (speed) of the particle's velocity, it does change the *direction* of the particle's velocity vector. The particle is continually deflected in a way which curves its trajectory into a circle of radius r.

The acceleration vector $\vec{a}$, according to Newton's second law ($\vec{F} = m\vec{a}$), is parallel to the magnetic force and is therefore perpendicular to the velocity vector and points directly toward the center of the circular trajectory. The acceleration is a centripetal acceleration (see Section 2.3) of magnitude v^2/r. Applied to this motion, Newton's second law,

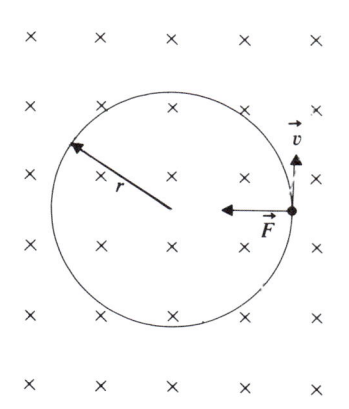

$$\vec{F} = m\vec{a}, \qquad (2.10)$$

Fig. 12-9 A positively charged particle moving perpendicular to a uniform magnetic field is deflected by the magnetic force $\vec{F}$. The trajectory is a circle of radius r. The magnetic field is directed into the paper, as indicated by the x's.

with $F = qvB$ and $a = v^2/r$, gives

$$qvB = mv^2/r. \tag{12.3}$$

Momentum Measurement

Solving Eq. (12.3) for the product mv gives

$$\text{momentum} = qBr. \tag{12.4}$$

This simple result shows that if a uniform magnetic field $\vec{B}$ is established over a track recorder like a cloud or bubble chamber (Section 1.3), then a particle's momentum can be determined from a knowledge of its charge and the radius of curvature r of its track. In this way, the momenta involved in elementary particle reactions are measured (Fig. 12-10).

Fig. 12-10 Charged particles moving in circular paths under the influence of a magnetic field. This picture shows electron-positron pair production as seen in a bubble chamber. (In the experiment in which this event occurred, the incident beam was composed of anti-protons. One of these anti-protons interacting with the liquid hydrogen in the bubble chamber has produced a neutral particle. This neutral particle, leaving no track, has decayed into an electron and a positron, which have left spiral tracks as they give up their initial energy. (Courtesy Brookhaven National Laboratory.)

When these circular orbits in a magnetic field are analyzed according to *relativistic* dynamics, one still finds Eq. (12.4) holds, but

$$\text{momentum} = \frac{mv}{\sqrt{1 - v^2/c^2}}. \tag{5.27}$$

Measurements of orbit radii of particles with speeds well over 99% of the speed of light are in complete agreement with this relativistic expression for the momentum.

Measurement of "Charge-to-Mass Ratio," Mass Spectrometers

The "charge-to-mass ratio" (q/m) of a particle can be accurately determined by measurement of the time T required for the particle to travel once around a circular orbit in a known magnetic field, B. Eq. (12.3) gives

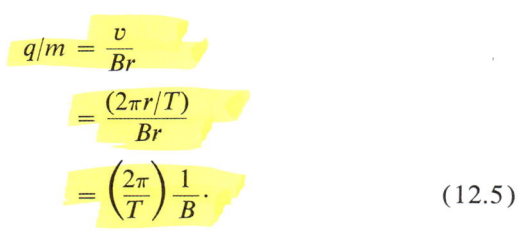

$$q/m = \frac{v}{Br}$$

$$= \frac{(2\pi r/T)}{Br}$$

$$= \left(\frac{2\pi}{T}\right)\frac{1}{B}. \qquad (12.5)$$

Remarkably, this relationship is independent of the orbit radius.

For ions of known charge, the ion's mass can be calculated after its charge to mass ratio has been measured. The apparatus which is used to determine the mass of an ion from observations of the ion orbit in a magnetic field is called a *mass spectrometer*. (In many mass spectrometers, instead of measuring T directly, the ion speed and orbit radius are measured.) Mass spectrometers have not only provided conclusive evidence of the existence of stable and radioactive isotopes of almost all elements but also furnished precise mass measurements of various isotopes with an accuracy of better than one part in a million.

Cyclotrons

In the cyclotron invented by Lawrence and Livingston (Section 4.6), a large electromagnet is used to establish an almost uniform magnetic field $\vec{B}$ over an evacuated region between its circular polepieces. This particle accelerator imparts to a charged particle a large kinetic energy by causing the particle to make repeated traversals of a region in which there is an accelerating electric field between two electrodes. It is instructive to see, in some detail, just how this is accomplished.

Consider a particle with charge q moving perpendicular to the magnetic field with a constant speed along a portion, *ab*, of its circular orbit (Fig. 12-11). In the region between the electrodes the particle experiences an electric force, $q\vec{E}$, exerted by the electric field. When this electric field has the direction indicated in Fig. 12-11, the particle speeds up during the trip, *bc*. The trajectory, *cd*, occurs in the region where the electric field is almost zero, so the particle moves with constant speed in a semicircle whose radius is determined by the momentum, according to Eq. (12.4). The time for this half circuit, from Eq. (12.5) is given by

$$T/2 = \frac{\pi m}{Bq}. \qquad (12.6)$$

This trip is completed in a time which is independent of the particle's speed or the size of the orbit. The polarity of the voltage applied across the electrodes is altered in successive intervals of $T/2$ so that the electric field between the electrodes reverses in time to again push this

Physics Today

E. O. Lawrence
(1901–1958)

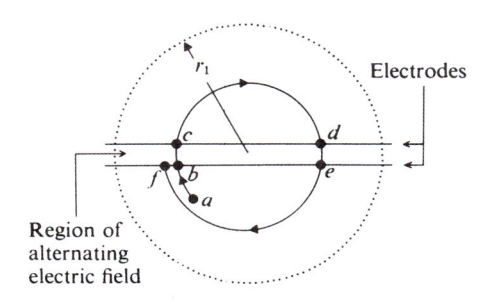

Fig. 12-11 Orbit of a charged particle in a cyclotron. A uniform magnetic field directed out of the paper is established over a circular region of radius r_1.

442 ELECTROMAGNETISM

particle to higher speeds during the traversal, *de*. The increased speed implies that the subsequent semicircular orbit, *ef*, will have a larger radius than the semicircle, *cd*.

A particle, projected from a central ion source, thus receives a succession of pushes from the electric field and moves in a sequence of semicircular orbits which increase in radius as the particle's speed increases, until the particle reaches the largest possible orbit (r_1 in Fig. 12-11) over which the guiding magnetic field B_1 has been established. The product, $B_1 r_1$, determines, from Eq. (12.4), the maximum momentum and consequently the maximum kinetic energy to which the particle can be accelerated in a given cyclotron:

$$\text{maximum momentum} = qB_1 r_1. \qquad (12.7)$$

This is the basic equation governing the design of all accelerators which employ a guiding magnetic field. To achieve high energy (and therefore high momentum), a strong magnetic field B_1 must be established over an orbit of large radius r_1. For this reason, high-energy particle accelerators are huge. For example, the orbit circumference of the accelerator at Weston, Illinois, is four miles.

A cyclotron, for which the time between successive reversals of the accelerating electric field is *constant*, cannot be used to accelerate particles to speeds close to the speed of light. The relativistically correct modification of Eq. (12.6),

$$\frac{T}{2} = \frac{1}{\sqrt{1 - v^2/c^2}} \times \frac{\pi m}{Bq}, \qquad (12.8)$$

shows that the time for the particle to complete a half circle in a given magnetic field in fact does depend on the particle's speed v but that this effect will become important only when the factor $\sqrt{1 - v^2/c^2}$ becomes appreciably different from the number 1, that is when the speed v is no longer much less than the speed of light, c.

Accelerators called synchrotrons do succeed in pushing particles to speeds extremely close to the speed of light. Modern synchrotrons, such as the Fermilab accelerator, keep the particle moving in the same circular orbit by increasing the magnetic field as the particle's speed builds up. The time between successive reversals of the accelerating electric field is correspondingly altered in accordance with Eq. (12.8).

12.4 MAGNETIC FORCE AND ELECTRIC CURRENT

Magnetic Force on a Wire Carrying a Current

In Fig. 12-12 a simple experiment is indicated which demonstrates that a magnetic field exerts a side thrust on a wire carrying an electric current.

Harvard News Office

M. S. Livingston
(1905–   )

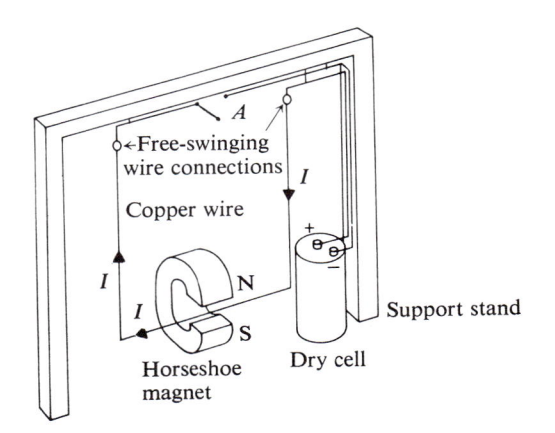

Fig. 12-12 Experimental setup demonstrating the force on a current-carrying conductor in a magnetic field. When the switch *A* is closed, the wire will swing out from the jaws of the magnet.

This is the type of force that was Faraday's first great discovery in electromagnetism, and which is exploited in all electric motors and many kinds of meters.

The origin of the force on the wire can be easily understood in terms of phenomena that we have already treated. The electric current in the wire consists of electric charges drifting along within the wire. When the wire is placed in a magnetic field $\vec{B}$,* each moving charge within the wire experiences a magnetic force as described in Section 12.2. Such forces on the moving charges are transmitted to the wire as a whole.

When the magnetic field $\vec{B}$ makes an angle θ with the wire, the resultant force $\vec{F}$, which is the vector sum of the magnetic forces exerted on all the moving charges in a length L of a conductor carrying a current I, can be very neatly expressed, after some straightforward analysis (that is outlined in Question S-11) by

$$F = BIL \sin \theta. \qquad (12.9)$$

The direction of the magnetic force $\vec{F}$ is determined by the direction of the force on moving positive charges, given by Eq. (12.2) and the right-hand rule of Fig. 12-8. $\vec{F}$ is perpendicular to the *wire* (a side thrust) and to the field $\vec{B}$. (See Fig. 12-13.)

When $\vec{B}$ is perpendicular to the wire, angle θ is 90° and $\sin \theta = 1$. Then Eq. (12.9) simplifies to $F = BIL$. On the other hand, if the direction of the wire happens to be parallel to $\vec{B}$ so that θ and $\sin \theta$ are both zero, then the magnetic force exerted by $\vec{B}$ is zero.

*To indicate the direction of the magnetic field produced by a magnet, one "pole" is labeled N (north) and the other S (south). Within the gap between the poles, $\vec{B}$ is directed away from the magnet's north pole toward its south pole.

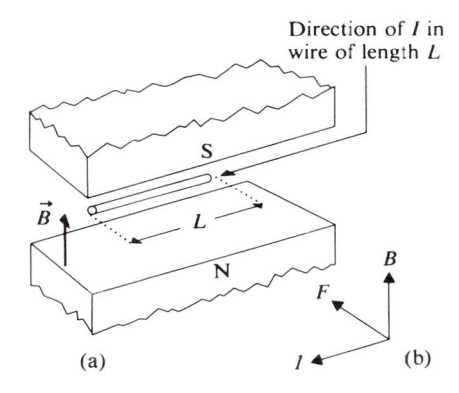

Fig. 12-13 (a) The physical setup in which a wire of length L carrying a current I would experience a force F (the connecting wires to the length L from some source of EMF are not shown). (b) The relative directions of the current I, the magnetic field B, and the force F on the wire.

Example 2 The magnetic field between the polepieces of an electromagnet (Fig. 12-14) is 0.080 weber/m² pointing horizontally eastward. A rectangular loop of wire is pivoted about a horizontal axis and placed in this magnetic field, orientated so that the plane of the coil is horizontal [Fig. 12-14(a)]. Each side of the loop parallel to the axis is 0.10 meter long.

(a) Find the magnitude and direction of the force exerted by the 0.080-weber/m² magnetic field on the left conductor of the coil when a current of 5.0 amp is passed through the coil in the direction indicated in Fig. 12-14(a).
(b) Find the magnetic force on the right conductor.
(c) Which way will the coil rotate?

Solution
(a) The left conductor is perpendicular to the magnetic field, so Eq. (12.9) with $\sin \theta = 1$ gives:

$$\begin{aligned} F &= BIL \\ &= (0.080 \text{ weber/m}^2) \times (5.0 \text{ amp}) \times (0.10 \text{ meter}) \\ &= 0.040 \text{ N}. \end{aligned}$$

The direction of this magnetic force is perpendicular to $\vec{B}$ and also perpendicular to the wire. The force is therefore vertical and the right-hand rule (Fig. 12-8) gives the direction as upward.

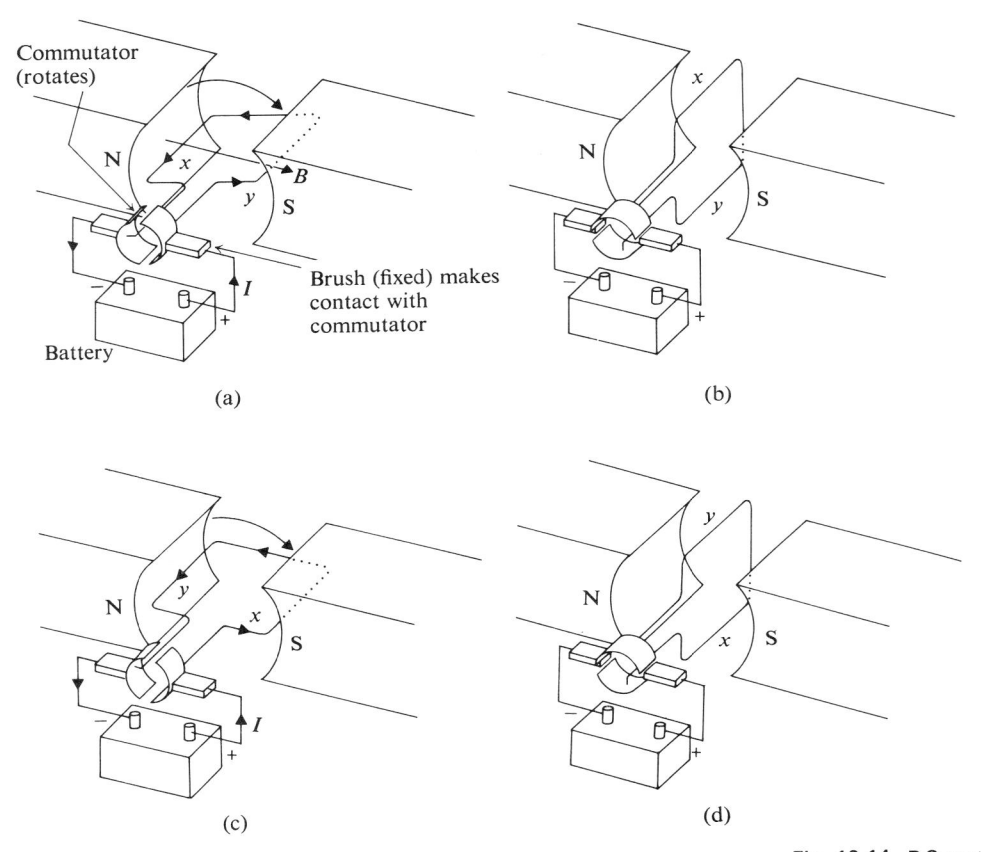

Fig. 12-14 DC motor. Diagram showing the basic elements. Parts (a), (b), (c), and (d) show successive positions involved in a complete rotation. In each case, x and y are the parts of the winding in the left and right positions, respectively, as shown in case (a).

(b) For the right conductor, we have also

$$F = BIL = 0.040 \text{ N}.$$

Since the current direction is opposite to that of the current in the left conductor, the magnetic force will be in the direction opposite to that of the magnetic force on the left conductor.

(c) These two forces exert a torque on the coil. The coil will rotate as indicated. When this coil is vertical, the torque of these forces is zero and, if the coil were at rest, it would remain in this vertical plane.

Meters and Motors

The magnetic force exerted by a magnetic field on a current-carrying wire is the force which makes electric motors and many meters function.

12.4 MAGNETIC FORCE AND ELECTRIC CURRENT 445

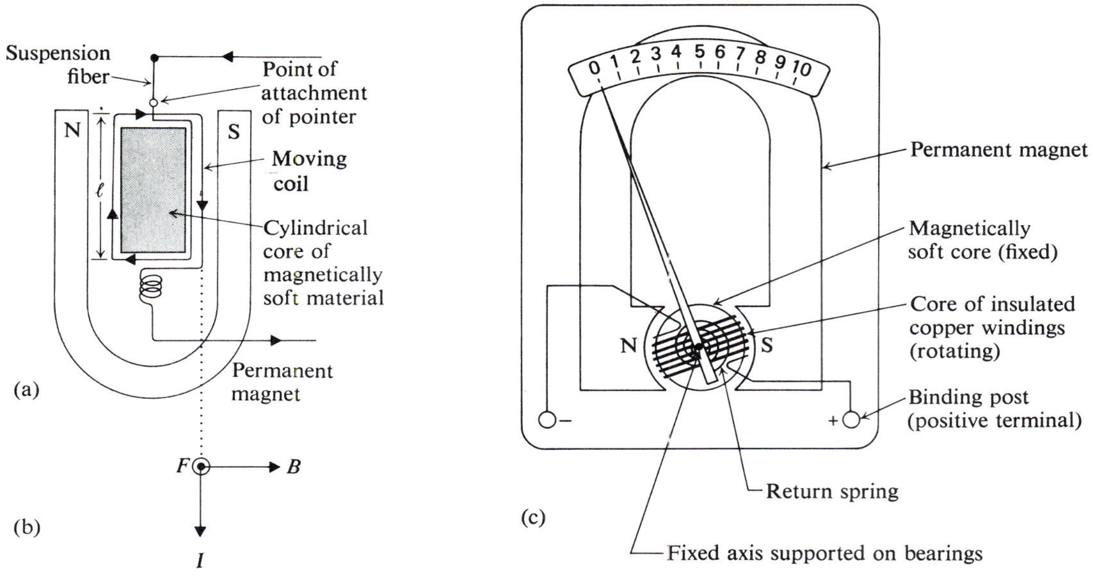

Fig. 12-15 D'Arsonval moving-coil galvanometer. (a) Diagram of the meter. (b) The force on the longitudinal part of the winding on the right, employing the directions associated with Eq. (12.9), is outward. (c) Galvanometer with the moving coil attached to an axis supported on bearings.

In a *galvanometer* (Fig. 12-15), a current under investigation is passed through a coil which is suspended in a magnetic field established by a permanent magnet. The coil then experiences current-dependent magnetic forces which cause the coil to rotate (as in Example 2) until their effect is balanced by restoring forces furnished by a hair spring. The amount of coil rotation, which is a measure of the coil current, is indicated by a pointer attached to the coil.

Galvanometers are generally very sensitive: a common type gives a rotation of 30° when the galvanometer current is merely 10^{-4} amp or 0.1 milliampere. Ammeters and voltmeters are simply galvanometers used with appropriate resistors.

An *ammeter* (Fig. 12-16) to measure currents ranging from, say, 0 to 10 amp, is obtained by using a very low resistance, called a *shunt* resistor, in *parallel* with a galvanometer. Most of the current through the ammeter passes through the shunt. The resistances involved can be so chosen that the small fraction of the current which *does* pass through the galvanometer coil produces full-scale deflection of the galvanometer needle when the total ammeter current is 10 amp.

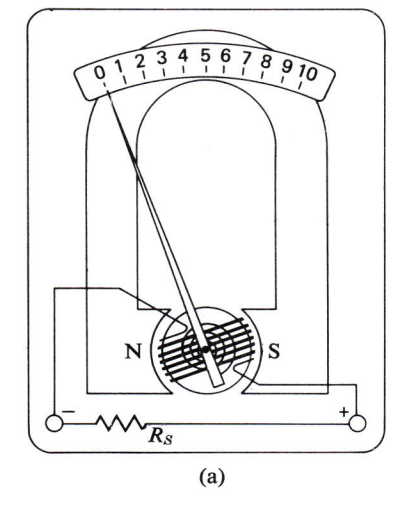

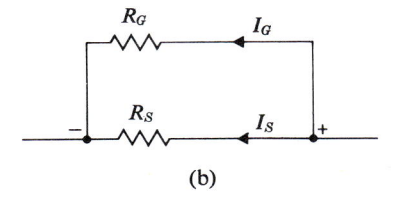

Fig. 12-16 Ammeter. (a) Diagram showing the moving-coil galvanometer connected in parallel with the very low resistance shunt, R_s. (b) Circuit diagram of the arrangement shown in (a).

446 ELECTROMAGNETISM

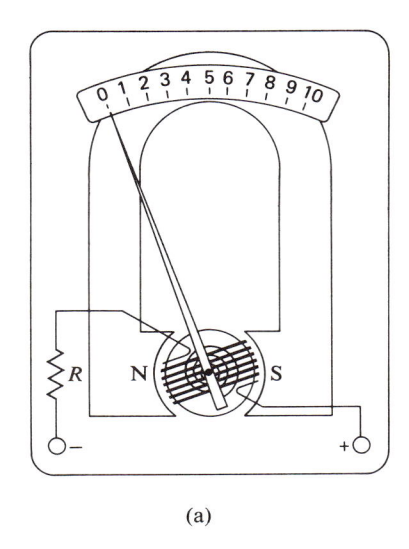

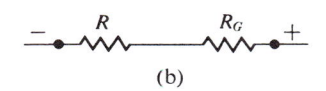

Fig. 12-17 (a) Voltmeter composed of a moving-coil galvanometer connected in series with a high resistance R. (b) Equivalent circuit diagram of this voltmeter. (R_G is the resistance of the wire of the galvanometer coil.)

A *voltmeter* (Fig. 12-17) consists of a high resistance connected in *series* with a galvanometer coil. To construct a voltmeter reading from 0 to 150 volts, the high resistance is selected so that the galvanometer current attains the value sufficient to produce full-scale galvanometer deflection when 150 volts are applied across the voltmeter terminals.

A prototype electric motor (Fig. 12-14) can be devised using the coil and magnetic field of Example 2. The magnetic forces cause this coil to rotate until its plane is vertical. Inertia causes the coil to continue this rotation, and now the same magnetic forces oppose continued rotation. In an electric motor, the force direction is therefore automatically reversed at this point by reversing the direction of the coil current.* You can easily verify that the magnetic forces will then be properly directed until the coil is again vertical. By reversing the coil current every half revolution, the coil is forced to continue rotating.

*The reversal is accomplished as fixed electrical contacts, called brushes, slide over different segments of the rotating armature (*the commutator of Fig. 12-14*).

Magnetic Field Produced by a Long, Straight Wire

As emphasized in Section 12.1, a wire carrying a current sets up, at all points of surrounding space, a magnetic field. In 1820, the French physicists, Biot and Savart, investigated in quantitative detail the field $\vec{B}$ produced by a long, straight wire carrying a current I, depicted in Fig. 12-3. Their experimental results were expressed by the formula

$$B = \frac{KI}{r}, \qquad (12.10)$$

which gives the magnitude B of the magnetic field at a distance r from the wire. The evaluation of the constant K will be discussed shortly.

12.4 MAGNETIC FORCE AND ELECTRIC CURRENT

The vector $\vec{B}$ has a direction at any point which is perpendicular to the plane containing the wire and the point (see Fig. 12-3). The right-hand rule of Fig. 12-3(c) can be used to determine the direction of the magnetic flux lines and consequently the direction of the magnetic field $\vec{B}$ at any point.

Magnetic Force Exerted by One Wire Upon Another

The assertion that one moving charge exerts a magnetic force on another moving charge was made at the beginning of this chapter. We are now in a position to examine such a magnetic interaction, at least for the simple case of two distinct steady streams of moving charged particles, constituting two currents, I_1 and I_2, in long, straight, parallel sections of wires which are a distance r apart (Fig. 12-18).

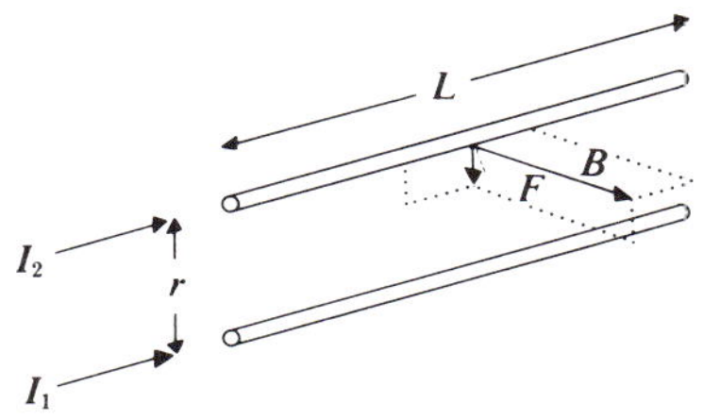

Proceeding by two distinct steps (see Section 12.1), we first evaluate the magnetic field produced by the current I_1 at the location of the second current. This field, from Eq. (12.10), has a magnitude

$$B = \frac{KI_1}{r},$$

and is perpendicular to the second wire, as shown in Fig. 12-18. Now, according to Eq. (12.9), this field exerts a force on the length L of the wire carrying the current I_2, given by

$$F = BI_2L$$
$$= \left(\frac{KI_1}{r}\right)I_2L. \qquad (12.11)$$

Fig. 12-18 Sections of two parallel long straight wires. The magnetic field B produced by current I_1 at the location of current I_2 exerts a force F on a length L of the wire carrying the current I_2.

This force, perpendicular to $\vec{B}$ and to the second wire, is, according to the right-hand rule of Fig. 12-8, directed toward the first wire. Evidently the magnetic force is attractive when the currents are in the same direction.

Calculation of the magnetic field produced by I_2 at location of I_1, and of the force then exerted by this field on I_1, leads again to an attractive force given by Eq. (12.11).

Definitions of the Ampere and the Coulomb

Measurement of the magnetic force between wires carrying a current gives a precise and reproducible laboratory method of establishing the size of the mks unit of current, the *ampere*. The size of the ampere is selected so that the constant K in Eq. (12.11) and Eq. (12.10) has a value of exactly 2×10^{-7} N/amp². In other words, from Eq. (12.11), one ampere is the current in two long, straight, parallel wires, one meter apart, that gives rise to a magnetic force of 2×10^{-7} N on each one-meter length of wire.

The mks unit of charge, the *coulomb*, is defined in terms of the ampere and the second:

$$1 \text{ coul} = (1 \text{ amp}) \times (1 \text{ sec}).$$

In words, a coulomb is defined to be the electric charge that passes a point in a circuit in one second when the circuit current is one ampere.

Having thus fixed the size of the coulomb, the constant of proportionality which occurred in Coulomb's law, Eq. (2.19), is determined experimentally to be 9.00×10^9 N m²/coul². The ratio of this constant to the constant in the magnetic force law [Eq. (12.11)] has the units of a velocity squared and a most interesting numerical value. This value is

$$\frac{k}{K} = \frac{9.00 \times 10^9 \text{ N m}^2/\text{coul}^2}{2 \times 10^{-7} \text{ N/amp}^2} = \frac{1}{2} \times (3.00 \times 10^8 \text{ meter/sec})^2.$$

The quantity in parentheses is the speed of light! Perhaps electricity and magnetism have something to do with light. This is a clue which contributed to a most brilliant work in theoretical physics, James Clerk Maxwell's electromagnetic theory of light, the topic of Chapter 14.

12.5 FARADAY'S LAW OF ELECTROMAGNETIC INDUCTION

The Electric Field Associated with a Changing Magnetic Field

In a region where a magnetic field is *changing*, there is an *electric field* which is related to the *rate of change* of the magnetic field.

Suppose, for instance, that an increasing current is driven through the coils of the electromagnet shown in Fig. 12-6. Then the magnetic field in the region between the polepieces increases. The associated electric field in a plane parallel to the polepieces is shown in Fig. 12-19. Here the electric field at any point is proportional to the rate of change of the magnetic field.

An electron, projected so as to travel through the perpendicular magnetic field in the circular orbit indicated in Fig. 12-19, will be continually accelerated by this *electric field*. Such an electric field is exploited in a betatron, a modern electron accelerator.* (See Fig. 12-20 on the following page.) In a betatron of the type used in medicine, shown in Fig. 12-20(a), the electrons circulate within the tube many thousands of times, gaining energies up to 50 MeV. If the betatron magnet is excited by an alternating voltage of 180 hertz, all this energy is gained in 1/720 sec, during which time the magnetic field increases from zero to its maximum value.

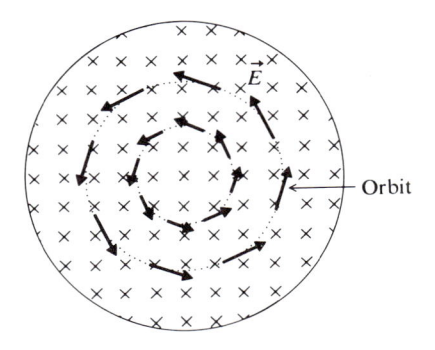

Fig. 12-19 The electric field $\vec{E}$ (represented by arrows) associated with a *changing* magnetic field $\vec{B}$. The magnetic field (represented by the x's) is directed into the paper and is of *increasing* magnitude.

*It can be shown that the electron gains kinetic energy at a rate sufficient to maintain an orbit of constant radius in the increasing magnetic field, if the magnetic field at the orbit at any instant is just half of the average at that instant of the nonuniform magnetic field occurring in the circular area within the orbit.

(a)

(b)

(c)

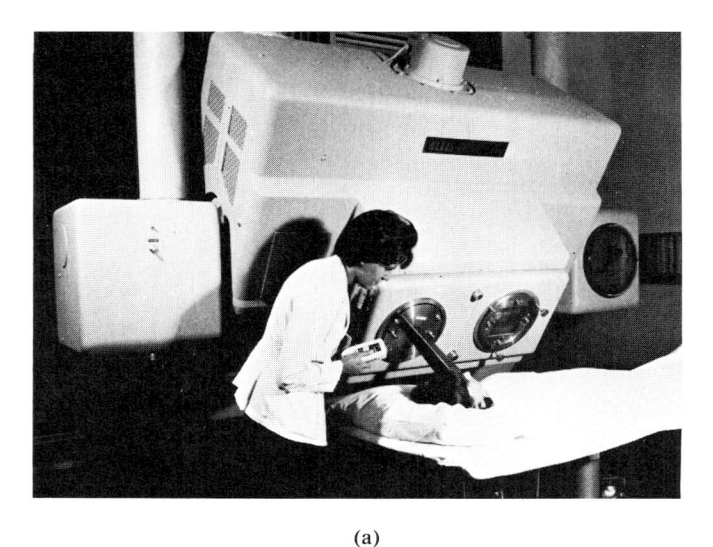

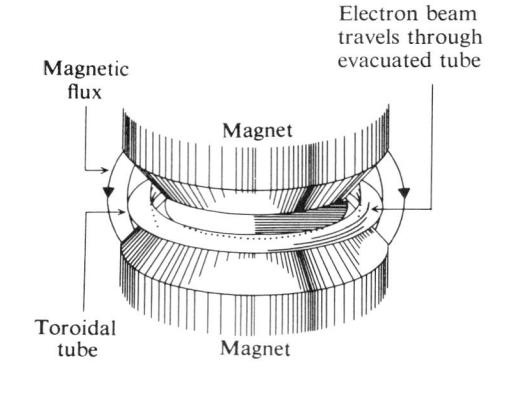

(d)

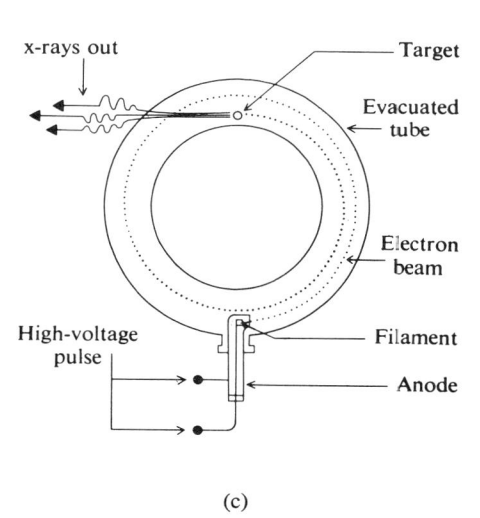

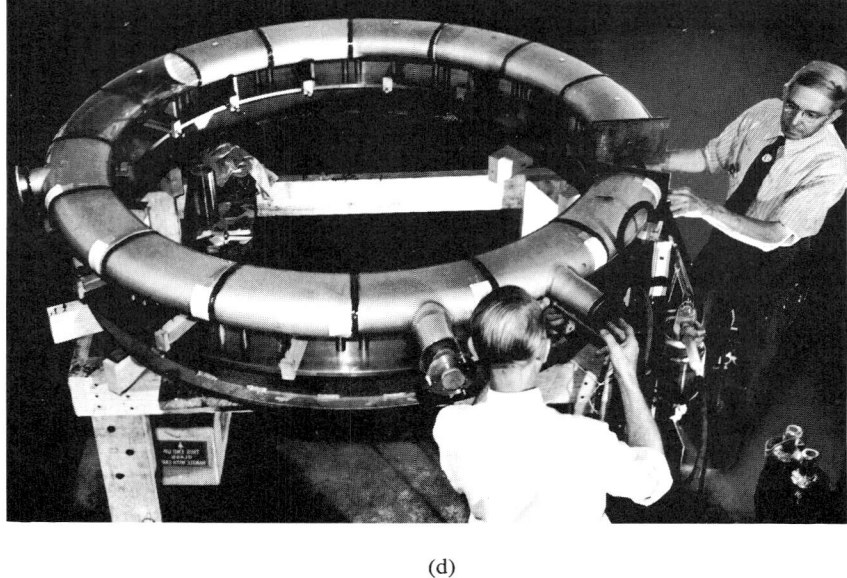

Fig. 12-20 The betatron. (a) A 25-million volt betatron in use for x-ray therapy. (Courtesy of ATC Betatron, formerly the Allis-Chalmers Betatron Department.) (b) Diagram of principal betatron components viewed from the side. (c) The toroidal tube showing electron beam as it might be used to produce x-rays by striking a solid target. (d) Assembling the toroidal tube in a research betatron. (Courtesy of General Electric Co.)

Because of the electric field, associated with a changing magnetic field, there is an EMF associated with a closed path like the electron orbit of Fig. 12-19. In other words, electric work, W, is done if a "test charge" q is taken around a closed path. The quotient W/q is the EMF. To discuss Faraday's great discovery which is summarized in a simple rule for the size of such an EMF, we must first define a new term, *magnetic flux*.

Magnetic Flux

Consider a closed path bounding a plane surface of area A (Fig. 12-21). Suppose there is a uniform magnetic field $\vec{B}$ perpendicular to this surface. The *magnetic flux*, Φ, through the surface of area A (or enclosed by the given path) is defined by the equation

$$\Phi = BA. \qquad (12.12)$$

The mks unit of flux is the *weber*; since $B = \Phi/A$, it can now be seen why the unit of B is named the weber/m².

When $\vec{B}$ is uniform but not necessarily perpendicular to the surface of area A, the definition for the flux through A is

$$\Phi = \text{(component of } \vec{B} \text{ which is perpendicular to } A) \times A. \qquad (12.13)$$

Finally, to define the flux through a surface which is so large that $\vec{B}$ has appreciably different values at different points on the surface, we simply subdivide the large surface into many small surfaces, small enough so that $\vec{B}$ changes only negligibly over each small surface. Then Eq. (12.13) defines the flux through each small surface, and the sum of all these fluxes is the total flux through the large surface.

Faraday, like many other students of nature, found it valuable to visualize even such intangible entities as magnetic flux. To do this, one maps out a magnetic field with a very large number of *flux lines* (Fig. 12-22) in such a way that *the flux through any surface is proportional to the number of flux lines going through the surface*. It is then an important property of the magnetic fields produced by currents that these flux lines are closed curves. A flux line never begins or ends.

Faraday's Law

Faraday discovered that when the magnetic flux Φ enclosed by a circuit is changing, an EMF $\mathcal{E}$ is "induced" in the circuit which is proportional to the rate of change of the flux Φ. In mks units, Faraday's law is

$$\mathcal{E} = \frac{\Delta \Phi}{\Delta t}, \qquad (12.14)$$

where $\Delta \Phi$ is the *change* in the flux enclosed by the circuit during the very short time interval Δt.

Example 3 The uniform magnetic field of Fig. 12-19 increases at a uniform rate from 0.10 weber/m² to 0.30 weber/m² in 5.0×10^{-2} sec. A single coil of wire, placed in the position of the circular orbit shown in Fig. 12-19, encloses an area of 0.40 m². Find the EMF induced in this coil while this flux change is occurring.

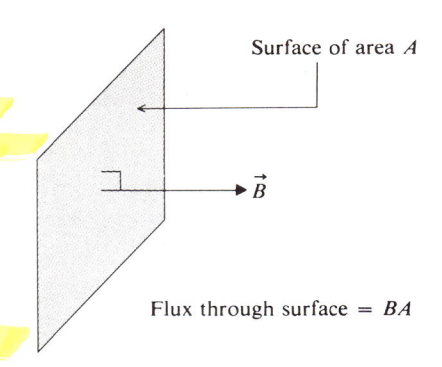

Fig. 12-21 The magnetic flux Φ through a plane surface of area A lying in a uniform magnetic field B which is perpendicular to the surface is given by: $\Phi = BA$.

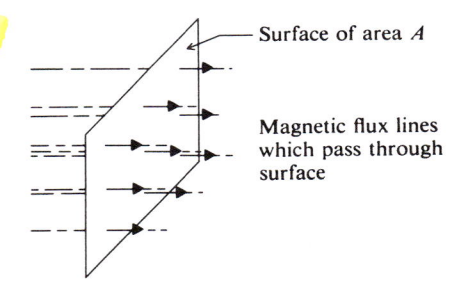

Fig. 12-22 The magnetic flux Φ through a surface may be visualized as proportional to the number of flux lines which pass through the surface.

12.5 FARADAY'S LAW OF ELECTROMAGNETIC INDUCTION

Solution The initial flux enclosed by the coil is, from Eq. (12.12),

$$(0.10 \text{ weber/m}^2) \times (0.40 \text{ m}^2) = 0.040 \text{ weber}.$$

After 5.0×10^{-2} sec have elapsed, the flux has increased to

$$(0.30 \text{ weber/m}^2) \times (0.40 \text{ m}^2) = 0.12 \text{ weber}.$$

The *change* in flux, $\Delta\Phi$, is given by

$$\Delta\Phi = 0.12 \text{ weber} - 0.04 \text{ weber} = 0.08 \text{ weber}.$$

This flux change occurs in 5.0×10^{-2} sec, so the rate of change of flux $\Delta\Phi/\Delta t$ is given by

$$\frac{\Delta\Phi}{\Delta t} = \frac{0.08 \text{ weber}}{5.0 \times 10^{-2} \text{ sec}} = 1.6 \text{ webers/sec}.$$

Faraday's law states that this is the EMF induced in the coil:

$$\mathcal{E} = \frac{\Delta\Phi}{\Delta t} = 1.6 \text{ volts}.$$

It is important to note that even if there is no coil of wire, there is still an electric field in space and there is an induced EMF associated with any closed path in space that encloses a changing magnetic flux. If the coil of Example 3 were removed, an electron orbiting at that very same location would acquire 1.6 eV additional kinetic energy on each revolution.

Lenz's Law

The polarity (or sense) of an induced EMF is determined by *Lenz's law: the sense of an induced EMF arising from a changing magnetic flux is such that the induced current is directed so as to create flux that opposes the flux change.* If the flux is *decreasing*, the EMF induced is such as to produce a current directed to create flux in the direction of the original flux. And if the flux is *increasing*, the flux created by the induced current will be opposite in direction to the original magnetic flux.

Thus, in Example 3, the downward flux through the coil is increasing. By Lenz's law, the induced current in the coil will be directed to create upward flux through the coil. This current direction and the flux produced by such a current is shown in Fig. 12-4(a). In Fig. 12-19 the direction of the induced current would be counterclockwise.

Transformers

EMFs associated with a changing magnetic flux are used to transform a varying voltage in one circuit into a larger or smaller voltage in another circuit. This is accomplished with a transformer (Fig. 12-23) consisting of two coils electrically insulated from each other and wound on the same ferromagnetic core.

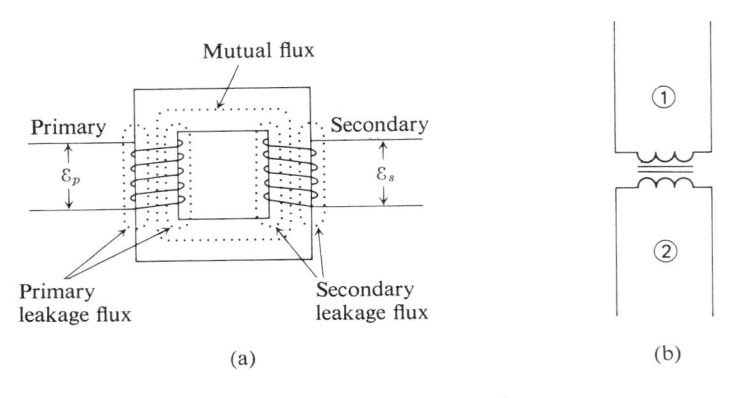

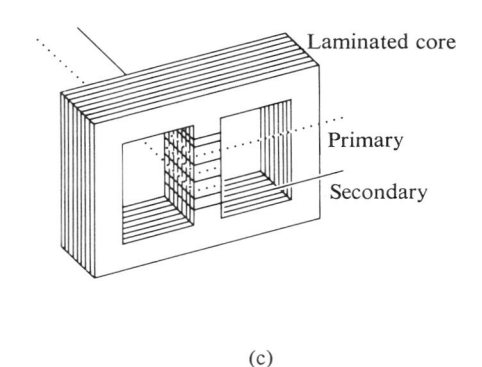

Fig. 12-23 Transformers. (a) Core-type transformer, showing magnetic flux leakage. (b) Circuit symbol for ferromagnetic core transformer. (c) Shell-type transformer. Most transformers are of this type, having concentric windings to minimize flux leakage. (d) Photograph of a common shell-type transformer.

Power is supplied to one coil named the *primary* coil. A varying current in this coil sets up a varying magnetic flux, largely confined to the ferromagnetic core. The other coil, called the *secondary* coil, thus encloses a varying flux. While this flux is changing at the rate $\Delta\Phi/\Delta t$ each of the turns in the secondary coil experiences, according to Faraday's law, an induced EMF equal to $\Delta\Phi/\Delta t$. With N_s turns in series, the EMF for the entire secondary coil is

$$\mathcal{E}_s = N_s \frac{\Delta\Phi}{\Delta t}. \tag{12.15}$$

The magnetic flux at any instant is almost the same through the primary and secondary coils. There is in the primary coil an EMF, $\Delta\Phi/\Delta t$, induced in each of the N_p turns. Therefore the EMF for the primary coil is given by

$$\mathcal{E}_p = N_p \frac{\Delta\Phi}{\Delta t}. \tag{12.16}$$

By dividing Eq. (12.15) by Eq. (12.16) we obtain

$$\frac{\mathcal{E}_s}{\mathcal{E}_p} = \frac{N_s}{N_p}. \tag{12.17}$$

According to this result, when there are varying currents, the ratio of the EMFs in the secondary and primary coil at any instant is equal to the "turns ratio," N_s/N_p. When N_s is greater than N_p, the secondary EMF exceeds the primary EMF, and the transformer is named a *step-up* transformer. In a *step-down* transformer like those used for toy electric cars and trains, N_s is less than N_p, and the household voltage applied across the primary is transformed into a safe low secondary voltage

12.5 FARADAY'S LAW OF ELECTROMAGNETIC INDUCTION

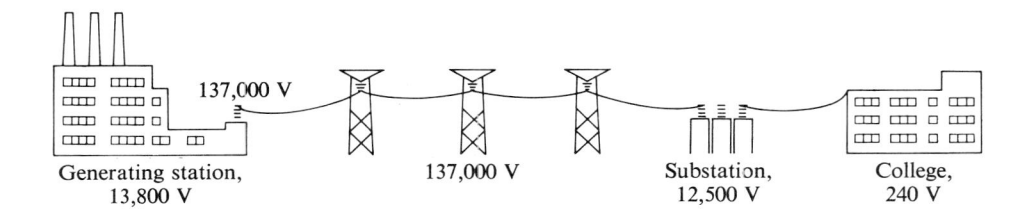

Fig. 12-24 Long-distance transmission of electrical energy. Transformers raise the voltage at the generator, lower it at city substations and again at residences or other points of usage. Large colleges and universities may have their own substations.

which delivers power to the toy. Step-up and step-down transformers find extensive use in the distribution of power in our electrical age. (See Fig. 12-24.)

It should be emphasized that only varying voltages ("AC" or pulses) should be applied across the primary terminals of a transformer. A steady DC voltage established across the primary coil does not produce a *changing* magnetic flux. No EMF is induced in the secondary. Moreover, since no opposing or "back EMF" is induced in the primary, the primary current will be limited only by the low primary coil resistance. The result is that the primary coil overheats and burns out!

Moving Circuits

A conductor like that of Fig. 12-25, which is orientated perpendicular to a magnetic field and pushed with a velocity $\vec{v}$ in a direction perpendicular to the field and to itself, can produce an induced current. Charges within the wire, since they are carried along with a velocity $\vec{v}$ perpendicular to $\vec{B}$, experience a magnetic force

$$F = qvB, \qquad (12.1)$$

which is directed along the wire. Mobile electrons will therefore move along the wire. If the conductor is part of a complete circuit, an induced current will be established.

It can be shown that the circuit EMF produced in this way is again given by Faraday's law:

$$\mathcal{E} = \frac{\Delta \Phi}{\Delta t} \qquad (12.14)$$

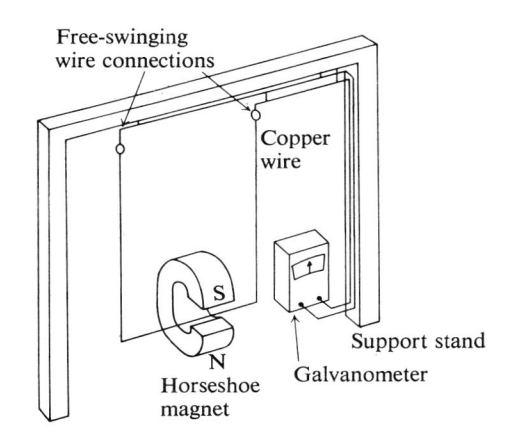

Fig. 12-25 Experimental setup to demonstrate the electric current induced in a conductor moving across a magnetic field.

even though the flux change, $\Delta\Phi$, is achieved, not by having a time-varying magnetic field but instead by moving all or part of the circuit.

In general, we find that an EMF equal to $\Delta\Phi/\Delta t$ is induced in a coil no matter how the flux change is produced. In some cases there is an electric field (Fig. 12-19), in other cases there may be only magnetic forces on moving charges (Fig. 12-25), but, rather amazingly, these very different physical situations are both embraced in the one statement,

$$\mathcal{E} = \frac{\Delta \Phi}{\Delta t}. \qquad (12.14)$$

454 ELECTROMAGNETISM

Generators

In a generator (or dynamo), moving conductors enclose a changing magnetic flux. Consequently an EMF is generated. The moving conductors experience a retarding magnetic force. The external force, which must be applied to maintain motion with constant speed, supplies mechanical energy. The generator serves to convert this mechanical energy into electrical potential energy, the rate of conversion being given by

Power converted = (generator EMF) × (current).

A prototype generator is obtained by rotating, with constant speed, a coil in a constant magnetic field (Fig. 12-26). As the coil rotates, the component of $\vec{B}$ which is perpendicular to the coil changes. Eq. (12.13) shows that the flux enclosed by the coil therefore changes. Consequently an EMF is induced:

$$\mathcal{E} = \frac{\Delta \Phi}{\Delta t}. \qquad (12.14)$$

The rate of change of flux alters as the coil rotates, so the EMF is not constant. A detailed investigation reveals that, in the course of one full revolution of the coil, this induced EMF builds up from zero to a maximum, decreases to zero, and then builds up again in the reverse sense to a maximum and again decreases to zero. Applied to a closed circuit, such an EMF produces what is called an alternating current (AC): a current whose magnitude is continually changing and whose direction reverses periodically. Such currents are the topic of the next chapter.

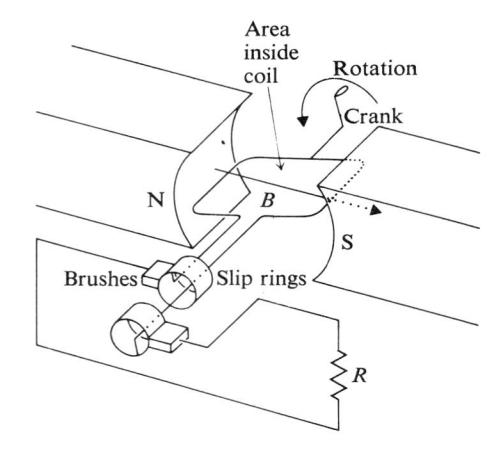

Fig. 12-26 Simple generator. Turning the crank causes an alternating EMF at the slip rings. The brushes make connection to an external circuit, giving an alternating current through R.

QUESTIONS

1. Is a stationary electric charge a source of an electric field? A magnetic field?

2. Draw a sketch showing the magnetic field vectors at several points in space near a long, straight wire carrying a current.

3. The north pole of a compass needle points in the direction of $\vec{B}$. Show the orientation of a compass needle at a point directly above a horizontal wire carrying a current.

4. What is the source of the magnetic field produced by a permanently magnetized iron bar?

5. Contrast a magnetized iron bar with an unmagnetized iron bar.

6. An electric current is passed through a coil of wire wound on a solid cylinder. The magnetic field within the cylinder is found to be much stronger when an iron cylinder is employed than when the cylinder is made of wood. Explain.

7. (a) Is there a magnetic force exerted by a magnetic field on a stationary electric charge?
   (b) Is the direction of the magnetic force exerted by a magnetic field on a moving charge in the same direction as the magnetic field?
   (c) What is the minimum amount of information needed to determine the direction of the magnetic force on a moving charge?

8. If the proton of Example 1, Section 12.2, moves vertically upward, what magnetic force does it experience?

9. Find the magnitude and direction of the magnetic force exerted on an electron moving vertically upward at a speed of 2.0×10^8 m/sec by a horizontal magnetic field of 0.50 weber/m² which is directed west.

10. From the defining equation for $\vec{B}$,

$$B = F/qv \sin \theta,$$

determine the dimensions of $\vec{B}$.

11. The equation $F = qvB \sin \theta$ involves the magnitudes of three vectors $\vec{F}$, $\vec{v}$, and $\vec{B}$. Of these three vectors, which pairs are always at right angles? Which pairs may have any angle between them?

12. What would be the magnitude and direction of the magnetic force on the electron in Question 9 if the westerly-directed magnetic field were angled at 30° above the horizontal?

13. What would be the magnitude and direction of the magnetic force on a proton moving in the same direction and through the same magnetic field as the electron in Question 12?

14. Crossed electric and magnetic fields are established over a certain region. The magnetic field is 0.10 weber/m² vertically downward. The electric field is 2.0×10^6 volt/m in a horizontal direction east. A proton, traveling horizontally northward experiences zero resultant force from these fields and so continues in a straight line. What is the proton's speed?

15. For a charged particle moving perpendicular to a uniform magnetic field, deduce, from Newton's second law, the result: $mv = qBr$.

16. Explain how magnetic fields can be used to measure
    (a) a charged particle's momentum;
    (b) the "charge-to-mass ratio" of a particle.

17. In a magnetron (designed to generate electromagnetic waves of radar frequencies) an electron moves perpendicular to a magnetic field of 1000 gauss. In a uniform field of this magnitude, what is the electron's frequency of revolution in a circular orbit?

18. A beam consisting of a mixture of protons, deuterons, and α-particles is accelerated (essentially from rest) through a difference of potential V, and it emerges into a region of uniform magnetic field $\vec{B}$. The particle velocities are perpendicular to $\vec{B}$. Express the radii of the orbits of the deuterons and the α-particles in terms of the radius r of a proton's orbit.

19. Show how the mass M of an ion of charge e can be determined from a knowledge of the proton's mass M_p and from measurement of the ratio T/T_p, where T is the time required for one revolution of the ion in a uniform magnetic field, and T_p is the time required for one revolution of a proton in the same magnetic field.

20. (a) What is the time required for a proton to complete one circular orbit in a uniform magnetic field of 1000 gauss? (1 gauss = 10^{-4} weber/m².)
    (b) What is the mass of a chlorine ion that takes 36.9 times as long as a proton to complete one revolution in the magnetic field of part (a)?

21. Give an example of an ordinary situation in which an object acquires a considerable kinetic energy, not from one large push, but from many properly timed successive small pushes. This is the idea of cyclotron acceleration. [*Hint:* Think of a playground.]

22. (a) When a magnetic field is employed in a particle accelerator, does the magnetic force increase the particle's speed?
    (b) For what purpose are magnetic fields used in particle accelerators?
    (c) What features of the magnet determine the maximum energy to which a particle can be accelerated in a cyclotron?

23. A cyclotron is to be designed to accelerate protons to an energy of 20 MeV, using a magnetic field of 1.5 weber/m². Determine:
    (a) the radius of the largest orbit over which the magnetic field must be established;
    (b) the period T and the frequency $f = 1/T$ of the voltage which is to be applied across the electrodes within the cyclotron in order to provide an accelerating electric field.

24. The cyclotron of the preceding question is to be converted to accelerate deuterons instead of protons.
    (a) By what factor must the frequency be changed?
    (b) What is the maximum energy that will be attained by the deuterons?

25. A 5.0-amp current is directed vertically upward in a straight section of a wire. The wire lies in a uniform

horizontal magnetic field of 0.020 weber/m² directed north. Find the magnitude and direction of the magnetic force exerted by this magnetic field on a 0.06-m section of the vertical wire.

26. A wire 0.60 m long, carrying 3.0 amp is located in a region in which there is a magnetic field of 0.015 weber/m². Find the magnitude of the magnetic force on this wire if the angle between the wire and the field $\vec{B}$ is: (a) 0°; (b) 30°; (c) 45°; (d) 90°.

27. (a) What information besides that given in Question 26 is required in order to be able to specify the direction of the magnetic force on the wire?
    (b) Is the answer to part (a) different for the case of 30° between the wire and $\vec{B}$ than it is for the case of 45° between the wire and $\vec{B}$? Explain.

28. What is the direction of the motion of the wire in Fig. 12-12 when the switch at A is closed?

29. In Example 2 [Fig. 12-14(a)], the side of the rectangular loop which is perpendicular to the axis of rotation is 0.070 m long.
    (a) Evaluate the torque exerted by the magnetic forces on this coil.
    (b) What is this torque when the coil has rotated to the position shown in Fig. 12-14(b)?

30. Fig. 12-15(a) shows the essential features of a moving-coil galvanometer. Give the rotation which would be caused by the indicated current.

31. (a) Describe the key parts of a moving-coil ammeter.
    (b) How would this instrument have to be modified to serve as a voltmeter?

32. A galvanometer gives full-scale deflection when a current of 1.0×10^{-4} amp passes through its coil. What is the total resistance of a voltmeter constructed using this galvanometer together with a series resistance, if the voltmeter is designed to measure from 0 to 150 volts?

33. Find the resistance of the shunt required to convert the galvanometer of the preceding question to an ammeter which reads 10 amp full scale. The resistance of the galvanometer coil is 30 Ω.

34. What are the dimensions of the constant K which appears in the Biot-Savart law, $B = KI/r$?

35. By considering the magnetic field produced by current in one wire at the location of another and the force exerted by this field, show that when the currents flow in opposite directions, the wires repel one another. (See Fig. 12-18 for wire orientation.)

36. How is the size of the mks current unit, the ampere, determined?

37. Two long, straight parallel wires, separated by a distance of 0.50 m, attract one another with a force of 0.10 N per meter of length when they carry currents of equal magnitude. What is the magnitude of these currents?

38. What are the dimensions of magnetic flux, Φ?

39. Find the magnetic flux passing through a horizontal surface of area 0.16 m² in a uniform vertical magnetic field of 0.30 weber/m².

40. What is the magnetic flux Φ through a circular loop antenna of radius 0.30 m, oriented in a plane perpendicular to the magnetic field $\vec{B}$ if the magnitude of $\vec{B}$ is 5.0×10^{-4} weber/m²?

41. A rectangular coil, 4.0×10^{-2} m by 5.0×10^{-2} m, is placed in a uniform magnetic field of 0.80 weber/m² directed vertically downward. What is the magnetic flux enclosed by the coil when
    (a) the plane of the coil is horizontal,
    (b) the plane of the coil is vertical?

42. (a) State Faraday's law of electromagnetic induction.
    (b) Describe any simple experiment which illustrates the above law.

43. Use Lenz's law to determine the sense of the EMF for the orbit shown in Fig. 12-19.

44. A bar magnet is thrust into a coil of wire, with the south pole entering first. Make a sketch indicating magnetic polarities and the direction of the induced current in the coil just as the south pole of the magnet enters the coil. (The magnetic flux emerges from the end of the magnet which is called the north pole and enters the magnet at the end called the south pole of the magnet.)

45. Find the magnitude of the EMF induced in a 200-turn coil with cross-sectional area of 0.15 m² if the magnetic field through the coil is caused to change from 0.10 weber/m² to 0.50 weber/m² at a uniform rate over a period of 0.02 sec.

46. With the coil of Question 41 maintained in a horizontal plane, the magnetic field is reduced at a uniform rate from 0.80 weber/m² to zero in 0.020 sec. Find the magnitude and sense of the induced EMF.

47. Two hundred turns of wire are wound on a cardboard cylinder 0.50 m in diameter. While the magnetic field through the coil is changed at a uniform rate from 0.15 weber/m² to 0.30 weber/m² in 1/50 sec, what is the EMF induced across the two ends of the wire?

48. When the EMF for the primary of a household transformer for an electric bell is 120 volts, the secondary EMF is 6.0 volts.
    (a) Is this a step-down transformer?
    (b) What is the transformer "turns ratio"?

49. Consider Fig. 12-24. What is the turns ratio of the transformer at the generating station? Is this a step-up or a step-down transformer?

50. Explain why there is a current induced in the circuit of Fig. 12-25 when the conductor is pushed out and away from the horseshoe magnet.

51. What will be the direction of the induced current in the preceding question?

52. Suppose a coil of wire were rotated in a uniform magnetic field.
    (a) Would the EMF induced be of the same magnitude throughout the rotation?
    (b) Would this EMF have the same polarity throughout the rotation?
    (c) What common device is described, in its essentials, by this type of arrangement?

53. What contributions to our knowledge of electromagnetism were made by Oersted? by Faraday?

SUPPLEMENTARY QUESTIONS

S-1. Consider a number of magnets lying on a table. The magnetic field at a certain point on the table is 0.07 weber/m² south, due to magnet M_1; 0.12 weber/m² east, due to magnet M_2; and 0.08 weber/m² west, due to magnet M_3.
   (a) What basic physical principle is used to find the magnetic field when magnets M_1, M_2 and M_3 are simultaneously present?
   (b) Find the magnitude and direction of the actual magnetic field at the point in question.

S-2. The magnets used by an ophthalmologist to remove ferromagnetic splinters from eyes range from small hand-held magnets to, in some cases, so-called "giant" magnets which have their own support structure.
   (a) Whatever the size of these magnets, they are electromagnets rather than permanent magnets. Why?
   (b) Explain why even the smaller hand magnets are quite heavy.

S-3. Fig. 12-27 shows an *electromagnetic relay*. When the switch in circuit P is closed, does the light in circuit Q go on or off? Explain how the relay works.

S-4. A protective device designed to prevent excessive currents from passing through a given circuit is represented schematically in Fig. 12-28. Describe the behavior of this *circuit breaker* when the current I reaches a potentially dangerous value.

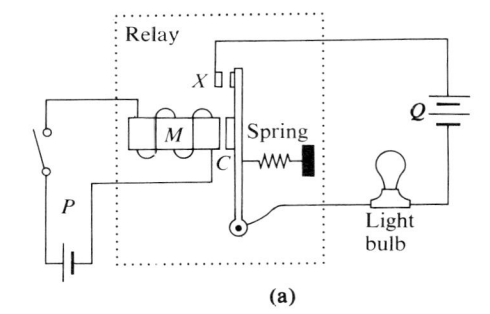

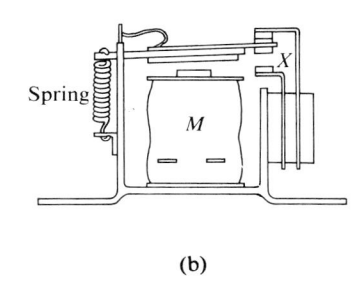

Fig. 12-27 Electromagnetic relay. (a) Schematic diagram of a simple relay. (b) A commercially produced relay. M and X are shown also in part (a) to indicate their function.

458 ELECTROMAGNETISM

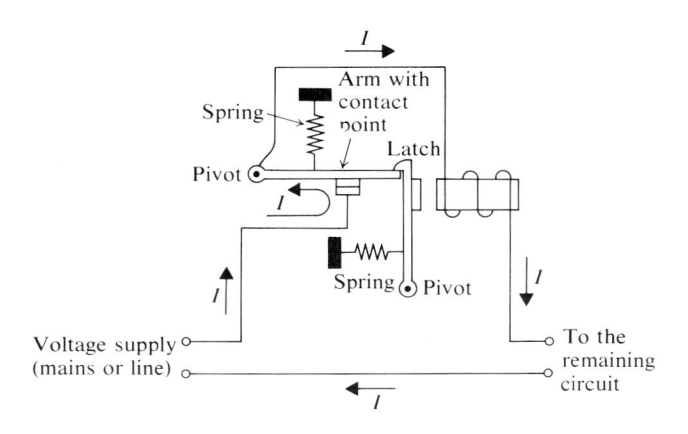

Fig. 12-28 Overload circuit breaker.

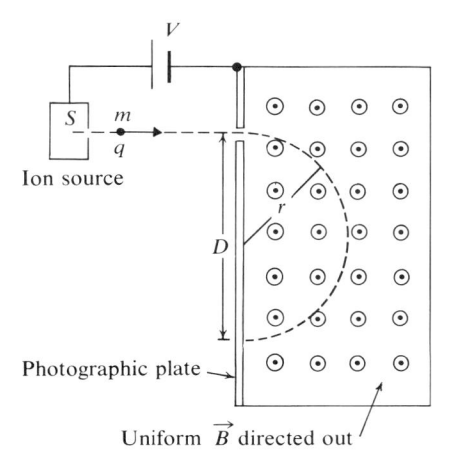

Fig. 12-29 Mass spectrometer.

S-5. In a mass spectrometer designed by Dempster (Fig. 12-29), ions of mass m and charge q are accelerated from rest through a difference of potential V and then encounter a uniform magnetic field $\vec{B}$. In the magnetic field, an ion moves in a semicircular path of radius r before striking a photographic plate. The ion's mass can be determined by measuring the distance D from the entry slit to the mark on the photographic plate. Show that

$$m = \frac{qB^2}{8V} D^2.$$

S-6. Consider two helium atoms, one singly ionized and the other doubly ionized, entering a mass spectrometer. If the magnetic field is uniform, and if both ions enter the field with the same kinetic energy while moving in a plane perpendicular to the field, compare the diameters of their orbits.

S-7. Electrons travel through a uniform magnetic field at right angles to the field and complete a circular orbit in 10^{-9} sec. What is the magnitude of the magnetic field? (See Appendix B for the charge-to-mass-ratio of the electron.)

S-8. The current density (current divided by cross-sectional area) in a horizontal copper wire is 2.0×10^6 amp/m². The current is directed from east to west. Find the magnitude and direction of the magnetic field which will exert a magnetic force on the wire sufficient to support the wire. (The density of the copper wire is 8.9×10^3 kg/m³.)

S-9. Parallel tracks, separated by 2.0×10^{-2} m, run north-south. There is a uniform magnetic field of 1.2 weber/m² pointing upward. A 4.0×10^{-2}-kg metal cylinder is placed across the tracks, and a battery connected between the tracks, with its positive terminal connected to the east track, drives a current of 3.0 amp through the cylinder. Find the magnitude and direction of the force acting on the cylinder and the acceleration of this cylinder assuming that it slides without friction.

S-10. Although a car has numerous conducting wires which may be carrying current, and although they are certainly moving through a magnetic field (the earth's), no cognizance is taken by automobile designers of the associated forces. Perform a calculation to show that these forces are of negligible magnitude compared to other forces, such as weight, which affect the car. Assume that the car is traveling horizontally due north, and that the earth's magnetic field at the car's location is 5.68×10^{-5} weber/m² due north at an angle of 65.0° below the horizontal. Also assume that there is a current of 10.0 amp in a wire 2.00 m long running right across the car under the windshield.

SUPPLEMENTARY QUESTIONS 459

S-11. The expression, $F = BIL \sin \theta$, for the magnetic force F exerted on a length L of a conducting wire which carries a current I can be deduced from the formula for the force on the moving charges within the conductor. To do this, consider that, within the length L of the conductor, there are N particles, each with charge q and moving with speed v along the length of the wire. By determining the time for the charge Nq to pass through one end of the conductor, show that the current I is given by

$$I = \frac{Nq}{(L/v)}.$$

Use this result, together with the expression

$$F = NqvB \sin \theta,$$

for the magnetic force on the N charges within the conductor, to show that

$$F = BIL \sin \theta.$$

S-12. Show that the torque exerted by the magnetic forces on the coil in Example 2 is given by

$$\tau = BIA \sin \alpha,$$

where A is the area of the coil and α is the angle between the normal to the plane of the coil and the direction of $\vec{B}$.

S-13. Determine the torque on a square coil (0.20 m long), consisting of 30 turns. The current in the coil is 0.25 amp, and the coil is in a region of uniform magnetic field of magnitude 2.0×10^{-2} weber/m². Consider the cases in which the angle between the normal to the plane of the coil and the magnetic field is 0°, 30°, 90°.

S-14. A galvanometer has an internal resistance of 1.0 ohm and gives maximum deflection for a current of 50 mA. Indicate with a sketch how this instrument can be changed into
(a) a voltmeter with a maximum reading of 2.5 volts;
(b) an ammeter with maximum reading of 2.5 amp.
In each case calculate the resistance of whatever resistors have to be used in conjunction with the galvanometer.

S-15. The box in Fig. 12-30 is placed in a uniform magnetic field of 0.40 weber/m² which is parallel to the 4.0-m edge of the box. Find the magnitude of the magnetic flux through each face of the box.

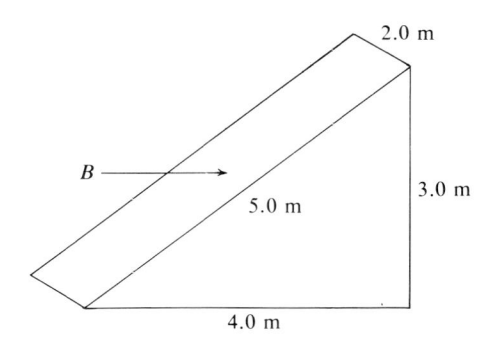

Fig. 12-30 Surfaces for the flux calculations of Question S-15.

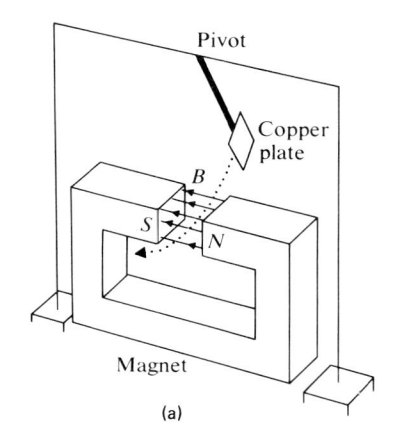

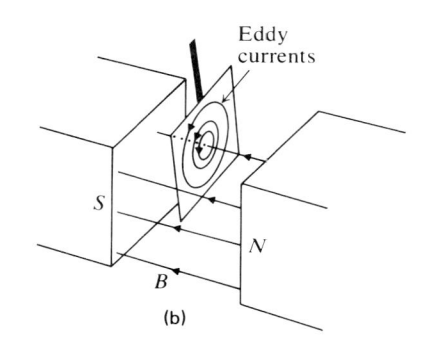

Fig. 12-31 Copper pendulum demonstrating eddy current damping. (a) The experimental setup. (b) Close-up of eddy currents in the plate.

S-16. Consider the coil and magnet in Question 44. With the aid of sketches, indicate what happens as the magnet is pulled all the way in, through, and out the other side of the coil.

S-17. Since a current will be induced in a loop of conducting material subjected to changing magnetic flux, and since one can think of a plate of this conducting material as a composite of many loops of the material, it is therefore not surprising that current loops are set up in any plate made of a conductor as long as there is a changing magnetic flux. Such loop currents are called *eddy currents*. Fig. 12-31 shows how the swing of a metallic pendulum can be damped by eddy currents. Many laboratory devices, such as sensitive chemical balances, employ eddy current damping.
   (a) Explain why the presence of eddy currents gives rise to a retarding force on the pendulum.
   (b) Indicate at least three ways in which the damping effect, as illustrated by Fig. 12-31, could be increased.
   (c) When the swinging motion (kinetic energy) is damped out where does this energy go?

S-18. The essential features of transformers are indicated in Fig. 12-23. Transformers generally have very high efficiency (90% or higher). However, there is some energy loss in transformers.
   (a) Indicate how, and in what part of the transformer, energy might be dissipated.
   (b) Explain what you think might be the purpose of the laminations in the core of the transformer [Fig. 12-23(c)].

S-19. The tracks of Question S-9 are joined by a copper wire at the south end. A metal cylinder is placed across the tracks and pushed so that it travels north with a constant speed. This cylinder starts from the south end and travels 0.20 m in the first second. Find
   (a) the magnitude of the induced EMF,
   (b) the sense of this EMF.

S-20. When a conducting bar which is perpendicular to a magnetic field $\vec{B}$ is moved with a velocity $\vec{v}$ perpendicular to the bar and to the magnetic field, the magnetic forces cause mobile charges to move until the charge distribution sets up an electric field $\vec{E}$ within the conductor. This electric field exerts on any charge q an electric force equal in magnitude but opposite in direction to the magnetic force.
   (a) Show that this electric field has a magnitude given by $E = vB$.
   (b) Show that the potential difference between the ends of the bar of length L is vBL.

S-21. Determine the magnitude of the EMF induced across a 2.0-m long bumper of a car if the car is going west at 21 m/sec (about 45 mph) in a region where the vertical component of the earth's magnetic field is 4.5×10^{-5} weber/m² downward.

S-22. The circuit of Fig. 12-32 encloses a magnetic flux Φ given by
$$\Phi = BA = BLx.$$
   (a) Show that when the segment of length L moves a distance $\Delta x = v\Delta t$ in a time interval Δt, the change in magnetic flux enclosed by the circuit is given by
$$\Delta\Phi = B\Delta A = BLv\Delta t.$$
   (b) Use Faraday's law to show that the EMF induced in the circuit is
$$\mathcal{E} = vBL.$$

S-23. Examine the magnetic force exerted by $\vec{B}$ on the moving conductor in Fig. 12-32 when the conductor carries an induced current I. Show that, to keep the conductor moving with a constant velocity, an external force F_{ext} must be applied to supply mechanical power
$$F_{ext}v = BILv.$$
Verify that the power converted from mechanical to electrical power is given by $\mathcal{E}I$ where $\mathcal{E} = vBL$.

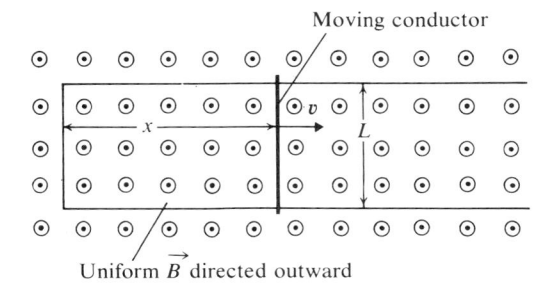

Fig. 12-32 Circuit with moving conductor of Question S-22.

S-24. Since an electric motor involves a conductor moving through a magnetic field, the motor, although driven by an external EMF, must itself generate an EMF. This EMF generated by the motor in motion is called the *back EMF*. The back EMF opposes the external EMF which drives the motor. Assume that a certain motor has a total resistance of 12.0 ohms associated with its various windings, and that under a certain load this motor requires a current of 6.00 amp from a 120-volt line.

(a) What is the back EMF?
(b) What power is developed in the motor?
(c) What is the motor's efficiency?

ADDITIONAL APPLICATIONS TO MEDICINE AND THE LIFE SCIENCES

Ophthalmological Hand Magnet

Magnets find use in ophthalmology in the removal of ferromagnetic materials from the eye. New pulsing techniques which result in momentary high currents and consequent high values of $\vec{B}$ are gradually retiring the so-called *giant magnet* which was mounted on a stand somewhat like a heavy telescope and aimed with considerable difficulty. *Pulsed magnets*, small enough to be held by the ophthalmologist in his two hands are now producing mean values of $\vec{B}$ of up to 0.5 weber/m², a value previously achieved in medical practice only with the cumbersome giant magnets. These hand magnets are turned on and off by a foot-switch, and they come supplied with various tips (of different shapes) which can be sterilized.

Berman Metal Locator: Faraday's Law

An interesting device that depends upon basic physics discovered by Faraday is the *Berman metal locator*, an electromagnetic device that quickly and accurately locates metallic foreign bodies, including magnetic intraocular foreign bodies (see Fig. 12-33). The distance of effectiveness in the case of magnetic materials is about 10 times the diameter of the foreign body, while in the case of nonmagnetic metals the distance is only about three times the diameter of the foreign body.

A foreign body, such as a metallic splinter, can be a very elusive object which may move even while the surgeon is probing for it. The Berman metal locator was developed in response to the expressed need of surgeons for a simple localizing device that keeps track of the foreign body as surgical probing proceeds. This localizing device takes up where radiography leaves off, giving guidance to the surgeon throughout the surgical removal.

This locating device literally passed its test under fire. After the first successful tryout in a New York hospital in November 1941, the locator was taken to Honolulu by a surgeon who had been invited to give a series of lectures before the Honolulu Medical Society in early December. The successful results obtained with the locator at the Tripler General Hospital following the December 7, 1941 Japanese attack showed this device to be an important contribution to the area of foreign-body surgery.

The detector probe is a thin pencil-like plastic-covered instrument which in turn can be covered by a sterilized rubber jacket. The probe itself likely consists of two sets of conductor wire windings wound on top of each other. When alternating current is passed through one set of windings there is, in accord with Faraday's law, an EMF induced in the second set, much as in the transformer described earlier. The magnitude of this induced EMF can be monitored, and the magnitude, which equals the rate of change of magnetic flux enclosed by the second set of windings, will depend not only on the alternating current in the first set of windings but also upon the effect of the

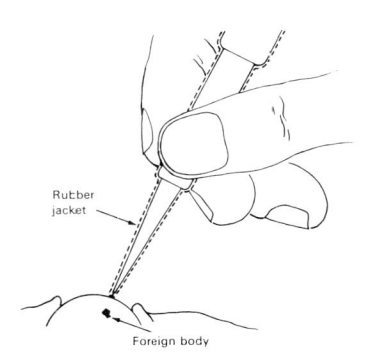

Fig. 12-33 Berman metal locator used in the localization of an intraocular metallic foreign body.

presence of ferromagnetic material on the magnitude of the variations of this magnetic flux.

The detection of nonmagnetic metals probably depends on the effect of magnetic flux changes produced by *eddy currents* (see Fig. 12-31) induced in the metal foreign body by the alternating magnetic flux due to the alternating current in the first set of windings.

Question

(a) Does covering the probe windings with a plastic covering and a rubber sleeve have any appreciable effect on the sensitivity of this device?

(b) What must be specified about the materials used to make the retractors which are employed to keep the incision open?

Answer

(a) No, because these materials are nonferromagnetic and nonmetallic, they have negligible effect on the magnetic flux generated by the instrument.

(b) Retractors must be made of nonmetallic material.

SUGGESTED READING

BITTER, FRANCIS, *Currents, Fields and Particles.* New York: Technology Press of Massachusetts Institute of Technology, and John Wiley, 1956. The author was a foremost authority on magnetism. (Bitter, and H. J. Williams of the Bell Telephone Laboratories, developed a method for outlining domains in an actual ferromagnetic sample by applying a colloidal suspension of iron oxide powder to the sample's surface.) Chapter 5 on magnetic fields is highly recommended to the advanced student. Diamagnetism is discussed in Chapter 6. Calculus is required for the quantitative part of the discussion, but a valuable insight into magnetic behavior may be gained without rigorously following the mathematical analysis.

EFRON, ALEXANDER, *Magnetic and Electrical Fundamentals.* New York: John F. Rider Publisher, Inc., 1959. Chapters 1 and 3 provide excellent brief supplementary reading for beginning physics students. The introduction of the left-hand motor rule on page 66, although quite correct, may be misleading to students of our text. The rule is equivalent to the right-hand rule stated in our Section 12.2.

FEYNMAN, RICHARD P., ROBERT B. LEIGHTON, and MATTHEW SANDS, *The Feynman Lectures on Physics*, Vol. II, Reading, Mass.: Addison-Wesley, 1964. The ten pages comprising Chapter 16 of these lectures by one of the foremost theoretical physicists of our time present an exceedingly lucid, nonmathematical treatment of induced currents, including the induction motor. Although the rest of this book is mathematically considerably beyond the level used in our text, this one chapter is well worth reading by anyone at any level of physics.

LEMON, HARVEY B., *What We Know and Don't Know about Magnetism*. Chicago: Museum of Science and Industry, 1946. This 60-page booklet is an excellent, thorough treatment of the nature of magnetic materials. Can be understood by anyone who has a basic idea of atomic structure such as outlined in our Chapter 1.

MARCUS, ABRAHAM, *Basic Electricity*. Englewood Cliffs, N.J.: Prentice-Hall, 1958. Chapter 6 is a detailed but elementary, nonmathematical treatment of DC meters. Chapter 17 on electrical motors and Chapter 11 on mechanical generators provide extra reading for those students who wish to look at these topics in more detail than we have provided. Practically no mathematics is employed, but nevertheless excellent descriptive passages and figures give a valuable insight.

SEARS, F. W., and M. W. ZEMANSKY, *University Physics (4th ed)*. Reading, Mass.: Addison-Wesley, 1970. Chapters 30 to 34 present the topics of our chapter at a higher level. Calculus is used. Recommended for the advanced student.

Charles Proteus Steinmetz

13 Alternating Currents

Considerably more than 90% of electrical energy consumed is in the form of alternating current (AC). In the United States alone something like a thousand billion kilowatt-hours of electrical energy are used in a year. It is therefore not hard to realize the importance of AC to users of electricity. Moreover, aside from the actual usage, it is safe to say that in the distribution of electrical power AC will predominate for years to come. In fact, even if DC appears more and more in future power transmission, DC's use will probably be for long-range transmission while the advantages of easy voltage changes, up or down, of AC seem to make it generally the more suitable way to distribute power to users.

Strangely enough, in spite of its obvious advantages in retrospect, it was not AC that was the first to be used in commercial power distribution. This distinction goes to DC, which was employed by the great American inventor, Thomas Alva Edison (1847–1931), when he arranged to light the main street of Menlo Park, New Jersey, on New Year's Eve, 1879.

The great difficulty facing the electrical industry in the late nineteenth century was the considerable loss that was encountered in the transport of electricity over wires. Since it was found that electricity could be transported more efficiently at high voltages, Nikola Tesla (1856–1943), a Croatian-American electrical engineer, developed transformers (Section 12.5) to boost electricity to high voltages for transmission and then to lower the voltages for relatively safe usage. But transformers, of course, do not work on steady DC; they are AC devices. There actually developed a bitter personal struggle between Edison and Tesla on the subject of DC versus AC. In the final analysis, the telling point was transport efficiency, and AC won out. A rather sad sidelight of this whole issue is that in 1912 the Nobel prize committee intended to award the Nobel prize in physics jointly to Edison and Tesla, but the latter refused to be associated with Edison. Consequently in the end the prize went to a Swedish inventor.

Before leaving the AC-DC controversy, we should note that this debate is far from finished. One of the very serious problems engendered by the use of AC for power transmission is that if a number of power

"In the gradual replacement of facts by the belief in theories more or less inadequately representing the facts, lies the chief danger to further scientific progress. Here is a field where splendid work in destructive criticism can and should be done by the younger generation, which, after leaving college with all the theoretical armament required, is not yet handicapped by personal relations with the men whose names are identified with the ruling theories."

Charles Proteus Steinmetz
(1865–1923)

At 5:16 p.m. Eastern Standard Time, November 9, 1965, the lights began to go out, and, from the northern mining country in Canada, around Timmins, to Boston and New York, everything depending on the public electrical supply stopped. The effects of a massive surge of power through interconnected electrical power grids were felt as far away as Montana and Florida. Some 30 million people were plunged into unexpected darkness. (Readers interested in more details of this event are referred to the article on the subject in the magazine Time, November 19, 1965.)

networks are linked together into one system (a so-called power *grid*), the frequencies of the AC of the different components of the system have to be precisely synchronized; if they are not, one part of the network that is supposed to be generating energy for the total network, can instead actually absorb energy from the rest of the system. Difficulties that arose from inadequately synchronized linkages contributed to the massive northeastern power blackout of 1965.*

The problems arising from such unplanned asynchronous linkages between power grids are, of course, obviated by the use of DC. It is for this reason in particular that considerable interest has been renewed in the transmission of power via DC. One of the factors which makes DC a more serious competitor of AC today is that the conversion equipment (from AC to DC and vice versa) required to permit the transmission and use of power at various voltages is now much superior to what it was only a decade or so ago. Moreover, the costs of this conversion equipment have been considerably reduced in recent years. At present there are several DC power transmission lines throughout the world. One interesting example is the power linkage in Europe between the United Kingdom and France. Other present-day practices in power transmission involve the reduction of resistive losses by the use of low temperatures in transmission. Cryogenic and superconducting transmission lines appear to be definite possibilities. However, as we pointed out earlier, there is little doubt that because of certain advantages inherent in AC it will continue to play a major role in electrical technology.

To speak of AC is in a sense to speak of Charles Proteus Steinmetz, a German who, because of his outspoken socialist views, emigrated to the United States in 1889. Steinmetz, a phenomenal electrical genius, worked out in mathematical detail the theory of AC circuitry, using complex numbers (a mathematical system in which appears that seemingly mysterious number, $\sqrt{-1}$). This work helped significantly in establishing the victory of AC over DC in that in Steinmetz' hands the analysis of AC circuits became elegant and almost as simple as that of DC. Steinmetz' book, *Theory and Calculation of AC Phenomena*, published in 1897, became the standard reference work of the electrical engineering profession for many years. This book and his subsequent eight others unquestionably served to educate that generation of engineers who pioneered the transition from Edison's DC power plants to the AC systems in general use today.

While to many of his generation Steinmetz was known only as an electrical wizard, he was much more than that. He was a philosopher, a sociologist, and above all a humanitarian. He wrote and spoke widely on social, educational, and political topics and by his writing as well as his *doing* exerted a pronounced beneficial influence, particularly in the city which became his home, Schenectady, N. Y.

Although for most household devices AC is eminently suitable, there are many devices from x-ray machines to television sets which require DC for their operation. So we change AC to DC. This process of changing AC to DC is called *rectification*. An important figure in this area is the Englishman, Sir John A. Fleming (1849–1945), who at one time served as a consultant to Edison. Fleming's *valve tube* (now called a vacuum diode) of 1904, based on an earlier discovery by Edison, is a device which, roughly speaking, allows AC to enter but permits only DC to leave.

Certain solid materials (some crystals, for example) have long been known to perform rectification functions similar to those of Fleming's tube. Barrier layer rectifiers, such as copper-copper oxide and iron-selenium have been developed. And in more recent years the work of W. B. Shockley, J. Bardeen, and W. H. Brattain of the Bell Telephone Laboratories showed that germanium crystals containing traces of certain impurities are better rectifiers than many of the crystals earlier employed. These three physicists were awarded the Nobel prize in physics in 1956 for their discovery that certain combinations of these new solid-state rectifiers could do all that a radio tube could do. That is, they discovered what is now called a *transistor*.

The transistor may be held largely responsible for the extraordinary rapidity with which the miniaturizing of electronic circuitry has swept the electronics industry. But the transistor itself, small compared to a radio tube, is enormous compared to the latest electronic development: tiny blobs of material which are formed into circuits occupying about the space of a pin head, known as *integrated circuits*. (See Fig. 16-17 in Chapter 16.)

13.1 ALTERNATING VOLTAGES AND CURRENTS

Advantages of AC

I^2R losses in conducting cables were mentioned in Section 11.5, where it was pointed out that it was advantageous to transmit power at high voltages. With step-up transformers this advantage of reducing I^2R heating losses in the cables can be achieved. However, transformers do not work with DC, but rather with AC. Therein lies a major advantage of AC.

Another point in favor of AC hinges on the induction motor, a very practical electric motor but one which requires AC. In general, it may also be said that most AC motors are probably cheaper and less subject to maintenance troubles than DC motors.

Finally, it should be added that AC, although at much higher frequency than for normal power requirements, is imperative for modern communication. Radio waves, for example, are generated by a rapidly alternating current in a transmission antenna.

AC Terminology

AC refers to a current which is periodically changing direction. One therefore cannot assign a fixed current direction in AC circuits such as was done in Chapter 11 in dealing with DC circuits. On the other hand, it now becomes imperative to distinguish between *instantaneous* and *average* values of current and voltage.

While in the true sense AC refers to alternating currents in general, without restriction as to how the current varies, we shall use the term AC to refer to a sinusoidally* varying current. In other words, we refer not to currents represented by the waveforms (b) and (c) of Fig. 13-1, but only to that labeled (a). We are concerned with sinusoidally varying current because the voltage supplied by the electric power company is essentially sinusoidal. This fact derives from the way conventional

*The current at the instant t is given by the expression $i = I_{max} \cos(2\pi ft + \theta_0)$, a function of the type studied in Section 3.10.

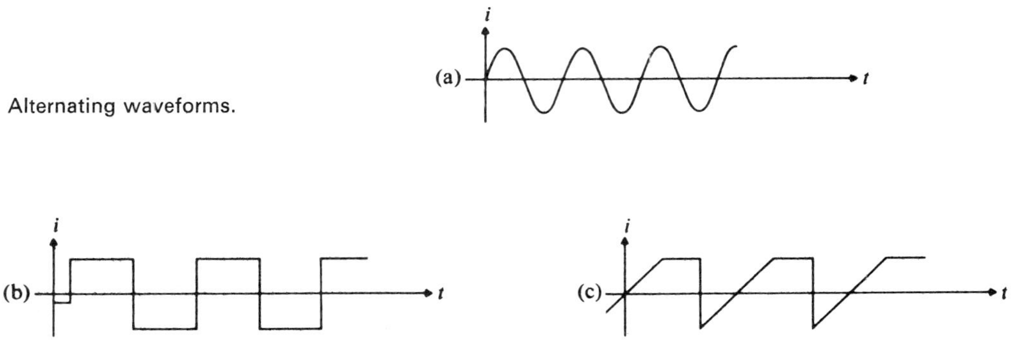

Fig. 13-1 Alternating waveforms.

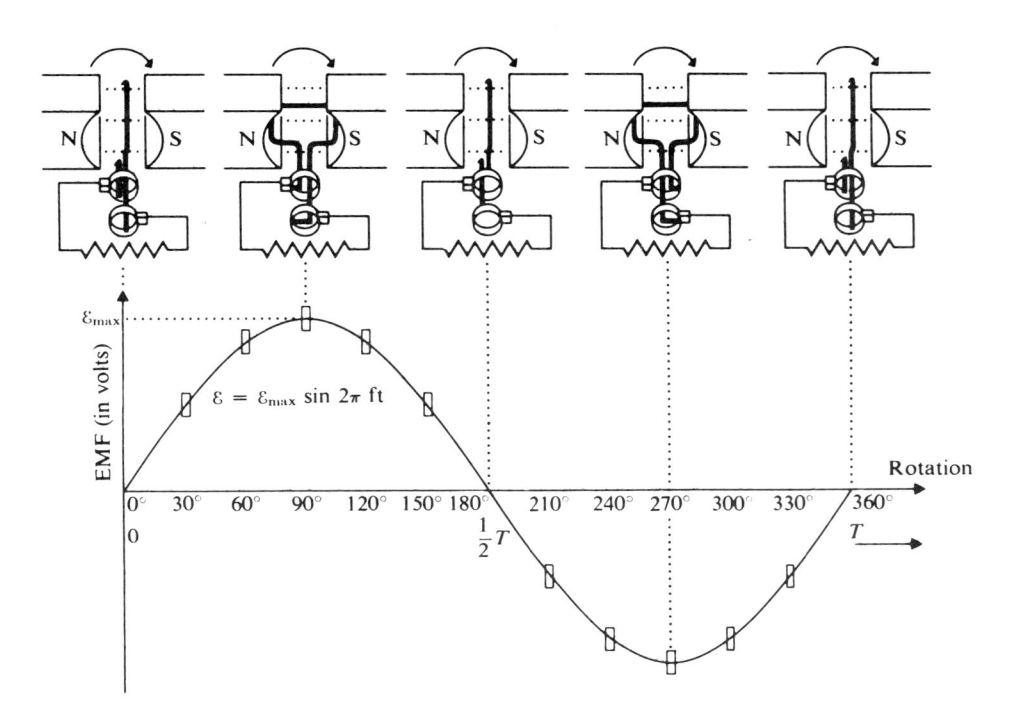

Fig. 13-2 Generation of a sinusoidal alternating EMF, showing a complete revolution of the armature (that is, one complete cycle). The abscissa shows how the degrees of rotation may be related to the period T, the time for a complete cycle.

electric generators operate. Fig. 13-2 shows how the rotation of a simple generator, such as that of Fig. 12-26, can be related to the alternating voltage induced.

Fig. 13-3 shows a sinusoidal voltage as might be produced by a conventional electric generator. With reference to this figure we now consider some of the relevant terms. First of all, the figure shows the value of the voltage at certain instants after an arbitrarily assigned time zero. For example, at t_1, the voltage has its maximum (or peak) value V_{max}, at t_2 the voltage is zero, and at t_3 the voltage is some negative value v_3, and at t_4 the voltage is again zero.

The value of the voltage at t_1, that is, its maximum or peak value, is also known as the *amplitude* of the voltage, quite in keeping with the usage of this term in the study of simple harmonic motion in Section 3.10.

The total time taken for the voltage to go from any given value to the succeeding same value at which the voltage is changing in the same direction, that is, the time required for one complete cycle of voltages, is naturally called the period, T. There is a definite relationship between

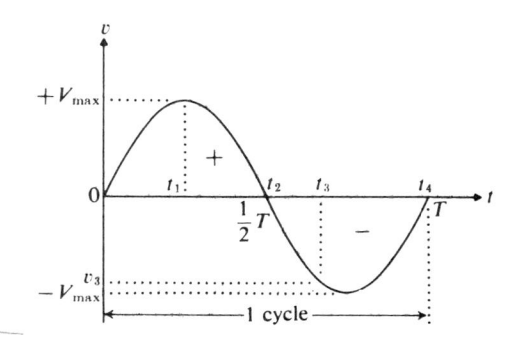

Fig. 13-3 Sinusoidally varying alternating voltage. The total time for one cycle is the period T.

13.1 ALTERNATING VOLTAGES AND CURRENTS 471

the frequency f and the period T. In the case of the voltage common in North America, the period for one complete cycle is 1/60 sec. The frequency in this case is 60 cycles per sec (or 60 hertz). We speak therefore of 60-cycle AC. In general, the relationship between frequency and period (Section 3.10) is

$$f = \frac{1}{T}$$

where f is the frequency and T is the period.

We emphasize: The important idea is that AC refers to current changing in a definite periodic fashion and that the EMF, responsible for the current, changes in the same fashion as does also the potential difference between various points in a circuit.* (Although AC literally means "alternating current" it has become common practice to speak of *AC current*, a redundancy, and *AC voltage*, in a way a contradiction in terms.)

Although in any AC circuit the voltage across part of a circuit and current in the circuit vary in the same way, that is, with the same frequency, they may not always be exactly in phase with each other. It may be, for example, that the voltage reaches its maximum value before the current (as in the case for an induction coil) or, vice versa, the current may reach its maximum value before the voltage (as is the case for a capacitor). In fact, only in the case of a resistor is the voltage simply proportional to the current. In this case one says the current and voltage are *in phase*, while in the previously mentioned cases the current and voltage are *out of phase*. But before making these facts plausible, we introduce one more set of AC terms having to do with voltage and current values, and also take a quick look at practical generators of AC.

Effective Current and Voltage Values

We have just mentioned a peculiar fact regarding AC. Unlike DC voltages and currents, AC voltages and currents can be out of phase. Another problem of AC, and one which is not met in steady DC, is that the question arises as to what really is meant by a given current, or voltage value. After all, over a cycle this value might be anything from zero to its maximum positive value to its maximum negative value and back to zero. There being an infinity of instantaneous current values (designated by i) between $-I_{max}$ and $+I_{max}$, the instantaneous value of a current is obviously of little general use. So one could consider taking the average value of the current over a complete cycle. But a moment's reflection on the nature of the AC waveform of Fig. 13-3 indicates that corresponding to each positive value of the current there exists in a cycle a negative value of equal magnitude. This means that the average over the cycle is zero. The average in this case is thus hardly very informative.

---

*We shall assume that all resistances are ohmic and the capacitances and inductances encountered are also constants, independent of v or i.

It is possible to employ some special averages which might serve some purpose. For example, if we took the average of the absolute values of the current, that is the magnitudes only, without being concerned with the concept of positive and negative, we would then get a non-zero average. But there is a more significant average. It is called the *effective value*, and it is this average which is most generally employed.

The *effective value* of an alternating current is that value which causes Joule heating of a resistor at the same rate as a direct current of the same numerical value. In other words, an AC current of 15 amp, effective value, leads to the same heating of a toaster as would a DC current of 15 amp applied to the toaster for the same period of time. Now it turns out that this effective value is a special kind of average of the alternating current. Recall the troublesome aspect of averaging over an AC cycle because of equal positive and negative values. One way of obviating this difficulty is to square the current. Then there are no negative values. Next we could take the average of the squared values of the current through one cycle and then take the square root of this average. What have we done? Taken the square root of the average value (or mean value) of the square of the current. This value is called the *RMS value*, the root-mean-square value. This special average, the RMS value, is also the effective value of the alternating current.

What the RMS, or effective, value of an alternating current is in terms of its maximum value, of course, depends on the waveform. In the case of the commonly employed sinusoidal AC, it is found that the effective value is given by

$$I_{\text{eff}} = I_{\text{RMS}} = 0.707\, I_{\text{max}}.*$$

If a certain type of average of current values is employed, then it seems reasonable that the same type of average value of voltage ought to be employed. We refer to the effective, or RMS, voltage given by

$$V_{\text{eff}} = V_{\text{RMS}} = 0.707\, V_{\text{max}}.$$

This means that the voltage at an electrical outlet in the home, if quoted at 120 volts, is 120 volts RMS. In other words, the actual maximum voltage appearing across the outlet is

$$V_{\text{max}} = \frac{V_{\text{RMS}}}{0.707} = \left(\frac{120}{0.707}\right) \text{ volts} = 170 \text{ volts}.$$

In the remainder of this chapter the symbols I and V when used without further qualification will always denote effective values. This is standard practice. Maximum values will be indicated, as in the past, with subscripts (e.g., I_{max}) and the instantaneous values by lower case letters (e.g., i).

---

*With $i = I_{\text{max}} \cos 2\pi ft$, the power supplied at the instant t to a resistor R is $Ri^2 = RI^2_{\text{max}} \cos^2 2\pi ft$. In Appendix A it is shown that the average value of the square of the cosine (or of the sine) during a complete half-cycle is the number $\frac{1}{2}$. Therefore the average power is

$$P = \tfrac{1}{2} R I^2_{\text{max}}.$$

Defining the effective current I to be such that $P = RI^2$, we obtain $RI^2 = \tfrac{1}{2} R I^2_{\text{max}}$, which implies $I = \dfrac{1}{\sqrt{2}} I_{\text{max}} = 0.707\, I_{\text{max}}$.

13.1 ALTERNATING VOLTAGES AND CURRENTS

The Production of AC

We now lean upon the basics of the electric generator (see Section 12.5) and review the essential ingredients in preparation for a look at practical generators.

The basic ingredients are a coil of wire (rotor) rotating in a constant magnetic field (produced by the stator), and a system of sliding contacts to lead off the induced current.

The induced EMF is, of course, increased by having a coil consisting not of a single loop but of many, and by increasing the speed of rotation; also, having the coil wound on a ferromagnetic core helps. This arrangement of wire wound on a core is referred to as the *armature*.

Practical Generators

In practice it is advantageous to achieve both a larger EMF and a higher frequency than is possible with a single coil rotating at a reasonable speed between a single pair of magnetic poles. One can achieve these ends by connecting several coils in series and providing additional pairs of magnetic poles. Fig. 13-4(a) shows the situation of two coils at right angles (each coil represented by one turn only) rotating between two pairs of magnetic poles. Notice what happens to the frequency of the alternations of the EMF in this case. Since a complete cycle of alternating EMF involves a coil passing by an N-pole to an S-pole and back again to an N-pole, it can be seen that in this case a complete EMF cycle is achieved with only half a revolution of the armature. To obtain a 60-cycle voltage this particular generator would have to rotate at only 30 revolutions per second. Obviously it is an advantage to reduce the speed of the rotating machinery as far as possible. Consequently, even more than four magnetic poles may often be employed.

Then, as the German electrical engineer and inventor, E. W. von Siemens (1816–1892), first proposed, it is advantageous to replace the various permanent magnets by electromagnets energized by current produced by the generator itself. You might wonder how a generator can ever get started under these conditions. The ferromagnetic cores of the electromagnets, once magnetized, retain a little residual magnetism, even after the magnetizing current is stopped. So once some outside exciter current has been utilized to start the generator in the first place, it is quite possible to restart it later without any outside current source.

In practice, armature winding is more complicated than the winding of the electromagnet, or field windings; and also the armature windings need better insulation than the field windings. Consequently the usual practice is to keep the armature stationary and to rotate the

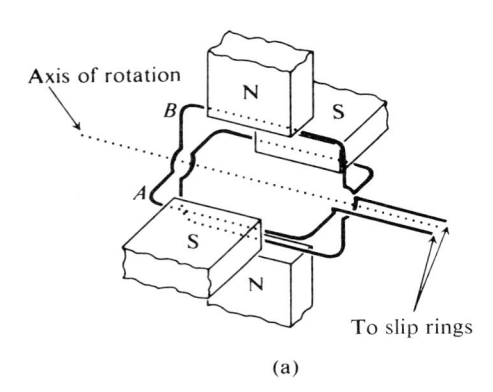

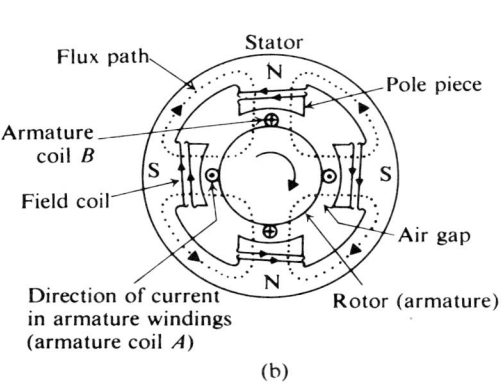

Fig. 13-4 Four-pole AC generator.
(a) Simplified version showing two coils, A and B, in series and at right angles.
(b) Schematic diagram of a four-pole commercial generator.

electromagnets or, as is commonly said, *rotate the field*. Such an arrangement is shown in Fig. 13-5. In this case, since the armature does not move, it is called the *stator*, and the rotating field coils are termed the *rotor*.

13.2 RESISTANCE, REACTANCE AND IMPEDANCE

Ohm's law, which proves so valuable in DC circuit analysis, is of much more limited value in AC.

For a rigorous analysis of AC, not only trigonometry but also calculus is required. Also complex numbers are used in the more elegant treatment of AC circuits, such as was devised by Steinmetz. In spite of these mathematical requirements *and without them*, we shall touch upon some of the key issues in AC—just enough to lead the reader to an expression analogous to Ohm's law for DC.

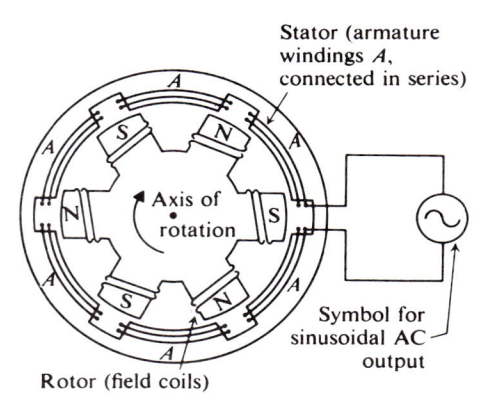

Fig. 13-5 Six-pole AC generator with the field coils as rotor. Direct current to energize the field coils is supplied through brushes and a split-ring commutator (not shown). The AC output comes directly from the ends of the armature windings.

Resistance

Consider an ohmic resistor in an AC circuit. Ohm's law relates the instantaneous voltage v across a resistor and the instantaneous current i through it:

$$v = iR. \tag{11.11}$$

This is indeed the Eq. (11.11) encountered earlier as Ohm's law, except that lower case symbols are used for voltage (v) and current (i) to denote their relation specifically at a given instant. (There is, of course, no concern with a distinction between instantaneous, maximum and average values when dealing with steady DC because in that case all these values are the same.)

Perusal of Eq. (11.11) shows that, since R is constant, as i changes so does v change correspondingly, and *at the same time*: We say v and i are *in phase*. From this fact it follows that one can also write

$$V_{max} = I_{max} R,$$

and that

$$V = IR \tag{13.1}$$

where V and I are effective values.

Further, from the definition of the effective value of an AC current it follows that the average power dissipated in a resistor can be written as in Eq. (11.14) for steady DC, namely

$$P = I^2 R \tag{13.2}$$

except that in this AC context the I refers to the effective current value.

Capacitors and Capacitance

In its simplest form a capacitor consists of two parallel metal plates (i.e., a conducting material) separated by a nonconductor (i.e., a *dielectric*), as shown in Fig. 13-6. Such a device (Fig. 2-25) was mentioned in the sections on electric forces (Section 2.7) and electric potential (Section 11.1; Fig. 11-1).

Capacitors in many respects may be considered as key components in our electronic way of life. Without capacitors our sophisticated communications systems would be impossible, since capacitors are used in both the generation and detection of radio waves and, in general, capacitors are an essential element in most AC circuits.

The description of the basic elements of a capacitor (Fig. 13-6) perhaps suggests what a capacitor does: It stores electric charge. But so does any insulated conductor, such as a metal plate supported on a glass stand. What makes a capacitor different? It can be shown that if we approach the one insulated metal plate with a second metal plate of an equal but opposite charge, then the amount of charge on each plate can be substantially increased, and the corresponding increase of potential difference between the plates will be relatively small if the plates are close together. A description that is perhaps not quite rigorous but at least plausible is that the positive charges on the one plate, due to Coulomb attraction, help to hold the charges on the other plate so that one can keep on loading up more and more charge.

In practice, capacitors may look quite different from the simple parallel plate "sandwich" described above. However, other forms of capacitors, although they seem quite different on the outside, must all have this same characteristic: They must be composed of two conducting materials separated by a dielectric. One common form is composed of

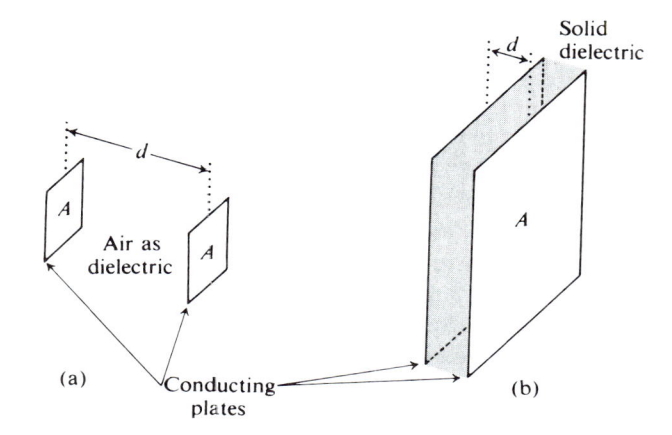

Fig. 13-6 Parallel plate capacitors. (a) Small capacitance: small plate area (A), large separation (d), low dielectric constant. (b) Large capacitance: large plate area (A), small separation (d), high dielectric constant.

476 ALTERNATING CURRENTS

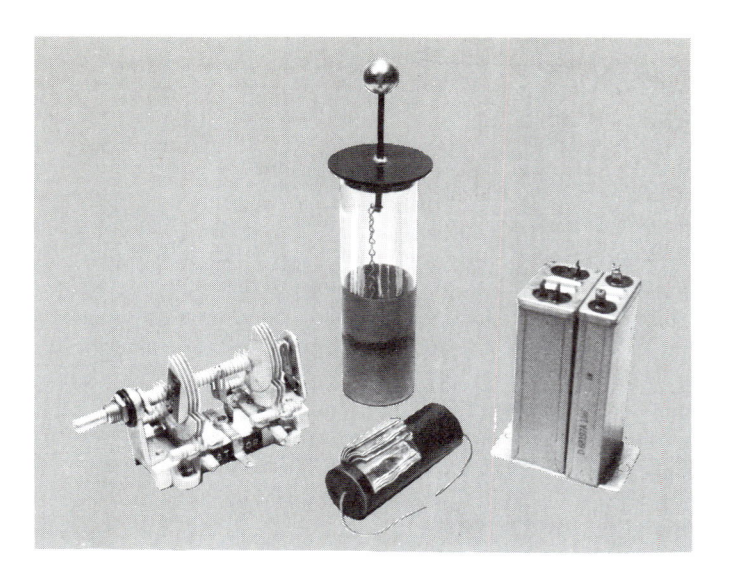

Fig. 13-7 Capacitors. The glass Leyden jar is shown in the center. To the left is a radio tuning capacitor, to the right an oil-filled capacitor, and in front center (partially cut open) a metal-foil capacitor.

metal foil and wax-paper strips rolled up. Such an arrangement effectively gives a larger capacitor area but uses very little space. This and other types of capacitors are shown in Fig. 13-7. One particularly interesting type seen in this figure is the variable capacitor used in radio tuning. When you select different stations, you are actually altering the *capacitance* of the capacitor by changing its effective area as the one set of vanes is moved in and out of the other. The dielectric in this case is, of course, simply air.

Note that, by reference to Fig. 2-25, there is to all intents and purposes no electric field outside the charged capacitor plates. This situation derives from the fact that, since the capacitor plates are conductors, the negative charges placed on one plate will move to be as close as possible to the positive charges placed on the other and that the electric field inside the conducting plates themselves is zero. The result is that outside the capacitor the contribution to the electric field of the positive charges more or less cancels the contribution due to the negative charges, and vice versa.

Note also that, except for the fringing at the ends, the electric field between the parallel plates is uniform (that is, the same everywhere between the plates). This characteristic of parallel-plate capacitors is a very useful attribute in electrical technology.

The word *capacitance* was used earlier without assigning a specific meaning to the term. We now define the capacitance, C, of a capacitor as follows:

$$C = \frac{q}{v}$$

13.2 RESISTANCE, REACTANCE, AND IMPEDANCE   477

where q is the magnitude of the charge on either of the plates and v is the potential difference between the plates. The unit for capacitance is thus coulomb/volt, a quantity known as the *farad*; that is

$$1 \text{ farad} = \frac{1 \text{ coulomb}}{1 \text{ volt}}.$$

In actual practice this turns out to be a rather large value of capacitance, and the microfarad (μf) and the micro-microfarad (or picofarad, abbreviated $\mu\mu$f)* are more frequently encountered.

*$1\mu f = 10^{-6}$ f; $1\mu\mu f = 10^{-12}$ f.

What affects the capacitance? The parameters involved are all quite plausible. Consider a parallel-plate capacitor. First of all, it seems quite reasonable that increasing the area of the plates would increase the capacitance. After all, there is then more place to store the charges. Secondly, the capacitance is increased by bringing the plates closer together. Thus one might expect

$$C \propto \frac{A}{d},$$

where A is the area of the plates and d is the distance between the plates. With the appropriate proportionality constant, the above proportionality turns into the equation

$$C = \epsilon \frac{A}{d},$$

where ϵ is a constant, characteristic of the dielectric material separating the capacitor plates. Thus it is possible to vary capacitances by changing the geometry (that is A and/or d) of the capacitor, or by inserting a different dielectric (see Fig. 13-6). Our presentation has served to make the above equation plausible; an exact analysis confirms that it is indeed correct.

Charging and Discharging the Capacitor

From the standpoint of Coulomb forces (charges on a particular plate will repel additional like charge as it is brought to that plate), it is expected that the higher the charge already on a capacitor, the more work would have to be done to add more of the same charges. In other words, when the capacitor is "empty" it is easy to add charge; and this process of adding charge becomes successively harder as the capacitor becomes more and more charged.

If a capacitor is charged (Fig. 13-8) by applying some kind of a constant voltage source such as an ordinary dry cell, the charging rate, that is, the current, will be large at first, and will finally go to zero as the voltage across the capacitor becomes the same as that of the dry cell. At this time, since there is no longer any potential difference between points a and b, and points c and d, no work can be done by the electric field in moving further charges; that is, the electric field within these

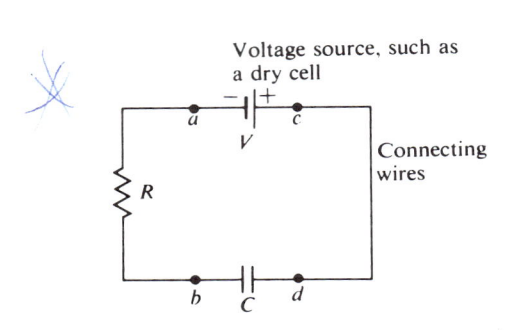

Fig. 13-8 Capacitor (C) connected to a constant voltage source (V). Figure shows usual circuit symbols for the capacitor, the voltage source, and the resistance (R).

478 ALTERNATING CURRENTS

wires is zero, and therefore the current is zero. It can be shown that the charging current falls to zero in an exponential fashion; that is, the "decay" of the current follows a law of the type encountered in Section 1.5 on radioactive decay.

Capacitive Reactance

The definition of capacitance, $C = q/v$, can be rearranged:

$$v = \frac{q}{C}. \tag{13.3}$$

(The charge on one plate of the capacitor at the instant in question is $+q$; the other plate has a charge $-q$.) Eq. 13.3 shows that as v changes so does q. But, according to what is meant by current, the current at any instant in the wire leading to a plate of the capacitor is determined by the change of electric charge (on that plate), Δq, which occurs in the short time interval Δt, that is

$$i = \frac{\Delta q}{\Delta t}. \tag{13.4}$$

Therefore in an AC circuit (see Fig. 13-9) there is a current i leading to one plate of a capacitor and away from the other plate, associated with the changing capacitor charge q. And in AC, q changes because v is continually changing. (This is, of course, quite different from the DC case in which the capacitor, once it is charged to a value such that the voltage across it is equal to the constant DC applied voltage, acts as an open circuit and there is no more current flow.)

All this is not to say that a capacitor in an AC circuit does not impede the current. It does. The impeding effect of a capacitor upon an alternating current is due to the reverse potential difference which appears across the capacitor as the electric charge builds up on the capacitor plates. This potential difference, or potential drop, affects the total circuit current just as the potential drop across a resistor does. The impeding effect of a capacitor upon the current is called the *capacitive reactance*. The voltage across the capacitor is given by

$$V_C = IX_C, \tag{13.5}$$

where V_C is the *effective voltage* across the capacitor, of capacitance C, I is the *effective current*, and X_C is the capacitive reactance which, like resistance, is measured in ohms.

The capacitive reactance in turn is given by

$$X_C = \frac{1}{2\pi f C}$$

where f is the frequency of the AC and C is the capacitance in farads.

Let us see if the inverse dependence of X_C upon f and C seems plausible. If C is large then Eq. (13.3) shows that for a given charge q

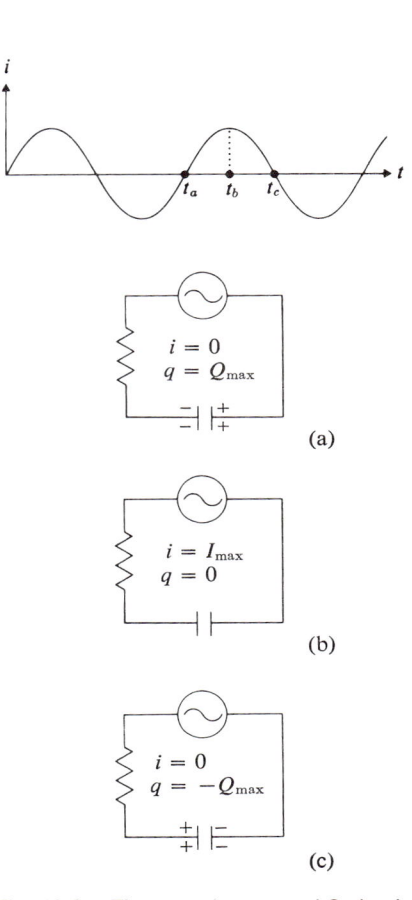

Fig. 13-9 The capacitor as an AC circuit element. (a) Capacitor charged, current at zero. (b) Capacitor fully discharged, current at maximum. (c) Capacitor now charged in opposite direction, current at zero, and about to start in direction opposite to that shown in (b).

13.2 RESISTANCE, REACTANCE, AND IMPEDANCE

on the capacitor plates, the voltage across the capacitor is smaller than in the case of a small C. And it is this voltage across the capacitor which gives rise to the current impeding effect known as capacitive reactance.

Further, if the frequency f is large, or high, then the period is short, each cycle is brief, and less charge q is deposited on the respective capacitor plates (for a given effective current) in each half of the cycle than if the frequency is low. Hence, according to Eq. (13.3), with a given capacitance C the implication is that the current-impeding voltage v is smaller at high frequencies, that is the opposition to current is less. Consequently the capacitive reactance is less at higher frequencies.

In the matter of phase relationship between the voltage across a capacitor and the circuit current, note that the reverse potential difference, the voltage opposing the current, is at a maximum when the capacitor has accumulated all the charge it will during the half cycle concerned. This fact can be seen by perusal of Eq. (13.3) which clearly indicates that the voltage has its maximum value when the charge q has its maximum value. But at the instant that q has reached its maximum value the current reaches zero. The current was at its maximum value or peak value earlier when the capacitor was uncharged and the voltage across the capacitor was consequently zero. In other words, the current is *out of phase* with the voltage across the capacitor. The current *leads* the voltage by a quarter of a period, as shown in Figs. 13-10 and 13-12.

This matter of phase difference leads to an interesting result regarding voltage addition in AC circuits. Consider a capacitor and a resistor together, in series, in the same circuit. The current and voltage in the case of a resistor are in phase. Furthermore, at any instant the current in the circuit is everywhere the same. Therefore the voltage across R and the voltage across C in the AC series circuit of Fig. 13-12 are out of phase. This is a new and different concept not encountered in our DC studies. And it means that *voltages in an AC circuit cannot be added by the straightforward arithmetic approach we used in DC*. The appropriate mathematical procedures for adding AC voltages are given as a supplementary topic at the end of this chapter.

The ideal capacitor, unlike a resistor, dissipates no energy. Energy is stored in the electric field of the capacitor during the quarter cycle in which the capacitor is being charged. This energy is returned to the circuit during the discharging process, that is, the following quarter cycle. In other words, the *average energy* and hence the *average power* delivered to a capacitor over a full cycle is zero.

Inductive Reactance

Now consider what happens if a coil of wire, such as shown in Fig. 13-11(b), is part of an AC circuit. Such a coil is called an *inductor* or

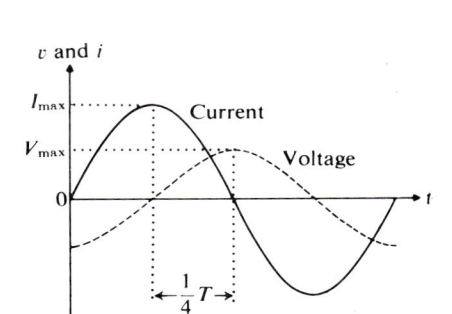

Fig. 13-10 Voltage and current phase relationship in a purely capacitive circuit. The current leads the voltage by a quarter of a period.

480   ALTERNATING CURRENTS

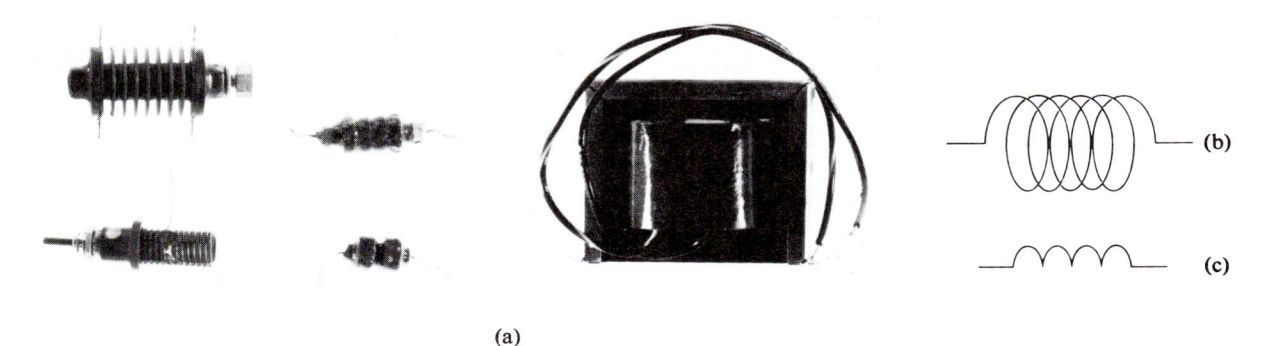

inductance coil. Inductors are important constituents of many electronic circuits, as well as electric motors, electromagnets, etc. For example, when you change the channel on your TV set you are likely switching from one inductance coil to another. The circuit symbol for an inductor, the electrical property of which is called *inductance*, is shown in Fig. 13-11(c).

The opposition to an AC current offered by an inductor stems from the self-induced *back EMF* associated with the changing current through the inductor. (The reader will recall that a changing current results in a changing magnetic flux and that a changing flux produces an EMF in accordance with Faraday's law.) This self-induced back EMF is what constitutes the voltage across an inductor, and is given by

$$V_L = IX_L, \qquad (13.6)$$

where V_L is the effective voltage across the inductor, of inductance L, I is the effective current through the inductor, and X_L is the *inductive reactance*, the opposition to current flow (analogous to R in the case of a resistor and X_C in the case of a capacitor) and, like R and X_C, is measured in ohms.

The inductive reactance in turn is given by

$$X_L = 2\pi f L,$$

where f is the frequency of the AC and L is the *inductance* of the inductor, a characteristic which we shall assume to be constant, i.e., not dependent upon the current through the inductor. The mks unit for L is the *henry*.

Increasing the number of turns of the induction coil increases L, and winding the wire coil around a core of magnetic material, such as iron, vastly increases L over its value with an air core. Since such a coil significantly impedes alternating current, the term *choke* or *choke coil* is sometimes used, in the sense that the inductor "chokes" the current.

So the inductor impedes an AC current. But like the capacitor it does not dissipate energy as does a resistor. However, it must be made

Fig. 13-11 Inductance coil. (a) A variety of inductors. (b) Schematic representation of an inductor. (c) Circuit symbol for an inductor.

clear that here we are talking about an ideal inductor, that is, one which has no resistance whatsoever. (In practice this is obviously not quite the case since even copper wire has some electrical resistance.)

As to the matter of energy, what happens in the case of the inductor is that, as the current in the circuit is increasing, energy is taken up by the inductor in establishing a magnetic field (see Chapter 12). But this energy is not dissipated at the rate i^2R as in a resistor, rather the energy is stored in the magnetic field. Then when the current decreases the magnetic field associated with the inductor in turn decreases and the energy goes back into the circuit. Since the inductor is thus seen to return to the circuit, on part of the cycle, the energy it took from the circuit during another part of the cycle, the *average energy* and hence *average power*, delivered to an inductor over a full cycle is zero. The measure of the ability of the inductor to store energy in its magnetic field is given by the inductance L.

Inductors, as non-energy-dissipating current limiters, are extensively employed in the starting circuits of fluorescent lights.

One other point regarding the inductance of an AC circuit has to do with phase difference, a matter touched on earlier in connection with the capacitor. Fig. 13-12 shows this voltage-current relationship. The current is seen to reach its maximum value $\frac{1}{240}$ sec later than the voltage across the coil, assuming, of course, 60-cycle AC. As in the case of a capacitor, the current and the voltage are *out of phase*. But the voltage across the inductor *leads* the current by a quarter of a period, while in the case of the capacitor the voltage across it lags the current by $\frac{1}{4}T$.

Fig. 13-12 AC series circuit. (a) The circuit consists of a resistance R, an inductance L, and a capacitance C. The impedance is determined by the resistance R, the inductive reactance X_L, and the capacitive reactance X_C.
(b) Voltage-time graph, showing the phase differences among the voltages across R, L, and C. (c) The current is the same throughout the circuit and is in phase with the voltage across R.

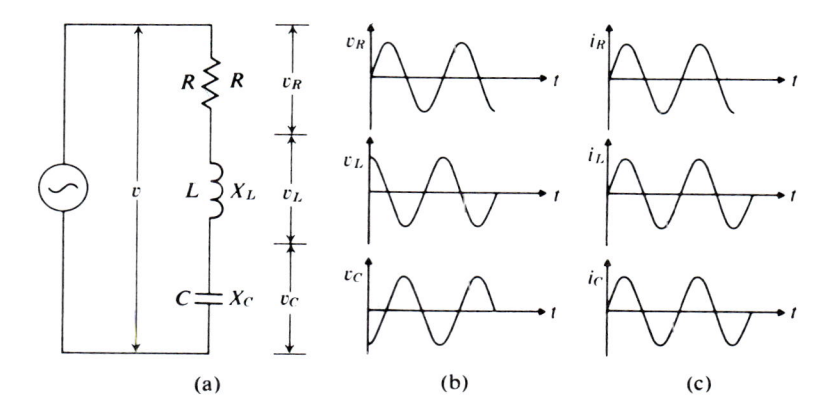

Impedance

An individual resistor, an inductor, or a capacitor each impede an alternating current. We now describe the current-impeding effects of a circuit element which is perhaps a complicated combination of these three types

482   ALTERNATING CURRENTS

of circuit elements. For any such circuit element we define a quantity known as the *impedance* of the circuit element in a fashion analogous to the definition of resistance, Eq. (11.9):

$$Z = \frac{V}{I} \qquad (13.7)$$

where Z is the circuit element's impedance in ohms, V the effective voltage across the circuit element in volts, and I the effective current through the element in amperes.

If the circuit element is the series combination of a resistance R, an inductor of reactance X_L, and a capacitor of reactance X_C, as shown in the circuit of Fig. 13-12, this impedance Z turns out to be given by

$$Z = \sqrt{R^2 + (X_L - X_C)^2}. \qquad (13.8)$$

The apparently peculiar addition format for these three current-impeding effects, R, X_C and X_L, arises from the fact that the voltages across R, L, and C are out of phase. The voltage across L leads by $\frac{1}{4}T$ that across R, which is in phase with the current, and the voltage across C lags by $\frac{1}{4}T$ as summarized in Fig. 13-12.

Eq. (13.8) shows that if the frequency is such that the term $(X_L - X_C)$ is zero, the separate terms X_L and X_C being equal in magnitude, then the impedance is at its minimum value and is simply equal to R. The circuit is purely resistive.

This condition of minimum impedance in a circuit including R, L, and C is referred to as the condition of *resonance*. The frequency at which this condition occurs is called the *resonant frequency*. In the ideal situation ($R = 0$, i.e., no energy dissipated) in a resonant circuit the energy would be continually exchanged between the inductor's magnetic field and the capacitor's electric field, as shown in Fig. 13-13.

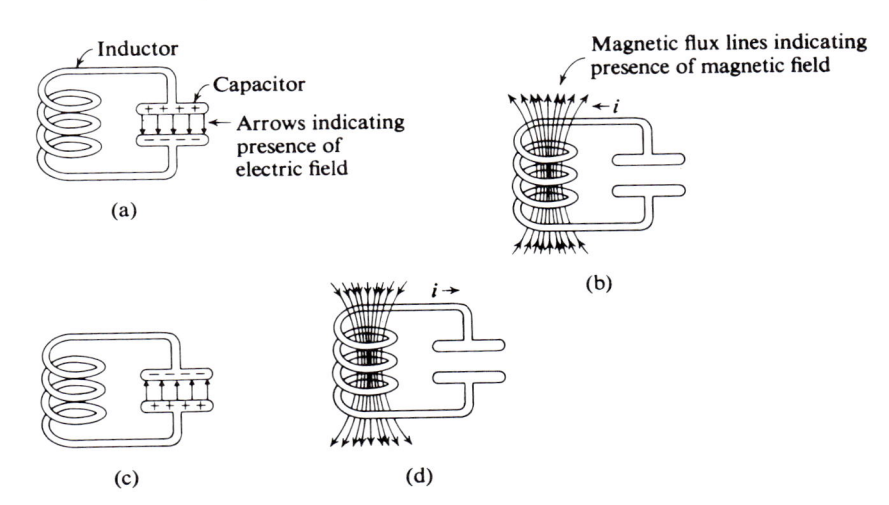

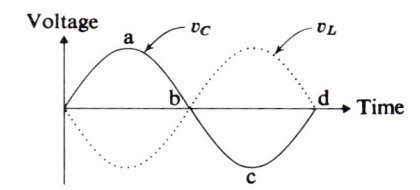

Fig. 13-13 Resonant circuit with a quarter cycle interval between successive diagrams (a), (b), (c), and (d). In (a) the capacitor is charged and the current is zero; the energy of the circuit is stored in the electric field of the capacitor. In (b) the capacitor is discharged and the current attains its maximum value; the energy is now stored in the magnetic field of the inductor. In (c) the capacitor is charged in the opposite direction and the current is again zero. The cycle is completed in (d) when the current again attains maximum magnitude but is in the direction opposite to that of the situation in (b). The energy is now back in the magnetic field. With reference to the voltage-time graph, note that at any instant $v_C + v_L = 0$, and that when $v_C = 0$, $v_L = 0$.

13.2 RESISTANCE, REACTANCE, AND IMPEDANCE

AC Ohm's Law Analogue

We now recall Eq. (13.7) and stipulate the impedance to be "constant," that is, independent of the voltage across the circuit component in question, and rewrite the equation in the form

$$V = IZ. \qquad (13.9)$$

The equation written in this form is sometimes known as "Ohm's law for AC."*

While this equation looks quite as harmless as its DC counterpart, Eq. (11.11) [and its AC adaptation, Eq. (13.1)], one must be considerably more careful in the application of this new equation. For example, it is emphasized that because of phase differences the effective voltages across various parts of an AC circuit do not add arithmetically to yield the total effective voltage. In fact, it is possible to have an effective voltage across an inductor or a capacitor much higher than the effective EMF or effective input voltage to the circuit.

*While Eq. (13.9) is similar in form to Ohm's law, it must be realized that this is really only a matter of the form of the equation and that Ohm's law itself, $v = iR$ [Eq. (11.11)] with i and v being instantaneous values, is valid in both DC and AC.

13.3 POWER IN AC CIRCUITS

AC Power

In previous paragraphs it was emphasized that the *average power* delivered to a capacitor and/or an inductor in an AC circuit is zero. However, in a resistor of resistance R the situation is different. No matter what the direction of the current through the resistor, energy is continually dissipated at the rate i^2R. So it is evident that the average power delivered to a resistor is not zero. In fact, as indicated earlier, the effective value of the current I is defined in such a way that the average power dissipated in a resistor is I^2R, as is the case for a DC current, I.

From this information about capacitors, inductors and resistors one can deduce a useful expression for the *average power P* dissipated as heat in any *impedance Z*.

Since *all* the power dissipation in an impedance occurs in its resistive component R, we have

$$P = I^2R. \qquad (13.2)$$

(This equation is entirely analogous to Eq. (11.14) except that now the I refers to the effective value of an AC current instead of a steady DC current.)

Using Eq. (13.7), the definition of impedance, we can write

$$I = V/Z,$$

where V is the voltage across the entire impedance in question.

We now enter this expression for I on the right-hand side of Eq. (13.2). Thus

$$P = I^2R = I(V/Z)R$$
$$P = IV(R/Z). \qquad (13.10)$$

The factor R/Z, by which the product IV must be multiplied to give the average power dissipated in an impedance, is called the *power factor* for the impedance. Clearly the power factor has the value *one* for an impedance which is just a resistance ($Z = R$), while the power factor is *zero* if the impedance has no resistive component ($R = 0$).*

*Our analysis is applicable only to a circuit element that contains no source of EMF (a passive circuit element). The general result (see Chapter 35 in the book by Sears and Zemansky listed in the Suggested Reading) is that $P = VI \cos \phi$, where ϕ is the phase difference between the current and the voltage across the circuit element. The term $\cos \phi$ is called the power factor. For a passive circuit element, $\cos \phi = R/Z$.

Transformers

Making use of our observations on AC power, it is feasible to further consider those circuit elements which have caused AC to triumph over DC in the matter of power distribution. (Earlier, in Fig. 12-24, a typical example of the various voltages employed in the distribution of electrical power from the supplier to the user was shown.)

Recall from Section 12.5 that the ratio of the EMF induced in the secondary coil to that of the primary coil is given by the turns ratio:

$$\frac{\mathcal{E}_s}{\mathcal{E}_p} = \frac{N_s}{N_p}. \qquad (12.17)$$

Using the present notation for effective values of voltage, we can rewrite this equation:

$$\frac{V_s}{V_p} = \frac{N_s}{N_p}, \qquad (13.11)$$

noting also at this point that the waveforms for V_s and V_p are identical.

It turns out that transformers are very efficient (90–99%).* It is therefore close to the truth to say that the *average power input* to the primary equals the *average power output* of the secondary. Looking at Eq. (13.10), quite the proper expression to use for AC, it can be seen that there is the problem of the power factor to be considered. Fortunately it can be shown by a thorough analysis that the power factor for the secondary is the same as for the primary. This fact yields the convenient result that, since $P_p = P_s$,

$$V_p I_p = V_s I_s.$$

And since

$$\frac{V_s}{V_p} = \frac{N_s}{N_p}, \qquad (13.11)$$

we get

$$\frac{I_p}{I_s} = \frac{N_s}{N_p}. \qquad (13.12)$$

*For a descriptive discussion of transformer power losses the reader is referred to the text by Ridgway and Thumm, listed in the Suggested Reading.

In other words, the currents in the primary and secondary windings are in the inverse ratio to the respective numbers of turns.

13.3 POWER IN AC CIRCUITS

13.4 ELECTRONICS

The advantages of AC notwithstanding, there are many electrical applications which require DC. In such an application, if the source of EMF is an AC source then it is necessary to effect a change. The changing of an alternating current to a pulsating uni-directional, or direct, current is known as *rectification*. The circuit elements which achieve this rectification are termed *rectifiers*.

Rectifiers are essentially of two types: vacuum or gas-filled tubes, and solid state devices. Of these, as yet the tube type is predominantly used wherever high voltages are involved. Consider the vacuum tube, or thermionic diode. This device is known as a *thermionic diode* because two separated conductors called *electrodes* are sealed into the tube, one at each end, and one of the electrodes, the *filament* (or *cathode*), is heated to give off electrons. Basically, this is the *valve tube* as originated by Fleming in 1904.

The study of the operation of these rectifier devices, as well as other vacuum and gas-filled tubes and certain solid state devices, falls under the general heading of electronics.

Vacuum-Tube Rectifier

Ideally a rectifier would be a device which would offer an infinite resistance when the voltage across it is applied in a given direction, and zero resistance if the polarity of the applied voltage is reversed. You have now studied enough physics not to expect such a device. However, to be a rectifier a device has to meet this criterion in good measure at least. And for our purposes we shall assume the ideal rectifier. In truth, certain rectifiers are close enough to ideal to warrant such approximate analysis.

Fig. 13-14 shows a vacuum diode and its schematic representation. When the filament of the tube is heated, electrons are given off. If now the other electrode is positive with respect to this filament, then the electrons will, under the influence of the electric field, migrate to the other electrode, called the *plate* (or *anode*): There will be a current in the

Fig. 13-14 Vacuum diode. (a) Photograph of a vacuum diode cut in half. Normally the cylindrical anode on the left surrounds the cathode spiral or filament seen in the central part of the tube on the right. This particular type of tube finds use in rectification in the high-voltage part of x-ray circuits. (b) Circuit symbol for a vacuum diode.

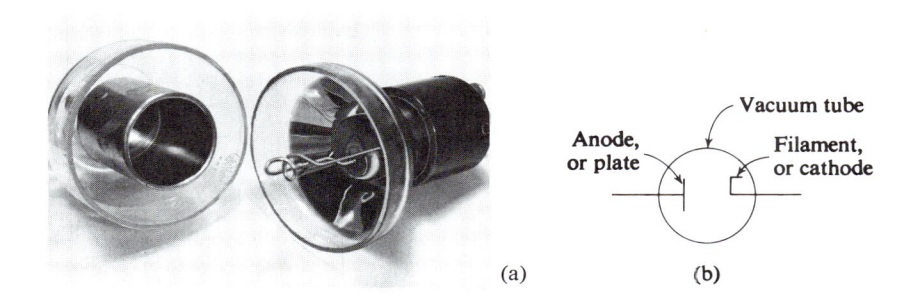

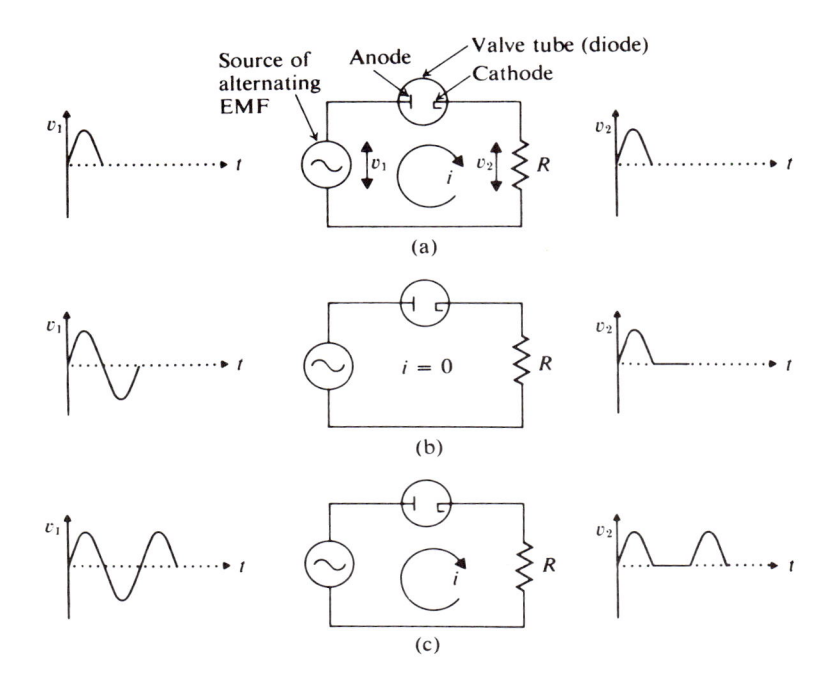

Fig. 13-15 Half-wave rectification employing a vacuum diode. (a) The first half-cycle, showing the input voltage v_1 supplied by the source and the output voltage v_2 across the resistor R. (b) The second half of the cycle, during which the tube does not conduct. (c) The first half of the next cycle.

tube, i.e., the tube conducts [see Fig. 13-15(a) and (c)]. If the plate is negative with respect to the filament (i.e., at a lower potential than the filament) the electric field in the tube is such as to drive any emitted electrons back to the filament. Then the tube will not conduct.

Basic Rectification Circuits

To pursue the essential details of rectification refer to Fig. 13-15. In part (a) of the figure we see the half cycle during which the plate of the vacuum diode is on the high potential side of the circuit: The tube conducts and consequently there is a current in the circuit. Since there is a current through the resistor R there will be an *iR-drop* voltage, $v_2 (=iR)$, across the resistor. This voltage will be like the input voltage, v_1.

Next [part (b)] the input voltage makes its negative excursion which thus places the plate on the low potential side of the circuit. The tube does not conduct. There is no current in the circuit. And so there is no iR-drop across the resistor: v_2 is zero, even though v_1 is not.

In part (c) there is the same situation as in part (a). The tube conducts, there is a current, and there is consequently a voltage across R.

Now notice two things. First of all, when there is a current in the circuit this current is always in the same direction. And as a consequence of this unidirectional aspect of the current, the output voltage across R

13.4 ELECTRONICS

(the iR-drop) always has the same polarity, despite the fact that the input voltage displays alternating polarity. So indeed AC has been changed to pulsating DC.

Second, it can be seen that although the input voltage is only momentarily zero, the current and consequently the output voltage is zero for a whole half-cycle, in each cycle. Therefore this particular rectification is called *half-wave rectification*.

Of course for most purposes it would be desirable not to lose a half-cycle of the input voltage. This desirable end can be achieved: The process is known as *full-wave rectification*. One way of achieving full-wave rectification is shown in Fig. 13-16. The transformer which is the source of alternating EMF is tapped at the center as well as at each end. When the top end of the transformer is the high-potential end then the upper tube conducts, with the current going through R from right to left. On the next half-cycle the lower end is at the high potential, the lower tube conducts, and the current again goes through R in the same direction as before.

Fig. 13-16 Full-wave rectification, using two vacuum diodes.

Note, however, that in the case shown, the voltage across R is, at best, half the entire voltage across the transformer. But this deficiency can be remedied. The design of such a circuit, using four rectifiers, is left to the reader (see Question 49).

Pulsating DC

Reference to Fig. 13-16 shows that the conversion of AC to DC does not produce a steady DC: Although the voltage is now no longer changing polarity it is continually changing in value; it is *pulsating*.

For many DC usages this pulsation is undesirable. To reduce the degree of pulsation appropriate circuitry can be devised. The process of reducing the pulsation is called *filtering*, and the reduced pulsation which remains after filtering is called the *ripple* (see Fig. 13-17).

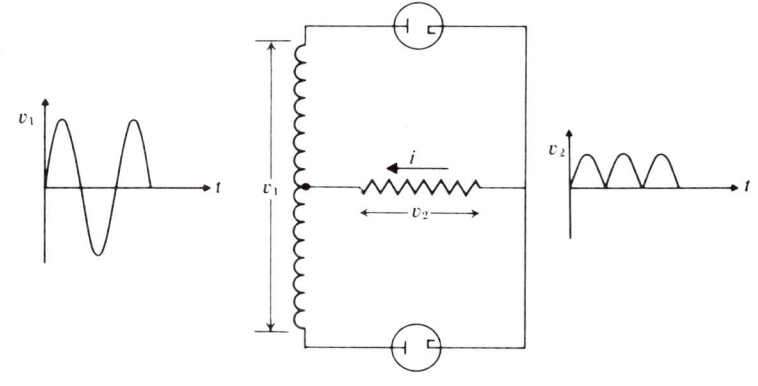

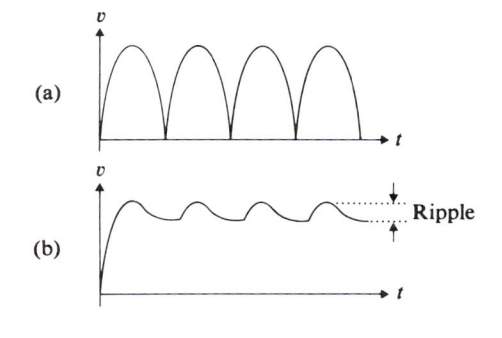

Fig. 13-17 Full-wave rectified AC. (a) The wave form upon rectification. (b) The wave form after filtering. Note the ripple. The wave form and amount of ripple depend on the details of the filtering circuit.

488 ALTERNATING CURRENTS

The above is not to be construed as implying that when DC is required it is necessarily always steady DC that is wanted. Often a DC pulsating type voltage is required in electronic applications. Moreover it is not unusual in such applications to start with a source of EMF, like a dry cell, which yields steady DC with no ripple at all, and then devise circuitry to give the kind of voltage or current pulses desired. Such circuits are called *pulse* or *pulsing* circuits.

One interesting example in which a DC voltage source (in the form of so-called mercury batteries) is used to yield periodic DC pulses of about 2 millisecond duration each, is the implantable *pacemaker*. The normal rhythmic functioning of the heart is triggered by a part of the heart structure known as the pacemaker. In certain ailments this pacemaker no longer provides adequate cardiac stimulation, and in some of these cases the patient's life can be sustained by implanting an electronic pacemaker which provides the necessary cardiac stimulation by means of two electrodes inserted in the heart muscle tissue. The entire pulsing circuit along with the source of EMF can be contained in a nonirritative material (e.g., epoxy resin). And the size of the entire arrangement being about that of a key case, it is possible (and has been done successfully) to implant the complete pacemaker unit adjacent to the heart and permit the recuperated patient to carry on diverse normal activities (which, according to medical reports, include having babies and water skiing). Fig. 13-18 shows the approximate nature of the electrical pulse delivered to the heart. A commonly employed frequency for this pulsed DC is around 70 min^{-1}, since the normal pulse rate in adults ranges from 60-80 beats per minute.*

*Recent laboratory experiments have indicated that the body's digested food and inhaled oxygen might be used to provide the electrical energy to drive a pacemaker.

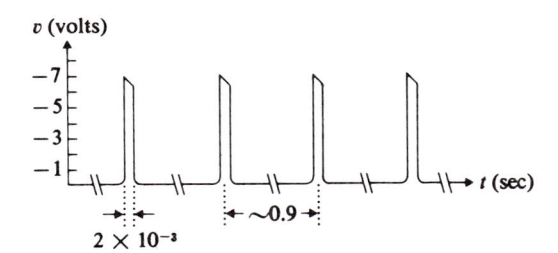

Fig. 13-18 Approximation of output voltage pulses of an implantable pacemaker. The current is of the order of 10 mA.

The Triode

A triode is a thermionic tube that has three electrodes. The provision of the third electrode, in addition to the filament and plate found in the diode, results in certain new and important tube characteristics.

The additional electrode in the triode is called a *grid*, and it is

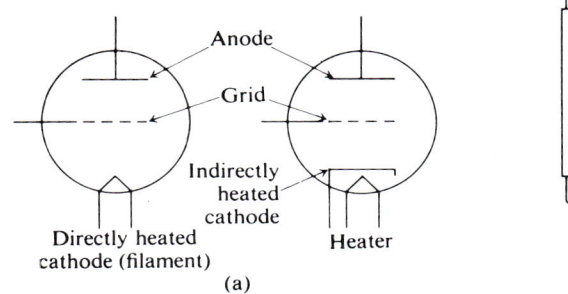

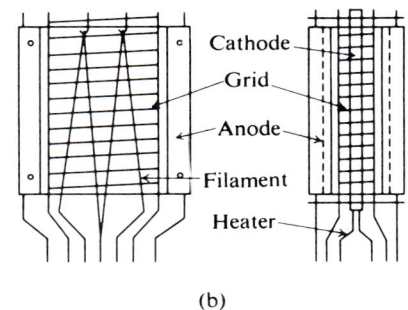

Fig. 13-19 The triode. (a) Diagrammatic representation. (b) Construction details.

usually schematically represented as shown in Fig. 13-19(a). Part (b) of the same figure shows one form of actual construction. The grid is seen to consist of an arrangement of parallel wires between the plate and the filament.

If the triode is used in a circuit with only the plate and filament connected, it will function more or less as a diode. Electrons will flow in one direction only, from the filament (or cathode) to the plate (or anode). But if the grid is made sufficiently negative with respect to the filament, there will be no electron flow through the tube. The negatively charged grid will repel the electrons emitted by the filament. These electrons will therefore be unable to reach the plate, even though the plate is positive with respect to the filament. This situation is shown in Fig. 13-20.

The voltage by which the grid is made negative is known as a *negative grid bias*. Because of the proximity of the grid to the filament, the amount of negative bias required to hold the tube current at zero is relatively small.

In order to permit electrons to flow through the tube, one reduces

Fig. 13-20 Negative grid bias. (a) The circuit diagram containing the triode. (b) The electric field (approximate) in the tube. The arrows show the direction of the field during negative grid biasing. Recall that the field direction is the direction of the force exerted on positive charges.

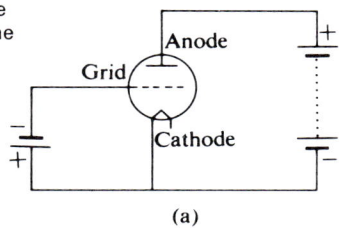

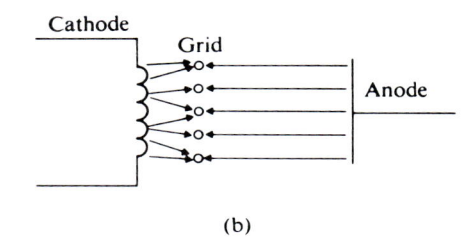

490 ALTERNATING CURRENTS

the negative grid bias. As the value of the grid voltage approaches zero, the effect of the distant but relatively large positive potential of the plate will predominate. Because electrons are emitted from the filament with a range of energies, some electrons will pass through the grid, which is now less negative than before. With a grid bias of zero, electron flow, that is the *plate current*, will be virtually unimpeded.

The significant relationship in the triode is that between the *plate current* and the cathode-to-grid voltage, or *grid bias*. For a fixed plate voltage (that is, the voltage between filament and plate), the plate current will vary greatly with small changes of grid voltage, as shown in Fig. 13-21. Therefore the application of a small input signal in terms of a small varying voltage between the grid and the filament will result in large changes of plate current. If this plate current is passed through a suitable resistor, the voltage across the resistor will then, quite in accordance with Ohm's law, also vary greatly: The output signal is said to be large. Such an arrangement, shown in Fig. 13-22(a), is called a *voltage amplifier*. (You may have one or more such amplifiers between the "pick-up" of your record player and the speaker.) Part (b) of the same figure shows graphically the relationships between the three voltages, input voltage superimposed on the grid bias voltage, resulting grid voltage, and output voltage.

Besides finding widespread application in radio and television equipment, the voltage amplifier is used to amplify small signals in all manner of industrial and medical electronic gear. In medicine, as an example, the faint signals picked up from the body in electrocardiography

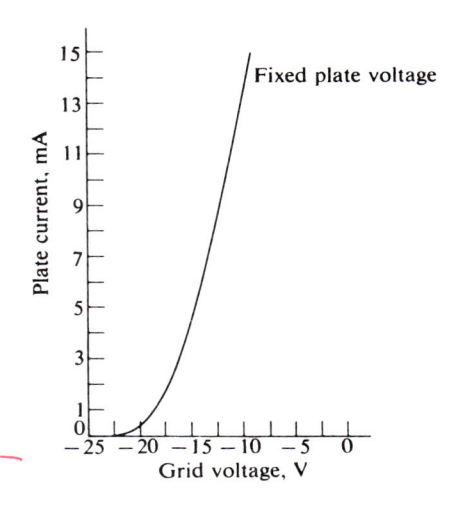

Fig. 13-21 The dependence of plate current on grid voltage in a triode. Small changes in grid voltage produce large changes in plate current. (The plate voltage has been kept constant.)

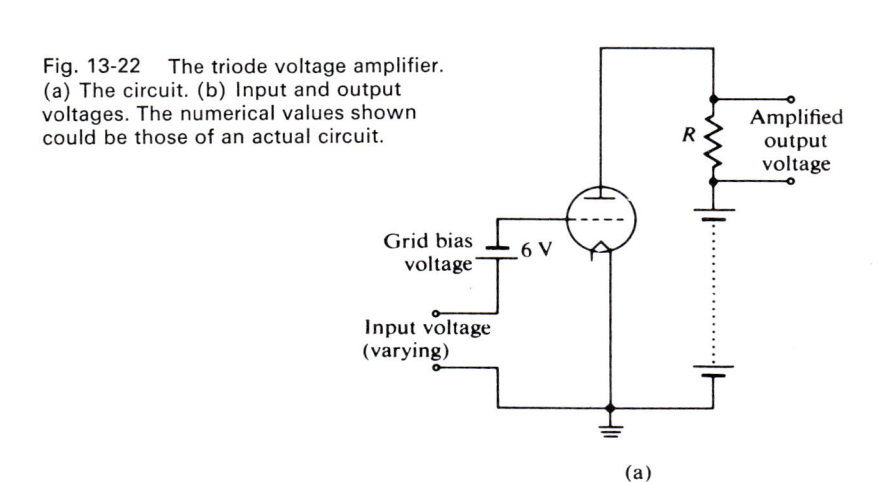

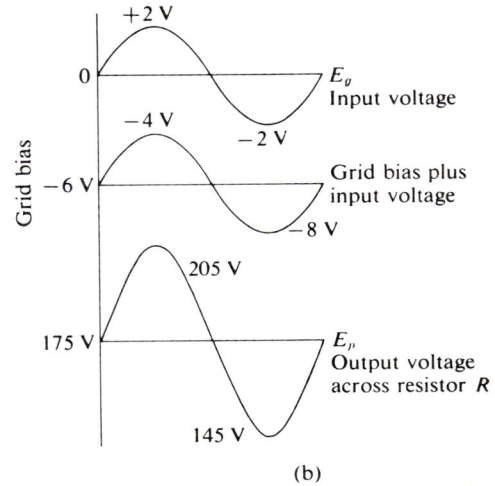

Fig. 13-22 The triode voltage amplifier. (a) The circuit. (b) Input and output voltages. The numerical values shown could be those of an actual circuit.

13.4 ELECTRONICS 491

and electroencephalography are very much dependent upon amplification before being useful.

The vacuum tube form of the triode, as described here, is rapidly being replaced by solid-state devices, *transistors*, the operation of which is analogous to that of the vacuum triode. The operation of solid-state diodes will be dealt with next. To extend that discussion to one on the transistor requires primarily the consideration of a third electrode which, like the grid of the vacuum triode, will control the movement of electric charge.

Solid-State Rectifiers

The circuitry mentioned earlier in connection with the pacemaker involves solid-state components, transistors. We shall not delve into the workings of transistors but will provide the reader with a start in the subject by considering the behavior of solid materials used to make rectifiers.

To achieve rectification the solid materials must obviously have non-ohmic characteristics. Semiconductors are such solids. In proper combination with another semiconductor or with a conductor, semiconductors achieve a high resistance to current in one direction and a low resistance in the opposite [see Fig. 11-10(b)].

The advantages of the solid-state rectifier over the thermionic valve tube are significant. The solid state device is more rugged, lasts longer with proper usage, requires no heating current, does not produce x-rays as may a vacuum diode under certain conditions, and finally, the solid-state rectifier takes up much less room. In the realm of high voltages, present silicon rectifiers about the size of a thick pencil can take the place of vacuum tubes usually somewhat larger than a beer mug. And in the area of low voltages the miniaturizing of circuitry which solid-state devices make feasible is truly staggering (see Fig. 16-17).

Actually there are three major different types of solid-state rectifiers, all of which depend in some measure on semiconductors: the metal-semiconductor combination, the p-n junction, and the conductivity modulation type. Examples of these types are the copper oxide rectifier, made of copper and copper oxide; the germanium rectifier; and the silicon rectifier, respectively. While the actual explanation of the behavior of these three types is different for each type, one can get some idea of the kind of behavior involved by considering what is probably the simplest type to describe, the *p-n junction rectifier*. Figure 13-23 gives a schematic representation of the charge transfer mechanisms in two types of semiconductors: n-type, with negative carriers, that is electrons; and the p-type with positive "holes" as carriers.* Next we note what happens when these two types are joined at some interface.

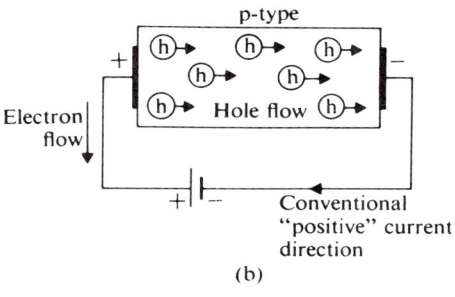

Fig. 13-23 Charge transfer in the two types of semi-conductor materials. (a) The n-type with negative carriers, that is, electrons. (b) The p-type with positive holes as carriers.

*While semiconductor conduction can be described in general terms at this stage, we must postpone to Chapter 16 a detailed discussion of the underlying physics.

492 ALTERNATING CURRENTS

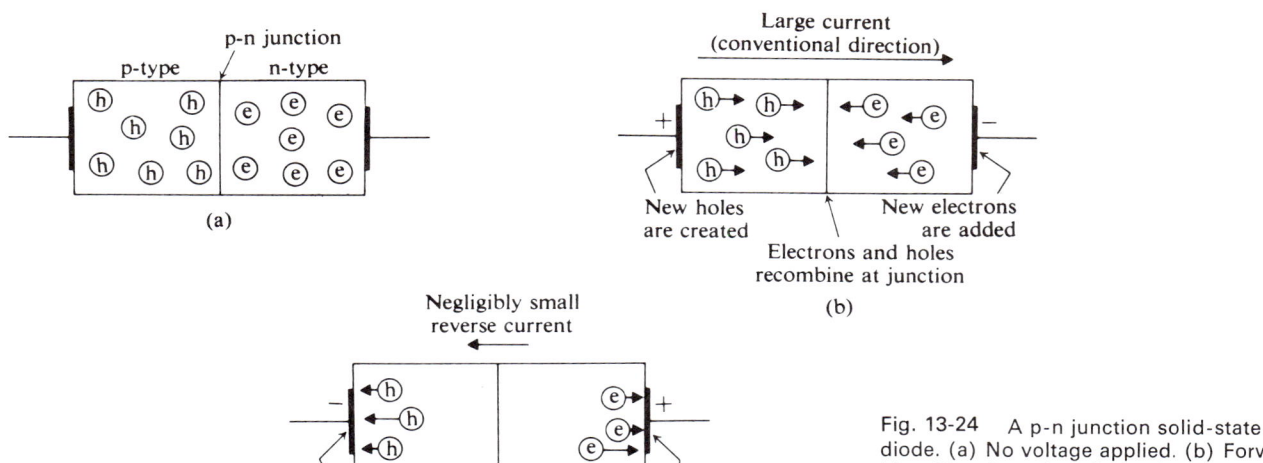

Fig. 13-24 A p-n junction solid-state diode. (a) No voltage applied. (b) Forward bias. Direction of applied voltage promotes a large current. This is the conducting direction. (c) Reverse bias. No appreciable conduction when voltage is applied in this direction.

Refer to Fig. 13-24. In (a) we see a p-n junction. In (b) the same junction is shown with a voltage applied such that the p-end of the junction device is positive and the n-end negative. This applied voltage results in a continual removal of electrons at the p-end and the consequent continual creation of holes at that end. In the meantime, electrons are being continually supplied to the n-end of the crystal. There are two things to note. First, the electrons and holes meet at the p-n junction and recombine there and again form part of the regular crystal structure, no longer acting as current carriers. Second, and most important, in spite of point one above, there is a continuous charge movement throughout the circuit. A current exists through a semiconductor in the direction of p to n.

Now let us reverse the applied voltage as in Fig. 13-24(c). The p-end is now negative, while the n-end is positive. The holes in the p-type material drift to the left and the electrons in the n-type material drift to the right. Soon there are hardly any holes in the p-type nor conduction electrons in the n-type. The current to all intents and purposes ceases. A significant current through a semiconductor in the direction of n to p does not exist. Consequently, a p-n junction has the property of a rectifier.

Before leaving the solid-state rectifier we first, in Fig. 13-25, redraw the circuit of Fig. 13-15, replacing the vacuum diode with a solid-state diode. Note that the arrowhead in the symbol for a solid-state diode points in the (conventional current) direction the diode conducts.

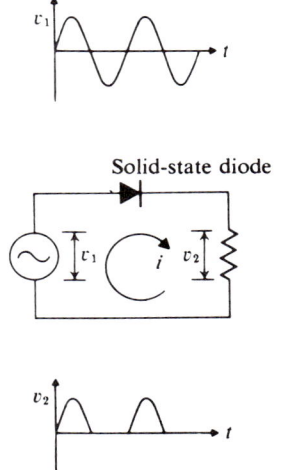

Fig. 13-25 Half-wave rectification employing a solid-state diode.

13.4 ELECTRONICS 493

13.5 ELECTRICITY AND SAFETY

Static Electricity Shocks

Since it has been established that it is primarily the current* rather than the voltage which is of major significance in physiological damage, there is usually little direct danger from static electricity* shocks. Of course there exists the hazard of even small electric discharges in an explosive atmosphere. And also even a mild static electric shock causes a person to recoil instinctively, thereby precipitating perhaps a serious accident.

Shocks Produced by 120-Volt AC

The AC encountered daily is another matter. Contact with the AC supply at the relatively low voltages employed in the home can, under certain circumstances, deliver enough current to be decidedly harmful. We have already indicated in Section 11.6 how a person might suffer such a passage of current in the course of using household appliances. The analysis pursued there holds also for AC and therefore the material presented now is limited to some details of the effects of current passage through the body.

The main factors which determine the extent of the harmful effect of a current passing through the body are (a) the path taken by the current, (b) the magnitude of the current, (c) the duration of the current.

The path of the current. A current flowing through vital organs is the most dangerous. In other words, a current path from the fingers of one hand to the elbow of the same arm is not nearly as perilous as a current from one hand through the chest to the other hand. Death from electric current is usually due either to cardiac or respiratory arrest, and the closer the current path is to the brain or to the critical organs in question, the greater the danger.

The magnitude of the current. Given any voltage, say the common 120-volt AC, the reader now realizes that the current will then depend upon the impedance of the body. The impedance will depend upon the two contact points selected as well as the area of contact and the condition of the skin at the contact points. Also this impedance will vary with the frequency of the AC and be different again for applied DC voltages. (In the latter case one would of course speak only of the resistance of the body rather than of its impedance.) Readers will find a wide range of values for body impedance (or resistance) quoted by various authors, but generally in the kilo-ohm range; values from 3000 to 20,000 ohms can be found depending on the details of voltage and contact conditions, etc. One study* employing an AC voltage regulated to yield a current of 1 mA through the body indicates that for 60-cycle AC the hand-to-hand impedance is of the order of 4000 Ω. Most of this is due to the resistance of the skin, dry skin being a poor conductor.

---

*More precisely it is the current density, that is, the current per unit body area, which is the crucial aspect in biological damage.

*Consider two different materials, both nonconductors. These materials will permit the accumulation of electric charge if some charge can in some way be transferred from one material to the other. In other words, the charge does not "flow" away readily; it accumulates and remains at rest. It is "static." Hence the term static electricity. The charge transfer from one body, say wool, to an ebonite rod or a hard rubber comb, although it requires only the contact of the two surfaces, is enhanced by rubbing the two materials together.

Reference to static electricity can be employed to arbitrarily define what we mean by negative and positive charge. For example, by definition negative electricity is the kind of electricity that predominates in a body composed of resin, such as an ebonite rod, after it has undergone electrification by rubbing with wool. From our current understanding this means, of course, that the wool, having given up some electrons to the resinous body, is now positively charged.

*Guest, P. G., V. W. Sikora, and Bernard Lewis, *Static Electricity in Hospital Operating Suites, Bulletin 520,* Bureau of Mines, U. S. Government Printing Office, Washington, D. C., 1953 (reprinted 1962).

Other body cells contain saline solutions and are thus fairly good conductors.

With a 4000-Ω impedance it can be seen from Eq. 13.7 that 120 volts would yield a current of 30 mA. Is this current unpleasant? Is it dangerous? The answer to both questions is unequivocably *Yes*, although the current is not necessarily fatal (likely lethal effects are estimated at 100 mA). Fortunately, such large area contacts as were used in the study which suggested the 4000-Ω impedance are seldom made with both hands; moreover there is usually a recoil as the slightest shock is felt. Probably in most cases the nature of the contact would lead to impedances of 10,000 to 20,000 Ω. In the case of 12,000 ohms this would give a current of 10 mA. How would 10 mA feel? Strong and painful cramps. Would you still be able to let go, control your muscles enough to break the contact? Yes, probably. Studies have indicated that a current of around 15 mA (not 15 amp!) was about the limit at which people can still voluntarily control muscles in the hand and wrist enough to let go.

What is really safe then is a debatable question and one which depends very much on the physical condition of the person involved. Persons with cardiac problems will naturally be in the danger zone at lower currents than persons with normal, strong hearts. One good answer is *Safe is zero current!* However, certain quantitative answers are suggested by some investigators.* They suggest that 6 to 9 mA is a reasonably safe current with 60-cycle 120-volt AC: safe but not pleasant. If contact is made with the hand, strong and painful cramps will be felt in the arm, and of course small holes (perhaps not even visible with the unaided eye) may be burned in the skin where the current enters and leaves the body.*

The duration of the current. Obviously, the shorter the contact time the better. In one of the papers referred to earlier, it is implied that the minimum current necessary to produce ventricular fibrillation (irregular movement of heart muscles, which is fatal if permitted to continue) in man is about 100 mA for shock durations of 1 sec or longer. Moreover, recent evidence suggests that this value is on the high side and that perhaps 70 mA is a better approximation.

Nowadays with medical diagnostic and therapeutic techniques probing the innermost depths of the body (witness organ transplants), it is probably worthwhile realizing that *internal* shock hazards are an entirely different matter from the *external* hazards we have been discussing. For example, when heart catheterization and cardiac pacemakers first became commonplace a rash of fatalities appeared, often defying immediate explanation. The cause of death in these cases was very likely ventricular fibrillation produced by extremely minute electric currents, perhaps a thousand times or so less than the threshold value for fibrillation initiated by currents entering at the surface of the body.

*Dalziel, C. F., and W. R. Lee, Lethal Electric Currents, *I.E.E.E. Spectrum*, Feb. 1969, pp. 44–50.

*It should be noted that the various current values given in this section refer to AC with a frequency of 60 hertz. In general, it has been found that to achieve comparable physiological effects higher current values are required in the case of DC, although as the current is increased to the fibrillation threshold, there appears to be no significant difference between AC and DC values.

It is definitely known that currents as low as 100 microamperes (i.e., 100×10^{-3} mA) and on occasion even as low as 20 microamperes have produced fibrillation in some cases in which electrodes made direct contact with the heart.

Such facts point to the need for an entirely new look at electrical safety measures in the hospital. It is therefore comforting to realize that these electrical problems have and are being seriously studied by the appropriate regulating bodies.

13.6 SUPPLEMENTARY TOPIC

Phasors and the Addition of Voltages in Series Circuits

At any instant, the voltage across the series combination of a resistor, an inductor, and a capacitor is the algebraic sum of the voltages indicated in Fig. 13-12(a) across each individual circuit element:

$$v = v_R + v_L + v_C.$$

The graph of this instantaneous voltage v as a function of time can be obtained by adding the three graphs of Fig. 13-12(b). We denote the amplitude of v by V_m and the effective value by V.

We want to determine the relationship between this effective voltage V and the effective voltages V_R, V_L, and V_C. Since effective voltages are proportional to amplitudes, our problem is solved if we can discover the relationship between the amplitude V_m and the amplitudes V_{Rm}, V_{Lm}, and V_{Cm} of the voltages across R, L, and C, respectively. The simplest and most illuminating method for finding the amplitude of the sum of sinusoidal functions is the method of *phasor addition*. For the benefit of those readers to whom AC circuit analysis is particularly important, we shall devote the remainder of this section to this rather advanced topic.

We proceed as in the study of simple harmonic motion and associate with a given sinusoidal function of time a certain *phasor*, represented by an arrow, on a phasor diagram which is constructed so that the horizontal component of this phasor is the sinusoidal function of interest. As the phasor rotates clockwise with a constant angular velocity $\omega = 2\pi f$, its horizontal component is a sinusoidal function of time with a frequency f and an amplitude which is the magnitude of the phasor. Phasors are added like vectors. Consequently, the amplitude of the sum of several sinusoidal functions is equal to the magnitude of the sum of the phasors which correspond to these functions. The utility of phasors comes from the fact that vector addition is much simpler and more readily visualized than is the direct addition of the sinusoidal functions themselves.

Phasors representing the voltages and current for the series circuit in Fig. 13-12 are shown in the phasor diagram in Fig. 13-26. In AC analysis, since we are more interested in effective values than in amplitudes, we take the magnitude of a phasor to be the effective value rather than the amplitude of the corresponding sinusoidal function. This choice merely changes the magnitudes of all phasors by the same factor, $1/\sqrt{2}$, and therefore will not affect relationships among the phasors.

The directions of the phasors are such that the angle between any two phasors is the phase difference (difference of phase angles) between the two corresponding sinusoidal functions. The current at any instant is the same in each of the circuit elements connected in series, so that there is only one current phasor for the circuit. It is convenient to show this current phasor in the horizontal position on the phasor diagram. Then the direction of each voltage phasor is determined from the phase difference between the current and the voltage in question. Since the voltage across a resistor is in phase with the current, the phasor for the resistor voltage is also horizontal, as shown in Fig. 13-26. For the capacitor, we have seen that the current leads the voltage by one-quarter of a period which corresponds to a phase angle of 90°. The phasor which represents the capacitor voltage therefore points vertically downward. And since the voltage across an inductor leads the current by one-quarter of a period, the phase difference is therefore 90° and the inductor-voltage phasor points vertically upward.

The addition of the phasors is performed by using the procedures for the addition of vectors. From Fig. 13-26(b), using Pythagoras' theorem, we find that the effective voltage across the series combination is given by

$$V = \sqrt{V_R^2 + (V_L - V_C)^2}.$$

The essential point is that, because the instantaneous voltages are not in phase, the effective voltage V across the combination is *not* simply the sum of the effective voltages V_R, V_L, and V_C.

Since $V_R = IR$, $V_L = IX_L$, $V_C = IX_C$, and $V = ZI$, the equation for V gives

$$ZI = \sqrt{(IR)^2 + (IX_L - IX_C)^2}.$$

A factor I can be canceled, and we reach the result that we quoted previously:

$$Z = \sqrt{R^2 + (X_L - X_C)^2}. \tag{13.8}$$

The voltage v across the series combination leads the current i by the angle ϕ shown in Fig. 13-26(b), and

$$\tan \phi = \frac{V_L - V_C}{V_R} = \frac{X_L - X_C}{R}.$$

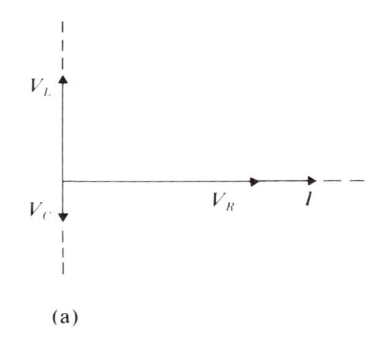

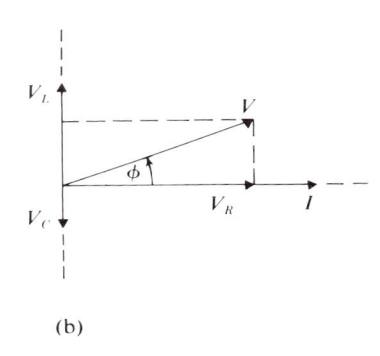

Fig. 13-26 Phasor diagram for a series circuit. (a) Phasors representing the circuit current and the voltages across individual circuit elements. (b) The addition of the phasors representing the voltages across L, C and R to give the phasor (of magnitude V) representing the voltage across the series combination.

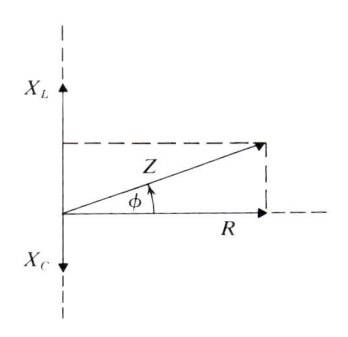

Fig. 13-27 Impedance diagram. The impedance phasor for the series combination of L, C and R has a resistive component R and a reactive component $X_L - X_C$. The relationships between these components and the magnitude Z and angle ϕ can be determined by inspection of this diagram.

For a given circuit element, the relationship between the sinusoidal voltage and the current is determined by the pair of numbers Z and ϕ or equivalently by the pair of numbers R and $(X_L - X_C)$. These quantities that characterize the circuit element can be displayed on an *impedance diagram* (Fig. 13-27) showing an impedance phasor of magnitude Z and angle ϕ with a (horizontal) resistive component R and a (vertical) reactive component $(X_L - X_C)$.

Example 1 Consider the circuit shown in Fig. 13-28(a).
(a) From the voltages across individual circuit elements, determine V_{ad}, the voltage across the entire part of the circuit that is external to the generator.
(b) Calculate the impedance Z of this external circuit and use this to determine V_{ad}.
(c) Determine the phase difference between V_{ad} and the current in the circuit.
(d) The value of C is adjusted until the circuit resonates. At what value of X_C will resonance occur?
(e) Give the impedance diagram for the R-L-C circuit element between the points a and d.

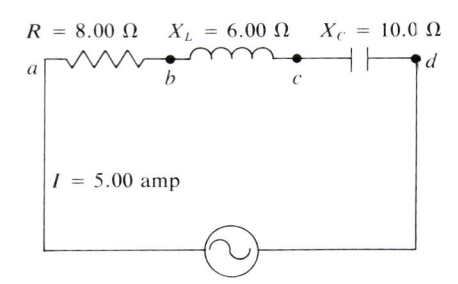

Fig. 13-28(a) The R-L-C series AC circuit of Example 1.

Solution

(a) $V_{ab} = V_R = IZ_R = IR = (5.00 \text{ amp})(8.00 \text{ ohm}) = 40.0$ volts,

and the phase angle $\phi_R = 0$.

$V_{bc} = V_L = IZ_L = IX_L = (5.00 \text{ amp})(6.00 \text{ ohm}) = 30.0$ volts,

and the phase angle $\phi_L = +90°$.

$V_{cd} = V_C = IZ_C = IX_C = (5.00 \text{ amp})(10.0 \text{ ohm}) = 50.0$ volts,

and the phase angle $\phi_C = -90°$.

Because of the 180° phase difference between the voltage across the inductor and the voltage across the capacitor, we see that the voltage across these two components, V_{bd}, has a magnitude given by

$$V_{L,C} = V_C - V_L = (50.0 - 30.0) \text{ volts} = 20.0 \text{ volts}.$$

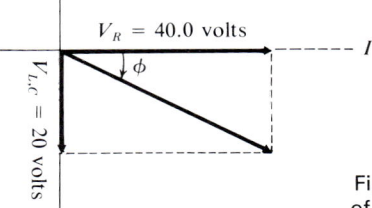

Fig. 13-28(b) Phasor diagram of voltages of Example 1.

498 ALTERNATING CURRENTS

Since V_C is greater than V_L, the phasor of magnitude $V_{L,C}$ points downward on a phasor diagram. Now, evaluating the phasor sum of the voltages $V_{L,C}$ and V_R [see phasor diagram, Fig. 13-28(b)], we find

$$V_{ad} = \sqrt{(40.0)^2 + (20.0)^2} \text{ volts} = 44.7 \text{ volts}.$$

(b)
$$Z = \sqrt{R^2 + (X_L - X_C)^2} \quad (13.8)$$
$$= \sqrt{(8.00)^2 + (6.00 - 10.0)^2} \text{ ohms}$$
$$= 8.94 \text{ ohms}.$$

Now using $V = IZ$, we obtain

$$V_{ad} = (5.00 \text{ amp})(8.94 \text{ ohms})$$
$$= 44.7 \text{ volts},$$

in agreement with the result of part (a).

(c) From Fig. 13-28(b) we see that

$$\tan \phi = \frac{-20.0 \text{ volts}}{40.0 \text{ volts}} = -0.500,$$

and
$$\phi = -26.6°.$$

(d) Resonance occurs when $X_L - X_C = 0$. This requires $X_C = X_L = 6.00$ ohms.

(e) The impedance diagram is shown in Fig. 13-29.

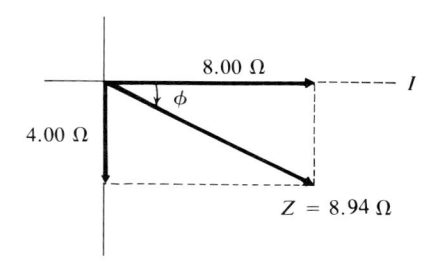

Fig. 13-29 Impedance diagram for the R-L-C circuit of Example 1.

QUESTIONS

1. Describe the difference between DC and AC.

2. What is an advantage of AC over DC?

3. Along with the fact that the peak voltage for AC is higher than for DC at the same power, there is another disadvantage of AC as compared to DC for long-range power transmission. What is this other disadvantage? Is there a power loss which is peculiar to AC? Comment. [*Hint:* The answer can be obtained by the implication of certain information given early in Section 13.1.]

4. Why does one not reduce power losses in transmission by going to even higher transmission voltages, say in the million-volt region?

5. Sketch the voltage-time graph for 60-hertz sinusoidal AC for two full cycles. Mark in the various times (in seconds) at which $v = 0$ (assuming $v = 0$ at $t = 0$) and at which v has its peak values.

6. In Europe AC in commonly supplied at 50 cycles per sec (i.e., 50 hertz). What is the period of this AC?

7. (a) What is the period of 15-hertz AC?
   (b) What disadvantage would the frequency above pose for illuminating purposes?

8. Would an AC current of 15 amp maximum (peak) result in the same temperature of a given heating element as a DC current of 15 amp applied for the same time? Explain.

9. If your kitchen stove works on 240 volts AC what is the maximum value of the voltage across the stove terminals in each cycle?

10. (a) Assuming that a sign which reads "Danger: 25,000 volts" refers to the common North American 60-hertz AC, what would be the maximum voltage in question?
    (b) How many times per second would this maximum potential difference occur?

11. (a) Why could an electric heater coil be used with either DC or AC?
    (b) If a given heater required a DC current of 9.5 amp, how large an AC current would be required to give the same heating effect? In answering this, should you specify the "instantaneous," the "RMS or effective," or the "maximum" value of the AC current?

12. If the speed of rotation of a generator were increased, what, if anything would happen to
    (a) the maximum EMF produced;
    (b) the frequency of the EMF?
    (c) What fundamental physical principle is employed to arrive at the answers above? [*Hint:* Refer also to Chap. 12.]

13. In a few lines indicate the difference between the **simple generator of Fig. 12-26 and a practical generator.**

14. (a) What would happen to the frequency of the AC output of a generator such as the one shown **in Fig. 13-4 if there were twice as many pairs** of magnetic poles?
    (b) How would the frequency of rotation of a generator with these extra poles compare with the frequency of rotation of the generator in **Fig. 13-4 if both generators produced 60-hertz AC?**

15. The alternating current in a 200-ohm resistor is 10 amp.
    (a) What is the voltage across this resistor?
    (b) What is the average power dissipated in the resistor?

16. The instantaneous current i of the preceding question is given by
    $$i = I_{max} \cos [(377 \text{ rad/sec})t].$$
    (a) What is the frequency?
    (b) Evaluate I_{max} from the data given in the preceding question.

(c) Find the voltage across the resistor at the following times: $t = 0$, $t = (1/240)$ sec, and $t = (1/120)$ sec.

17. Define capacitance and give the unit in which capacitance is usually stated.

18. Describe the structure and electrical property of a capacitor.

19. What charge is required to produce a potential difference of 1200 volts across a capacitor of capacitance 12 μf?

20. A parallel-plate capacitor with plates of area 0.10 meter^2 and separated by a distance of 0.010 meter, with air as a dielectric, has a capacitance of 8.9×10^{-12} farad.
    (a) What is the capacitance of a capacitor exactly like the one described above except for one difference: the plate separation is merely 0.0010 meter?
    (b) What is the capacitance of a capacitor which has plates of area 1.00 meter^2 and a separation of 0.010 meter, with air serving as the dielectric?

21. **The capacitor shown on the left in Fig. 13-7 is one** that might be used to tune a radio receiver to a given frequency. By what means is the capacitance of this capacitor changed?

22. It can be shown that when a capacitor is being charged **in a circuit, such as in Fig. 13-8, the charging rate** is rapid at first and then slows progressively. In this connection, a commonly used term is the *time constant* of the R-C circuit. The time constant is the time it takes the capacitor charge to reach 63% of its ultimate value. Show that the time constant, which is given by the product RC, has the dimensions of time.

23. At what frequency will a 5.00-μf capacitor have a reactance of 4000 ohms?

24. The 60-hertz current in a 4.00-μf capacitor is 20.0 amp. Find the voltage across the capacitor.

25. Sketch a graph showing the reactance of a capacitor as a function of the frequency. Examine and comment on the values assumed by this reactance at very low and at very high frequencies.

26. At what frequency would the magnitude of the reactance of a 5.00-henry inductance be the same as that of the capacitor of Question 23 (4000 ohms)?

27. What is the current in a 1.50-millihenry inductor when the 60-hertz alternating voltage across the inductor is 120 volts?
28. Evaluate the current in the 1.50-millihenry inductor of the preceding question when the alternating voltage is maintained at 120 volts, but the frequency is:
    (a) 6.0 cycles/sec;
    (b) 600 cycles/sec.
29. (a) Sketch a graph showing the reactance of an inductor as a function of the frequency. Examine and comment on the values assumed at very low and at very high frequencies.
    (b) Comment on these inductive reactance values as compared to capacitive reactance values over a similar frequency range.
30. The instantaneous voltage across a 5.0-millihenry inductor is given by

    $v = (170 \text{ volts}) \cos 377t$.

    Find the effective value of the current in this inductor.
31. Refer to Fig. 13-30.
    (a) Suppose R represented a light bulb. What would you note as you increased the inductance of the variable inductance coil shown? Explain.
    (b) By what means might one devise a variable inductance?
32. A resistor, inductor, and capacitor connected in series have a resistance of 4.0 Ω, an inductive reactance of 12 Ω, and a capacitive reactance of 15 Ω respectively when the frequency is 60 cycles/sec.
    (a) Sketch this circuit showing an AC voltage source.
    (b) What is the impedance of the circuit?
    (c) What current would there be in this circuit if a 60-hertz AC EMF of 10 volts were applied?
    (d) What value of the capacitive reactance would make the impedance a minimum?
33. (a) Show directly from Eq. (13.7) defining impedance, $Z = V/I$, and from the information given earlier in Section 13.2, that for a circuit element which is a
    (i) pure resistance R, $Z = R$;
    (ii) pure capacitance C, $Z = \dfrac{1}{2\pi f C}$;
    (iii) pure inductance L, $Z = 2\pi f L$.
    (b) Show that the impedances in cases (i), (ii), and (iii) in part (a) above can all be obtained as special cases of Eq. (13.8).
34. (a) What must be the inductive reactance of an inductance coil in order for it to be used with a capacitor of capacitive reactance of 100 ohms to form a resonant circuit?
    (b) If the coil and the capacitor are connected in series, and if the coil has a resistance of 100 ohms, what current will pass through the circuit if the applied EMF is 10 volts?
35. What is the capacitance of the capacitor required in series with a 40-millihenry inductance coil to provide a circuit which resonates at a frequency of $(500/\pi)$ hertz?
36. A 25.0-μf capacitor, a 0.100-henry inductor and a 25.0-ohm resistor are connected in series with a 120-volt 60-cycle/sec power line. Find
    (a) the inductive reactance;
    (b) the capacitive reactance;
    (c) the impedance of the circuit;
    (d) the current in the circuit.
37. A 0.2-henry pure inductor, in series with a capacitor, carries a current of 3.0 amp when the combination is placed across a 120-volt, 50-cycle/sec line.
    (a) What is the reactance of the capacitor?
    (b) What is the voltage across the inductor?
38. (a) Assume a certain carbon arc lamp has a resistance of 12.0 ohms when the current is 5.00 amp. What resistance should a rheostat in series with this lamp have if the arc is to be operated on a 120-volt DC power line?
    (b) Assuming that the arc lamp is operated on a 120-volt 60-cycle/sec AC line, and that an inductance coil is used in place of a rheostat, what must be the value of the inductance in order to maintain the current at 5.00 amp?

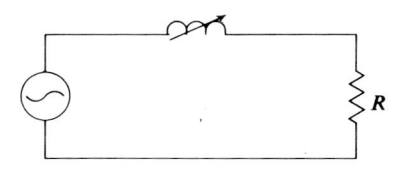

Fig. 13-30 AC circuit with variable inductance, L.

39. In Fig. 13-31 *R* has a resistance of 250 ohms, *L* has an inductance of 0.50 henry and no resistance, and *C* has a capacitance of 0.020 μf.
   (a) What is the resonant frequency of the circuit?
   (b) The capacitor can withstand a *peak* voltage of 350 volts. What maximum *effective* terminal voltage can the generator have at resonant frequency?

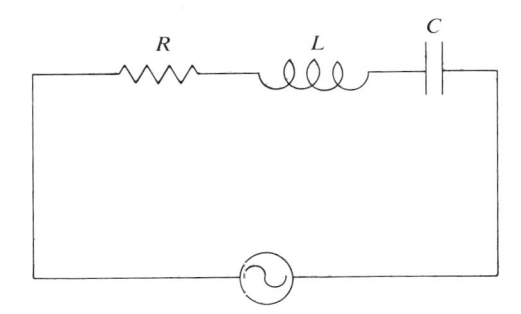

Fig. 13-31 Circuit of Question 39.

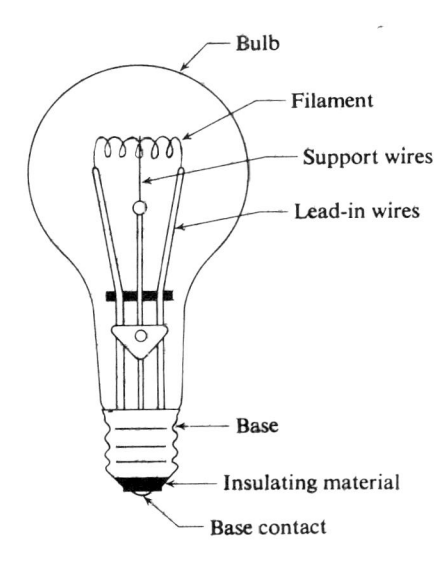

Fig. 13-32 Incandescent light bulb.

40. Consider AC power.
   (a) State the general expression for the average AC power delivered to a circuit element carrying an effective current *I* and across which there is an effective voltage *V*. (Remember the power factor.)
   (b) What is the unit of power?
   (c) Give the expression for the average electrical power dissipated in a circuit element, in a form which is applicable to both AC and DC.

41. If in an AC circuit a current of 20 amp at a voltage of 110 volts yields an average power dissipation of 2.2 kilowatts, what can be said about the nature of the circuit impedance?

42. Refer to Fig. 13-32 depicting a light bulb. Assume the bulb shown is a 60-watt, 120-volt bulb such as one would buy for use in the AC circuits in the home.
   (a) To what base parts must the lead-in wires of the light bulb be connected to provide for a complete circuit when the bulb is screwed into a socket?
   (b) Would this bulb work or not work on DC? Explain.
   (c) In the case of such a bulb does the 120 volts refer to the instantaneous voltage, the RMS value of the voltage, or the maximum voltage of the applied AC voltage?
   (d) What is the value of the current this bulb will draw?
   (e) Since the bulb is working on AC what assumption have you made in solving part (d)?
   (f) What would happen if the bulb were connected to a 12-volt source, such as a car battery? Explain.
   (g) What would happen if the bulb were connected to a 240-volt source? Explain.

43. Suppose that in an AC series circuit such as is shown in Fig. 13-12(a) the voltage across the resistor is 100 volts, across the inductor 40 volts, and across the capacitor 80 volts, while the current is 15 amp.
   (a) What current and voltage values (instantaneous, average, or RMS) are presumably meant?
   (b) What are the individual average power dissipations of the capacitor, resistor, and inductor, respectively?
   (c) What is the average power dissipation of the circuit?

(d) Is the input voltage, V, larger, smaller, or the same as the voltage across the resistor? Explain.

44. Using a sketch, show the difference between a step-up transformer and a step-down transformer.

45. A step-up transformer has a turns ratio of 500 to 1, and 100 volts are applied to the primary side of this transformer.
    (a) What is the output voltage?
    (b) If the secondary current is 100 mA, what is the primary current in amperes?
    (c) What is the power output of the transformer in watts?
    (d) What assumptions were made in your solutions to parts (a), (b), and (c) above?

46. What would happen if a DC voltage were applied to a transformer?

47. A transformer has 400 turns on the primary winding and 1600 turns on the secondary. The secondary winding is connected across a load which is a pure resistance of 1000 ohms. What is the value of the ratio of the primary voltage to the primary current?

48. What is the number of current pulses per second passing through an x-ray tube in a half-wave rectified x-ray machine operating on 60-hertz AC? (This means that the voltage across the x-ray tube is half-wave rectified AC.) Draw the appropriate current-time graph for the x-ray tube.

49. Sketch a circuit using four vacuum diodes to yield full-wave rectification.

50. If in the case of full-wave rectification of an AC input voltage one rectifying tube were burnt out, what would be the result on the rectified output voltage?

51. Consider the triode amplifier of Fig. 13-22.
    (a) Between what two electrodes of a triode is the input signal voltage applied?
    (b) If we define the voltage amplification by the ratio:
    $$\frac{\text{output voltage change}}{\text{input voltage change}},$$
    what is the voltage amplification in this case?
    (c) If the input voltage change applied were only 1 volt, what might you expect the amplitude of the output voltage to be?

52. (a) What is meant by a "solid state rectifier"?
    (b) Describe how such a rectifier works, by making reference to "electrons" and "holes."

53. Why can birds perch without harm on electric wires?

54. (a) Which is more dangerous: to accidentally ground an electrical appliance through your body when touching it with a wet hand or when touching it with a dry hand? Explain.
    (b) In the case of a wet hand, which is more dangerous: one wet with perspiration or one simply wet after being rinsed with water? When giving your answer suggest a possible explanation.

55. (a) What is the approximate maximum current passing through a person's hand in contact with a "live" electrical wire which would still permit the person to control his muscles to voluntarily let go of the wire?
    (b) Is this "let go" current larger or smaller than that which one might normally expect in accidental contact with the 120-volt household AC?

SUPPLEMENTARY QUESTIONS

S-1. Sketch the voltage time graph of the AC signal one might see displayed on an oscilloscope when a tuning fork is vibrating at 256 hertz near a microphone connected to the vertical input of the oscilloscope. Indicate how this trace would differ if, instead of a tuning fork, various musical instruments were used which were all sounding at the fundamental frequency of 256 hertz.

S-2. (a) Recall that the definition of capacitance is $C = q/v$. Keeping this definition in mind, show that when several capacitors are connected in parallel the total capacitance C of the combination of capacitors C_1, C_2, C_3, etc., is given by the general expression
$$C = C_1 + C_2 + C_3 + \ldots + C_n.$$

(b) The foregoing result shows the sense in constructing certain commercial capacitors of many interleaved plates of conductor and dielectric in order to achieve a high capacitance in a small unit. Show in a diagram how the plates of such a capacitor must be connected.

S-3. Show that the equivalent (or total) capacitance of n capacitors connected in series is given by

$$\frac{1}{C} = \frac{1}{C_1} + \frac{1}{C_2} + \frac{1}{C_3} + \ldots + \frac{1}{C_n}.$$

S-4. Assume that the voltage input to the R-C circuit of Fig. 13-33 is supplied by an oscillator which supplies 10.0 volts AC at 10.0 kilohertz. Under this condition the milliammeter reads 2.50 mA. When the oscillator frequency is changed to 20.0 kilohertz without changing the magnitude of the oscillator voltage, the milliammeter indicates 3.50 mA. Determine
(a) the impedance of the circuit when the frequency is 10.0 kHz;
(b) the impedance at 20.0 kHz.
(c) Use the results of these calculations to determine both the resistance and the capacitance of the circuit components. [*Hint:* You have two simultaneous equations (a) and (b), and you have two unknowns, R and C.]

S-5. Because the impedance in an AC circuit depends upon the frequency, it turns out that it is crucial in modern high-speed medical x-ray apparatus to compensate automatically for voltage changes across the x-ray tube resulting from frequency-dependent impedance changes. For example, at 60 hertz a frequency change to 57 hertz can result in the reduction of the voltage from 140 volts to 132 volts across a certain circuit component in the low voltage circuitry of the x-ray machine.
(a) Is the impedance of this circuit component in the low voltage circuitry essentially inductive or capacitive?
(b) Suppose the current in this low voltage circuitry is 10.0 amp at 140 volts, what is its impedance?
(c) If we make the oversimplification that the resistive component of the impedance of (b) is negligible and the other impedance component is either capacitive or inductive as determined in (a), then what is the inductance (or capacitance) of this low voltage circuitry?

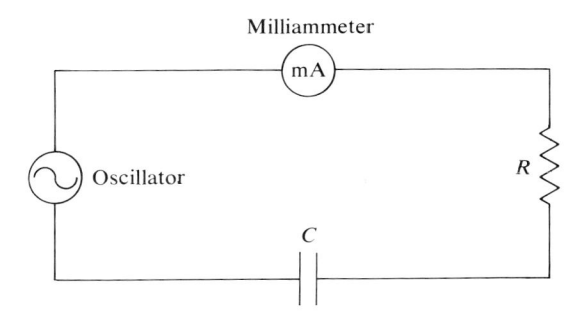

Fig. 13-33 Circuit of Question S-4.

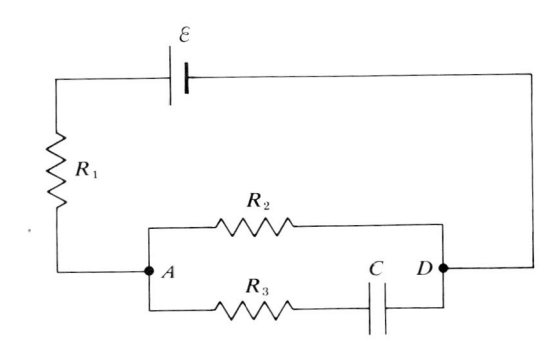

Fig. 13-34 Circuit of Question S-8.

S-6. A series circuit consists of a 25-ohm resistor, a 10-μf capacitor, and an inductor having a resistance of 12 ohms and inductance of 0.1 henry. Determine at frequencies of 100 Hz and 1000 Hz,
(a) the impedance of the circuit;
(b) the impedance of the inductor.

S-7. It is desired to deliver 1000 kilowatts over a transmission line with a resistance of 100 ohms, and with the resistive loss in the line of no more than 1% of the input power. What is the minimum input voltage at the generator end of the line that will satisfy these conditions?

S-8. Consider the circuit in Fig. 13-34 with $\mathcal{E} = 12.0$ volts, $R_1 = 30.0$ ohms, $R_2 = 16.2$ ohms, $R_3 = 5.0$ ohms, and $C = 1.0$ μf.

(a) Calculate the steady state current I through the source of EMF.
(b) What is the voltage drop V_{AD} between the points A and D in the circuit?
(c) What is the charge on the condenser in the steady state?
(d) What is the power supplied by the EMF source in the steady state?
(e) What happens to the power calculated in (d) above?

S-9. Suggest a reason, or reasons, why the ferromagnetic core of an electromagnet gets warmer when the magnet's coils carry AC current than when DC current is used?

S-10. The AC electric motor used to drive a certain type of bread dough mixer in a large commercial bakery draws a current of 70.0 amp at an input voltage of 220 volts. If the power is transmitted to the plant at 12.8×10^3 volts, find:
(a) the turns ratio of the transformer used at the plant;
(b) the current drawn from the transmission line.

S-11. A tone control circuit in a home stereo set consists of a 100-ohm resistor in series with a capacitor. The circuit is designed to have twice the impedance at 100 Hz that it has at 300 Hz. How large a capacitance is required?

S-12. The resistance to DC of a certain electric clock is 200 ohms. The impedance of the clock with 60-cycle/sec current is 2000 ohms. What must be the capacitance of a capacitor inserted in series with the clock in order for the clock's resistance to DC to be infinite, and for its impedance at 60 cycles/sec to be unchanged?

S-13. Assume that the efficiency of a transformer is 100%. When 150 volts is applied across the primary winding, the current in the primary is 2.00 amp. What is the voltage across the secondary winding when the current in the secondary is 0.100 amp?

S-14. Nikola Tesla (who was mentioned in the introduction to this chapter) devised a circuit that produces very high voltage pulses at high frequencies. The essential features of his device are represented schematically in Fig. 13-35. Explain how this device produces the high voltage pulses across the gap A-B.

S-15. Describe how the battery charger represented schematically in Fig. 13-36 works. First indicate the functions of the various circuit components in circuit A and then explain the purpose of circuit B.

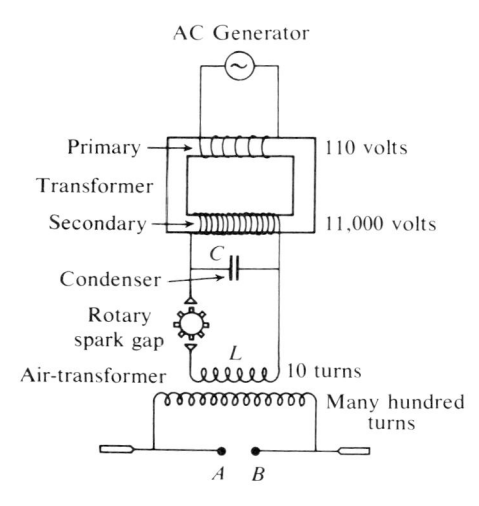

Fig. 13-35 Tesla apparatus for generating high-frequency high voltage (see Question S-14).

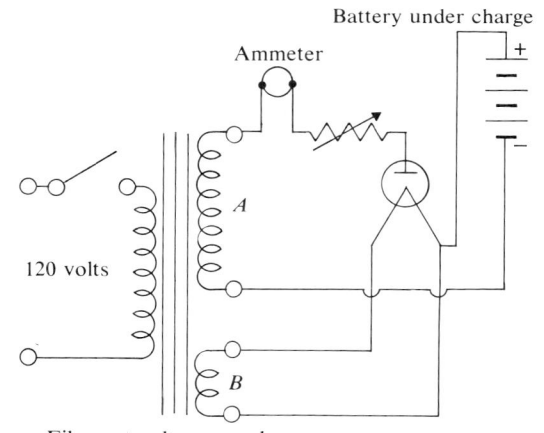

Fig. 13-36 Schematic representation of a simple battery charger operating from a 120-volt AC supply.

S-16. Fig. 13-37 shows a so-called characteristic curve of a diode, such as that of the diode in the circuit of Fig. 13-36. This curve shows the current (plate current) through the tube for a given voltage across the tube (plate voltage) as a function of the heating current in the filament. Outline the information that can be obtained from this characteristic curve, and comment on the significance of the "saturation current."

S-17. In addition to stepping a voltage up or down, transformers may also be used to insulate one circuit from another. A transformer that is used for this purpose is called an *isolation transformer*. Explain why it makes sense to have an isolation transformer between the house electrical supply and an electric razor.

S-18. Prove that, as stated in Section 13.1, the effective value of a sinusoidal voltage is $1/\sqrt{2}$ times the peak voltage. In other words, show that when a voltage

$$v = V_{max} \cos 2\pi f t$$

is applied across a resistor, the average power dissipated in the resistor is given by

$$P = V^2/R,$$

where $V = V_{max}/\sqrt{2} = 0.707 V_{max}$.

S-19. A 30-ohm resistor and a capacitor which has a capacitive reactance of 40 ohms are connected in series to a source of alternating potential difference. The current is 3.0 amp.
(a) What is the voltage across the resistor?
(b) What is the voltage across the capacitor?
(c) Sketch these voltages on a phasor diagram.
(d) Determine the voltage across the series combination of the capacitor and the resistor.
(e) Does the voltage across the capacitor-resistor combination lead or lag the current through it?

S-20. A resistance of 20 Ω, an inductive reactance of 80 Ω, and a capacitive reactance of 50 Ω are connected in series to an AC source of 60 hertz.
(a) Draw the circuit diagram.
(b) Find the total impedance of the circuit.
(c) Find the EMF required to produce a current of 4.0 amp in the circuit.
(d) Draw the phasor diagram for this situation, and indicate the phase angle by which the voltage across the series combination leads the current.

S-21. The phasor diagram in Fig. 13-38 shows the phasors which correspond to the voltages across two components connected in series in an AC circuit, as well as the total voltage V across these circuit components.

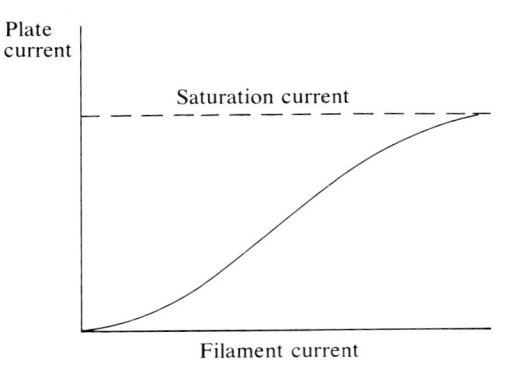

Fig. 13-37 Characteristic curve for a vacuum diode (see Question S-16).

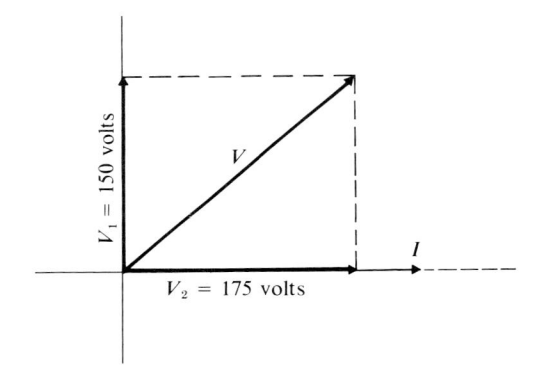

Fig. 13-38 Phasor diagram for Question S-21.

(a) Is the circuit reactance primarily inductive or capacitive?
(b) Is there any resistive component in the circuit impedance?
(c) Does the voltage lag or lead the current in the circuit?
(d) If the current in the circuit is 11 amp, what is the circuit impedance Z?

S-22. For the circuit described in Question S-21, calculate
(a) the total voltage across the circuit;
(b) the phase angle between the voltage across the circuit and the current in the circuit.

S-23. Consider two impedances Z_1 and Z_2 in series. $Z_1 = 80\ \Omega$ with phase angle $\theta_1 = 20°$, and $Z_2 = 120\ \Omega$ with $\theta_2 = -45°$. The current in the circuit is 1.2 amp. Determine the voltage across
(a) Z_1;
(b) Z_2;
(c) the series combination of Z_1 and Z_2.

S-24. An alternating voltage of 110 volts at 60 hertz, is applied to a series circuit having an inductance L of 1.00 henry, a capacitance C of 2.00 μf, and a resistance R of 400 ohms. Calculate the
(a) inductive reactance;
(b) capacitive reactance;
(c) total impedance;
(d) current in the circuit;
(e) power factor of the circuit;
(f) voltage across R, V_R;
(g) voltage across L, V_L;
(h) voltage across C, V_C sketching approximately to scale a phasor diagram showing the phasors corresponding to V_R, V_L and V_C;
(i) power dissipation in the circuit;
(j) peak voltage across the circuit;
(k) peak current.

S-25. An alternating voltage of 110 volts at 180 hertz, is applied to a series circuit consisting of an inductance L of 1.00 henry, a capacitance C of 2.00 μf, and a resistance R of 400 ohms. Calculate
(a) the inductive reactance, capacitive reactance, impedance, current, and power factor of the circuit;
(b) the potential differences across R, L, and C, sketching approximately to scale the phasor diagram of these voltages;
(c) the power dissipation in the circuit.

S-26. A capacitor C and a resistor R are connected in series to a source of sinusoidal EMF (as in Fig. 13-9). The capacitive reactance is 37.0 ohms, the resistance is 45.0 ohms, and the current (effective value) is 2.00 amp.
(a) What is the voltage across C?
(b) What is the voltage across R?
(c) Sketch these voltages on a phasor diagram.
(d) Determine the total voltage across both C and R.
(e) Does the voltage across the RC combination lead or lag the current through it?

S-27. Suppose that in the circuit in Question S-26, the capacitor is replaced by an inductor L with an inductive reactance of 37.0 ohms.
(a) What is the voltage across the inductor?
(b) What is the voltage across the resistor?
(c) What is the voltage V across both L and R?
(d) How, if at all, would the phasor diagram for these voltages differ from that of Question S-26?
(e) Does the voltage across L lead or lag the current through the circuit?
(f) Does the voltage across L lead or lag the voltage across R?
(g) What is the phase difference (in degrees) between the voltage V across the series combination of L and R, and the current through them?

S-28. Consider the circuit of Fig. 13-39. Assume the frequency is 60 cycles/sec. Calculate
(a) the current in the circuit;
(b) the inductive reactance of the inductance L_1;
(c) the impedance Z_1 of the components L_1 and R_1.
(d) Determine the magnitude of the voltage across Z_1 and that across Z_2, drawing the appropriate phasor diagram.
(e) What is the phase difference between the 208 volts across the circuit and the current in the circuit? Does the voltage lead or lag the current?

S-29. A 60-hertz 10-volt potential difference is applied to a circuit, and the result is a current of 5.0 amp lagging the voltage by 60°.
(a) Represent the current and the voltage on a phasor diagram.
(b) Determine the inductive reactance.
(c) Determine the resistance.

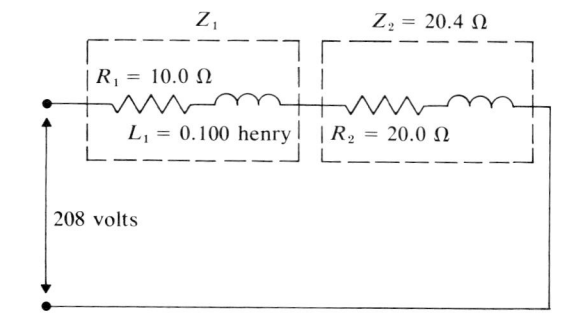

Fig. 13-39 Circuit of Question S-28.

S-30. An AC voltage generator yielding a 2.00×10^2 volt output is connected to a circuit which includes a resistance $R = 1.50 \times 10^2$ ohm, an inductance $L = 5.25 \times 10^{-1}$ henry and a capacitor $C = 2.00$ μf in series. The frequency of the generator voltage is 1.00×10^2 hertz.

(a) Draw the schematic diagram of this electrical circuit.
(b) Calculate the inductive reactance X_L of the inductor at the frequency of the generator.
(c) Calculate the capacitive reactance X_C of the capacitance at the frequency of the generator.
(d) Calculate the impedance of the series R-L-C circuit at the frequency of the generator.
(e) What is the current in the circuit?
(f) What is the power dissipated in the circuit?
(g) What would be the impedance of the circuit at its resonant frequency?

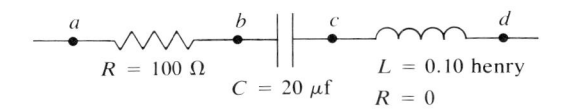

Fig. 13-40 Circuit of Question S-32.

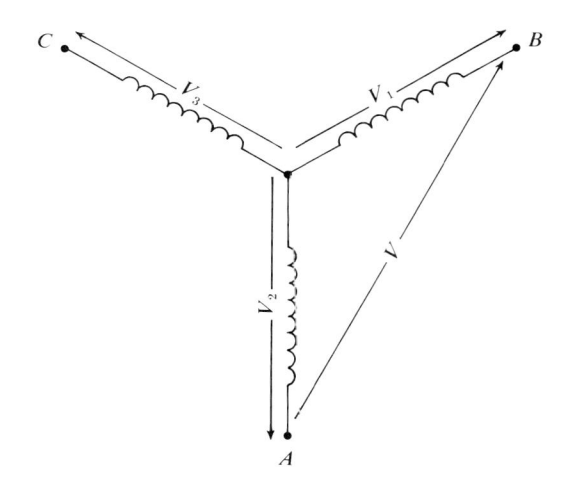

Fig. 13-41 Three-phase AC Y (or star) connection of Question S-33.

S-31. (a) In a DC circuit, would it be possible to have a voltage across any circuit component larger than the EMF applied to the whole circuit? In an AC circuit, would it be possible to have an effective voltage across some circuit component higher than the effective EMF applied to the entire circuit? Explain.
(b) One of the characteristics of a resonant L-R-C series circuit is the large "output" voltage, across either the inductor or the capacitor, obtained with a relatively small input voltage across the entire L-R-C series circuit. This "gain," given by (output voltage)/(input voltage), can be shown to be equal to the ratio X_L/R, called the "Q" of the circuit. Determine the Q of the circuit of Question 39.

S-32. For the circuit in Fig. 13-40 draw the following curves, using the same set of axes, over a frequency range from 50 to 200 cycles/sec:
(a) reactance versus frequency for the capacitor;
(b) reactance versus frequency for the inductor;
(c) resistance versus frequency for the resistor.
(d) Construct $X_L - X_C$ graphically by subtracting ordinates. Indicate the part of this curve where voltage leads, the part where the voltage lags, and the point of resonance.

S-33. Most electrical energy is produced and distributed as far as local electrical substations in the form of **three-phase AC. Fig. 13-41 shows a so-called Y** (or *star*) connection to a three-phase supply. The voltages v_1, v_2, v_3 across each of the windings 1, 2, 3, respectively, are (for a balanced user's load) the same in magnitude but 120° out of phase with the other two voltages. These voltages are called *phase voltages*.
(a) On the same axes, draw a graph of these three voltages as a function of time, assuming 60-hertz three-phase. (Plot voltage on the ordinate and time on the abscissa.)
(b) In addition to making connection to any or all of these phase voltages, it is possible to connect a load across two phases (see points A and B). Such a connection yields the so-called *line* voltage, V. Using a phasor diagram, make the calculations necessary to show that the line voltage (RMS or peak value) is $\sqrt{3}$ times the magnitude of the corresponding phase voltage value, and that the line voltage is 30° out of phase with a phase voltage.

ADDITIONAL APPLICATIONS TO MEDICINE AND THE LIFE SCIENCES

Coolidge X-Ray Tube

Somewhat akin to the vacuum diode is the hot cathode vacuum x-ray tube first reported by W.D. Coolidge in the *Physical Review* of Dec. 1913 (see Fig. 13-42). Almost all x-ray tubes in use today are some modification of the tube invented by Coolidge at the General Electric Laboratory in Schenectady. The invention of this type of x-ray tube marked a significant turning point in most x-ray work and particularly in medical radiography.

Prior to Coolidge's invention, gas-filled tubes were used for radiography. These partially evacuated tubes suffered from two main disadvantages for radiographic work. First of all the production of x-rays was very variable, varying widely from minute to minute, and secondly, it was impossible to vary the tube current (and consequently the x-ray intensity) without varying the voltage across the tube (and consequently the penetrability of the radiation). Or, as we say today, one could not change the *quantity* of the radiation without changing its *quality*. The latter limitation in particular severely restricted the gas tube's flexibility. Insofar as possible, what the radiographer wants is a tube which permits the variation of the tube current (called mA, standing for the average value of the current through the tube), and hence the radiation quantity, independently of the voltage across the tube (called the kV, being the maximum value of the voltage across the tube). This desirable feature for radiography can be achieved with the Coolidge tube.

In the Coolidge tube the electrons which are accelerated toward the target by the voltage across the tube are "boiled off" the cathode just as in the case of the vacuum rectifier diode. There is a distinct difference, however, in that the voltage in the x-ray tube is sufficiently high that electrons are propelled toward the target as soon as they are emitted. Increasing the voltage will not increase the current because no more electrons are available. The tube is said to be operated in the *temperature-limited* region, i.e., the upper limit of the tube current is determined by the filament temperature (see Fig. 13-43). (Rectifier diodes on the other hand are operated in the *space-charge limited* region, i.e., many more electrons are boiled off than drift over to the anode so there is always an excess of conduction electrons available in a "cloud" about the filament.)

Modern medical x-ray circuits and tubes (see Fig. 13-44) are designed to operate (depending on the make and model of the x-ray machine) approximately in the range of 40-140 kV, and 10-1000 mA.

Question

(a) Reasonable exposure factors for a chest x-ray are 70 kV, 100 mA, and 1/10 sec. X-ray production at these

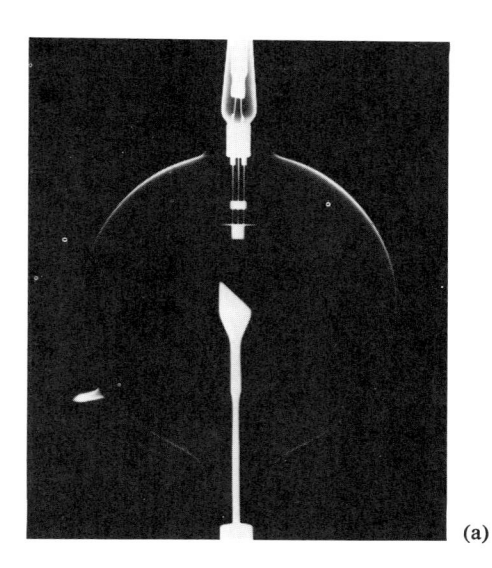

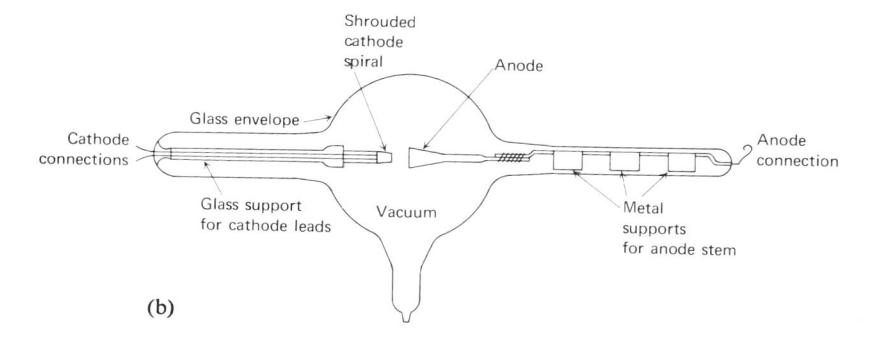

Fig. 13-42 (a) Radiograph of the Coolidge tube, circa 1915. (b) The original Coolidge tube (adapted from W. D. Coolidge's paper in *Physical Review*, December 1913). (Courtesy A. Ridgway and W. Thumm, *The Physics of Medical Radiography*, Reading, Mass.: Addison-Wesley, 1968.)

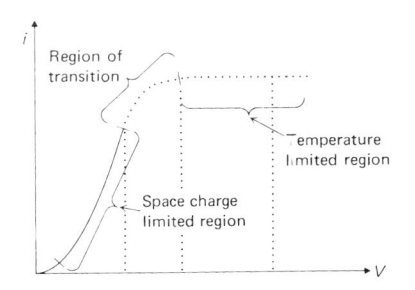

Fig. 13-43 X-ray tube current as a function of the voltage across the tube. Note the temperature-limited region.

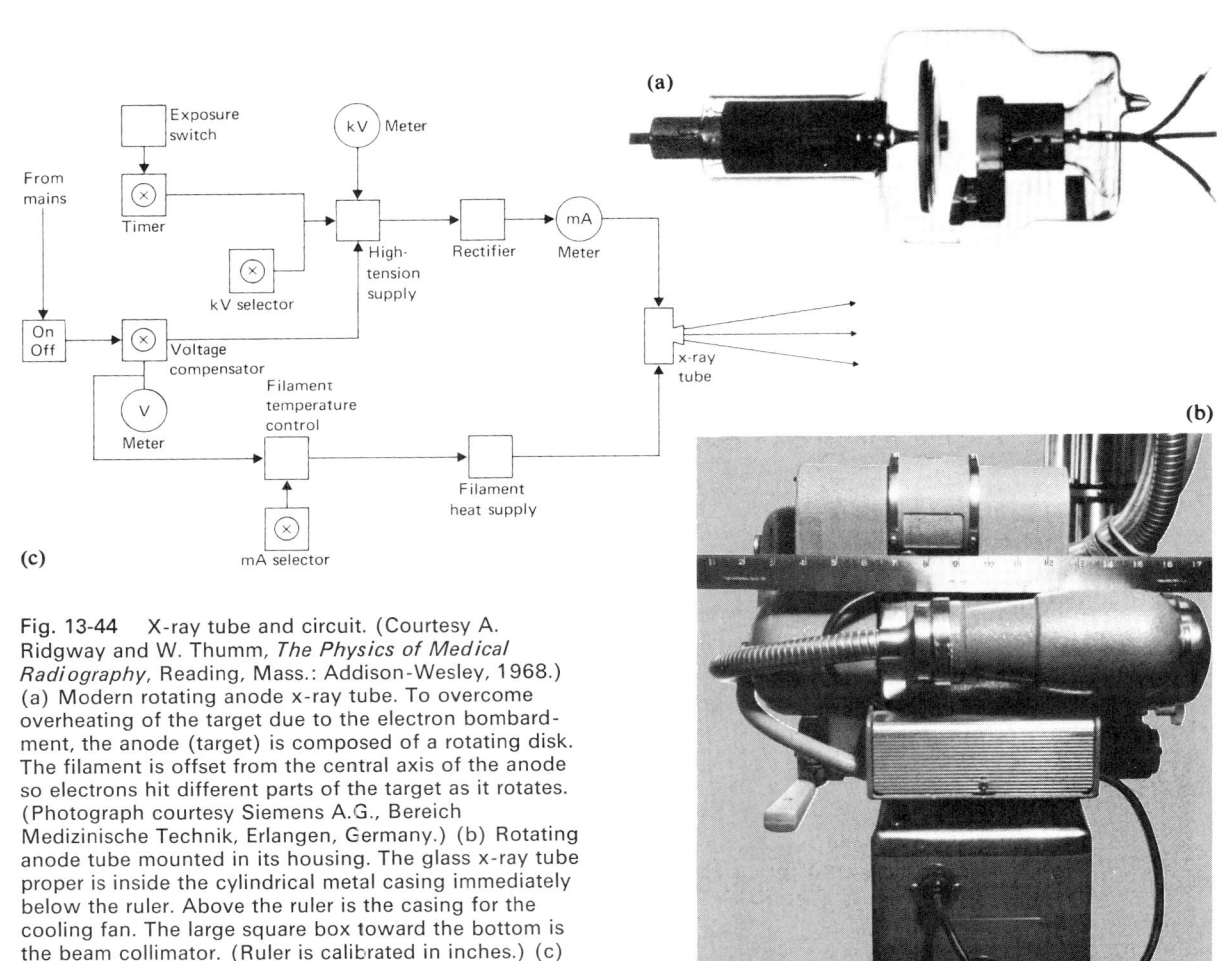

Fig. 13-44 X-ray tube and circuit. (Courtesy A. Ridgway and W. Thumm, *The Physics of Medical Radiography*, Reading, Mass.: Addison-Wesley, 1968.) (a) Modern rotating anode x-ray tube. To overcome overheating of the target due to the electron bombardment, the anode (target) is composed of a rotating disk. The filament is offset from the central axis of the anode so electrons hit different parts of the target as it rotates. (Photograph courtesy Siemens A.G., Bereich Medizinische Technik, Erlangen, Germany.) (b) Rotating anode tube mounted in its housing. The glass x-ray tube proper is inside the cylindrical metal casing immediately below the ruler. Above the ruler is the casing for the cooling fan. The large square box toward the bottom is the beam collimator. (Ruler is calibrated in inches.) (c) Block diagram of basic x-ray machine components, shown in order from the electric mains supply on the left to the x-ray tube on the right. Circles denote meters on the control panel of the x-ray machine. Crossed circles denote control knobs on the control panel.

510 ALTERNATING CURRENTS

voltages is very inefficient, over 99% of the input energy going into heating the target and less than 1% going into x-ray energy. How much thermal energy is developed in the x-ray tube target in such an exposure?

(b) Suppose in an x-ray procedure such as described above, the radiographer wanted to reduce the tube current from 100 mA to 90 mA. How is such an adjustment achieved?

Answer

(a) Assume total input energy goes into thermal energy of target, then

$$W = VIt = (70 \text{ kV})(100 \text{ mA})(0.10 \text{ sec}) = 700 \text{ joules}.$$

(Actually the energy is a little less than 700 joules. The reason is that in our calculation we have assumed RMS (or effective values) of current and voltage while in x-ray practice it is customary to express the voltage in peak values and the current in average values for reasons we shall not pursue here. Suffice it to say that the correct answer would be 700 H.U., where H.U. is known as a *radiographic heat unit:*

$$1 \text{ H.U.} = 0.785 \text{ joule}$$
$$= 0.188 \text{ calorie.})$$

(b) By adjusting a control (the mA selector, see Fig. 13-44c) which does not vary the *tube* current directly, but rather varies the current through the *filament* and thus, in this case, reduces the temperature of the filament. (Filament currents vary typically in the range of 3-5 amp at 6 to 12 volts.) Reducing the temperature of the filament results in fewer electrons available to traverse the tube and therefore a lower tube current.

Further Reading

For more on medical x-ray work including simple physics problems see Ridgway, A., and W. Thumm, *The Physics of Medical Radiography*, Addison-Wesley, Reading, Mass., 1968. For an overview of the field of radiology from history to equipment see Etter, L.E., ed., *The Science of Ionizing Radiation*, Thomas, Springfield, Illinois, 1965.

Operating Theater Hazards: Static Electricity Shocks

Many of the volatile agents used in anesthesia form highly explosive mixtures with air. Although the trend is to employ less explosive anesthetics, ether, because of many of its other desirable characteristics may continue to be used on occasion.

It is true that mobile x-ray units and other electric and electronic gear for use in operating rooms may be quite safe, in that dangers from the electricity involved in their operation have been eliminated by having the units hermetically sealed and flash-proofed for use in explosive atmospheres. However, the dangers arising from the accumulation of static charge still exist. It is quite possible to acquire relatively small charges and yet achieve relatively high potential differences. And static electricity of high voltage on an insulated conductor such as a metal stretcher with rubber casters, or a mobile x-ray unit on rubber wheels, can discharge through the air if another object (such as a person) at different potential comes near. The spark involved in such a discharge may contain more than enough energy to ignite an explosive anesthetic mixture even though the spark may be much too small to be seen.

Tests have shown that potential differences between the floor and an operating table of as much as several thousand volts can be generated by the mere act of stripping a wool blanket from a rubber mattress. All these effects are worse in places of low relative humidity. The more moist the air, the more readily the accumulated charge leaks off before a high potential difference sufficient to cause a spark may be built up. To aid this charge leakage, it is not uncommon practice to drape a wet towel around the base of an operating table and thereby ensure better electrical contact with the floor. Also the relative humidity of the theater is appropriately controlled to minimize the buildup of static electric charge.

The above comments are not to be construed as implying that the moisture itself makes the air electrically more conductive. Actually, pure water is more or less an insulator, although not a good one because it is usually very slightly ionized. However, if the air is relatively moist, condensation on surfaces such as clothing, floors, and equipment of all kinds takes place. The thin film of moisture on these articles, like the water on the wet towel referred to above, usually dissolves a certain amount of material, thereby rendering itself conductive.

Question

Would it be better to wear leather-soled or rubber-soled shoes while performing duties in a hospital operating theater?

Answer

Leather-soled shoes because leather is a better conductor than rubber. Actually, however, since leather tends to be slippery and leather shoes cannot be cleaned readily, one now uses cotton overshoes or sometimes shoes with special conductive rubber.

Further Reading

More details on operating room hazards and the related physics can be found in MacIntosh, Sir Robert, W.W. Mushin, and H.G. Epstein, *Physics for the Anaesthetist*, F.A. Davis, Philadelphia, 1968.

More on Electrical Hazards in Patient Care Areas

The problem. Hospital electrical hazards fall into two general categories: electrical shock to the body, and electrostatic spark generation in the presence of an explosive atmosphere.

The existence of these two categories presents a dilemma: procedures to eliminate the accumulation of static charge (conductive flooring, conductive clothing, etc.) in turn lead to current flow through people using defective electrical appliances. Obviously *fault currents* can more readily pass through persons to ground if the person is standing on conductive flooring, wearing conductive shoes. But in most cases, fortunately, such currents are relatively small.

Fault or *leakage* currents are currents taking paths other than those intended. The present electrical codes in the U.S.A. and in Canada permit a leakage current of 1 mA.

Compounding this electrical current hazard is the fact that electrodes applied internally rather than on the surface of the body give rise to far more complex safety problems than those associated with externally applied electrodes. A further complication of the problem arises from the increasing use of electrical apparatus in hospitals (for treatment, for diagnosis, and for patient monitoring).

The patient, often readily susceptible to dangerous electrical stimulation because of his disease or medication, is generally grounded. Bedside plumbing, motorized beds, lamps, and electrical devices for signaling and for entertainment are so commonplace that their possible hazard may often be ignored. Nevertheless, usually these conveniences provide unavoidable access to "ground." This grounding may well make part of the patient the path of leakage currents that would otherwise flow elsewhere. It is, of course, also

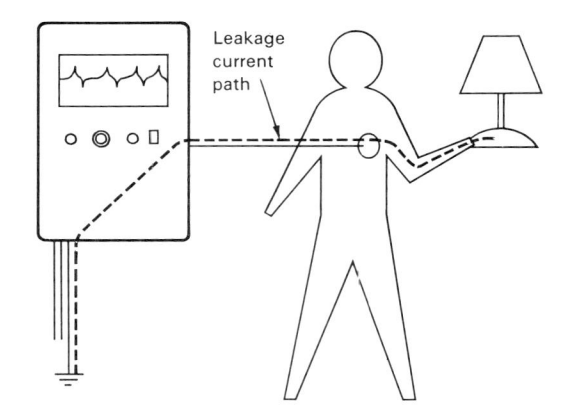

Fig. 13-45 Patient, connected to a grounded physiological monitor, turning on a lamp which has a leakage current. Possible pathway of leakage current to ground is shown by dotted line.

possible that defective wiring actually permits contact with electric power at above ground potential.

A normally harmless leakage current in a lamp can become dangerous if the patient has implanted electrodes. Take the case of a patient who reaches over to turn on his bedlamp, a two-wire lamp that has a leakage current of 1 mA and thus still passes present safety requirements. If the patient (see Fig. 13-45) is connected to a monitor, this 1 mA current could pass through the patient's arm, then through his trunk to the grounded monitor, with a portion of the current going through the patient's heart. In fact, if a grounded electrode were attached to his heart, all of the 1 mA might flow through his heart, causing either heart standstill or perhaps ventricular fibrillation. The existence of this leakage path would go undetected, unless perhaps a doctor or a nurse, accidentally touching both lamp and monitor, also noticed a shock. Fortunately, as a rule, when a patient is monitored, an alarm sounds if his heart starts fibrillating or standstill is noted, so that resuscitation may be effected at once.

A most important consideration in this field of electrical hazards is the interplay of any given electrical appliance with other electrical equipment being used. Compatibility of any piece of electrical gear with other equipment being used on the patient is imperative.

Dangerous currents. Before we consider one specific example of the hazard of interaction of different equipment items, we review briefly those current magnitudes which constitute danger.* Actual current values, of course, depend on various variables such as AC or DC, frequency of the AC, duration of contact, where on the body, and how contact is made, etc. It is essentially the current, and more specifically, the current density, which determines the damage resulting from electric current through living tissue. Small currents are relatively innocuous. That is why one can withstand a static shock of several kV with only minor discomfort. (This not infrequently happens when one touches a door handle after walking across a carpet when the humidity is low.) The current then is very small.

However, when the current increases, the situation changes. One can certainly feel a shock of two or three mA. Above 10 mA the shock becomes painful. Between 15 and 30 mA one cannot relax contracted muscles. Above this level one goes through stages of muscular paralysis, respiratory interference, and finally death, in the region of 70 mA to 200 mA. Curiously, shocks of 200 mA or higher may not cause death but rather inflict severe burns and arrest respiration. This strange phenomenon is explained by the fact that small currents above a certain threshold level may cause ventricular fibrillation while larger currents may produce momentary heart standstill which usually reverts to normal rhythm.

As a rough guide, one may say that the biological damage follows a rule somewhat like this,

$$\text{Damage} \propto \left(\frac{I}{A}\right)^2 T.$$

where I is the magnitude of the current through region of contact, A is the surface area of contact, and T is the length of time of the contact.

It is important to realize that the previously mentioned current thresholds were determined for shocks with electrical contact made on the surfaces of the body. And it is these current values which were used to establish conventional safety standards in hospitals. But, as we know, for internal body connections the situation is different. For example, pacemaker leads placed on the myocardium provide a direct route to the heart for accidental currents. So do diagnositc catheters which may be conductive or filled with conductive fluid, or both. Therefore, it is worth noting that the threshold of fibrillation in such instances may be 1000 times lower than when the electrodes make contact at the surface of the body, and that, in fact, currents even as low as 0.02 mA have been known to produce fibrillation.

Example of equipment interaction hazard. To the general public, and probably to most hospital workers, all seems safe if there is no resistive coupling, that is, no actual conductive connection between a voltage source and a user. This placid assumption is naive in light of the fact that most of our electrical gear works with AC. A very obvious example: Any transformer works without a resistive coupling between the primary and the secondary circuit. In other words, if AC is employed, then leakage currents may arise due to inductive coupling or, as is more common, capacitive coupling. The capacitor represents an open circuit for DC but in AC anything which qualifies as a capacitor, i.e., two conductors separated by some insulator, will permit AC passage.

Consider a particular case (see Fig. 13-46): a 26-year-old man underwent heart catheterization because of a congenital pulmonary ailment. Equipment used included an ECG monitor and a motorized injection apparatus for cineangiography (x-ray movie-making of flow in blood vessels). The injection syringe was primed with Hypaque (a radiopaque saline medium to provide the necessary contrast between the blood and other body tissues). When the saline-

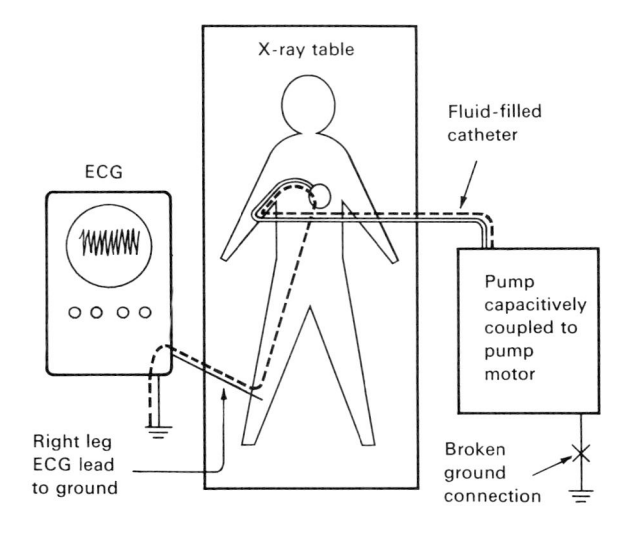

Fig. 13-46 Leakage current flows through the fluid-filled catheter, through the patient's heart to ground. See dotted line.

---

* For the convenience of the reader we reiterate here certain of the pertinent information cited earlier in the chapter.

filled catheter was coupled to the syringe, i.e., to the injection apparatus, the ECG display showed 60 Hz interference. Decoupling the catheter restored the ECG trace which then showed the heart was in fibrillation. The chest was opened, cardiac massage performed, and the heart was then electrically defibrillated. (The patient recovered and underwent uneventful surgery 6 weeks later—quick thinking and quick action countermeasures saved his life.)

What had gone wrong? Apparently the three-pin power plug of the motorized injection apparatus had been replaced by a two-prong plug, with no provision made for the independent ground. A 79-volt leakage potential (capacitively coupled) then existed between the casing of the injector and ground. At the moment of connection of the catheter, with the aid of the conductive properties of the saline solution, there appeared a current from the injection apparatus casing through the patient's heart, through the right-leg ECG lead to the ground. (These facts were established by reconstructing the events using the self-same equipment and a dog.)

Question

Suppose a patient in an intensive care unit is connected by internal electrodes to an ECG monitor and that the intracardiac blood pressure is being monitored by a conductive catheter inserted into the heart. The blood pressure measuring transducer and related electronic monitoring device is also "grounded" through a different outlet and therefore at a different point in the hospital wiring system. Could a situation arise in which a current could flow through the patient's heart between the ECG probe (grounded) and the blood pressure measuring catheter (also grounded but at a different point)?

Answer

Yes (see Fig. 13-47). Suppose the ECG outlet is grounded at the main distribution panel some 100 feet away, and it is there that the ground wire from the blood pressure monitor is also connected. The patient is therefore connected between two "grounds" which are indeed the same, providing no leakage current flows in either ground wire. But

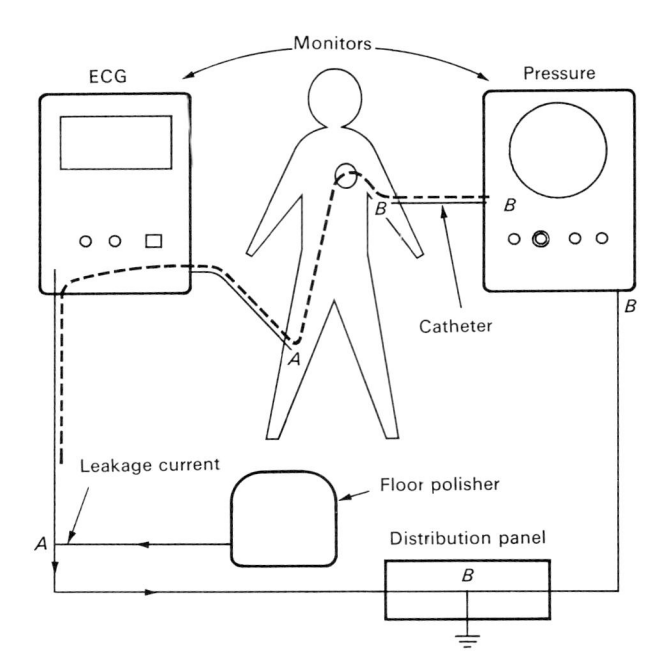

Fig. 13-47 Patient is connected to two grounded monitors, with the common ground at a distant electrical distribution panel. The leakage current from the floor polisher yields a potential difference between points *A* and *B* and thus precipitates current (dotted line) through the patient's heart.

suppose someone plugged an electric floor polisher or some such electrical device into an outlet near the ECG outlet, and further suppose this resulted in a leakage current of 1 amp through a 100-ft 12-gauge copper ground wire having a total resistance of about 160 milliohms. This yields a potential difference of 160 mV between the "ground" of the ECG monitor probe and the "ground" of the blood pressure monitor probe. If we assume that the impedance of the heart between the two probes is about 3200 ohms (a plausible figure), then the patient's heart could be subjected to a current of 0.05 mA.

SUGGESTED READING

LATHAM, D. C., *Transistors and Integrated Circuits*. Philadelphia: Lippincott, 1966. This book is for the reader who wants to unravel the mystique of these modern electronic developments but who does not necessarily have any previous mathematics or physics background to bring to the subject. It includes many good pictures and a glossary of the terminology of solid-state physics.

LURCH, E. NORMAN, *Electric Circuits*. New York: John Wiley, 1963. A reference for those readers wishing to further analyze electric circuits, including polyphase systems.

RIDGWAY, A., and W. THUMM, *The Physics of Medical Radiography*. Reading, Mass.: Addison-Wesley, 1968. AC meters, transformer losses and other such matters are treated. This book is of particular interest if the reader wants to learn a little about three-phase AC, since this topic is seldom discussed elsewhere in elementary terms.

SEARS, F. W. and M. W. ZEMANSKY, *University Physics* (4th ed.). Reading, Mass.: Addison-Wesley, 1970. Chapters 27, 33, and 35 in particular of this well-known text provide excellent collateral reading.

James Clerk Maxwell

14 Electromagnetic Waves

Having completed a look at electromagnetism in the last chapter, we now turn to the subject of light, a transition which is most natural in that light, after all, turns out to be an electromagnetic phenomenon.

The study of light has extended over thousands of years. The question of the nature of light is undoubtedly one of the first to arise in an inquiring mind with a scientific bent. After all, humans are equipped with marvelous light detectors, and the information that these detectors garner profoundly influences human activity. Moreover, since the eighteenth century, our advancing knowledge of light has had a tremendous impact on the entire sphere of physics and particularly on fundamental theories.

The Greeks utilized the concept of "rays" of light, which were straight lines in a uniform medium, to formulate the law of reflection and to describe refraction, that is, the bending of a light ray encountered when it passes from one transparent medium to another. In the eleventh century, the Arabian scholar, Alhazen, recorded his knowledge of optics in a book which remained a standard authority for 600 years. (Alhazen was the first physicist to give a detailed description of the human eye. Some of our names for the various parts of the eye, e.g., *retina*, derive from the Latin translation of his work.)

The telescope was pioneered by the Dutch spectacle maker, Hans Lippershey, and the compound microscope by his countryman, Z. Janssen, near the beginning of the seventeenth century. These inventions stimulated the great work of that century: the discovery in 1621 of the law of refraction by the Dutch mathematics professor, Willebrord Snell (1591–1626); the investigation, the significance of which was not realized until the nineteenth century, of Francesco Maria Grimaldi (1618–1663), professor of mathematics at the University of Bologna, into the phenomenon of the bending of a light beam (diffraction) encountered with small obstacles or apertures; the dispersion of white light into a spectrum of colors and other experiments reported by Newton in his *Opticks* (1704); the establishment by the Danish astronomer, Olaf Roemer (1644–1710), that light has a definite speed and does not move through space instantaneously, and the

"*This velocity is so nearly the velocity of light, that it seems we have strong reason to conclude that light itself (including radiant heat and other radiations if any) is an electromagnetic disturbance in the form of waves propagated through the electromagnetic field according to electromagnetic laws.*"

James Clerk Maxwell
(1831–1879)

demonstration by the great Dutch scientist, Christiaan Huygens (1629–1695), that a wave theory of light could embrace the laws of reflection and refraction.

Thus arose the vexed question: Are there particles of light or is light some kind of wave motion? To Newton it seemed most likely that a light beam was a stream of particles emitted by the light source rather than purely a wave motion in some hypothetical, invisible, all-pervading, jelly-like, imperceptible substance called the *ether*. Newton and his contemporaries were familiar with the way in which water waves and sound waves spread out after passing through apertures and how these waves bend around obstacles (diffraction phenomena). No one during the seventeenth and eighteenth centuries was able to show how a wave theory of light could be reconciled with the everyday observation that light apparently travels in straight lines so that obstacles cast sharp shadows and so that behind an aperture there is a well-defined beam of light.

The viewpoint that light was a stream of particles therefore held the day until a beautiful series of experiments in the early nineteenth century drove physicists to investigate a wave theory more thoroughly. An erudite Englishman, Thomas Young (1773–1829), trained as a physician, published in 1802 experimental results showing that light exhibited the phenomenon of interference: a typical wave phenomenon that involves fixed regions in space of high light intensity, where crests of one wave coincide with crests of another (constructive interference), separated by regions of low intensity where crests of one wave are canceled out by troughs of another wave (destructive interference). Young's experiment allowed a calculation of the wavelength of the "light waves": an exceedingly short length, indeed (approximately 5×10^{-7} meter, or 5000 angstrom units). Young's conclusion as to the wave nature of light earned him a decade of ridicule which subsided only in the face of overwhelming evidence of the wave-like behavior amassed by men such as Augustin Fresnel and Dominique Arago in France, and Joseph von Fraunhofer in Bavaria.

Careful mathematical investigation of the predictions of a wave theory showed that diffraction effects were very small and approximately straight line propagation would occur unless the dimensions of obstacles and apertures were roughly as small as the wavelength. Everyday objects like a window have dimensions which are a million times bigger than the wavelength of light, and so diffraction is not apparent. However, diffraction effects with light were indeed evident in experiments using very small apertures and obstacles.

Another triumph of the wave theory of light came in 1850 when J. B. L. Foucault (1819–1868) succeeded in showing that, in accordance with Huygens' wave theory of refraction, light traveled more slowly in water than in air.

Physics Today

Thomas Young
(1773–1829)

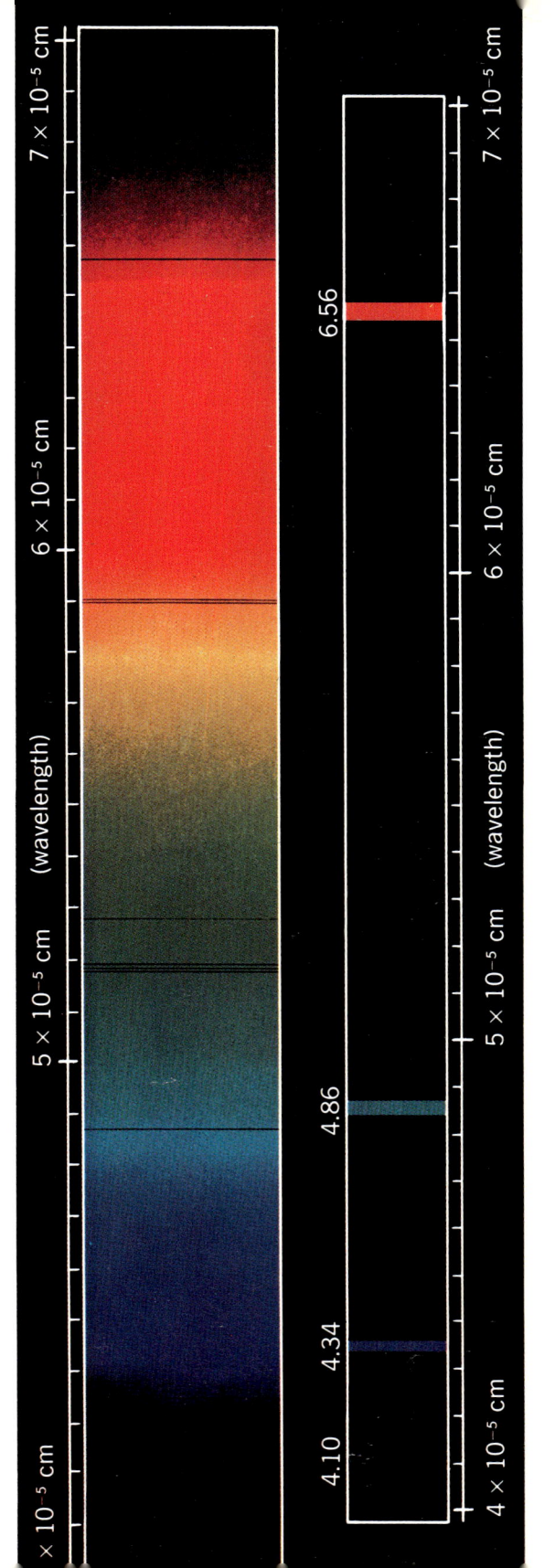

Lower spectrum shows the visible portion of the *line* spectrum emitted by hydrogen atoms. Upper spectrum is the continuous spectrum emitted by the sun. The dark lines appear because cooler gases surrounding the sun absorb light of certain discrete wavelengths. Courtesy of Eastman Kodak

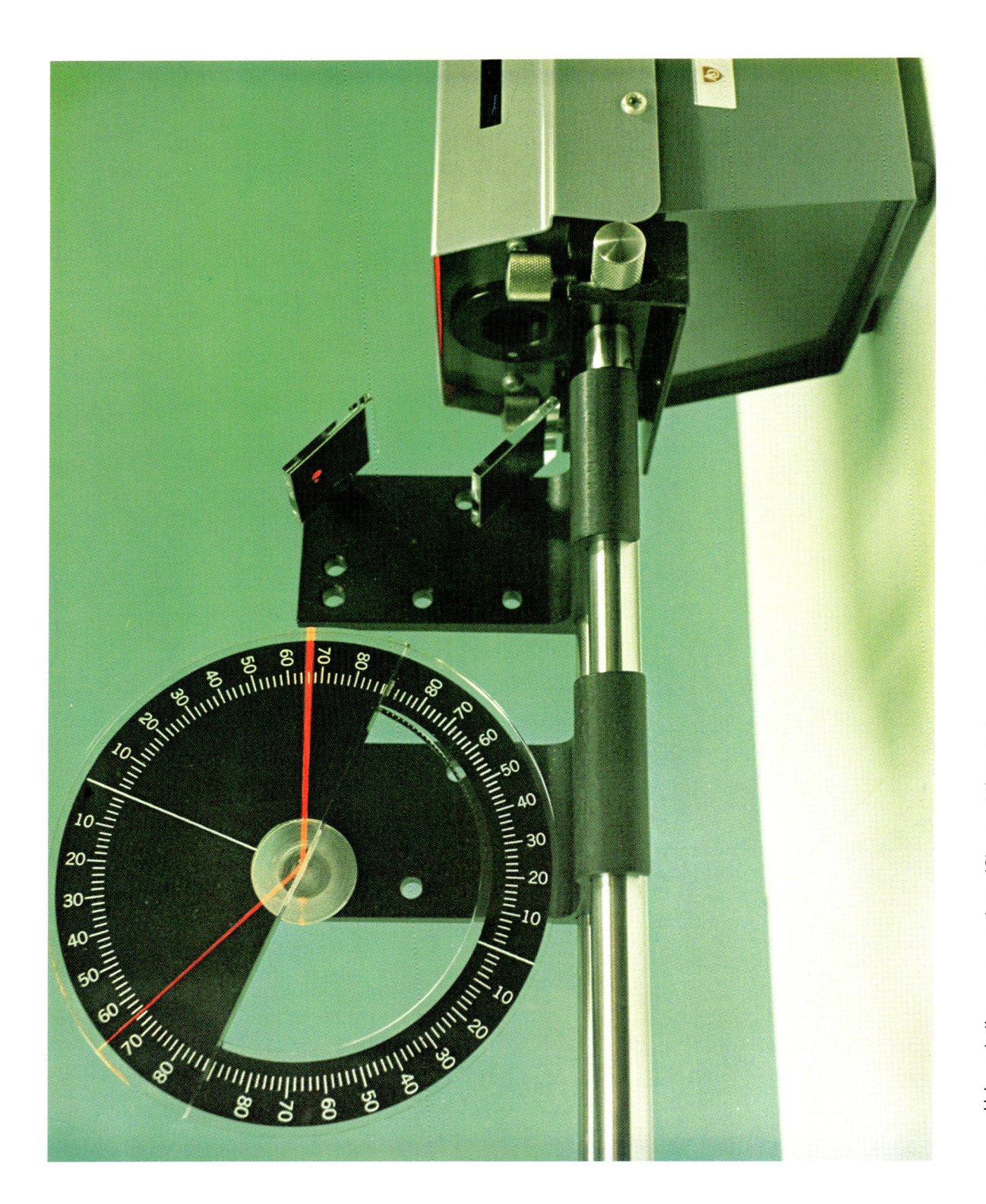

Using a helium-neon gas laser (Chapter 16) to study geometrical optics (Chapter 15). The red light beam from the laser is raised by means of two angled mirrors and is then directed into a transparent semicircular plastic disk placed at an angle to show the phenomenon of total internal reflection. (Queen's University at Kingston, Ontario.)

Just what it is in a light wave that is waving was elucidated in the 1860s by the British physicist, James Clerk Maxwell. The famous American physicist, Richard P. Feynman (1918–), who won the 1965 Nobel prize for his contributions to quantum electrodynamics, gives the following appreciation of Maxwell's work. "From a long view of the history of mankind—seen from, say, ten thousand years from now—there can be little doubt that the most significant event of the nineteenth century will be judged as Maxwell's discovery of the laws of electrodynamics. The American Civil War will pale into provincial insignificance in comparison with this important scientific event of the same decade."

What Maxwell had done was to write down, in four succinct equations, laws that embraced all the previously known behavior of electric and magnetic fields. He found that a new term had to be added to make these laws consistent, and this new term implied that there should be such things as *electromagnetic waves* emitted whenever an electric charge is accelerated. The velocity of these electromagnetic waves, calculated from Maxwell's equations, agreed with the measured velocity of light. This immediately suggested what has since been abundantly confirmed: light waves are electromagnetic waves. Maxwell had synthesized the fields of electricity, magnetism, and light. Over 20 years later, the gifted experimentalist, Heinrich Rudolf Hertz (1857–1894) was able to generate and detect in the laboratory electromagnetic waves of the wavelengths now employed in radio broadcasting. Today's physicist, surveying modern radio and television developments, views the existence of these industries as a direct consequence of the paper and pencil work of James Clerk Maxwell.

In this chapter we examine the wave nature of electromagnetic radiation, particularly light. Additional facts and theory, bearing on the twentieth century discovery that this radiation consists of particles (photons), are postponed to Chapter 16. The nineteenth century theory of electromagnetic waves, although an incomplete account of the phenomena involved, nevertheless is enormously successful in many situations and is a necessary preliminary to quantum theory which embraces both the idea of photons and that of the electromagnetic field.

14.1 CHARACTERISTICS OF ELECTROMAGNETIC WAVES

Much more abstract than waves on a duck pond are the electromagnetic waves that have such an influence on our daily lives. Although radio waves and x-rays beat upon us, we would remain as unaware of their existence as is a savage, were it not for the curiosity, imagination, and creative efforts of a few highly intelligent people whose first object was simply to understand nature.

Michael Faraday's great experimental work in electricity and magnetism had been helped and guided by his picture of electric and magnetic fields in space. Impressed by the aging Faraday's fruitful use of the concept of a field, Maxwell applied himself to the problem of the mathematical formulation, in terms of fields, of the experimentally discovered laws of electricity and magnetism. He was able to write equations relating the spatial and time rates of change of electric and magnetic fields.

From Faraday's law of electromagnetic induction, Maxwell inferred that at any point in space, an electric field was associated with a changing magnetic field (as in Fig. 12-19). Then Maxwell found that, in order to have mathematically consistent relations between the fields, he would have to assume that a magnetic field was associated with a changing electric field. This in turn implied that it would be possible to have a self-sustaining electromagnetic wave, consisting of fluctuating electric and magnetic fields, propagating through space. The speed of propagation of this electromagnetic wave was calculated from Maxwell's equations to be $\sqrt{2\,k/K}$. This quantity, mentioned in Section 12.4, has the value 3.00×10^8 meters/sec, the experimental result for the speed of light. A comparison of other properties of electromagnetic waves with experiments involving light piled up convincing evidence that light waves were electromagnetic waves. Maxwell had united the subjects of electricity, magnetism, and light.

So much physics has emerged from electromagnetic theory that Maxwell's reputation increases with time. The basic physical laws used by today's communication engineers are precisely Maxwell's equations. The most successful part of theoretical physicists' present efforts to understand elementary particles utilizes Maxwell's equations for the electromagnetic field, albeit with a more abstract interpretation.

The special theory of relativity created by Einstein has historical roots in Maxwell's electromagnetic theory. Einstein realized that, if all inertial frames of reference are equivalent for the description of nature, then Maxwell's theory and Newton's mechanics could not both be correct. Einstein decided in favor of Maxwell and proceeded to modify Newton's mechanics. Special relativity demands no modification of Maxwell's electromagnetic theory, which has the distinction of being the first relativistically correct theory, even though it was created by Maxwell more than forty years before Einstein's work.

The Basic Nature of the Waves

We now describe, in some detail, an electromagnetic wave. An electromagnetic wave is created by the *acceleration* of an electric charge. The accelerated charge loses energy which is carried away by the electromagnetic wave moving with a velocity c. Let us assume, as is usually the case in radio broadcasting, that electric charge is accelerated periodically by being forced to oscillate up and down in a vertical antenna with a frequency f and a period $T \, (= 1/f)$. Suppose we now station ourselves at a distant point and measure the electric field $\vec{E}$ at this point. We find that this electric field fluctuates with the same frequency f and period T as that of the accelerated charge in the remote antenna. Some values of $\vec{E}$ which occur at this one point in space at different times are depicted in Fig. 14-1(a).

There is also a magnetic field $\vec{B}$ which is perpendicular to $\vec{E}$ and proportional to it. Thus, as shown in Fig. 14-1(b), the magnetic field also fluctuates reaching its maximum value at the same instant as does the electric field and becoming zero when the electric field is zero.

We now investigate the fields at different points far from the antenna along an axis OZ which points away from the antenna in the direction of propagation of the wave. The electric and magnetic fields encountered at different points along OZ at one instant of time, a mental "snapshot" of the arrows representing these field vectors, is shown in Fig. 14-1(c). At points on the axis OZ separated by a distance of one wavelength, λ, one finds electric fields (and similarly magnetic fields) have the same value. If, after the lapse of a short time interval, one took another "snapshot," one would find the wave form of Fig. 14-1(c) displaced to the right along the axis OZ.

Fig. 14-1 Electromagnetic waves. (a) Variation with time of the electric field $\vec{E}$ at a given point in space. The field is a sinusoidal function of time when the charges in the antenna execute simple harmonic motion. (b) The changing magnetic field, at right angles to the changing electric field, is added to the diagram of part (a). $\vec{B}$ is in phase with $\vec{E}$. (c) Representation of (sinusoidal) electromagnetic waves traveling to the right.

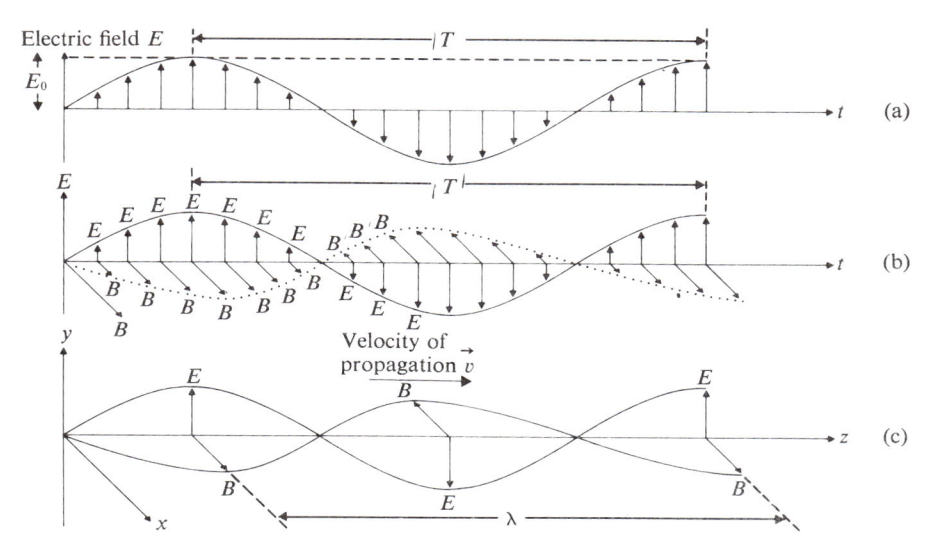

14.1 CHARACTERISTICS OF ELECTROMAGNETIC WAVES

The relative orientation of the vectors $\vec{E}$, $\vec{B}$, and the wave velocity vector $\vec{v}$ at any point in space, shown in Fig. 14-1 (c), should be noted. In an electromagnetic wave, these three vectors are always mutually perpendicular. The orientation of the electric field $\vec{E}$ in space is such that it lies in the plane containing both the wave velocity $\vec{v}$ and also the straight line along which the charge is being accelerated, that is, the antenna.

Intensity

A traveling electromagnetic wave transports energy. The *intensity*, I, of an electromagnetic wave is defined as the average power crossing a small area A, divided by A:

$$I = (\text{average power})/A, \tag{14.1}$$

where the small surface of area A is oriented so that it is perpendicular to the direction in which the wave is traveling. The mks unit of intensity is the watt/m². From this definition one can derive the fact that the intensity of an electromagnetic wave is proportional to the square of the amplitude of the wave, as indicated in the following equation:

$$I = (1.33 \times 10^{-3} \text{ watt/volt}^2) E_0^2, \tag{14.2}$$

where E_0 is the amplitude of the electric field of the electromagnetic wave [see Fig. 14-1(a)].*

*Although an excellent approximation in many situations, this equation is exactly true only for what are called plane waves (Fig. 14-7) traveling in a vacuum.

Example 1 The light from a certain lamp is directed so that 10.0 watts of electromagnetic radiation are uniformly distributed over a table top whose area is 2.00 m². Find the intensity of the electromagnetic waves striking the table and the amplitude E_0 of this wave.

Solution From Eq. (14.1) we calculate the intensity:

$$I = \frac{10.0 \text{ watts}}{2.00 \text{ m}^2}$$
$$= 5.00 \text{ watts/m}^2.$$

Now, from the relationship between the intensity and the wave amplitude, E_0, given by Eq. (14.2), we obtain

$$E_0^2 = \frac{5.00 \text{ watts/m}^2}{1.33 \times 10^{-3} \text{ watt/volt}^2}$$
$$= 3.76 \times 10^3 \text{ (volts/m)}^2.$$

Therefore,
$$E_0 = 61.3 \text{ volts/m}.$$

Example 2 Lasers, to be discussed in Chapter 16, are light sources which provide a narrow beam of high intensity. Suppose a 3.0×10^3-watt laser beam were concentrated by a lens into a cross-sectional area of about 10^{-10} m². Find the corresponding intensity and the amplitude of the electric field.

Solution The intensity is, from Eq. (14.1),

$$I = \frac{3.0 \times 10^3 \text{ watts}}{10^{-10} \text{ m}^2}$$

$$= 3.0 \times 10^{13} \text{ watts/m}^2.$$

From Eq. (14.2), we find

$$E_0^2 = \frac{3.0 \times 10^{13} \text{ watts/m}^2}{1.33 \times 10^{-3} \text{ watt/volt}^2}$$

$$= 2.3 \times 10^{16} \text{ (volts/m)}^2.$$

Therefore,

$$E_0 = 1.5 \times 10^8 \text{ volts/m}.$$

Because a laser beam has such a high power and can be brought to a very sharp focus, it can be used as an exceedingly effective "drill" to burn through a target.

Electromagnetic Spectrum

All electromagnetic waves travel with the same speed c in a vacuum. Any wavelength λ is possible. And for a given electromagnetic wave, its wavelength λ and its frequency f are related by Eq. (10.2), which for this case gives

$$c = \lambda f. \qquad (14.3)$$

In Maxwell's day, the only well known electromagnetic waves were those corresponding to visible light, although some knowledge had been acquired concerning the ultraviolet and infrared. In Fig. 14-2, on the following page, we show the vast range of electromagnetic radiation wavelengths now familiar. Moreover, Fig. 14-2 shows the various radiations placed in an orderly sequence.* Such an orderly arrangement according to wavelength or frequency is termed a *spectrum*. Note that for half of the spectrum in Fig. 14-2, wavelengths are specified in angstrom units (Å). Let us emphasize how small this unit is: one meter is ten billion angstrom units long.

The very short wavelengths of visible light range from approximately 7000 Å ($= 7 \times 10^{-7}$ m), which corresponds to the color red, through decreasing wavelengths corresponding to the colors orange,

*The reader is not to infer from Fig. 14-2 that there are arbitrary or well-defined cutoff points along this spectrum at which one "type" of radiation stops and another begins. The purpose is simply to indicate regions in which the energies of the electromagnetic radiations are such as to exhibit some particular characteristic behavior with respect to interactions with matter. Nor are there upper or lower bounds to the electromagnetic spectrum, as perhaps is implied in the figure.

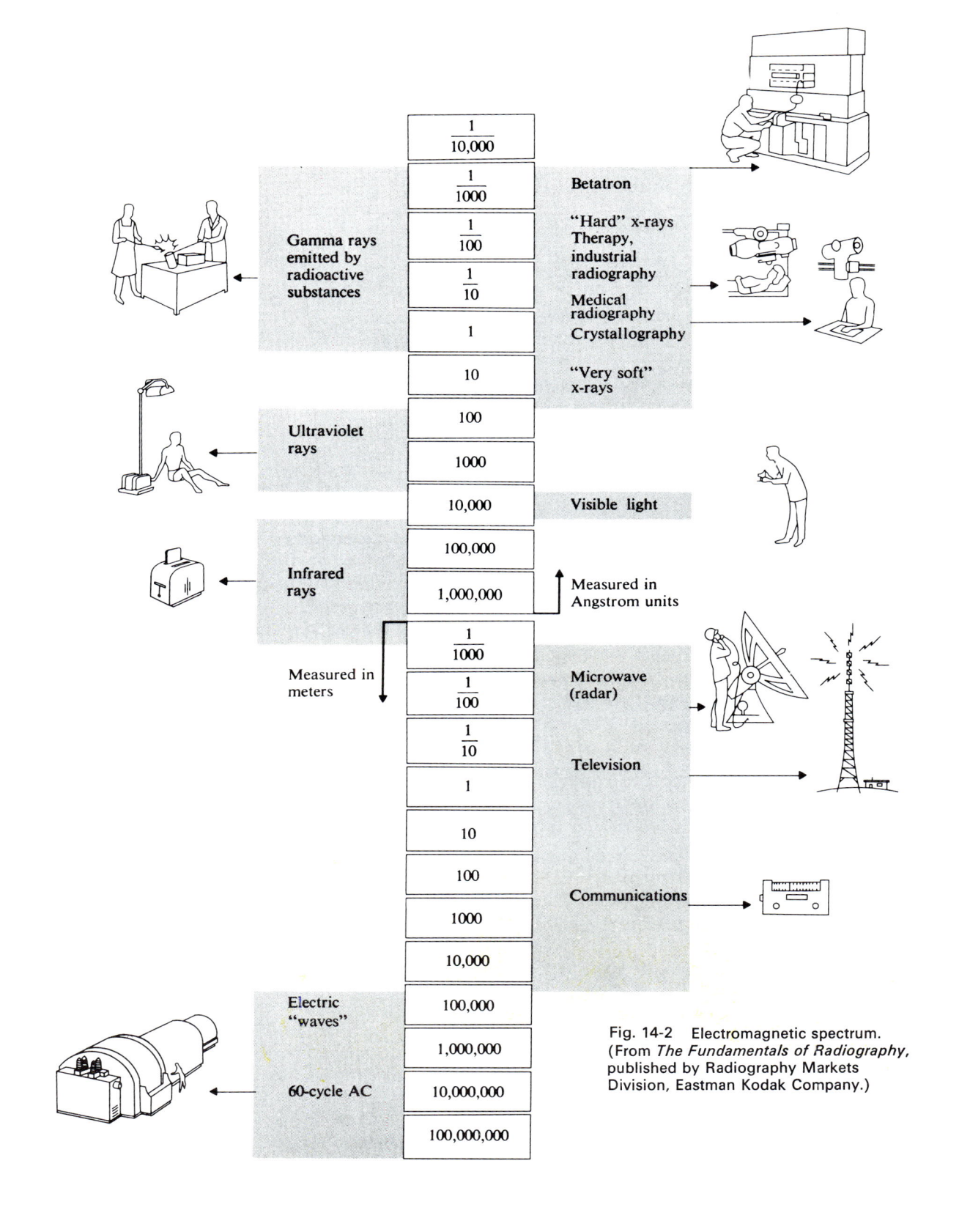

Fig. 14-2 Electromagnetic spectrum. (From *The Fundamentals of Radiography*, published by Radiography Markets Division, Eastman Kodak Company.)

yellow, green, blue, and finally violet at approximately 4000 Å. Just how such short wavelengths are measured will be described when we discuss Young's experiment involving the interference of light beams. The narrow band of wavelengths which can be detected by the human eye is bounded on the short wavelength side by what are called ultraviolet waves, easily detected by photographic films or by utilizing the photoelectric effect to be discussed in Section 16.2. On the long wavelength side of visible light are infrared waves copiously emitted by any hot body.

Although, as shown in Fig. 14-2, electromagnetic waves of different wavelengths have acquired different names such as x-rays and radio waves, it must be stressed that such waves are of exactly the same nature. However, the emission, absorption, scattering, and transmission through and around obstacles are decidedly different for electromagnetic waves of different wavelengths. Therefore, investigations of the characteristics of waves of different parts of the spectrum require distinctly different experimental techniques. And these differences among the various electromagnetic radiations give rise to vastly different applications in technology.

A striking step in the exploration of the electromagnetic spectrum was made by Heinrich Hertz in 1887 in Germany: this was after the appearance of Maxwell's theory, of course. Hertz generated what we call *radio* waves by the oscillatory discharge across a spark gap of separated electrodes which had been given opposite charges. To detect the electromagnetic waves emitted by the accelerated charges, he used a loop with its two ends slightly separated. A spark across this gap would indicate an oscillatory electric field in the vicinity. Sure enough, his detector sparked: Maxwell was right! Hertz then showed that these waves were transverse waves and were reflected by a metal sheet but that they would pass through insulators such as wood, undergoing a change of direction at the surface (refraction). The measured wavelength was 9.6 m when the frequency was 3×10^7 hertz.

From this data, the relationship $c = \lambda f$ [Eq. (14.3)] allowed calculation of the wave speed:

$$c = (9.6 \text{ meters}) \times (3 \times 10^7 \text{ hertz})$$
$$= 3 \times 10^8 \text{ m/sec}.$$

The long wavelength electromagnetic waves discovered by Hertz became of great practical importance with the development of electronic oscillators and amplifiers in the twentieth century. With such oscillators we now generate a spectrum ranging from radar waves or "microwaves" (usually 0.3-cm to 10-cm wavelength) through the television broadcast band (about 0.3-m to 5-m wavelength) and the wavelengths of short and long wave radio. Commercial broadcasting extends to wavelengths of 600 meters (a frequency of 500 kilohertz). Aircraft and marine broadcasting use wavelengths as long as 3000 meters (100 kilohertz).

Deutsches Museum, courtesy of AIP Niels Bohr Library

Heinrich Hertz
(1857–1894)

In 1895, a few years after Hertz' experiments with radio waves, Röntgen's sensational discovery of *x-rays* precipitated the investigation of a range of wavelengths much shorter than those of visible light. However, it was not until 1912 that the wave nature of x-rays was confirmed and their wavelengths measured to be of the order of 1 Å.

Of course, we must not forget the γ-rays we discussed in Chapter 1. Nuclear γ-rays were discovered in 1900 by Paul Villard. Much shorter wavelength γ-rays (10^{-13} Å) produced by cosmic radiation have since been found.

Polarization

The interaction of an electromagnetic wave with matter often depends on the orientation in space of the vibrating electric vector of the wave. A specification of the direction of the electric vector determines what is called the *polarization* of the wave. The wave shown in Fig. 14-1, with the vertical axis assumed to represent a vertical direction in space, is said to be *linearly polarized in a vertical direction*, meaning simply that at any given point in space, the electric vector $\vec{E}$ fluctuates along a *vertical line*, sometimes pointing straight up and sometimes straight down. Recall that electromagnetic waves are transverse, that is, that $\vec{E}$ (and $\vec{B}$) are always perpendicular to the direction in which the wave is traveling as indicated by the wave velocity vector $\vec{v}$.

In Fig. 14-3 we illustrate the results of an experiment in which light, linearly polarized in a vertical direction, is incident upon a thin sheet of material known as polaroid which can be used as a polarization analyzer. The polaroid sheet has a direction known as the *transmission axis*. When the polaroid is positioned so that the transmission axis is parallel to the electric vector of the incident linearly polarized light, it is found that this light is transmitted through the polaroid with almost undiminished intensity. Now, if the polaroid is rotated through 90°, practically no light is transmitted! At intermediate rotations light is transmitted with a diminished intensity corresponding to transmission of only the component of the vector $\vec{E}$ along the transmission axis. The component of the vector $\vec{E}$ which is perpendicular to the transmission axis is *absorbed*. The transmitted light is therefore linearly polarized in the direction of the transmission axis.

These facts about the transmission of polarized light through a polarization analyzer were discovered early in the nineteenth century by Etienne Malus.

Now polarization cannot arise in a longitudinal wave—a wave motion in which the vibrations are along the direction of the wave velocity vector $\vec{v}$. Why? Because in this case there is nothing to distinguish any one transverse direction from another transverse direction. *The phenomenon of polarization therefore showed that light involved transverse rather than longitudinal vibrations.* This was established half a

Courtesy of the American Institute of Physics, Meggers Gallery of Nobel Laureates

W. C. Röntgen
(1845–1923)

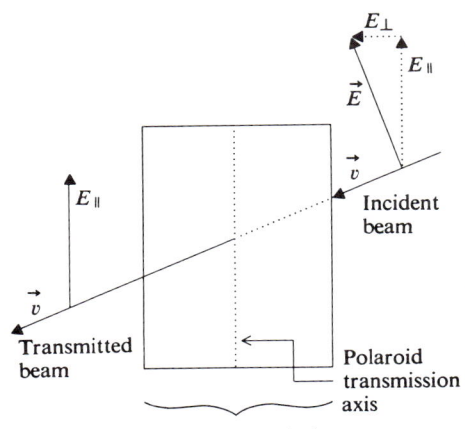

Fig. 14-3 Polaroid. The component of the incident $\vec{E}$ that is parallel to the transmission axis, $E_\parallel$, is transmitted. The other component, $E_\perp$, is absorbed in the polaroid.

The edge of a polarizer such as the edge of a polaroid sunglass lens

526 ELECTROMAGNETIC WAVES

century before Maxwell identified the transverse vibrating quantities as electric and magnetic fields.

Large sheets of a material with a transmission axis were not available until 1938 when Edwin H. Land invented polaroid. This is manufactured by stretching a plastic sheet, containing long hydrocarbon chains, to align the molecules. Next the sheet is dipped in an iodine solution. The iodine attaches to the long hydrocarbon chains, providing each chain with mobile electrons. In a fluctuating electric field, the component of the electric field along the chain drives these electrons back and forth along the length of the chain and the energy associated with such a field component is absorbed. The transmission axis is the direction perpendicular to the line of aligned molecules because in this direction the absorption mechanism cannot function.

Most light sources emit what we call *unpolarized light*. Picturing a radiating atom as containing an oscillating electron which radiates as it does in an antenna, we expect that the radiated wave will be linearly polarized. However, a given atom radiates only for about 10^{-8} sec. With a lamp, for one 10^{-8} sec time interval, we receive waves from one particular collection of atoms and this wave will be polarized in a certain direction. But in the next 10^{-8} sec interval we will get waves from a different set of atoms with different orientations, and the polarization of the wave will generally be in a different direction. With the atoms randomly orientated and radiating independently, we will therefore find that the wave at a given observation point has a *polarization which jumps about* assuming many million different transverse directions in a second, with all transverse directions being equally likely. For detectors like the human eye, which responds to the average intensity over intervals like 1/10 sec, we say such light is unpolarized. Unpolarized light incident on a polaroid will emerge linearly polarized in the direction of the transmission axis. The transmitted intensity is not quite 50% of the intensity of the incident unpolarized light.

We shall mention only one of several interesting applications of our knowledge of polarized light. A most common source of light that is at least partially polarized is that which is reflected from any smooth surface, such as a road or a lake. This glare, mostly linearly polarized in a horizontal direction, can be discriminated against by sunglasses made of polaroids with their transmission axes vertical.

Wave Fronts and Rays

In the case of waves on the surface of a duck pond, the eye can literally keep track of the motion of crests and troughs. With electromagnetic waves we similarly focus our attention on crests and troughs but our concepts must be extended to describe waves in three-dimensional space and, of course, we do not actually see the motion of these waves.

Physics Today

Edwin H. Land
(1909–)

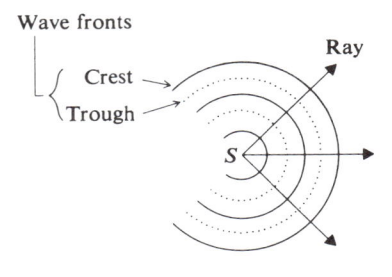

Fig. 14-4 The spherical wave fronts emitted by an electron executing small oscillations at S. Rays are perpendicular to the wave fronts.

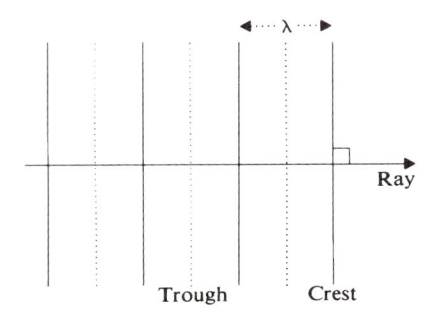

Fig. 14-5 Plane wave fronts.

*The present theory known as quantum electrodynamics predicts that scattering of light by light does occur but the effect is completely negligible with beams of ordinary intensities.

To picture an electromagnetic wave in space at a certain instant we draw a succession of surfaces through points where there are crests, that is through points where the vibrating quantity, the electric field in an electromagnetic wave, is at its maximum value, its *amplitude*. The intersection of such surfaces with a two-dimensional plane gives the circular solid lines shown in Fig. 14-4. The dotted lines between the crests show the location at this instant of the troughs, that is, the places where the electric vector has a magnitude equal to the amplitude but points in a direction opposite to that corresponding to a crest. Such *surfaces* indicating the location of crests and troughs are called *wave fronts*. The wave fronts in the electromagnetic wave radiated by an electron executing small oscillations (at S in Fig. 14-4) are spherical. A small portion of a large spherical surface is approximately a plane. Therefore, with a remote source the wave fronts are planes, as shown in Fig. 14-5.

A *ray* is a line in the direction of the wave velocity vector $\vec{v}$ and intersects the wave fronts at right angles. Geometrical optics (which deals with the analysis of mirrors, lenses, optical instruments, etc.) is developed entirely in terms of light rays. (See Chapter 15.)

14.2 INTERFERENCE

Superposition and Interference

The superposition principle for electric and magnetic fields states that, at any instant, the field at any point in space arising from several sources is the vector sum of the contributions that each source would furnish if it were acting alone. This principle is a consequence of Maxwell's equations. According to the superposition principle, the field contributed by one source does not affect the contribution of another source; the two fields just add up vectorially. For electromagnetic waves this implies that different radio signals, light beams, and x-rays can travel through the same region of space without influencing each other in the slightest. Intersecting light beams are undeviated and unchanged by the encounter with one another.*

However, in the regions where there is a superposition of two or more waves *there are variations of intensity* which, under certain conditions, are easily detectable and yield interesting information. Consider as sources of electromagnetic waves, two parallel antennas, S_1 and S_2 in Fig. 14-6, each driven by the same oscillator in such a way that the oscillating charges in each antenna are doing the same thing at the same time: we then say the *antennas are in phase*. Now let us consider the resultant electric field $\vec{E}$ at a distant observation point P. The electromagnetic wave from antenna S_1 contributes a fluctuating field $\vec{E}_1$ at the point P, and S_2 contributes a field $\vec{E}_2$. The superposition principle then

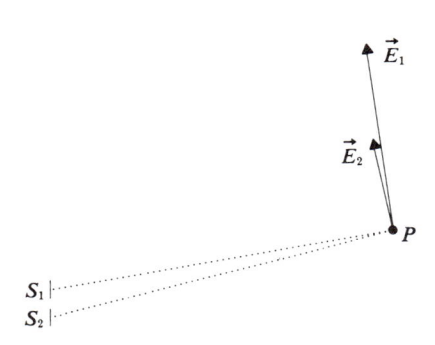

Fig. 14-6 At a certain instant, antenna S_1 contributes a field $\vec{E}_1$ at the observation point P, and antenna S_2 contributes a field $\vec{E}_2$.

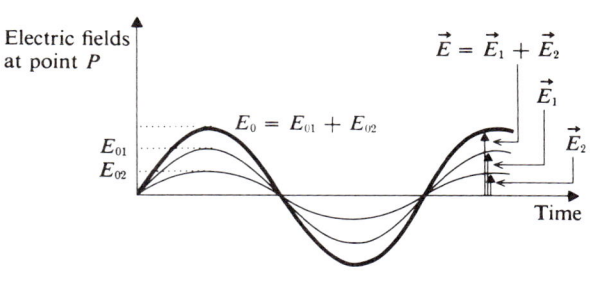

Fig. 14-7 Constructive interference. Graph of the fluctuating fields $\vec{E}_1$, $\vec{E}_2$, and their superposition $\vec{E}\ (=\vec{E}_1+\vec{E}_2)$ at different times at an observation point P where crests from each source arrive at the same time. Here the fields $\vec{E}_1$ and $\vec{E}_2$ are *in phase* and the amplitude E_0 of the superposition is the sum of the amplitudes E_{01} and E_{02}.

states that, at any instant,

$$\vec{E} = \vec{E}_1 + \vec{E}_2.* \tag{14.4}$$

The resultant electric field $\vec{E}$ fluctuates with the same frequency as that of the contributions $\vec{E}_1$ and $\vec{E}_2$. In Fig. 14-7 we show the values of $\vec{E}_1$ and $\vec{E}_2$ at different times, and also the value of their vector sum, $\vec{E}$. The amplitude E_0 of the fluctuations of the resultant field $\vec{E}$ is of particular interest because E_0^2 determines the intensity I of the resultant wave at the observation point P. Recall that

$$I = (1.33 \times 10^{-3}\ \text{watt/volt}^2)\ E_0^2. \tag{14.2}$$

The relationship between the amplitude E_0 and the amplitudes E_{01} and E_{02} of two overlapping waves is different at different observation points. At a location where crests from one source arrive at the same time as crests from the other source, the fluctuating fields add up as shown in Fig. 14-7 to give a resultant fluctuating field $\vec{E}$ whose amplitude E_0 is simply the sum of the amplitudes (E_{01} and E_{02}) of the waves from each source. That is,

$$E_0 = E_{01} + E_{02}, \tag{14.5}$$

at locations where crests overlap.

However, at observation points where a crest from one source arrives at the same time as a trough from the other source, the fluctuating fields add as shown in Fig. 14-8 to produce a fluctuating resultant field $\vec{E}$ whose amplitude is the difference of the amplitudes of the waves from each source. That is,

$$E_0 = E_{01} - E_{02}, \tag{14.6}$$

*Since the point P is very far from the antennas, to a good approximation we can say that the vectors $\vec{E}_1$ and $\vec{E}_2$ lie along the same line, that is, the waves have the same polarization.

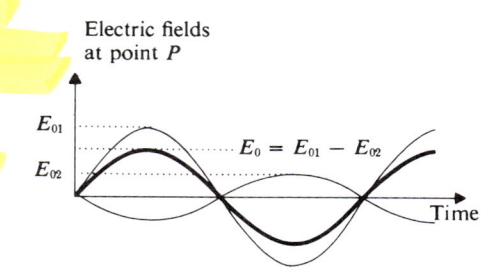

Fig. 14-8 Destructive interference. At an observation point P where crests from one source arrive with troughs from the other source, the amplitude E_0 of the superposition is the *difference* of the amplitudes E_{01} and E_{02} of the waves contributed by each source.

14.2 INTERFERENCE

at locations where crests and troughs overlap. At intermediate locations the amplitude E_0 has a value somewhere between its minimum value, $E_{01} - E_{02}$, and its maximum value, $E_{01} + E_{02}$.

The intensity I at any location is determined by E_0^2, the square of the amplitude of the fluctuating field. The general result turns out to be

$$I = I_1 + I_2 + \text{(interference term)} \tag{14.7}$$

where I_1 [$= 1.33 \times 10^{-3} E_{01}^2$ from Eq.(14.2)] is the intensity that source S_1 would contribute if it *alone* were radiating. Similarly I_2 is the intensity that S_2 alone would contribute. The value of the interference term is different at different locations. The phenomenon *that the intensity I is not everywhere simply the sum of the intensities I_1 and I_2* is termed *interference*. We say the two overlapping waves *interfere* and the pattern in space of regions of high and low intensity is called an *interference pattern*.

The interference term in Eq. (14.7) is at some locations positive so that I is greater than the sum of I_1 and I_2, and here we say the interference is *constructive*. Maximum constructive interference occurs at locations where crest meets crest so that the amplitude of the superposition is the sum of the amplitudes of the two interfering waves, as in Fig. 14-7.

At other locations the intensity I is less than the sum of I_1 and I_2, and we say that there is *destructive interference*. Minimum intensity occurs where crests meet troughs producing a superposition whose amplitude is the difference of the amplitudes of the interfering waves (Fig. 14-8).

The case when the amplitudes of the two interfering waves are equal ($E_{01} = E_{02}$) is particularly simple and gives rise to striking interference patterns. When the amplitudes are equal, the intensities that each wave alone would produce, I_1 and I_2, are also equal. In this case, at locations of constructive interference where crest meets crest, Eq. (14.5) gives for the amplitude

$$E_0 = 2E_{01}.$$

The reader is asked, in Question 18, to show that this result, combined with Eq. (14.2), implies

$$I = 4I_1. \tag{14.8}$$

At the locations of destructive interference, where crest meets trough, we now see, from Eq. (14.6), that the amplitude is

$$E_0 = E_{01} - E_{02} = 0.$$

Then, from Eq. (14.2),

$$I = 0. \tag{14.9}$$

Thus the intensity, in a region with two interfering waves of equal amplitude, ranges from zero at some locations to $4I_1$ at other places.

Interference and the Measurement of the Wavelength of Light

If the wavelength of interfering waves of equal amplitude is such that the electromagnetic waves are visible light, then we find total darkness where $I = 0$ and brightness where $I = 4I_1$. Such a situation was achieved by Young in 1802 using the arrangement shown in Fig. 14-9. The double slits S_1 and S_2 are illuminated by light from the same source coming through a single slit. The double slits now serve (instead of antennas) as the two radiation sources which are always "in phase." While one naturally cannot see the crests and troughs which are drawn in Fig. 14-9,

Fig. 14-9 Schematic plan of Young's double-slit experiment; not to scale. For the waves from either slit, S_1 or S_2, the distance between adjacent crests is grossly exaggerated. This distance, the wavelength, is actually about 5×10^{-5} cm for green light. Also, the distance *l* to the viewing screen has to be much larger than the separation between the double slits. In (a) the dashed semicircles represent troughs, while in both (a) and (b) the solid semicircles represent crests. (Courtesy Holton and Roller, *Foundations of Modern Physical Science*, Addison-Wesley, 1958.)

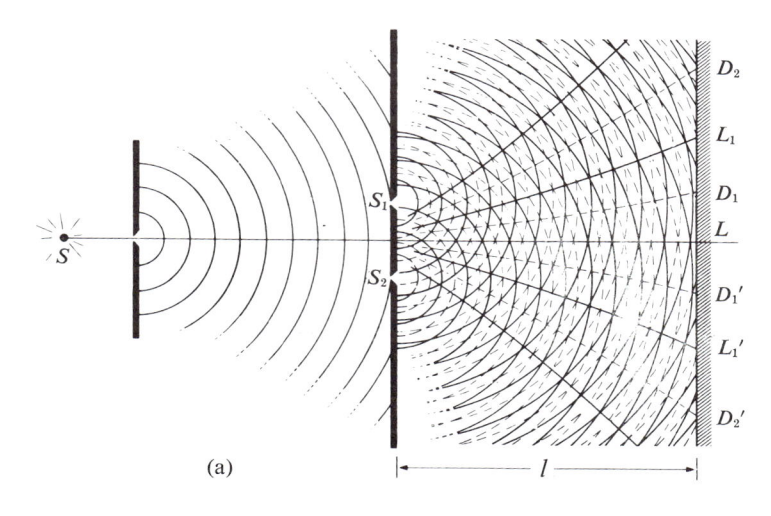

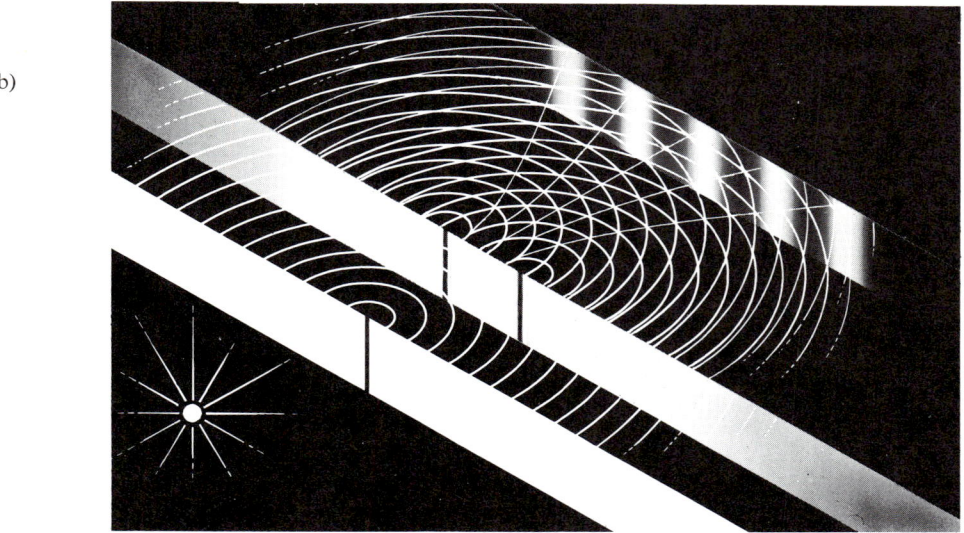

14.2 INTERFERENCE

one does observe the interference pattern. The regions of brightness, called interference fringes, which alternate with regions of darkness, are readily seen by allowing the light waves to illuminate the screen shown at the right of the figure. Such an interference pattern recorded on a photographic plate (used instead of a screen) is shown in Fig. 14-10.

An understanding of the appearance of the interference patterns like those of Fig. 14-9 and Fig. 14-10 enabled Young to do a remarkable thing: to measure the wavelength λ of the light even though this wavelength was obviously much too small to be measured directly.

Fig. 14-10 The interference pattern produced on a photographic plate by two parallel long, narrow slits. (By permission, from Alonso and Finn, *Fields and Waves*, Addison-Wesley, 1967.)

His idea is not hard to understand. Let us return to Fig. 14-9. First, the point L, which is the same distance from each slit, will be a brightly lighted maximum of intensity [$I = 4I_1$ from Eq.(14.8)], simply because crests, leaving at the same time from S_1 and S_2 and having the same distance to go, arrive at the same time at L. Further up on the screen at the point D_1, where the distance from S_2 is $\frac{1}{2}\lambda$ further than the distance from S_1, there will be darkness [$I = 0$, from Eq.(14.9)] because crests from S_1 arrive at D_1 at the same time as troughs from S_2. Still further up on the screen at the point L_1, which is a distance λ further from S_2 than from S_1, crests from both sources will arrive simultaneously. So at L_1 the interference is again constructive ($I = 4I_1$) and we find brightness. We emphasize that L_1 is a point such that*

$$S_2L_1 - S_1L_1 = \lambda. \tag{14.10}$$

*In the following equations S_2L_1 denotes the distance between points S_2 and L_1, and similarly S_1L_1 denotes the distance between points S_1 and L_1.

To determine λ we must know $(S_2L_1 - S_1L_1)$ and, although this small difference of two large distances would be extremely difficult to measure directly, it can be related to distances which are easy to measure.

After some purely geometric labor with the distances involved in Fig. 14-9, the following approximate geometric relationship can be deduced:

$$S_2L_1 - S_1L_1 = \frac{(LL_1) \times (S_1S_2)}{l}. \tag{14.11}$$

532 ELECTROMAGNETIC WAVES

The approximation is good if the distance l from the slits to the screen is much greater than both the slit separation (S_1S_2) and the distance (LL_1) on the screen between two adjacent intensity maxima. The left-hand side of Eq. (14.11), according to Eq. (14.10), is just one wavelength, so we have

$$\lambda = \frac{(LL_1) \times (S_1S_2)}{l} \qquad (14.12)$$

from which the wavelength can be calculated.

Example 3 In the double-slit experiment shown in Fig. 14-9, the yellow light from a sodium vapor lamp falls upon two parallel slits which are 1.0×10^{-4} meter apart. The distance between the central bright fringe and the adjacent fringe is 0.47×10^{-2} meter when the screen is placed 0.80 meter from the slits. Determine the wavelength λ of the yellow light.

Solution

$$\lambda = \frac{(LL_1) \times (S_1S_2)}{l} \qquad (14.12)$$

$$= \left(\frac{0.47 \times 10^{-2} \times 1.0 \times 10^{-4}}{0.80} \right) \text{meter}$$

$$= 0.59 \times 10^{-6} \text{ meter}$$

$$= 5.9 \times 10^3 \text{ Å}.$$

To understand what determines the separation (LL_1) of the bright fringes in the interference patterns of Fig. 14-9, we solve Eq. (14.12) for LL_1:

$$LL_1 = l\lambda/(S_1S_2). \qquad (14.13)$$

This result shows that the closer together we have the slits, that is, the smaller we make S_1S_2, the greater will be the distance LL_1 between the interference fringes. Also increasing the wavelength of the light increases the fringe separation. For very short wavelengths the entire interference pattern will be squeezed together and we will have difficulty resolving the individual maxima.

Young's experiment was of great importance because it brought the wave nature of light to the fore and at the same time provided a method of measuring the wavelength. A modification of Young's method, which exploited interference from not two but *many* evenly spaced slits, was developed by Fraunhofer in the 1820s. The interference from a regular array of many slits (the array is known as a *diffraction grating**) produces intensity maxima which are very bright narrow lines. The distance from the central bright fringe to the next fringe for a given color can then be determined with considerable precision, allowing extremely accurate

*The result of a large number of sources interfering is often called a diffraction pattern instead of an interference pattern. There is no fundamental distinction to be made between interference and diffraction.

measurements of wavelengths. The American physicist, Henry A. Rowland (1848–1901), succeeded in making excellent diffraction gratings with up to 1000 ruled parallel lines per millimeter of a glass surface, each line playing the role of S_1 or S_2 in Young's double-slit experiment. The tremendous accuracy with which wavelengths can be measured with such an instrument is indicated by quoting a typical result for one of the yellow wavelengths emitted by a sodium atom:

$$\lambda = 5889.95 \pm 0.04 \text{ Å}.$$

Measurement of X-ray Wavelengths

After Röntgen's discovery of x-rays in 1895, it was suspected by many physicists that these radiations were electromagnetic waves of a shorter wavelength than any that had been encountered up to that time. However, they soon found this wave nature of x-rays was no easy matter to prove with the optical apparatus at hand. One sort of difficulty that was encountered can be appreciated by trying to repeat the Young double-slit experiment with the geometry of Example 3, but using waves of wavelength 1 Å. Then Eq. (14.13) gives the fringe separation LL_1:

$$LL_1 = \left(\frac{0.80 \times 10^{-10}}{1.0 \times 10^{-4}}\right) \text{ meter}$$

$$= 0.80 \times 10^{-6} \text{ meter} \quad \text{(that is, less than 1/1000 of a millimeter!)}.$$

The fringes are too close together!

Clearly what is needed in order to increase the separation of the intensity maxima is some arrangement with "slits" separated by a very small distance: S_1S_2 must not be too much greater than the wavelength under investigation. In 1912, it occurred to Max T. F. von Laue (1879–1960) that the regular array of atoms in a crystal (Fig. 14-11), which have interatomic spacings of a few angstrom units, would be useful in looking for interference effects with very short wavelength electromagnetic radiation. Each atom would reradiate the electromagnetic waves. Intensity maxima arising from constructive interference in the superposition of the radiations from all the atoms in the crystal would arise only for certain wavelengths and certain directions.

Experiments confirmed these expectations in a beautiful fashion. Not only was the wave nature of x-rays clearly demonstrated but also the hypothesis that the crystals were regular arrays of atoms was verified. A typical Laue diffraction pattern is shown in Fig. 14-12 on the following page. The spots are intensity maxima with different spots generally corresponding to constructive interference for different wavelengths.

In England, a father and son team of physicists, Sir William Henry Bragg (1862–1942) and Sir William Lawrence Bragg (1890–   ), rapidly discovered a straightforward method of relating these diffraction patterns

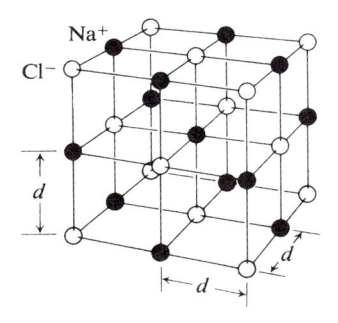

Fig. 14-11 Crystal lattice. Model of the arrangement of ions in NaCl. The lattice constant d is about 3 Å.

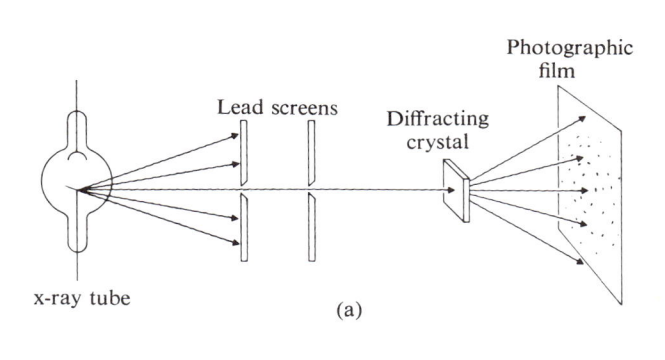

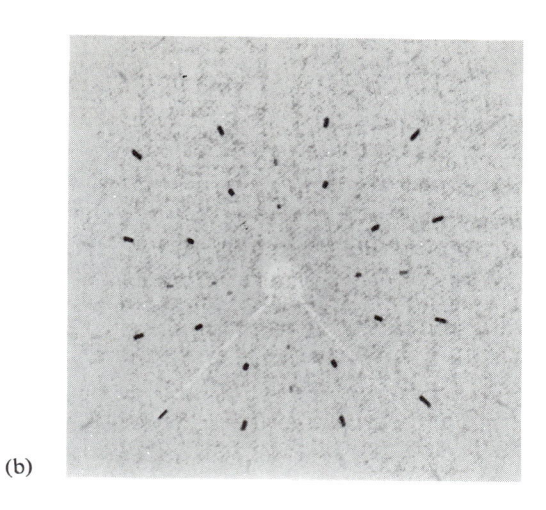

to the crystal structure. For crystals of known structure, the x-ray wavelength could be calculated. With x-rays of known wavelength, the structure of other crystals could be determined. Many variations of the original experimental techniques are still in use today to determine the structure not only of crystals but also of molecules. For example, x-ray diffraction techniques have been used to discover the atomic structure of such complex molecules as $C_{22}H_{23}O_8N_2Cl$, known as Aureomycin.

These achievements were recognized by the award of Nobel prizes to von Laue in 1914, and to the two Braggs in 1915.

Fig. 14-12 Von Laue pattern. (a) The experimental setup. (b) Laue diffraction by a crystal. The white central disk is due to a small metal disk held in front of the film by two wires (leading at an angle down toward the bottom of the film) to prevent extensive blackening of the film by the undeviated incident x-ray beam.

Coherence and Incoherence

If we illuminate a region with light from two independent sodium lamps and measure the intensity I by eye or with a photographic plate, we find that

$$I = I_1 + I_2, \qquad (14.14)$$

where I_1 is the intensity when the first lamp alone is operating and I_2 is similarly defined. *No interference is observed!*

If, in the Young double-slit experiment, slit S_1 is illuminated by one sodium lamp and S_2 by a similar sodium lamp, the eye or a photographic plate records merely a uniform intensity across the screen,

$$I = I_1 + I_2 = 2I_1.$$

The pattern of alternate bright and dark bands is no longer apparent. Again *no interference is observed.*

In such cases, when no interference is observed, we say the sources are *incoherent*. With *independent* identical sources S_1 and S_2, the sources are no longer always in phase, nor is there even what is called a fixed phase relationship between the oscillating charges in the different sources.

Except in the case of lasers (Chapter 16), there is no control over the instant at which an individual atom starts and stops radiating. Individual atoms radiate at random. Perhaps for one 10^{-8}-sec interval, there will be electron oscillations in the atoms of the two sources such that a crest leaves one source as a trough leaves the other source. For this short time interval, there will be a certain interference pattern in space. But in succeeding intervals, unrelated electron oscillations in the two sources take over with the result that perhaps crests leave one source a quarter of a period before crests leave the other source. While these oscillations persist there will be a definite interference pattern in space but it will be different from the interference pattern which preceded it. The regions of high and low intensity will occur at different locations. Thus we obtain a succession of different interference patterns, shifting after intervals of approximately 10^{-8} sec. A detector which responds only to averages over time intervals like 1/10 sec, will observe the average of some ten million different patterns, and, in this average, no interference will be apparent.

At a given observation point, for each 10^{-8}-sec interval, Eq. (14.7) gives

$$I = I_1 + I_2 + \text{(interference term)}. \quad (14.7)$$

However, as time goes by and electron oscillations in one source shift in a random fashion relative to the oscillations in the other source, the interference term changes value. After ten million such shifts (in 10^{-1} sec) with positive and negative values equally likely, the *average value of the interference term will be zero* and the *average intensities* will then satisfy,

$$I = I_1 + I_2. \quad (14.14)$$

No interference is observed by detectors whose response is determined by this *average* intensity. We say the two sources are *incoherent*, as far as this method of detection is concerned.

Nevertheless, in spite of what many textbooks say, it *is* possible to detect interference between two independent light sources. To do this one needs detectors that can "read" the interference pattern in a time less than 10^{-8} sec. Modern photomultiplier tubes can do the job. In a beautiful experiment (1956), an interference pattern produced by light coming from different parts of the star Sirius was detected by R. H. Brown and R. O. Twiss. Measurement of the distance between the bright fringes in this interference pattern allowed calculation of the separation of the sources of light. In this way, for the first time, the diameter of Sirius was determined.

Now let us return to Young's experiment. Usually, to observe interference phenomena with an ordinary light source, as in this experiment, the wave radiated by a given atom in the source is split into *two parts*. Each part of the wave travels a different path, and these two parts eventually meet at some point on the screen.

In Young's experiment, the sources S_1 and S_2 are really two different parts of the same wave. Changes in the oscillation of charges in the original light source will affect both S_1 and S_2 in the same way, and they will always be in phase. Sources such as S_1 and S_2 are *coherent* (for any method of detection), and produce interference patterns.

One of the most important novel features of the *laser* is the *coherence* of the light emitted by its different atoms. The laser exploits *stimulated emission*. Instead of individual atoms radiating at random, the light from one atom is used to stimulate the emission of light from another atom. The emitted light is locked in phase with the stimulating light.

Interference of light beams coming from different portions of the same laser can be easily observed by eye. Indeed two *independent* lasers emitting the same wavelength light will produce fringes that can be detected using high-speed television techniques. The difficulty here is not with phase shifts between the two sources, but rather that the wavelengths cannot be made precisely equal.

Diffraction

Waves spread out after passing through narrow apertures. In passing obstacles, waves bend and travel into the region behind the obstacle. Such departure from propagation in a straight line is termed *diffraction*.

Everyone is familiar with the diffraction of sound waves. You can speak to someone who is behind a tree, because the sound waves will bend around the tree and fall upon his ear. The phenomenon of diffraction is clearly displayed by water waves, as in Fig. 14-13.

Newton, and the majority of physicists up to the 19th century, felt that light could not be a wave motion because of the apparent absence of diffraction phenomena with light. True, Grimaldi, in Newton's century, had observed some departure from straight-line propagation when light passed through a small aperture or when shadows were cast by small objects. But the observed bending of the light beam was small and was interpreted as due to the deflection of light particles as they passed near matter.

It was not until Young showed with his double-slit interference experiment that a wave theory gave detailed mathematical agreement with experimental results that physicists looked into the subject of diffraction of light with care and persistence. Fresnel and Fraunhofer found that large diffraction effects indeed could be obtained with light waves (see Fig. 14-14). Moreover, the patterns of intensity maxima and minima (called diffraction patterns) that were produced in these experiments could be understood in quantitative detail using a wave theory of light. In the analysis of diffraction effects from the point of view of electromagnetic theory, one is concerned with the superposition of the electromagnetic waves radiated by many sources: the oscillating electric charge

Fig. 14-13 Diffraction of a water wave which passes through a small aperture. (From Alonso and Finn, *Physics*, Addison-Wesley, 1970.)

in an original light source and the charges in obstacles and screens that are forced to oscillate by the presence of fluctuating electric fields. Without doing the analysis, we quote an instructive result for diffraction of plane waves of wavelength λ which pass through a single circular hole of diameter D. The beam spreads out behind the hole, to give, on a distant screen, the diffraction pattern shown in Fig. 14-14. Speaking roughly, the light behind the hole fans out into a cone with a vertex angle (see Fig. 14-15) whose order of magnitude, when λ is less than D, is given by

$$\theta \text{ (in degrees)} \approx 10^2 \, (\lambda/D). \tag{14.15}$$

Fig. 14-14 Fraunhofer diffraction pattern. Light comes through a circular aperture and then spreads out to produce this pattern on a photographic plate.

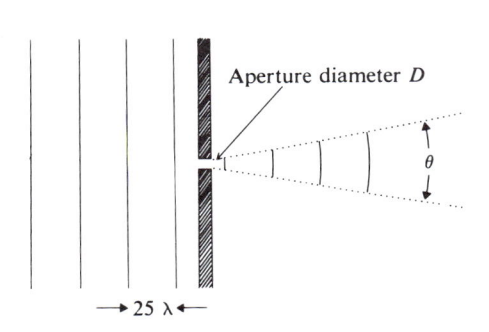

Fig. 14-15 Diffraction spreading behind an aperture when plane waves are incident. In order to have as little spreading as this, the wavelength would have to be about 1/25 of the distance between the indicated wavefronts.

According to this result, the spreading of the beam behind the hole will be significant if the wavelength λ is not too short compared to the hole diameter D. For example, if $\lambda = D/10$, we compute from Eq. (14.15) that $\theta \approx 10°$. However, if the wavelength is much less than the hole diameter the spreading is very small. Thus if $\lambda = 5000 \text{ Å} = 5 \times 10^{-7}$ meter, and $D = 5 \times 10^{-3}$ meter so that $\lambda/D = 10^{-4}$ we find, again from Eq. (14.15), that $\theta = 10^{-2}$ degree.

These results illustrate something that is true in general. *Diffraction effects become difficult to detect, and apparent straight-line propagation results when the wavelength is so small that the ratio of λ to the size of the aperture is very much less than one.* For light waves and apertures like house windows, this ratio is about 10^{-6}! This is the answer to Newton's objection to the wave theory. Notice the vital role that *quantitative information*, the order of magnitude of the wavelength of light, plays in reconciling the wave nature of light with our everyday experience of the propagation of light in a straight line.

Eq. (14.15) has an interesting application to lasers. One of the most important advantages of a laser is that its energy is not spread out in all directions. The energy is concentrated in a narrow beam of high intensity. The beam *stays* narrow! Laser light from the earth has already been shone on the moon, the beam spreading out over an area on the moon of only a few miles in diameter. The best optical searchlight beam would spread out to be larger than the moon itself, yielding a hopelessly low intensity.

Now why does a laser beam spread at all? The answer is diffraction. The unavoidable diffraction spreading of any beam of initial diameter D is given by Eq. (14.15). For lasers, λ/D can be made of the order of 10^{-5}, so the spreading is small. Having made a comparison of the laser with a searchlight, we must add that the relatively large searchlight beam-spread is not due to diffraction. Searchlights are unable to achieve plane wave fronts with great precision. Diverging rays result, causing a beam-spread much larger than the diffraction spreading.

14.3 REFLECTION, REFRACTION, AND DISPERSION

Reflection

When light is incident on a smooth plane surface of a material like glass in Fig. 14-16, part of the light is *reflected*, and part is passed into the glass, that is, *refracted*. The directions of the *incident*, *reflected*, and *refracted* beams are specified by giving the angles θ_1, θ_r, θ_2 that these rays make with the *perpendicular* to the surface. Experiments, dating back to the time of the ancient Greek civilization, lead to the *law of regular reflection:*

1. The reflected ray lies in the same plane as that containing the incident ray and the perpendicular to the surface.
2. The angle of reflection, θ_r, equals the angle of incidence, θ_1. That is,

$$\theta_r = \theta_1. \qquad (14.16)$$

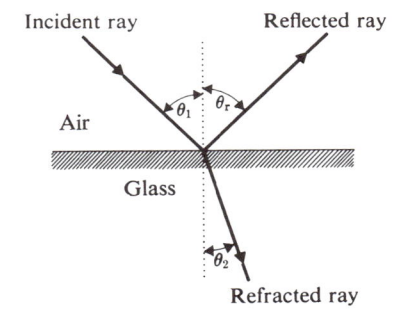

Fig. 14-16 Incident, reflected, and refracted rays.

Most of the objects that we see have on their surfaces irregularities which are large compared to the wavelengths of light. Such objects reflect the light in all directions, as shown in Fig. 14-17(b). This is called *diffuse reflection*.

(a) Regular reflection

(b) Diffuse reflection

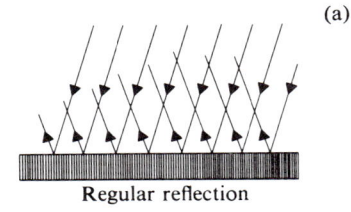

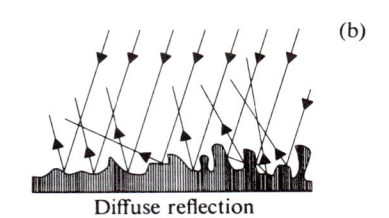

Fig. 14-17 (a) Regular reflection. (b) Diffuse reflection.

Refraction

A plane wave of frequency f, traveling at a speed v_1 in air, will be "bent" as shown in Fig. 14-18, as it is transmitted into glass where its speed is changed to a value v_2. This abrupt bending at the interface between two different transparent media is called *refraction*, a phenomenon which can be understood by considering the wave nature of light. The positions of two successive wave crests are shown in Fig. 14-18, in which a ray is drawn perpendicular to these wave fronts. The key point is that the frequency of the light is the same in the glass as it is in the air (and this is the frequency of the light as emitted from its source). Thus, denoting the wavelength of the light in the glass by λ_2 and in the air by λ_1, we can use Eq.(10.2) to write $v_1 = \lambda_1 f$ and $v_2 = \lambda_2 f$. And from Fig. 14-18(b) we see that $\sin \theta_1 = \lambda_1/D$ and $\sin \theta_2 = \lambda_2/D$. Therefore, we obtain

$$\frac{\sin \theta_1}{\sin \theta_2} = \frac{(\lambda_1/D)}{(\lambda_2/D)} = \frac{\lambda_1}{\lambda_2} = \frac{(v_1/f)}{(v_2/f)} = \frac{v_1}{v_2}, \qquad (14.17)$$

which implies

$$(1/v_1) \sin \theta_1 = (1/v_2) \sin \theta_2. \qquad (14.18)$$

It is convenient to express this relationship in terms of the *refractive indices*, $n_1 = c/v_1$ and $n_2 = c/v_2$, of the two different media. Multiplication of both sides of Eq. (14.18) by c gives

$$n_1 \sin \theta_1 = n_2 \sin \theta_2, \qquad (14.19)$$

a result which is known as *Snell's law of refraction*. This relationship determines the bending of a light ray at the interface of any two media with refractive indices n_1 and n_2; it is valid for light traveling from the second to the first medium as well as for light traveling in the opposite direction.

Snell's law was originally stated simply as an empirical law—an equation in agreement with experimental measurements of the angles θ_1 and θ_2. From such measurements, the ratio n_1/n_2 of the refractive indices of the two media can be calculated. By definition, the refractive index of a vacuum is exactly the number *one* [$n_{\text{vacuum}} = c/v_{\text{vacuum}} = c/c = 1$]. The refractive index of any material medium depends upon the nature of the medium and also upon the wavelength of the light. Typical values for visible light are as follows: crown glass, $n = 1.52$; water, $n = 1.33$; air, under standard conditions, $n = 1.0003$.

Let us return for a moment to the "waves" versus "particle" arguments regarding light. In the 17th century, Huygens showed, as we have, that waves would satisfy Snell's law of refraction. And since, for water,

$$n = c/v = 1.33,$$

the speed, v, of light waves in water must therefore be less than the speed of light in a vacuum, c.

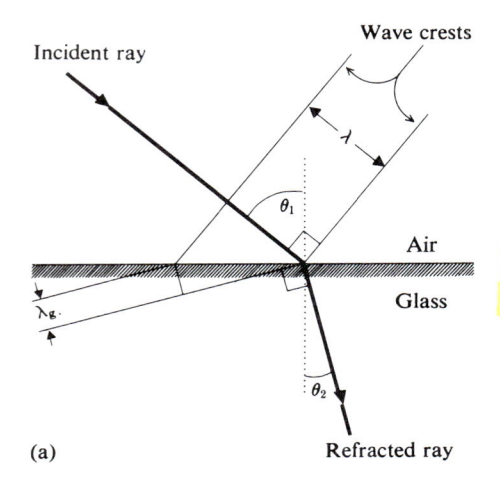

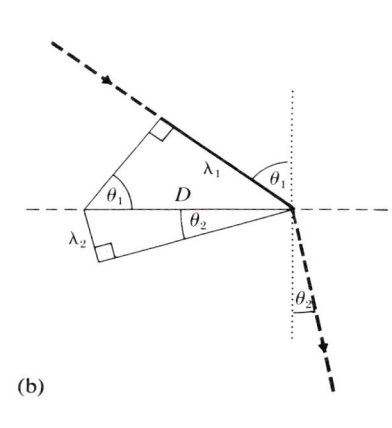

Fig. 14-18 Refraction. (a) The wave travels more slowly in glass than in air. (b) Enlargement which shows that $\sin \theta_1 = \lambda_1/D$ and $\sin \theta_2 = \lambda_2/D$.

Now, if light were a beam of particles *obeying Newton's mechanics*, a light beam would be bent, as observed, at a vacuum-water interface if the water exerted an attractive force on the particles. A particle's component of velocity perpendicular to the interface then would be increased inside the water, and this would explain why its velocity vector points closer toward the perpendicular. But then the particles would travel faster in water than in a vacuum!

Here was an opportunity to decide between Huygens' wave theory of light and the particle theory of light advanced by the great French mathematician and philosopher, René Descartes (1596–1650), and considered more plausible by Newton. One simply had to find out whether light travels more slowly or more rapidly in water than in a vacuum. But this crucial experiment was beyond the capabilities of physicists until 1850, when Jean Léon Foucault devised a successful laboratory method. The experimental result was that light travels more slowly in water than in a vacuum. The prediction of the wave theory was upheld.

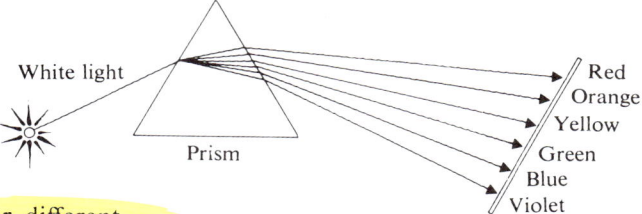

Dispersion and Spectra

The speed v of light waves inside a material is different for different wavelengths. Consequently, the index of refraction ($n = c/v$) of the material has different values for different wavelengths or colors. This phenomenon is known as *dispersion*. It is involved in one of Newton's greatest experimental discoveries, the fact that a beam of white light is *dispersed* into a spectrum of separated colors by a prism (Fig. 14-19).

Fig. 14-19 Ordinary white light (such as from the sun) passing through a glass prism is separated into the spectral colors indicated.

The deflection of a ray by a prism involves *refraction* at two surfaces. The greater the refractive index, the greater the deflection. The value of the refractive index of the prism glass for violet light is greater than its value for red light, and so the violet light in the incident beam is deflected more than the red.

The spectrum of thermal radiation emitted from a hot substance, such as the glowing tungsten wire in an electric light bulb, shows a continuous distribution of wavelengths. One color shades smoothly into another as in the upper portion of the color plate facing page 526. Such a spectrum is termed a *continuous spectrum*.

In contrast, the spectrum of light from an arc or from an electric discharge through a tube of gas like a neon sign is a *line spectrum*, in that only certain separated lines occur, corresponding to a set of discrete wavelengths. (See color plate and also Fig. 14-20 on the following page.) Each type of atom radiates its own characteristic spectrum, a point that was established in the middle of the 19th century by Gustav Kirchhoff

14.3 REFLECTION, REFRACTION, AND DISPERSION

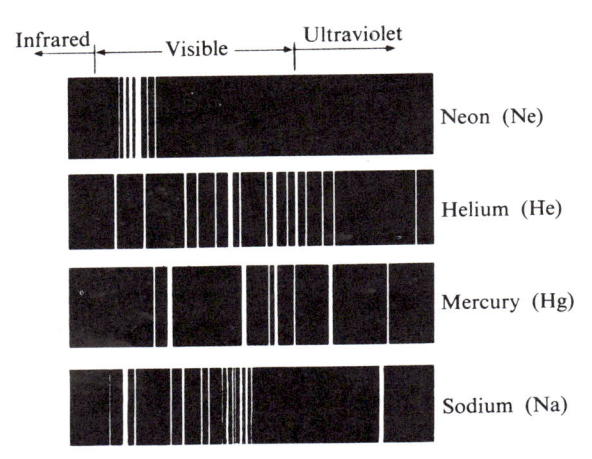

Fig. 14-20 Portions of the bright-line spectra of neon, helium, mercury, and sodium. (Courtesy A. B. Arons, *Development of Concepts of Physics*, Addison-Wesley, 1965.)

(1824–1887) and Robert Bunsen (1811–1899). Consequently, spectroscopy became a primary method of chemical analysis. Stellar compositions are revealed by the spectra of starlight. One element, helium, was first found in the sun! Its spectral lines were detected in sunlight.

After the appearance of Maxwell's electromagnetic theory, it was realized that the emission or absorption of light by an atom was associated with some sort of oscillatory motion of charge within the atom. But until the notion of the *photon* and of *atomic energy levels* was exploited by Bohr, no one saw how to interpret the information about the mechanics of atoms that was contained in their spectra. Bohr's work, mentioned in Chapters 1 and 2, is one of the topics developed in Chapter 16, where we continue the account of man's efforts to understand the nature of light.

14.4 DOPPLER EFFECT IN ASTRONOMY

Doppler Shift in Light

When a light source is moving relative to an observer, there is a shift in the observed frequency analogous to the Doppler shifts of sound waves that were discussed in Section 10.6. For light waves the effect has particular significance because astronomers' measurements of Doppler shifts have yielded so much information about the universe—information that is as astounding as anything encountered in science.

The special theory of relativity must be used to determine the Doppler effect for light waves. We consider a light source which produces waves of frequency f and wavelength λ in a frame of reference in which the source is at rest. If the light source is approaching an observer, then the observer will measure a higher frequency f_o and a shorter wavelength λ_o. The exact mathematical relationship between f_o and f has a quite different form than that for the case of sound waves because there are significant differences in the underlying physics. For light waves in a vacuum, there is *no* "transmitting medium" which provides a special frame of reference. Light travels with a speed c relative to *any* inertial frame, while sound travels with a speed v relative to the transmitting medium but has a different speed relative to an observer moving through the medium. Only the velocity of the light source *relative* to the observer has physical significance, in contrast to the case for sound waves where source and observer velocities relative to the transmitting medium determine the precise magnitude of the frequency shift.

In spite of these differences between light and sound waves, *approximate* formulas for their Doppler frequency shifts have the same mathematical form. The frequency shift of light waves, when the relative velocity is much less than c, is given to a good approximation by

$$\Delta f / f = \text{(velocity of light source relative to observer)}/c$$

where $\Delta f = f_o - f$.

The corresponding approximate relationship for the change in the wavelength of light waves is

$$\Delta \lambda / \lambda = \text{(light source recession velocity)}/c, \qquad (14.20)$$

where $\Delta \lambda = \lambda_o - \lambda$. When the light source is moving away, the observer measures a longer wavelength.

$$\lambda_o = \lambda + \Delta \lambda.$$

Visible light is thus shifted toward the red end of the visible spectrum.

Example 4 Spectrographic analysis (Fig. 14-21) of light coming from a galaxy in the constellation Ursa Major reveals a line in the calcium absorption spectrum with a wavelength of 4.17×10^3 Å. In the laboratory, with the absorbing calcium atoms more or less at rest, this absorption line corresponds to a wavelength of 3.97×10^3 Å. Assuming that the difference in wavelength is a Doppler shift, what is the velocity of this galaxy relative to the earth?

Solution The "red shift" is

$$\begin{aligned}\Delta \lambda &= (4.17 \times 10^3 \text{ Å}) - (3.97 \times 10^3 \text{ Å}) \\ &= 0.20 \times 10^3 \text{ Å}.\end{aligned}$$

Then from Eq. (14.20),

$$\text{Galaxy recession velocity}/c = \frac{0.20 \times 10^3 \text{ Å}}{3.97 \times 10^3 \text{ Å}} = 0.050.$$

Therefore this red shift corresponds to the velocity of recession which is 5.0% of the speed of light.

From similar observations of red shifts in the spectrum of the light coming from different galaxies (Fig. 14-21), an extraordinary conclusion

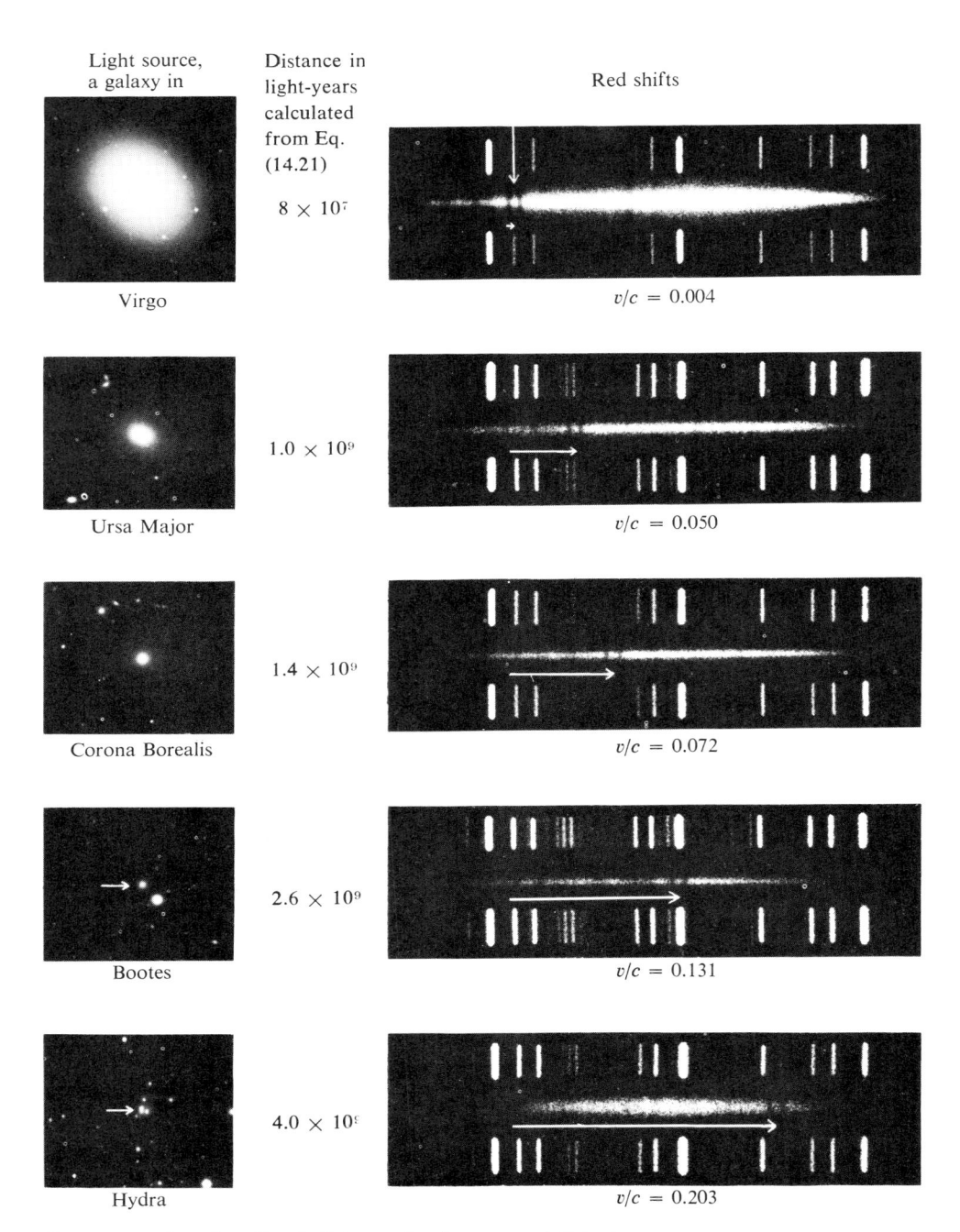

Light source, a galaxy in	Distance in light-years calculated from Eq. (14.21)	Red shifts
Virgo	8×10^7	$v/c = 0.004$
Ursa Major	1.0×10^9	$v/c = 0.050$
Corona Borealis	1.4×10^9	$v/c = 0.072$
Bootes	2.6×10^9	$v/c = 0.131$
Hydra	4.0×10^9	$v/c = 0.203$

Fig. 14-21 Doppler effect in light coming from several distant galaxies. The pair of dark lines (indicated by the arrow) in each photograph on the right is caused by absorption of light by calcium atoms. (Photograph courtesy Hale Observatories.)

emerges. Practically all galaxies seem to be receding from the earth. The most distant galaxies recede with the greatest velocities. This remarkable discovery was made in the 1920s by E. E. Hubble, an astronomer at the Mount Wilson Observatory in California. In 1931 Hubble and his colleague, M. L. Humason, presented evidence that the recessional velocity v_r of a galaxy was approximately proportional to its distance r from us. With modern data an approximate expression of this Hubble-Humason law is

$$r = (20 \times 10^9 \text{ light-years})(v_r/c). \tag{14.21}$$

(One light-year is the distance that light travels in one year: 1 light-year = 9.46×10^{15} meters = 5.88×10^{12} miles.)

These observations imply that the universe is in a state of expansion. The number 20×10^9 years, called the Hubble time, is the time needed for galaxies to reach their present distances if they had always been receding from us with their present velocities. There are many different cosmological models which are consistent with the Hubble-Humason law. One increasingly plausible theory postulates that the present state of the universe has evolved from a "big bang" in the past, at which time all matter was crowded together with the same density as the nucleus of an atom. The time back to the start of the expansion of the universe, as determined from the evolution of the stars and the elements, is approximately 10×10^9 years.

E. Hubble (1889–1953)

Quasars

The largest shifts of wavelengths that astronomers have discovered correspond to wavelengths slightly more than three times their usual value. This occurs in the spectra of light from a *quasar*. Interpreted as a Doppler shift, this red shift corresponds to a recessional speed of $0.8c$.

Quasi-stellar objects (called quasars for short), discovered in 1961, are among the most amazing and puzzling objects in the universe. Their luminosity is up to 100 times that of the most luminous normal galaxy. Some are also strong sources of radiation with radio wavelengths. Although a quasar radiates a power which is perhaps a million million times that of the sun, the apparent size of a quasar is very small compared to a normal galaxy. At the time of writing, although the nature of quasars is unknown, many astronomers are tending toward the opinion that quasars are an unusual phase of normal galaxies.

QUESTIONS

1. Sketch the electric and magnetic field vectors at one instant of time at different points along a line in the direction of propagation of an electromagnetic wave. Indicate the significance of the wavelength λ.

2. A helium-neon laser used in a student laboratory continuously emits 1.0×10^{-3} watt of red light in a beam with a diameter of 1.4 millimeters. Find
   (a) the beam intensity,
   (b) the amplitude of the electric field in the beam.

3. A suitable light intensity at the actual point of operation in surgery is of the order of 17 watts/m² (about 1000 foot-candles, or 1000 lumens/ft² in the parlance of illumination).
   (a) If this light is uniformly spread out over an area of 0.5 m², how many watts of power illuminate this area?
   (b) Calculate the amplitude of these electromagnetic waves.

4. The frequencies of radio waves in the "AM broadcast band" range from 0.55×10^6 hertz to 1.60×10^6 hertz. What are the longest and shortest wavelengths in this band?

5. Channels for FM (frequency modulation) broadcast stations begin at about 88.1 megahertz and continue in successive steps of 200 kilohertz to over 100 megahertz.
   (a) Express 88 megahertz in terms of hertz using exponential notation, i.e., powers of ten.
   (b) Are the electromagnetic waves used for FM longer or shorter than those used for AM (amplitude modulation) broadcasting? [*Hint:* See Question 4.]
   (c) What wavelengths are used by a station broadcasting in the 88.3 megahertz band, that is, between 88.2 and 88.4 megahertz?

6. Use the data of Fig. 14-2 to estimate the wavelength of x-rays used in medical radiography (such as in a chest x-ray, for example).

7. A certain atom radiates orange light of wavelength 6.0×10^3 Å for a time interval of 10^{-8} sec. How long is the emitted wave train? How many wavelengths does it contain?

8. In what way is unpolarized light different from polarized light?

9. Explain how it is that polaroid sunglasses may eliminate some of the glare from a smooth road.

10. The electric vector $\vec{E}$ of a linearly polarized electromagnetic wave has, at maximum amplitude, a horizontal component of 3.0 volts/meter and a vertical component of 4.0 volts/meter. Find the amplitude of the wave transmitted by a polaroid when the transmission axis is
    (a) parallel to $\vec{E}$,
    (b) horizontal,
    (c) vertical.

11. In Fig. 14-3, the intensity I_t of the beam transmitted by the polaroid is related to the intensity I_i of the incident beam by

    $$I_t = I_i \cos^2 \theta,$$

    where θ is the angle between the transmission axis and the electric vector. This equation is known as the *law of Malus*. Deduce this result, using Eq. (14.2) to express I_t in terms of $E_\parallel$, I_i in terms of E, and then noting that $E_\parallel = E \cos \theta$.

12. A light beam has an intensity of 10 watts/m² and is linearly polarized in a vertical direction. Use the law of Malus (see preceding question) to find the intensity transmitted by a polaroid when the polaroid's transmission axis makes the following angles with the vertical: 30°, 45°, 60°, 90°.

13. The beam in the preceding question passes through a polaroid with its transmission axis at 30° with the vertical. The transmitted beam passes through a second polaroid with transmission axis at 90° to the vertical.
    (a) What is the intensity of the beam emerging from the second polaroid?
    (b) What would be this intensity if the first polaroid were removed?

14. A beam of unpolarized light with an intensity of 10 watts/m² is incident upon a stack of two polaroids with an angle θ between their transmission axes. Find the intensity of the beam transmitted by this stack when θ has the following values: 0°, 30°, 45°, 60°, 90°. (Assume that when unpolarized light is incident on a single polaroid, the transmitted beam has an intensity which is 50% of the intensity of the incident beam.)

15. A beam of unpolarized light of intensity I_0 is incident upon a stack of four polaroids, each with its transmission axis rotated 30° clockwise with respect to the preceding polaroid. Express the transmitted intensity in terms of I_0.

16. If, in the preceding question, the angle between successive polaroids were increased to 60°, what would be the intensity of the transmitted beam?

17. The electric field is measured at different points in a certain region which is receiving a wave of amplitude 0.50 volt/meter from one antenna and also a wave of amplitude 0.30 volt/meter from another antenna. The polarization of both waves is the same, and the two antennas are in phase.
    (a) What will be the maximum amplitude encountered and the intensity at such a location?
    (b) What will be the minimum amplitude encountered and the corresponding intensity?

18. Deduce Eq. (14.8). That is, show that if one antenna contributes an intensity
$$I_1 = 1.33 \times 10^{-3} \text{ (watt/volt}^2\text{)} E_{01}^2,$$
then, at a location where the superposition of the contributions from two antennas has an amplitude $E = 2E_{01}$, the intensity is
$$I = 4I_1.$$

19. Show that the intensity at a point where two waves are in phase is given by
$$I = I_1 + I_2 + 2\sqrt{I_1 I_2}.$$
(Notation is that used in Section 14.2)

20. Show that the intensity at a point where two waves are 180° out of phase is given by
$$I = I_1 + I_2 - 2\sqrt{I_1 I_2}.$$

21. Two antennas are 40 meters apart and radiate waves of wavelength 2.0 meters. The antennas are in phase. The intensity is measured 4.0×10^3 meters away at a point which is equidistant from each antenna. What is the minimum distance the detection apparatus will have to be moved parallel to the line joining the antennas to encounter another intensity maximum?

22. In Young's double-slit interference experiment, how does the interference pattern on the screen change when

(a) the wavelength is increased?
(b) the slit separation is decreased?

23. In the double-slit experiment in Fig. 14-9, the parallel slits are 2.0×10^{-4} m apart. The distance between two adjacent bright fringes is 0.94×10^{-2} m, and the screen is placed 3.20 m from the slits. What is the wavelength of the light?

24. If one tripled the distance between the slits S_1 and S_2 in the preceding question, what would be the distance between adjacent bright fringes?

25. Suppose each atom of the NaCl crystal in Fig. 14-11 radiates electromagnetic waves. The vertical distance between adjacent horizontal layers of atoms in the crystal is 2.8 Å. For the case when the oscillations of the radiating atoms are in phase, consider the intensity of the electromagnetic wave at a distant observation point vertically above the crystal. Find the longest wavelength for which the waves radiated from the layer of atoms in the upper horizontal plane will be in phase with the waves radiated from the atoms located in the next horizontal layer. In what region of the electromagnetic spectrum (Fig. 14-2) will this wavelength lie?

26. Give all the wavelengths for which there will be an intensity maximum at the observation point in the preceding question.

27. Give an example of a pair of sources which are
    (a) coherent,
    (b) incoherent.

28. With incoherent sources, $I = I_1 + I_2$. How is it that there is no contribution of the interference term which appears in Eq. (14.7)?

29. Estimate the diffraction spreading of the laser beam of Question 2 by calculating the order of magnitude of the vertex angle of the beam cone.

30. Is the diffraction spreading of a beam behind a slit important for the success of Young's double-slit experiment? Explain.

31. A pinhole of radius 1.0×10^{-5} m is made in the end of a box which is 0.40 m long. The box is aligned so that light from a small distant source passes through the pinhole and strikes the opposite end of the box. What is the approximate diameter of the illuminated area at the end of the box? Assume that the light has a wavelength of 6000 Å.

32. Why is it that we can hear someone who is hidden behind a tree but cannot see him?

33. In many situations a large diffraction spreading is desired in order that a beam will "bend around corners."
   (a) Estimate the diameter of the aperture for which there will be diffraction spreading into a cone with a vertex angle of at least 30° for radio waves of frequency 1.0×10^6 hertz and then for light waves with $\lambda = 5000$ Å.
   (b) Use the answers to part (a) to indicate why radio waves diffract around buildings while light waves do not.

34. What are the dimensions of the refractive index, $n = c/v$, of a medium?

35. Show that the definition of the refractive index of a medium ($n = c/v$), implies that the refractive index of a vacuum is exactly one.

36. The index of refraction of a certain type of glass is 1.55 for green light. What is the speed of light waves in this glass?

37. What is the index of refraction of water for light waves which travel in water at a speed of 2.25×10^8 m/sec?

38. Show that
$$n_1/n_2 = \lambda_2/\lambda_1 ,$$
where λ_1 is the wavelength of light of frequency f in a medium of refractive index n_1, and λ_2 is the wavelength of light of the *same frequency* f in a medium of refractive index n_2.

39. A certain light wave has a wavelength in a vacuum of 5000 Å. If this wave passes into water, what is the wavelength within the water? (Assume that the water has a refractive index of 1.33).

40. A ray of light traveling through a vacuum makes an angle of 53.1° with the vertical. This ray is refracted as it passes into still water, where it makes an angle of 36.9° with the vertical. Use Snell's law to determine the index of refraction of water from these measurements.

41. Consider refraction at an interface between water and glass when a light ray within the glass makes an angle of 30° with the normal to the interface. Within the water, what is the angle between the ray and the normal? Does the answer depend on which way the light is traveling? (Use $n_{glass} = 1.52$, $n_{water} = 1.33$.)

42. A lamp on the bottom of a pool 5.0 m deep is a horizontal distance of 3.0 m from the pool's edge. At this edge, what angle does a ray from the lamp make with the vertical after it has emerged into the air? (Assume $n_{air} = 1.00$, $n_{water} = 1.33$.)

43. Foucault's experimental result that light travels more slowly in water than in a vacuum was interpreted as strong evidence in favor of the wave nature of light. Why?

44. A line in the spectrum of light coming from a galaxy in the constellation Virgo is red shifted to a value of 3984 Å. In laboratory experiments on earth this line has a wavelength of 3968 Å. Find the velocity of recession of this galaxy.

45. Use the Hubble-Humason law to estimate the distance from the earth to the galaxy in Ursa Major with a recessional velocity of 0.050c (Example 4).

46. A spaceship returning to the earth at a speed of 11.0×10^3 m/sec broadcasts a frequency of 50×10^6 hertz. Will receivers on the earth detect a higher frequency than 50×10^6 hertz? What is the magnitude of the frequency shift?

SUPPLEMENTARY QUESTIONS

S-1. Show that, when *unpolarized* light is incident on a polarizer such as polaroid, the law of Malus implies that the intensity of the transmitted beam is one-half the intensity of the incident beam. [*Hint:* The average value of $\cos^2 \theta$ is evaluated in Appendix A.]

S-2. Unpolarized light of intensity I_0 is incident upon a stack of two polaroids whose transmission axes make an angle θ with each other. Express the intensity I of the emerging beam as a function of I_0 and θ, and sketch a graph showing I as a function of θ for values of θ ranging from 0° to 180°.

S-3. It can be shown that the superposition of two sinusoidal waves (of amplitudes E_{01} and E_{02}, respectively) is a sinusoidal wave of amplitude E_0 such that

$$E_0^2 = E_{01}^2 + E_{02}^2 + 2E_{01} E_{02} \cos(\theta_2 - \theta_1),$$

where $(\theta_2 - \theta_1)$ is the phase of difference between the two waves at the observation point. Show that this general result implies that:
(a) when $\theta_2 - \theta_1 = 0$, $E_0 = E_{01} + E_{02}$;
(b) when $\theta_2 - \theta_1 = 180°$, $E_0 = E_{01} - E_{02}$ (where it is assumed that E_{01} is greater than E_{02}).

S-4. From Eq. (14.2), and from the general result quoted in the preceding question, deduce Eq. (14.7) and evaluate the interference term. That is, show that

$$I = I_1 + I_2 + 2\sqrt{I_1 I_2} \cos(\theta_2 - \theta_1).$$

S-5. From the geometry of Fig. 14-9, show that

$$(S_2 L_1 - S_1 L_1)(S_2 L_1 + S_1 L_1) = 2(L\ L_1)(S_1 S_2),$$

and consequently that Eq. (14.11) is a good approximation if $S_2 L_1$ and $S_2 L_2$ are approximately the same as the distance l in Fig. 14-9. [Hint: Pythagoras' theorem gives $(S_1 L_1)^2 = l^2 + (LL_1 - S_1 S_2/2)^2$, and there is a similar equation for $(S_2 L_1)^2$.]

S-6. Consider refraction at an interface when the incident ray is in the medium of higher refractive index (say n_2).
(a) Show that Snell's law implies that the refracted ray just grazes the interface ($\theta_1 = 90°$) if θ_2 has the value $\theta_{critical}$ given by

$$\sin \theta_{critical} = n_1/n_2.$$

(b) The angle of incidence $\theta_{critical}$ is called the *critical angle for total reflection* because as the angle of incidence increases from 0, the intensity of the reflected ray increases and that of the refracted ray decreases until, at the value $\theta_{critical}$, the intensity of the refracted ray is zero and the light is totally reflected. Total reflection occurs for all angles of incidence greater than $\theta_{critical}$. (See Section 15.3 for further discussion.) What is the critical angle for total reflection for light traveling in water toward an air-water interface?

S-7. A lamp is at the bottom of a pool which is 3.0 m deep. When the water is still, what is the radius of the largest circle at the water's surface through which the light can emerge into the air?

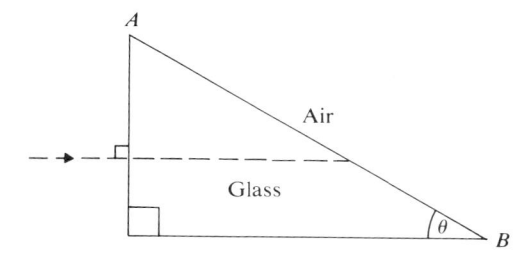

Fig. 14-22 Prism of Question S-8.

S-8. Light is incident normally on a face of a glass prism ($n = 1.52$), as shown in Fig. 14-22. Find the largest value of the angle θ such that this light will be totally reflected from the face AB.

S-9. The difference Δf between the frequency of an incident microwave beam and the beam reflected from a car moving at speed v is measured to monitor highway speeds.
(a) Show that the magnitude of this frequency shift is $2f(v/c)$.
(b) When $v = 30$ m/sec and the incident microwaves have a frequency of 2.5×10^9 hertz, what is the magnitude of the frequency shift?

S-10. The spectrum of a nearby star, Arcturus, gradually changes in a six-month interval from a small shift toward the blue to a larger red shift. In the following six months, the spectrum regains its shift toward the blue. Give a plausible interpretation of this observation.

S-11. Derive the exact formula for the Doppler effect for light for the case when the source recedes from the observer at a speed v. That is, show that the observed frequency is given by

$$f_0 = \sqrt{\frac{1 - v/c}{1 + v/c}}\ f,$$

where f is the frequency of the light source as measured in the frame in which the light source is at rest. One way of proceeding is to do as follows. Take the observer as fixed in the frame OX and the light source as fixed in $O'X'$ which has a velocity v relative to OX.

(a) Show that the distance between successive crests (the wavelength λ) measured in OX is given by

$$\lambda = (c + v)T$$

where T is the source period measured in OX.

(b) Show that

$$T = \frac{T'}{\sqrt{1 - v^2/c^2}},$$

where $T' = 1/f$ is the source period measured in $O'X'$.

(c) Use the above result in calculating the observed frequency from

$$f_0 = \frac{c}{\lambda}.$$

S-12. Show that when v is much less than c, the Doppler effect formula for light, $f_0 = f\sqrt{1 - v/c}/\sqrt{1 + v/c}$, can be approximated by $\Delta f/f = -v/c$, where v is the velocity of recession of the light source relative to the observer. [When x is small, the binomial theorem implies that $(1 + x)^{-\frac{1}{2}}$ is approximately $(1 - \frac{1}{2}x)$ and $(1 - x)^{\frac{1}{2}}$ is approximately $(1 - \frac{1}{2}x)$.]

ADDITIONAL APPLICATIONS TO MEDICINE AND THE LIFE SCIENCES

Diathermy: Microwaves

If a part of the body is placed between two microwave electrodes (or antennas), considerable thermal energy can be generated at subcutaneous levels. The clinical procedure for producing such body heating is called *diathermy*. In general, medical microwave diathermy is indicated in any condition in which heating of subcutaneous and muscular tissues to a relatively greater depth than possible by the direct application of heat would be beneficial; for example, the localized heating of inflammatory regions, such as in certain types of arthritis. Even in ophthalmology, in which considerable microwave hazard to the eye exists, there have been reports of success in the treatment of inflammatory lesions of the eyelid. In these applications relatively large areas are subjected to the radiation and it is customary to use two electrodes. There is, however, use for much more localized heating of very small regions. In this case only one electrode, perhaps as fine as a toothpick, may be employed. An example is in the use of diathermy in the treatment of detached retinas.

In essence, medical diathermy equipment is apparatus which utilizes a radio frequency generator. The operation of medical diathermy equipment may thus constitute a serious source of interference to authorized radio communications services, unless appropriate steps are taken to prevent such interference. The most common frequency employed is 2450 MHz. The power is generally around 125 watts (as compared to megawatts used in communications networks). However, the relatively lower power units used in ophthalmology in the case of retinal detachments do not constitute any problem in communications interference.

The problem of dosage in microwave diathermy has not yet been adequately solved. The total energy applied to the body surface can be determined, but not the actual amount of energy effectively absorbed in the underlying tissue. Studies are proceeding in this field, of course, because of interest in more efficacious treatment and because of the possible dangers inherent in microwave exposure. Some studies have revealed that temperatures of 40°C (104°F) or higher can be consistently obtained in human muscles in some 20 minutes of exposure.

The potential hazard of microwaves was dramatically revealed in 1957 when the death of a patient was reported due to exposure of the patient to high-intensity radar radiation for less than one minute at a distance of 3 meters from the antenna. In another instance an industrial worker who developed lenticular opacities was found to have done bench work near a microwave antenna installation.

Question

(a) A common frequency for medical diathermy generators is 2450 MHz. What is the wavelength in air of this radiation?
(b) What might be expected to happen if a patient receiving diathermy treatment has a metallic implant?

Answer

(a)
$$\lambda = \frac{c}{f} = \frac{3 \times 10^8 \, \text{m/sec}}{2450 \times 10^6 \, \text{sec}^{-1}}$$
$$= 0.12 \, \text{m}.$$

(b) The induced currents in the metal would be expected to cause a pronounced Joule heating effect, larger than that in surrounding tissue. It has, in fact, been demonstrated by investigators that in some cases metallic implants can increase sufficiently in temperature to cause damage to adjacent tissue.

Ultraviolet Radiation

Ultraviolet radiation is commonly produced by using mercury vapor lamps. Radiation of the ultraviolet (UV) wavelength is emitted by mercury atoms following their electrical excitation. The natural supply of UV radiation is, of course, from the sun. While a certain amount of UV radiation is desirable in that it precipitates the formation of vitamin D (a deficiency of vitamin D causes rickets), UV radiation in excess can be harmful. While UV radiation can produce tanning it can also produce severe burning and, on occasion, skin cancer. The eyes are particularly sensitive to UV radiation, a fact to be remembered by those administering UV radiation treatment.

Certain microscopes are built to employ UV radiation for studies in cellular biology. Here UV has the advantageous characteristic of being strongly absorbed by certain components such as protein and nucleic acid. These components therefore show strong image contrast in UV light without the need for staining. This desirable feature of UV light has resulted in the unraveling of the roles of DNA and RNA in living cells before, during, and after cell division, by means of the *ultraviolet microscope.*

Besides direct UV treatment for vitamin D deficiency, lamps emitting UV radiation find additional use in the medical realm. Such lamps may be used to sterilize air in critical areas as the shorter UV wavelengths particularly are germicidal.

Even in diagnosis UV is found useful. For example, in dermatology a high-intensity UV source of 3360 Å permits detection of ringworm of the scalp. Infected hairs fluoresce, while clean healthy hairs do not.

Question

In the diagnostic use of UV radiation the lamp usually has a red-purple filter (composed of a material which transmits UV, of course). What is the purpose of this filter?

Answer

To filter out most of the visible light produced along with the UV because this visible light would mask the visible light due to the fluorescence produced by the UV

SUGGESTED READING

American Scientist. November-December 1974. Sigma Xi, The Scientific Research Society of North America. The article by J. A. Wheeler, "The Universe as Home for Man," gives the reader a stimulating overview of central problems in science and contains particularly valuable sections on astronomy and cosmology.

CRAWFORD, F. S., *Waves and Oscillations, Berkeley Physics Course,* Vol. III. New York: McGraw-Hill, 1968. Interference, diffraction, and coherence are discussed in Chapter 9 of this volume. Recommended for students who wish to look at a more thorough and advanced treatment than is offered in our text.

HODGE, P. W., *Concepts of the Universe.* New York: McGraw-Hill, 1969. Red shifts, quasars, and cosmology are treated in this entirely descriptive little book. There are many excellent illustrations.

HOLTON, G., and D. H. ROLLER, *Foundations of Modern Physical Science.* Reading, Mass.: Addison-Wesley, 1958. Chapters 29 and 30 give an elementary and very readable account of electromagnetic waves and light, replete with historical detail.

Johannes Kepler

15 Optics

Although both Chapters 14 and 16 are concerned with light, the roots of this chapter on optics and the fundamental physical principles involved are to be found primarily in Chapter 14. Those principles are extended in this chapter into some of the more practical aspects of optics. Particular emphasis will be on the study known as *geometrical optics*. In geometrical optics, which is developed in terms of light rays, ray diagrams and related analysis ultimately lead to an understanding of optical devices.

It is difficult to specify a historical genesis of the subject of optics, for it is probably fair to say that man has been keenly interested in the subject of light from the beginning of intellectual awareness. From the technological viewpoint we are better equipped to be specific historically. Many historians suggest that *optometry*, a very practical branch of optics, was practiced at least as early as the thirteenth century.

Here are some of the milestones in the development of optics:

130 AD Ptolemy of Alexandria, a Greek astronomer, discussed the laws of reflection and made a systematic study of refraction. He believed that the eye emits visual rays which are propagated with great speed in straight lines.

1000 AD Alhazen, an Arabian scholar, wrote a book which was the authoritative treatise on optics for several centuries and invented the *camera obscura* (pin-hole camera).

1589 Giambattista della Porta wrote the first systematic treatise on lenses, *Magia Naturalis* (Natural Magic), in which he proposed an optical system of two lenses. He may well have been first to conceive the principle of the telescope.

1590 Z. Janssen of Holland made a compound microscope.

1604 Johannes Kepler, a German mathematician, physicist, and astronomer, summarized and synthesized in his *Ad Vitellionem Paralipomena* the propositions governing the rectilinear propagation of light.

1608 Hans Lippershey, a Dutch spectaclemaker, pioneered the field of lens and telescope manufacture.

"The die is cast; I have written my book; it will be read either in the present age or by posterity, it matters not which; . . ."

Johannes Kepler
(1571–1630)

1610   Galileo discovered Jupiter's satellites with a telescope he had constructed himself.

1621   Willebrord Snell, a Dutch mathematics professor, discovered the law of refraction (but did not publish it).

1650   Pierre de Fermat, a French magistrate whose hobby was mathematics and science, explained reflection and refraction in terms of the "principle of least time."

1660   Advances in glass-polishing techniques led to the manufacture of larger lenses. This in turn sparked ever-increasing interest in optical instruments.

1674   Antony van Leeuwenhoek, a Dutch microscopist employing a single-lens microscope which he built himself, gave an accurate description of the disc-shaped red blood corpuscles of man and other mammals. By the time of his death in 1723, Leeuwenhoek had built 247 microscopes, with magnifying powers ranging from 50 to 200.

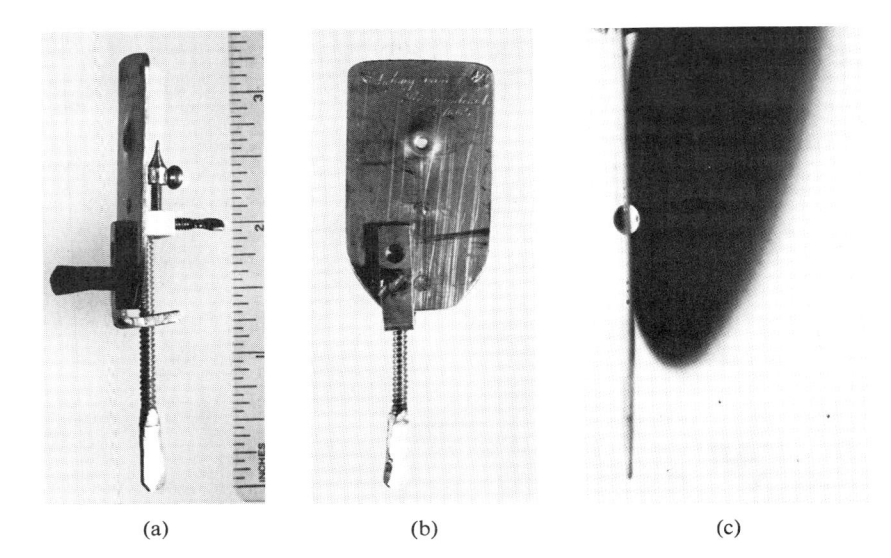

(a) Facsimile of one of van Leeuwenhoek's microscopes. The movable mechanism at the bottom arranges the position of the specimen to be viewed through the fluid drop lens.
(b) Microscope seen from the top. Note the hole filled with the fluid drop that acts as a lens.
(c) Fluid lens (drop of glycerin) in an aperture $\frac{1}{16}$ inch in diameter. This is the kind of lens employed with the van Leeuwenhoek microscope shown in (a) and (b).

(a)   (b)   (c)

1676   O. Roemer, a Danish astronomer, from observations of the periods of revolution of one of Jupiter's moons, established that light has a finite velocity.

1690   Christiaan Huygens of Holland showed in his *Traité de la Lumiere* (Treatise on Light) that a wave theory of light could embrace both the laws of reflection and refraction.

1704   Isaac Newton discovered the dispersion of white light into spectral colors. This and other optical experiments are recorded in his *Opticks*.

1850 Leon Foucault, a physicist at the Paris observatory, demonstrated that, in accordance with Huygens' wave theory of refraction, light traveled more slowly in water than in air.

1873 James Clerk Maxwell, in his *Treatise on Electricity and Magnetism,* identified light as an electromagnetic disturbance in space.

1926 Albert Michelson, the first American scientist to be awarded a Nobel prize (1907), measured the speed of light over a distance of 22 miles between Mount Wilson and Mount San Antonio in California, using rotating octagonal mirrors.

1951 The 200-inch (diameter of concave mirror) reflecting telescope, still the largest in the world, became operational at Mount Palomar, California.

With this bird's-eye view behind us, we now take a closer look at a man encountered in Chapter 2 in connection with his celebrated laws of planetary motion: Johannes Kepler, the scientist who managed to identify the basic physical principles from which the laws of geometrical optics were ultimately derived.

Kepler was the son of a man best described as a "soldier of fortune" who abandoned his family. The self-reliance that young Johannes acquired must have been particularly helpful later when, as a mathematics teacher at the Protestant Seminary in Graz, he had to flee the country when Archduke Ferdinand of Württemburg issued an edict against Protestant teachers. For Kepler this enforced move was fortunate in that it brought him to Prague and into contact with the astronomer, Tycho Brahe. Ultimately, Kepler made his major scientific contributions in astronomy.

As well as laying sound foundations for geometrical optics for the first time, Kepler's optical work of 1604 also clearly separated the physical and physiological problems of optics.

15.1 STRAIGHT LINES

The Rectilinear Propagation of Light

Our treatment of geometrical optics will be based on the assumption that light travels in straight lines; that is, we assume the *rectilinear propagation* of light, as may be inferred from the straight rays shown in Figs. 14-4 and 14-5 on page 528. As has been amply indicated in Chapter 14, this assumption is valid if the dimensions of the apertures through which light passes and the obstacles it encounters are large compared to the wavelength of the light.

While the idea of the rectilinear propagation of light is commonly taken for granted, there may be merit in considering for a moment an easily performed demonstration to support this contention.

Let's look at what happens when some smoke or chalk dust is blown in the way of a beam from a slide projector. You are probably familiar with what is observed in such a situation (Fig. 15-1); your impression is that light travels in a straight line. You might ask why the beam from such a projector lamp is not visible from the side prior to the infusion of smoke or dust. The answer is that the dust scatters the light in the beam, and thus sufficient light is reflected to our eyes to make the beam visible. If there is nothing around to reflect light towards our eyes, we don't see anything. This is why outer space, where there is hardly anything to reflect light, appears essentially dark even though the sunlight is traveling through this space.

One interesting aspect of this rectilinear propagation of light is that by and large we expect to see fairly sharp outlines of objects viewed against a background of light. An intriguing example is the sharply defined outline of the moon seen in a solar eclipse (Fig. 15-2).

Umbra and Penumbra

Another phenomenon explicable in terms of the rectilinear propagation of light is observed when the source of light is relatively large compared

Fig. 15-1 Light beam from a slide projector.

Fig. 15-2 Solar eclipse as seen about noon on March 7, 1970, at Elizabeth City, N. C. The photograph was taken with the aid of an optical system, employing a two-element achromatic lens of focal length 50 inches, devised by R. E. Baran. (Photo courtesy R. E. Baran and M. J. Kesteven, Astronomy Group, Department of Physics, Queen's University at Kingston, Ontario.)

556 OPTICS

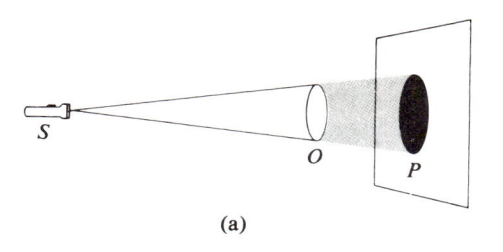

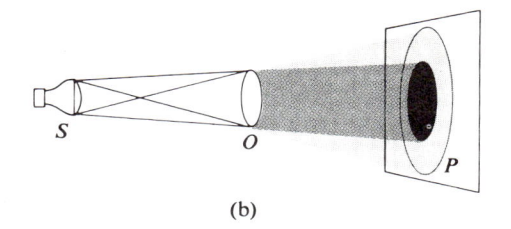

to the object casting the shadow. Note first [Fig. 15-3(a)] that if there is a source of light small compared to the obstacle in the way, we have a fairly well-defined shadow. But when the source is not small compared with the shadow-throwing object [Fig. 15-3(b)], two distinct types of shadows are cast: a dark shadow, the *umbra*, and a lighter shadow, the *penumbra*. Again this phenomenon, evidenced on a large scale by the shadows thrown on the earth when the moon eclipses the sun (Fig. 15-2), is clearly explicable in terms of straight lines, that is, in terms of the rectilinear propagation of light (see Fig. 15-4).

Fig. 15-3 The effect of the size of a light source S on the shadow cast by the object O on the screen P. (a) The "point" source casts a sharply defined shadow. (b) The large source results in a dark shadow (umbra) and a lighter shadow (penumbra).

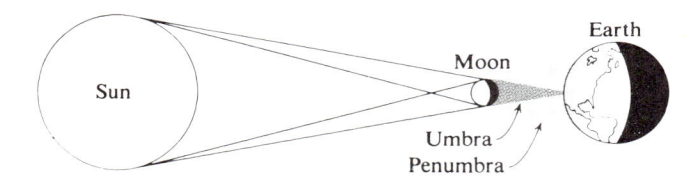

Fig. 15-4 A solar eclipse, showing the region of total darkness, the umbra, and the region of partial darkness, the penumbra.

15.2 REFLECTION

As was emphasized in Chapter 14, the physics of light is the physics of electromagnetic waves. But in much of optics, we do not have to take into account the wave nature of light. A sufficiently accurate description can be given in terms of light rays.

Fermat's Principle

One principle of physics underlies the phenomena of geometrical optics: rectilinear propagation, reflection, and refraction. We shall use this principle, *Fermat's principle*, to make plausible both the rectilinear propagation concept and the law of reflection (Section 14.3): the angle of reflection equals the angle of incidence. That is,

$$\theta_r = \theta_1 \qquad (14.16)$$

where θ_r and θ_1 are the angles shown in Fig. 14-16, which is reproduced here for convenience as Fig. 15-5. We shall then simply proceed with

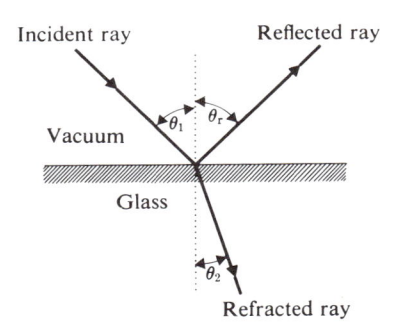

Fig. 15-5 Reflection and refraction at a vacuum-glass interface. This situation is essentially the same as that at an air-glass interface. The angles θ_1, θ_r, and θ_2 are the angles of incidence, of reflection, and of refraction respectively.

refraction on the understanding that those students wishing to pursue this branch of physics further will find that refraction too can be explained in terms of Fermat's principle. Herein lies the true beauty of a scientific idea: a simple, uncomplicated proposition, in some respects no more than a way of thinking, can lead to formulas and then to the actual numbers which are found to describe the workings of nature. Fermat's principle* is simple. To say what it is, or what it means, is really just to give the principle its other name: the *principle of least time*. Fully stated, the idea is that of all possible paths light might take to get from one point to another, light takes that path which requires the shortest time.* Consider Fig. 15-6. If light (traveling in one medium, say air of constant density) is to travel from point A to point B in the shortest time, it must go by the shortest route, which in this case is a straight line from A to B. So immediately we account for the rectilinear propagation of light which Kepler dwelt upon some 50 years before Fermat's generalization.

Now what happens if we require the light to go from A to B but also require that, although remaining in the same medium, it go by way of some reflecting surface? Again the path of least time would be the path of the shortest distance. Fig. 15-6 shows many alternative paths. How is the shortest path to be determined?

Various mathematical tricks could be employed to establish the correct path, but a mechanical analogue will suffice to demonstrate the pertinent geometrical facts. Refer to Fig. 15-7. A string is tied at point B and goes down through a pulley at the "reflecting surface." This pulley can be slid back and forth along the reflecting surface. From this pulley the string goes over a second pulley fixed at point A. At the end of the string a weight is tied to keep the string tight. Now the path of least time from A to B by way of the "reflecting surface" would be the path of the shortest distance, that is, the path for which the length of string from A to B is least. If the pulley is slid back and forth along the "reflecting surface," we see the weight under A going up and down.

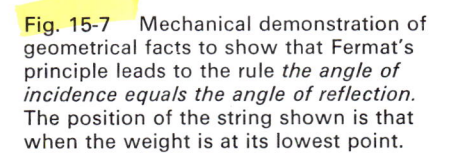

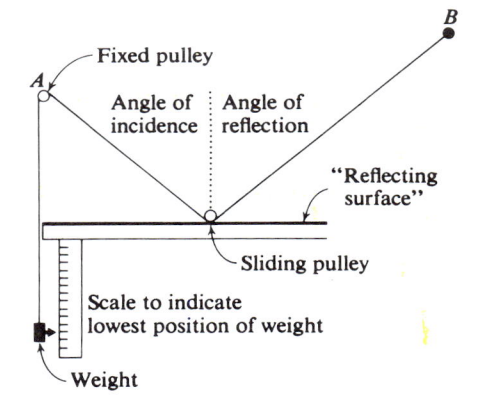

---

*This principle goes a long way back in history. According to some historians, Hero of Alexandria stated a somewhat similar, but more restrictive proposition, some 2000 years ago.

*Fermat's principle can be shown to be a consequence of the wave nature of light. Along the path selected in accordance with Fermat's principle, the waves contributed by all the oscillating electric charges are in phase and the superposition of these waves results in a large amplitude (high light intensity). Along paths which are appreciably different from this "least time path," one encounters out-of-phase contributions which produce a superposition of negligible amplitude (light intensity nil to all intents and purposes).

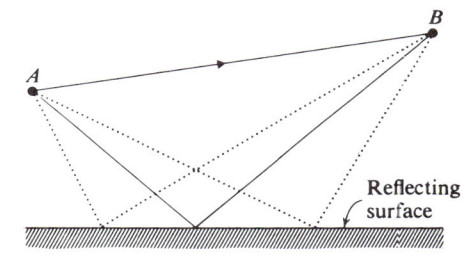

Fig. 15-6 Which path does light take? Directly from A to B it follows the straight line joining A and B. From A to B via the reflecting surface it follows the solid line.

Fig. 15-7 Mechanical demonstration of geometrical facts to show that Fermat's principle leads to the rule *the angle of incidence equals the angle of reflection*. The position of the string shown is that when the weight is at its lowest point.

558 OPTICS

Obviously when the sliding pulley has been moved into the position at which the weight hangs down the lowest under A, we have the least string between A and B. According to Fermat's principle, the string then indicates the path that light rays would take from A to B by way of a flat, or plane, mirror. When this demonstration is carried out, it is found that when the least string is between A and B (when the weight hangs down lowest), the pulley on the reflecting surface is at a position such that the angle of the string on one side of a line perpendicular to the reflecting surface at this position is the same as the angle between this perpendicular line and the string on the other side. The angle of incidence equals the angle of reflection.

Image Formation by a Plane Mirror

The face that greets you when you look in a plane mirror is what is called the *image* of your face. The formation of such an image by reflection in a mirror is easy to understand after one has considered the reflection of different light rays which originate from a single point object such as the point O in Fig. 15-8(a). When these reflected rays are extended backward they intersect at a single point I located behind the mirror. Therefore, rays which actually originate at O, *appear* to emanate from I. The point I is called the image of the object O.

Fig. 15-8(b) shows how the image I can be located by considering just two light rays which emanate from a point object O. One light ray is incident perpendicular to the surface and will be reflected back upon itself. Another light ray which strikes the mirror at an arbitrary point P is reflected in accord with the law of reflection:

$$\theta_r = \theta_1. \qquad (14.16)$$

Elementary geometry now implies that in Fig. 15-8(b) the angles MIP and MOP must be equal and consequently that the triangles MIP and MOP are congruent. Therefore the distances MI and MO are equal; that is,

Distance of image from mirror = distance of object from mirror.

Notice that since the argument above is valid no matter what the location of the point P on the mirror, *all* rays from O which are reflected by the mirror must pass through I when extended backward, as indicated in Fig. 15-8(a).

Images formed by systems of lenses and mirrors are categorized as follows:

1. *Real image*. Different light rays originating from a point object are bent so as to *actually pass through* the image point.
2. *Virtual image*. Different light rays originating from a point object are bent so as to *appear* to diverge from the image point because their backward extensions intersect at this point.

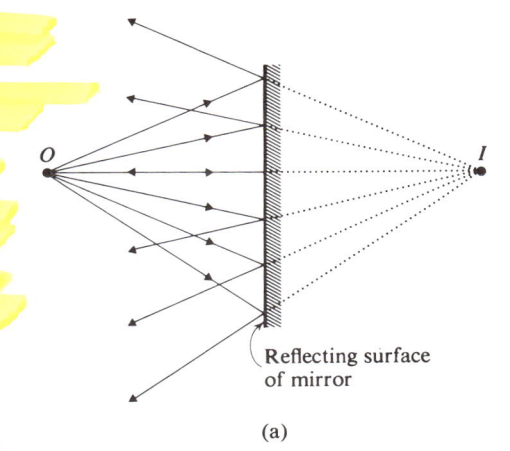

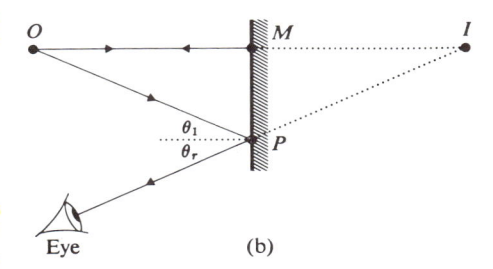

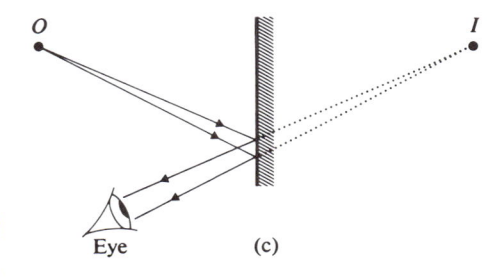

Fig. 15-8 Plane reflection. (a) All rays from point O that are reflected by the mirror *appear* to emanate from a point I. Point I is called the *virtual image* of the point O. (b) Two rays which emanate from the object O and are reflected by the mirror. The extensions of these rays behind the mirror intersect at point I, and the distance $MO = MI$. (c) A small bundle of rays from O enters the eye upon reflection from a small portion of the mirror.

15.2 REFLECTION 559

Evidently an image in a plane mirror is a *virtual* image.

If the eye in Fig. 15-8(c) were turned toward O then light from O would come directly to the eye and the eye-brain system, interpreting this signal, would tell us something about O. In this case we would say that we saw O the *object*. If we now turn the eye toward the mirror, then light comes to the eye as if the object, point O, were at the position of I. Now if the eye-brain system did not know that the eye were facing a mirror, the brain would simply interpret what was being seen at the location of I as an object. But there is no object at the location of I. It is this fact that the object O appears to be at the location of I that we describe by saying I is the *image* of O.

Not only is the light signal to the eye-brain system not coming directly from O, the signal is not even coming from I. The mirror is in the way and no light gets behind or through the mirror. If we were to hold a screen at the location of I, there would be no light on the screen. If we took the eye-brain system around to the back of the mirror there would be no sign of O. So, in fact, the image of O is created by the reflection at and near the point P of a small bundle of light rays [Fig. 15-8(c)]. What we really see is light coming from points near P, and the eye-brain system makes the necessary backward projection to create the illusion that the light signal is coming from point I. The fact that there is actually no light at I due to light from O, but that it appears as though there were, is emphasized by saying I is a *virtual* image.

With an extended object such as a human face, each point on the object has a corresponding image point so that an extended image is produced which is a point-by-point reproduction of the object. The virtual image formed by reflection in a plane mirror is upright and of the same size as the object. As we have stressed, this virtual image is as far behind the mirror as the object is in front.

Left-to-Right Reversal

When you look into an ordinary plane mirror you see an image of yourself (Fig. 15-9) with all the characteristics of the mirror image described earlier: That is, you appear to be as far behind the mirror as you actually are in front of it, and you appear to be as tall and wide as you actually are. You wave to yourself and the mirror image responds instantaneously with a similar greeting. But there is a difference: You waved with your right hand and your mirror alter ego waved with his left. Since you are standing upright and your image is also upright, perhaps there is something different about reflections in the vertical plane as opposed to the horizontal plane. So you lie on your side in front of the mirror. No solution. Your image head is still opposite your head and even the feet are still in the right place. So you make a slight motion with your left foot and your image wiggles its right foot.

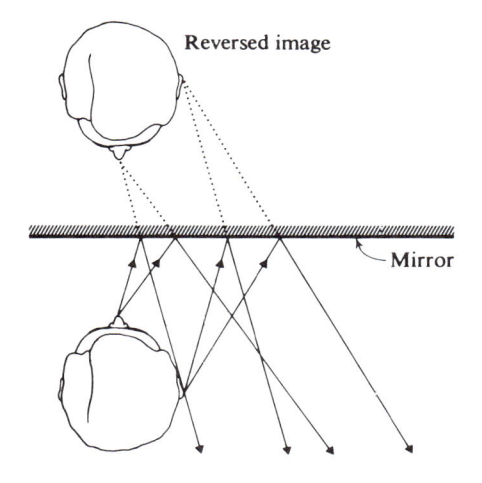

Fig. 15-9 "Left-to-right" reversal of mirror image. Note hair parted on left side while the reflected image shows hair parted on right side.

560 OPTICS

The answer to this peculiar apparent left-to-right reversal without a corresponding up-down reversal of image to object is easily explained in mathematical terms by saying it is not a question of left-right reversal but a question of front-to-back reversal. Let us see if this can be clarified. Imagine standing in front of a large mirror oriented in an east-west direction and you are facing the mirror and looking north. The image is upright. So there is no problem in this regard. The up-down axis seems all right. Then you clench your west fist and the image obliges by clenching its west fist. You kick with your east leg and the image kicks with its east leg. So the east-west axis, like the up-down axis has kept its orientation in our three-dimensional space. But what about the third dimension, the north-south axis? You are looking north and the image is looking south. You take a step closer to the mirror. You took a step north, but the image took a step south. So what has been reversed by the mirror is the north-south axis or, in the more general sense, simply the front-back axis. What the mirror really does is reverse point for point, the pattern or figure in front of it along an axis perpendicular to the mirror. And it is this front-to-back reversal which achieves the apparent left-right reversal without a corresponding up-and-down reversal. The mirror shows no preference for up and down as opposed to right and left, but does produce a front-to-back reversal.

Curved Mirrors

In an analysis of image formation in curved mirrors, the same law holds for reflection as for plane surfaces. However, it is necessary to note that the perpendicular (called the *normal*) to the surface is obviously not parallel to adjacent normals as in a plane mirror. But given any point on a curved surface, a normal can be constructed, and with respect to that normal the angle of incidence equals the angle of reflection.

Curved mirrors are basically of two types according to whether incoming parallel light rays are diverged, or whether these light rays are converged. The mirrors shown in Fig. 15-10 are called *convex* and *concave* respectively. The particular examples shown are parts of spherical surfaces. In such cases some convenient ray geometry pertains. In the convex mirror the rays parallel to the axis perpendicular to the center of the mirror (the *principal axis*) are all reflected in such a way as to appear to come from one point on the far side of the mirror. This point, F', in Fig. 15-10(a) is called the *virtual focus* of the mirror. In the case of the concave mirror these parallel rays are all reflected through a common point, F, on the principal axis. This is the *focus, focal point*, or *principal focus* of the mirror. It is a *real* focus because the light rays not only *appear* to come to or from that point (as in the case of the convex mirror), but they actually do so.

Another useful fact about reflection in spherical mirrors is that if a

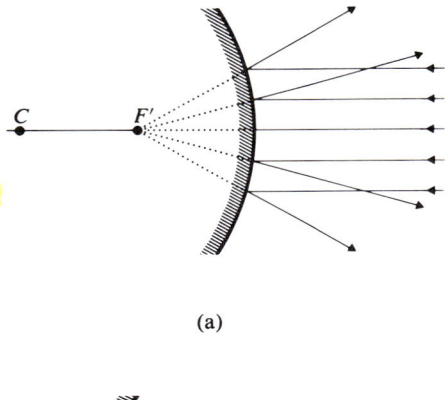

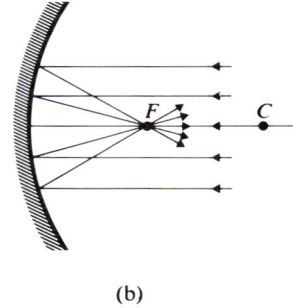

Fig. 15-10 Curved mirrors. These diagrams show cross sections of spherical mirrors. (a) Convex mirror. (b) Concave mirror.

15.2 REFLECTION 561

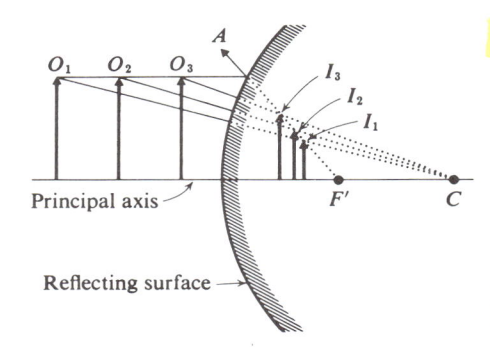

Fig. 15-11 Object-image representation in the case of the convex mirror. Note for each object position O_1, O_2, O_3 two rays are employed to locate the virtual images I_1, I_2, I_3: one ray directed through the center of curvature C and one ray which is incident parallel to the principal axis and is reflected in a direction such that its backward extension passes through the virtual focus, F'.

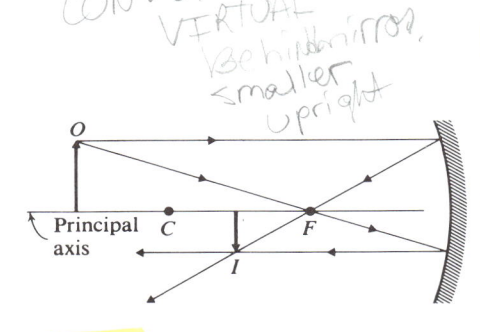

Fig. 15-12 Concave mirror. Position of image formed by the mirror when the object is between infinity and the center of curvature.

ray approaches the mirror along a direction through the center of the sphere of which the mirror surface is a part, then because the ray strikes the surface perpendicularly, it is reflected back upon itself. Construction of ray diagrams is easy when you know this about rays approaching parallel to the principal axis and those approaching through the center of the sphere, called the *center of curvature*, C. It is also useful to know that in spherical mirrors the distance from the mirror surface to F, a distance called the *focal length* of the mirror, is just half the radius of the sphere, that is, half the distance from the center of curvature to the mirror surface.

These ideas can be used to construct an object-image diagram for the case of a convex mirror (Fig. 15-11). Note that, as in the case of the plane mirror, the image is virtual. Regardless of the distance of the object from the mirror, one will always get a virtual, upright, smaller-than-object image. This feature makes convex mirrors useful as rear-view mirrors in vehicles and as observation mirrors in stores. Since images are always smaller than the object, the mirror can cover a wide region of observation.

Consideration of the concave mirror reveals that, unlike the single object-image relationship of the convex mirror, there are six different possibilities (the image being sometimes smaller, sometimes larger, and so on) depending on the distance of the object from the mirror and the mirror's focal length. Only one case is shown in Fig. 15-12, and the reader is asked to use his knowledge of ray optics gained to this point to draw other cases in Questions 16 and 17.

Concave mirrors find wide use as components of optical systems including those of astronomical telescopes. However, in many of these applications the mirror's shape is not exactly that of part of a sphere. Rather the shape is what is called a *paraboloid* (a cross section of such a mirror surface does not look like the arc of a circle; it is a parabola). The reason for this slight departure from the spherical surface is that if the spherical mirror is more than a very shallow dish, parallel light rays striking the mirror do not converge precisely at the same point. This defect is called *spherical aberration* and is prevented by constructing mirrors in a parabolic shape.

Reversibility of Light Rays

So far we have focused our attention primarily on light rays coming from some distance away from the mirror and being reflected. In the extreme we have looked at rays parallel to the principal axis and pointed out that these rays, in the case of a concave mirror, are brought to focus at a single point, the principal focus. Now imagine the light originating at the principal focus. Light rays from the source at F are reflected by the mirror and leave the mirror as a parallel beam. This is shown in Fig. 15-13 for the case of a parabolic mirror. This path is the same as that of the parallel incoming beam. In other words, either way the path is the same; only the direction, indicated by the arrows, is changed. This seemingly obvious and yet very significant fact is referred to as the *reversibility* of light rays. If we can trace rays in one direction, then we also know how they would travel in the reverse direction. This fact is less obvious in refraction, which we are about to discuss, but is still true and, from the standpoint of optical analysis, often very useful.

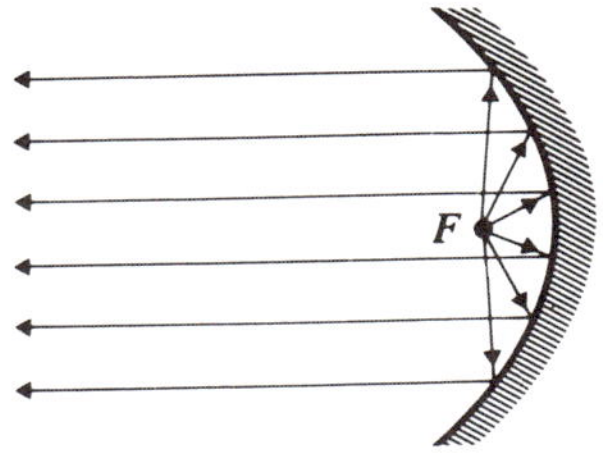

Fig. 15-13 The "reversibility" of light rays: a parabolic mirror such as might be employed in a searchlight. The light source is at the principal focus, F.

15.3 REFRACTION

As noted in the previous section on reflection, the basic law of refraction, Snell's law, can be derived from Fermat's principle of least time, keeping in mind that the speed of light is different in different media. In this section the discussion of refraction that was presented in Section 14.3 will be amplified to describe the role that refraction plays in some practical devices and in certain natural phenomena.

Figure 15-5 indicates that as light passes from one material into another material in which the speed of light is less, the light is refracted toward the normal. In terms of the refractive index,

$$n = \frac{c}{v},$$

light is bent toward the normal when passing into a substance of greater refractive index, and away from the normal in passing into a substance of lesser n.

Glass

Of the various transparent substances of greater refractive index than air, glass is by far the most widely used material in optical instruments. Not always was there available such a great variety of glass so homogeneous in constitution as modern optical glass. Notable progress toward the state of the art today was made by Professor E. Abbe of Jena, Germany. He, in collaboration with Dr. O. Schott, carried out research in this field in the late 1800s to the point that Germany was the acknowledged leader in the field of optical glass until the end of World War II. Modern lens designers are able to choose from more than 100 different varieties of glass, all of which can conform to the following requirements: uniformity of chemical composition, physical homogeneity, absence of color, high degree of chemical and physical stability, freedom from internal strain, and a high degree of transmission of all the wavelengths of visible light. Optical glass is usually made from a very pure (that is, free from iron) form of sand to avoid giving a greenish tinge. This sand is mixed with one or more of a variety of certain metallic oxides to impart the desired properties, both physical and optical.

Table 15-1

Glass*	Refractive index	Chromatic dispersion
Light crown	1.52	Relatively low
Dense crown	1.61	Medium
Borate flint	1.55	Medium
Dense flint	1.96	Fairly high

*The chief chemical difference between crown and flint glass is that calcium oxide is used in the former and lead oxide in the latter. The refractive indices given are those for green light. For our purposes they suffice for calculations involving the mixture of wavelengths known as white light.

Along with the refractive index, another very important characteristic of optical glass is its *chromatic dispersion*. This term refers to the extent to which, in a given glass, the refractive index for, say, orange light is different from the index for blue light; the bigger the difference in these refractive indices, the bigger the chromatic dispersion. In other words, the chromatic dispersion is an indicator of the lens's characteristic as a color separator. Refractive indices for a few types of optical glass are shown in Table 15-1. Ordinary eye glasses are usually made of crown glass.

Total Internal Reflection

In refraction the relation between the angle of incidence and the angle of refraction, the angles θ_1 and θ_2 in Figs. 14-18 and 15-5, is dependent upon the index of refraction. This fact was presented in Chapter 14, when it was also pointed out that the exact relation between these angles is expressed by Snell's law. One of the features of Snell's law is that it indicates that the larger the angle θ_1 the larger the angle θ_2.

Let us now consider light originating in a material such as glass and look at the refraction as the light passes into a medium of lesser refractive index, air. We can construct a ray diagram in which the arrows of the rays involved in refraction, as shown in Fig. 15-5, are simply reversed, in accord with the notion of the reversibility of light rays (see Fig. 15-14). Note that as well as the refracted ray emanating from the glass there is also a ray reflected back into the glass at the glass-air interface. There is always some reflection at any interface between materials of different refractive indices. Fig. 15-15 gives an indication of the approximate relative intensities of the refracted and the reflected light.

Let us imagine what happens as we increase the angle θ_2, the angle of incidence in such a case. The angle θ_1 becomes larger. Not only does this angle of refraction, θ_1, become larger but also the percentage of the intensity of the incident light which is reflected becomes larger. If the angle θ_2 is further increased, the angle of refraction will become so large

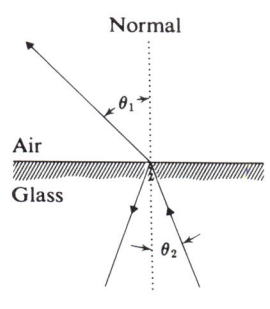

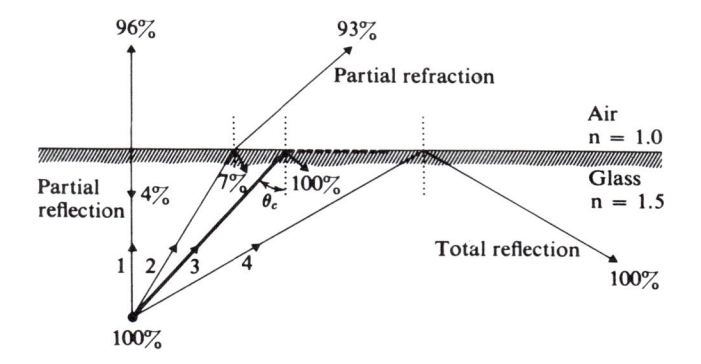

Fig. 15-14 Light ray refracted when going from glass into air. In this case the incident angle is θ_2 and the angle of refraction is θ_1. Notice the refracted ray is bent *away* from the normal.

Fig. 15-15 Light going from glass to air, a medium with a lower refractive index. (1) Angle of incidence equals angle of refraction (both zero); (2) refracted ray; (3) critical angle (θ_c); (4) total internal reflection. Note that as well as refraction reflection occurs. In case 1 the reflected light intensity will be only a few percent of the incident intensity. The percentage of reflected light increases with increasing angle of incidence until at incident angles greater than the critical angle (case 4) 100% of the incident light is reflected.

that the refracted ray will graze the glass-air interface. At a value of θ_2 greater than this, no refracted light is possible. All the incident light is reflected.

This angle of incidence which, if exceeded, will result in *total internal reflection* is called the critical angle. In the situation pictured in Fig. 15-15 the critical angle is about 42°. In the case of a water-air interface the critical angle is about 49°. The smaller the difference between the refractive indices of the two media in question the larger the critical angle. Figure 15-16 shows a simple experiment the reader could perform to show the phenomenon of total internal reflection.

Total internal reflection comes into wide play in optical instruments because, among other reasons, one can achieve 100% reflection of the incident beam. This efficiency cannot be reached with mirrors. A very familiar example is the case of the reflecting prisms of binoculars. Another example is illustrated in Fig. 15-17. Here is a device in which a succession of total internal reflections causes light to "bend" around corners. Various instruments based on this principle are employed in medicine for internal body examinations.

Fiber Optics

This "bending" of light can be taken one step further to transmit pictures from one point to another by means of a bundle of thousands of thin glass fibers, each about 0.01 mm or less in diameter, formed into a flexible cable (see Fig. 15-18). A whole technology, known as *fiber optics,* has blossomed forth to embrace the multitudinous possible applications. It has been found that for the transmission of pictures under certain conditions, the transmission utilizing fiber optics is considerably

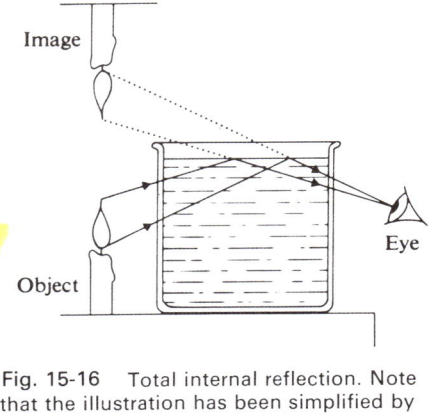

Fig. 15-16 Total internal reflection. Note that the illustration has been simplified by omitting the refractions which occur at the air-glass and the glass-water interfaces. The image in this case is a virtual image.

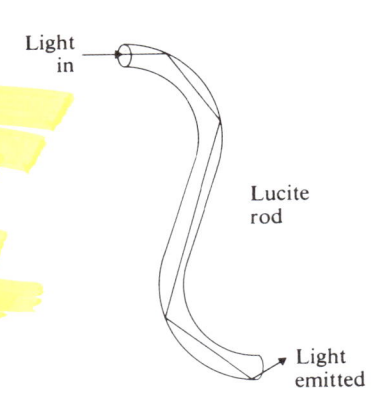

Fig. 15-17 "Light pipe." Light follows the bends in the lucite rod through the process of total internal reflection.

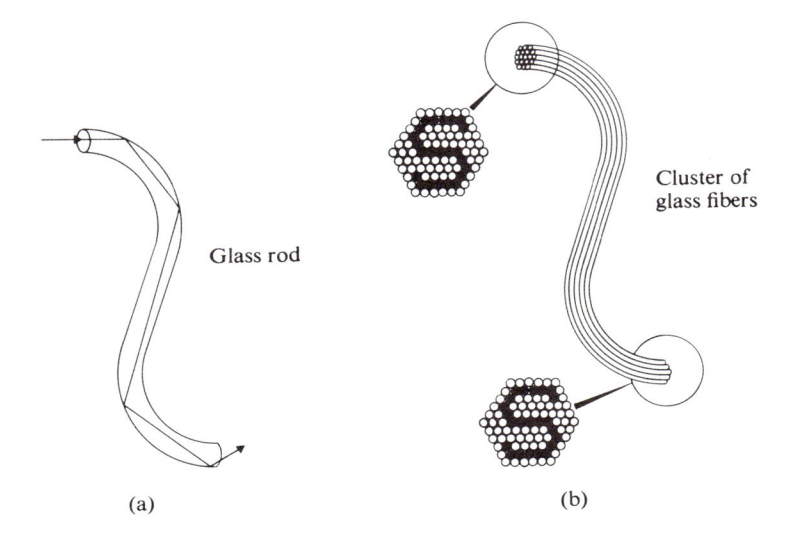

Fig. 15-18 Fiber optics. (a) Total internal reflection as employed in each single fiber. (b) A bunch of fibers capable of transmitting an optical signal, the letter *S* in this case.

15.3 REFRACTION   565

more efficient and more versatile than conventional optical systems (lenses, prisms, and mirrors).

A particularly fascinating possibility made feasible by fiber optics is the study of deep blood vessels. In this application the imaging system is a bundle of fibers contained in a hypodermic needle. The needle and its bundle is attached to a microscope. When the needle is inserted through the skin, a small light bulb at the point where the fiber bundle is coupled to the microscope supplies the light for visualization of tissues beneath the skin of living beings. The complete instrument is called a *fiber-optics hypodermic microscope.*

Atmospheric Refraction

At an interface which separates two media of different indices of refraction, the incident ray undergoes an abrupt change of direction (unless it approaches the interface as a normal). But if there is no well-defined interface, if the index of refraction of a medium changes gradually, as is the case with air which is not all at the same temperature, the incident light ray suffers a continual and gradual refraction. The path of light becomes curved. This is the essence of atmospheric refraction phenomena such as looming, mirages, and "heat waves."

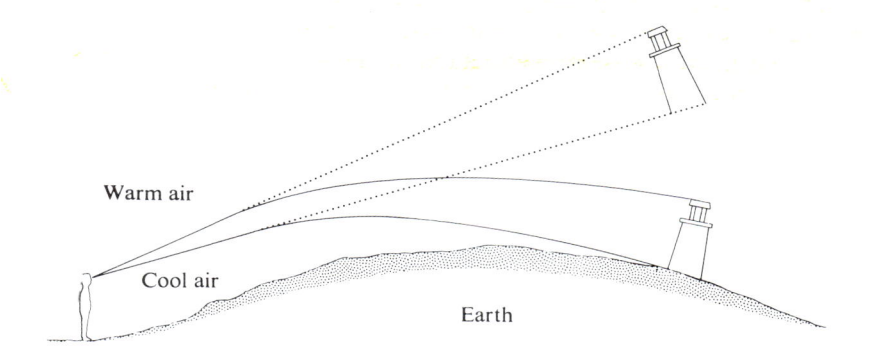

Fig. 15-19 Looming. On some occasions this atmospheric refraction results in "stretching" the appearance of an object which is in the direct line of vision—that is, rather than appearing in a different position, the object appears elongated.

In *looming*, objects such as ships, lighthouses and buildings appear to be suspended in the sky (see Fig. 15-19). On some bodies of water, such as the Gulf of California and Chesapeake Bay, looming is fairly common. The warm air is cooled at the water's surface so that the light rays passing into warmer, less dense air—which has a lower index of refraction than cold, dense air—are continually refracted away from the normal into a downward curved path.

The reverse conditions to those producing looming, that is, warm and less dense air near the surface of the earth with cooler air lying above it, result in a *mirage* (Fig. 15-20). Light from the top part of an object can in such a case reach the observer by the two routes indicated in Fig. 15-20. So the object is seen in its actual position simultaneously with its inverted image below it. If this happens on a desert the unfortunate interpretation of the observer is that there is a reflecting surface, a body of water between him and the object.

A related observation is that of "heat waves." The masses of heated air which rise from a highway on a hot day are of different densities than the surrounding air. The result is an irregular and continuously changing pattern of refraction. This mildly turbulent condition gives the impression that an object, viewed through this moving mix of air of different densities, is dancing. And the turbulent change in refractive indices gives the impression of "heat waves" rising from the surface.

On a larger scale, consider the entire layer of the earth's atmosphere. The density of this air is greater at lower altitudes and in effect fades to more or less zero when we get far enough away from the earth. This variation in refractive index results in the bending of light coming to the earth from outer space (Fig. 15-21). At sunrise and sunset, the sun can be seen when it is actually below the horizon.

As a last point dealing with atmospheric refraction, we consider the *rainbow*. We need to recall that the refractive index for a given medium has different values for different wavelengths of light. We thus get color dispersion (Fig. 14-19). It is this fact that gives rise to the concern over chromatic dispersion in optical glass (Table 15-1). If we keep in mind that long wavelengths are refracted least and short wavelengths are refracted most, then we can understand an important process in the formation of the rainbow in terms of Fig. 15-22, which shows what is

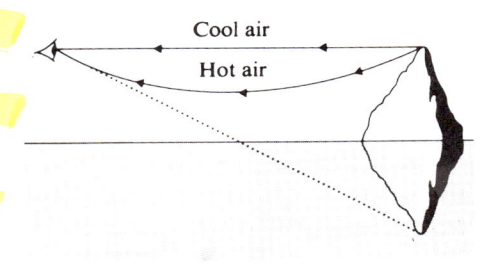

Fig. 15-20 A mirage.

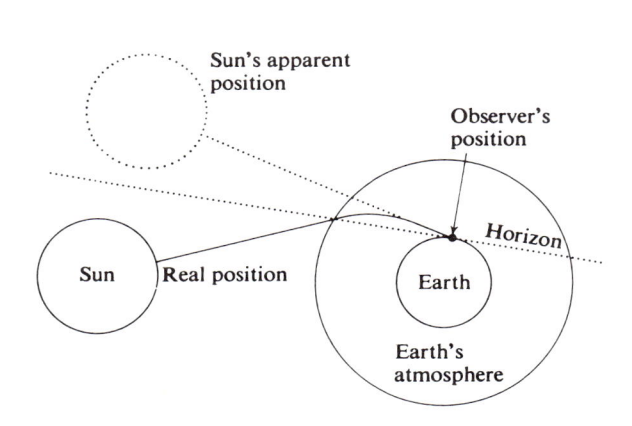

Fig. 15-21 Bending of light rays from the sun by the earth's atmosphere.

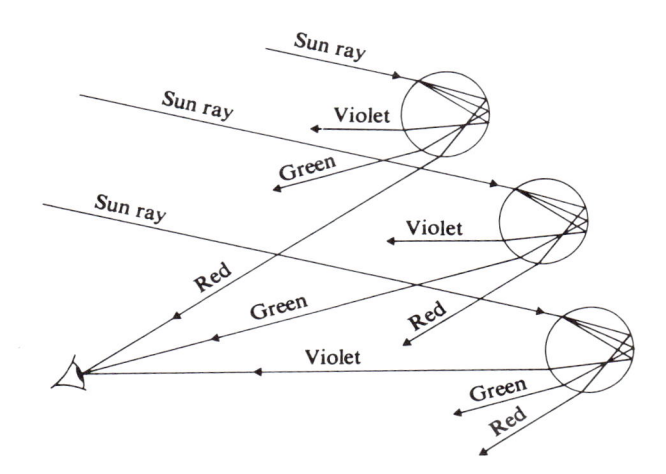

Fig. 15-22 Rainbow formation.

15.3 REFRACTION 567

called a *primary rainbow*. A *secondary rainbow* can be formed from rays which enter near the bottom of a drop and undergo two internal reflections before emerging from the upper part of the drop. In this secondary bow the observer sees the colors in an order reverse to that of the primary bow. (A complete theory of rainbows is rather involved. For further details the reader is referred to books in the Suggested Reading.)

15.4 LENSES

Of the major optical instrument components, mirrors, prisms, and lenses, the last of these are utilized in the most widespread applications.

Basically a lens may be considered simply as a combination of prisms (Fig. 15-23). In practice, lenses are usually ground in spherical symmetry. This produces better, more clearly defined images than would a series of disjointed prisms. Moreover, this shape is very practical for the grinding of lenses. Our analysis will be restricted to that of thin lenses (that is, not more than a few millimeters thick). This makes the analysis simple and yet widely applicable because optical devices commonly employ what one would term a *thin lens*.

Fig. 15-23 The lens as a combination of prisms. Note that parallel rays are brought more nearly to bear at a sharp focal point when many prisms are used. The greatest deviation is produced by the outermost prisms because they have the greatest angle between their refracting surfaces. If we use many thin slices of prisms, we approach the continuously curved surface of a lens.

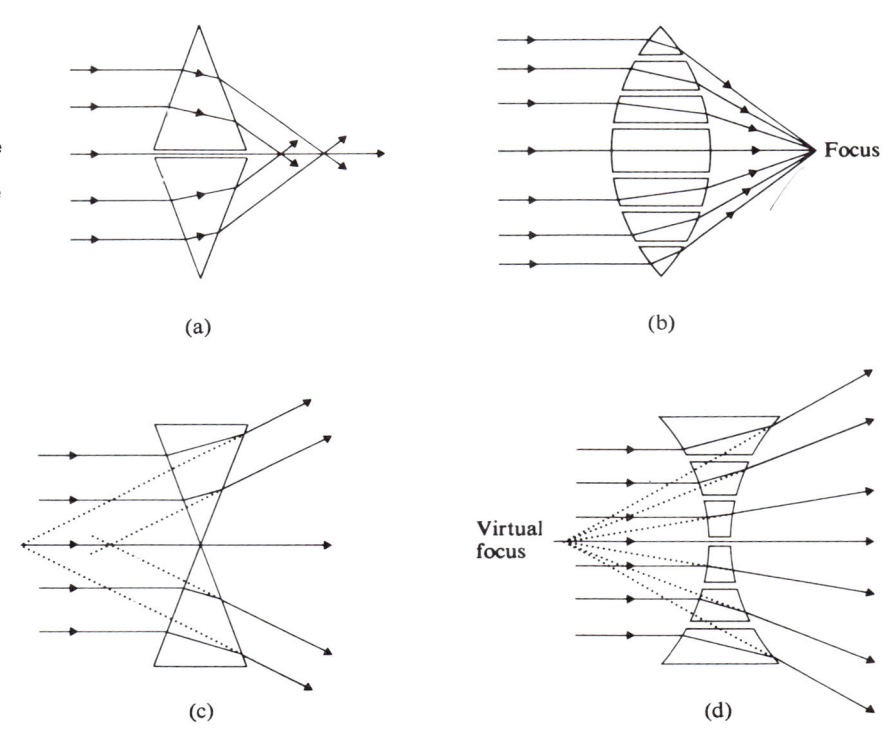

568 OPTICS

Converging Lenses

Converging (or convex, or positive) lenses have the basic characteristic that they are thicker in the middle than they are at the outer edge [Fig. 15-24(a)]. This property causes them to deviate incident parallel light rays so that the rays converge upon a single point.

We introduce some terms used in this chapter. These terms may be identified with reference to Fig. 15-25.

Principal axis. The axis which passes through the center of the lens and is perpendicular to the two faces of the lens at the point of intersection.
Optical center. The point in the lens, on the principal axis, through which rays pass without deviation. A ray through the optical center is undeviated because the surfaces of the lens are parallel at the lens axis. The transverse displacement of such a ray is negligible if the lens is thin.
Principal focus. The point on the principal axis at which incident light rays parallel to the principal axis are brought to a focus.
Focal length. The distance along the principal axis between the center (optical center) of the lens and the principal focus.

The location of images formed by convex lenses is straightforward. The intersection of two light rays from a point on the object is required to locate the corresponding point of the image. The defining properties of the optical center and the principal focus provide the open sesame to

Fig. 15-24 (a) Converging lenses. (b) Diverging lenses.

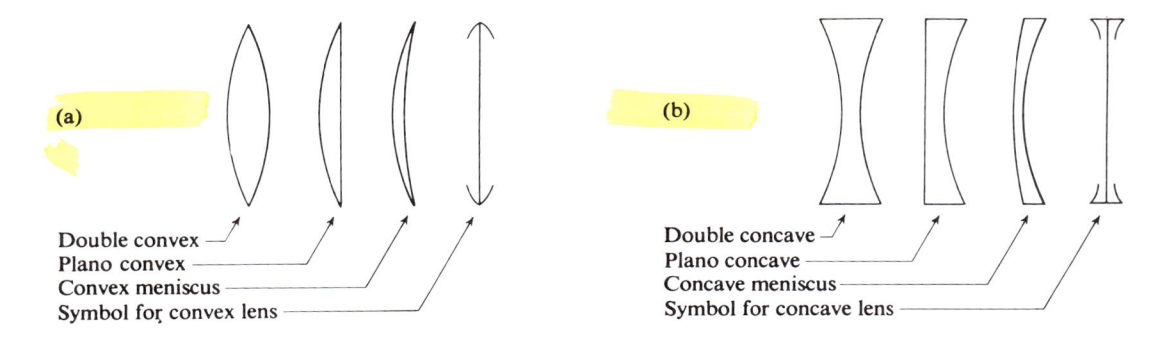

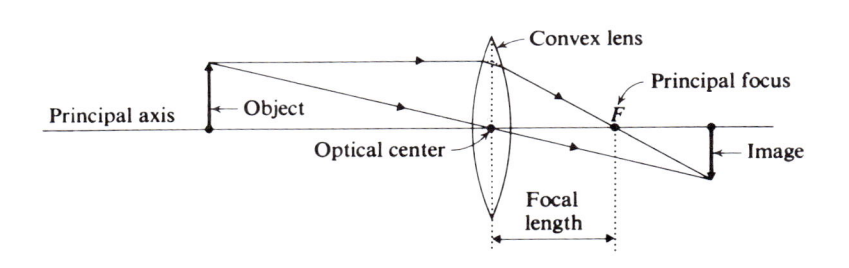

(a) Double convex
Plano convex
Convex meniscus
Symbol for convex lens

(b) Double concave
Plano concave
Concave meniscus
Symbol for concave lens

Fig. 15-25 Ray diagram for a thin lens. Note that although the upper ray is refracted twice, once on entry to the lens and once on leaving the lens, it is common practice (and to all intents and purposes correct for thin lenses) to draw ray diagrams in which the ray follows the dotted line and is bent just once on the central plane of the lens. In subsequent figures we shall follow this simpler schematic approach.

Fig. 15-26 The six possible cases of the thin convex lens. *O* refers to the object, *I* to the image. The image in the last case is dotted to indicate that it is a virtual image, as opposed to all the others which are real images. (The points marked 2*F* indicate a distance twice the focal length.)

the appropriate ray diagrams. Refer to the procedure illustrated in Fig. 15-25. As one ray, a line is drawn through the optical center of the lens (it is not deviated), and as the other ray, a line is drawn parallel to the principal axis (this ray is refracted through the principal focus).

The location and the nature of the image is determined by the distance of the object from the lens compared to the focal length of the lens. It turns out that with *convex* lenses there are six distinct cases (as there are for *concave* mirrors). These various cases are shown in Fig. 15-26 along with a listing of some practical applications.

Cases of convex lens	Position of image	Description of image	Example of application	Optical diagram
Object very far from lens; rays entering lens almost parallel	Almost at *F*	Real, almost a point, inverted	Telescope	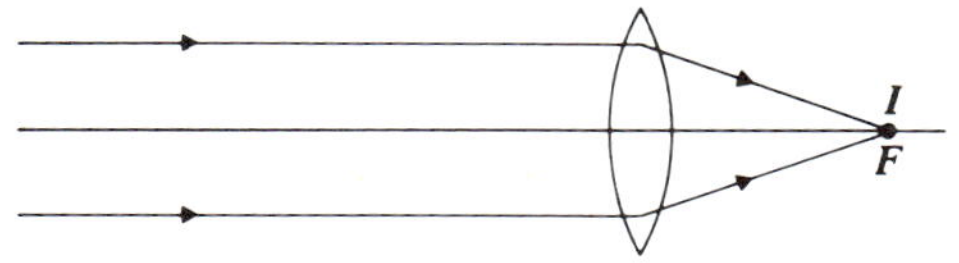
Object farther than twice the focal length	Between *F* and 2*F*	Real, smaller, inverted	Snapshot camera, Movie camera	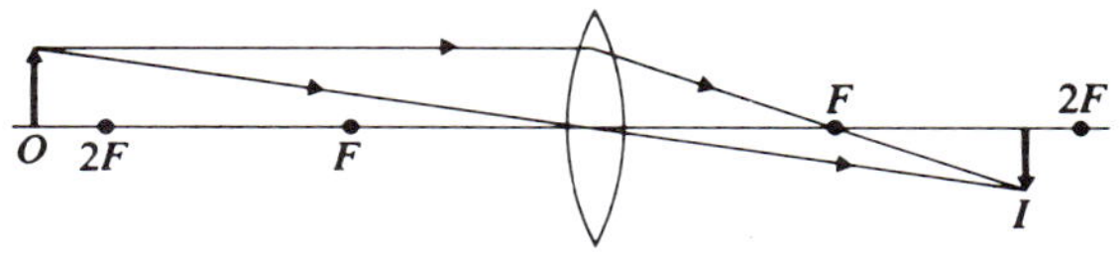
Object at 2*F*	At 2*F*	Real, same size, inverted	Copying camera	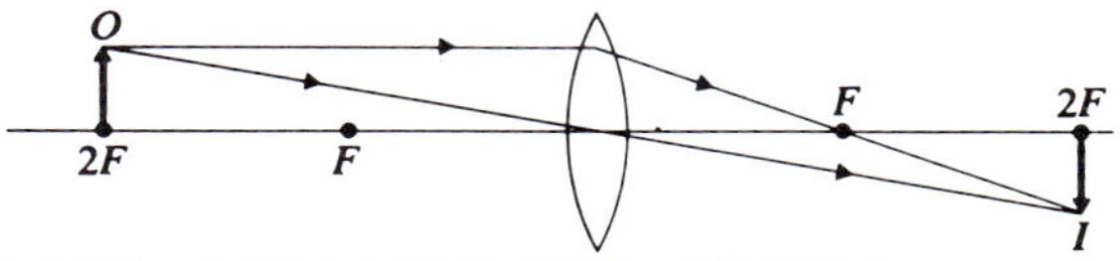
Object between *F* and 2*F*	At greater than 2*F*	Real, larger, inverted	Enlarging camera, Slide projector	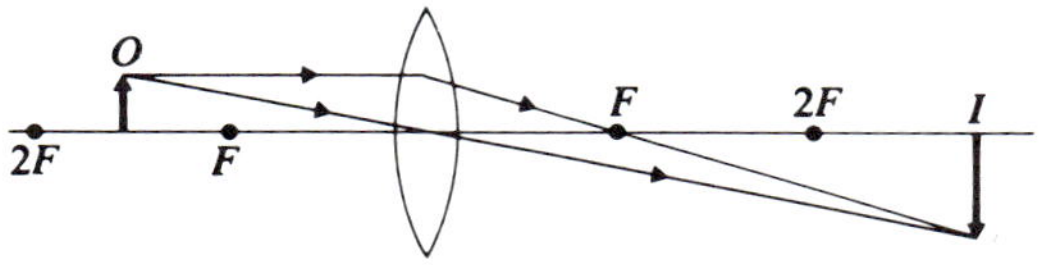
Object at the principal focus	No image, refracted rays parallel		Light source for parallel beam	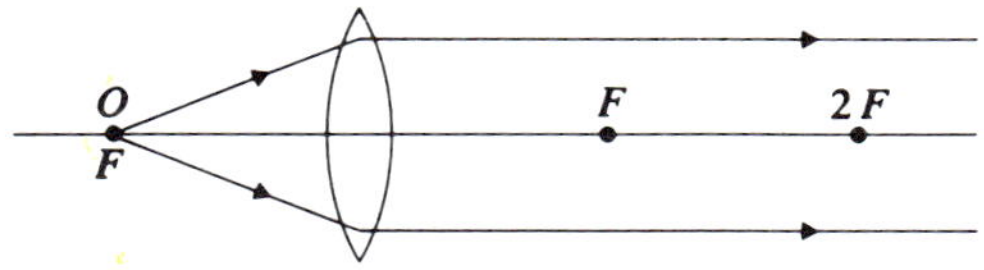
Object between *F* and lens	On same side of lens as object	Virtual, erect, larger	Magnifying glass	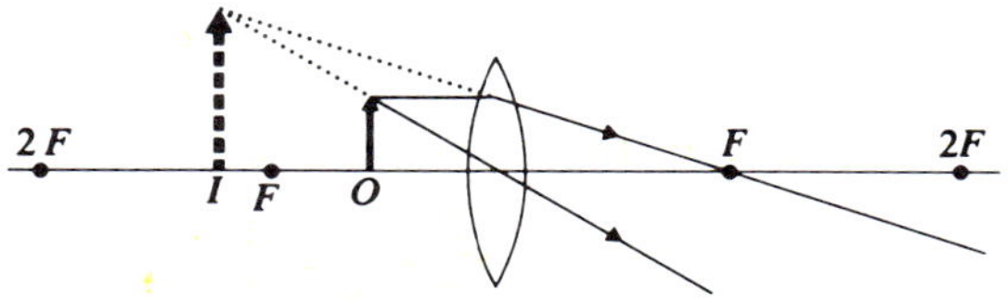

570 OPTICS

The Lens Equation

As an alternative to drawing ray diagrams such as the one shown in Fig. 15-25, the relative positions and sizes of objects and images can be determined by numerical computation when the focal length of the lens in question is known. Using plane geometry, we now derive the necessary equation with reference to Fig. 15-27.

First note that the triangles $O'XO$ and $I'XI$ are similar. Therefore

$$\frac{II'}{OO'} = \frac{XI}{XO}.$$

Also the triangles YXF and FII' are similar. Therefore

$$\frac{II'}{XY} = \frac{FI}{FX}.$$

But $OO' = XY$ because the ray $O'Y$ is parallel to the principal axis, so we can rewrite the last equation of ratios as

$$\frac{II'}{OO'} = \frac{FI}{FX}.$$

But from our very first equation we have

$$\frac{II'}{OO'} = \frac{XI}{XO},$$

so that if we use this relation in the last equation we have

$$\frac{XI}{XO} = \frac{FI}{FX}.$$

At this point, for convenience in subsequent algebra as well as later calculations, we introduce the letter q for the distance XI, the *image distance*; the letter p for the distance XO, the *object distance*; and f for FX, the *focal length*. Using these new symbols our last equation becomes

$$\frac{q}{p} = \frac{q-f}{f}.$$

Division by q yields

$$\frac{1}{p} = \frac{q-f}{qf}$$

$$= \frac{q}{qf} - \frac{f}{qf}$$

$$= \frac{1}{f} - \frac{1}{q}.$$

Transposing the term $1/q$, we obtain what is called the *lens equation*:

$$\frac{1}{p} + \frac{1}{q} = \frac{1}{f}. \tag{15.1}$$

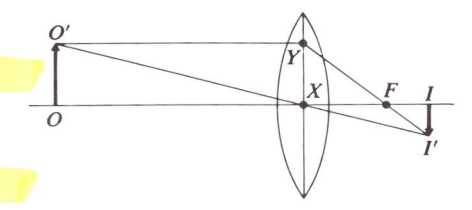

Fig. 15-27 Ray diagram showing image II' formed of object OO'. Optical center of lens is point X. Focal length is the distance XF.

*Different sign conventions are possible. Before using the equations presented in any book on optics, the reader is cautioned to examine the sign convention that the author has adopted.

Although we derived Eq. (15.1) for a special case of a convex lens, it can be shown that this lens equation can be applied to both converging and diverging thin lenses, regardless of object position, provided a certain sign convention* is followed:

1. f is positive for a converging lens and negative for a diverging lens,
2. q is positive for a real image and negative for a virtual image,
3. p is positive for a real object.

Example 1 An object is located 0.20 m to the left of a thin convex lens which has a focal length of 0.040 m. Where will the image be formed?

Solution To avoid working in fractions, we change the respective distances to centimeters. Then

$$\frac{1}{p} + \frac{1}{q} = \frac{1}{f},\qquad(15.1)$$

or

$$\frac{1}{q} = \frac{1}{f} - \frac{1}{p},$$

so

$$\frac{1}{q} = \frac{1}{4.0 \text{ cm}} - \frac{1}{20 \text{ cm}}$$

$$= \frac{4}{20 \text{ cm}}$$

$$= \frac{1}{5.0 \text{ cm}}.$$

Therefore $q = 5.0$ cm and the image is real and formed to the right of the lens (see the second case in Fig. 15-26).

Example 2 If the object of the previous example were placed further to the right, so that it was just 3.0 cm from the lens, where would the image be formed?

Solution

$$\frac{1}{q} = \frac{1}{f} - \frac{1}{p}$$

$$= \frac{1}{4.0 \text{ cm}} - \frac{1}{3.0 \text{ cm}}$$

$$= -\frac{1}{12 \text{ cm}},$$

so that

$$q = -12 \text{ cm}.$$

This means the image is virtual and is on the same side of the lens as the object. (See the sixth case in Fig. 15-26.)

One more useful expression can be obtained from the geometrical relationships portrayed in Fig. 15-27. From the fact that triangles $O'XO$ and $I'XI$ are similar, we get

$$\frac{I'I}{O'O} = \frac{XI}{XO},$$

which in terms of the p and q notation gives us

$$\frac{I'I}{O'O} = \frac{q}{p}.$$

We rewrite this equation in the form

$$\frac{\text{Image size}}{\text{Object size}} = -\frac{q}{p}, \qquad (15.2)$$

introducing a minus sign to permit distinction between upright and inverted objects and images according to the following rule: *Upright sizes are positive, and inverted sizes are negative.* The ratio on the left of Eq. (15.2) is called the *magnification*.

Example 3 In the case of Example 1, if the object were 6.0 cm long, how long would the image be? What is the magnification?

Solution

$$\frac{\text{Image size}}{\text{Object size}} = -\frac{q}{p} \qquad (15.2)$$

so

$$\text{Image size} = -\frac{q}{p} \times \text{Object size}$$

$$= -\frac{5.0 \text{ cm}}{20 \text{ cm}} \times 6.0 \text{ cm}$$

$$= -1.5 \text{ cm}.$$

$$\text{Magnification} = \frac{\text{Image size}}{\text{Object size}}$$

$$= \frac{-1.5 \text{ cm}}{6.0 \text{ cm}}$$

$$= -0.25.$$

In summary, we see that the image is inverted and one-quarter the size of the object.

The reader has now become acquainted with some quite useful mathematical tools for the design of optical systems employing thin lenses.

Diverging Lenses

We now apply Eqs. (15.1) and (15.2) to concave lenses. *Concave* or *diverging* lenses are always thinner in the middle than at the edges

Case of concave lens	Position of image	Description of image	Application	Optical diagram
All the same, regardless of position of object	Same side of lens as object	Virtual, smaller, erect	Telescopic lens combinations Camera viewers Corrective lenses for nearsightedness	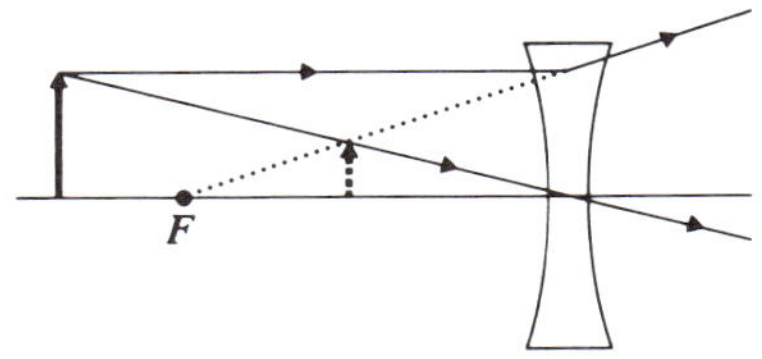

Fig. 15-28 Image formation by a concave or diverging lens.

[Fig. 15-24(b)]. The optical center of these lenses is at the thinnest point of the lens; incident parallel light rays are caused to diverge; the image is always virtual, smaller, and located on the same side of the lens as the object. (See Fig. 15-28.)

Example 4 A concave lens has a focal length of magnitude 4.0 cm. At what distance from the lens will the image of an object 12.0 cm away be formed?

Solution

$$\frac{1}{f} = \frac{1}{p} + \frac{1}{q}. \tag{15.1}$$

Hence

$$\frac{1}{q} = \frac{1}{f} - \frac{1}{p},$$

where $f = -4.0$ cm (because parallel light rays are diverged, there is no real focus; the divergent light only appears to come from a single point, a *virtual focus*; and therefore in accord with the sign convention associated with Eq. (15.1) the focal length is negative), and $p = 12.0$ cm. Substituting these values in the above equation, we get

$$\begin{aligned}\frac{1}{q} &= \frac{1}{-4.0 \text{ cm}} - \frac{1}{12.0 \text{ cm}} \\ &= -\frac{1}{4.0 \text{ cm}} - \frac{1}{12.0 \text{ cm}} \\ &= \frac{-3.0 - 1.0}{12.0 \text{ cm}} \\ &= -\frac{4.0}{12.0 \text{ cm}}.\end{aligned}$$

Therefore

$$\begin{aligned}q &= -\frac{12.0 \text{ cm}}{4.0} \\ &= -3.0 \text{ cm},\end{aligned}$$

where the minus sign indicates the image is virtual, a fact which is in accord with Fig. 15-28.

Example 5 If the object in Example 4 were 8.0 cm tall, how tall would the image be?

Solution

$$\frac{\text{Image size}}{\text{Object size}} = -\frac{q}{p}. \tag{15.2}$$

So

$$\begin{aligned}
\text{Image size} &= -\left(\frac{q}{p}\right) \times \text{Object size} \\
&= -\left\{\frac{-3.0 \text{ cm}}{12.0 \text{ cm}}\right\} \times 8.0 \text{ cm} \\
&= \frac{1}{4} \times 8.0 \text{ cm} \\
&= 2.0 \text{ cm}.
\end{aligned}$$

Again this agrees with Fig. 15-28 which shows the image must be erect and smaller than the object in the case of a diverging lens.

Focal Length

Eq. (15.2) gives us the relative sizes of the image and the object in terms of q and p and thus indicates the *magnification* (= image size/object size) achieved with a given lens and object position. However the ratio q/p employed in Eq. (15.2) is in turn dependent upon the focal length f of the lens, as can be seen in Eq. (15.1). So the crucial issue as regards possible magnification by a lens is the focal length of the lens. It is this characteristic of a lens, therefore, with which one is most concerned when selecting a lens for some specific application.

The focal length of a lens is determined both by the index of refraction of the material of which the lens is composed* and the curvature of the lens surfaces. If we consider the extreme case of a flat disk of glass, we realize that it yields no magnification, that parallel rays incident on one side do not come to a focus at a point on the other side. Now one can say that the radius of curvature of a flat surface is infinite. If one then mentally works upon this glass disk to give it curvature so that its central part is thicker, we visualize that incident parallel rays do come to a focus on the other side, the curvature is "less than infinite," we have a definite focal length, and the lens can be used to magnify. The shorter the radius of curvature, the shorter the focal length and the greater the magnification possible. Lenses vary widely in their focal lengths, which can be very short (say a few millimeters) or quite long (tens of meters or more).

---

*Also of importance is the index of refraction of the material in which the lens is immersed. For example, the focal length of a lens immersed, say, in clear oil is not the same as the focal length of the same lens in air.

The Lensmaker's Formula

Because of the importance of the focal length of a lens and its dependence upon the curvature of the lens surfaces, it would be useful to have a relationship which could be used to calculate the required curvature for a specific lens. There is such an equation: it is called the *lensmaker's formula*:

$$\frac{1}{f} = (n - 1)\left(\frac{1}{R_1} + \frac{1}{R_2}\right) \tag{15.3}$$

where f is the focal length, n the index of refraction (for this purpose usually specified for yellow light emitted by a sodium lamp), and R_1 and R_2 are the radii of curvature of the two lens surfaces (Fig. 15-29). Again a sign convention is required: R is positive for convex surfaces.

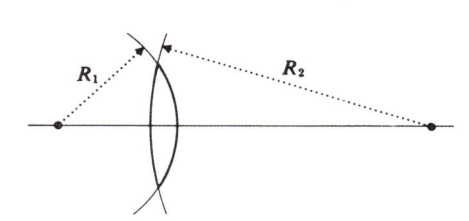

Fig. 15-29 Radii of curvature of the lens surfaces of a double convex lens.

Example 6 A double convex lens (Fig. 15-29) is made of crown glass. If the radii of curvature of both surfaces are to be the same and the desired focal length is 5.0 cm, what must be the radius of curvature to which the lens surfaces are ground and polished?

Solution

$$\frac{1}{f} = (n - 1)\left(\frac{1}{R_1} + \frac{1}{R_2}\right), \tag{15.3}$$

and since $R_1 = R_2$, we get

$$\frac{1}{f} = (n - 1)\left(\frac{2}{R_1}\right).$$

Rearranging this equation to get R_1 on the left, that is, multiplying both sides by $R_1 f$, yields

$$R_1 = (n - 1)(2f).$$

Using the index of refraction for crown glass, from Table 15-1, i.e., 1.52, we get

$$\begin{aligned} R_1 &= (1.52 - 1)(2 \times 5.0 \text{ cm}) \\ &= (0.52)(10.0 \text{ cm}) \\ &= 5.2 \text{ cm}. \end{aligned}$$

Diopters

Because the focal length is such an important characteristic of a lens and because it is frequently simpler to perform calculations if reciprocals like the $1/f$ in Eq. (15.1) are avoided, it is common practice to use a unit called the *diopter* rather than f, the focal length. The *power of a lens in diopters* is defined as the reciprocal of the numerical value of the focal length in meters (Fig. 15-30). (In this case "power" is used in a rather

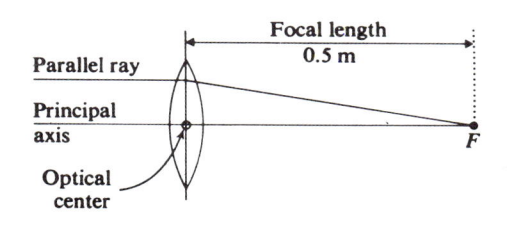

Fig. 15-30 Lens with refractive power of two diopters $\left(\frac{1}{0.5} = 2\right)$.

576 OPTICS

colloquial sense, and not as defined in earlier chapters dealing with energy and power.) So we see that a "powerful" lens, which would have a short focal length, would have a large power expressed in diopters. The human eye has quite a powerful lens system of some 60 diopters.

Example 7 What is the refractive power in diopters of a lens which has a focal length of 5.0 cm?

Solution

$$D \text{ (diopters)} = \frac{1}{\text{focal length in meters}}$$
$$= \frac{1}{0.050}$$
$$= 20.$$

The refractive power of the lens is said to be 20 diopters.

Lenses in Combination

Most optical instruments employ a combination of optical devices: mirrors, prisms, and lenses. The combination of several lenses is very common (microscopes, telescopes, achromatic lenses, overhead projectors, and so on). While we shall not pursue any mathematical examples involving such combinations, we do wish to point out that the reader has by now the basic information required to carry on such further analysis himself. All one needs to realize in this combination of lenses is that if light passes successively through two lenses, the combined effect of the two lenses can be deduced by taking the image which is (or would be) formed by the first lens as the object for the second lens (Fig. 15-31). Optical diagrams can then be drawn and calculations can be made using Eqs. (15.1) and (15.2).

Fig. 15-31 Ray diagram to locate the image of a two-lens system. The *heavy lines* show the formation of image I_1, formed by the lens L_1, and also the final virtual image I_2. This final image is located with *light lines* by using I_1 as the object for lens L_2 and again drawing the two convenient rays (one parallel to optic axis, refracted through principal focus, F_2 in this case; the other, not refracted, through the optical center of the lens). The two lens system shown is essentially that of a microscope.

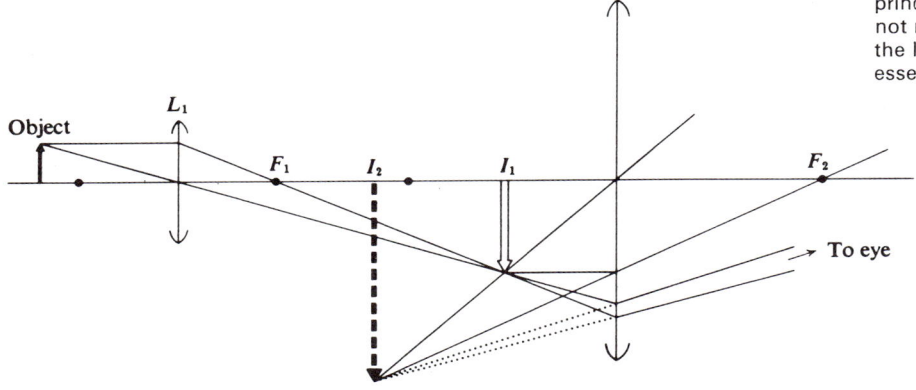

A final point on combinations of lenses: if two (thin) lenses of focal length f_1 and f_2 respectively are placed in contact, it can be shown, using the ideas expressed in the previous paragraph, that the focal length of the combination is given by

$$\frac{1}{f} = \frac{1}{f_1} + \frac{1}{f_2}. \qquad (15.4)$$

This fact is of considerable interest since, as will be indicated shortly, certain lens defects can be remedied by constructing compound lenses consisting of two different lenses joined together. In terms of refractive power in diopters, Eq. (15.4) takes the convenient form

$$D = D_1 + D_2, \qquad (15.5)$$

where D is the power in diopters of the lens combination.

All of these comments regarding the combination of two lenses can be extended step by step to combinations of three or more lenses.

Lens Defects

The relatively simple equations we have presented for lenses were based on a number of assumptions, not always explicitly stated: the lens is thin, all incident rays make small angles with the principal axis, light is monochromatic. These assumptions when not met fairly closely result in images somewhat at variance with those predicted by the simple theory. These image defects are referred to as *aberrations* or, more usually, *lens aberrations*. These aberrations may be classified into six categories: (1) spherical aberration, (2) chromatic aberration, (3) astigmatism, (4) coma, (5) curvature of field, (6) distortion. We shall restrict our comments to only the first two of the above. (Readers who wish to delve further into the subject of lens aberrations can do so in a nonmathematical manner by consulting the book by Gluck cited in the Suggested Reading.)

Spherical aberration results in a blurred image because in a lens with spherical surfaces the rays passing through the edge of the lens are deviated through too great an angle compared with those that pass near the center (see Fig. 15-32). This aberration will increase with an increase in lens diameter. Therefore, at the expense of light intensity, one can reduce the aberration by masking off the outer edge area of the lens, making a small aperture in order to allow light to pass through the lens only near its central region.

Another way of reducing spherical aberration is to grind the lens in a nonspherical way. Modifying the lens curvature into a parabolic shape helps (see Fig. 15-33). But grinding lenses with nonspherical surfaces is a difficult task. Fortunately there is another solution: making a

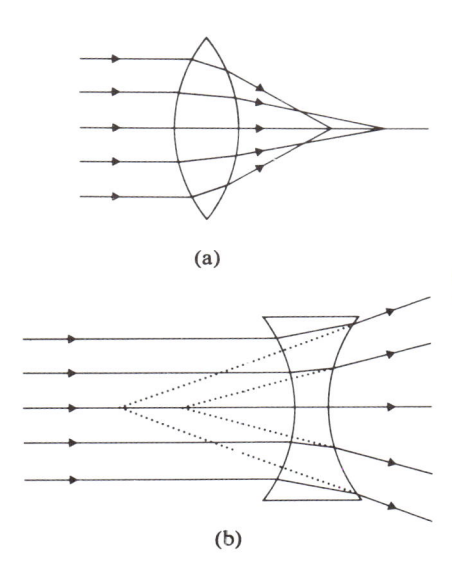

Fig. 15-32 Spherical aberration in the case of (a) a converging lens, and (b) a diverging lens.

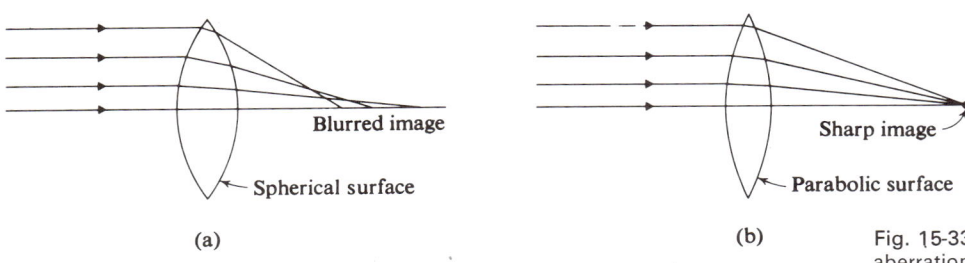

Fig. 15-33 Correction of spherical aberration [shown in (a)] by using a parabolic lens shape as in (b).

compound lens, as shown in Fig. 15-34. Here in what is called a *two-element lens*, the concave curvature of the one element, that is, the diverging lens, will result in spherical aberration in the opposite direction to that of the convex curvature of the other element, the converging lens. In this manner the net aberration can be minimized. Of course the power (diopter values) of the two lenses in combination must be different, or else the power of the combination will be zero.

Because light proceeding from an object is not monochromatic, a lens forms a number of colored images. This defect in image formation is called *chromatic aberration*. It arises from the fact that when white light passes through any single lens, the violet light is deviated more than the red. That chromatic aberration should be expected is to be inferred from our discussion of dispersion in Chapter 14. Moreover, since the refractive index, n, is different for different colors of light the lensmaker's formula

$$\frac{1}{f} = (n-1)\left(\frac{1}{R_1} + \frac{1}{R_2}\right) \qquad (15.3)$$

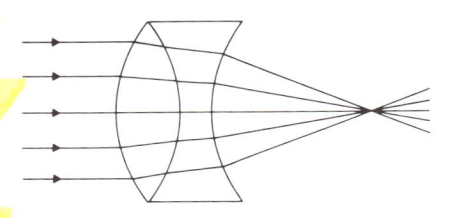

Fig. 15-34 Elimination of spherical aberration by use of a compound lens composed of a converging lens on the left and a diverging lens (of lesser power) on the right.

implies that a lens has a different focal length for each different color.

Chromatic aberration can be corrected to a large extent by constructing multiple element lenses. In this case the combined elements need each to be lenses of materials of different chromatic dispersion (Table 15-1) as well as of different powers. Complete correction of this defect can be obtained only for selected colors; there always remains a small aberration for the remaining colors.

As well as the various aberrations, there is another problem to be considered with lenses. Just as was shown in Fig. 15-5, aside from refracted rays one must also consider reflection at the glass surface. In some cases this reflection actually produces an undesirable visual effect and, in any event, the reflection results in a loss of light intensity through the lens. This loss is no problem with an ordinary magnifying glass, as the loss involved is only a small fraction of the intensity striking an air-glass interface. However, in devices such as cameras, telescopes, and microscopes many lenses and prisms may be involved so that the total loss of intensity becomes appreciable. Consider, for example, an ordinary pair of binoculars. There may be ten or more air-glass interfaces. If the intensity loss at each interface is only 5% then almost half the incoming light never gets to the viewer.

Fortunately this problem is easily remedied. Recall the subject of interference discussed in Chapter 14. With this in mind, we coat a lens with a thin layer of a transparent substance, such as magnesium fluoride, having an index of refraction between that of air and that of glass. Then there are two reflected beams of approximately equal intensity: one reflected at the air-coating interface and the other reflected at the coating-glass interface. If this transparent coating has a thickness of $\frac{1}{4}$ of the wavelength of light waves in this medium, then light reflected at the interface between the coating and the glass will have traveled $\frac{1}{2}\lambda$ by the time it joins the light reflected at the interface between the air and the coating. This path difference gives us destructive interference between these two reflected light beams because now at the surface of the layer a trough and a crest combine. In this circumstance, the superposition of these two reflected waves is practically zero. There is no reflected beam. Almost all the energy in the incident beam is now transmitted through the coated lens. Since most optical equipment is used with white light, it is reasonable to ask what wavelength one chooses for determining the thickness of the nonreflecting coating. A compromise is in order. A wavelength near the middle of the visible spectrum is chosen, corresponding to the color green. This means red and violet are least affected by the destructive interference and therefore these colors give coated lenses a slightly purplish appearance.

15.5 THE EYE AND SEEING

It seems almost traditional to describe the camera and then show how the eye can be compared to a camera. We follow a somewhat different procedure using a very simple camera, the pinhole camera, to single out some of the problems the eye has to solve. Then after discussing the eye, we shall lean upon that discussion to abbreviate the matter of the camera when it comes up in Section 15.6 on optical instruments.

Camera Obscura: The Pinhole Camera

Camera obscura literally means a "dark chamber." If light is permitted to enter such a dark chamber through a tiny hole in one wall, an image of the light source will be formed on the opposite wall. This idea is exactly what is exploited on a smaller scale in the *pinhole camera* of Fig. 15-35. An even simpler arrangement (Fig. 15-36) is one which could be used to observe a solar eclipse.

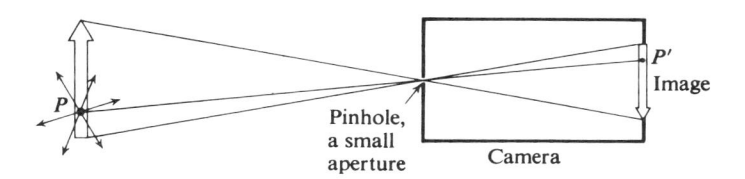

Fig. 15-35 The pinhole camera. If the object is illuminated, light is reflected from all points of it and, as shown at P, is reflected in all directions. Only one "ray" from each point, however, can reach the screen through the pinhole (providing the pinhole is very small), thus forming a well-defined image of P at P'.

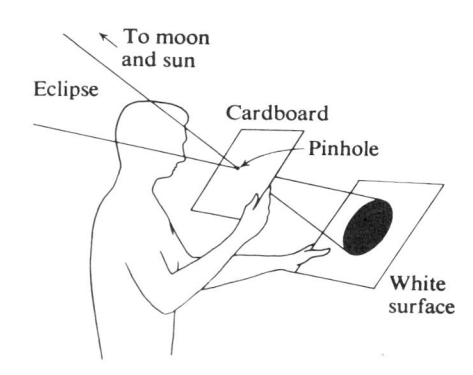

Fig. 15-36 Using the pinhole camera idea to see an eclipse of the sun. Watching a solar eclipse is dangerous because looking at the sun can cause irreparable damage to the eye. Thus a viewing procedure of the sort suggested above is recommended.

Only if the pinhole of the camera is very small will the image be sharply defined. If the hole is made large (Fig. 15-37), more light will gain admission to the camera. But, as is shown in Fig. 15-37, divergent rays emanating from a single point on the object will be admitted. This results in the image of any particular point on the object being spread out, causing the full image to be composed of many overlapping regions: A general blurring of the image occurs and the blurring will be more severe the larger the hole. It seems therefore advisable to keep the pinhole small. But if the hole is kept small there is a decided limitation on the intensity of the light in the camera. The image will not be very bright. For this reason, if one actually wants to "take a picture" with such a camera the film will have to be exposed for a long time. A dilemma: We may either have a sharp, dim image or a bright, blurred image. This dilemma can be resolved by using a lens, as is shown in Fig. 15-38.

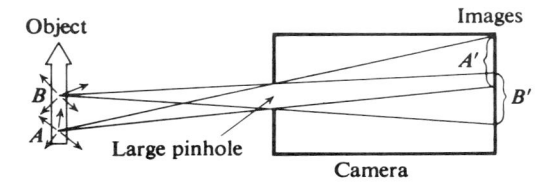

Fig. 15-37 The effect of a large hole in a pinhole camera. Rays of light from any point A enter the camera and produce an image A', which is larger than a point. Rays from any other point B on the object produce another region of image, B'. Two adjacent points on the object are shown as overlapping regions of image. Since this happens for all points on the object, the resulting image is blurred.

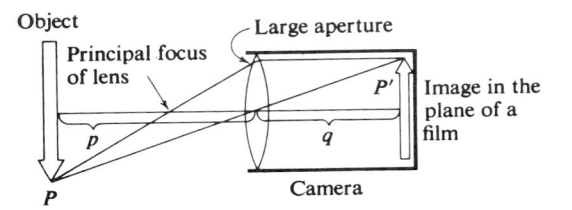

Fig. 15-38 The lens camera. All rays of light which enter the lens from any point P on the object reach the screen at a point P'. The image is therefore sharply defined, while nevertheless a large opening or aperture (as large as the lens) can be employed. Thus much light enters the camera.

15.5 THE EYE AND SEEING

But there is one more difficulty. Examination of Fig. 15-38 in conjunction with Eq. (15.1) indicates that for a given lens (i.e., fixed focal length) the image will be focused clearly at a specific image distance q for a given object distance p. The problem engendered by the desirability of a system capable of focusing a sharp image for various object distances thus requires either a variable focal length f or a variable camera lens-to-film distance, or both.

Optics of the Eye

All the previously cited difficulties, and more, are resolved automatically by the human eye.

As an optical system the eye is unique in the incredible intricacy with which it performs functions which range from activities like that of a stereoscopic movie camera (in color, of course) to that of a computer. In simplified terms the eye can be considered as a lens system with a refractive power of about 60 diopters. Of this power, some 45 diopters is due primarily to the cornea (Fig. 15-39). The remaining refractive power is provided by the lens, which has the happy faculty of being able to change its curvature. Thus the eye has a lens system of variable power, that is, of variable focal length, and so the difficulty of focusing to get sharp images with a fixed distance q while having variable distances p is overcome. This property of the eye to deal adequately with various object distances p is called *accommodation*. (See Fig. 15-40.)

In young people accommodation may extend from infinity to 10 cm or less. As one gets older accommodation usually gets progressively more restricted. With many people a close-range focusing problem begins at about age 40.

To control the amount of light entering the eye, the *pupil* is made larger or smaller. As an aside we might note that the pupil size is not influenced by light intensity alone. Emotion and drugs also have an effect. In fact, strong *miotics*, that is, drugs which make the pupil small, find use in medicine. Such drugs can reduce the pupil of the eye to the point where the eye behaves like a pinhole camera. This allows clarity of vision despite refractive defects in the eye. Of course there is the obvious price to pay: Decreasing the pupil to this extent limits the light entering the eye; the patient cannot function normally in dim light.

Adjusting the size of the pupil is achieved by muscles which are capable of contracting or expanding the *iris*. This control of the aperture performs another useful function. By preventing light from going through the edges of the eye lens system, spherical aberration is reduced.

Light which has passed through the lens system is focused on the *retina*, which is the screen (or the film),* that is, the optical detector of the eye system. The light signal is then transmitted by nerve fibers from the retina. These fibers come together to form the optic nerve which transmits the light signals to the brain.

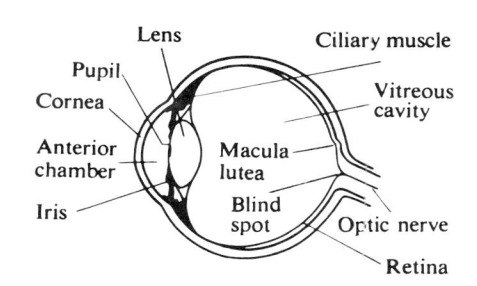

Fig. 15-39 The eye.

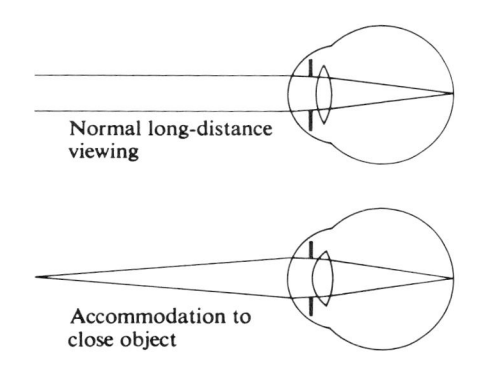

Fig. 15-40 Accommodation.

*Present theory suggests that the function of the retina is more complex: The retina is not simply a signal receptor, such as a screen or film, but is involved, along with the brain, in the actual analysis of the light signals received.

In different parts of the retina there are different structures. *Rods* are the structures whose density relative to the other visual cells, the *cone* cells, is greatest near the periphery of the retina. While rods are a type of visual cell, they are not as capable of sharp vision as are the cone cells. Cone cells are most numerous in a depression in the retina known as the *macula lutea* (Fig. 15-39) which in turn has a tiny pit, the *fovea centralis*. This is the region of most acute vision. Acuteness of vision falls off the greater the distance of the image from the foveal region of the macula lutea. In the normal act of looking, quite without realizing it, we direct our line of sight such that the image will be formed on the macula lutea. It follows that in order to see large objects clearly we need to continuously scan the object with the macula lutea in "quantum" jumps. Reading is an example. We see and digest a few words and then jump to the adjacent group of words.

All this also explains the relative inefficiency of peripheral vision. When we see something "out of the corner of our eye" we are in effect obtaining an unsharp image *on* the retina but *off* the macula lutea area; that is, rods rather than cones are the predominant signal receptors. It should be noted that when we speak of inefficiency of vision in the above sense, we refer to the inefficiency with respect to observing clarity of form. On the other hand, the area peripheral to the macula lutea is much more efficient than the actual foveal region in detecting very faint light. This fact has long been exploited by those interested in observing stars. Focusing on a star and then glancing just slightly to one side of it makes it possible to note if the star has a fainter neighbor.

Each eye has a blind spot. There is a point on the retina where the nerves that carry the information to the brain leave the eye (Fig. 15-39), and at this point there are no visual cells.

There are other limitations to vision imposed by various factors such as, for example, the fact that the visual cells, rods and cones, are spaced apart in the retina. While this spacing is only of the order of a few thousandths of a millimeter, it is nevertheless significant. Because of the various limitations, the best the eye can do under ideal laboratory conditions is distinguish between images which are of the order of 0.003 mm apart on the retina. This corresponds to two points 1 cm apart at a distance of 50 to 60 meters. These visual limitations are obviously more restrictive with small images on the retina than with larger ones. Objects that are either very small or very far away give us trouble, and so for centuries many men have been preoccupied with devising and improving microscopes and telescopes.

Binocular Vision

Not only in automatic adjustment to varying object positions and varying light intensities, but also in the matter of depth perception is the eye the prototype of optical instrumentation. Our appreciation of the

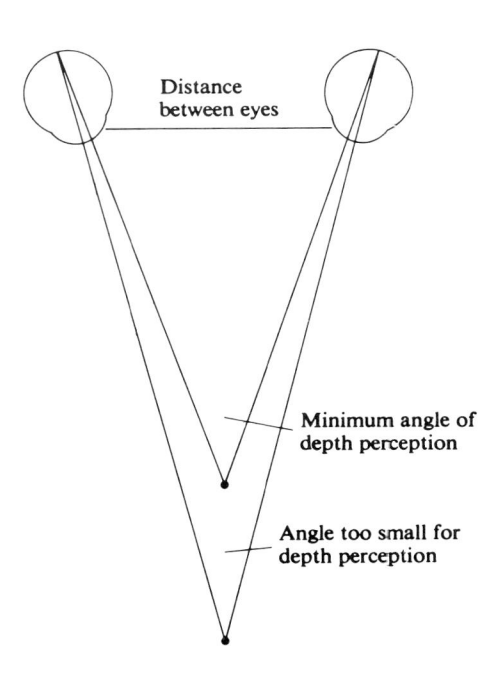

Fig. 15-41 Binocular vision: there is a limit to the stereoscopic acuity of one's eyes.

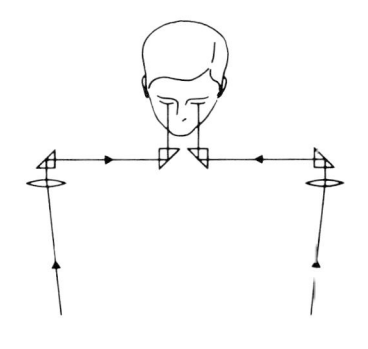

Fig. 15-42 "Separating the eyes": use of a range finder to extend the limit of stereoscopic acuity. (Notice the principle of total internal reflection employed in the two sets of prisms.)

relative size and distance of objects is in large measure due to the distance of about 6 cm separating the eyes. Having two separated eyes corresponds to having a built-in range finder. To make this human range finder work, it is of course necessary to use both eyes to look at the object. It is for this reason that infants do not display binocular vision; a newborn baby's eyes move about with relatively uncoordinated control. This is probably largely due to the fact that in the infant the fovea centralis is not fully developed, and therefore the requisite fixation on small objects is not to be expected. Slowly, over the months, the fovea develops, as do requisite muscular control and related mental processes, coincident with experience and with the aid of touch, until depth perception and three-dimensional interpretations as described below become as natural as breathing.

The depth or distance sensation derives somewhat like this. The two eyes are fixed on some point (Fig. 15-41). Light rays from this point are sent to each retina and, through the optic nerve, signals are sent to the brain to be computerized. In the computer language of the day, these light signals on the retina are the input. Our sensation of distance is the output of the brain, a magnificent computer. In the language of physiological optics, this ability of depth perception is called *stereoscopic acuity*. As indicated in Fig. 15-41, there are limits to this visual acuity. Depth perception based on the binocular aspect is most acute at short range, say about 1 meter, and is of significance up to about 10 meters. Naturally this ability varies widely with individuals, depending on a multitude of factors, including experience as well as physiological and psychological aspects. One's ability to determine greater distances appears to be largely dependent on experience with respect to relative brightness, size, shading, etc. For this reason one-eyed persons are less handicapped in judging long distances than short ones. Of course it is possible to extend the advantages of binocular depth perception by "increasing the separation of the eyes." One achieves this effect by employing optical instruments such as binoculars and range finders (Fig. 15-42).

Why an object is perceived in a three-dimensional aspect may be explained as follows. Consider again both eyes fixed on a given point on some object. The rays are focused on the retina of each eye. But other points on the object, nearer and farther, also send light signals to the retinas. Because of the separation of the eyes, if the rays from our first point are focused on corresponding points on the two retinas, then the rays from these other points will be focused on different and not corresponding points on the two retinas. This produces a slight doubling of the image. The brain unscrambles these doubled images to give us a *stereoscopic* three-dimensional impression. If there is some malfunction in this complex process the unfortunate individual is subject to *double vision*, a distressing symptom which may be due to minor malfunction of the eye muscles or to more serious neurological disorders.

Eye Defects

From our earlier discussion of lenses and from the previous paragraphs dealing with the eye we can recognize that if images on the retina are to be sharp, then the relationship of Eq. (15.1) must be satisfied. This means there is some limit to the image distance q, for the eye, since the focal length f of its lens system does not have unrestricted capabilities for variation. The distance from the cornea to the retina must be within the range of image distances to which the lens system of the eye can adapt (around 25 mm), if a sharp image is to result. Unfortunately, for a considerable portion of the North American population such is not the case. For these people the eye is either too long or too short for the focal length of its lens system. The image tends to focus in front of, or behind, the retina. Either way a blurred image results. People with these eye defects are said to be myopic (nearsighted) or hyperopic (farsighted) respectively. Applying our earlier considerations of converging and diverging lenses, the reader can understand the correction of these eye defects as indicated in Fig. 15-43.

A third common defect of the eye is astigmatism, which also may be corrected by eyeglasses. In this case the corrective lens must have some asymmetry of shape to counteract the lack of spherical symmetry in the eye's lens system, for the defect is that the cornea is not spherical but rather curved more sharply in one plane than in another. This trouble, *astigmatism*, makes it impossible, for example, to focus clearly on both horizontal and vertical lines on squared paper at the same time.

Rather than correcting defects such as myopia, hyperopia, and astigmatism by wearing ordinary eyeglasses (spectacles), many people use contact lenses. Although contact lenses date back to the last century, only since the middle of the present century has it been possible to produce such lenses small and comfortable enough to permit steady wearing. At present, contact lenses usually are made from a transparent plastic having about the same index of refraction as crown glass.

Contact lenses, 10 mm or less in diameter and somewhat less than 1 mm thick, are made to fit the cornea. Adhesion probably involves both air pressure* (since the minute space between the lens and the cornea is devoid of air, being filled by tears) and the adhesive force between the tears and the contact lens material. The shaping of contact lenses to give corrective action is similar to that employed in the more common eyeglasses (Fig. 15-43). However, the details are a little different. The actual shape of the cornea is measured, and then the contact lens is ground to correct the total eye defect. So in a sense the wearer has an "artificial cornea." A final point of interest: Astigmatism is

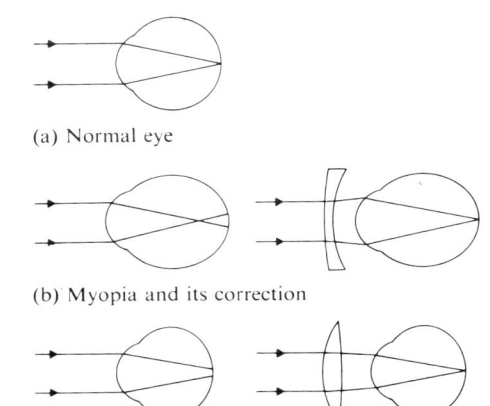

(a) Normal eye

(b) Myopia and its correction

(c) Hyperopia and its correction

Fig. 15-43 The normal eye, two common defects, and their correction.

*See Chapter 6.

frequently overcome by the film of tears trapped between the contact lens and the cornea. The index of refraction of tears is similar to that of the cornea, and so in a sense a corrective "tear lens" may be formed.

15.6 OPTICAL INSTRUMENTS

Overhead Projector

The overhead projector has proved to be a versatile and most useful component of the arsenal of modern visual teaching aids. As can be seen from Fig. 15-44, the optics is relatively simple. Basically only two lenses, L_1 and L_2, as well as the mirror M_1 (contained in the projector head) would suffice. However, in such a case much of the light passing through the stage and the projectual to be displayed would not reach the first image forming lens L_2. Thus letters, figures, etc., at the outer edge of a projectual, if it is the full size of the stage, would not form distinct shadow images on the screen, since the light passing close to figures at the outer edges is not directed at L_2. One therefore introduces a third lens, called a *condensing lens*, between the light source and the stage in order to focus the light from the stage onto L_2.

In most modern overhead projectors the condensing lens consists of a *Fresnel lens*. A Fresnel lens may appear to be only a thin plate of transparent plastic, but it actually consists of a series of concentric circular prisms (see Fig. 15-45). The angles of the prism faces are graded from the plate's center to its outer edge so that the plate has the refracting effect of a thick converging lens, but with these distinct advantages: Since the plate, that is, the Fresnel lens, is thin neither has it the bulk of a conventional lens nor does it absorb nearly as much light as would a large convex lens of comparable focal length.

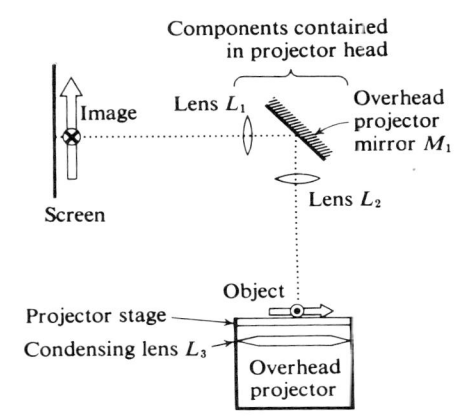

Fig. 15-44 Overhead projector optics. The circled cross indicates that the image on the screen is inverted left to right horizontally with respect to the object (the projectual) on the projector surface (the stage).

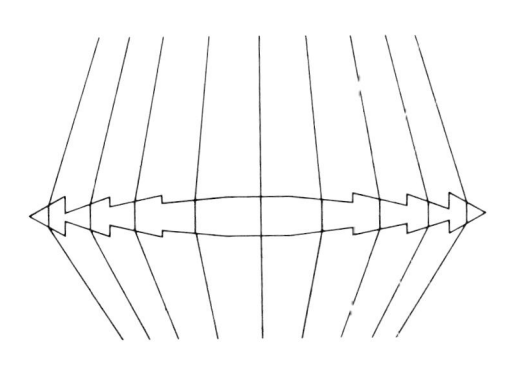

Fig. 15-45 Cross section of Fresnel lens.

The Camera

The basic operation of the camera can be inferred from the description of the eye in Section 15.5 and the comparison shown in Fig. 15-46. The type of camera shown in this figure has adjustable bellows to permit varying the film distance, because the image distance q has different values for different object distances p. This is necessary because, unlike the eye, the camera is saddled with a lens of fixed focal length f. On the commonly employed 35 mm cameras, adjustment of the distance q is achieved by rotating the lens in and out on a helical mounting. [There is, of course, a simpler type of camera, the box camera (see Question 58) which does not have the capability of varying the lens-to-film distance.]

The camera is also equipped with a diaphragm which acts like the iris of the eye. This diaphragm performs three functions: (1) controls the amount of light entering the camera; (2) controls the depth of focus; (3) decreases lens aberrations. Of these functions the latter is probably

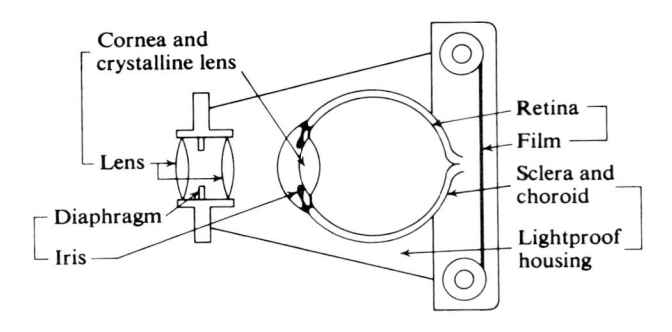

Fig. 15-46 The camera and the eye, showing comparable parts.

the least significant because of the great advance in the construction of multiple-element lenses to reduce aberrations. Even a relatively cheap camera is likely to have a triplet (three-element) lens, while seven-element lenses are not uncommon in medium-priced cameras.

The second function, *depth of focus* control, is important. Depth of focus means the region either side (in front or behind) of the finely focused image in which an acceptably clear image is still obtained. In other words, the depth of focus in the camera is the region of fairly sharp focus. Reducing the size of the lens aperture by closing in the diaphragm increases the depth of focus. This in turn increases the depth of field, that is the region in the landscape, or whatever, in which both near and far objects will be in reasonable focus on the film. Obviously the advantage of increased depth of focus is purchased at the price of light entering the camera, that is, at the price of increasing the exposure times.

Over the diaphragm is a shutter, control of which permits various ranges of exposure times commonly *up to* (in photographers' language) *maximum shutter speeds* of perhaps $\frac{1}{300}$ sec or $\frac{1}{500}$ sec. Ardent amateurs and professionals prefer even higher shutter speeds, up to say $\frac{1}{1000}$ sec.

The prime consideration in selecting a camera is a good lens, one as free as possible of the aberrations discussed in Section 15.4. The focal length of such a lens in the case of a folding camera is of the order of 10 cm. The focal length of most standard lenses is usually equal to the diagonal of the film area, and in the case of most folding cameras this is usually about 10 cm. This approach to focal length selection yields an angle of view approximately equal to that of the eye and therefore gives pictures with the most natural perspective in most conditions. For special purposes supplementary lenses are available for cameras with a fixed lens: telephoto lenses and wide-angle lenses are the most common. Better than attachments to a fixed lens are interchangeable lenses because the addition of a supplementary lens to a normal lens introduces further aberrations.

It is important for a photographer to be able to control the quantity of light that enters the camera and reaches the film. At a given lens diaphragm opening, the light intensity at the film with a lens of long focal length is less than with a lens of short focal length. Obviously the other factor which determines the amount of light at the film for a given exposure time is the size of the diaphragm opening, the aperture. Both these contributions to brightness at the film can be readily understood by imagining the camera as a closed room with only one window. The light intensity on the wall opposite the window will be greater when the distance from the window to the wall is less (small focal length), and also when the window is large (large aperture). A single number is employed to yield the necessary information regarding light intensity at the film of the camera. This is the f-number, which is defined as follows:

$$f\text{-number} = \frac{\text{focal length}}{\text{diameter of opening}}.$$

Example 8 A camera lens has an aperture 2.5 cm in diameter and a focal length of 5.0 cm. What is the f-number?

Solution

$$\begin{aligned} f\text{-number} &= \frac{\text{focal length}}{\text{diameter of opening}} \\ &= \frac{5.0 \text{ cm}}{2.5 \text{ cm}} \\ &= 2. \end{aligned}$$

In photographic practice the aperture setting corresponding to an f-number of 2 is indicated by $f/2$, which means that the aperture diameter is $\frac{1}{2}$ the focal length.

It can be seen from the definition of the f-number that the smaller the f-number the greater the light intensity at the film. And of course the f-number is adjustable by changing the opening, so if a camera is designed for very fast shutter speeds, that is, extremely short exposures, it should have a small f-number. Since the small f-number implies either a large aperture or a short focal length (short relative to film size) or, as is more generally the case, both, such lens systems tend to be expensive because of satisfactorily correcting the lens defects. The range of f-numbers may be from 1.4 to 16.

Probably the most popular camera with amateur photographers is the so-called 35 mm camera, a compact, versatile and relatively economic piece of optical equipment. Such a camera is designed for 35 mm film, has a lens of short focal length (4–5 cm), and may be equipped to accommodate various accessories. Some of these cameras also have automatic exposure timing mechanisms, eliminating decisions regarding f-numbers and exposure times. The photoelectric effect (Section 16.2) is the crucial element here. The light intensity at a given time and place triggers a circuit which, with a little hearing-aid type mercury dry cell, automatically adjusts the diaphragm to the proper size.

The Spectroscope

Spectra as shown in Fig. 14-20 on page 542 may be observed visually by means of a *spectroscope*. If a camera attachment is used to record the spectra, the instrument is referred to as a *spectrograph*.

Fig. 15-47 shows a spectroscope. In part (b) of this figure the schematic diagram of the optical system is presented. The light to be investigated enters a narrow, adjustable slit S and then is formed into a narrow, almost parallel beam by the achromatic lens L_1. The light beam leaving L_1 is said to be collimated. All the rays of the beam will strike the prism at essentially the same angle of incidence. (*Note*: In place of the prism one might use a diffraction grating; see p. 533.) In going through the prism, the light undergoes deviation and dispersion. The objective lens L_2 of the telescope now focuses the dispersed beam components at different points so that with a white light source at S one sees, through the telescope, a continuous spectrum of color like that of a rainbow.

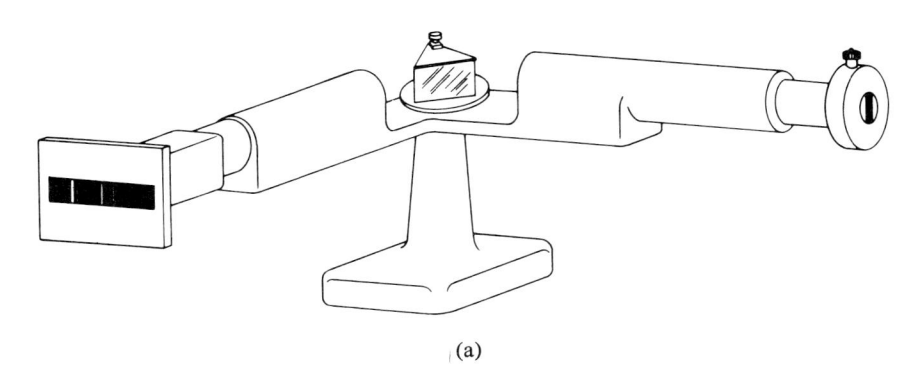

Fig. 15-47 A spectroscope. (a) The instrument with a spectrum displayed on a translucent screen. (b) The optical diagram of the instrument shown in (a).

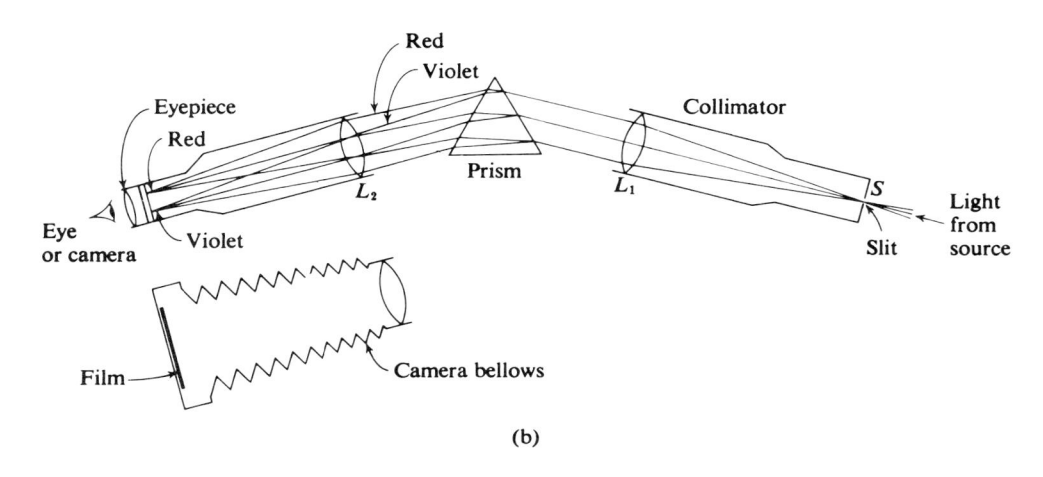

If, on the other hand, the light source were emitting only certain distinct wavelengths, as would be the case in an electric discharge through sodium vapor, only certain narrow *lines* of different colors would be seen. (See Fig. 14-20). Each of these lines can be considered to be the image of the slit S formed by light of a particular wavelength.

If such an instrument is provided with a mechanism for accurately measuring the angle through which light of a given wavelength is deviated, one can calculate the wavelength of the light (provided the instrument has been calibrated using a spectrum the wavelengths of which have been measured with a diffraction grating).

The Microscope

Discovery is the inevitable consequence of people resisting the restriction imposed by the narrowness of the normal human scale in both time and space. Man is ever attempting to extend the realm of his sensory perception in an effort to escape from an imprisonment which obscures much of the vast panorama of reality. In many instances he ultimately must take refuge in mathematical abstractions supported not by direct but by secondary (and interpreted) observations as is the case in quantum mechanics. But in the end we still tend to adhere to the dictum "to see is to believe." And here optics comes into its own.

As noted in the preamble of this chapter, the earliest and simplest of microscopes consisted of only a single convex lens. This lens may be no more than an ordinary drop of water, as in Fig. 15-48.

This small microscope may be improved by using glycerin instead of water (because glycerin has a greater refractive index), or by using a glass lens (because glass has a greater refractive index than water and can be ground to yield specific focal lengths).

The next step up in sophistication involves an instrument with two or more glass lenses. A relatively simple version of such a *compound microscope* is shown in Fig. 15-49. (Recall that the manner in which one can draw the optical diagram, Fig. 15-49(b), was shown earlier with reference to Fig. 15-31.) The two lenses are fairly widely separated and the one closer to the object is called the *objective* while that closer to the eye is called the eyepiece or *ocular*. In more complex instruments the objective and the ocular may each consist of a system of lenses rather than a single lens, and these lenses in turn may be compound lenses to minimize lens aberrations.

The optical system of the compound microscope may be so arranged that the ocular forms a greatly enlarged virtual image about 25 cm from the lens. This is because the eye is placed close to the ocular and with a normal eye 25 cm is a good distance for distinct vision. On the other hand, the optical system may also be arranged so that the image is formed at infinity.

Fig. 15-48 Water-drop microscope. A drop of water has been placed on a thin sheet of transparent plastic over a newspaper. Note the enlargement of the first three letters of the word "action."

(a)

(b)

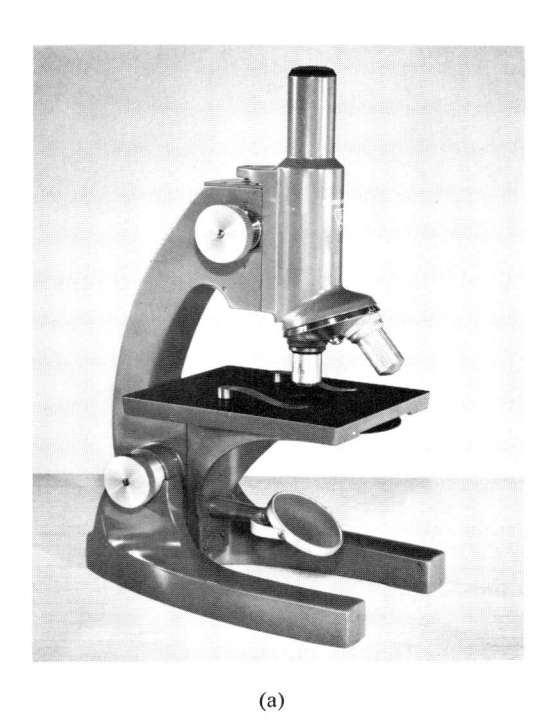

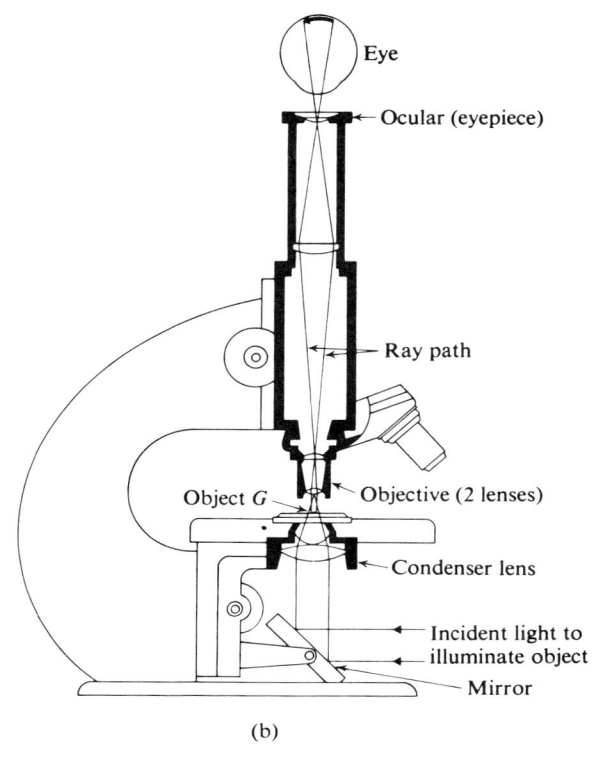

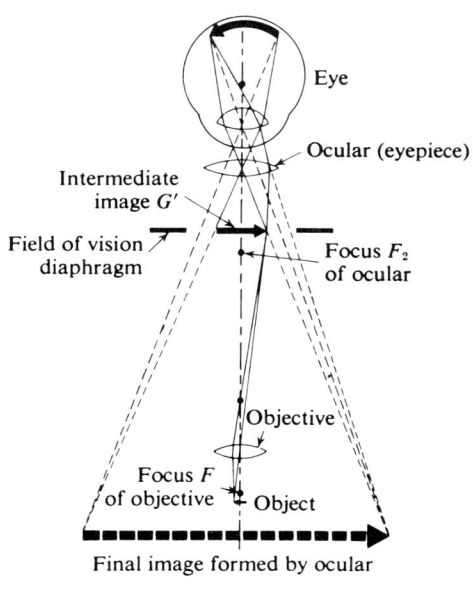

(c)

Fig. 15-49 Compound microscope. (a) Microscope with two objective lens systems, permitting two ranges of magnification. (Courtesy Bausch and Lomb.) (b) Schematic diagram of the microscope in (a). (c) Optical diagram showing the formation of the enlarged virtual final image.

15.6 OPTICAL INSTRUMENTS

In the case of a compound microscope, the total magnification is the product of the magnification achieved by the objective alone times the magnification of the ocular alone. This is so because the second of these lenses employs the image of the first as its object, as can be seen in Figs. 15-31 and 15-49. In other words, if the objective of a given microscope has a magnification of 40 times (written 40×, and referred to as 40 *times*, or 40 *diameters*) and the ocular has a magnification of 10×, the overall magnification achieved is 40 × 10 = 400×.

No matter what the degree of complexity of an optical microscope, diffraction makes it impossible to do better than distinguish details about the size of the wavelength of the light that is used (~5000 Å). It can be shown that this fact implies that magnifications much in excess of about 400 do not increase detail. Higher magnifications are however sometimes employed because a larger image may lend itself to more thorough study even if no further detail can be discerned. In general, the practical upper limit of magnification by optical microscopes is considered to be about 2000×.

Other Microscopes

Much higher magnifications are attainable with an *electron microscope*. [Using the deBroglie wavelength (Section 16.4), it is found that electrons accelerated through a potential difference of 50,000 volts have a wavelength about 10^5 smaller than the 5000 Å mentioned above for the optical microscope. This implies that it is possible for an electron microscope to distinguish very small details that would be blurred by diffraction if the wavelength involved were the relatively long wavelength of visible light.] Electron microscopes with magnifications of 200,000 diameters and more have been constructed. (See Fig. 15-50).

In Fig. 15-51 we see a comparison between the basic elements of the optical microscope and an electron microscope. In the electron microscope, electrons are thermionically emitted at the cathode (as in an x-ray or radio vacuum tube). The electrons are accelerated through application of a high voltage between the emitting filament and an anode with a hole in it to permit passage of the electrons. The electron beam is focused either by magnetic or electric fields and then impinges upon the specimen to be examined. Depending on the thickness and composition of this object, the electrons undergo various amounts of attenuation so that an image due to differing electron intensity is formed on a fluorescent screen or photographic plate sensitive to electrons.

Only relatively recently the electron microscope has been used to gain knowledge about the early history of the solar system. Moon dust brought home by the Apollo astronauts, and magnified several thousand times, reveals minute glassy spheres. Some reports suggest these spheres may be construed as an indication that lunar rock was at one time

Fig. 15-50 Electron micrograph of some virus molecules. The smaller spheres are the viruses; the larger sphere, used for calibration, has a diameter of 2000 Å. (Courtesy R. P. Feynman, R. B. Leighton, and M. Sands, *The Feynman Lectures on Physics*, Addison-Wesley, 1963.)

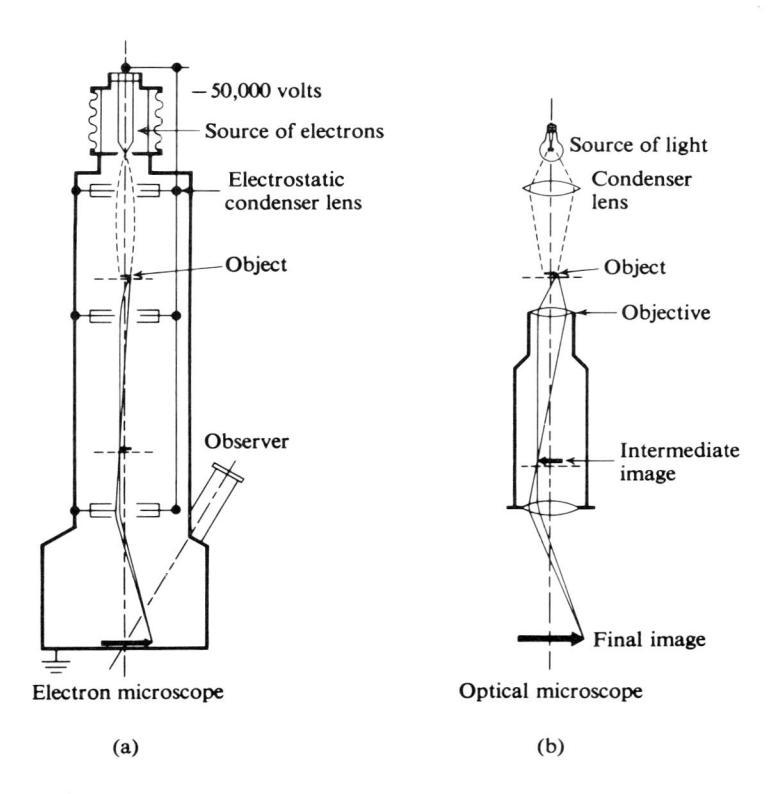

Fig. 15-51 Electron microscope (a) as compared to optical microscope (b). The focusing of the electron beam in the electron microscope is achieved using electric fields; magnetic fields can also be employed for this purpose.

melted, perhaps by volcanoes, crashing meteorites, or possibly exceedingly intense radioactivity.

Because of its ability to permit seeing large molecules (and small crystals), the electron microscope is now standard equipment in many medical, biological, and metallurgical research laboratories. There is also a device which magnifies not just a few hundred thousand times but several million times. This is the *field ion microscope*, invented by E. W. Müller (see Fig. 1-3). In this remarkable device a very fine needle point of the material to be "photographed" is placed at the center of a glass sphere which contains no other gases but a small amount of helium. The inner surface of the glass sphere is coated with a thin conducting layer of fluorescent material and a very high voltage is applied between the needle and this coating. In this situation the electric field at the needle tip is extremely intense and the needle is positive with respect to the fluorescent screen. When a helium atom makes contact with the needle tip the strong electric field rips off an electron. The helium is now in the form of a positively charged ion which is propelled along the direction of the electric field from its point of contact at the needle tip to the fluorescent screen where its impact causes the screen to fluoresce. Since the center of an atom in the needle ionizes helium at a somewhat different rate than do the regions between the atoms, the density of helium ions striking the screen is different depending upon from where on the needle the ions came. In this way the ions "paint" a picture of the atomic configuration of the needle tip upon the fluorescent screen.

Telescopes

The Dutch spectaclemaker, Hans Lippershey, in 1608 applied for a patent giving him for 30 years the exclusive right to make an "instrument for seeing at a distance." Although he was not granted a patent, on the grounds that others had known of the device, he apparently developed a successful business constructing telescopes for his government, neglecting however to make any astronomical use of the instrument himself.

The first astronomical use of the telescope was made by Simus Marius in 1609. However, because Galileo so outshone Marius' discoveries only a few months later, it is Galileo who is given the credit of being the founder of telescopic astronomy, his first discovery being three of the four bright satellites of Jupiter.

Two of the telescopes actually constructed by Galileo are preserved in the Galileo Museum in Florence. Both instruments are made with a paper tube and two lenses, a plano-convex and a plano-concave. The larger of the two instruments is about 4 ft long, has a diameter of less than 2 in., and magnifies 32 times.

The optical system of these telescopes, as shown in Fig. 15-52(a), is well adapted to such low-power instruments as opera glasses but is not at all practical for observations demanding high magnification because the magnification is achieved at the expense of the size of the field of observation. In fact, with Galileo's larger telescopes it is not possible to see more than a quarter of the diameter of the moon at once. However, this optical system has some distinct advantages such as producing an erect image, which is particularly convenient for terrestrial observations, and a sharp field, improved by the partial compensation of the aberrations of the convex objective by the concave ocular.

Fig. 15-52 Refracting telescopes. Objects O are shown close to objective lens. In astronomical usage the object is usually so far away that the rays reaching the objective are very nearly parallel.

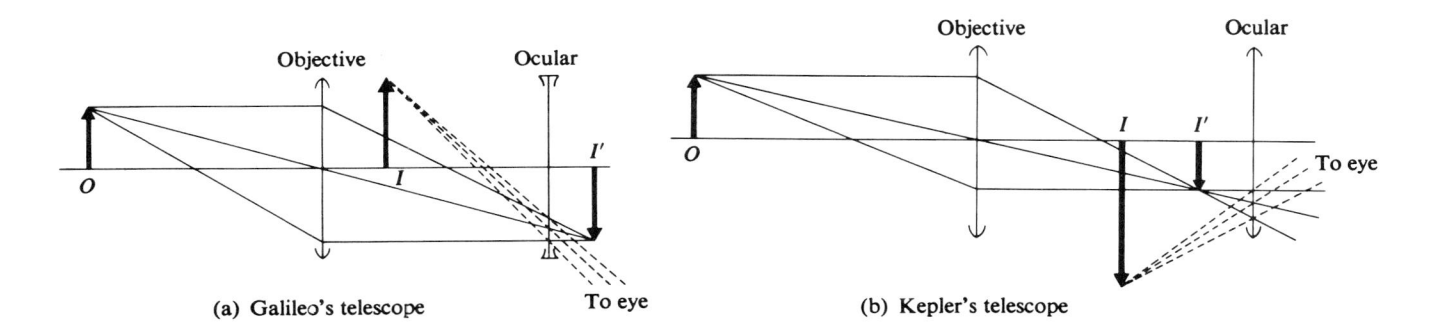

(a) Galileo's telescope

(b) Kepler's telescope

594 OPTICS

The difficulty of being unable to combine high magnification with a field of vision of reasonable size finally led to Kepler's construction of a telescope with two convex lenses [Fig. 15-52(b)].

This new telescopic system, however, still suffered from the two chief aberrations of a single lens: chromatic aberration and spherical aberration. The first steps towards the elimination of these faults were proposed by Descartes in his *Study of Dioptrics* of 1637, wherein he also indicated the advantage of a very long objective focal length for increased magnification. In the following decades there then appeared an extreme lengthening of focal distances, the Huygens brothers constructing instruments of up to 210 ft in focal length.

While the optical diagrams of Fig. 15-52 do illustrate a way of focusing a refracting telescope (the diagrams were drawn in this manner to clearly demonstrate the final image I), in actual practice the intermediate image I' is focused to be at, or very near, the principal focus of the ocular. This means that the rays from each point of the object are rendered parallel by the ocular, and so the eye sees the final image at infinity. In other words, the eye is relatively relaxed in the way it normally is when looking at distant objects.

A telescope does not really "magnify" an object in the sense that a microscope does nor does it "bring objects closer" if the final image is at infinity. What the telescope does do is increase the visual angle (Fig. 15-53), thereby increasing the image formed on the retina of the eye. So the magnification formula, Eq. (15.2), will not do for telescopes. Obviously we don't see an image of the moon larger than the object, the moon itself! If we express magnification of a telescope as

$$\text{magnification} = \frac{\text{angle object subtends at eye with aid of instrument}}{\text{angle object subtends at unaided eye}},$$

then this means the magnification in Fig. 15-33 is θ_2/θ_1.* Using plane geometry, it can be shown that the above definition yields to a good approximation the useful fact that for refracting telescopes

$$\text{magnification} = \frac{\text{focal length of objective}}{\text{focal length of ocular}}.$$

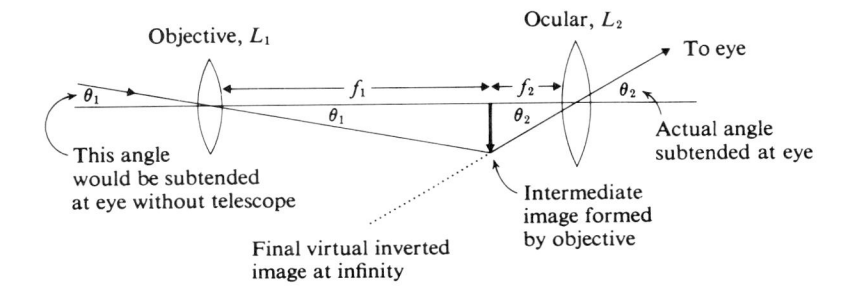

*This is known as the angular magnification, *as distinct from the* linear, or lateral *magnification of Eq. (15.2).*

Fig. 15-53 Magnification of a telescope is given by the ratio θ_2/θ_1, where θ_1 is seen to be the angle that the rays from a point of a distant object would subtend at the eye if no telescope were employed, and θ_2 is the angle subtended at the eye because of the properties of the telescope.

The next great change in the construction of telescopes was initiated in 1663 by James Gregory, a Scottish mathematician. He was first to use a parabolic mirror and so obtain an erect image. Further improvement of the reflector type instrument was made by Newton. This turn of events was apparently precipitated by Newton's erroneous assumption that the dispersion of a light beam through a prism varied as its deviation. If this were so then a type of glass that would deviate (refract) a beam more would also spread the beam out into a more pronounced color spectrum. Newton therefore concluded that it would be impossible to design a combination of lenses which would not suffer from what is now called chromatic aberration, so he looked for a means of forming an image without employing lenses. He spent his time improving Gregory's idea of the mirror system until in 1688 he constructed a telescope which was the first successful reflecting type.

Work on the reflecting principle was carried on by various astronomers, and other types of reflecting telescopes were developed. Fig. 15-54 shows the optical diagram of a typical reflecting instrument. The advantages of the reflecting telescopes are that they are devoid of chromatic aberration and, insofar as it is possible to construct much larger mirrors than lenses, give a better image because they can "gather more light." Also there is the very practical advantage that the largest and heaviest optical component, the mirror in this case, can be placed at the bottom of the instrument rather than extended far out, away from the mounting and the observer's position. Modern telescopes of this type have been built with increasingly large mirrors. The culmination of efforts in this direction to date is the telescope on Mount Palomar in Southern California. The diameter of the objective mirror of this instrument is 200 in.

In the broadest sense, telescopes are not necessarily restricted to use with visible light. Special telescopes, *radiotelescopes*, which are essentially really large focused radio antennas, have been used extensively in *radioastronomy*. The radiotelescopes are of great value to astronomers because various stars, planets, nebulae, and other bodies and regions of the universe are sources of electromagnetic radiations much longer in wavelength than visible light. The radiations in question are in the wavelength region of radio communication waves. This accounts for the term *radioastronomy*. In a way, these telescopes have opened another window

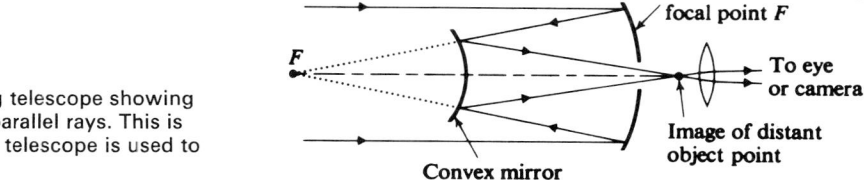

Fig. 15-54 Reflecting telescope showing the case of incoming parallel rays. This is the situation when the telescope is used to view stars.

upon our universe. And on some occasions unusual and especially interesting stars and other more distant celestial objects have been detected by optical telescopes only after the radioastronomer informed his "optical colleague" where to look.

Binoculars

To take advantage of binocular vision (stereoscopic acuity) in distance viewing, field glasses and opera glasses were constructed. Early models consisted essentially of two Galilean telescopes, one for each eye. Later on the desirable features of the Keplerian telescope were employed. This required that not only the image be brought upright but also that the problem of the relatively long distance between objective and ocular be solved. The solution to both these problems is attained by prism binoculars (Fig. 15-55).

The magnification of such binoculars is commonly in the range of $6\times$, $7\times$, or $8\times$. Magnifications of less than $4\times$ are of questionable merit in that attendant lens aberrations offset the magnification advantage to the point that the average person can usually see better with the unaided eyes. If magnifications much in excess of $15\times$ are required, normal hand jitter is so magnified in the field of view that the instrument requires some sort of support to hold it steady.

Aside from magnification, the other characteristic of prime interest in binoculars is the extent to which they bring light to bear on the eye. This factor which depends in part on the diameter of the objective lens is of particular importance in the case of night vision. The specification

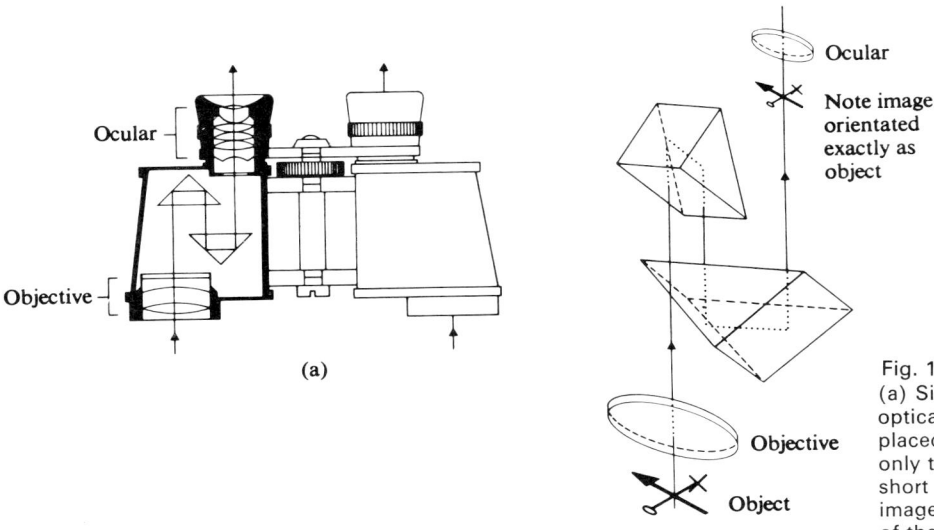

Fig. 15-55 Prism binoculars. (a) Simplified cutaway section showing optical path. (b) Details of how prisms placed at 90° to each other are used not only to provide for a long optical path in a short space, but also to obtain an upright image of the objective. This image, like that of the Kepler telescope, would otherwise be inverted.

15.6 OPTICAL INSTRUMENTS 597

of binoculars is therefore quoted in terms like 6 × 30. This means the magnification is 6 and the effective diameter of the objective is 30 mm.

Sometimes the term *brightness* is employed. Numerically the *relative brightness* of glasses is expressed by dividing the objective diameter by the magnification and squaring the result. The reason for squaring this ratio derives from the fact that, while the effective diameter of the objective can be construed as the "entrance pupil" for light, this ratio yields the diameter of the "exit pupil" for light actually impinging on the eye. And the light available to the eye depends on the area of this "exit pupil." So squaring the ratio gives a better indication of the intensity of light at the eye. For example, if the glasses were 6 × 30 then relative brightness would be $(30/6)^2 = 25$.

Smog Chamber

Optical instrumentation is often an important part of the apparatus used by teams of scientists from different disciplines in their approach to problems arising in our technological society. An interesting illustration is provided by the smog chamber of Fig. 15-56.

The smog chamber is used to provide data on the smog-forming tendencies of different test compounds which are mixed with pure air and subjected to the action of artificial sunlight at a controlled temperature. An array of ultraviolet lamps, sunlamps, blue and daylight fluorescent lamps provide the artificial sunlight. This light is transmitted to the chamber through borosilicate glass walls which are transparent to electromagnetic radiation down to 3000 Å. As the smog components

Fig. 15-56 Smog chamber. (Courtesy Stanford Research Institute.)

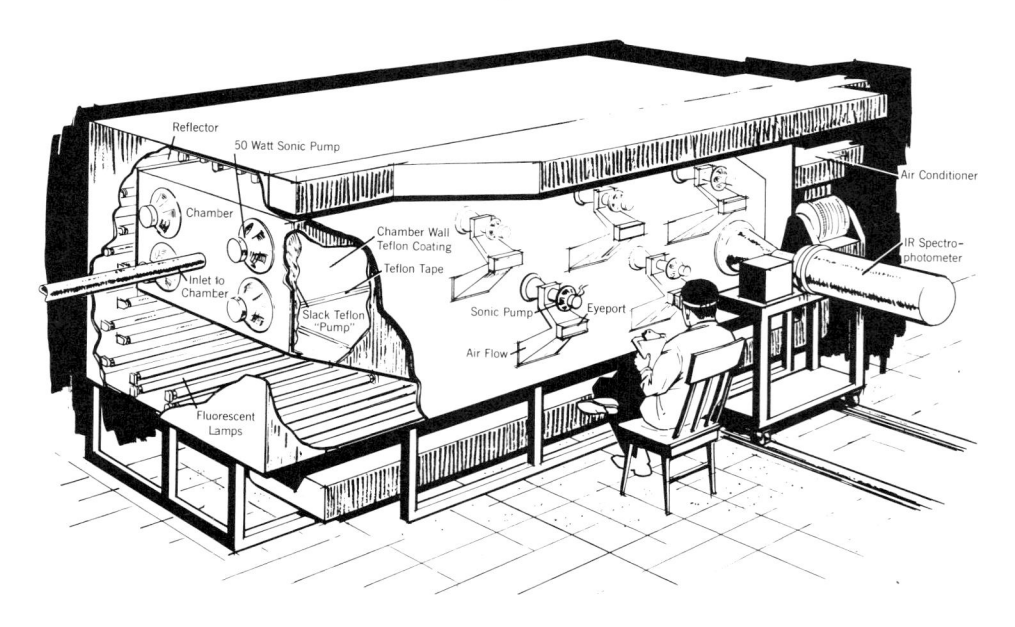

develop, information about their chemical compositions is desired. This is obtained, not only by the methods of "wet chemistry," but also by using an infrared spectrophotometer to determine the infrared wavelengths that the smog absorbs. The eye irritation produced by this smog is evaluated by humans stationed at various eyeports past which the smog is circulated by sonic wind pumps.

Such experiments yield the scientific information necessary to establish reasonable standards for smog control.

QUESTIONS

1. Give an example demonstrating the rectilinear propagation of light, other than the examples cited in the text.

2. If we extend our contention that light travels in straight lines to the propagation of electromagnetic waves of even longer wavelength, speculate on how it is that powerful radio transmitters can send signals to virtually all parts of our spherical earth?

3. (a) In the case of an eclipse of the sun by the moon, what does one term the region of partial eclipse on the earth?
   (b) Why is there both a region of total and of partial eclipse?

4. Fig. 15-57 shows the "reflection" of billiard balls. Does this behavior support a "particle" concept of light? Does it refute a "wave" concept of light?

5. What stipulation must be made so that Fermat's principle of *least time* can be interpreted as a principle of *least distance*?

6. What is meant by the term "image"?

7. How far in front of a plane mirror is a person whose image appears to be 2.00 m behind the mirror? Draw the appropriate ray diagram using a single point as the object.

8. Use the law of plane reflection to sketch a ray diagram showing the various images of an object placed in front of two plane mirrors at right angles. How many images will there be? Try the experiment yourself.

9. Employ your knowledge of reflection to determine the shortest possible plane mirror in which a girl 1.6 m (5 ft 3 in) tall could see a complete image of herself. Draw the appropriate ray diagram, and for the sake of simplicity assume her eyes are at the top of her head!

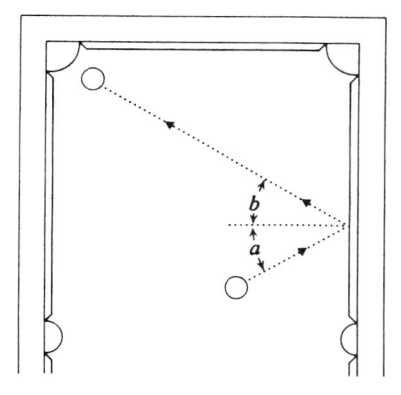

Fig. 15-57 Reflection in billiards.

10. Repeat the appropriate sketch for Question 9 but this time let the girl's eyes be below the top of her head. This sketch of the actual situation should nevertheless yield the same answer as was arrived at in Question 9.

11. State the major difference between a *real* and *virtual* image.

12. All the discussions in this chapter have related to *regular* reflection. Do you think that the law of reflection (angle of incidence equals angle of reflection) can be used to explain the diffuse reflection shown in Fig. 14-17(b) on page 539. Explain.

13. (a) Sketch a ray diagram to show the location of the image of an object in the case of a concave mirror if the object is closer to the mirror than the focal point (principal focus) of the mirror.
    (b) What is the nature of such an image? Could the image be captured on a screen? How is it that an observer can see such an image?

14. The radius of curvature of a convex (spherical) mirror is 2.0 meters.
    (a) What is the mirror's focal length?
    (b) Is the focal point *real* or *virtual*?
    (c) Sketch a diagram showing the location of the mirror, its center of curvature and its focal point.

15. The radius of curvature of a concave (spherical) mirror is 0.50 meter. Draw a diagram showing the location of the mirror, its center of curvature, and its focal point.

16. For the case of the concave mirror, sketch the ray diagram for an object at
    (a) infinity;
    (b) the center of curvature.

17. For the case of the concave mirror, sketch the ray diagram for an object at
    (a) a location between C (center of curvature) and F (focal point);
    (b) F.

18. Make a sketch to indicate the path of a light ray through a
    (a) thick plate of glass with parallel sides;
    (b) triangular prism;
    (c) piece of glass with a circular cross section.

19. (a) Sketch a ray diagram to show that a person looking down at an angle at a fish in the water gets the impression the fish is closer to the surface than it actually is [*Hint:* Simply represent the fish by a point under the water and then trace two rays to the eye of the observer above the water surface (one ray normal to the air-water interface and one ray with a small angle of incidence.)]
    (b) State a general conclusion about "apparent depth" with respect to underwater objects as seen from above the water surface.

20. If you were to construct a magnifying glass with the largest possible magnification but a reasonably limited thickness of glass, would you choose crown glass or flint glass? Explain.

21. Explain what is meant by the term "total internal reflection." Use a diagram.

22. Will total internal reflection take place more readily (i.e., will the critical angle be smaller) if the refractive index of the material of a "light pipe" is large or if it is small? Explain.

23. Sketch a ray diagram to show that when looking in at the side of an aquarium it is possible to see the fish directly, and also by total internal reflection.

24. What is meant by the term "fiber optics"? Suggest a practical application of fiber optics.

25. Trace a ray entering a sandwich of three layers of glass with successively higher indices of refraction. (This shows you how air with a gradually changing density, and hence changing index of refraction, can curve light rays.)

26. (a) Distinguish between "looming" and a mirage.
    (b) What is the principle common to the explanation of both those atmospheric effects?

27. What effect, if any, does atmospheric refraction have upon the time the sun is first seen in the morning?

28. Sketch a diagram to show how the dispersion of white light into its constituent colors when the light is refracted by water droplets can produce a rainbow.

29. **In the refraction shown in Fig. 15-22, aside from dispersion, what other interesting optical phenomenon is indicated?**

30. An object is placed 12.0 cm from a converging lens of focal length 8.0 cm.
    (a) Locate the image by drawing a ray diagram.
    (b) **Use Eq. (15.1) to determine the position of the image.**

31. An object is placed a distance from a converging lens such that an image of exactly the same size as the object is formed. The lens has a focal length of 15 cm.
    (a) Use Eq. (15.1) to determine the location of the object and the image. [*Hint:* First use Eq. (15.2) to establish p in terms of q.]
    (b) Describe the image (inverted or upright? real or virtual?).
    (c) Sketch the ray diagram.

32. Two hills whose summits are 1.0 kilometer apart are photographed from a point 4.0 kilometers removed along the perpendicular bisector of the line joining them. What is the distance separating the

images of the two summits on the film if the camera's converging lens has a focal length of 20.0 cm? [*Hint:* The distance between the summits can be taken as the object size *O* and their separation on the film, the image size *I*.]

33. A lamp and a screen are 3.0 m apart. If the image of the lamp on the screen is to be twice the size of the lamp, what should be the focal length of the converging lens employed?

34. A diverging lens has $f = -15.0$ cm. An object is placed on the optical axis 12.0 cm from the lens.
    (a) Sketch the optical diagram.
    (b) Use Eq. (15.1) to determine the position of the image, and describe the image.
    (c) If the object is 1.8 cm tall, how tall would the image be?
    (d) What is the magnification in this case?

35. Repeat Question 34, but let $f = -8.0$ cm.

36. What happens to the focal length of a lens immersed in transparent oil? Is the focal length increased or decreased? Explain. [*Hint:* A diagram, keeping in mind the concept of refractive indices, helps.]

37. It is possible to specify for spherical mirrors that the focal length is half the radius of curvature. Is it possible to make a similar general statement for lenses? Explain.

38. A double concave lens is made of glass with $n = 1.50$. If the radii of curvature of the two lens surfaces are both 30.0 cm, what is the focal length of the lens?

39. What must be the radius of curvature of the curved surface of a plano-convex lens if it is to be made of glass of refractive index 1.52 and is to have a focal length of 30.0 cm? [*Hint:* The radius of curvature of the flat side is infinite so that $1/R_1 = 0$.]

40. What is the refractive power in diopters of a convex lens of focal length 10 cm?

41. The power of a human eye is 60 diopters. Calculate its focal length.

42. Show that when two thin lenses of focal lengths f_1 and f_2 are placed *in contact*, the focal length f of the combination is given by
$$\frac{1}{f} = \frac{1}{f_1} + \frac{1}{f_2}.$$
Start by drawing the appropriate ray diagram, using the image of the first lens as the object of the second. Assume the lenses are in exactly the same position, that is, their optical axes coincide.

43. Show that Eq. (15.5), $D = D_1 + D_2$, follows from Eq. (15.4) and the definition of the diopter.

44. Assume that an achromatic lens is made of two lenses in contact. One is a converging crown-glass lens of focal length 4.0 cm, and the other lens, a divergent one composed of a type of glass with different chromatic dispersion properties, has a focal length of -8.0 cm. Find the focal length of this compound lens.

45. A certain achromatic lens consists of a converging crown-glass lens with $f = 8.0$ cm and a diverging flint-glass lens with $f = -10.0$ cm. What is the focal length of this achromatic lens?

46. An optical system consists of two converging lenses separated by 1.00 m along their common optical axis. Each lens has a focal length of 3.00 m. An object is placed 1.50 m in front of the first lens.
    (a) Sketch the optical diagram.
    (b) Where is the final image formed? [*Hint:* First find the image due to the first lens, then use this image as the object for the second lens.]

47. Sketch a diagram of a convex and a concave converging lens combination to show how the concave component reduces the chromatic aberration of the converging combination.

48. In the case of chromatic aberration of a converging lens, which color of the visible spectrum will be focused nearest to the lens?

49. Explain the action of "nonreflective" lens coatings in terms of the interference effects discussed in Chapter 14.

50. How thick should the "nonreflecting" coating on binocular prisms be in order that light whose wavelength in air is 6000 Å be completely transmitted? The index of refraction of the coating is 1.25. [*Hint:* First determine the wavelength of the light in this "nonreflecting" coating.]

51. Consider "nonreflective" coatings on a lens (n = 1.61).
    (a) About how thick should the coating with n = 1.50 be so that it is effective for light in the middle of the visible spectrum (p. 523).
    (b) What would happen if the lens coating were made twice as thick?

52. In the case of a pinhole camera, what does the size of the image depend upon?

53. If the refractive power of the eye is 60 diopters and that of the cornea is 45 diopters,
    (a) what is the power of the lens?
    (b) what is the focal length of the lens? (For simplicity assume the cornea and lens act as a simple combination of thin lenses in contact.)

54. The eye has a certain limit of visual discrimination which prohibits unlimited precision of definition imposed upon it by the finite separation of rods and cones. Suggest the comparable limiting factor of a photographic film.

55. The distance from the front of the eye to the retina is about 25 mm. Suppose one gets a clear image of an object 50 centimeters away.
    (a) What is the focal length of the combination lens system of the eye in this case?
    (b) What is the refractive power in diopters?

56. Refer to the preceding question. Suppose the object were now removed to a distance of 2.00 meters.
    (a) Would the muscles controlling the shape of the lens now cause the lens to become flatter or rounder?
    (b) What is the power in diopters now?

57. As one goes farther away from an object, "angular size" at the eye diminishes, that is the retinal image becomes smaller.
    (a) How, in your opinion, does one then know anything about the size of distant objects since their "apparent size" is obviously a variable?
    (b) What other aspect of vision (other than size perception) may be aided by a valid appreciation of real sizes?

58. Consider a box camera which consists basically of a convex lens, a box of fixed dimensions, and a shutter mechanism.
    (a) Sketch an optical diagram for such a camera.
    (b) Could this camera be used for taking pictures of distant objects? Explain.
    (c) Could it be used for taking close-ups? Explain.

59. A "universal-focus" camera lens is simply one of fairly short focal length, say 5.00 cm.
    (a) How far from the optical center of the lens would the image of an infinitely distant object be focused?
    (b) How far from the optical center of the lens would the image of an object 6.00 m distant be formed?
    (c) In what sense does your answer to (b) support the term "universal lens"?
    (d) Would a camera with such a lens lend itself to close-ups? Under what condition, if any?

60. Show that the image size on the film in a camera is proportional to the focal length of the camera lens, providing the object distance is much greater than the focal length. [*Hint:* Use Eqs. (15.1) and (15.2) and simplify Eq. (15.1) by using the fact that p is very much greater than f.]

61. The moon has a diameter of 3.5×10^6 m (about 2000 miles) and is at a distance from the earth of 3.8×10^8 m (about 240,000 miles). Calculate the size of an image of the moon produced on the film of a special camera with a lens of 1.00 m in focal length. [*Hint:* Use the relation between image size and focal length stated in Question 60.]

62. A widely used type of Polaroid camera has its widest aperture of 2.50 cm at the setting $f/4.5$. What is the focal length of the lens system?

63. Is a lens with $f/4$ "faster" or "slower" than one with $f/11$? (Faster of course, means that a given photograph can be taken with a shorter exposure time.) Explain.

64. (a) The aperture of a camera lens is set at $f/8$. What is the focal length of the lens if the aperture is 0.50 cm in diameter?
    (b) If the camera above had aperture settings ranging from $f/2.8$ to $f/11$, which of these two would you select in order to make as short an exposure as possible?
    (c) Which of the two f-numbers above would you select on a very bright day?

65. A certain box camera has a single lens, while a particular portrait photographer's camera employs a multi-element lens system of 5 lenses. In which of these two cameras is it more important that the lenses have coatings to reduce reflection? Explain.

66. When the slit of a spectroscope is illuminated by the light from a mercury arc lamp, certain spectral lines of visible light are evident.
    (a) Of these lines, which are spatially separated the most by the spectroscope: violet and blue; yellow and orange; green and yellow; yellow and red; or red and blue?

(b) What information about mercury can be gained by such a spectroscopic investigation?

67. What can be seen when a white light source illuminates the slit of a spectroscope?

68. Consider a simple microscope, that is, a single lens magnifying glass.
    (a) Draw a ray diagram showing the formation of the image.
    (b) **Using Eqs. (15.1), (15.2), and the fact that a** convenient distance for distinct vision is 25 cm, show that the magnification of such a magnifying glass is given by $1 + 25\text{cm}/f$ where f is the focal length of the lens in cm.

69. Given that the magnification of a single lens can be expressed as indicated in Question 68 (b), determine the magnification that van Leeuwenhoek achieved with his simple microscope with lens having $f = 1.25$ mm.

70. The ocular of a microscope has a magnification of $12\times$ and the objective of $50\times$. What is the total magnification?

71. (a) If ultraviolet light were used in place of white light to illuminate an object under a microscope, could the resolution (that is, the useful detail) be increased; in other words, would higher magnification be possible? Explain.
    (b) What problems would be encountered in using a microscope with ultraviolet light in place of white light?

72. What is the magnification produced by a refracting telescope employing an objective with a focal length of 1.0 m and an ocular with a focal length of 2.0 cm?

73. A refracting telescope is constructed with an objective with a power of 1.5 diopters and an ocular with a power of 4.5 diopters. What is the magnification?

74. The reflecting telescope in which the image is "extracted" through a hole in the big mirror (shown **in Fig. 15-54) is called a *Cassegrain type*. Show** by means of an optical diagram how the image from a reflecting telescope might be extracted by means of a plane mirror or a prism (total internal reflection) without having to have a hole through the main concave mirror. Such a system is called a *Newtonian telescope*.

75. (a) What is the physical phenomenon employed in prism binoculars to contain a long optical path between objective and ocular in a relatively short instrument length?
    (b) What is achieved by employing this phenomenon, aside from shortening the length of the instrument?

76. The specification of binoculars is given as 8×30. What does this mean?

77. Which binoculars would be better for night vision: 8×30s, or 7×50s? Explain.

SUPPLEMENTARY QUESTIONS

S-1. The moon subtends an angle of approximately $0.467°$, and its light takes 1.28 sec to reach the earth. Calculate the moon's diameter and compare it to the earth's diameter.

S-2. A beam of light is reflected from a plane mirror. Show that when the mirror is rotated through an angle θ, the reflected beam rotates through an angle 2θ.

S-3. Sketch a ray diagram and indicate the positions of the images formed when a point object is placed in front of two plane mirrors forming a right angle. How many images will there be?

Fig. 15-58 Plane mirrors forming an angle of 45°. See Question S-4.

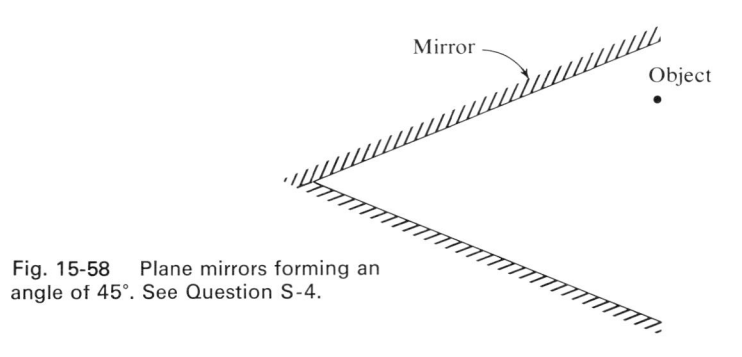

S-4. A point object is placed in front of two plane mirrors which form an acute angle of 45° (Fig. 15-58). How many images will there be?

S-5. The rotation of an object (for example, the deflection of the needle of a galvanometer) is often measured by attaching a mirror to the rotating body and observing the scale by means of the light reflected from the mirror (see Fig. 15-59). Such a device is called an *optical lever*.

(a) Show that for the optical lever in Fig. 15-59

$$\frac{x}{d} = \tan(\alpha + 2\theta) - \tan \alpha.$$

(Note that the angle α is a constant which depends partly on the initial position of the mirror and partly on the direction of the axis of the telescope: $\tan \alpha = a/d$.) It can be seen from the derived expression that, since x can be read on the scale, it is possible to determine the angle θ through which the mirror turns.

(b) Show that (as is the case in many practical situations), if the angles α and θ are both small so that one may substitute the angles (in radians) for their tangents, the expression of part (a) simplifies to

$$\theta = \frac{x}{2d}.$$

S-6. A spherical mirror gives an image 4.00 cm behind the mirror when a 3.00-cm high object is located 6.00 cm in front of the mirror.
(a) Determine the nature of the mirror.
(b) Calculate the mirror's radius of curvature.
(c) What is the height of the image?

S-7. In a laboratory experiment designed to determine the focal length of a convex mirror, the first step was to use other optical apparatus to produce a real image on a screen. The next step was to place the mirror 12.0 cm in front of the screen, whereupon a new image was obtained on another screen 60.0 cm from the mirror. From these observations, determine the focal length of the mirror.

S-8. The "black box" in Fig. 15-60 contains some optical component or components and is open on one side. On the basis of the light rays shown, suggest what would be in the "black box."

S-9. A ray of light strikes the surface of crown glass ($n = 1.50$) at an incident angle of 50°. Determine the directions of the reflected and refracted rays.

S-10. A flat-bottomed glass vessel has an index of refraction 1.50. In the vessel there is a 10.0-cm depth of

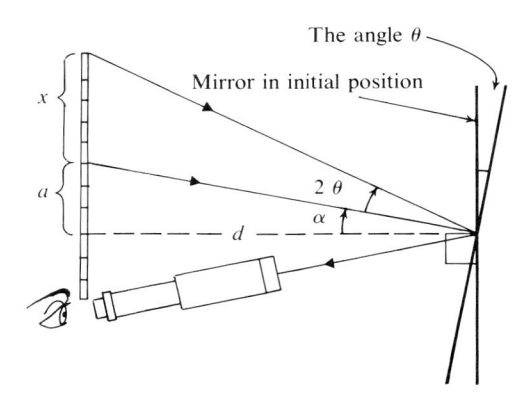

Fig. 15-59 Optical lever of Question S-5.

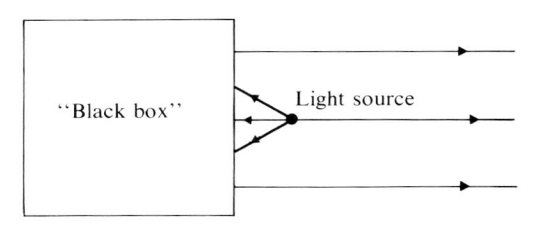

Fig. 15-60 The "black box" of Question S-8.

water of index of refraction 1.33. A point source of light is embedded in the bottom of the vessel, 3.00 cm below the glass-water interface. Considering first the refraction of light proceeding from glass to water, and then the further refraction as the light goes from water to air, determine the apparent depth of the light source below the top surface of the water, as estimated by an observer looking down at an angle from above. Proceed in two main steps:
(a) Derive the appropriate formula, clearly defining all symbols.
(b) Use the fomula derived in (a) to obtain a numerical answer.

S-11. A common "science trick" is to place an opaque container, such as a pie plate, on a table at which a person is sitting. A coin is dropped into the container which is then moved so that the spectator at the table is not able to see the coin in the bottom of the container. The "magician" then pours water into the container, and the coin becomes visible to the spectator. Explain this "magic," using a suitable ray diagram.

S-12. (a) Starting with Snell's law, $n_1 \sin \theta_1 = n_2 \sin \theta_2$, show that the critical angle, mentioned in connection with total internal reflection, is given by

$$\sin \theta_c = \frac{n_1}{n_2},$$

where $\sin \theta_c$ is the critical angle for light originating in the medium of index of refraction n_2 being refracted at the interface between that medium and a medium having a lower index of refraction n_1.

(b) Show that if air is the medium with the lower refractive index (index of refraction = 1.0), then the critical angle is given by

$$\sin \theta_c = \frac{1}{n},$$

where n is the index of refraction of the material in which the light originates.

S-13. Describe how the index of refraction of a substance might be determined experimentally. [*Hint:* Consider the critical angle.]

S-14. Sketch the ray diagram in Fig. 15-61 and complete it by continuing the rays, assuming the critical angle is 48°.

S-15. (a) Suppose you wanted to hide a transparent object in a bowl of water. What would the index of refraction of the object have to be?

(b) It is sometimes difficult to see contact lenses immersed in a transparent sterilizing solution. What can be said about the refractive indices of the lens material and the solution?

S-16. (a) Sketch a ray diagram showing how light is refracted as it passes through a triangular glass prism with prism angle A. (The angle between the two refracting sides of the prism is called the *prism angle*.)

(b) The angle between the incident ray and the ray emerging from the prism is called the *angle of deviation, D*. It can be shown that, when the angle of incidence of the incident ray equals the angle of refraction of the emerging ray, the angle D attains a minimum value, the *angle of minimum deviation, D_m*. (If the cross-section of the prism is isoceles, with the two equal angles at the base, then D_m is attained with the ray inside the prism parallel to the base of the prism.) Show that the angle of minimum deviation D_m is related to the prism angle A and to the index of refraction n by

$$n = \frac{\sin \frac{1}{2}(A + D_m)}{\sin \frac{1}{2} A}.$$

S-17. Show that if the prism angle is small then the angle of minimum deviation referred to in Question S-16 is given approximately by

$$D_m \approx (n - 1)A,$$

where D_m is the angle of minimum deviation, and A is the prism angle.

S-18. Describe how one might determine experimentally the angle of minimum deviation of a prism. (See Question S-16 for definition of angle of minimum deviation.)

S-19. A recent development in fiber optics is a light pipe that really is a pipe. It is an optical fiber composed of a liquid filled capillary. Light loss (that is, loss in light intensity) is about 10 decibels per kilometer. In communications engineering power ratios are

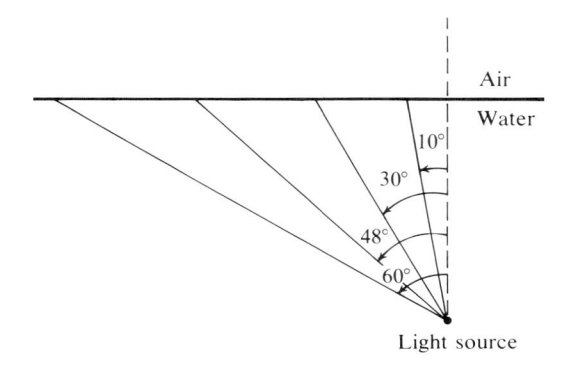

Fig. 15-61 Ray diagram for Question S-14.

measured in decibels. The gain or loss of power expressed in decibels is ten times the logarithm to base 10 of the power ratio. If a fiber, such as described above, were used for moderately long-range transmission of information, what would be the ratio of the input to output light power when transmitting a signal over 3.0 kilometers?

S-20. In fiber optics the so-called *acceptance angle* (see Fig. 15-62) is important; it defines the cone from within which a light guide (or fiber) can accept light for transmission.
(a) What is the acceptance angle for a plastic optical fiber the index of refraction of which is 1.48? [*Hint:* First determine the critical angle.]
(b) If a bundle of thin fibers is intended for the transmission of a visual image (for example printing on a page) then the acceptance angle should be relatively small. Explain.
(c) Consider a fiber such as described in (a). The acceptance angle of such a fiber is decreased in practice by sheathing the fiber in a transparent polymer of lower refractive index. Calculate the refractive index of the sheathing if the acceptance angle of the fiber in (a) is to be reduced to 70°.

S-21. As mentioned in Chapter 13, the problems of electrical hazards and patient safety are of considerable concern in modern hospitals. In a short paragraph, discuss the patient-monitoring arrangement pictured in Fig. 15-63 of the section on fiber optics at the end of this chapter; that is, propose, on the basis of the physics studied so far, an explanation of the components shown and of their linkage. Indicate in what manner the fiber optical coupling contributes to patient safety.

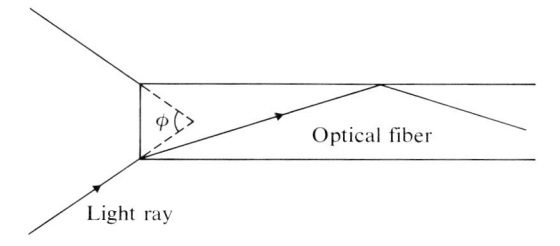

Fig. 15-62 The angle ϕ is the *acceptance angle* of a fiber optics light guide. (Question S-20).

S-22. The prominent nineteenth-century British physicist, John Tyndall (who succeeded Michael Faraday at the Royal Institution) discovered that light could be transmitted by a stream of water. He demonstrated this effect by illuminating water in a tank and then opening a tap. Water poured out carrying light into a darkened room. Using a suitable ray diagram, explain this phenomenon.

S-23. A lamp and a screen are 160 cm apart.
(a) What must be the focal length of a lens that will produce on this screen a real image of the lamp three times as large as the actual lamp?
(b) Draw the appropriate ray diagram.

S-24. Determine by two different means the focal length of a converging lens which produces a real image three times as large as the object placed 10.0 cm from the lens.
(a) Use the graphical method.
(b) Use the analytical method; that is, perform the calculation using the appropriate algebraic relation.

S-25. Draw a ray diagram for a single convex lens with an object placed between F and $2F$. Use this diagram to derive the relation

$$s_0 s_i = f^2,$$

where s_0 is the distance between the object and F on one side of the lens, and s_i is the distance between the image and F on the other side of the lens. [The relation $s_0 s_i = f^2$, with a suitable sign convention, turns out to be general for all lenses and is called the *Newtonian* form of the lens equation. The lens equation we have chosen to use in the text $(1/f = 1/p + 1/q)$ is known as the *Gaussian* form, after the great mathematician Karl F. Gauss.]

S-26. (a) What is the power in diopters of a converging crown glass lens ($n = 1.50$) each surface of which has a radius of curvature of 25 cm?
(b) What are the position and size of the image of an object located 15 cm from this lens?
(c) What are the characteristics of this image?

S-27. (a) Assuming the radii of curvature of the curved surfaces to be the same, which of these two lenses has the shorter focal length: a thin plano convex lens, or a double convex lens?
(b) Verify your answer to (a) by deriving a relationship between the focal lengths of these two lenses.

S-28. The lensmaker's formula (Eq. 15.3) as presented in Section 15.4 is based on the assumption that the lens is in air (with $n = 1.00$). A more general analysis shows that the lensmaker's formula is actually

$$\frac{1}{f} = \frac{n - n'}{n'} \left(\frac{1}{R_1} + \frac{1}{R_2} \right).$$

where n is the index of refraction of the lens material, and n' is the index of refraction of the substance in which the lens is immersed.
(a) Show that this lensmaker's formula reduces to Eq. (15.3) if the lens is immersed in air (which has a refractive index of one, to all intents and purposes).
(b) Show that the focal length of a lens ($n = 1.50$) immersed in water ($n = 1.33$) is about four times the focal length of the same lens in air.

S-29. (a) What is the focal length of a combination lens composed of 2 thin lenses, one of which is a concave flint glass lens with $f = 12.0$ cm, and the other a convex crown glass lens of $f = 8.00$ cm?
(b) What is one of the advantages of such a combination lens?

S-30. Two converging lenses, each of focal length 40.0 cm, are arranged 20.0 cm apart. A point source of light is located 30.0 cm in front of the first lens.
(a) Sketch the ray diagram showing the location of the final image of the point source with this 2-lens system.
(b) Calculate the location of the image of the point source.

S-31. (a) How thick would the nonreflecting coating on a lens have to be if the coating material's refractive index were 1.25 for light of wavelength in air of 5000 Å?
(b) For what wavelength (in air) is this a truly nonreflecting coating?

S-32. Assume that the refractive power of the cornea-lens combination of an eye is 50.0 diopters.
(a) What is the focal length of this system?
(b) If the print on a newspaper is held 30.0 cm from the eye, what is the height on the retina of the image of a 1.40-mm high newsprint letter?

S-33. Use the lensmaker's formula as stated in Question S-28 to calculate the focal length of the lens of the eye. Assume that the radius of curvature of the anterior lens surface is 11.0 mm and that that of the posterior surface is 5.7 mm, while the refractive index of the lens is 1.4 and that of the material in the vitreous cavity and in the anterior chamber (see Fig. 15-39) is about that of water.

S-34. A man wearing eyeglasses with a refractive power of 1.3 diopters wears a second pair on top of the first pair in order to obtain a 3.0 diopter power for reading. What is the focal length of the lenses in his second pair?

S-35. Normally, if one attempts to look into another person's eye, that person's pupil appears black, and no area of the retina is seen. (This is a significant problem in *ophthalmoscopy*, that is, examination of the eye for diagnositc purposes. Readers interested in this subject are referred to the section on Ophthalmoscopes at the end of this chapter.)
(a) Explain the reason for the problem described above. [*Hint:* Consider the relative positions of the eye to be inspected, the viewer's head and possible light sources.]
(b) This difficulty of seeing the retina of another person's eye is obviated by means of an ophthalmoscope. Fig. 15-64 shows schematically one type of direct ophthalmoscope. Explain how this optical instrument overcomes the difficulty, and indicate the function of the glass plate of the ophthalmoscope.
(c) What effect does the patient's own eye have when its interior is examined by an ophthalmologist?

S-36. A Fresnel lens is used to construct a "sunlight cooker."
(a) If a cooking pot is located 40 cm below the lens, what should the approximate focal length of this lens be?
(b) In what ways is a Fresnel lens better than a conventional lens for this purpose?

S-37. In judging exposure times in photography, one may use the relation that the required exposure time varies as the square of the f-number. Suppose a camera had a range of f-numbers from 1.4 to 16. What is the ratio for a given photo of the maximum to the minimum exposure times?

S-38. A converging lens of 15 diopters is used as a simple magnifying glass. What is the angular magnification? Sketch the ray diagram. (Angular magnification is described in Section 15.6 in the discussion of the telescope.)

S-39. The main concern in ophthalmology is generally with angular magnification (as described in Section 15.6 in connection with the telescope).
(a) Assuming 25 cm as the convenient distance for distinct viewing, show that the angular magnification of a lens is given by $M = D/4$, where D is the refractive power of the lens.
(b) If a patient's eye has $D = 59$, what is the angular magnification of this patient's retina as seen by an examiner?

S-40. Regarding the compound microscope discussed in Section 15.6, it was stated that the overall magnification is obtained by multiplying the magnification of the objective by the magnification of the ocular. More precisely, this product is that of the linear magnification of the objective and the angular magnification of the ocular.
(a) Draw the ray diagram of a compound microscope.
(b) Use this ray diagram and the appropriate symbols to show that the overall magnification M (angular magnification as described in Section 15.6) is given by

$$M = mm',$$

where m is the linear (or lateral) magnification of the objective, and m' is the angular magnification of the ocular.
[Hint: Let y represent the size of the object and y' the height of the image formed by the objective. Then $\tan \theta_1 = y/25$, where θ_1 is the angle subtended by y at the unaided eye at an object distance of 25 cm; and $\tan \theta_2 = y'/f_2$, where y' is the size of the image formed by the objective, θ_2 is the angle subtended at the eye aided by the microscope, and f_2 is the ocular's focal length.]

S-41. Determine the total angular magnification (as described in Question S-40) of a compound microscope in the following case. The linear magnification of the objective is 50.0, and the (angular) magnification of the ocular is 13.6.

S-42. (a) Show how two convex lenses can be arranged to make an astronomical telescope. Draw the appropriate ray diagram.
(b) Show that to a good approximation the (angular) magnification of a telescope (as defined in Section 15.6) can be expressed by the relation

$$\text{magnification} = \frac{\text{focal length of objective}}{\text{focal length of ocular}}.$$

(c) If the two lenses used in such a telescope have focal lengths of 200 cm and 20 cm, which of the two lenses is the objective lens, and what is the magnification of the telescope?

S-43. The concept of *angular magnification* is extensively used in optometry and ophthalmology. Consider the case of a converging lens of power 5.00 diopters, placed 15 mm from the eye, so that print can be read 40 cm (the usual reading distance) from the lens. The distance from the front of the eye to the retina is 25 mm.
(a) What is the nature of the eye defect being corrected?
(b) Sketch the appropriate ray diagram. (For the sake of simplicity assume that all the refraction of the eye takes place at the surface of the cornea.)
(c) What is the angular magnification of the print as seen through this lens?

ADDITIONAL APPLICATIONS TO MEDICINE AND THE LIFE SCIENCES

More on Fiber Optics in Medicine

As indicated in this chapter total internal reflection lends itself readily to a wide variety of medical applications. One such application, which has a bearing on "electrical hazards in patient care areas" (see Chapter 13) is the optical coupling of the amplified electrical signal from a patient to the readout monitor (see Fig. 15-63). Electrical isolation of the patient is thus achieved because the amplifier which amplifies the patient's recorded potential differences can be battery operated. The amplified electrical signal is next changed to a light signal and then transmitted to the monitor (which is connected to the hospital electrical supply) by means of a flexible "light pipe." This light pipe could consist of a single large diameter fiber, but for the sake of flexibility may consist of a bundle of fibers of say 0.05 to 0.10 mm in diameter. Since in this case the fiber optics system is used simply to conduct light, such relatively large diameter fibers are quite adequate; moreover they need no special alignment with one another. All that is required is that the fibers be clad with a material of lower refractive index than that of the fiber (in the case of a single fiber not touching another one, the lower refractive index material is often simply air).

Similar applications of fiber optics are the conduction of the output of a laser directly to the tissue to be coagulated and the illumination of obscure surgical fields.

If the fiber bundle is designed for viewing, that is, for the transmission of an image, then, as indicated earlier in the chapter, smaller fibers of 0.01 mm or less are employed. And in this case the fibers must be aligned so that their relative positions at the two ends of the bundle are the same. In medical applications involving direct viewing, such as in endoscopes, for example, one bundle of aligned fibers is used for viewing while generally another, perhaps central, group of fibers is used to conduct light to the field to be viewed. The illuminating fibers obviate the need for complex illuminating systems or for a light source and its attendant heat being inside the body structure to be viewed.

Recently the "viewing" aspect has found application also in the medical x-ray department where the fiber optical system has in certain cases replaced conventional optics (mirrors and lenses) for coupling the image produced by an x-ray image intensifier to a television camera.

While image transmission with fiber optics is very efficient, it is not to be construed that like the total internal reflection process itself this image transfer is 100% efficient. The efficiency of image transmission along a fiber bundle is influenced by losses at each end of the bundle as well as attenuation along the fibers. Attenuation along the fibers is due largely to absorption by the core glass (or plastic) of the fiber. In the range of visible light wavelengths this absorption may be as much as 50% for every 2 meters of fiber length.

Question

Could fiber optics lend-itself to direct magnification of an internal body structure?

Answer

Yes. Say the bundle is composed of fibers 0.05 mm diameter at the viewing end and that these fibers have each been drawn to a smaller diameter, 0.005 mm, at the opposite end. This results in a 10× magnification.

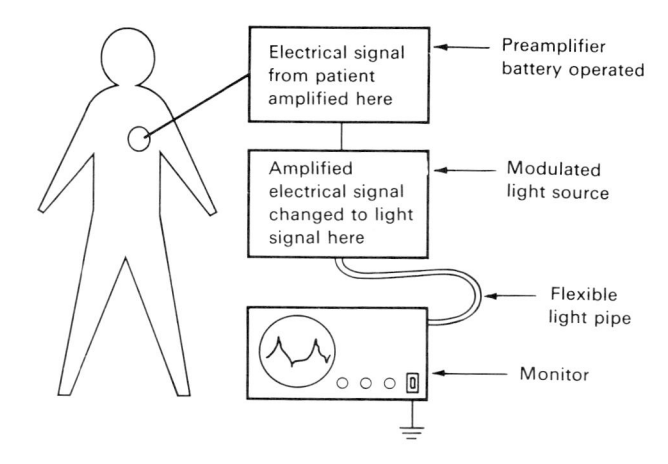

Fig. 15-63 Optical coupling. Electrical isolation of the patient is achieved by changing the electrical signal from the patient to a light signal. The light signal is transmitted through a flexible "light pipe" which works on the principle of total internal reflection. The light signal is then fed into an electronic optical demodulator which displays the information in the usual fashion of a physiological monitor.

Ophthalmoscopes

An *ophthalmoscope* is an instrument for observing the interior of the eye. Basically there are two types of these instruments, the direct ophthalmoscope and the indirect ophthalmoscope.

The direct ophthalmoscope allows the patient's eye to act as a simple magnifying lens with the retina being examined placed near the principal focus (of the optical system—lens and cornea—of the eye being examined). Thus the image of the retina appears at a great distance behind the eye being observed and the examiner is able to view the image with relatively relaxed accommodation. These facts are not to imply that a direct ophthalmoscope consists simply of the patient's own optical system. Normally, as we attempt to look into another person's eye, the pupil appears black—no area of the retina is seen, even though there may be high general illumination. The reason for this fact may be readily adduced by sketching a suitable ray diagram: those rays of light which would be reflected from the retina such as to enter the examiner's eyes must follow approximately the same paths as those rays which need to enter the eye to illuminate the retina. These particular

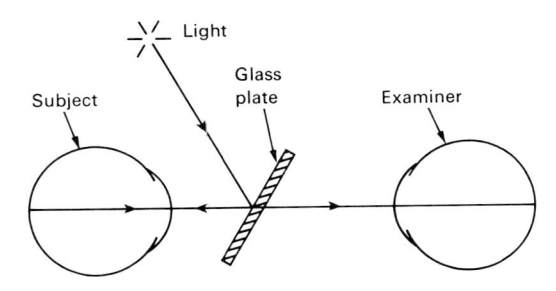

Fig. 15-64 Direct ophthalmoscope.

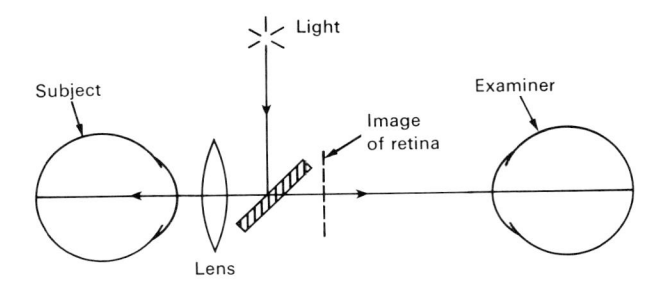

Fig. 15-65 Indirect ophthalmoscope.

incident rays are, of course, usually blocked in a large measure by the head of the examiner. One way to overcome this difficulty is to use a partially reflecting glass surface held at an angle to the eye to be examined (see Fig. 15-64). A light source from above the head of the examiner can thus throw light at the retina to be examined while the examiner can see the retina through the same glass. In the modern ophthalmoscope the function of such a partially reflecting glass plate is usually performed by a total reflecting prism (with the examiner looking just over the edge of this prism) or by a small mirror (with the examiner viewing through a small hole in the mirror).

The *indirect ophthalmoscope* (see Fig. 15-65) uses the patient's eye as a simple magnifying glass, just as does the direct ophthalmoscope. But in addition the indirect ophthalmoscope uses a convex glass lens (about 13 diopters) which performs much the same function as the ocular of an astronomic telescope [see Fig. 15-52(b)]. The advantage of indirect ophthalmoscopy is increased field size. This is gained at the expense of lower magnification than in direct ophthalmoscopy. Usually indirect ophthalmoscopes are constructed as binocular systems. In this case there is also the advantage of stereoscopic viewing.

Question

(a) In ophthalmology one generally concerns oneself with angular magnification. Using trigonometry it can be shown that (assuming the convenient distance of distinct viewing of 25 cm) the angular magnification is given by $M = D/4$ where D is the power of the patient's eye in diopters. If the patient's eye has $D = 59$, what is the magnification of his retina as seen by the examiner? Is this answer significantly different from the one obtained for linear magnification using the expression derived in Question 68?

(b) In indirect ophthalmoscopy, is the image real or virtual?

(c) What does an examiner need to do in direct ophthalmoscopy if he is looking into an ametropic eye (ametropia = condition of eye in which incident parallel rays fail to come to focus on the retina)?

Answer

(a) $M = D/4 = \dfrac{59}{4} = 15$.

Using $M = 1 + \dfrac{25\,\text{cm}}{f}$, with

$f = \dfrac{100}{59}\,\text{cm} = 1.69\,\text{cm}$, we get

$M = \dfrac{1.69 + 25}{1.69} = 16$,

which suggests that to a good approximation the magnifying glass formula $M = 1 + 25\,\text{cm}/f$ is satisfactory even though in the eye the relative position of "lens" and object cannot be adjusted.

(b) Sketching the ray diagram shows the image to be inverted and real. In direct ophthalmoscopy the image is, of course, virtual.

(c) The examiner must insert an appropriate correcting lens in front of his own eye.

Further Reading

Geometrical optics specifically for ophthalmologists is treated in Ogle, K.N., *Optics, An Introduction for Ophthalmologists*, Thomas, Springfield, Illinois, 1968.

Biomicroscopy of the Ocular Fundus

Reference to Fig. 15-26 and Fig. 15-49(c) shows that in the application of the microscope the object to be viewed must be placed inside the focal length of the objective lens. This fact has to be borne in mind in *biomicroscopy of the ocular fundus* (fundus = remotest part of a hollow organ; in ophthalmology this usually refers to the retina). So when microscopy in clinical ophthalmology is indicated, one has to use the image of the fundus, as formed by the eye's optical system, as the "object" for viewing with the microscope. It is necessary to cause the image of the fundus to be formed within the focusing range of the microscope. Since the objective lens of a microscope has a short focal length, it is necessary to bring the image of the fundus forward (in a direction toward the microscope). One way of fulfilling this requirement is to place a lens of high minus power (concave lens of short focal length) next to the cornea. A so-called *Hruby lens* is a plano-concave lens which has a radius of curvature approximately equal to that of the first surface of the cornea so that it can be placed in contact with the cornea and thus provide the fundus image within the focal length of the microscope (see Fig. 15-66).

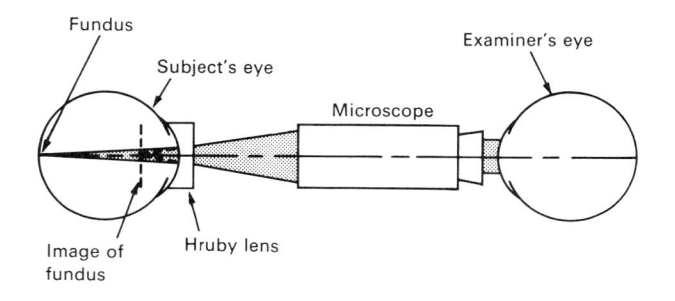

Fig. 15-66   Biomicroscopy of the ocular fundus. Hruby lens in contact with the cornea.

Question

Sketch the ray diagram which shows the image formed by a Hruby lens. Is this image real or virtual?

Answer

Virtual.

SUGGESTED READING

BASFORD, S., and J. PICK, *The Rays of Light, Foundations of Optics.* London: Samson Low, Marston and Co., 1966. Light reading; a very elementary descriptive treatise of about 125 pages. Worth consulting for the numerous excellent colored explanatory diagrams. Matters of particular interest include an explanation of the phase-contrast microscope and both additive and subtractive color mixing.

BURKHARDT, D., W. SCHLEIDT, and H. ALTNER, *Signals in the Animal World.* New York: McGraw-Hill, 1967. How an insect's motion is based on its vision, how owls depend upon acoustics, and how data processing is applied in vision are some of the many topics examined in detail. There are magnificent color photographs, including that of the eye of a gad-fly, showing how the multitudinous lenses are arranged in regular hexagonal arrays. Research in the behavioral sciences is integrated with recent findings in biochemistry and biophysics. Written in collaboration with a dozen prominent scientists in their respective fields, the style is nevertheless of a down-to-earth variety which will appeal to most readers. Medical students as well as systems engineers should find this book stimulating. Highly recommended as collateral reading.

FEYNMAN, R. P., R. B. LEIGHTON, and M. SANDS. *The Feynman Lectures on Physics*. Reading, Mass.: Addison-Wesley, 1963. Chapter 26, "Optics: The Principle of Least Time," provides a more exhaustive treatment of Fermat's principle than we did; the principle is employed to explain refraction as well as reflection. The mathematics of this chapter is very elementary and hardly beyond that employed in our book. Chapters 35 and 36 treat the human eye and various aspects of vision, including color vision. In these lectures, Professor Feynman took particular pains to point out that subject boundaries such as those between physics and physiology are man-made for convenience and should thus not be used to avoid making a most natural bridge between fields.

GLUCK, I. D. *Optics, The Nature and Application of Light*. New York: Holt, Rinehart and Winston, 1964. A 150-page paperback intended for a wide audience: students, amateur astronomers, camera fans, and hobbyists. This nonmathematical treatment deals with the basic principles of light and follows this introduction with the essentials of modern optical instruments starting with an interesting early history of glass and lens grinding. A chapter on the eye is intended to point out that nature is still a formidable competitor of man in the ingenuity with which she (nature) "uses the laws of science."

NUSSBAUM, A., *Geometric Optics: An Introduction*. Reading, Mass.: Addison-Wesley, 1968. An interesting short (130 pages) book which covers, in its first few chapters, elementary lens systems in the traditional manner, using only algebra and trigonometry. This part of the book makes useful supplementary reading to Chapter 15 of this text. The object of the latter part of Nussbaum's book is to show that the analysis of a lens system—no matter how complex—can be simplified and made amenable to computer solution. That part of the book would be of interest to students who have studied matrix algebra and particularly to those who want practice in using the computer to solve physics problems. A brief Appendix deals with FORTRAN programming.

SEARS, F. W., *Optics*. Reading, Mass.: Addison-Wesley, 1958. This standard text by a well-known physicist-teacher-author covers in some depth some of the topics of our Chapters 14, 15, and 16. Recommended for the student who would pursue the subject of optics. While calculus is used occasionally, most of the mathematical analysis can be mastered if the reader has some acquaintance with trigonometry.

VAN AMERONGEN, C., (adaptation and translation), *The Way Things Work*. New York: Simon and Schuster, 1967. This book, a translation of the German *Wie funktioniert das?*, is not a reference in the ordinary sense. Rather it has been designed "to give the layman an understanding of how things work, from the simplest mechanical functions of modern life to the most basic scientific principles and complex industrial processes that affect our well-being." Pages 133 to 164, which encompass, with plentiful illustrations, topics ranging from lenses and camera shutter speeds to the electron microscope, are recommended for those with practical minds.

Niels Bohr

16 Quantum Physics

The story of efforts to understand the nature of light, begun in the preceding chapters, now leads us to the realm of quantum physics where we meet again some of the ideas introduced in Chapters 1 and 4.

Nature's greatest surprises were in store for twentieth century physicists. The wave theory of light had been verified in great detail in myriad situations concerned with the transmission of light from one point to another. Then Einstein in 1905 suggested that in a process involving the absorption of light (the photoelectric effect first noticed by Hertz) there was evidence that the energy was not continuously distributed over a wave front, but rather localized into bundles of energy which are now called *photons*. Within the next twenty years, the evidence accumulated for the existence of photons, particles of light which were manifest in the emission, absorption, or scattering of light.

What now then? One has to face the fact that one particle, a photon, shows interference effects (a typical wave phenomenon). Following the suggestion in 1924, of Louis de Broglie (1892–   ) of France, and the experiments of G. P. Thomson (1892–1975), C. J. Davisson (1881–1958), and L. H. Germer (1896–   ), we now recognize that this behavior is not confined to photons: *All particles show interference effects*.

As we shall describe in the course of this chapter, this astounding fact concerning particles is coped with in quantum mechanics by associating with the state of any mechanical system, a "wave function." In this theory, the wave functions allow calculation of the probability of a given process taking place; one no longer determines certainties as in Newtonian mechanics. And abstruse as it may appear, the fact is that the interference which occurs is an interference of probabilities.

It is this theory, quantum mechanics, now approximately half a century old, which is *the* successful theory of atoms and molecules. It is this theory which is our guide as we seek to understand the nucleus and the many elementary particles which have been discovered during recent decades.

"The task of science is both to extend the range of our experience and to reduce it to order."

Niels Bohr
(1885–1962)

A renowned precursor to the quantum mechanics of today was Bohr's theory of the hydrogen atom. At the age of 26, Niels Bohr, having received his doctorate in Denmark, joined Rutherford's group at Manchester University in 1911 and was caught up in the excitement aroused by the discovery that atoms have a charged massive central nucleus. He pondered the puzzles posed by a nuclear model of the atom.

What are some of these puzzles? An electron orbiting about a nucleus experiences a large centripetal acceleration. Therefore, according to Maxwell's electromagnetic theory, the electron should continually radiate electromagnetic waves, and, as its energy is thus radiated away, collapse into the nucleus. Besides having this grave defect of predicting a rapid collapse of the atom's electronic structure to nuclear dimensions, this model also suggested that the radiated spectrum should be continuous with frequencies determined by the electron's frequencies of revolution. But experiment shows that the atom has a stable structure some 10,000 times larger than its nucleus and emits only the discrete frequencies of light that constitute its *line* spectrum.

To produce a theory reconciled with these experimental facts, Bohr tried modifications of Newtonian mechanics and radiation theory. He came to the realization that if one thought of the light radiated by an atom in terms of photons, the line spectrum was evidence that there were restrictions on the possible changes of an atom's energy and that an atom probably could possess only certain discrete values of energy (now called the *energy levels* of the atom). The frequency of every line in the hydrogen atom spectrum was given with remarkable precision in terms of a simple formula for the atom's energy levels. And this formula for the energy levels was found to follow from the assumption that, for some reason foreign to Newtonian mechanics, the orbital angular momentum of the electron in a hydrogen atom was restricted to be an integral multiple of $\hbar$ (Section 3.9).

Bohr's theory was published in 1913. For a decade, experiments stimulated by this work confirmed the existence of discrete energy levels of atoms and lent credence to Bohr's picture of photon emission or absorption as associated with an atomic transition from one energy level to another. But Bohr realized that his theory was only a guide toward some more comprehensive theory which would embrace, in some quite natural way, the growing array of peculiarities encountered in atomic behavior.

In 1925, as young men like W. K. Heisenberg (1901–) and P. A. M. Dirac (1902–) created the first outlines of quantum mechanics, they congregated at Copenhagen around Bohr. Exposed to Bohr's probing and stimulated by his eagerness to consider new ideas, this Copenhagen school was for years the center of activity in quantum theory. The work of this group may well come to be recognized as an unsurpassed flowering of the human mind.

16.1 PROPERTIES OF PHOTONS

As we have emphasized in Chapter 14, the nineteenth century saw the accumulation of evidence of the wave nature of light. This evidence still stands today and must be faced by any acceptable theory. But in Chapter 14 we have presented only part of nature's lessons on light. In our century, physicists have met incontrovertible evidence for the existence of *photons*. As we stated in Chapters 1 and 4, photons are particles of light, or, in general, particles of electromagnetic radiation. Photons are also called the *quanta of the electromagnetic field*. Before examining the relevant experimental evidence, we shall summarize some of the present knowledge regarding photons.

Photon Energy, Momentum, and Wavelength

A beam of electromagnetic radiation, considered as an electromagnetic wave, is characterized by its frequency f or its wavelength λ which are related, according to Eq. (14.3), by

$$f = c/\lambda.$$

The same beam, considered as a stream of photons, is characterized by the energy E* or the momentum p of the individual photons. A photon has no mass and travels at a speed, $c = 3.00 \times 10^8$ m/sec. Its energy and momentum are related by

$$E = cp. \qquad (4.23)$$

The vital connecting link between these two descriptions of the *same beam of radiation*, proposed by Einstein, is that the photon energy E is proportional to the frequency f of the electromagnetic wave:

$$E = hf. \qquad (16.1)$$

The constant of proportionality h is named *Planck's constant* for reasons that will be mentioned in the following section. Modern experiments yield the value, $h = 6.626 \times 10^{-34}$ joule-sec. (Nature's unit of angular momentum, introduced in Section 3.9, is Planck's constant divided by 2π: i.e., $\hbar = h/2\pi$.)

Evidently electromagnetic radiation can be classified according to the energy of its photons, or the wavelength, or the frequency, whichever is most convenient. For example, from $f = c\lambda$, Eq. (16.1) can be written to show the relationship between photon energy E and wavelength:

$$E = hc/\lambda. \qquad (16.2)$$

Photon energies are usually specified in electron volts, and wavelengths in angstrom units. Inserting the numerical values of h and c and the required conversion factors, Eq. (16.2) can be written in a form which is

*Note that here the symbol E denotes the energy of a photon, while in Chapter 14 the same symbol was used to denote a completely different quantity, an electric field. The reader is cautioned to interpret symbols according to the context.

convenient for calculations:

$$E = \frac{(12.4 \times 10^3 \text{ Å-eV})}{\lambda} \quad (16.3)$$

where E is the photon energy in *electron volts* and λ is the wavelength in *angstrom units*.

Example 1 Find the energy of the photons in a beam whose wavelength is
(a) 6.2×10^3 Å (orange light);
(b) 4.13×10^3 Å (violet light);
(c) 1.0 Å (x-ray);
(d) 10.0 meters (radio wave).

Solution
(a) The orange photon has an energy which is, from Eq. (16.3),

$$E = \frac{(12.4 \times 10^3 \text{ Å-eV})}{(6.2 \times 10^3 \text{ Å})}$$
$$= 2.0 \text{ eV}.$$

(b) The violet photon has an energy

$$E = \frac{(12.4 \times 10^3 \text{ Å-eV})}{(4.13 \times 10^3 \text{ Å})}$$
$$= 3.0 \text{ eV}.$$

Notice that the photons of visible light have energies of a few electron volts. These are typical of energies involved in atomic and chemical processes.

(c) This x-ray photon has an energy

$$E = \frac{(12.4 \times 10^3 \text{ Å-eV})}{(1.0 \text{ Å})}$$
$$= 12.4 \times 10^3 \text{ eV}.$$

In medical practice, most diagnostic work employs x-rays with wavelengths from 0.1 Å to 1.0 Å, that is photons of energies from approximately 10^4 to 10^5 electron volts. Thus, typical x-ray photon energies are ten thousand times *greater* than the energy of photons of visible light.

(d) The wavelength of this radio wave = 10.0 meters = 10.0×10^{10} Å. Therefore, the radio wave photon has an energy

$$E = \frac{(12.4 \times 10^3 \text{ Å-eV})}{(10.0 \times 10^{10} \text{ Å})}$$
$$= 12.4 \times 10^{-8} \text{ eV}.$$

A single radio wave photon has so little energy that it is undetectable. Many such photons are required to build up a detectable effect.

Next we turn to an important relationship between the photon's wavelength λ and its momentum p. Eq. (14.3) implies

$$\lambda = c/f.$$

Substituting $f = E/h$ from Eq. (16.1), we get

$$\lambda = hc/E.$$

To express E in terms of p we use $E = cp$, Eq. (4.23), and obtain

$$\lambda = h/p. \qquad (16.4)$$

This equation, which associates a wavelength with a photon's momentum, we shall meet again. It turns out to have a universal significance!

Beam Intensity and the Number of Photons

The *intensity*, I, of a beam of radiation is determined by both the number of photons in the beam and the energy of each photon. Suppose that, when a light beam shines on a surface perpendicular to the beam, there are N identical photons per second per unit area striking the surface. The beam intensity, being defined as the energy per second per unit area, is therefore given by

$$I = N \times \text{(energy of a single photon)}. \qquad (16.5)$$

Now when we describe the same light beam as an electromagnetic wave with an amplitude E_0 (recall that the amplitude E_0 is the maximum value attained by the fluctuating electric field), the *intensity of the beam is determined by the square of the wave amplitude*:

$$I = (1.33 \times 10^{-3} \text{ watt/volt}^2) E_0^2. \qquad (14.2)$$

Here we see a connection between the two ways of thinking about a light beam: as a wave or as a collection of photons. The *square of the wave amplitude* is related to the *number of photons in the beam*. Bright parts of an illuminated screen are struck by many photons per second (N is large) and here the electromagnetic wave has a large amplitude E_0.

As a light wave spreads out from a candle, the amplitude E_0 and the intensity I diminish. However, the energy radiated remains concentrated in "lumps" that we call photons. The energy of a single photon does *not* get spread out across a broad wave front.

Example 2 A desk is illuminated with violet light of wavelength 4.13×10^3 Å. The amplitude of this electromagnetic wave is 61.3 volts/m. Find the number of photons per sec per meter^2 striking the table top.

Solution In mks units the intensity is given by Eq. (14.2), which yields

$$I = (1.33 \times 10^{-3} \text{ watt/volt}^2) \times (61.3 \text{ volts/m})^2$$
$$= 5.00 \text{ watts/m}^2.$$

Using the definition of the electron volt (1 eV = 1.60×10^{-19} joule), we get

$$I = \frac{5.00 \text{ (joules/sec m}^2)}{1.60 \times 10^{-19} \text{ (joule/eV)}}$$
$$= 3.12 \times 10^{19} \text{ eV/sec m}^2.$$

Now the violet photon with $\lambda = 4.13 \times 10^3$ Å, according to Eq. (16.3) has an energy

$$E = \frac{(12.4 \times 10^3 \text{ Å-eV})}{(4.13 \times 10^3 \text{ Å})}$$
$$= 3.00 \text{ eV}.$$

With N of these photons striking the desk per second per square meter the intensity, from Eq. (16.5), is

$$I = N \times (3.00 \text{ eV}).$$

Therefore,

$$N \times (3.00 \text{ eV}) = 3.12 \times 10^{19} \text{ eV/sec m}^2,$$

and

$$N = 1.04 \times 10^{19} \text{ photons/sec m}^2.$$

Evidently some billion billion photons strike this page of the book every second, as you read it.

16.2 EXPERIMENTAL EVIDENCE OF PHOTONS

The Photoelectric Effect

In a modern physics laboratory equipped with the particle detectors described in Chapter 1, it is a commonplace fact that electromagnetic energy in light, x-rays, and γ-rays is always found to be emitted or absorbed in localized lumps each containing an energy $E = hf$. This discovery, which caused a great shock in early twentieth century physics, has an interesting history.

Einstein, again in 1905, the annus mirabilis, thinking about the ejection of electrons from a metal surface which was illuminated with light, can be said to have "discovered" the photon. The phenomenon in question is called the *photoelectric effect* (Fig. 16-1). A clean metal surface, illuminated by a light beam of sufficiently high frequency, f, immediately (the time delay is less than 10^{-9} sec) ejects electrons with a

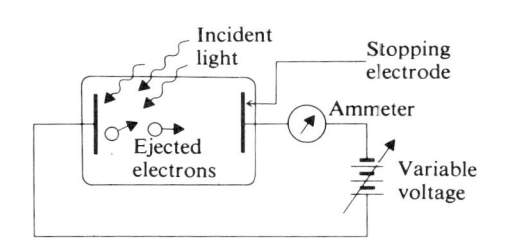

Fig. 16-1 Photoelectric effect.

variety of kinetic energies which range up to a maximum value, K. Increasing the intensity of the light beam does not affect the kinetic energy of the ejected electrons in the slightest. The only effect of an intensity increase is that the number of electrons ejected per second increases proportionately. But the maximum kinetic energy K of the ejected electrons depends only upon the frequency f of the light according to what is called the Einstein photoelectric equation

$$K = hf - W, \qquad (16.6)$$

where W (named the work function) is a constant whose value depends on the nature of the surface.

Einstein deduced this result by postulating that the light beam consists of photons, each of energy hf. An electron inside the metal absorbs all the energy of a photon and then may escape from the metal, after losing at least an energy W in this escape. The work function W is the binding energy of the most energetic electrons in the metal. Energy conservation then immediately requires that the largest kinetic energy K of the ejected electrons is given by the photoelectric equation, Eq. (16.6).

It should be pointed out that experimental knowledge of the photoelectric effect was very fragmentary at the time that Einstein leaped to the right answer. It was not until 1916 that this photoelectric equation was experimentally verified in detail by Millikan. The maximum kinetic energy K of the ejected electrons was determined by measuring the voltage required to prevent any electrons from reaching a stopping electrode (see Fig. 16-1) which faced the metal surface from which the electrons were emitted.

Example 3 A tungsten surface is illuminated with ultraviolet light of wavelength 2.0×10^3 Å. Electrons are ejected from the tungsten, and it is found that a retarding voltage of 1.6 volts applied between the tungsten and another (stopping) electrode is just enough to prevent any electrons from reaching the stopping electrode. What is the work function W for this surface?

Solution Applying conservation of mechanical energy, as discussed in Example 3 of Chapter 11, we conclude that an electron, which is reduced to rest after moving against the electric forces through a difference of potential of 1.6 volts, started with a kinetic energy of 1.6 eV. Thus, $K = 1.6$ eV. Calculating the energy of the incident photons from Eq. (16.3), we obtain

$$E = \frac{(12.4 \times 10^3 \text{ Å-eV})}{(2.0 \times 10^3 \text{ Å})}$$
$$= 6.2 \text{ eV}.$$

Max Planck (1858–1947)

Physics Today

Now the photoelectric equation,

$$K = hf - W, \qquad (16.6)$$

gives

$$1.6 \text{ eV} = 6.2 \text{ eV} - W.$$

Therefore,

$$W = 4.6 \text{ eV}.$$

That is, the work function for the tungsten surface is 4.6 eV.

Einstein's bold postulate, that light had a particle nature after all, even though physicists had successfully pictured light as a wave for a hundred years, was suggested to him by the prior work of Max Planck (1858–1947). In 1900, Planck had found that the different amounts of electromagnetic energy radiated at different wavelengths in the continuous spectrum emitted by a hot substance could be understood only if one assumed that energy transfers from matter to the radiation always occurred in bundles of the size $E = hf$. (Planck attributed this limitation to the nature of the oscillating matter, not to the idea that the radiation consists of photons.)

So far we have spoken only of Einstein's successful way of interpreting the photoelectric effect. Of course Einstein, and many others, tried first to understand the phenomenon in terms of the way the electric vector of an electromagnetic wave would be expected to interact with the electrons in the metal. This "classical" theory failed in three important respects.

1. The kinetic energy acquired by an electron should increase as the field amplitude E_0 increases, and consequently as the light intensity increases. *But this does not happen.*
2. The electron's energy should not depend particularly on the light frequency f. *But it does.*
3. With a low-intensity light, it would take many seconds for an electron to accumulate sufficient energy to escape if the incident energy in the electromagnetic wave were continuously distributed across the wave fronts. *But electron ejection commences without delay.*

Clearly, old ideas had to be amended to explain the photoelectric effect. The success of the photon hypothesis was striking in this instance. If true, photons should abound in our world and the thing to do was to examine other processes for evidence of their existence. A promising region to investigate would be at shorter wavelengths because here the energy of each photon would be larger and the presence of photons should thus be easier to detect.

Bremsstrahlung X-rays

The photoelectric effect involves the *absorption* of light. *Emission* processes also contribute evidence of photons.

Consider a specific case. Say electrons in an x-ray tube are accelerated through a difference of potential of 25×10^3 volts; they acquire a kinetic energy of 25×10^3 electron volts before striking the anode. Within the anode, some of these electrons passing close to a highly charged nucleus are subjected to a large acceleration and radiate away energy as electromagnetic waves of x-ray wavelength (see Fig. 16-2). X-rays produced in this way are called Bremsstrahlung, the German word for "braking radiation." Measurement of the wavelengths of Bremsstrahlung x-rays reveals a continuous spectrum with a minimum wavelength (Fig. 16-3).

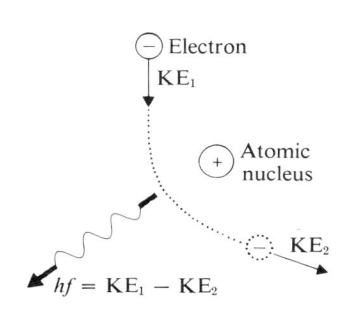

Fig. 16-2 Bremsstrahlung, a radiative "collision" of a high-speed electron of original kinetic energy KE_1 with a target nucleus.

Fig. 16-3 Bremsstrahlung gives a continuous x-ray spectrum with a minimum wavelength.

The reason for a minimum wavelength can be easily understood in terms of photons. In this process of an electron radiating a photon, the maximum energy that the photon can acquire is the entire kinetic energy of the electron. And this maximum photon energy, $E = hf$, corresponds to a maximum frequency f or a minimum wavelength, $\lambda \; (= c/f)$.

Example 4 A potential difference of 50×10^3 volts is maintained between the cathode and anode of an x-ray tube. Find the shortest wavelength λ in the spectrum of the emitted x-rays.

Solution Electrons acquire a kinetic energy of 50×10^3 eV before they hit the anode. The maximum photon energy is therefore 50×10^3 eV. From Eq. (16.3), a photon with this energy has a wavelength given by

$$\lambda = \frac{(12.4 \times 10^3 \text{ Å-eV})}{(50 \times 10^3 \text{ eV})}$$
$$= 0.25 \text{ Å}.$$

Photons with a lower energy will have a longer wavelength than this.

*Experiments which revealed a longer wavelength present in the scattered radiation had been reported previously (1913 and 1920) by the Canadian physicist, J. A. Gray.

Courtesy of Washington University, St. Louis, Missouri

A. H. Compton (1892–1962)

*Compton's equation for the wavelength λ' of the scattered photon when the incident photon has a wavelength λ is

$$\lambda' - \lambda = (h/mc)(1 - \cos\theta),$$

where θ is the scattering angle (see Fig. 16-4) and m is the mass of the recoiling electron. The quantity h/mc has the value 0.024 Å.

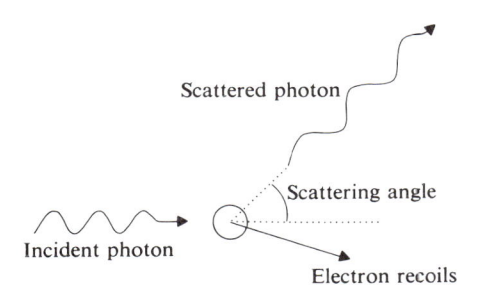

Fig. 16-4 Compton effect. The scattered photon has less energy and therefore a longer wavelength than the incident photon.

Compton Scattering of X-rays

The particle nature of x-rays is very evident when they are scattered by free or loosely bound electrons. The American physicist, A. H. Compton (1892–1962), in an experimental investigation of x-ray scattering, found that part of the scattered radiation had a longer wavelength than that of the incident radiation.* On the basis of classical physics, the scattered radiation results from electron oscillations produced by the incident radiation and therefore has the same frequency (and wavelength) as this incident radiation. Here again, classical physics gives a wrong answer. But by analyzing the scattering process as a photon-electron collision (Fig. 16-4), Compton gave a clear and precise account of his experimental observations.

In an interaction or "collision" between a photon and an electron, momentum and energy are conserved. Therefore, after the collision, the electron recoils. Since the recoiling electron has acquired kinetic energy, the scattered photon must have less energy (and therefore a lower frequency and a longer wavelength) than the incident photon.

Compton calculated the difference between the wavelength of the scattered photon and that of the incident photon from the equations expressing conservation of momentum and energy in a photon-electron collision. His result is that, when the scattering angle (see Fig. 16-4) happens to be 90°, the scattered x-rays should have a wavelength 0.024 Å longer than the wavelength of the incident x-rays. Larger increases in wavelength are predicted for larger scattering angles.*

The agreement between Compton's theoretical predictions and his experimental results was impressive. This particularly direct and convincing evidence of the existence of photons, particles of light which carry both energy and momentum, led to the award of the 1927 Nobel prize in physics.

16.3 ENERGY LEVELS

Photons and Energy Levels of Atoms

Photons were the key to understanding the emission and absorption of light by atoms.

Voluminous tables of precise measurements of the wavelengths of the lines in atomic spectra were accumulated throughout the nineteenth century. No one, before Bohr, had seen a way of reading the message that nature was broadcasting. It was apparent that there was order in the different spectra, and painstaking efforts were made to find an empirical formula that would fit the facts. The first success came in 1885 to a Swiss schoolteacher, Johann Balmer (1825–1898), who found a formula for certain lines in the visible spectrum of the simplest atom, hydrogen.

624 QUANTUM PHYSICS

In the early twentieth century, all the lines in the hydrogen atom's spectrum were known to be given by the then seemingly peculiar empirical rule: Calculate a sequence of terms proportional to the reciprocal of the square of an integer n; then each frequency in the spectrum is the difference of two of these terms.

Bohr saw the physical laws underlying this rule. When an atom emits a photon, conservation of energy implies that *the energy of the atom must change* from an initial value E_u (the subscript u denotes the upper energy level, as in Fig. 16-5) to a lower value E_ℓ such that

$$E_{photon} = E_u - E_\ell. \qquad (16.7)$$

This "Bohr frequency condition" determines the photon frequency f, since $E_{photon} = hf$. The empirically determined sequence of terms whose differences determine the frequencies evident in the hydrogen atom spectrum must then be the possible values of the energy of the hydrogen atom. These energies,* called the *energy levels* of the hydrogen atom, are given by

$$E_n = \frac{-(13.6 \text{ eV})}{n^2} \qquad (16.8)$$

where n is any positive integer and is called the *principal quantum number*. That is: $E_1 = -13.6$ eV, $E_2 = -3.40$ eV, $E_3 = -1.51$ eV, etc., as displayed in Fig. 16-6.

Thus, for example, a hydrogen atom can exist for a short while ($\sim 10^{-8}$ sec) in a state with energy $E_3 = -1.51$ eV. If, after the emission of a photon, the atom is left in the state with the lower energy, $E_2 = -3.40$ eV (see Fig. 16-6), the photon emitted must have an energy, according to Eq. (16.7), given by

$$E_{photon} = E_3 - E_2 = (-1.51 \text{ eV}) - (-3.40 \text{ eV})$$
$$= 1.89 \text{ eV}.$$

The wavelength λ of this photon is, by Eq. (16.3),

$$\lambda = \frac{(12.4 \times 10^3 \text{ Å-eV})}{(1.89 \text{ eV})}$$

$$\lambda = 6.56 \times 10^3 \text{ Å}.$$

The novel idea thus advanced by Bohr was that the energy of the hydrogen atom (and in fact all atoms, molecules, and any bound system) can have only certain *discrete values* (instead of a continuous range of values) in its bound states. That is, there exist discrete *energy levels*.

The lowest possible of these energy levels is called the *ground state* and all higher levels are called *excited states*. Given sufficiently high energies the electron can escape from the atom. Then the separated parts can have any amount of kinetic energy so there then exists a continuum (Fig. 16-6), that is, a continuous range of possible energies, corresponding to such an ionized atom.

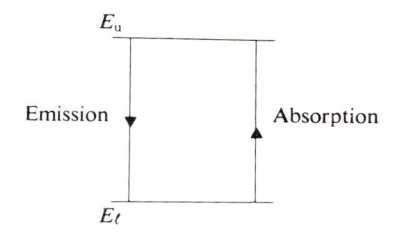

Fig. 16-5 Two energy levels, E_u and E_ℓ, and the transitions which accompany either emission or absorption of energy.

*The total energy of the hydrogen atom, including the rest energy of the proton ($M_p c^2$) and the rest energy of the electron (mc^2) is $M_p c^2 + mc^2 + E_n$. The quantity, $-E_n$, is therefore just the binding energy of the atom, and E_n is a negative number (so $-E_n$ is a positive number) because this proton-electron system with attractive Coulomb forces is stable against decomposition into a separated electron and proton.

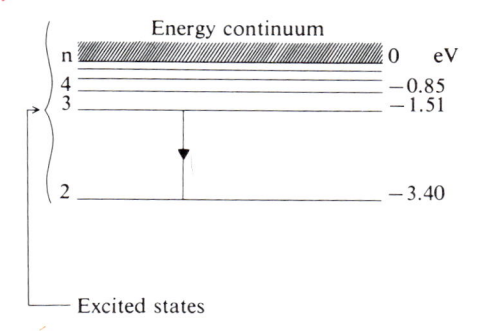

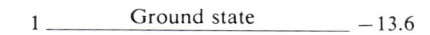

Fig. 16-6 Energy levels of the hydrogen atom.

16.3 ENERGY LEVELS 625

When an atom gives off energy it passes from an upper to a lower energy level. When an atom absorbs energy it passes from a lower to a higher level (Fig. 16-5). For the absorption of a photon, the Bohr frequency condition, Eq. (16.7), still applies, but now the lower energy E_l is the initial energy of the atom.

The excited states of an atom are unstable. Each state has its own characteristic half-life, usually of about 10^{-8} sec. For an isolated atom, the transition from an excited state to a lower energy level occurs spontaneously and is accompanied by the emission of a photon. Such a decay or transition is entirely analogous to the radioactive decay of nuclei, discussed in Chapter 1. Apparently there is a definite probability per unit time for any given decay, but one cannot say at what instant a given atom in an excited state will decay. The timing of such events follows the laws of chance!

The existence of energy levels in all atoms and the truth of the preceding statements were abundantly verified in experiments performed during the decade following the publication of Bohr's great paper in 1913 in which he both used and laid aside parts of the mechanics of Newton.

Newtonian mechanics, applied to a system of a proton and an orbiting electron, does *not* lead to discrete energy levels. Bohr realized that Newtonian mechanics would have to be modified in some way. His preliminary efforts met only partial success. The extent of the drastic revision of thought required to build a mechanics that would encompass the experimental knowledge of atoms was not apparent until the next decade.

Nuclear Energy Levels

After recognizing that a hydrogen atom had discrete energy levels, Bohr soon realized that this was a particular instance of a very general quantum mechanical phenomenon. The energy of *any bound system* is restricted to certain discrete values which are called the energy levels of the system. For instance, the nucleus $_{28}Ni^{60}$ has a stable ground state and many discrete energy levels corresponding to excited states.

Fig. 16-7 shows the nuclear energy levels involved in the β-decay of radioactive cobalt-60 (Section 1.7):

$$_{27}Co^{60} \rightarrow {_{28}Ni^{60}} + \beta^- + \bar{\nu}.$$

The cobalt nucleus makes a transition from its ground state to an excited state of the nickel nucleus as an electron and an antineutrino are created which share a total of 0.3 MeV available as kinetic energy. This excited state of the nickel nucleus is unstable and decays (for all practical purposes immediately) to another excited state which is 1.2 MeV lower, as it emits a 1.2-MeV γ-ray. Finally the nucleus in this state, which is 1.3 MeV above the ground state, suddenly changes to the ground state configuration and simultaneously emits a 1.3-MeV γ-ray.

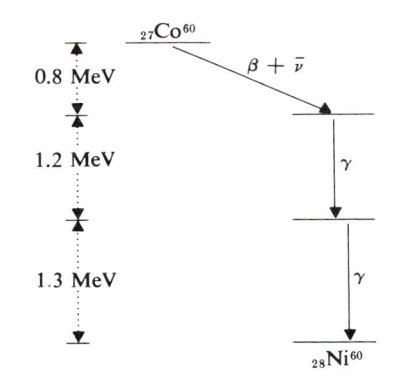

Fig. 16-7 Nuclear energy level diagram showing the β-decay of cobalt-60.

The β-decay is therefore accompanied by two γ-rays of energies 1.2 MeV and 1.3 MeV. It is these γ-rays that are so widely used in hospitals throughout the world in the treatment of cancer.

16.4 SOME QUANTUM MECHANICS

De Broglie Wavelength of a Particle

An electromagnetic wave of wavelength λ is associated with photons of momentum p such that

$$\lambda = h/p. \tag{16.4}$$

In 1924, a young French student, Louis de Broglie, submitted a Ph.D. thesis in which he speculated that, with *any particle* of momentum p, there was in some way associated a wave with wavelength λ given by

$$\lambda = h/p. \tag{16.9}$$

It soon became apparent that he was right! The formula $\lambda = h/p$ does have a universal significance.

This "de Broglie wavelength" in the case of most particles is very small if the energy of the particle is high enough to make its detection easy. As we have learned with light and x-rays, typical wave effects like diffraction are not evident if the wavelength is much shorter than the significant geometrical dimensions in an experiment. For electrons with a kinetic energy of 150 eV, the de Broglie wavelength calculated from $\lambda = h/p$ is 1.00 Å. This is in the same range as x-ray wavelengths and suggests that, as with x-rays, crystals could be used to produce diffraction patterns consisting of well-defined intensity maxima at locations of constructive interference of the waves scattered from the regularly spaced atoms.

An English physicist, G. P. Thomson (the son of the discoverer of the electron, J. J. Thomson), thus set out to obtain diffraction patterns with electrons like those obtained with the transmission of x-rays through thin films. By 1928, he had succeeded in a very convincing fashion.

In the meantime, in the Bell Telephone Laboratories in the United States, C. J. Davisson and L. H. Germer had already (1926) come across clear evidence of constructive interference of electrons reflected from the surface of a single metal crystal. Moreover, they had verified the formula, $\lambda = h/p$, to better than 1%. Since that time, diffraction patterns, confirming always that $\lambda = h/p$, have been obtained for neutrons and even for a structure as large as an α-particle. The evidence is overwhelming that $\lambda = h/p$ refers to a universal connection between wavelength and momentum, regardless of the nature of the object involved. The consequences of this discovery have been great.

De Broglie's insight was recognized with the award of the 1929 Nobel prize. Davisson and Thomson shared the 1937 Nobel prize for their discoveries.

Physics Today

Louis de Broglie (1892–   )

The Mystery of Particle Interference

To bring out the startling implications of the fact that particles produce interference patterns, we shall consider a simple idealized interference experiment with electrons: the Young double-slit (Fig. 14-9) experiment but with the light source replaced by an electron gun (Fig. 16-8) which shoots out electrons with a well-defined energy.

A photographic plate will detect the arrival of electrons on the screen. With both slits open we find the photographic plate shows exactly the distribution of intensity I at the screen depicted in Fig. 14-9. This is an interference pattern with intensity maxima $[I = 4I_1$, Eq. (14.8)] and intensity minima $[I = 0$, Eq.(14.9)]. The separation of interference maxima on the screen is that expected from Eq. (14.13), $LL_1 = l\lambda/S_1S_2$, where now λ is given by de Broglie's relation $\lambda = h/p$ and p is the momentum of each electron.

But what is going on? No one has ever been able to explain these observations in purely classical terms. And it is freely predicted that no one ever will.

Let us investigate more closely. We decrease the intensity by having the gun fire so few electrons per second that there is only one electron in transit at a time. Now we have to increase the exposure time of the film on the screen but the *interference pattern is unchanged*. This shows that interference does *not* arise from the interaction of one electron in transit with another in transit. *An electron interferes with itself!* What now?

Interference therefore has to be understood by thinking about a *single particle* in transit and two alternatives, slit 1 or slit 2. Certain points are clarified by replacing the photographic plate of Fig. 14-9 by the very small idealized counter of Fig. 16-8 that counts the arrival of individual electrons. It is found that this counter clicks erratically as though the timing of the arrival of an electron were governed by chance. At intensity maxima, the counting rate is large, and where the intensity is low, the counting rate is small. In fact, the intensity is simply proportional to the counting rate.

Evidently the *probability P for a single particle* to reach a given small region on the screen will determine the intensity I that will be observed at that point after many particles, one after the other, have gone from the electron gun to the screen. Since P and I are proportional, for each relationship between intensities, we can write a corresponding relationship between probabilities. Thus, corresponding to the intensity equation, $I = I_1 + I_2 +$ (interference term) given as Eq. (14.7), we have for the probability:

$$P = P_1 + P_2 + \text{(interference term)} \qquad (16.10)$$

where the interference term has different values at different locations.

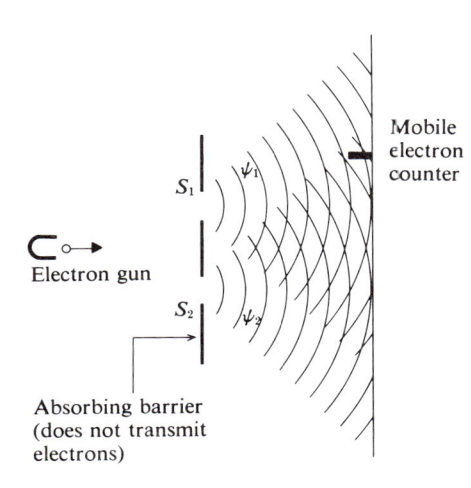

Fig. 16-8 Double-slit experiment with electrons.

628 QUANTUM PHYSICS

At intensity maxima where the interference is constructive, corresponding to the equation $I = 4I_1$ [Eq (14.8)], we have

$$P = 4P_1. \qquad (16.11)$$

Where the interference is destructive, giving intensity minima corresponding to the equation $I = 0$ [Eq. 14.9)], we have

$$P = 0. \qquad (16.12)$$

We see that nature has presented us with *interfering probabilities*.

The probability P_1 refers to the situation where slit 1 is open and slit 2 is closed. We can consider P_1, and consequently the intensity I_1, as uniform across the region of interest on the screen.* That is, there is no alteration of maxima or minima on the screen with just one slit open. The probability P_2 has a similar interpretation.

Now we focus on the situation with both slits open so that there are two alternatives. A single particle is in transit. Surely one would think (erroneously, as it turns out) that one could assume the particle goes through either slit 1 or slit 2 and then correctly deduce some consequences. One logical consequence of this assumption would be that the presence of the slit that the particle does not go through would not affect the probability of the particle reaching a given point on the screen. But it does! For instance, when both slits are open the particle has zero probability [Eq. (16.12)] of arriving at the location of the intensity minima. How does the slit that the particle does *not* go through prevent the particle from reaching such a location on the screen? And how is it that, at the location of an intensity maximum, the probability that an electron will reach such a region is not doubled but quadrupled [Eq. (16.11)] by having two slits open instead of one? The classical physics of particles has no answer.

What about closing slit 1 for a day and then opening it and closing slit 2? Under these circumstances it is certain that the particles in transit go through one slit or the other, and at any instant it can be determined which slit it is. But after this experiment we observe a completely uniform distribution. At every point on the screen

$$I = I_1 + I_2,$$

and correspondingly,

$$P = P_1 + P_2.$$

The interference term is zero. In general, we find that whatever we do, if one can ascertain by some experimental method which of the two alternatives (slit 1 or slit 2) is taken, the interference disappears.

How can things be so different when both slits are open for a day (interference occurs), than when one slit at a time is open for a day (no interference)? When both are open, does an electron split into two parts and have a piece go through each slit? Definitely not! We always detect

*This could be achieved to a good approximation with a slit so narrow that its width is the size of the de Broglie wavelength of the electrons.

entire electrons, never a part of an electron. Moreover, the entire story is not contingent upon using electrons. All the evidence suggests that *any particle will interfere with itself*. We are dealing with a *universal behavior of matter*, not with the peculiarities of some particular particle.

Quantum Mechanical Superposition Principle

The more we thus examine the interference manifested by a particle, the stranger it seems. Faced with particle interference, classical physics is at an impasse. However, we need not be reduced to a scientific paralysis as were some Greek philosophers by Zeno's paradoxes. The thing to do is to face the experimental facts and build a new mechanics in accordance with these facts. This has been done, and the resulting theory is called *quantum mechanics*.

An encouraging feature in the face of the mystery of particle interference is that one has on hand a very simple way of correctly predicting all details of the interference pattern. Just think of waves of wavelength $\lambda = h/p$. The square of the amplitude of the waves at a certain location on the screen is associated with the intensity or the probability that the particle will hit the screen at that location. Following this clue, physicists were led to the most important postulates upon which quantum mechanics is founded.

Without attempting a precise account, we shall merely indicate some salient features of this theory, for a system consisting of a single particle. With each possible state of affairs at any instant is associated some wave denoted by ψ which is called a wave function. The wave function indirectly determines probabilities. In particular, the square* of the wave function ψ at a certain point in space is proportional to the probability of the particle being detected within a given small volume enclosing the point in question.

A postulate of central importance is the *quantum mechanical superposition principle*: If ψ_1 is one wave function corresponding to a possible physical state and ψ_2 is a different wave function corresponding to a different possible physical state, the sum,

$$\psi = \psi_1 + \psi_2, \qquad (12.13)$$

is itself the wave function of still another possible physical state. The state ψ is called the superposition of the states ψ_1 and ψ_2.

For instance, in the double-slit experiment, let ψ_1 be the wave function for a state which corresponds to the particle going through slit 1 as shown in Fig. 16-8. Similarly, let ψ_2 be the wave function corresponding to the particle going through slit 2. Then according to the superposition principle, the superposition, $\psi = \psi_1 + \psi_2$, is the wave function of another possible state. If we are asked to describe this new state ψ in words, we would have to say that in this state the particle (without breaking up)

*For readers who are familiar with complex numbers, it should be pointed out that ψ at any point is a complex number. Its magnitude squared *determines a probability*.

goes through both slits! You can now appreciate that the quantum mechanical superposition principle, which seems so innocent and is such a simple postulate to state mathematically, introduces radically new possibilities. Many of these new situations seem to be nonsensical when one attempts a verbal description in terms of everyday ideas. But there is no difficulty in giving a quantum mechanical description. One is simply concerned with a superposition of states.

Now when the square of the magnitude of ψ ($= \psi_1 + \psi_2$) is computed to determine a probability, we obtain

$$P = P_1 + P_2 + \text{(interference term)},$$

as we had found in the double-slit interference experiment and expressed in Eq. (16.10). The point to notice is that the superposition principle together with the rule for computing probabilities leads to something different from $P = P_1 + P_2$. There is also an *interference term*. We have *interfering probabilities*. This remarkable fact, unnoticed in our everyday world, is of common occurrence in many circumstances in the domain of molecules, atoms, nuclei, and elementary particles. It is a typically quantum mechanical phenomenon.

Heisenberg Uncertainty Relation

Consider a superposition, $\psi' = \psi_1' + \psi_2'$, of two states in which the particle has different momentum values along some direction OX; say for example that the state ψ_1' corresponds to a momentum p_1 (that is, ψ_1' has a wavelength $\lambda_1 = h/p_1$), and state ψ_2' corresponds to momentum p_2. Then the superposition is a state ψ' in which the *particle does not have a single definite value of momentum*. A measurement could yield either the value p_1 or p_2. Proceeding in this way to build the superposition of many states with different momentum values, we are led to consider a state ψ which corresponds to a whole range of momentum values. This spread of momentum values is aptly named the *uncertainty in momentum* and is usually denoted by Δp. (In this instance, the symbol Δp refers to the *spread* of possible momentum values which range from approximately p to $p + \Delta p$, not to a change in momentum.)

We now examine the same wave function ψ at different points in space along the axis OX. We find that, even though the wave may be very concentrated in one region and nearly zero everywhere else, there is a certain range of x values between x and $x + \Delta x$ where the wave function ψ has an appreciable size. A position measurement would be most likely to give a value of x within this range Δx. This range of x values, Δx, is named the *uncertainty in x*.

In Newtonian mechanics, a mechanical state is specified by giving a precise value of momentum and position. In quantum mechanics, a typical state ψ corresponds, not to definite momentum and position, but

*The symbol $\gtrsim$ means "is of the order of or is greater than."

rather to ranges of momentum and position values measured by Δp and Δx. Now, from the postulates of quantum mechanics, it can be shown that for a given state ψ, the uncertainties Δp and Δx are related by the celebrated Heisenberg uncertainty relation:

$$\Delta x \Delta p \gtrsim h.*$$

So small Δx implies large Δp and vice versa! This means that if a state is such that ψ is well localized in space so that Δx is small, then ψ is a superposition of states with a large range (Δp) of different momentum values. On the other hand, if we have a state which is a superposition of states which have only a small spread (Δp) of momentum values, then the wave function ψ is widely spread out in space (Δx is large). In particular, there is no state possible in quantum theory for which *both* Δx and Δp are zero. The precise simultaneous specification of position and momentum (or velocity) characteristic of Newtonian mechanics does not occur in quantum theory.

A picture, mental or otherwise, of a particle trajectory with arbitrarily well defined position and momentum at any instant is therefore not consistent with a quantum theory description. There must always be a smudging or blur, a lack of definition, consistent with $\Delta x \Delta p \gtrsim h$. For macroscopic objects like a football (see Question 33), both Δx and Δp can be so close to zero that for all practical purposes there is no uncertainty, because h is so small. But for an electron in a hydrogen atom, the uncertainty in position is of the order of the size of the atom itself. Bohr's picture of electron orbits (Section 2.9) thus disappears in quantum theory.

At this point the reader may well say, "Put *theory* aside." Suppose one makes a simultaneous measurement of position and momentum with such accuracy that $\Delta x \Delta p$ is smaller than Planck's constant h. What then?

Then the particle would be in no state consistent with quantum theory and this theory would be wrong on a most fundamental point. A major theoretical revision would be dictated.

But are such simultaneously accurate measurements possible, even in principle? It seems the answer is "No." Heisenberg and many others have analyzed many "thought experiments" and found no situation which beats the uncertainty principle. For instance, if our particle is an electron and a position measurement is to be made using light, then we are concerned with a photon scattered by an electron which recoils. The scattered photon enters a detecting instrument through some aperture. Because of the diffraction of light, there will be uncertainties associated with this photon's momentum and with the location of the scattering event. There will be a related uncertainty in the measurement of the position and momentum of the electron. The analysis can be made quantitative and the result is in agreement with the uncertainty principle.

Physics Today

W. K. Heisenberg
(1901–   )

632 QUANTUM PHYSICS

Remarks on Quantum Mechanics

The theory that is known as quantum mechanics was developed in a few years from 1924 to 1928. This golden age of theoretical physics, its heroes, their reminiscences, successes, and foibles are charmingly described in the book by Barbara Cline listed as suggested reading for Chapter 1. In contrast with the history of relativity, the quantum theory was the work of many men: Heisenberg, Schrödinger, Dirac, Born, and Pauli all won Nobel prizes for their great contributions.

The theory is abstract and, at first sight, strange. But so are the experimental facts of atomic physics. Our experience in the macroscopic world does little to prepare our intuition for successful responses in this new realm of the very, very small. In any case, quantum mechanics is the most quantitatively successful theory that science has seen. It is *the* underlying theory in atomic physics, chemistry, solid state physics, and nuclear physics. Studies of interactions of elementary particles that are so copiously created and destroyed have not yet led to a comprehensive theoretical understanding of this branch of physics. Nevertheless the main principles of quantum mechanics seem to stand up. As far as we know, we live in a relativistic, quantum mechanical universe.

Before continuing this important topic, we shall summarize several features of quantum mechanics, many of which we have already encountered. Each *state* of a mechanical system is described by a *wave function* ψ. The laws of quantum mechanics allow one to determine the state at some future time in terms of the present state. A knowledge of ψ enables one to compute *probabilities* for the results of measurements on the system. It is characteristic of both experimental and theoretical quantum physics that we deal with *probabilities rather than certainties*. The superposition, $\psi = \psi_1 + \psi_2$, of two possible states, ψ_1 and ψ_2, is another possible state and in such a state we encounter interfering probabilities. A typical state ψ corresponds to many possible values of a dynamical variable, such as the momentum p or the position x. The spread of values, or the uncertainties, Δx and Δp are related by $\Delta x \Delta p \geq h$, the Heisenberg uncertainty relation.

Corresponding to the bound states of a system are discrete energy levels. For the hydrogen atom, quantum theory gives $E_n = -(13.6/n^2)$ eV, where the "principal quantum number" n is an integer. Each excited state ($n \geq 2$) is unstable and has a characteristic average life τ (= half-life/0.69). An excited state does not correspond to a single definite value of the energy. There is a spread of energy values, ranging roughly from some value E to $E + \Delta E$. The spread or width ΔE of the energy level is related to the average lifetime of the level τ by the uncertainty relation:

$$\tau \Delta E \geq h.$$

According to this result, the shorter this lifetime of a state, the greater its width ΔE.

Physics Today

E. Schrödinger
(1887–1961)

Physics Today

Max Born
(1882–1970)

In the macroscopic world, where the predictions of Newtonian mechanics are in accord with experiments, quantum mechanics and Newtonian mechanics agree. This comes about because, for large masses, the de Broglie wavelength $\lambda = h/p$ is so short compared to the size of macroscopic objects that the peculiar wave effects, characteristic of the microscopic world, are no longer apparent. The uncertainty relations pose no significant restriction for a description of macroscopic motion.

16.5 THE HYDROGEN ATOM

The first field to be triumphantly put into order by quantum mechanics was that of atomic physics. Almost all the details of atomic spectra and atomic structure became at least comprehensible, if not calculable. In this program, the theory of the simplest atom—the hydrogen atom—was the key.

For the hydrogen atom, the mathematical complexities of quantum mechanics are not too formidable. Exact theoretical predictions can be calculated. Fortunately, certain general features of the hydrogen atom theory are applicable to electrons in *any* atom. Consequently, an understanding of the hydrogen atom is a big step in understanding atomic structure in general. Without concerning ourselves with the mathematics involved we will survey in the following sections these modern notions of atomic structure. This is the historical route followed by the pioneers of quantum theory, and it remains a most instructive approach to quantum physics.

The electromagnetic interaction between electrically charged particles is the only significant interaction in atomic physics, that is, in the physics of the atom outside the nucleus. The values of many physical quantities of importance in atomic physics that are determined by electromagnetic interactions can be conveniently expressed in terms of one constant α defined by

$$\alpha = \frac{ke^2}{\hbar c} \qquad (16.14)$$

where k is the constant of proportionality in Coulomb's law. Substitution of the experimental values of ke^2, $\hbar$ and c gives

$$\alpha = \frac{1}{137},$$

almost exactly. This quantity is a dimensionless number and therefore has the same value no matter what system of units is used for the measurement of charge (e), speed (c) and angular momentum ($\hbar$). The constant α acquired the name "the fine structure constant" in early studies of hydrogen atom energy levels and this name has persisted, even though it is not particularly apt.

States of the Electron in a Hydrogen Atom

The hydrogen atom consists of an electron and a proton. These two particles interact by exerting Coulomb forces on each other. Since the proton's mass is 1836 times the electron's mass, the proton's acceleration, according to classical mechanics, is merely 1/1836 of the electron's acceleration. Then, to a good approximation, one can visualize the proton as fixed and attribute all the motion to the electron. This picture carries over to quantum mechanics. The problem then is to determine all possibilities for an electron which finds itself in the electric field established by a charged fixed nucleus.

Quantum mechanics now prescribes a set of different wave functions, each corresponding to a different possible state for the electron in the type of atom under consideration. From the wave function for a given state, one can determine the probability of finding an electron in any given region near the atom. This probability can be visualized by imagining a cloud of electric charge with a density in a given region proportional to this probability. Such *charge clouds* are displayed in Fig. 16-9.

State with
$n = 2, l = 0, m_l = 0$

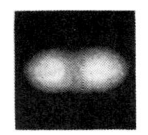

State with
$n = 2, l = 1, m_l = 1$

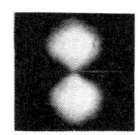

State with
$n = 2, l = 1, m_l = 0$

Fig. 16-9 Charge clouds for three different possible states of the electron in a hydrogen atom. (From *Introduction to Atomic Spectra*, H. White; New York, McGraw-Hill, 1934.)

The charge cloud indicates the "size" of the *atom* in different states. In quantum mechanics, objects do *not* have a precise and well-defined size, just as a cloud does not have a distinct bounding surface.

It is the size of the atom, *not the size of the electron*, that is determined by the spatial extension of the charge cloud. Scattering experiments with high energy electrons show that, relative to the size of the atom itself, the upper limit to the electron's size is less than that of any dot in these charge cloud pictures. As far as the electron is concerned, charge clouds are simply a useful way of picturing probabilities: it is more likely that the electron be found where the cloud is dense than where it is tenuous.

Ground State

For the hydrogen atom *ground state*, the electron charge cloud is distributed in the space in such a way that the most probable distance of the electron from the nucleus is given by*

$$a = \frac{\hbar^2}{mke^2}$$
$$= 0.53 \text{ Å},$$

where a is called the *Bohr radius*. (According to the early "Bohr model" of the hydrogen atom, the electron in the ground state orbited about the nucleus in a circle of this radius.) In the ground state, the charge cloud thus extends over a spherical region somewhat larger than 1 Å in diameter. Quantum mechanics thus predicts that the "size" of a hydrogen atom in its ground state is roughly that of a sphere 1 Å in diameter (Fig. 1-10). This is in reasonable agreement with the estimate of atomic size which is furnished by measuring the volume occupied by a known number of hydrogen atoms in a chunk of hydrogen which has been solidified by cooling.

The ground state wave function corresponds to many different values of the component of the electron's momentum in a given direction. These momentum values range from about $m\alpha c$ corresponding to motion in one direction through 0 to about $-m\alpha c$ corresponding to motion in the opposite direction. The range Δp of momentum values encountered in the ground state is therefore about $2m\alpha c$. It is interesting to relate this to the size of the charge cloud. The bulk of the charge cloud is confined to a region which is two or three times the Bohr radius in diameter, so we can write for the uncertainty in position

$$\Delta x \approx 3a.$$

Then the product of these uncertainties is

$$\Delta x \Delta p \approx (3a)(2m\alpha c)$$
$$\approx 6\left(\frac{\hbar^2}{mke^2}\right)\left(\frac{mke^2}{\hbar}\right)$$
$$\approx 6\hbar$$
$$\approx 6 h/2\pi$$
$$\approx h,$$

in accord with the Heisenberg uncertainty relation.

A momentum of $m\alpha c$ corresponds to an electron speed $\alpha c\ (=c/137)$ which is less than 1% of the speed of light. Therefore relativistic effects are small in the hydrogen atom.

---

*Throughout this chapter, the symbol m denotes the rest mass of an electron. An electron has a rest energy (Section 4.4) given by $mc^2 = 0.511$ MeV.

The electron's total energy in the ground state, according to quantum mechanics, is given by

$$E_1 = -\tfrac{1}{2}\alpha^2 mc^2$$
$$= -\tfrac{1}{2}(1/137)^2 (0.511 \times 10^6 \text{ eV})$$
$$= -13.6 \text{ eV},$$

in agreement with Bohr's formula, Eq. (16.8). Here zero total energy has been chosen to correspond to the situation when the electron is completely removed from the vicinity of the nucleus. The atom is then said to be ionized. When a hydrogen atom is in its ground state, an energy of at least 13.6 eV must be supplied to the atom for ionization to be possible.

Quantum Numbers

There are many possible bound states of the electron in a hydrogen atom. Each possible state is specified by giving the values of four *quantum numbers*: n, ℓ, m_ℓ, and m_s. The rules which emerge from quantum mechanics for the possible numerical values of these quantum numbers are summarized in Table 16-1.

It is a peculiarity of the hydrogen atom that the energy of a state depends only on the value of the "principal quantum number" n, according to Bohr's formula,

$$E_n = -13.6 \text{ eV}/n^2, \tag{16.8}$$

the validity of which is confirmed by quantum mechanics. This quantum number n also determines the size of the atom in the state with principal quantum number n: For a state with $\ell = n - 1$, the most probable distance of the electron from the nucleus is given by

$$r_n = n^2 a$$

where $a\ (=\hbar^2/mke^2)$ is the Bohr radius.

Table 16-1. QUANTUM NUMBERS FOR STATES OF AN ELECTRON IN AN ATOM

Quantum number		Values
n	principal quantum number	Can be any positive integer: 1, 2, 3, ...
ℓ	orbital angular momentum quantum number	Given n, there are different states corresponding to $\ell = 0, 1, 2, 3, 4, 5, \ldots, n - 1$.
m_ℓ	magnetic quantum number	Given ℓ, there are $2\ell + 1$ different states corresponding to $m_\ell = \ell, \ell - 1, \ldots 1, 0, -1, \ldots -(\ell - 1), -\ell$.
m_s	electron spin quantum number	For each set of values of n, ℓ, m_ℓ, there are two different states, one for which $m_s = +\tfrac{1}{2}$ and another for which $m_s = -\tfrac{1}{2}$.

Example 5 Use the rules given in Table 16-1 to give the quantum numbers of all the possible states for which $n = 2$.

Solution The maximum value of ℓ is $n - 1 = 2 - 1 = 1$. For $\ell = 1$ there are different states corresponding to $m_\ell = 1$ or 0 or -1. The value $\ell = 0$ is also possible and corresponding to this ℓ value, m_ℓ can have only the value 0. For each possible set of values of n, ℓ, m_ℓ, the quantum number m_s can be either $+\tfrac{1}{2}$ or $-\tfrac{1}{2}$.

There are therefore the following 8 different states (each labeled by the values of n, ℓ, m_ℓ, m_s):
$(2, 1, 1, +\tfrac{1}{2})$, $(2, 1, 1, -\tfrac{1}{2})$; $(2, 1, 0, +\tfrac{1}{2})$, $(2, 1, 0, -\tfrac{1}{2})$;
$(2, 1, -1, +\tfrac{1}{2})$, $(2, 1, -1, -\tfrac{1}{2})$; and $(2, 0, 0, +\tfrac{1}{2})$, $(2, 0, 0, -\tfrac{1}{2})$.

For our purposes, it is not necessary to dwell on the physical significance of the quantum numbers ℓ and m_ℓ which justifies the names shown in Table 16-1. As a guide for future studies of readers who wish to delve more deeply into this interesting topic, we mention the following points. The ℓ value of a state determines the rotational symmetry properties of its charge cloud. For instance, the charge cloud of any state with $\ell = 0$ can be rotated any amount about any axis through its center without producing any change. These $\ell = 0$ states are said to be spherically symmetric. Rotational symmetry and its associated ℓ value determines the magnitude of the quantum mechanical analogue of the *angular momentum* that was defined for Newtonian mechanics in Section 3.9. The "magnetic quantum number" determines the orientation of the electron charge cloud in space. When a magnetic field is applied, states with the same n and same ℓ values have different energies in the different possible orientations corresponding to $m_\ell = \ell, \ell - 1, \ldots, 1, 0, -1, \ldots, -(\ell - 1), -\ell$. In a state with quantum number m_ℓ, the electron's orbital angular momentum has a component $m_\ell \hbar$ in the direction of the applied magnetic field. The electron itself has an intrinsic angular momentum called "spin" whose component in the direction of an applied magnetic field is restricted to two values, either $+\tfrac{1}{2}\hbar$ or $-\tfrac{1}{2}\hbar$. We denote this component of spin angular momentum by $m_s \hbar$ and conclude that the spin quantum number m_s is restricted to the values $+\tfrac{1}{2}$ or $-\tfrac{1}{2}$.

Such details about the possible states of an electron in a hydrogen atom enable one to understand all the complexities of the line spectra emitted by these atoms under various excitations and in circumstances when a magnetic field or an electric field has been applied to the light source. Quantum mechanics allows calculation of not only the wavelengths but also of the intensities of each spectral line. Rather than describing these achievements, we shall show how the structure of different atoms is related to the facts listed in Table 16-1 concerning electron states in hydrogen.

16.6 THE ELECTRONIC STRUCTURE OF ATOMS

An incredible wealth of chemical and spectrographic detail becomes comprehensible when the electronic structure of each type of atom is known. Theory of this structure rests on the facts that have been presented about the electron states in hydrogen.

Pauli Exclusion Principle

Consider an atom with atomic number Z. The nucleus has a positive charge Ze which produces an electric field strong enough to bind Z electrons. Each electron moves in the electric field created by the positive nuclear charge Ze and by the negative charges of the other $Z - 1$ electrons in the atom. Analogously to the case of the hydrogen atom, the possible states for an individual electron can be specified by giving the values of the four quantum numbers: n, ℓ, m_ℓ and m_s.

The state of the entire atom depends on just which of the possible electron states are occupied. And here we come to a simple but peculiar rule of quantum mechanics, the *Pauli exclusion principle: In a system containing several electrons, no two electrons can occupy states with the same values of all four quantum numbers*, n, ℓ, m_ℓ, m_s. The possibility of occupancy of a given state by more than one electron is excluded. This basic law, which has profound implications in many fields of physics, was formulated in 1925 by the German physicist, Wolfgang Pauli, as a key to an understanding of atomic structure. In the ground state of an atom, the electron energy levels are filled from the lowest energy level upward with precisely one electron in each possible state until the atom's supply of electrons is exhausted. The properties of these occupied states determine the electronic structure of the atom.

Subshells of Electrons

The order in which successive electron energy levels are filled is shown in Fig. 16-10. In many-electron atoms, although the energy of an electron state depends principally on the quantum number n, there is also a dependence on the value of ℓ with higher energies corresponding to higher ℓ values. An $n\ell$ level in Fig. 16-10 is labeled by a number which gives the value of the principal quantum number n, followed by a lowercase letter which corresponds to the value of ℓ according to the following convention:

ℓ value	0	1	2	3
letter	s	p	d	f

Thus a state with $n = 3$ and $\ell = 1$ is called a 3p state.

The set of electron states associated with a given $n\ell$ energy level is called a *subshell*. The number of different states in a subshell is just the number of different sets of values of m_ℓ, m_s that are possible according

Physics Today

Wolfgang Pauli
(1900–1958)

Fig. 16-10 Shell structure of atoms. Different $n\ell$ states are filled from the bottom up in this order.

Table 16-2. ELECTRON SUBSHELLS

Subshell ℓ value	0	1	2	3
Subshell letter	s	p	d	f
Number of states $2(2\ell + 1)$	2	6	10	14

to the rules of Table 16-1. Results are given in the last row of Table 16-2. According to the Pauli exclusion principle, the number of states in a subshell is also the maximum number of electrons that can occupy the subshell.

The *electron configuration* of an atom in any state is the name given to a specification of the number of electrons in each subshell. The electron configuration for the ground state of any atom can be read off from the diagram of Fig. 16-10. For example, sodium with 11 electrons has 2 electrons filling the 1s subshell, 2 electrons filling the 2s subshell, 6 electrons filling the 2p subshell, and one electron remaining which will occupy the 3s subshell. Placing the number of electrons in a shell as a superscript, we designate this electron configuration by $1s^2\ 2s^2\ 2p^6\ 3s^1$.

A study of electron charge clouds reveals that the lowest energy level corresponds to the innermost subshell and that subshells of increasing energy usually have charge clouds that are increasingly distant from the nucleus. Bohr's early notion of atomic structure which pictured the atom as built up from concentric shells of electrons thus retains considerable validity in quantum mechanics. The highest energy subshell to be occupied usually corresponds to the outermost electrons. The physical and chemical properties of an atom are determined by these outermost electrons.

Except for helium, the inert gases listed at the right of Fig. 16-10 are characterized by having a full p-subshell as their outermost subshell. This is evidently a particularly stable structure. The energy gap in Fig. 16-10 between each of the p levels and the next highest level is abnormally large, indicating that an unusually large energy must be provided to an inert gas atom in order for it to reach its first excited state. For helium this excitation energy is about 20 eV.

The alkali atoms such as lithium, sodium, and potassium have just one electron outside a closed p-subshell. This electron is in an s-state and a rather small excitation energy (about 2 eV) is required to promote the electron to the next highest energy level. The metallic behavior of solids composed of atoms of any elements of this type is associated with this low excitation energy.

These remarks are merely a first indication of the wealth of chemical information that can be inferred from the electron configurations dictated by Fig. 16-10. But enough has been said to make clear the reason for the recurrence of the same chemical properties over and over again as we proceed through Mendeleeff's celebrated *periodic table* of the chemical elements (Table 16-3).

Inner Electrons and X-rays

The emission or absorption of photons by an atom is associated with a transition from one state of the atom to another. When an *outer* electron of an atom passes from one possible electron state to another, its energy

Table 16-3. PERIODIC TABLE OF THE ELEMENTS

(Numbers in parentheses indicate the mass number of the longest-lived isotope of radioactive elements.)

Outer electrons are in the	I	II	III	IV	V	VI	VII	VIII			O	Electrons per shell
First or K-shell	1 H 1.00797										2 He 4.0026	2
Second or L-shell	3 Li 6.939	4 Be 9.0122	5 B 10.811	6 C 12.01115	7 N 14.0067	8 O 15.9994	9 F 18.9984				10 Ne 20.183	2,8
Third or M-shell	11 Na 22.9898	12 Mg 24.312	13 Al 26.9815	14 Si 28.086	15 P 30.9738	16 S 32.064	17 Cl 35.453				18 Ar 39.948	2,8,8
Fourth or N-shell	19 K 39.102	20 Ca 40.08	21 Sc 44.956	22 Ti 47.90	23 V 50.942	24 Cr 51.996	25 Mn 54.9380	26 Fe 55.847	27 Co 58.9332	28 Ni 58.71		
	29 Cu 63.54	30 Zn 65.37	31 Ga 69.72	32 Ge 72.59	33 As 74.9216	34 Se 78.96	35 Br 79.909				36 Kr 83.80	2,8,18,8
Fifth or O-shell	37 Rb 85.47	38 Sr 87.62	39 Y 88.905	40 Zr 91.22	41 Nb 92.906	42 Mo 95.94	43 Tc (99)	44 Ru 101.07	45 Rh 102.905	46 Pd 106.4		
	47 Ag 107.870	48 Cd 112.40	49 In 114.82	50 Sn 118.69	51 Sb 121.75	52 Te 127.60	53 I 126.9044				54 Xe 131.30	2,8,18, 18,8
Sixth or P-shell	55 Cs 132.905	56 Ba 137.34	57-71 La series*	72 Hf 178.49	73 Ta 180.948	74 W 183.85	75 Re 186.2	76 Os 190.2	77 Ir 192.2	78 Pt 195.09		
	79 Au 196.967	80 Hg 200.59	81 Tl 204.37	82 Pb 207.19	83 Bi 208.980	84 Po (210)	85 At (210)				86 Rn (222)	2,8,18, 32,18,8
Seventh or Q-shell	87 Fr (223)	88 Ra (226)	89-103 Ac series**	104 Ku (260)	105							

| *Lanthanide series: | 57 La 138.91 | 58 Ce 140.12 | 59 Pr 140.907 | 60 Nd 144.24 | 61 Pm (145) | 62 Sm 150.35 | 63 Eu 151.96 | 64 Gd 157.25 | 65 Tb 158.924 | 66 Dy 162.50 | 67 Ho 164.930 | 68 Er 167.26 | 69 Tm 168.934 | 70 Yb 173.04 | 71 Lu 174.97 | 2,8,18, 32,9,2 |
| **Actinide series: | 89 Ac (227) | 90 Th 232.038 | 91 Pa (231) | 92 U 238.03 | 93 Np (237) | 94 Pu (244) | 95 Am (243) | 96 Cm (247) | 97 Bk (247) | 98 Cf (251) | 99 Es (254) | 100 Fm (257) | 101 Md (256) | 102 No (255) | 103 Lw (257) | 2,8,18, 32,32,9,2 |

16.6 THE ELECTRONIC STRUCTURE OF ATOMS   641

change is only a few electron volts. The photon emitted in a downward transition or absorbed in an upward transition may be visible, or perhaps infrared, or ultraviolet. But if the opportunity is provided for an *inner* electron of an atom of atomic number greater than about 12 to undergo a transition, the change in electron energy is so large that the photon emitted or absorbed is an x-ray photon.

As a typical example of an x-ray emission process involving inner electrons, we consider a tungsten atom in the anode of an x-ray tube. The anode is bombarded by a beam of energetic electrons. When a tungsten atom is struck by one of these electrons, it sometimes happens that an electron occupying the 1s shell is knocked right out of the tungsten atom. The electrons of this atom which are residing in higher energy levels now have "Pauli's permission" to make a transition to this vacant 1s state. A very common occurrence in this situation is that an electron in a 2p state tumbles into the 1s state. In this process the electron decreases its energy by 59.3×10^3 eV, and an x-ray photon of this energy is emitted. Another electron can then drop into the vacated 2p state emitting another x-ray. Processes of this type lead to the emission of photons of several different energies that are characteristic of the electronic structure of the atom. Each element therefore has its own *characteristic spectrum* (Fig. 16-11), a set of discrete x-ray wavelengths superimposed on the continuous spectrum of *bremsstrahlung* x-rays described in Section 16.2.

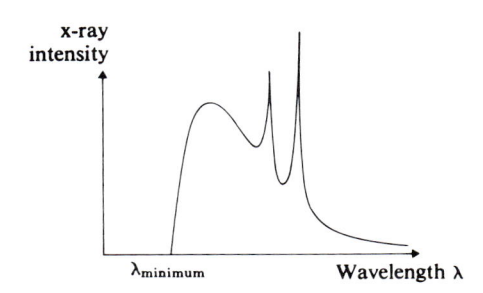

Fig. 16-11 The characteristic x-ray spectrum (spikes) is shown superimposed on the continuous bremsstrahlung spectrum of Fig. 16-3.

16.7 ENERGY BANDS IN SOLIDS

The quantum mechanical description of electron states in crystalline solids leads to a simple characterization of electrical conductors, insulators, and semiconductors and provides the theory underlying modern semiconductor technology.

Within a crystal there is a three-dimensional array of atomic nuclei. Each atom furnishes to the crystal its complement of electrons. The quantum mechanical problem is to determine the possible states for an electron confronted by this assembly of interacting particles. Fortunately, a few general features of the solution to this formidable problem are sufficient to shed considerable light on the behavior of electrons in crystals.

An instructive example is provided by metallic sodium. The electron energy levels for a single isolated atom are illustrated in Fig. 16-12(a). If we now consider N sodium atoms, so far apart that their interaction is negligible, the electron energy level diagram for this aggregate of particles [Fig. 16-12(b)] will be identical to that of a single atom except for the fact that the number of states corresponding to each level is increased by a factor of N. Finally, we consider the electron energy levels when these N atoms are brought into the positions they occupy

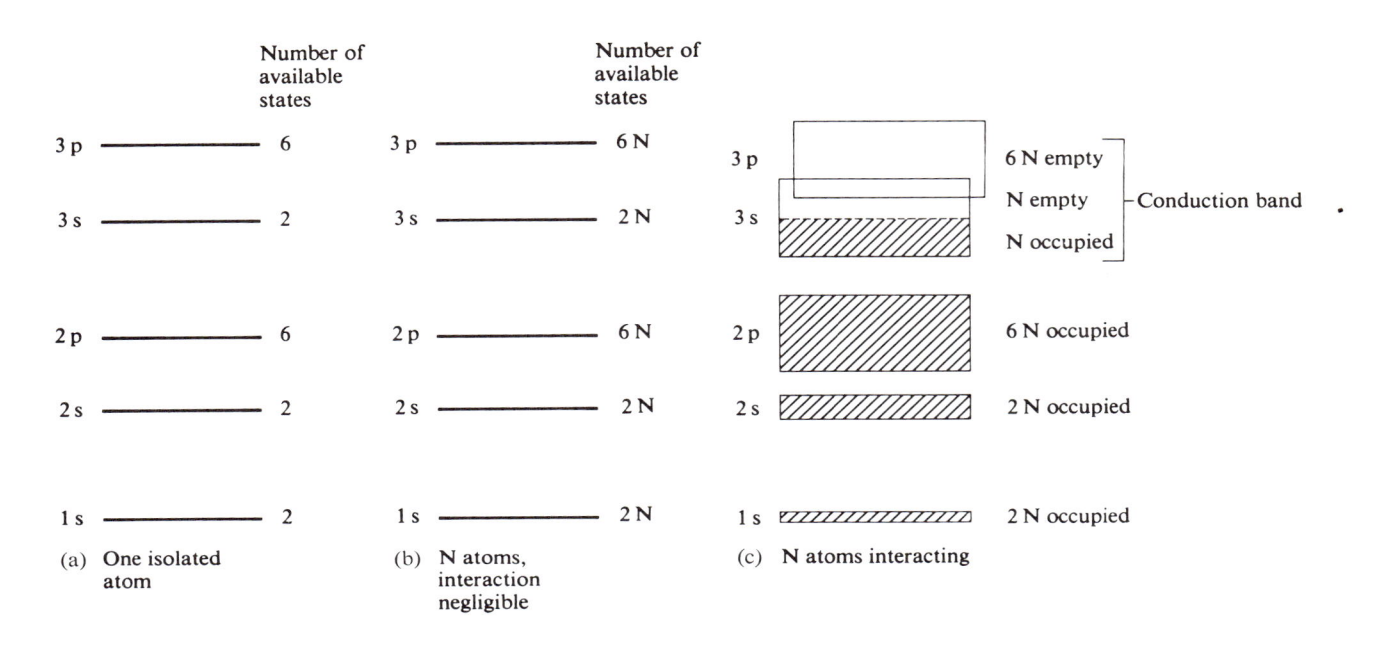

	Number of available states		Number of available states		
3 p ———	6	3 p ———	6 N	3 p	6 N empty
3 s ———	2	3 s ———	2 N	3 s	N empty / N occupied — Conduction band
2 p ———	6	2 p ———	6 N	2 p	6 N occupied
2 s ———	2	2 s ———	2 N	2 s	2 N occupied
1 s ———	2	1 s ———	2 N	1 s	2 N occupied
(a) One isolated atom		(b) N atoms, interaction negligible		(c) N atoms interacting	

Fig. 16-12 Schematic representation of sodium energy levels for (a) one isolated atom; (b) N atoms so far apart that their interaction is negligible; (c) N atoms interacting as they do in a metallic crystal. Here the levels are grouped into energy *bands*.

within a crystal of metallic sodium. Because of the *interaction*, each set of levels that were coincident becomes *separated into many distinct levels*. When N is large, these levels are so closely spaced that we can consider them as forming a *band* of energy. Thus in Fig. 16-12(c), the 2s levels have spread out to become the 2s band containing 2N distinct states, the 2p levels have spread out to become the 2p band with 6N distinct states, and so on. The interaction has the greatest effect and leads to the greatest band widths for electron states whose charge clouds range furthest from the nucleus. For example, the 2s band is wider than the 1s band. Bands at higher levels often overlap in the fashion illustrated in Fig. 16-12(c) by the 3p and 3s bands of sodium.

The gaps between bands are called *forbidden bands* since they correspond to no possible value of electron energy.

The *occupation* of the available energy levels is governed by the Pauli exclusion principle. Each state can be occupied by at most one electron. In the lowest energy state for the 11N electrons of a sodium crystal, 2N electrons fill the 1s band, 2N electrons fill the 2s band, 6N electrons fill the 2p band, and the remaining N electrons will occupy the lowest levels in the 3s band. Since the 3s band contains 2N different states, the upper half of this band will be empty [Fig. 16-12(c)].

16.7 ENERGY BANDS IN SOLIDS

The electrons in this band that occupy levels just below vacant energy levels are in a unique position. They are the only electrons that can make a transition to a slightly higher level without violating the exclusion principle. For this reason, these are the electrons that are responsible for the high electrical conductivity of sodium metal. When an electric field is applied within the material, these electrons can move readily, and so acquire a modest kinetic energy associated with a transition to a nearby higher energy level. The electrons in the lower filled bands have such an opportunity denied by the exclusion principle. We see that a *partially filled* upper band is characteristic of a *conductor*. Such a band is named the *conduction band*.

Energy bands for a typical *insulator*, diamond, have lower bands completely filled with electrons (Fig. 16-13), and then a gap of 6 eV to a band which is *empty*. Small energy increments of electrons in the filled bands are impossible so these electrons cannot respond to an applied electric field (unless the field is formidably large). This is why the material is an insulator. Evidently insulators are characterized by a sizable energy gap between the last filled band, called the *valence band*, and the first empty band, called the *conduction band*.

We have been considering the situation when all electrons occupy the lowest available states. There are several ways for an electron to absorb energy and make an upward transition to an unoccupied state. The energy of a particle participating in thermal agitation at room temperature averages about $\frac{1}{40}$ eV. Electrons near empty levels in the conduction band will undergo frequent thermal excitation to levels about $\frac{1}{40}$ eV higher. However, thermal excitation through about 1 eV will be a rare event.

When a beam of light shines on a crystal, the photons will be absorbed by the crystal's electrons if, and only if, appropriate excited states are available for the electrons. Diamond is transparent to visible light because a visible photon does not furnish enough energy for an electron to make a 6 eV jump from the full valence band to the empty conduction band. On the other hand conductors are opaque because they have many empty energy levels into which electrons can be promoted from the same band.

16.8 SEMICONDUCTORS

Intrinsic Semiconductors

Certain crystalline solids such as silicon and germanium are similar to diamond in that they have a full valence band and an empty conduction band at low temperatures. But the energy gap between the valence and conduction bands is merely 1.1 eV for silicon and 0.7 eV for germanium, in contrast to the 6 eV gap for diamond. In the materials with a small

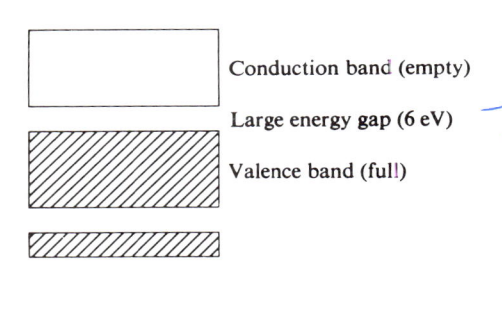

Fig. 16-13 Energy bands for diamond, an insulator. In the ground state, the conduction band is empty and the valence band is full of electrons.

gap, thermal agitation at room temperature will impart to a few electrons in the valence band sufficient energy for them to jump up to the conduction band. While in this almost empty band, these electrons can acquire kinetic energy in an applied electric field because there is an abundance of nearby vacant energy levels. The material thus conducts electricity. It is called a *semiconductor* because it has a conductivity greater than that of insulators with a large gap above the valence band but less than that of conductors with many electrons in the conduction band.

An electron which has been promoted from the full valence band to the conduction band leaves behind an empty state or *hole* in the valence band. When an electric field is applied, an electron in the valence band can make a transition to the empty state, with the result that the hole is elsewhere. Transitions in the valence band are most easily described by keeping track of the hole. Since the hole moves in the direction opposite to that of an electron, the *hole* behaves as a *positive charge*.

Thermal excitation in a semiconductor thus promotes electrons to the conduction band and leaves holes in the valence band (Fig. 16-14). In an electric field the holes move one way and the electrons in the conduction band move the other way. The conductivity is due to both motions. Such a semiconductor is called an *intrinsic semiconductor* to distinguish it from the important class of semiconductors whose conductivity arises largely from impurities.

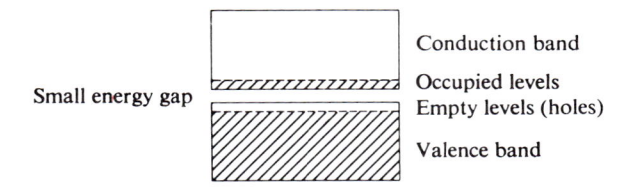

Fig. 16-14 Energy bands in a semiconductor. The conductivity of an intrinsic semiconductor arises from the motion of electrons which have been excited up to the conduction band and also from the motion of the holes which have been left in the valence band.

Impurity Semiconductors

The addition of certain impurities to a semiconductor drastically affects its conductivity. This phenomenon is exploited throughout the semiconductor technology which has revolutionized the electronics industry. By incorporating minute amounts of appropriate impurities, it is possible to make either

1. An *n-type* semiconductor in which electric current is carried by negative charges: the electrons donated to the conduction band by *donor* impurities; or
2. A *p-type* semiconductor in which electric current is carried by the motion of "positive charges": the holes in the valence band created when a valence band electron becomes bound to an *acceptor* impurity.

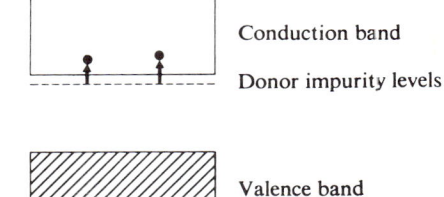

Fig. 16-15 Energy levels of an n-type semiconductor (donor impurities).

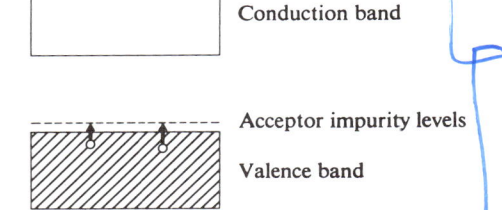

Fig. 16-16 Energy levels of a p-type semiconductor (acceptor impurities).

First we consider an example of a donor impurity. Suppose an arsenic atom replaces a germanium atom in a crystal of germanium atoms. The arsenic atom differs from the germanium atom in having one more electron in its outer shell. This extra electron is excluded from the full valence band. It occupies, instead, an energy level called a *donor level* just below the conduction band (Fig. 16-15). A mere 0.01 eV is sufficient to detach this electron from the arsenic atom, that is, to promote the electron from a donor level up to the conduction band. At room temperature thermal agitation provides ample energy, and almost all these extra electrons are donated by the arsenic impurities to the conduction band. Then most of the electron population in the conduction band comes from these "donor" arsenic impurities, even when the impurity concentration is as little as 1 part in 10^{10}. An electric current through such a crystal involves the motion of these conduction band electrons. Because the charge carriers are negative, the semiconductor is said to be of the *n-type*.

Semiconductors of the *p-type* are created by introducing *acceptor* impurities that will give rise to holes by accepting electrons from the valence band. Suppose, for example, that some of the atoms of a germanium crystal be replaced by gallium atoms. In its outer subshell, a gallium atom has one less electron than germanium. The electron deficits associated with the presence of the gallium impurity atoms appear as vacant energy levels just above the valence band (Fig. 16-16). Thermal excitation then promotes electrons into these acceptor levels, leaving vacant states, or holes, in the valence band. If an external electric field is applied, these holes will move like positive charges and effect a transfer of electric charge through the crystal. The conductivity of this p-type semiconductor will be determined by the concentration of acceptor impurities.

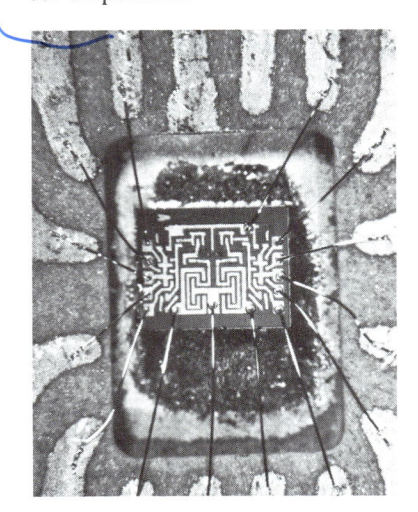

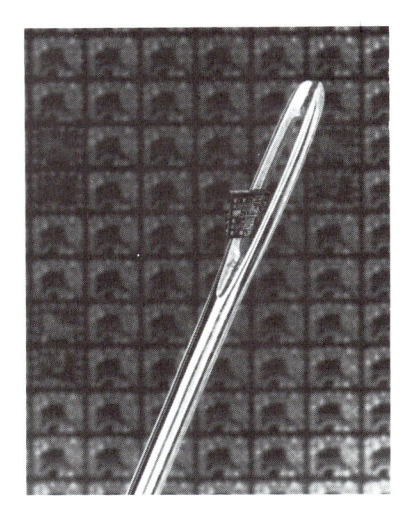

Fig 16-17 (a) Dual nand silicon microcircuit. This dual nand microcircuit performs a logic operation in new electronic switching systems. Integrated into the 60 by 80 thousandths of an inch silicon slice are 30 components: 4 transistors, 22 diodes, and 4 resistors. Microcircuits are more reliable, less expensive, and smaller in size than conventional circuits. (b) Eye of needle indicates the size of a typical microcircuit. (Courtesy Bell-Northern Research Ltd., Ottawa, Canada.)

Combinations of semiconductors find a host of applications in solid-state devices such as: rectifiers which allow a one-way passage of current, transistors which permit a weak alternating signal to be amplified, and integrated circuits or microcircuits. Miniaturization marvels like the microcircuit of Fig. 16-17 involve thin film deposits (a few molecular layers only) and diffusion of materials upon a thin wafer. Since it is possible to construct circuit elements with the desired electrical characteristics using only "molecular amounts" of matter, complex circuits can be accommodated within the eye of a needle.

16.9 THE LASER

Much of the physics discussed in the last two chapters enters into the operation of a *laser*, an acronym for "light amplification by stimulated emission of radiation."* This is a light source which produces an intense beam (Section 14.2) of coherent light (Section 14.3) with beam spread which can be reduced to the diffraction limit [Eq. (14.15)]. The light is confined to essentially one wavelength.

The key physical process, *stimulated emission*, was predicted by Einstein in 1917. This is different from spontaneous emission (discussed in Section 16.3) which involves an atom in an excited state that *spontaneously* emits a photon and makes a transition to a state of lower energy. In stimulated emission a photon of the right energy, incident from elsewhere, interacts with an atom in an excited state and *stimulates* the emission of a photon *identical* to the one already present. The decrease in the atom's energy equals the energy of the emitted photon, and this matches the energy of the original photon. At the conclusion of the process there are two photons with the same energy proceeding in the same direction. These two photons are coherent.

By stimulated emission, one photon gives rise to two. A chain reaction producing light amplification is possible if more than half of these photons go on to stimulate further emission. Some of these photons are absorbed by atoms which happen to be in the lower of the two energy states involved in the transition which produces these photons. Light amplification will be achieved only if there are more atoms in the higher than in the lower energy state. This desirable situation is called a *population inversion* because it is the reverse of the population of the atoms in different energy states in equilibrium (Section 7.3) at any temperature. In different types of lasers the problem of obtaining a population inversion is solved in different ways.

The first laser, constructed in 1960 by T. H. Maiman at the Hughes Aircraft Company, was made from a single cylindrical crystal of ruby with its ends ground flat and silvered (Fig. 16-18). Ruby consists of aluminum oxide with a few aluminum atoms replaced by chromium. The chromium atoms produce the laser light. Relevant energy levels of a

C. H. Townes (1915–)

*Invention of the laser followed the development of the *maser (a device which amplifies microwaves rather than light)* in 1954 by a group led by C. H. Townes at Columbia University. For "fundamental work in the field of quantum electronics, which has led to the construction of oscillators and amplifiers based on the maser-laser principle," Professor Townes shared the 1964 Nobel prize in physics with the Russian physicists N. G. Basov and A. M. Prokhorov.

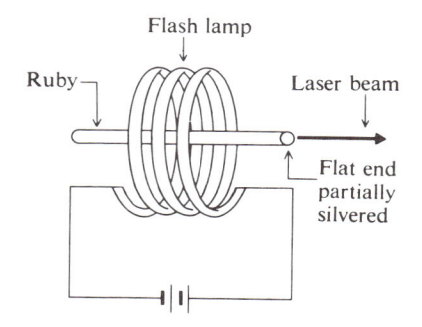

Fig. 16-18 Ruby laser. After the spiral flash lamp provides optical pumping, a bright beam of red light emerges from the partially silvered end of the ruby crystal.

chromium atom are shown in Fig. 16-19. If only spontaneous decay occurs, the excited state of energy E_M has a half-life of 3×10^{-3} sec, almost a million times greater than the half-life of most excited states of atoms. Such a long-lived state is termed *metastable*. In the ruby laser, atoms in this metastable state are stimulated by red photons (with $\lambda = 6943$ Å) to make a transition to the ground state and emit another identical red photon.

The population of chromium atoms in the metastable state is made momentarily much greater than the population in the ground state by a method known as *optical pumping*. The ruby is illuminated by a bright flash of yellow-green light ($\lambda = 5500$ Å). Absorption of these photons by chromium atoms in the ground state produces transitions to an excited state such as E_u in Fig. 16-19. After some 10^{-8} sec, many excited atoms will have made downward transitions from E_u to the metastable state E_M.

Fig. 16-19 Energy levels of a chromium atom involved in a ruby laser. Absorption of pump photons indirectly increases the population of atoms in the metastable state. A stimulated transition from this metastable state to the ground state produces an additional red photon identical to the stimulating photon (6943 Å).

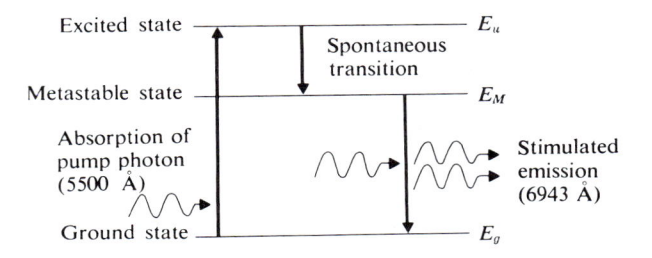

With a majority of the chromium atoms in their metastable state, the time is ripe for laser action. A photon with a wavelength 6943 Å, emitted by one atom in a spontaneous transition to its ground state, will stimulate similar transitions in other chromium atoms.

To enhance the probability of photon interaction and to achieve a unidirectional beam, reflection from flat parallel ends is exploited. Only light that is perpendicular to these ends will make repeated traversals of the crystal and cause substantial amplification. At the partially silvered end about 1% of the incident photons escape and constitute the emerging laser beam. With Maiman's laser this was a flash of red light lasting 0.3×10^{-3} sec with a peak power of 10^4 watts.

The manifold applications of lasers derive from the fact that the laser beam is unidirectional and intense, consisting of coherent light with a very sharply defined wavelength. When a high-power laser beam is brought to a sharp focus (Example 2, Chapter 14) the intensity is sufficient to vaporize rapidly any substance. Minute holes in diamond dies used in making fine wires are drilled by laser beams. In medicine lasers are employed in a variety of ways, such as to "weld" on detached

Fig. 16-20 Surgical laser. The surgeon fires the laser with a foot switch upon directing the laser at the tissue to be destroyed. The aiming of the laser beam involves the use of an incandescent light source coincident with the laser beam's path. This type of laser is finding use as a tool for cancer surgery. (Designed and constructed by the Biomedical Engineering and Instrumentation Branch, Division of Research Services, National Institutes of Health, U.S.A.)

retinas or to destroy inaccessible eye tumors (Fig. 16-20). In a way the eye is a fairly obvious target for laser surgery in that the transparent outer regions allow light of appropriate wavelength to pass through to the more opaque tissues at the back.

Because of the lack of spreading of a laser beam, surveyors find lasers a most useful device to locate objects relative to fixed landmarks and by means of interference patterns to detect minute changes in relatively large distances very accurately. In Japan, for example, an engineering firm used the laser technique to measure the movement of the top of a dam as the water flowed in.

Also, interesting studies of the interaction of matter with large electric fields have been opened to research by the availability of the tremendous oscillating electric field which is present in a high-intensity laser beam.

16.9 THE LASER

16.10 INTERACTION OF ELECTROMAGNETIC RADIATION WITH MATTER

When a beam of electromagnetic radiation passes through matter the photons of the beam may be absorbed or scattered from the beam by various processes. If the incident photon has an energy of a few MeV, the production of electron-positron pairs by the photon near a nucleus is the predominant process of removal of photons from the beam. As was discussed in Section 4.3, this process cannot occur unless the photon energy is over 1.02 MeV. For photon energies between a few MeV and about 0.05 MeV, Compton scattering is usually the most important mechanism removing photons from the beam.

The ejection of an electron from an atom by absorption of the incident photon is always possible if the photon can furnish an energy at least as great as the binding energy of the electron. This "atomic photoelectric effect" becomes very probable when the energy of the incident photon almost matches a binding energy. For incident photons with energies ranging from 2×10^5 eV to 5 eV, the photoelectric effect is often the most probable absorption event.

Now let us consider photon energies that are low enough so that the atom can absorb the photon and make a transition to a discrete excited state, a bound state. At normal temperatures almost all the atoms in an absorbing material will be in their ground states. We find that a photon with an energy which is close to the energy difference

$$hf_{1n} = E_n - E_1 \qquad (16.15)$$

between an excited state E_n and the ground state E_1 has a high probability of being absorbed. We say that light of such a frequency is in *resonance* with the atom and that the frequencies f_{1n} are the resonant frequencies of the atom.

Actually, to some extent the atom reacts to light of any frequency and nonresonant processes are also significant. In fact, these processes are often responsible for the visual appearance of an object. To bring out the essential features of such processes, it is necessary to do more than talk about quantum jumps or transitions of the atom from the ground state to excited energy levels. We require a more complete quantum mechanical description of the interaction.

From the wave function for the ground state of an atom one can determine the probability of finding an electron in any given region near the atom. This probability can be visualized by imagining a cloud of electric charge with a density in a given region proportional to this probability (Fig. 1-10). If an atom is in its ground state and is undisturbed, its charge cloud is stationary.

When a beam of electromagnetic radiation falls upon an atom, the atom's *charge cloud undergoes forced vibrations* at the frequency of the incident radiation. The amplitude of the charge cloud oscillations is minute, roughly 10^{-17} meter, which is only one ten millionth of the size of the atom. (The situation is analogous to the forced vibrations of a solid exposed to sound waves. Most of us have felt the room shake when the radio or record player volume was turned up very high.) The amplitude of the charge cloud oscillations is largest at the *resonant frequencies* f_{1n} given by Eq.(16.15), which states that $hf_{1n} = E_n - E_1$. When incident radiation has a frequency far above or below a resonant frequency of the atom's charge cloud, the amplitude of charge cloud oscillation will be much less than at a resonant frequency.

Whatever its amplitude, the oscillating charge cloud can radiate electromagnetic waves of the *same frequency* as the incident radiation and coherent with it. Indeed it is this small vibration that re-emits the light by which we see the objects around us.

It is also possible that the excited atom get rid of its excess energy by some other process. An atom may collide with another atom and the excitation energy may be transformed into kinetic energy of the atoms emerging from the collision. The net result of such a process is the *absorption of an incident photon* and an increase in the thermal motion of the atoms. This absorption process is particularly important in liquids or solids at or near a *resonant frequency* of the atoms of the material.

With the above ideas in mind, we now describe some features of the interaction of light of various frequencies with water (H_2O) molecules. First, what are the resonant frequencies of the charge cloud of the molecules? Like most atoms and molecules, the resonant frequencies of the electron charge cloud for a water molecule are higher than the frequencies of visible light; they lie in the ultraviolet. *Molecules*, however, can perform oscillations in which the *atoms* move with respect to one another within the molecule. Because of the larger mass of the moving atoms, the frequency of such vibrations is low. The resonant frequencies corresponding to *motions of atoms* within the molecules lie in the infrared region: a water molecule has strong resonances in the infrared.

Now when sunlight falls on the surface of the ocean, what happens? The water molecule's resonances in the infrared and ultraviolet lead to a strong absorption in these regions of the spectrum. And with no resonances for frequencies of visible light, water is quite transparent. The infrared resonance, however, is so strong that the absorptive effect extends even into the visible red. At an ocean depth of 30 meters, practically all red light has been absorbed. This fact has an interesting consequence, as far as the color of marine life is concerned. At great depths red looks black and therefore has the same survival value. This explains why deep-sea crustaceans can be red and not be discriminated against by the selection mechanisms of evolution.

The small forced oscillations of the molecule's electron charge cloud which lead to re-emission of light of the same frequency as the incident beam, account for both the reflected beam and the refracted beam. Within the bulk of the water, the incident beam and the light which is re-emitted coherently from the various molecules give a superposition which is the *refracted wave.** Backward radiation is cancelled by interference, except for a layer of molecules (roughly $\lambda/2$ thick) near the surface, whose contributions constitute the *reflected wave.*

X-ray photography, that is, radiography, is based on the fact that different types of atoms interact differently with a given x-ray photon. For example, a photon whose energy is 90×10^3 eV has a frequency well above any of the resonant frequencies of the oxygen atoms in a water molecule. However, lead atoms have resonant frequencies in the vicinity of the frequency of 90×10^3-eV photons. Consequently, for such photons lead is opaque while water is relatively transparent.

16.11 QUANTUM ELECTRODYNAMICS

The description of atomic physics that has been outlined in the preceding sections is based on a quantum mechanical but nonrelativistic theory of electron behavior. At low energies such a theory is a useful approximation but at high energies we know that major changes are required. A theory in accord with the special theory of relativity must have the electron's energy E related to its momentum p and its rest mass m by

$$E^2 = (mc^2)^2 + (cp)^2. \qquad (5.28)$$

And at high energies the number of electrons in a system is not constant. Creation and annihilation of electron-positron pairs may occur (Section 4.4).

The construction of a theory of the interaction of photons, electrons, and positrons that fits the verified facts of both relativity and quantum mechanics has proved to be an arduous task but well worth the effort. This theory is called *quantum electrodynamics.* In a certain sense, quantum electrodynamics is the prize gem of all scientific theories because it embraces atomic physics and chemistry and indeed all phenomena governed by the electromagnetic interaction, furnishing results that in some instances have the unprecedented precision of one part in 10^9. Nobel prizes were earned by the most celebrated architects of this theory: Dirac in 1933; J. Schwinger of Harvard University, R. P. Feynman of the California Institute of Technology, and the Japanese theoretical physicist, S. I. Tomonaga in 1965.

Feynman Diagrams

In quantum electrodynamics, the state of a system is specified by giving the number of electrons, positrons, and photons as well as their energies, momenta, and spins. Any transition or change from one state to another

---

**It is rather surprising that, although the wave contributed by each molecule travels at the same velocity c, the superposition of the waves radiated from molecules in different locations can and does produce a refracted wave which travels at a lesser velocity $v = c/n$, where n is the refractive index of the water.*

is described in terms of the annihilation of particles in the original state and the creation of particles in the new state. Every change is a catastrophic change.

Thanks to Feynman's work, it is possible to "picture" the creation and annihilation events that are associated with any change. More precisely, Feynman showed that to each abstruse mathematical expression in the theoretical description of a transition from one state to another, there corresponds a diagram with a straightforward physical interpretation. In the *Feynman diagrams* of Fig. 16-21, a spatial dimension, such as an x-axis, is plotted horizontally. The vertical direction is that of a time axis with low regions corresponding to early times and higher regions to later times. A photon is represented by a broken line and an electron by a solid line.

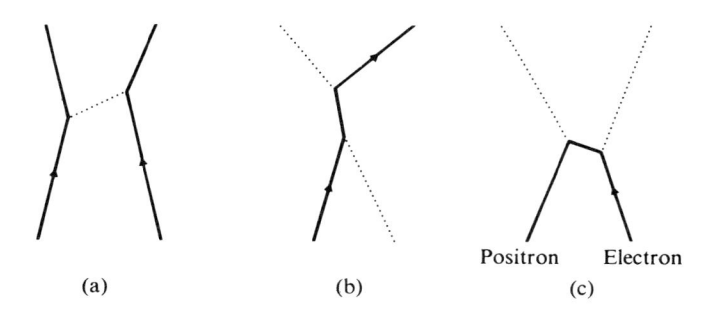

Fig. 16-21 Three Feynman diagrams. Solid lines represent electrons or positrons, and broken lines represent photons. (In these schematic diagrams no attempt has been made to relate the direction of a line with particle velocity.) The bottom of the figure represents the initial state and the top represents the final state. (a) Electron-electron scattering by exchange of a virtual photon, (b) Compton scattering, (c) Electron-positron annihilation with the production of two photons.

The first diagram, Fig. 16-21(a), depicts the basic phenomenon underlying the interaction of two charged particles. The initial state with two electrons is represented by the two solid lines at the lower part of the diagram. At the left vertex a photon is created, the incident electron is annihilated, and an electron with different momentum and energy is created. This new electron is represented by the line emerging toward the upper left-hand corner. The photon travels a short distance and is absorbed at a vertex where the other incident electron is annihilated and another electron is created. In the final state there are again just two electrons and no photon. According to this diagram a "collision" of two electrons involves the *exchange* of a *photon*. The photon is the mechanism by which energy and momentum are transferred from one electron to another. The interaction whose electric part we had described by simply postulating a Coulomb force,

$$F = \frac{kQq}{r^2}, \qquad (2.19)$$

which in this case of two electrons is simply

$$F = ke^2/r^2,$$

can thus be interpreted as arising from the emission and absorption of photons by the two charged particles.

A Feynman diagram for the Compton effect is shown in Fig. 16-21(b). The initial photon is absorbed at one vertex and the final photon is emitted at another vertex.

The annihilation of an electron-positron pair (Example 6, p. 172) with the subsequent production of two photons is depicted in the Feynman diagram of Fig. 16-21(c).

The key ingredient in every diagram is a vertex with one photon line and two lines representing charged particles. Using such vertices, much more complicated processes can be diagrammed, and their relative importance in linking the initial and final states can be evaluated by the mathematical procedures of quantum electrodynamics. In such a calculation a factor α (the fine structure constant introduced in Section 16.5) appears for each vertex in the diagram. This constant has the value

$$\alpha = ke^2/\hbar c \qquad (16.14)$$
$$= 1/137.$$

Since high powers of $\frac{1}{137}$ are very small, diagrams with many vertices are unimportant compared to the two-vertex diagrams of Fig. 16-21.

There is a subtlety associated with these diagrams. The simple process depicted at any vertex is never observed. Such a process cannot satisfy the conservation laws for both energy and momentum. For all observed processes, such as the three processes of Fig. 16-21, the total energy and momentum in the final state are the same as in the initial state—that is, energy and momentum are conserved. Nevertheless, for an *intermediate* state, processes which violate these conservation laws are possible in quantum mechanics, provided the intermediate state is of sufficiently short duration. This possibility is associated with the Heisenberg uncertainty relation for energy and time: A state which lasts for some limited time τ does not correspond to a single definite value of the energy, but instead is a superposition of different energy values ranging roughly from some value E to $E + \Delta E$, where the spread or uncertainty ΔE in the value of the energy cannot be reduced below the value given by

$$\tau \, \Delta E \approx h.$$

Therefore, a state of short duration τ may involve a process in which energy conservation is violated by any amount which does not exceed that given by
$$\Delta E \approx h/\tau.$$

Alternatively we can assert that for a state which involves a creation or an annihilation of a given amount of energy ΔE, the duration of the state must not be longer than a time given by

$$\tau \approx h/\Delta E.$$

The intermediate states represented in the diagrams of Fig. 16-21 involve the creation of an undetected particle at one vertex and its absorption at another vertex. Such particles are called *virtual* particles. Since the presence of a virtual particle implies a violation of energy conservation by a certain amount ΔE, the virtual particle can have only a transitory existence, not exceeding a time τ given by

$$\tau \approx h/\Delta E.$$

In this terminology, the electron-electron collision of Fig. 16-21(a) involves the exchange of a virtual photon. The word *virtual* serves to emphasize that the Feynman diagrams, although extremely useful, are by no means equivalent to a classical picture of an interaction. The strangeness of quantum mechanics persists.

The theoretical predictions of quantum electrodynamics accord with experiment not only for a wide variety of scattering phenomena (Fig. 16-21) but also for certain atomic energy level details that were not predicted by nonrelativistic quantum mechanics: A small separation of hydrogen atom energy levels which have the same value of the principal quantum number n but different values of ℓ is a relativistic effect, as is the very existence of electron spin.

Also from the postulates of quantum electrodynamics it can be deduced that electrons must obey the Pauli exclusion principle. In nonrelativistic quantum theory the exclusion principle was an unexplained addition to the basic postulates.

In spite of its many successes, quantum electrodynamics is still plagued with mathematical and conceptual difficulties. As one illustration of these problems, let us consider the energy of the virtual photon exchanged between the electrons of Fig. 16-21(a). The shorter the path of this photon, the less its lifetime τ and the greater its maximum allowable energy,

$$\Delta E \approx h/\tau.$$

Therefore the allowed energy of exchanged photons increases as the separation of the electrons decreases, becoming infinite for zero separation. An infinite photon energy entails a host of problems. This train of thought leads one to suspect that the electron-photon vertices, instead of being mathematical points, must have some structure, and then the electron itself would have a structure. If this is so, quantum electrodynamics in its present formulation will fail whenever interactions at very small distances are important. So far no such evidence of electron structure has been detected even though experiments at high energies have tested the electromagnetic interaction down to particle separations as small as 10^{-16} m.

Strong Interactions

Attempts have been made to develop a theory of strong interactions, such as those which lead to the formation of the nucleus, using ideas similar to those that have been so successful in the description of the weaker electromagnetic interaction. The interaction between two nucleons (neutrons or protons) is attributed to the exchange of a virtual particle. The range R of the interaction is determined by the mass M of this virtual particle according to the following argument. An upper limit to the distance R that the virtual particle can travel during its lifetime τ is given by the distance it can travel at a speed c:

$$R = c\tau.$$

The existence of a virtual particle of mass M implies the existence of at least a rest energy Mc^2. Taking this energy as the order of magnitude of the uncertainty ΔE in the energy of the intermediate state, we obtain

$$Mc^2 = \Delta E = h/\tau = h/(R/c) = hc/R.$$

This yields
$$M = h/cR.$$

Use of the experimentally determined range of nuclear forces, about 10^{-15} m, gives M equal to several hundred times an electron's mass.

The π-mesons discovered by Powell in 1947 (Section 4.4) have masses of this order of magnitude (actually 273 m) and interact strongly with neutrons and protons. Following this discovery it has been assumed that the exchange of *virtual π-mesons* is a most important contribution to the force between strongly interacting particles, such as the constituents of a nucleus. The existence of particles like the π-mesons, which could be exchanged between nucleons and give rise to the short-range nuclear force, had been brilliantly conjectured by the Japanese theoretical physicist, H. Yukawa, twelve years before their discovery.

After this promising start, subsequent development of strong interaction theory has been disappointing. One trouble is that in the Feynman diagram for any strong interaction process, each vertex involves a factor of about 1 instead of the factor of 1/137 encountered in the electromagnetic interaction. Consequently there is no reason to think that processes represented by complex diagrams with many vertices are unimportant.

Much of the progress that has been made in theories of strong interactions (and also the weak interactions mentioned in Table 2-2, p. 80) is based on studies of "symmetries." A discussion of this topic, and many others in the vast field of quantum physics, can be found in the books listed in the Suggested Reading.

Physics Today

H. Yukawa (1907–   )

QUESTIONS

1. Arrange light beams of the colors red, green, violet, in order of increasing
   (a) wavelength,
   (b) frequency,
   (c) photon energy.

2. Find the energy in electron volts of a photon of green light for which $\lambda = 5.0 \times 10^3$ Å.

3. What is the wavelength of a beam of light consisting of photons which have an energy of 2.0 eV each? What color is this light?

4. The γ-ray photon emitted when a proton captures a slow neutron has an energy of 2.3 MeV. Find the wavelength of this photon and compare it with a typical size for a nucleus of low atomic number, say 10^{-15} m.

5. A table top of area 2.0 m² is uniformly illuminated by 10 watts/m² of light with wavelength 6.2×10^3 Å. How many photons per second strike the table?

6. A radio station with an average power output of 4.0×10^3 watts broadcasts at a frequency of 1.5×10^6 hertz.
   (a) What is the energy of a photon in electron volts?
   (b) How many photons per second are emitted?

7. What is the number N (photons per sec per unit cross-sectional area) in the laser beam described in **Question 2 of Chapter 14? (The wavelength of this** red laser light is 6.3×10^3 Å.)

8. (a) What is the photoelectric effect?
   (b) State and explain the relationship between the maximum kinetic energy of the electrons ejected from a surface and the frequency of the light used to irradiate the surface.

9. Green light ejects electrons from a certain surface. Yellow light does not. Do you expect electrons to be ejected when the surface is illuminated with
   (a) red light? Why?
   (b) violet light? Why?

10. The work function for a cesium surface is 2.0 eV.
    (a) Find the maximum kinetic energy of the ejected electrons when the surface is illuminated by violet light with $\lambda = 4.13 \times 10^3$ Å.
    (b) What is the "threshold wavelength" (the largest λ) for the photoelectric effect to occur with this metal?

11. When a clean zinc surface is irradiated with ultraviolet light it is found that no electrons are ejected from the surface unless the light has a wavelength less than 2.93×10^3 Å.
    (a) What is the work function (in eV) of the zinc surface?
    (b) What is the maximum kinetic energy of the electrons that are ejected by light of wavelength 1.24×10^3 Å?

12. In what ways are the experimental facts regarding the photoelectric effect in disagreement with the predictions of classical electromagnetic theory?

13. (a) Describe the mechanism for production of Bremsstrahlung x-rays.
    (b) What is the shortest wavelength that will emerge from an x-ray tube when the difference of potential between the cathode and anode is 100×10^3 volts?

14. (a) What is the Compton effect?
    (b) Why does the photon, scattered from a free electron at rest, have a longer wavelength than the incident photon?

15. Using Compton's equation, $\lambda' - \lambda = (0.024$ Å$)(1 - \cos \theta)$, calculate the increases in wavelength, $\Delta \lambda = \lambda' - \lambda$, which correspond to scattering angles of 0°, 30°, 90°, 180°.

16. A photon with a wavelength of 0.049 Å is scattered backwards ($\theta = 180°$) by an electron initially at rest.
    (a) Find the initial energy and the final energy of **the photon.**
    (b) What is the kinetic energy of the electron after this event? [*Hint:* Use the result of the previous question for $\theta = 180°$.]

17. Find the magnitude (in eV/c) and the direction of the momentum vectors of the incident photon, the scattered photon, and the recoiling electron of the preceding question.

18. What is the wavelength of the incident photons if Compton scattering at an angle of 90° gives photons with energies equal to one-half the energies of incident photons?

19. A hydrogen atom in the excited state, $E_2 = -3.40$ eV, makes a transition to its ground state, $E_1 = -13.6$ eV, and emits a photon. Find the photon energy and wavelength. Is this photon in the ultraviolet, the visible, or the infrared portion of the spectrum of electromagnetic waves?

20. The color plate in Section 14.3 shows four lines of the hydrogen atom spectrum. From the wavelengths given on this plate, determine the corresponding photon energies and identify on an energy-level diagram the transitions associated with the emission of these photons.

21. The four lines referred to in the preceding question are the first four lines of the Balmer series. Find the photon energy and the wavelength for the next line in this series.

22. A collection of hydrogen atoms in their ground states is irradiated with ultraviolet photons which each have an energy of 12.1 eV. Some atoms each absorb a photon and are excited to the state with energy $E_3 = -1.51$ eV.
    (a) Draw an energy level diagram and label with arrows all the possible transitions to lower energy levels.
    (b) Calculate the energies and wavelengths of all the photons that will be emitted by the hydrogen atoms.

23. A hydrogen atom in its ground state is ionized by the absorption of an 800-Å photon. (This is an "atomic photoelectric effect.") Find the kinetic energy of the ejected electron.

24. After a collision with some other particle, an atom must be left either in its ground state or in an excited state, with an energy corresponding to one of its possible energy levels. Consider hydrogen that is bombarded by a beam of electrons, each of which has a kinetic energy of 12.5 eV.
    (a) What excited states are possible for a hydrogen atom after interaction with an electron in this beam?
    (b) What wavelengths will be emitted by hydrogen atoms after collision with electrons of this beam?

25. When hydrogen is bombarded by a beam of electrons of kinetic energy K, what is the lowest value of K for which *visible* light will be emitted by the hydrogen atoms? What color is this light?

26. What is the change in the energy of a nucleus that emits a γ-ray with a wavelength of 3.1×10^{-3} Å?

27. Show that, for an electron that moves at a speed well below the speed of light, and whose kinetic energy K can thus be evaluated from $K = \frac{1}{2}mv^2 = p^2/2m$, the de Broglie wavelength of the electron can be expressed in terms of its kinetic energy by

$$\lambda = \sqrt{\frac{150 \text{ eV} - \text{Å}^2}{K}}$$

28. Find the de Broglie wavelengths that correspond to the following electron kinetic energies: 1.5 eV, 150 eV, 15×10^3 eV.

29. Describe, in terms of the arrival of photons, the formation of the intensity maxima and minima shown on the screen in Fig. 14-9 and Fig. 14-10. Assume the light source is extremely dim.

30. The electron gun of Fig. 16-8 is adjusted so that, with just one slit open in the barrier, the small counter records an average of 100 counts per hour. With one slit open, this counting rate does not change appreciably as the counter is moved about to explore different regions on the screen. When the second slit is opened, what is the counting rate at the location of
    (a) an intensity maximum?
    (b) an intensity minimum?

31. Suppose one were to try to perform a laboratory experiment using a beam of electrons instead of a light beam in a double-slit interference experiment. Assume that the geometry of the slits and of the screen are to be exactly as described in Example 3 of Chapter 14. If, in the interference pattern on the screen, the distance between locations of maximum intensity is to be the same as that which was observed when the yellow light from the sodium vapor lamp was used, what must be the kinetic energy of the electrons in the electron beam?

32. (a) Compute the de Broglie wavelength of a football of mass 0.40 kg when it is moving at a speed of 10 m/sec.
    (b) Have any of the experiments discussed in this book measured wavelengths this short?

33. If the momentum of the football of Question 32 has an uncertainty of merely about one part in a million (that is, $\Delta p = 4 \times 10^{-6}$ kg m/sec), find the minimum uncertainty in position Δx consistent with the Heisenberg uncertainty relation.

34. Verify that the fine structure constant, $\alpha = ke^2/\hbar c$, is a dimensionless number.

35. What is meant by the term "charge cloud" in atomic theory?

36. Compare the de Broglie wavelength of an electron moving at a speed αc with the circumference of the Bohr orbit of radius $a = \hbar^2/mke^2$.

37. Describe the uncertainty in position and momentum for an electron in the hydrogen atom ground state. Verify that these uncertainties are consistent with the Heisenberg uncertainty relation.

38. Evaluate the ground state energy of the hydrogen atom from the formula $E_1 = -\frac{1}{2}\alpha^2 mc^2$, using $\alpha = 1/137$ and $mc^2 = 0.511 \times 10^6$ eV.

39. (a) Show that each of the following three constants has the dimensions of a length: $\hbar^2/mke^2$, called the radius of the Bohr orbit and denoted by a; $\hbar/mc$, called the reduced Compton wavelength of an electron and denoted by λ_c; and ke^2/mc^2, called the classical electron radius and denoted by r_e.
    (b) Show that $r_e = \alpha \lambda_c = \alpha^2 a$ where $\alpha = ke^2/\hbar c = 1/137$.

40. Using the rules of Table 16-1, give the possible values of the quantum numbers n, l, m_l and m_s for (a) a state with $n = 1$; (b) a state with $n = 3$.

41. List the quantum numbers of all states for which $n = 4$. Verify that there are 2×4^2 different states.

42. State the Pauli exclusion principle.

43. (a) What is a subshell?
    (b) Verify that the numbers in the last row of Table 16-2 do follow from the rules of Table 16-1.

44. From the diagram of Fig. 16-10, work out the electron configuration for the ground states of each atom from $Z = 1$ to $Z = 9$.

45. What is the electron configuration for the ground state of magnesium ($Z = 12$)?

46. What would be the electron configuration of magnesium if no restrictions were imposed by the Pauli exclusion principle?

47. Give the electron configuration of the outer subshell for each inert gas.

48. (a) Give the electron configuration of the outer subshell for lithium ($Z = 3$), sodium ($Z = 11$), and potassium ($Z = 19$).
    (b) Give the electron configuration of the outer subshell for fluorine ($Z = 9$) and chlorine ($Z = 17$).

49. Relate your answers to Questions 47 and 48 to the chemical behavior of these elements and their position in the periodic table (Table 16-3).

50. Describe a sequence of events that can lead to the emission of x-ray photons in the characteristic spectrum of an element.

51. Sketch typical energy level diagrams showing the occupancy of various bands for a conductor and an insulator. Label the conduction band for each case.

52. Explain why the presence of many empty levels in a partially filled band leads to a high conductivity.

53. Why is diamond transparent?

54. What is the significant difference between an insulator and a semiconductor?

55. Draw typical energy level diagrams showing the occupancy of the valence and conduction bands for n- and p-type semiconductors.

56. In what ways is a laser light beam different from a beam from an ordinary flashlight?

57. Explain laser action making reference to stimulated emission, population inversion, and optical pumping.

58. A certain laser using carbon dioxide produces a beam with a total energy of 150 joules which is confined to a pulse lasting approximately 10^{-7} sec.
    (a) What is the power output of the laser during the emission of this pulse?
    (b) It should be possible to achieve controlled nuclear fusion reactions (Section 4.5) by irradiating heavy water pellets with a sufficiently powerful laser beam. If an energy of 10^6 joules is needed for pulse duration of 10^{-7} sec, what is the required power output of such a laser?

59. Name and describe the three different interactions of photons with matter that are shown schematically in Fig. 16-22.

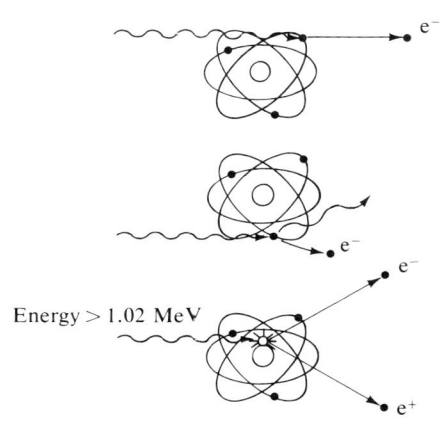

Fig. 16-22 Summary of x-ray and gamma-ray interactions with matter.

60. Show that the resonant frequencies of the ground state of the hydrogen atom lie in the ultraviolet region. Demonstrate this by showing that the lowest photon energy obtained from Eq. (16.15) [used in conjunction with Eq. (16.8)] corresponds to an ultraviolet wavelength.

61. Why is water transparent to visible light but strongly absorptive for ultraviolet and infrared?

62. Calculate the distance that the virtual photon of Fig. 16-21(a) can travel in a time $\tau = h/\Delta E$, for the case where ΔE equals the rest energy (mc^2) of an electron.

63. Does the range of an interaction depend on the mass of the virtual particle that is exchanged between the interacting particles? Explain your answer.

SUPPLEMENTARY QUESTIONS

S-1. One method of determining the value of Planck's constant is based on observations of the photoelectric effect. Suppose that a metallic surface is irradiated by light and that measurement is made of the retarding voltage required to stop the most energetic of the electrons ejected from the surface. Suppose also that for light with a wavelength of 4.00×10^3 Å the stopping potential is 2.00 volts, while for light with a wavelength of 6.00×10^3 Å the stopping potential is 1.00 volt. Use this data to determine the value of Planck's constant and the work function of the surface.

S-2. When the potential difference between the cathode and the anode of an x-ray tube is maintained at 62.5×10^3 volts, it is found that 0.20 Å is the shortest wavelength in the spectrum of the emitted x-rays. Calculate the value of Planck's constant from this data.

S-3. X-rays with a wavelength of 0.036 Å emerge from a metal target at an angle of 90° with the direction of the incident beam. Assume that these x-rays are scattered from free electrons. Find
   (a) the wavelength of the x-rays in the incident beam;
   (b) the momentum vectors of an incident photon and a scattered photon;
   (c) the magnitude and direction of the momentum vector of a recoiling electron that participates in such a scattering event.

S-4. In the description of the Compton effect given in Section 16.2, it is mentioned that *part* of the scattered radiation has a longer wavelength than that of the incident radiation. It should also be pointed out that the other part of the scattered radiation (called the unmodified radiation) has essentially the same wavelength as the incident beam. The origin of this "unmodified" scattered radiation can be understood by considering the change in photon wavelength, $\lambda'' - \lambda$, which occurs in scattering events in which the photon interacts with an atom of mass M and the entire atom recoils. The appropriate form of Compton's equation is then $\lambda'' - \lambda = (h/Mc)(1 - \cos\theta)$. Show that

$$\frac{\lambda'' - \lambda}{\lambda' - \lambda} = \frac{m}{M},$$

where $\lambda' - \lambda$ is the change in photon wavelength when the recoiling particle is an electron of mass m. Then calculate the wavelengths λ' and λ'' that will be observed at a scattering angle of 90°, when x-rays with $\lambda = 0.100$ Å are incident on carbon.

S-5. Following Bohr's line of thought, apply the principles of Newtonian mechanics to describe the motion of an electron (of mass m and charge e) acted upon by a Coulomb force ke^2/r exerted by a proton of charge e. To a good approximation, the acceleration of the relatively massive proton can be ignored and the proton can be regarded as stationary.
(a) Show that when the electron moves in a circular orbit of radius r about the proton, Newton's second law gives
$$\frac{ke^2}{r^2} = \frac{mv^2}{r}.$$
(b) Show that when an electron is moving in a circular orbit of radius r, its total mechanical energy E is given by
$$E = \tfrac{1}{2}mv^2 - \frac{ke^2}{r} = \frac{-ke^2}{2r}.$$

S-6. As described in Section 16.3, Bohr concluded that the line spectrum of the hydrogen atom was evidence that the energy of the atom could have only certain discrete values (called energy levels) in its bound states. From the result of the previous question, $E = -ke^2/2r$, it is apparent that the possible values of the electron's total energy E will be restricted to certain discrete values, $E_1, E_2, E_3, \ldots$, only if the radius of the electron's orbit is correspondingly restricted to certain discrete values, $r_1, r_2, r_3, \ldots$. At this point Bohr made the brilliant guess that, for some reason foreign to Newtonian mechanics, the electron's orbital *angular momentum* (mvr from Fig. 3-34) is restricted to values which are an integral multiple of $\hbar$. That is, Bohr made the following *quantum hypothesis*:
$$mvr = n\hbar,$$
where the possible values of the *quantum number* n are the integers $1, 2, 3, \ldots$.
(a) Show that this assumption, together with our previous result, $ke^2/r^2 = mv^2/r$, implies that the radii of the possible circular orbits are given by
$$r_n = n^2 a,$$
where $a = \hbar^2/mke^2$ (the so-called Bohr radius).
(b) Next, using $E = -ke^2/2r$, with r restricted to the values $r_n = an^2$, show that possible values of E are given by
$$E = \frac{E_1}{n^2},$$
where $E_1 = -\tfrac{1}{2}\alpha^2 mc^2$ and $\alpha = ke^2/\hbar c$. This is the famous Bohr formula for the energy levels of the hydrogen atom.

S-7. Experimental measurement of the wavelengths in the line spectrum of the hydrogen atom shows that all the observed wavelengths are given by the formula
$$\frac{1}{\lambda} = R\left(\frac{1}{n_l^2} - \frac{1}{n_u^2}\right),$$
where $n_l = 1, 2, 3, \ldots$ and n_u is an integer greater than n_l. The experimental value of the constant R, called the Rydberg constant, is 1.09677×10^7 m^{-1}. Using Bohr's formula for the energy levels of the hydrogen atom, show that Bohr's theory implies that
$$R = \frac{\alpha}{4\pi a},$$
where $\alpha = ke^2/\hbar c$ and $a = \hbar^2/mke^2$. Compare the predicted value $\alpha/4\pi a$ with the experimental value of R.

S-8. Show that the shortest wavelength in the line spectrum of hydrogen is much larger than the Bohr radius a.

S-9. Show that, according to Newtonian mechanics, the electron's speed in the nth Bohr orbit (of radius $r_n = n^2 a$) is given by
$$v = \frac{v_1}{n},$$
where $v_1 = \alpha c$ and $\alpha = ke^2/\hbar c$. [Hint: Use the result, $ke^2/r^2 = mv^2/r$, together with $r_n = n^2 a$.]

S-10. Use the postulates of Bohr's theory of the hydrogen atom to deduce an expression for the energy levels of singly ionized helium (a helium atom from which one electron has been removed).

S-11. According to the Bohr theory, what is the radius of the electron orbit for a singly ionized helium ion in its ground state?

S-12. (a) Consider the frequency f of radiation emitted by a hydrogen atom in a transition from an energy level with quantum number $n + 1$ to a level with a quantum number n. Using the expression $E_n = -\tfrac{1}{2}\alpha^2 mc^2/n^2$, show that if n is large, the radiated frequency is given approximately by
$$hf = \frac{\alpha^2 mc^2}{n^3}.$$

SUPPLEMENTARY QUESTIONS 661

(b) Classical theory predicts that an electron moving in a circular orbit should radiate electromagnetic waves with a frequency equal to that of the electron's revolution in its orbit. What is this frequency for an electron traveling in an orbit of radius $r_n = n^2 a$ at a speed $v = \alpha c/n$?

(c) Bohr's thoughts were guided by what he called the *Correspondence Principle*, which may be stated as follows: Predictions of quantum theory must agree with predictions of classical theory in the limit where quantum discontinuities may be treated as negligibly small (as is the case when the quantum numbers are very large). Are the results of parts (a) and (b) consistent with the correspondence principle?

S-13. For an electron orbit in a hydrogen atom, show that if Bohr's quantum hypothesis (given in Question S-6) is satisfied, then the circumference of the orbit contains an integral number of de Broglie wavelengths. In other words, show that the equation

$$mvr = n\hbar$$

implies that

$$2\pi r = n\lambda,$$

where $\lambda = h/mv$.

S-14. Starting from the relativistically correct relationship between energy E and momentum p,

$$E = K + mc^2 = \sqrt{(mc^2)^2 + (cp)^2}\ ,$$

show that the de Broglie wavelength of a particle of mass m can be expressed in terms of its kinetic energy K by:

$$\lambda = \frac{h}{mc}\ \frac{1}{\sqrt{(2K/mc^2) + (K/mc^2)^2}}\ .$$

S-15. Show that, in an extreme relativistic range of kinetic energies, where K is much greater than mc^2, the de Broglie wavelength is *independent* of the mass of the particle being given by

$$\lambda = hc/K,$$

after making the approximation that mc^2 is negligible compared to K.

S-16. Using the appropriate results of preceding questions (27, S-14, S-15), evaluate the de Broglie wavelength of an electron and of a proton for the following values of the particles' kinetic energies: 30 eV, 30 keV, 30 MeV, 30 GeV. (For an electron $mc^2 = 0.511$ MeV and $h/mc = 0.024$ Å. The mass of a proton is 1836 times the mass of an electron.)

S-17. If an electron were confined to a region the size of a nucleus then the uncertainty in one of its position coordinates would be approximately 10^{-14} m.

(a) Evaluate the corresponding uncertainty in momentum.

(b) Since the possible momentum values must be at least as large as the uncertainty in momentum, the value Δp_x calculated in part (a), can be used as a typical value of the momentum p. Calculate the corresponding value of the electron's kinetic energy. (Relativistic expressions must be used.)

S-18. (a) A dramatic demonstration using a ruby laser is performed by placing two rubber balloons, one red, one blue, in the path of the laser beam. The beam is then switched on. One of the balloons bursts, the other does not. Which one bursts? Explain.

(b) Would the red light from a ruby laser be effective on blood, and therefore useful in the medical treatment of vascular abnormalities? Explain.

S-19. Argon ion lasers can be used in ophthalmology to cause what is known as photocoagulation, by means of which splits and holes in the retina of the eye may be healed. The principal output of the argon laser is at two wavelengths, 5145 Å and 4880 Å. These lasers are usually operated in the 0.50–2.0-watt power range.

(a) What are the photon energies associated with the two wavelengths cited?

(b) What are the colors?

(c) Would one expect blood vessels to absorb much of this laser energy? Explain.

(d) Suppose a given treatment was carried out at 0.95 watt and lasted for 30 milliseconds. How much energy is involved?

(e) If the total energy in (d) were split evenly between the main two photons [see (a)], how many of each type of photon would there be?

S-20. In medical radiography, the chest "x-ray" is used as a standard check for tuberculosis. What might you expect to happen to the x-ray film contrast if a radioactive γ-ray source were used rather than a conventional x-ray generator? [*Hint:* Most gamma ray emitters, but not all, yield gamma rays in the MeV energy region, while medical x-rays have energies in the region of 40 keV to 140 keV.]

ADDITIONAL APPLICATIONS TO MEDICINE AND THE LIFE SCIENCES

Phototubes: The Photoelectric Effect

The photoelectric effect finds application in many different fields. A very common ingredient of many control circuits today is the *phototube* which can activate a circuit to turn off a light, open a door, trigger an alarm, etc.

In the medical x-ray department the phototube is used in *automatic exposure timers*. These timers operate to terminate an x-ray exposure when the film has received enough radiation to produce a density suitable for the purpose involved. At the heart of this *phototimer* is the phototube, or photoelectric tube (see **Fig. 16-23**). This tube may consist of a glass envelope from which the air has been evacuated. Inside the tube is a small curved piece of metal coated with a photoemissive surface having an appropriate work function. This piece of metal is the cathode of the tube (see **Fig. 16-1**); near the cathode is the anode. In use, these two electrodes are connected to an external source of EMF, but when they are in relative darkness, no current passes through the tube. When light falls on the tube, electrons are ejected from the photoemissive surface, and so under the influence of the applied EMF, a current is produced. In x-ray equipment the phototube with a small fluorescent screen in front of it is positioned underneath (or behind) the film (see **Fig. 16-24**). X-rays pass through the patient to the film, and some of the x-rays will continue through the film and strike the fluorescent screen. (X-rays are not used directly for precipitating photoemission, since the efficiency of this process is much lower than that for visible light.) The photo current may now be employed to charge a capacitor. The charged capacitor is in turn used to trigger another circuit, which in turn terminates the x-ray exposure. The required x-ray film density can be achieved simply by selecting the appropriate value of the capacitance of the capacitor. The charge on the capacitor, which is the charge transferred by the photocurrent, is proportional to the total x-ray energy applied.

Another very fruitful application of the photoelectric effect is in the photomultiplier tube, a device which now forms part of much radiation detection equipment, not only in medicine and biology, but also in many other aspects of physical research and engineering. For example, since in the x-ray machine phototimer the light intensities are quite low, the simple photoelectric tube has been superseded in today's x-ray equipment by the *photomultiplier* tube (see **Fig. 16-25**). Between the anode and the cathode of this tube there are several intermediate electrodes called *dynodes*.

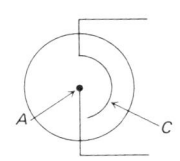

Fig. 16-23 A photoelectric tube. *A* is the anode and *C* is the photocathode, which emits electrons when illuminated. Both electrodes are enclosed in the evacuated glass tube represented by the outer circle.

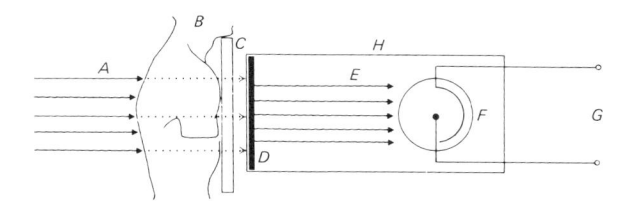

Fig. 16-24 A phototimer unit. *A* is the x-ray beam; *B*, the patient; *C*, the cassette which holds the x-ray film; *D*, a fluorescent screen facing *F*, the photoelectric tube which is inside the light-tight box, *H*. *E* is the light coming from the fluorescent screen, and *G* represents the connections to an external circuit. The different components are not to scale; an actual phototimer is quite small compared with the film cassette.

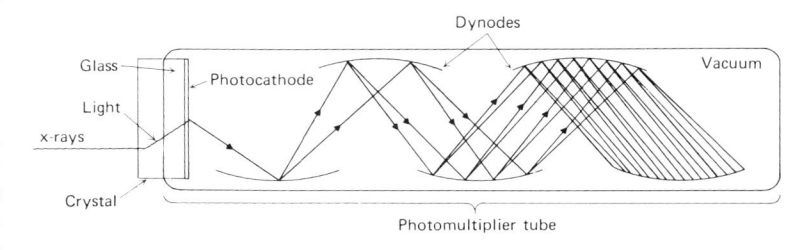

Fig. 16-25 Photomultiplier tube as used for x-rays.

Each of these is maintained at a voltage higher than the one preceding and each is coated with a material which emits secondary electrons when bombarded with electrons from the dynode before. Although the primary electrons are emitted as a result of light energy, the secondary electrons are emitted as the result of the energy imparted to them by the primary electrons, a phenomenon known as *secondary emission*.

When light falls on the cathode, electrons are emitted. They are then accelerated to the first dynode, where they produce secondary electrons. These electrons are accelerated to the second dynode, where each produces more than one new secondary electron. As this process is repeated through the photomultiplier tube, the final electron flow is many times—perhaps 10^8 times—greater than the original. This flow is therefore large enough to be easily and directly utilized to operate an x-ray exposure circuit, whereas using only the simple phototube would make it necessary to have additional amplifying circuits.

A gamma camera, such as mentioned in Chapter 1 makes use of many such photomultiplier tubes. Use of these tubes permits radioisotope scanning with much lower patient radiation dose than would otherwise be possible.

Question

In the notes above, two types of electron emission were mentioned. List all types of electron emission processes you can think of and indicate the type of energy supplied to cause the emission.

Answer

1. Thermionic emission—thermal energy (see **Chapter 13**).
2. Secondary emission—kinetic energy of impinging primary electrons.
3. Photoemission—electromagnetic radiation.
4. Field emission—large electric field. This process has not been mentioned earlier in the text nor in these notes. In field emission, if the electric field E at the emitter surface has a sufficiently high value, say 10^6 kV/m or more, then electrons are emitted and the number emitted per unit time increases approximately exponentially with the applied electric field. For certain medical applications, x-ray machines using field emission rather than thermionic emission have been found effective (see **Fig. 16-26**).

X-Ray Energies in Medical Radiography: Bremsstrahlung

When a medical radiographer (x-ray technician, or radiologic technologist) sets a voltage of 100 kV on his machine this does not mean that all the x-ray photons have an energy of 100 keV. As can be seen from **Fig. 16-3**, the maximum intensity (energy per unit area per unit time) occurs for x-rays having an energy 2/3 of the maximum photon energy or a wavelength of about 1.5 times the minimum wavelength (maximum energy).

In addition to this Bremsstrahlung radiation, if the voltage is sufficiently high (60 kV or higher for a tungsten x-ray target), a certain amount of energy in the form of characteristic x-rays also appears (see **Fig. 16-27**). In diagnostic x-ray work the energy contribution by characteristic x-rays may be as much as 10% or more of the total x-ray energy.

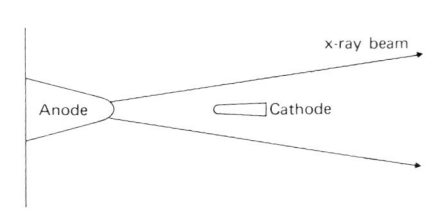

Fig. 16-26 The field-emission x-ray tube. A high voltage is applied between the electrodes. Because of the small area of the cathode the current density is large.

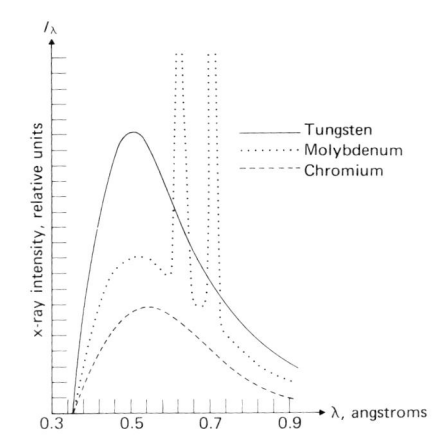

Fig. 16-27 X-ray spectra of three target metals at 35 kV. [C. T. Ulrey, *Phys. Rev. 11*, 405 (1918)]. Note characteristic radiation evidenced by intensity peaks.

From all this one can see the x-ray beam is far from being homogeneous (or monochromatic); a wide variety of x-ray photon energies appears, having an upper limit set by the maximum voltage across the x-ray tube. A reasonable approximation is that the effect of the heterogeneous beam produced is about the same as that of a homogeneous beam having an energy of from 30% to 50% of the maximum beam energy of the heterogeneous beam. The point of consequence is then that an x-ray beam at 90 kV does not produce photons all of 90 keV energy, but rather produces a beam which to a certain extent resembles a homogeneous beam of say 30 or 40 keV photons.

The penetrability of an x-ray beam, usually expressed as the *quality* of the beam, is indicated in terms of the half-value layer (the HVL is that thickness of a given material required to reduce the intensity of the beam to half its original value).

Question

(a) What is the most common photon energy in an x-ray beam when the tube voltage is set at 70 kV?
(b) The half-value layer of an x-ray beam is given as 1.5 cm of Al. What does this mean?

Answer

(a) Maximum photon energy

$= 70$ keV.

Most common photon energy

$= 2/3 \times 70$ keV

$= 46$ keV.

(b) The beam is of such a quality that when incident on Al of thickness 1.5 cm the emerging beam has only half the intensity of the incident beam.

Xeroradiography (or Xerography)

Knowing something about semiconductors allows us to consider a relatively old technique now being revitalized: xerography. This is essentially an electrostatic process in which x-rays are employed in the conventional way to obtain an image of internal body structures. It is different from conventional radiography in that the image is not recorded on a film. Rather the image is obtained by using a thin layer of a semiconductor called a photoconductor (a substance which normally has very few charge carriers but in which radiation quanta can release many new carriers by raising electrons into the conduction band). One material which is used is selenium, which is a good insulator in the dark and will also retain a superficially-applied electric charge for some hours.

The selenium-covered recording plate, which has been given a surface electric charge, is placed under the patient in the same way as conventional film and is then subjected to the x-radiation which has passed through the patient. When x-rays strike the selenium layer the selenium becomes conductive. The surface charge now "leaks" away in those areas which have been exposed to x-rays and thus an electric charge image of the incident radiation pattern is left. This charge image may be rendered visible by exposing the selenium plate to an aerosol of electrically charged powder particles. These powder particles adhere to the surface of the selenium plate principally in the region of strong electric field and thus clearly delineate variations in the charge density on the plate. This pattern may be viewed directly or recorded permanently by several different methods.

The xerographic process is becoming more practical largely through benefits resulting from technological advances made in photocopying. In medicine, the xerographic image offers several important advantages, including improved resolution and a longer contrast scale as compared to the conventional film technique. There is also an effect known as "edge enhancement." All these factors tend to make xerography particularly valuable in cases where fine distinctions between fibrous and vascular structures within soft tissue are desirable. The xerographic process is thus receiving renewed attention, particularly in the field of mammography, which is used for early radiographic detection of malignant breast tumors.

To read more about xerography, we recommend to the student interested in medicine, "Xerographic Recording of Mammograms" by Boag, Stacey and Davis in *The British Journal of Radiology, 45*, 633-640, 1972. For the physicist: "Xeroradiography" by J.W. Boag in *Physics in Medicine and Biology, 18*, 3-37, 1973.

More on the Laser as a Surgical Tool

Because the laser shows much promise as a medical tool it is becoming, and in many medical centers has already become, a standard treatment device. Work is proceeding on laser technology and its medical application in such fields as cancer surgery, vision research, retinal coagulation, dental application, dermatological studies, neurological surgery and the study of the effect of the laser on the nervous system, controlled coagulation and cutting and microsurgery, including the use of microsurgery for research on a cellular and subcellular level.

Recently the development of the first *self-contained laser operating microscope system* has been announced by a team working at the University of Cincinnati Medical Center. Reports from that center, and the Boston University School of Medicine, have indicated that laser radiation may be of significant value for various techniques of microsurgery when the procedure can be viewed and controlled under the magnification of the standard Zeiss operating microscope.

The advantages of the laser in such surgical procedures include:

1. Ability to cause complete thermal necrosis within a small, precisely controlled tissue volume.
2. Ability to have a selective response in tissue structures which have natural (or artificially induced) pigmentation.
3. Ability to make incisions in moderately vascular tissues with a minimum of bleeding.
4. Since laser-induced lesions are well circumscribed, coagulation of tiny bleeding points can be effected with a minimum of peripheral tissue trauma. Also, because there is nothing (except the laser light beam) which touches the tissue, there are no problems of reopening bleeding points because of the mechanical trauma of a probe.

In the realm of cancer treatment some successful laser work on malignant tumors in humans, employing some 300-500 joules of energy per cubic centimeter of tumor has been reported. Tumor destruction is workable because, providing the energy per unit volume per unit time is not too high, the effect of laser energy on normal tissue can be kept relatively minimal, and healing is rapid. But laser energy does have a greater effect; that is, a selective destructive effect, on certain malignant tumors in which it produces tumor regression and/or dissolution. While the biological processes in this connection are possibly not yet fully understood, the basic physical principle seems clear: tumors, having a greater vascular structure than surrounding tissue, absorb more energy. This selective destructive effect is, of course, particularly pronounced if the wavelength of the laser light is suitably chosen. For example, in such cases the argon laser (principal output at green, $\lambda = 5145$ Å, and blue, $\lambda = 4880$ Å) is much more effective than the ruby laser ($\lambda = 6943$ Å).

In ophthalmology, photocoagulation achieved with the laser, operated usually at power levels between 0.5 to 2 watts, (see Fig. 16-28) is now a common treatment for a variety of ocular disorders and diseases. (In this field of medicine, the argon laser and the ruby laser are the ones most frequently used.) The basic reaction in photocoagulation is a thermally induced retinal lesion resulting directly from the conversion of the radiation energy of the laser beam into thermal energy of the tissue. The absorbed energy produces a highly localized temperature rise (15-20C°), which ultimately results in the target tissue and adja-

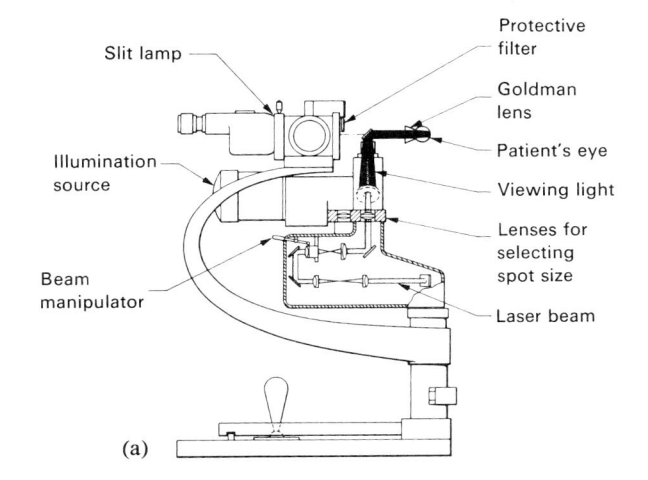

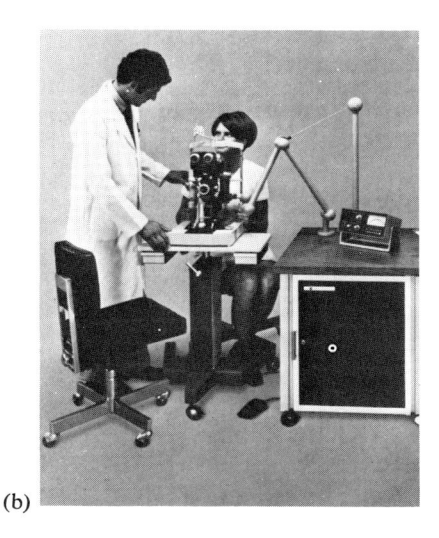

Fig. 16-28 Argon ion laser photocoagulator. (Courtesy Coherent Radiation.) (a) Schematic diagram. (b) The apparatus ready for use. The laser itself is housed in the console at the right.

cent protein substances becoming denatured and/or coagulated. The extent of the "lesion spreading" from the target material to the adjacent tissue is determined by a number of factors which include the laser beam intensity, wavelength, and exposure time.

While, as implied above, the laser has been found successful in the treatment of retinal detachments, laser treatment is by no means always indicated. One of the difficulties in using the laser in the case of retinal detachments is due to the possible presence of fluid which elevates the retina from the underlying tissue. Hence, even if a scar is produced by means of the laser, the retina may be prevented by this fluid from adhering to the underlying tissue and thus sealing the break. However, if no such pronounced retinal elevation is present then the laser is an excellent treatment device.

At the moment, cryosurgery and to a lesser extent diathermy are still the major methods in the treatment of retinal detachments. No doubt the picture will change somewhat as new lasers and associated techniques are developed.

Question

(a) Would ruby laser light be effective on blood (and hence for vascular abnormalities)?
(b) Should the operating theater in which laser surgery is done have overall high general illumination, or should the illumination be kept to a minimum for adequate vision?

Answer

(a) No. Ruby laser light is red. Red light is minimally absorbed by red arterial blood and is therefore not found effective against many blood and vascular abnormalities. The argon laser with its blue-green light proves much more effective in the case of vascular disorders.
(b) Laser establishments should probably have high general illumination (white light) to keep the pupils of attending personnel constricted. This is not to say that potential injuries to the retinas of attendant personnel would be thereby so reduced that the surgeon could safely wave the laser beam around the room. Rather, the potential damage to the eyes due to unexpected momentary reflections of the beam would be reduced.

Further Reading

For some of the latest authoritative information on lasers in medicine, the reader is referred to *Laser Applications in Medicine and Biology*, Vol. I, M.L. Wolbarsht, Ed., Plenum Press, New York, 1971.

The Laser for Sterilization of Instruments

While the lasers discussed previously have been ones which emit radiation primarily in the visible region of the electromagnetic spectrum, the carbon dioxide laser emits radiation primarily in the infrared region. It has been found that this radiation of the CO_2 laser is exceedingly effective for sterilizing procedures. For example, heat resistant spores which required some three hours at 170°C before they were destroyed by conventional steam sterilization methods have apparently been rendered completely inactive with a 0.01-sec pulse of a 50-watt CO_2 laser.

One particularly attractive feature of this type of sterilization is that surface contamination can be rendered sterile on materials which are themselves quite sensitive to increases in temperature. Materials such as certain sutures, plastic and rubber catheters, and wound dressings come to mind. In this connection Pratt* has determined that a 20 watt/cm² beam of a CO_2 laser pulsed for (1/20) sec which delivers 1 joule to the target would raise the temperature of 1 cm² of paper 0.01 cm (10^6 Å) thick by about 50°C. This temperature increase is below that required to ignite the paper under normal conditions. Yet this same 1 joule of infrared energy delivered to a 1 cm² layer of spores 10^5 Å thick would raise the spore temperature by some 500°C.

X-Ray Attenuation Processes in Medical Radiography

The interaction between x-rays and matter involves various processes (see Fig. 16-29).

The principal attenuation processes in medical diagnostic x-ray work (maximum photon energies from say 30-150 keV; i.e., tube voltage range 30-150 kV) are *unmodified scattering* (in which the incident and emitted photons have the same energy—essentially the incident photon has simply changed direction upon interaction with an entire atom within body components), *photoelectric attenuation*, and *Compton scattering*.

Of these processes the most significant one in medical radiography is photoelectric attenuation which in the region of x-ray energies commonly used may account for 50% or more of the total energy removed from the incident x-ray beam. To review briefly, photoelectric absorption takes

---

* George W. Pratt, "Effect of Infrared Laser Radiation on Biological Systems," in Stanley, H.E. (ed), *Biomedical Physics and Biomaterials Science*, Cambridge, Mass.: The MIT Press, 1972.

Process	Products
Unmodified scattering	Scattered photons
Photoelectric absorption	a) Photoelectrons b) Characteristic photons
Compton effect	a) Compton electrons b) Scattered photons
Pair production	a) Electrons b) Positrons c) Annihilation photons

Incident x-ray photons →

Fig. 16-29 Summary of processes of interaction between x-rays and matter. This represents only the first-order processes but, of course, may be extended to the nth degree. Photons and particles from the "products" column may then become incident agents in further interactions with matter.

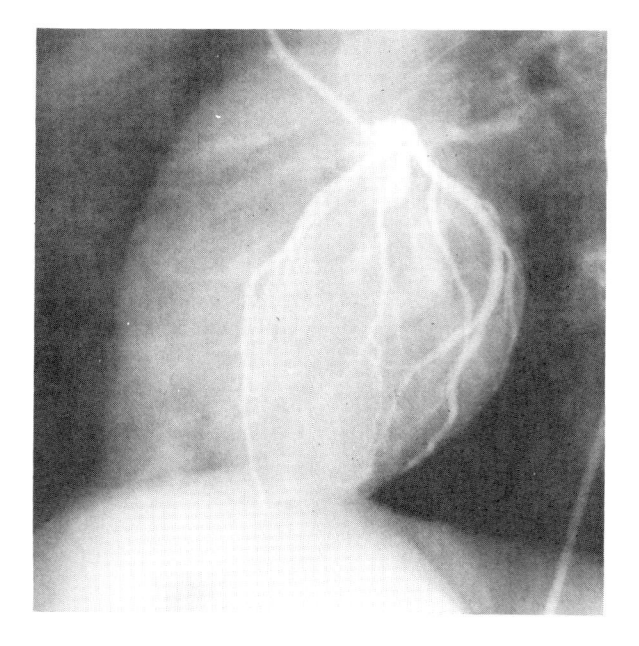

Fig. 16-30 Radiograph of coronary arteries outlined by use of a radiopaque contrast medium (Renografin). The contrast medium is introduced into the arteries by means of a catheter. (This radiograph was taken with a tube current of 1000 mA and an exposure time of 1/200 sec.) Courtesy Dr. Melvin P. Judkins, Department of Radiology, Loma Linda University, California.

place when an incident x-ray photon interacts with an inner atomic electron of some body constituent and has sufficient energy to eject the electron from its shell. All the x-ray photon energy, which must equal or exceed the binding energy of the particular electron, is transferred to the electron: quite in accord with the Einstein equation

$$K = hf - W.$$

The probability of photoelectric attenuation per unit thickness of attenuator varies approximately as the cube of the wavelength of the x-ray and as the cube of the atomic number, Z, of the attenuator.

From the above information one can see the rationale of medical radiography: due to certain interactions between x-rays and matter, the transmitted beam contains information. For example, the distinction between bone (with effective $Z \approx 13.8$) and the soft tissue (with effective $Z \approx 7.4$) is readily understood.

A problem comes to light in that a certain part of the body, for example the stomach, may attenuate x-rays to no greater or lesser extent than the tissue surrounding it (all this tissue having approximately the same effective Z). To overcome this problem the body cavity of interest (kidney, stomach, or blood vessel) may be filled with a contrast medium before the x-ray exposure is made. Such a medium consists of a material with an atomic number Z and a density different from that of the adjacent tissues (see Fig. 16-30).

Question

(a) Does a radiographic contrast medium necessarily have to have an atomic number higher than that of the surrounding tissue?
(b) As one gets into the higher voltages in medical radiography (say average photon energy from 60 to 100 keV), the relative probability of Compton scattering to photoelectric attenuation increases slightly. What effect might this phenomenon have on the x-ray photographs produced?

Answer

(a) No. Contrast media may have atomic numbers less than that of surrounding structures. Air and carbon dioxide are sometimes used. (Of the contrast media of higher Z the preference is for salts of barium and iodine.)
(b) The Compton scattered photon may in some instances be nearly as energetic as the incident photon. The scattered photon may thus have sufficient energy to reach the x-ray film from a variety of random directions. This causes "fog" on the film, a problem which is consider-

ably reduced by using *grids*. A commonly used type of grid is the *Potter-Bucky grid* which consists of a series of parallel lead strips (permitting only photons parallel to the incident beam to strike the film) which move across the film during the x-ray exposure. This type of grid thus blocks scattered radiation in a large measure but by moving across the film does not leave its own image (series of parallel strips) on the film.

SUGGESTED READING

American Scientist. January-February 1975. Sigma Xi, The Scientific Research Society of North America. The article by N. Bloembergen, "Lasers: A Renaissance in Optic Research," gives an easily understood summary of the principles of laser operation and a description of modern applications.

FEYNMAN, RICHARD P., ROBERT B. LEIGHTON, and MATTHEW SANDS, *The Feynman Lectures on Physics*, Vol. 1. Reading, Mass.: Addison-Wesley, 1963. The chapters on quantum mechanics, Chapters 37 and 38, are particularly recommended. In fact, for those with an interest in modern physical theories, one is tempted to say that if you were to choose only one of our Suggested Readings, pick Chapter 37 of this book.

HOFFMAN, B., *The Strange Story of the Quantum*. New York: Dover, 1959 (second edition). A completely nonmathematical account for the layman of the growth of ideas leading to our present knowledge of "the atom." The 1959 postscript itself is worth reading in that it takes the reader, in Hoffman's delightfully lighthearted yet accurate prose, through the quantum mechanical developments of relatively recent times.

Physics Today. June 1975. New York: The American Institute of Physics, Inc. The article "Light as a Fundamental Particle" by Steven Weinberg describes current ideas in particle physics.

Scientific American. September 1968. New York: Scientific American Inc. This issue is devoted to light. The article by V. F. Weisskopf on the interaction of light with matter gives a very readable and extensive discussion of the topics introduced in our Section 16.10.

WEIDNER, R. T., and R. L. SELLS, *Elementary Modern Physics*. Boston: Allyn and Bacon, 1968 (second edition). The laser and energy bands are described in the last chapter. Electronic structure of atoms is presented in Chapter 7 after a detailed semi-classical introduction to the various quantum numbers.

WICHMANN, E. H., *Berkeley Physics Course, Volume 4, Quantum Physics*. New York: McGraw-Hill, 1971. Chapter 1 and Chapter 3 are not too advanced to be read with great profit by our readers.

Richard P. Feynman

Epilogue

Surveying the many ramifications of science, Feynman has remarked: "I would like not to underestimate the value of the world view which is the result of scientific effort. We have been led to imagine all sorts of things infinitely more marvelous than the imaginings of poets and dreamers of the past."

This world view is to a large extent contained in the five great theories of physics upon which our attention has been concentrated:

Newtonian mechanics   The theory that describes the motion of tangible objects such as stones, spaceships, and planets with a more than adequate accuracy and with a simplicity that ensures its continued use.

Special relativity   A theory that includes modifications of Newtonian mechanics. These modifications become important at speeds so high that they are a significant fraction of the speed of light.

Electromagnetism   The theory of electricity, magnetism, and electromagnetic radiation.

Quantum mechanics   The theory that has been developed to fit the facts of the atomic and subatomic world.

Statistical physics and thermodynamics   The theory of collections of many atoms. It describes the behavior of macroscopic systems in terms of such quantities as temperature, thermal energy, and entropy.

Each of these theories has arisen from the interplay of experiment and the mathematical expression of human imagination. Within those domains that have been emphasized in this book, these theories have been so thoroughly tested and have proved so useful that it seems almost certain that, during our century, they will remain the theoretical foundation for much of science and technology.

Present-day physics is an impressive human creation. As Bertrand Russell has said, "We know very little, and yet it is astonishing that we know so much, and still more astonishing that so little knowledge can give us so much power."

". . . a man cannot live beyond the grave. Each generation that discovers something from its experience must pass that on, but it must pass that on with a delicate balance of respect and disrespect, so that the race does not inflict its errors too rigidly on its youth, but it does pass on the accumulated wisdom, plus the wisdom that it may not be wisdom."

Richard P. Feynman
(1918–)

It is clear that physics is far from finished. The extremes, represented by quarks and quasars, are enshrouded in mystery. There is every reason to expect discoveries of new objects—large and small—as well as novel theories which will bring out relationships between the chaos of facts that are emerging on the frontiers. And there is a suspicion among many physicists that their quest for an understanding will never end, that the frontier of the unknown can be rolled back indefinitely.

Appendix A

MATHEMATICAL REVIEW

A knowledge of elementary algebra and some plane geometry on the part of the reader has been assumed. A brief review of some relevant mathematics is given here for the benefit of those readers whose mathematics needs a little refreshing.

A.1 SOME BASICS OF ALGEBRA

Symbolism

In a way, algebra may be considered as arithmetic in which letters of the alphabet (or other symbols) are used in place of numbers like 2, 67, and so on. One great advantage of algebra is that a given problem can be solved in a general way such that complicated calculations involving numbers can be reduced or avoided. Also, a general solution allows the same result to be used in different cases, that is, cases with different numbers. If and when a specific numerical answer is sought, it is necessary only to replace the symbols in the algebraic solution with the appropriate numbers and perform the indicated arithmetic.

Basic Manipulations

The basic manipulations of algebra are those of *addition* (and its inverse, *subtraction*), and *multiplication* (and its inverse, *division*).

Say we have two quantities, x and y. If we add them, we write

$$x + y.$$

For convenience, we generally give another name for the result, say z:

$$x + y = z,$$

where z is the sum of x and y. By the same token, the difference obtained by subtracting y from x could be written

$$x - y = w,$$

where w is the difference.

There is nothing magical about x and y (or a and b, etc.). We just arbitrarily pick symbols that suit us. For example, one generally uses g to represent gravitational acceleration. But we could just as well have chosen p instead of g—just as long as we remain consistent in our usage throughout a given calculation.

Mulplication of x and y may be written a variety of ways, e.g.,

$$x \times y = p,$$
$$x \cdot y = p,$$

or simply

$$xy = p.$$

The latter is the briefest and usually the most commonly employed designation.

Division may be written

$$x \div y = g, \quad \frac{x}{y} = g, \quad \text{or} \quad x/y = g.$$

All three systems of noting division of x by y mean exactly the same thing.

If more than one operation is to be performed, it is sometimes necessary to indicate, by means of parentheses, the order in which the separate operations are to be performed. This use of parentheses is really no more than an indication of which ones of several symbols belong together. For example, $(p + q) c$ in terms of the operation to be performed simply means that we add p to q and then multiply this sum by c. If the parentheses were not there, i.e., if we had $p + qc$, this would mean that first q is to be multiplied by c and then this product is to be added to p.

Let us take one more example: xy^2 means we multiply x by the square of y. $x(y^2)$ also means we multiply x by y^2—in this case the parentheses, although not necessary, emphasize that one squares y and multiplies the result by x. However, $(xy)^2$ is a different matter. This means doing what is inside the brackets first, i.e., multiplying x times y, and then squaring the product of x and y. (Another way of expressing this case is simply $x^2 \times y^2$, because it can be shown that $(xy)^2 = x^2 \times y^2$.)

The Commutative Property of Multiplication

One seemingly obvious property of multiplication of ordinary algebra is that

$$a \times b = b \times a.$$

This indeed seems obvious in that we are quite familiar with the fact, for example, that

$$2 \times 4 = 8 = 4 \times 2.$$

However, it is worth spelling out this property, the *commutative* property of multiplication in the algebra we use (especially since there are other algebras which have multiplication operations which are not commutative, i.e., in which $a \times b \neq b \times a$; an example of the latter is a certain type of vector multiplication which, however, we shall avoid). The object of this emphasis on the commutative property of algebraic multiplication is, in our case, simply to emphasize that many of the physical laws expressed in symbols might well be expressed in different ways and yet obviously be the same law. Take, for example, Newton's second law. We could write

$$F = ma \quad \text{(the usual form)},$$

but we could equally as well write

$$F = am,$$

because

$$ma = am.$$

A.2 ALGEBRAIC SUMS

When we say *algebraic sum* we mean a sum in which due account is taken of the signs of the quantities added. For example, the algebraic sum of the quantities (2) and (3) is (5), but the sum of (-2) and (3) is (1). We encounter this matter of algebraic summation in dealing with quantities such as electric charge. Suppose the charge on a nucleus to be $(5e)$ and the charge of the orbital electrons of an atom to be $(-5e)$, then the algebraic sum of the electric charge (i.e., the net charge) is zero. We can write

$$(5e) + (-5e) = 5e - 5e = 0.$$

A.3 EXPONENTS

Nomenclature

Because it is frequently necessary to multiply a quantity by itself one or more times, it is worth employing a "shorthand" notation for this process. Take the simplest case: $x \times x$. This can be written x^2, which is usually read as "x squared." Another example: $r \times r \times r$ is written r^3, and usually read "r cubed."

The general situation can be expressed this way: $x \times x \times x \ldots \times x$ where x is multiplied by itself n times is written x^n and read "x to the nth" or "x to the nth power." Thus b^6 means $b \times b \times b \times b \times b \times b$, which is read "$b$ to the sixth," and, of course, q^1 is simply q. In other words, "q raised to the first power" is simply q.

In the general case of x^n the "n" is called an *exponent*. The exponent is the *power* to which "x is raised." In b^7 we say "b is raised to the seventh power."

Calculating with Exponents

It can be shown that a very simple and convenient rule holds for multiplying a given quantity raised to some power by the same quantity also raised to some power: the product is that same quantity raised to the sum of the powers. That is,

$$x^m x^n = x^{m+n}. \tag{A.1}$$

Consider these examples:

$$x^4 x^2 = x^6;$$
$$y \times y^7 = y^1 y^7 = y^8.$$

Another convenient rule to recall in calculating with exponents is

$$(x^m)^n = x^{mn}, \tag{A.2}$$

that is,

$$(x^3)^4 = x^{3 \times 4} = x^{12}.$$

We define negative and fractional exponents in such a way that the above rules remain valid. Consider the negative exponent first: We define x^{-1} as $1/x$, and in general by x^{-n} one means $\dfrac{1}{x^n}$. That being the case, we see from the rule expressed by Eq. (A.1) that

$$x^3 x^{-2} = x^{3-2} = x,$$

or in general,

$$\frac{x^m}{x^n} = x^m x^{-n} = x^{m-n}.$$

The rule expressed by Eq. (A.2) applied to a fractional exponent, say $1/n$, gives

$$(x^{1/n})^n = x^{n/n} = x^1 = x.$$

Therefore $x^{1/n}$ is a number which, when multiplied by itself n times, produces x. Consequently, one recognizes $x^{1/n}$ as the nth root of x. For example, $x^{1/2}$ is $\sqrt{x}$, and $(x^{1/2})^2 = x$.

Further illustrations are

$$a^{-2} a^7 = a^5.$$
$$y^{1/2} y^5 = y^{11/2}.$$
$$(p^{4.2})^2 = p^{8.4}.$$
$$(q^2)^{1/2} = q^{2 \times 1/2} = q.$$

From the above considerations it follows that any number at all raised to the *zeroth power* equals one. To show this fact, we point out that

$$\frac{x^n}{x^n} = x^n x^{-n} = x^0,$$

while

$$\frac{x^n}{x^n} = 1.$$

Therefore

$$x^0 = 1.$$

Exponents and Units

All this dealing with exponents is equally true in the case of units (e.g., meters) as it is in the case of some number x. For example, consider a lawn 12.0 m long and 12.0 m wide. Its area, A, is written

$$\begin{aligned} A &= (12.0 \text{ m})(12.0 \text{ m}) \\ &= (144)\,(\text{m} \times \text{m}) \\ &= 144 \text{ m}^2. \end{aligned}$$

Exponential Notation

Now that we have recalled the essential principles of dealing with exponents, we note that exponential notation is extensively used in physics and other mathematical disciplines for expressing very large and very small numbers. This convenient custom derives from the fact that all numbers can be expressed in terms of some number between 1 and 10 multiplied by 10 raised to the appropriate power. Reference to Table A.1 will help in the consideration of the following examples.

$$500 = 5 \times 100 = 5 \times 10^2.$$
$$829 = 8.29 \times 100 = 8.29 \times 10^2.$$
$$2370 = 2.37 \times 1000 = 2.37 \times 10^3.$$
$$\text{three million} = 3{,}000{,}000 = 3 \times 10^6.$$
$$\text{one three millionth} = \frac{1}{3{,}000{,}000} = \frac{1}{3} \times \frac{1}{10^6} = \frac{1}{3} \times 10^{-6}$$
$$= 0.333 \times 10^{-6}$$
$$\text{or } 3.33 \times 10^{-7}.$$
$$0.0056 = 5.6 \times 0.001 = 5.6 \times 10^{-3}.$$

Table A-1. POWERS OF 10 FROM 10^{-6} TO 10^6

10^{-6}	= 0.000,001
10^{-5}	= 0.000,01
10^{-4}	= 0.000,1
10^{-3}	= 0.001
10^{-2}	= 0.01
10^{-1}	= 0.1
10^{0}	= 1
10^{1}	= 10
10^{2}	= 100
10^{3}	= 1000
10^{4}	= 10,000
10^{5}	= 100,000
10^{6}	= 1,000,000

This method of writing very large and very small numbers has a great advantage in that it makes numerical calculations less cumbersome and therefore less prone to errors in many cases. Examples that obviously support this contention are calculations involving such quantities as the speed of light and/or Planck's constant. Imagine calculating with Planck's constant expressed as

0.000000000000000000000000000000000663 joule-sec

instead of simply

$$6.63 \times 10^{-34} \text{ joule-sec.}$$

A.4 SIGNIFICANT FIGURES

There is always some *uncertainty* in the value of any number that is obtained as a result of experimental measurement. Scientists write such numbers in a conventional way which displays not only the best estimate of the value of the number but also gives some information about the uncertainty in this estimate. For example, when we write for a rocket's speed, v, that

$$v = 1.52 \times 10^3 \text{ m/sec},$$

we assert that our best estimate of the speed is 1520 m/sec but that we are uncertain about the digit 2, the last digit displayed in 1.52, and that we know nothing about the value of digits to the right of 2.

The digits 1, 5, and 2 in the number 1.52×10^3 m/sec are called *significant figures* or *significant digits*: they are the digits in our estimate that are believed closer to the actual value than any others. And, as indicated above, the last significant digits of a series of significant digits are the ones about which we are in doubt. Look again at the above example of the rocket speed; there are three significant figures in

1.52×10^3 m/sec, as well as in 0.152×10^4 m/sec, 15.2×10^2 m/sec, or 152×10 m/sec, and in each case there is doubt about the digit 2.

Zeros are significant only when preceded by another significant figure. Thus the number 1.520×10^3 (or equivalently, 1520) has *four* significant figures, and if we write

$$v = 1.520 \times 10^3 \text{ m/sec},$$

we are asserting that, although there is a measure of doubt, we do have some grounds for believing that the fourth digit is 0 instead of 1 or 2 or 3, etc.

By the same token, if we write

$$v = 1.5200 \times 10^3 \text{ m/sec},$$

the last zero here is also significant since it follows a zero which is itself significant. In other words, we assert in this way that both the fourth and the fifth digits are each 0 and not 1, 2, or 3, etc.

A still more accurate measurement might yield the result,

$$v = 1.5200418 \times 10^3 \text{ m/sec},$$

which has eight significant figures. The same information is given by

$$v = 1520.0418 \text{ m/sec},$$

which also displays eight significant figures. Notice that the number of significant figures is not changed by merely moving the decimal point and correspondingly changing the power of ten.

When then is a zero not a significant figure? When it does not follow a significant figure, as was stated before. And this situation occurs when the zero is employed simply as a decimal place holder. As an example, consider 0.0089. This number has only two significant figures. The zero before the decimal point is employed conventionally simply to emphasize that this is not a whole number but a fraction (i.e., that there is no digit like 1, 2, 3, etc., in front of the decimal point). The two zeros after the decimal point are decimal place holders only. They tell us that the decimal fraction is not 89/100, nor 89/1000, but rather 89/10,000. This expression could equally well be written

$$8.9 \times 10^{-3},$$

which notation clearly indicates that there are only two significant figures.

(It is worth realizing that what has been described here is a fairly common convention regarding significant figures. But some authors have adopted slightly different conventions, none of which we shall pursue here.)

In this text, in the examples and questions, we usually present data with two or three significant figures and, after calculating with these data, we retain in the results only the number of significant figures consistent with the uncertainty of the original data.

A.5 EQUATIONS

A statement which says that a certain quantity is equal to another one is an *equation*. In that sense,

$$5 + 8 = 13 \text{ is an equation.}$$

And so is

$$x + 8 = 13 \text{ an equation.}$$

This latter equation tells us that $x = 5$.

The process of finding out what x is in such a case is called *solving* the equation. We say $x = 5$ is a *solution* of this equation. Equations may have more than one solution. For example, consider

$$r^2 = 4.$$

We could say $r = 2$, or $r = -2$, because both $(2) \times (2)$ and $(-2) \times (-2)$ equal 4.

Equations need not have any numbers in them. For example, the earlier example of Newton's second law is an equation,

$$F = ma.$$

Basic Manipulations

There is only one point to remember in performing manipulations with, or upon, equations. And that is to remember *equal* ($=$) means exactly that, i.e., "equal." In other words, we cannot disturb this sign's meaning. The obvious implication therefore is that whatever operation is performed on the right-hand side of an equal sign must also be performed on the left.

Let us take as a first example an equation employed earlier:

$$x + 8 = 13.$$

If we subtract 8 from the left side we must also subtract 8 from the right side:

$$x + 8 - 8 = 13 - 8; \text{ that is, } x = 5.$$

We see here the basic reason why one needs to manipulate equations at all: to find an explicit answer—in this case to find out exactly what the value of x has to be to satisfy the equality sign.

Equations and Units

The equations encountered in physics inevitably have some unit associated with them because these equations deal with physical quantities (or dimensions) with which units are associated. For example, the meter (m) is a unit of length, the second (sec), a unit of time. What to do about units in equations? These units abide by all the previously stated and implied rules, i.e., they handle algebraically.

Here are some examples of equations from physics which involve units.

1. Determine I, if $V = IR$, and $V = 12$ volts while $R = 6$ ohms.
$$V = IR$$
is the same as
$$IR = V,$$
and now dividing both sides by R, we get
$$I = \frac{V}{R}$$
$$= \frac{12 \text{ volts}}{6 \text{ ohms}}$$
$$= 2 \; \frac{\text{volts}}{\text{ohms}} \quad \left(\text{the unit } \frac{\text{volt}}{\text{ohm}} \text{ is called an ampere}\right).$$

2. If $v = v_0 + at$ (where v is the speed in m/sec, v_0 is the speed at some arbitrary starting time zero from which the time t is measured in seconds, and a is the acceleration in meter/sec^2), find the time it takes for a falling stone to reach the speed of 39.2 m/sec if the stone is dropped from rest (i.e., $v_0 = 0$), and the acceleration due to gravity is 9.8 m/sec^2.
$$v = v_0 + at$$
$$v = 0 + at$$
$$v = at,$$
or,
$$at = v \qquad \text{(it is simply a matter of custom to place the unknown quantity, } t \text{ in this case, on the left).}$$

Dividing both sides of the equation by a, we get
$$t = v/a.$$
Replacing v and a with the appropriate numerical values and units, we have
$$t = \frac{39.2 \text{ m/sec}}{9.8 \text{ m/sec}^2} = \frac{39.2}{9.8} \; \frac{\text{m/sec}}{\text{m/sec}^2},$$
which (if the reader will recall from his arithmetic background that dividing by a fraction is the same as multiplying by that fraction inverted) is
$$t = \frac{39.2}{9.8} \; \frac{\text{m}}{\text{sec}} \; \frac{\text{sec}^2}{\text{m}};$$
whereupon performing the division with both the numerical quantities and the units, we see that
$$t = 4.0 \text{ sec.}$$

Dividing One Equation by Another

Suppose we have the two equations

$$a = b$$

and

$$c = d.$$

If we now divide the left-hand side of the first equation by c and the right-hand side of the first equation by d (which is equal to c), we obtain another equation:

$$\frac{a}{c} = \frac{b}{d}.$$

This process of obtaining the third equation from the first two is often called dividing the first equation by the second. For example, if

$$F_1 = \frac{GMm}{r_1^2}$$

and

$$F_2 = \frac{GMm}{r_2^2}$$

and we wish to find the ratio F_1/F_2, we proceed as follows: Divide the first equation by the second, and thus get

$$\frac{F_1}{F_2} = \frac{\left(\dfrac{GMm}{r_1^2}\right)}{\left(\dfrac{GMm}{r_2^2}\right)} = \left(\frac{GMm}{r_1^2}\right)\left(\frac{r_2^2}{GMm}\right) = \frac{r_2^2}{r_1^2}.$$

A.6 PROPORTIONALITY

Direct Proportions

Many equations in physics express a *proportional relationship*. For example, one finds that the extension of a spring is proportional to the force exerted in stretching this spring. What this word "proportional" means here is simply that if we exert twice the force, we get twice the extension of the spring.

In more general terms, we say that when the value of a certain quantity y depends upon the value of another quantity x in the particular way that doubling x means y also doubles, tripling x results in y tripling, and so on, then y is said to be directly proportional to x. Suppose then that y is directly proportional to x. Let us take an example in which y

is twice as large as x:

$$y = 0 \text{ when } x = 0$$
$$y = 1 \text{ when } x = \tfrac{1}{2}$$
$$y = 2 \text{ when } x = 1$$
$$y = 4 \text{ when } x = 2$$
$$y = 6 \text{ when } x = 3$$

and so on. Whatever the value of x, the ratio of y to x remains constant:

$$\frac{y}{x} = 2 = \text{constant} = k.$$

The fact that x is proportional to y is usually expressed in the following way:

$$y = kx,$$

where k is termed the *proportionality constant*. In our particular case, the proportionality constant is 2.

If we were to graph such a proportionality, we would get a straight line (see Fig. A-1). In the parlance of mathematics, we say such a graph shows y as a function of x. And in this particular case we say y is *directly* proportional to x. Moreover, since the graph is a straight line, one can also say:

1. y is a linear function of x, or
2. $y = 2x$ is a linear equation.

A practical example of such a proportionality is Ohm's law:

$$V = RI.$$

Here the two variables analogous to x and y in our example are I, the current, and V, the voltage. The proportionality constant is R, the resistance.

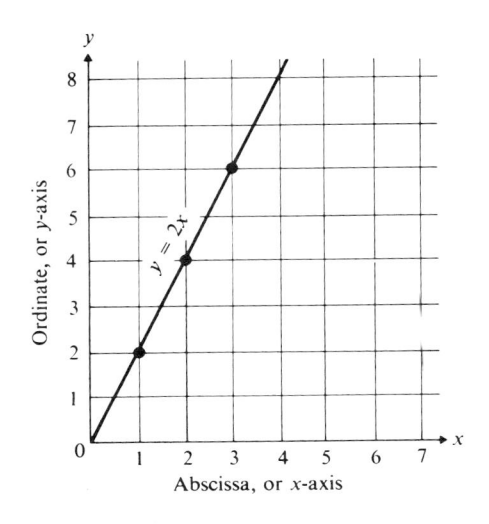

Fig. A-1 Graph of a linear equation: $y = 2x$.

Proportions Involving Higher Powers of Numbers

Proportionalities other than direct proportionalities are, of course, also found in physics. We might find, for example,

$$A = 0 = 0 \times 3 \text{ when } B = 0$$
$$A = 12 = 4 \times 3 \text{ when } B = 2$$
$$A = 27 = 9 \times 3 \text{ when } B = 3$$
$$A = 48 = 16 \times 3 \text{ when } B = 4.$$

In this case, $A = 3B^2$. We say A is directly proportional to B^2. An example is in the heating effect of an electric current:

$$P = I^2 R,$$

where P is the power, or the rate that thermal energy is developed in a heating element of electrical resistance R if the current in the element is I. We say that P varies as I^2. (With reference to our A and B example, P is analogous to A, I to B, and R to 3.) See also Fig. A-2.

Another possibility is an inverse proportionality, that is

$$A = c\frac{1}{B}, \text{ or } AB = c.$$

Such a situation exists in the case of gases. In a given mass of a gas, at a constant temperature, the volume (V) varies inversely as the pressure (P), that is, $PV = $ constant.

Inverse Square Law

Of course, the inverse proportion like the direct proportion could be to higher powers. That is, we could have

$$y = c\frac{1}{x^2}.$$

Coulomb's law in electricity provides an important example of such a relationship. This law states: The force, F, of one electrical charge on another varies inversely as the square of the distance, s, between them, that is,

$$F = \frac{\text{constant}}{s^2}.$$

A.7 GEOMETRY

Pythagorean Theorem

A very useful geometric relationship in a right (or right-angled) triangle is the *Pythagorean theorem*, which states that the sum of the squares of the sides adjacent to the right angle is equal to the square of the hypotenuse. For the triangle of Fig. A-3

$$a^2 + b^2 = c^2.$$

Hence we can always express the length of any of the sides of a right triangle in terms of the other sides:

$$a = \sqrt{c^2 - b^2}$$
$$b = \sqrt{c^2 - a^2}$$
$$c = \sqrt{a^2 + b^2}.$$

Similar Triangles

Another useful point to recall relates to *similar triangles*. Stating it very loosely, one might say that similar triangles have the same shape but

x	x^2	y
0.0	0.0	0.0
1.0	1.0	0.5
2.0	4.0	2.0
3.0	9.0	4.5
4.0	16.0	8.0

(a)

(b)

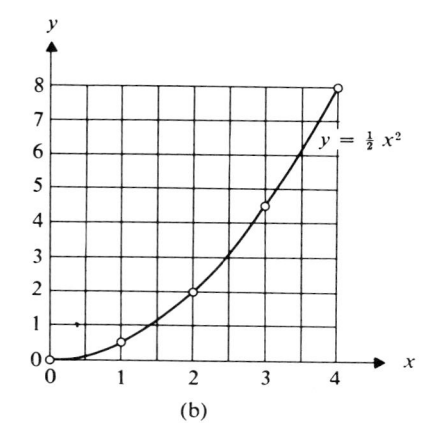

Fig. A-2 A nonlinear relation. (a) The tabulation of some values of the relation $y = \frac{1}{2}x^2$. (b) The graph using the tabulated values of (a).

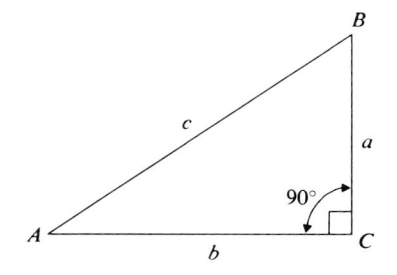

Fig. A-3 A right-angled triangle. The angle at C is a right angle, that is, the angle is 90°.

not necessarily the same size, and certainly not necessarily the same orientation in space. Fig. A-4 shows two similar triangles. By definition, for these triangles:

$$\theta_1 = \theta_2, \phi_1 = \phi_2, \text{ and } \alpha_1 = \alpha_2.$$

For these triangles it can be shown that the ratios of the corresponding sides are equal. That is

$$\frac{a_1}{a_2} = \frac{b_1}{b_2} = \frac{c_1}{c_2}.$$

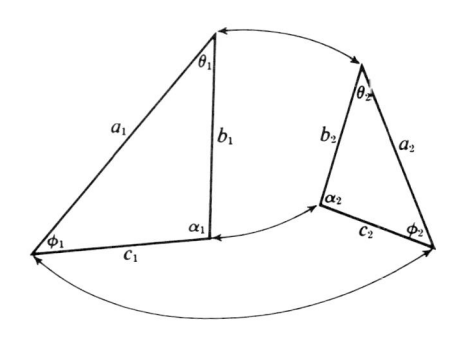

Fig. A-4 Similar triangles.

Congruent Triangles

Two triangles which have the three sides of one respectively equal to the three sides of the other are said to be *congruent*. Congruent triangles also have their corresponding angles equal.

A.8 TRIGONOMETRY

Trigonometric Functions

It is often necessary to know the relationships among the various sides and angles of a right triangle. The three basic trigonometric functions can be defined in terms of the triangle shown in Fig. A-5.

We shall label θ the angle at A included between the side b of the triangle and its hypotenuse c. The sine of this angle, which is abbreviated $\sin \theta$, is the ratio of the side a opposite θ to the hypotenuse c:

$$\sin \theta = \frac{a}{c} = \frac{\text{opposite side}}{\text{hypotenuse}}.$$

The cosine of the angle θ, abbreviated $\cos \theta$, is the ratio of the side b adjacent to θ to the hypotenuse c:

$$\cos \theta = \frac{b}{c} = \frac{\text{adjacent side}}{\text{hypotenuse}}.$$

$\sin \theta = a/c$
$\cos \theta = b/c$
$\tan \theta = a/b$

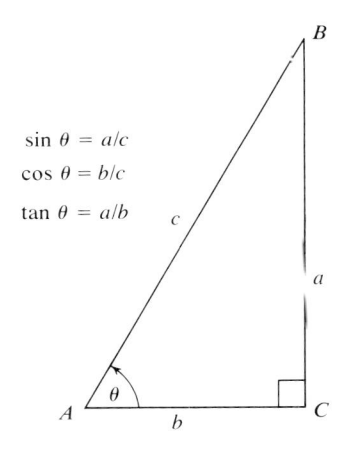

Fig. A-5 A right triangle with trigonometric functions defined.

The tangent of the angle θ, abbreviated $\tan \theta$, is the ratio of the side a opposite to θ to the side b adjacent to θ:

$$\tan \theta = \frac{a}{b} = \frac{\text{opposite side}}{\text{adjacent side}}.$$

From these definitions we can obtain a useful result:

$$\frac{\sin \theta}{\cos \theta} = \frac{\text{opposite side/hypotenuse}}{\text{adjacent side/hypotenuse}}$$
$$= \frac{\text{opposite side}}{\text{adjacent side}} = \tan \theta.$$

684 MATHEMATICAL REVIEW

The tangent of an angle is equal to its sine divided by its cosine.

Appendix E contains numerical tables of sin θ, cos θ, and tan θ, for angles from 0° to 90°. These tables can be used for angles from 90° to 180° with the help of the following formulas:

$$\sin(90° + \theta) = \cos\theta,$$
$$\cos(90° + \theta) = -\sin\theta,$$
$$\tan(90° + \theta) = -\frac{1}{\tan\theta}.$$

One side of a right triangle can be expressed in terms of another side and the appropriate trigonometric function. For example, suppose that we know the side c and the angle θ in the triangle of Fig. A-5, and we want to find side b. Since $\cos\theta = b/c$, we find

$$b = c\cos\theta.$$

The Quadrants

The general definitions of trigonometric functions for any value of an angle θ are given in terms of the coordinate system shown in Fig. A-6. The hypotenuse c is considered always to be positive, but a and b, taken to be the coordinates of the end of the hypotenuse, can be either positive or negative. Consequently, a given trigonometric function is positive or negative depending upon the quadrant in which the hypotenuse lies. For example, in the second quadrant $\sin\theta$ is positive because a is positive, while $\cos\theta$ is negative because b is negative in that quadrant.

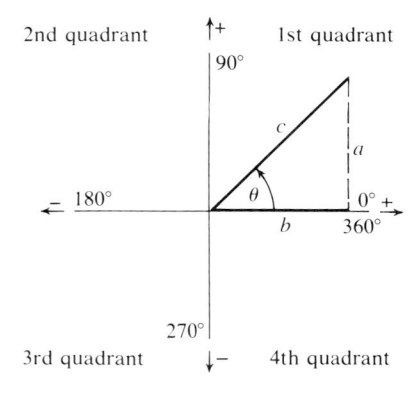

Fig. A-6 The four quadrants.

Radian Measure

Another common unit of measure of an angle, besides the degree, is the *radian*. The measure of an angle in radians is given by dividing the arc s, which subtends the angle θ at the center of a circle, by the circle's radius r (see Fig. A-7); that is

$$\theta = s/r.$$

It is clear from this definition that for a full circle we get

$$\text{angle in radians} = \frac{2\pi r}{r} = 2\pi.$$

In other words, 360° equals 2π radians, 180° equals π radians and 90° equals $\pi/2$ radians. (Also, since π is approximately 3.14159, one degree is about 0.01745 radian.)

A very interesting fact emerges when we look at a figure such as Fig. A-8, keeping in mind the angular measure in radians. In part (a) of Fig. A-8 we see that $\sin\theta = p/r$. The interesting thing is that, as we make θ very small [as in part (b) of the figure], the length s of the arc, and p become very nearly the same length. And since θ (in radians) is

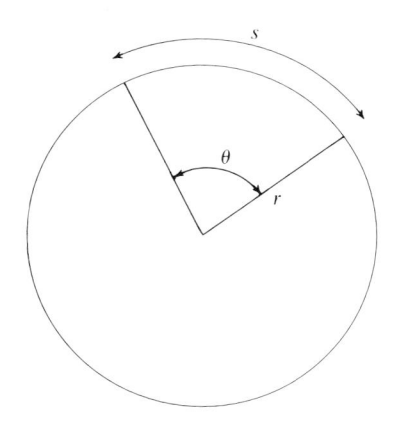

Fig. A-7 Radian measure: θ (in radians) $= s/r$.

A.8 TRIGONOMETRY   685

s/r, and $\sin\theta$ is p/r, one sees that as θ gets smaller and smaller, the value of $\sin\theta$ approaches that of θ. In other words, for small angles one may use the approximation,

$$\sin\theta = \theta \text{ (in radians)}.$$

Graphing a Trigonometric Function

Keeping in mind the previous comments on the positive and negative values of a trigonometric function, let us examine a graphical representation of a trigonometric function as the angle varies continuously. In Fig. A-9, we plot a graph of $\cos\theta$ for values of θ from $-360°$ to $360°$.

Note that the maximum value for $\cos\theta$ is $+1$, that the minimum value is -1, and that these extreme values occur at $\theta = 0$ and $\theta = \pm 180°$, respectively. Fig. A-5 and Fig. A-6 show that these extreme values are correct, for as θ approaches zero, say as in Fig. A-5, we can visualize side b closing up on c and, since the angle $C = 90°$, becoming closer to c in length. At $\theta = 0$, we have $b = c$ and

$$\cos 0° = \frac{b}{c} = \frac{c}{c} = 1.$$

Also to be noted from the graph (Fig. A-9) is that $\cos\theta = 0$ for $\theta = \pm 90°$ and $\pm 270°$. The zero value is also readily apparent from the triangle of Fig. A-5. If we imagine θ approaching $90°$, then we see that b gets smaller as it approaches the value 0. And at $\theta = 90°$ we have $b = 0$. Hence

$$\cos 90° = \frac{b}{c} = \frac{0}{c} = 0.$$

The negative values of cosine and of θ, shown in Fig. A-9, arise from the general definition of trigonometric functions given in terms of the rotating "hypotenuse" in Fig. A-6. We see that the values of the

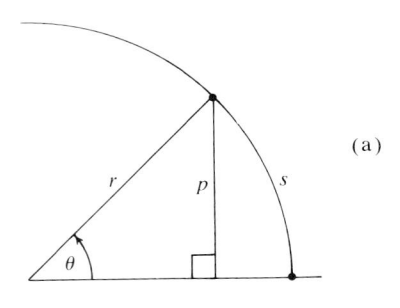

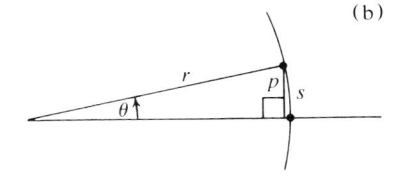

Fig. A-8   For small θ, θ in radians is approximately equal to $\sin\theta$.

Fig. A-9   Graph of $\cos\theta$ versus θ.

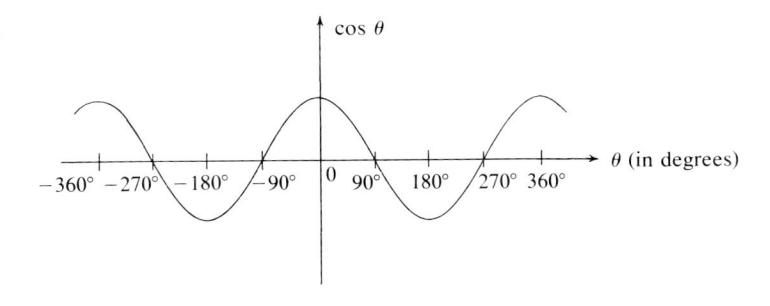

686   MATHEMATICAL REVIEW

cosine function, for example, go from a maximum through zero to a minimum, through zero again and back to a maximum as the "hypotenuse" c rotates through a full circle.

Components

The reader will find it convenient to solve many physics problems involving vectors by means of the *Method of Components* described in Chapter 3. Here we simply repeat the general rule (which holds without exception): To get the component of a vector, multiply the magnitude of the vector by the cosine of the angle between the vector and the direction in which you want the component.

Some Useful Trigonometric Relations

Among many useful trigonometric relationships is the *cosine law*: In any triangle (not just a right triangle), if two sides and the included angle are given, the third side can be found by using the relation

$$c^2 = a^2 + b^2 - 2ab\cos\theta,$$

where c is the side opposite the angle θ (see Fig. A-10).

Note that if $\theta = 90°$, that is, if the triangle is a right triangle, this relation reduces to the Pythagorean Theorem of Section A.7:

$$c^2 = a^2 + b^2,$$

because $\cos 90° = 0$.

A form of the cosine law which is often convenient is

$$c^2 = a^2 + b^2 + 2ab\cos\phi,$$

where ϕ is the angle shown in Fig. A-10.

A useful relationship between $\sin\theta$ and $\cos\theta$ follows from the Pythagorean theorem. In the right triangle in Fig. A-5, Pythagoras' theorem gives

$$c^2 = a^2 + b^2,$$

which yields

$$(a/c)^2 + (b/c)^2 = 1.$$

Since $\sin\theta = a/c$ and $\cos\theta = b/c$, we have the important result

$$\sin^2\theta + \cos^2\theta = 1.$$

We are confronted in several physical situations with the problem of evaluating the average value of $\cos^2\theta$ over a range of values of θ from $0°$ to $180°$ or over any other $180°$ range of values (called a half-cycle of the functions $\cos\theta$ and $\sin\theta$). Now, except for a $90°$ shift along the θ-axis, the graph of $\sin\theta$ is the same as the graph of $\cos\theta$ (Fig. A-9).

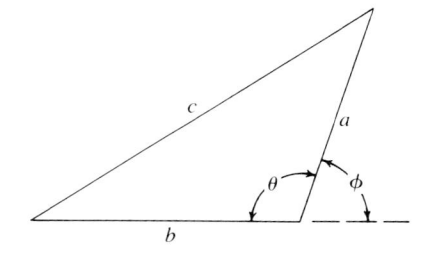

Fig. A-10 Reference figure for the cosine law.

Therefore, over a half-cycle,

$$\text{average of } \sin^2\theta = \text{average of } \cos^2\theta.$$

We now use

$$\sin^2\theta + \cos^2\theta = 1,$$

which implies

$$(\text{average of } \sin^2\theta) + (\text{average of } \cos^2\theta) = 1.$$

Consequently, over a half-cycle,

$$2(\text{average of } \cos^2\theta) = 1$$

which yields the simple answer,

$$\text{average of } \cos^2\theta = \tfrac{1}{2}.$$

SUGGESTED READING

CRANE, H. R., *Programmed Math Reviews*. New York: Appleton-Century-Crofts, 1966. This is a set of five programmed reviews on angles and triangles, trigonometry, vectors, powers of ten, and algebra. Each set is composed of about 8 pages embracing some 50 frames each. A frame consists essentially of a statement followed by a question the reader must answer before proceeding. The answers to each frame are available for immediate confirmation of the correct response. These individual reviews would be excellent for one who desired a quick refresher in elementary mathematical techniques.

Handbook of Chemistry and Physics, Cleveland: Chemical Rubber Publishing Co. Any edition of this well-known handbook provides exhaustive trigonometric and other mathematical tables as well as numerous compilations of various physical and chemical properties.

MARION, J. B., and R. C. DAVIDSON, *Mathematical Preparation for General Physics*. Philadelphia: W. B. Saunders, 1972. This 300-page paperback presents mathematical ideas from the standpoint of the physicist. Mathematical examples are drawn from the realm of physics, and for this reason the text is particularly valuable for physics students who need more review and practice in mathematics than can be gained from our Appendix.

SWARTZ, C. E., *Used Math*. Englewood Cliffs, N.J.: Prentice-Hall, 1973. A very practical review of mathematics actually used in science. Although some of the book deals with topics in calculus, several chapters, such as *Units and Dimensions, Graphs, Simple Functions of Applied Math, Geometry*, and *Vectors*, are recommended for readers who wish to reinforce the material in this review.

WASHINGTON, A. J., *Basic Technical Mathematics* (2nd ed.). Menlo Park, California: Cummings, 1970. The first half of this well-known text deals with the mathematics employed in our book. The reader looking for more help with mathematics than supplied in our Appendix will find many examples worked out in detail.

Appendix B

VARIOUS PHYSICAL CONSTANTS AND OTHER PHYSICAL DATA

Air pressure, normal at sea level	14.7 lb/in.^{2} = 1.01 × 10^5 N/m^2
Alpha particle, charge	$+2e$ = 3.204 × 10^{-19} coul
Alpha particle, mass	M_α = 4.00260 amu
	= 6.6443 × 10^{-27} kg
Avogadro's number	N_0 = 6.0225 × 10^{23} particles/mole
Boltzmann's constant	k = 1.3805 × 10^{-23} joule/°K
Coulomb's law constant	$k = 9.0 \times 10^9 \dfrac{Nm^2}{coul^2}$
Earth, acceleration of gravity near surface	g = 9.8 m/sec^2 = 32 ft/sec^2
Earth, equatorial circumference	24.9 × 10^3 miles = 4.008 × 10^7 m
Earth, average distance to sun	9.3 × 10^7 miles = 1.5 × 10^{11} m
Earth, average orbital speed	2.98 × 10^4 m/sec
Earth, mass	5.975 × 10^{24} kg
Earth, average radius	3959 miles = 6.371 × 10^6 m
Electron, charge	$-e$ = $-$1.602 × 10^{-19} coul
Electron, mass	M_e = 5.4860 × 10^{-4} amu
	= 9.109 × 10^{-31} kg
Electron charge/mass ratio	e/m = 1.76 × 10^{11} coul/kg
Gravitational constant	G = 6.67 × 10^{-11} Nm^2/kg^2
Hydrogen atom radius (Bohr radius)	5.29 × 10^{-11} meter
Light, speed in vacuum	c = 3.00 × 10^8 m/sec
	= 186 × 10^3 miles/sec
Moon, acceleration of gravity at surface	1.67 m/sec^2
Moon, distance from earth (center to center)	2.389 × 10^5 miles
	= 3.844 × 10^8 m
Moon, mass	7.34 × 10^{22} kg
Moon, radius	1.738 × 10^6 m
Moon, period of revolution	27.3 days = 2.36 × 10^6 sec
Neutron, mass	M_n = 1.00866 amu
	= 1.6748 × 10^{-27} kg
Planck's constant	h = 6.63 × 10^{-34} joule-sec
Proton, charge	$+e$ = +1.602 × 10^{-19} coul
Proton, mass	M_p = 1.00728 amu
	= 1.6725 × 10^{-27} kg
Sun, mass	M_s = 1.99 × 10^{30} kg
Sun, radius	7.0 × 10^8 m

Appendix C

THE METRIC SYSTEM AND SOME EQUIVALENTS

GENERAL SYSTEM OF METRIC MULTIPLES

Multiple	Prefix	Symbol
10^9	giga	G
10^6	mega	M
10^3	kilo	k

(i.e., 1 kilogram = 10^3 grams)

10^2	hecto	h
10	deca	da
10^{-1}	deci	d
10^{-2}	centi	c
10^{-3}	milli	m
10^{-6}	micro	μ
10^{-9}	nano	n

LENGTH

1 angstrom = 1 Å
    = 10^{-8} cm (centimeter)
    = 10^{-10} m (meter)
1 micron = 1 μ
    = 10^{-3} mm (millimeter)
    = 10^{-6} m
1 cm = 0.3937 in. (inch)
1 m = 3.281 ft (foot)
1 kilometer = 1000 m
    = 0.621 mile
1 in. = 2.540 cm (exactly, by definition)
1 ft = 30.48 cm

SPEED

1 m/sec = 3.60 km/hr
    = 3.28 foot/sec
    = 2.24 miles/hr

HEAT

1 calorie = 4.185 joules
    $\approx 30 \times 10^{18}$ eV

VOLUME

1 cm^3 = 0.0610 cu. in. (or in.^{3})
1 liter = 0.2642 U.S. gal (gallon)
1 liter = 0.220 Imperial gal

MASS

1 kg = 1000 grams
1 atomic mass unit (amu)
    = 1.66×10^{-27} kg
    = 1.49×10^{-10} joule
    = 931 MeV

WEIGHT

Metric (gravitational to absolute units)
1 gram-force = 980 dynes
1 kg-force = 9.8 newtons
English (avoirdupois) to metric (gravitational)
1 oz = 28.35 grams-force
1 lb = 16 oz = 0.4536 kg-force

TEMPERATURE

Celsius to Fahrenheit: degrees Fahrenheit: $t_F = (t_C \times \frac{9}{5}) + 32°\text{F}$
Fahrenheit to Celsius: degrees Celsius: $t_C = \frac{5}{9}(t_F - 32°\text{F})$
Kelvin (Celsius absolute) to Celsius: $t_C = T - 273.15°\text{K}$

ENERGY

1 joule = 1 watt-sec
    = 0.738 foot-pound
    = 0.239 calorie
    = 6.24×10^{18} eV
1 eV (electron volt)
    = 10^{-6} MeV
    = 1.60×10^{-19} joule
1 kilowatt-hour
    = 3.6×10^6 joules

Appendix D

ALPHABETICAL LIST OF THE ELEMENTS

Element	Symbol	Atomic number Z	Element	Symbol	Atomic number Z	Element	Symbol	Atomic number Z
Actinium	Ac	89	Hafnium	Hf	72	Praseodymium	Pr	59
Aluminum	Al	13	Helium	He	2	Promethium	Pm	61
Americium	Am	95	Holmium	Ho	67	Protractinium	Pa	91
Antimony	Sb	51	Hydrogen	H	1	Radium	Ra	88
Argon	A	18	Indium	In	49	Radon	Rn	86
Arsenic	As	33	Iodine	I	53	Rhenium	Re	75
Astatine	At	85	Iridium	Ir	77	Rhodium	Rh	45
Barium	Ba	56	Iron	Fe	26	Rubidium	Rb	37
Berkelium	Bk	97	Krypton	Kr	36	Ruthenium	Ru	44
Beryllium	Be	4	Kurchatovium	Ku	104	Samarium	Sm	62
Bismuth	Bi	83	Lanthanum	La	57	Scandium	Sc	21
Boron	B	5	Lawrencium	Lr	103	Selenium	Se	34
Bromine	Br	35	Lead	Pb	82	Silicon	Si	14
Cadmium	Cd	48	Lithium	Li	3	Silver	Ag	47
Calcium	Ca	20	Lutetium	Lu	71	Sodium	Na	11
Californium	Cf	98	Magnesium	Mg	12	Strontium	Sr	38
Carbon	C	6	Manganese	Mn	25	Sulfur	S	16
Cerium	Ce	58	Mendelevium	Md	101	Tantalum	Ta	73
Cesium	Cs	55	Mercury	Hg	80	Technetium	Tc	43
Chlorine	Cl	17	Molybdenum	Mo	42	Tellurium	Te	52
Chromium	Cr	24	Neodymium	Nd	60	Terbium	Tb	65
Cobalt	Co	27	Neon	Ne	10	Thallium	Tl	81
Copper	Cu	29	Neptunium	Np	93	Thorium	Th	90
Curium	Cm	96	Nickel	Ni	28	Thulium	Tm	69
Dysprosium	Dy	66	Niobium	Nb	41	Tin	Sn	50
Einsteinium	Es	99	Nitrogen	N	7	Titanium	Ti	22
Erbium	Er	68	Nobelium	No	102	Tungsten	W	74
Europium	Eu	63	Osmium	Os	76	Uranium	U	92
Fermium	Fm	100	Oxygen	O	8	Vanadium	V	23
Fluorine	F	9	Palladium	Pd	46	Xenon	Xe	54
Francium	Fr	87	Phosphorus	P	15	Ytterbium	Yb	70
Gadolinium	Gd	64	Platinum	Pt	78	Yttrium	Y	39
Gallium	Ga	31	Plutonium	Pu	94	Zinc	Zn	30
Germanium	Ge	32	Polonium	Po	84	Zirconium	Zr	40
Gold	Au	79	Potassium	K	19			

NATURAL TRIGONOMETRIC FUNCTIONS

Appendix E

Angle Degree	Radian	Sine	Cosine	Tangent
0°	.000	0.000	1.000	0.000
1°	.017	.017	1.000	.017
2°	.035	.035	0.999	.035
3°	.052	.052	.999	.052
4°	.070	.070	.998	.070
5°	.087	.087	.996	.087
6°	.105	.104	.994	.105
7°	.122	.122	.992	.123
8°	.140	.139	.990	.140
9°	.157	.156	.988	.158
10°	.174	.174	.985	.176
11°	.192	.191	.982	.194
12°	.209	.208	.978	.212
13°	.227	.225	.974	.231
14°	.244	.242	.970	.249
15°	.262	.259	.966	.268
16°	.279	.276	.961	.287
17°	.297	.292	.956	.306
18°	.314	.309	.951	.325
19°	.332	.326	.946	.344
20°	.349	.342	.940	.364
21°	.366	.358	.934	.384
22°	.384	.375	.927	.404
23°	.401	.391	.920	.424
24°	.419	.407	.914	.445
25°	.436	.423	.906	.466
26°	.454	.438	.899	.488
27°	.471	.454	.891	.510
28°	.489	.470	.883	.532
29°	.506	.485	.875	.554
30°	.524	.500	.866	.577
31°	.541	.515	.857	.601
32°	.558	.530	.848	.625
33°	.576	.545	.839	.649
34°	.593	.559	.829	.674
35°	.611	.574	.819	.700
36°	.628	.588	.809	.726
37°	.646	.602	.799	.754
38°	.663	.616	.788	.781
39°	.681	.629	.777	.810
40°	.698	.643	.766	.839
41°	.716	.656	.755	.869
42°	.733	.669	.743	.900
43°	.750	.682	.731	.933
44°	.768	.695	.719	.966
45°	.785	.707	.707	1.000

Angle Degree	Radian	Sine	Cosine	Tangent
46°	0.803	0.719	0.695	1.036
47°	.820	.731	.682	1.072
48°	.838	.743	.669	1.111
49°	.855	.755	.656	1.150
50°	.873	.766	.643	1.192
51°	.890	.777	.629	1.235
52°	.908	.788	.616	1.280
53°	.925	.799	.602	1.327
54°	.942	.809	.588	1.376
55°	.960	.819	.574	1.428
56°	.977	.829	.559	1.483
57°	.995	.839	.545	1.540
58°	1.012	.848	.530	1.600
59°	1.030	.857	.515	1.664
60°	1.047	.866	.500	1.732
61°	1.065	.875	.485	1.804
62°	1.082	.883	.470	1.881
63°	1.100	.891	.454	1.963
64°	1.117	.899	.438	2.050
65°	1.134	.906	.423	2.145
66°	1.152	.914	.407	2.246
67°	1.169	.920	.391	2.356
68°	1.187	.927	.375	2.475
69°	1.204	.934	.358	2.605
70°	1.222	.940	.342	2.747
71°	1.239	.946	.326	2.904
72°	1.257	.951	.309	3.078
73°	1.274	.956	.292	3.271
74°	1.292	.961	.276	3.487
75°	1.309	.966	.259	3.732
76°	1.326	.970	.242	4.011
77°	1.344	.974	.225	4.331
78°	1.361	.978	.208	4.705
79°	1.379	.982	.191	5.145
80°	1.396	.985	.174	5.671
81°	1.414	.988	.156	6.314
82°	1.431	.990	.139	7.115
83°	1.449	.992	.122	8.144
84°	1.466	.994	.104	9.514
85°	1.484	.996	.087	11.43
86°	1.501	.998	.070	14.30
87°	1.518	.999	.052	19.08
88°	1.536	.999	.035	28.64
89°	1.553	1.000	.017	57.29
90°	1.571	1.000	.000	∞

Appendix F

ANSWERS AND HINTS FOR ODD-NUMBERED QUESTIONS

CHAPTER 1

1. (a) No. (b) Arrow points away from B.
3. $-e, +e, 0, 0$.
5. Infrared, visible light, ultraviolet, x-rays.
7. Photon, electron or β-particle, proton, neutron.
9. (a) 13. (b) 27. (c) 14.
11. (a) The nucleus of a helium atom. The constituents of an α-particle are two protons and two neutrons.
    (b) $+2e$. (c) 2. (d) 4.
13. The atom contains as much negative charge as positive charge.
15. Deuterium is an atom with one orbital electron and a nucleus which is a deuteron. The deuteron's constituents are a neutron and a proton. Nuclear forces have a short range and are much stronger than the electrical forces involved in the formation of an atom.
17. (a) In some encounters between an α-particle and a molecule of the air, an electron is ripped off from the molecule.
    (b) A trail of tiny water drops which have condensed around ions.
19. An α-particle which experiences a head-on collision with a gold nucleus will be scattered backwards.
21. A radioactive series, like the uranium series, is a sequence of radioactive nuclei, each nucleus of the series resulting from the decay of the preceding nucleus.
25. (b) $_{88}Ra^{226} \longrightarrow {}_{86}Rn^{222} + {}_2He^4$, $_{86}Rn^{222} \longrightarrow {}_{84}Po^{218} + {}_2He^4$, $_{84}Po^{218} \longrightarrow {}_{82}Pb^{214} + {}_2He^4$, $_{82}Pb^{214} \longrightarrow {}_{83}Bi^{214} + {}_{-1}\beta^0$,
    (c) β-particles.
    (d) A source containing the inert gas, radon-226, would soon contain, as daughter products, the useful lead-214 and bismuth-214. Therefore, the same useful γ-rays would emerge from this source as from a radium-226 source. However, since the half-life of radon-222 is merely 3.5 days, the source activity would decrease rapidly.
27. 1 millicurie.
29. 1.0×10^{16} particles per second.
31. I-131; I-131.
33. A proton or an α-particle experiences an electrical repulsive force as it approaches a nucleus but there is no such force acting on a neutron. A neutron therefore can easily penetrate to the nucleus where it interacts strongly with the particles of the nucleus.
35. $_0n^1 + {}_{15}P^{31} \rightarrow {}_{15}P^{32} + {}_0\gamma^0$.   37. 40 days.

39. (a) The present ratio of C-14 to C-12 in the atmosphere is the same as it was in the past.
    (b) A million years is about 200 half-lives so the amount of C-14 remaining would be too small to be detected.

S-3. $_{42}Mo^{99} \rightarrow {}_{43}Tc^{99} + {}_{-1}\beta^{0}$.   S-5. 20 hr.

CHAPTER 2

1. (a) 4.0 inches/year.   (b) 12 inches.
   (c) 6.0 ft.   (d) No. Growth rate will change.
3. Velocity constant.   5. 44.5 m.
7. Acceleration is the rate of change of velocity.
9. (a) 6 m/sec.   (b) 16 m/sec.
11. (a) 94 m, 136 m, 166 m, 184 m, 190 m.
    (b) Graph is a straight line with a negative slope (-3.0 m/sec^2). At $t = 10.00$ sec, the velocity is zero.
13. 1080 m.
15. 35 m/sec, 58 sec.
17. (a) 2.83 sec.   (b) 27.7 m/sec.   (c) 14.7 m/sec, 14.7 m.
19. (a) 40 m/sec.   (b) No.   (c) Yes.
21. Magnitude is 50 m/sec.
23. 30 units, south.
25. $v_x = 52$ m/sec, $v_y = 30$ m/sec.
27. $x = -6.4$ m, $y = 6.4$ m.
29. $a_x = 0$, $a_y = -9.8$ m/sec.^{2}
31. $r = 10.0$ m, direction $36.9°$ above X-axis.

33. The position vector is a vector directed from the origin of the frame of reference to location of the object. The magnitude of the position vector is the distance from the origin to the object.
    The velocity vector is the rate of change of the position vector.
    The acceleration vector is the rate of change of the velocity vector.
35. (a) $\vec{v}$ and $\vec{a}$ both directed north.
    (b) $\vec{v}$ directed north, $\vec{a}$ directed south.
37. Velocity vector is tangent to the path. Acceleration vector points inward and makes an acute angle with the velocity vector.
39. Acceleration vector has a centripetal component 4.0 m/sec^2 south, and a tangential component 3.0 m/sec^2 east. Magnitude of acceleration is 5.0 m/sec^2.
41. (a) 5 N, east.   (b) 1 N, east.   (c) 0.
    (d) 5 N, about $37°$ east of north.   (e) 4 N, east.

43. 600 N, downward.
45. 1000 N directed upward.
47. 0.167 m/sec².
49. 10 N in the direction opposite to that of the velocity vector.
51. 4.0 N, downward.
53. (a) The weight of an object is the gravitational force exerted on it.
    (b) If an object of mass m placed at the point experiences a gravitational force $\vec{F}$, then $\vec{g} = \vec{F}/m$.
55. (a) 9.00 N/kg.   (b) 54.0 N.
57. 6.7×10^{-11} N/kg; 6.7×10^{-10} N.
59. (a) 2.71×10^2 N/kg.   (b) 5.90×10^{-3} N/kg.
63. (a) 27×10^{-11} N/kg directed toward 2.0-kg mass.
    (b) 15×10^{-11} N/kg directed toward masses.
65. 5 × (earth's radius).
67. 6.7×10^{-10} N.
69. 3.53×10^{22} N.
71. 2.18.
73. $\Delta \vec{r}$ is not necessarily in the direction of $\vec{F}$ but $\Delta \vec{v}$ is in the direction of $\vec{F}$ for a short time interval.
75. Gravitational field produced by the earth is weaker at points more distant from the earth.
    An astronaut's weight is not zero in orbit because, at the location of the orbit, there is still a gravitational field produced by the earth and this field exerts a gravitational force on the astronaut (weight).
77. Velocity vector is tangent to orbit. Acceleration vector points toward earth's center. Mass m is slowing down because the acceleration vector has a component in the direction opposite to the velocity vector.
79. 4.2×10^7 m.
81. 1.90×10^{27} kg.
83. 2.01×10^{30} kg.
85. 1.87 years.   87. No. No.
89. (b) 33.9 m/sec.   (c) 2.86 sec.   91. 35 m/sec.
93. Electric forces can be attractive or repulsive but gravitational forces are always attractive. Within an atom electric forces are much stronger than gravitational forces.
95. (a) 5.0×10^3 N/coul.
    (b) Plates horizontal with upper plate charged negatively and lower plate positively.

97. (a) 26×10^3 N/coul directed toward the negative charge.
    (b) 4.2×10^{-15} N.
103. (a) 3.20×10^{-18} N in the direction of the electron's velocity.
     (b) 3.51×10^{12} m/sec^2 in the direction of the electron's velocity.
     (c) 1.96×10^{-2} m, 7.1×10^5 m/sec.
105. (a) Ellipse, parabola, hyperbola.   (b) Ellipse.
S-3. (d) 30 m/sec.
S-5. (a) 268.2 m.   (b) 2.40 sec.
S-7. Automobile was speeding at 17 m/sec.
S-9. 24 sec.
S-13. (a) 3.84 N/kg.
      (b) 980 N on earth, 384 N on Mars.
S-15. (a) $4\pi^2 r^3/GT^2$.   (b) $10^4 \pi^2 r/T^2$.
S-17. 20.4 m.
S-19. (a) 1.0×10^{-9} sec.
      (b) 3.52×10^{15} m/sec^2, upward.
      (c) 3.52×10^6 m/sec, upward; 1.76×10^{-3} m, upward.

CHAPTER 3

1. 1000 N directed toward man.
3. 800 N forward, 200 N backward. Acceleration = 4.00 m/sec^2.
5. (a) 326 N.   (b) 366 N, upward.   (c) 366 N, downward.
7. (a) Initial momentum = 8.0 kg-m/sec,
       final momentum = 20 kg-m/sec.
   (b) 4.0 kg-m/sec^2.
   (c) 4.0 N in the direction of the velocity.
9. 1.4 m/sec west.
11. $(0.01 \text{ kg}) \times (1000 \text{ m/sec}) = -(0.01 \text{ kg}) \times (200 \text{ m/sec}) + (100 \text{ kg})v$
    $(100 \text{ kg})v = 12$ kg m/sec.
    $v = 0.12$ m/sec.
13. $mv_1 + 0 = 0 + mv_2$. Therefore, $v_1 = v_2$.
15. 2.0 m/sec, south.
17. 13.3 m.
19. Law of conservation of momentum.
21. 2.4 m/sec, east.

25. $2m\, \vec{r}_c = m\vec{r}_1 + m\vec{r}_2$
    $\vec{r}_c = \frac{1}{2}(\vec{r}_1 + \vec{r}_2)$.

27. (a) The center of mass of a system moves like a particle of mass equal to the total mass of the system, acted upon by the vector sum of the external forces acting on the system.

29. $a = 5.00$ N, $36.9°$ below the direction of X-axis; $b = 13.00$ N, $67.4°$ above the negative x-direction.

31. 8.83 N, $14.4°$ west of north.

33. 544 N, $36.0°$ north of east.

35. (a) 4.5×10^3 N. (b) 1.5×10^3 N.

37. (a) 4.50×10^3 m. (b) 1.50×10^3 N.

39. $g \sin \theta$.

41. (a) 1.67 N to the right. (b) 3.33 N to the right.
    (c) The acceleration of the blocks is the same in parts (a) and (b) and therefore the same resultant force acts on the 10-kg block. In part (b), the force exerted by the 20-kg block is not the only force contributing to this resultant.

45. (a) 78.4 N, upward.
    (b) 15.7 N in the direction opposite to the block's velocity.
    (c) 1.96 m/sec² in the direction opposite to the block's velocity.
    (d) 2.04 sec.

47. 4.7 N.

49. 0.19.

51. 11.8 N, $\mu_s = 1.00$.

53. Rope 115 N, string 58 N.

55. Torque $= FD$, where the moment arm D is the perpendicular distance from the line of action of $\vec{F}$ to the axis.

57. In a uniform gravitational field, gravitational torque can be computed as if the entire weight were located at the center of mass.

59. 9.8×10^3 m-N.

61. With the XY-plane vertical:
    (a) Algebraic sum of x-components of external forces $= 0$,
    (b) Algebraic sum of y-components of external forces $= 0$.
    (c) For any axis:
        Sum of clockwise torques $=$ sum of counterclockwise torques.

63. At 40-cm mark, 16.7 N; at 100-cm mark, 3.3 N.

65. $T = 400$ N, $F_h = 400$ N, $F_v = 600$ N.

67. (a) $H = 10$ N. (b) $F_h = 10$ N, $F_v = 60$ N.

69. $f = 62.5$ N, $N' = 62.5$ N, $N = 300$ N.

71. $\mu_s = 0.292$.

73. $\omega_{sec} = 0.105$ rad/sec; $\omega_{min} = 1.74 \times 10^{-3}$ rad/sec.

75. 15.7 rad/sec.
77. (a) 1.6×10^3 m/sec, the upper tip moves to the left and the lower tip moves to the right.
    (b) 3.2×10^6 m/sec², directed inward toward the propeller shaft.
79. (a) Mass moves from $x = -A$ through the origin to $x = +A$ and then returns through the origin back to $x = -A$.
    (b) 0.25 sec.
81. 125 sec.
83. The amplitude of a simple harmonic motion is the magnitude of the maximum displacement from the equilibrium position.
85. (a) $T = 15.7$ sec, $A = 0.30$ m.
    (b) $T/4 = 3.9$ sec.
87. Arrow is 2.00 m long and is directed upward.
89. (a) 0.12 m/sec.   (b) 0.12 m/sec.
    (c) At the equilibrium position ($x = 0$).
    (d) The minimum speed, zero, is attained at $x = \pm 0.30$ m.
91. (a) 4.8×10^{-2} m/sec² directed toward center of circle.
    (b) 4.8×10^{-2} m/sec².
    (c) At $x = \pm 0.30$ m.
    (d) The minimum magnitude of the acceleration, zero, occurs at the equilibrium position ($x = 0$).
93. $a_x = 2.6$ m/sec².
95. (a) 200 N/m.   (b) 0.40 sec.
99. (a) 0.10 m.   (b) 0.10 m.
    (c) 1.6 hertz, 0.10 m.
101. 2.01 sec.
103. 9.79 m/sec².

S-1. (a) 1.41×10^{-4} c.
     (b) 2.96×10^{-2} c at an angle of 89.4° to the direction of motion of the incident neutron.
S-3. $sm/(m + M)$, forward.
S-5. (a) On the line joining the spheres, 9.4×10^{-2} m from the 5.0-kg sphere.
     (b) 5.0 m/sec² in the direction of the applied force.
S-7. (a) 4.0 m/sec².
     (b) 8.0 m/sec east.
        $(5.0 \text{ kg}) \times v_1 + (10 \text{ kg}) \times (6.0 \text{ m/sec}) = (15 \text{ kg}) \times (8.0 \text{ m/sec})$
        $v_1 = 12$ m/sec east.
S-9. (a) 3.0 m/sec².
     (b) Tension = 54 N, $M = 8.0$ kg.

S-11. 100.

S-13. If the applied force is applied smoothly and the 2.0-kg mass has zero acceleration, the resultant force acting on this mass is zero. Consequently, the tension in thread *A* equals the tension in thread *C plus* the weight of the 2.0-kg mass. In this case thread *A* will break, rather than thread *C*. If a sudden jerk is applied at *C*, because of the inertia of the 2.0-kg mass, there is a small time delay before this mass acquires a downward displacement sufficient to cause the tension in thread *A* to be increased significantly. In the meantime the tension in thread *B* exceeds that in thread *A* and thread *B* may break.

S-15. (a) 98 N.   (b) 29.4 N.
      (c) 2.0 m/sec².   (d) 2.72 m.

S-23. 346 N.

S-27. $\omega = 1.99 \times 10^{-7}$ radians/sec, $v = 2.98 \times 10^4$ m/sec.

S-29. (a) 1.05×10^3 rad/sec.
      (b) Magnitude of angular acceleration is 1.74 rad/sec².

S-31. Magnitude of the angular acceleration is 0.23 rad/sec². 4.1 revolutions.

S-33. 50 kg-m².

S-35. 13 m/sec.

S-37. (a) $\omega = 1.57$ rad/sec, $f = 0.25$ hertz, $T = 4.0$ sec.
      (b) Initial phase $= \pi/3$ radians, Amplitude $= 5.0$ m.
      (c) $x = -2.5$ m, $v_x = 6.8$ m/sec, $a_x = 6.16$ m/sec².

S-39. 0.038.

S-41. (a) The tension reaches its maximum value when the pendulum is vertical.

CHAPTER 4

1. Work is done by girl in part (a) since there is motion in the direction of the force the girl exerts. No motion and therefore no work in part (b).

3. (a) 600 joules.   (b) 0 joule.   (c) -200 joules.

5. -1.04×10^3 joules.

7. 1.25×10^3 joules.

9. $MV = mv$
$V/v = m/M$
$$\frac{\tfrac{1}{2}MV^2}{\tfrac{1}{2}mv^2} = \left(\frac{M}{m}\right)\left(\frac{m}{M}\right)^2 = \frac{m}{M}.$$

11. 10^2 watts.

ANSWERS AND HINTS

13. 1.3×10^4 watts.
17. 3.6×10^6 watts.
19. (a) There are different expressions for the potential energies arising from the work done by different types of forces.
    (b) No. There is no definite amount of frictional work that can be assigned to a given change in position.
21. (a) ML^2T^{-2}.
23. To hold two positive charges at rest, the giant would have to pull to balance the repulsive Coulomb forces between like charges. He would do negative work as their separation increased. The potential energy of the two charges would decrease. The initial potential energy would be positive.
25. (a) 71×10^2 joules. (b) 71×10^2 joules. (c) 71×10^2 joules.
    (d) Work done by man in (a) is increased; increase in potential energy (b) is unchanged; kinetic energy at bottom (c) is less.
29. $+13.6$ eV.
31. 1.12×10^4 m/sec.
35. (b) $\left(\dfrac{m}{m+M}\right)^2 \dfrac{v^2}{2\mu_k g}$.
37. (a) $x = \pm A$. (b) $x = 0$.
    (c) $x = \pm \tfrac{1}{2}A$. (d) $x = \pm A/\sqrt{2}$
39. 16.
41. 1.1×10^2 sec.
43. (a) 1.3 kilocalories. (b) 0.66 C°.
45. (a) 3.2×10^3 N. (b) 4.
47. (a) The ideal mechanical advantage of a machine is the mechanical advantage of an idealized replica which would operate without friction or energy storage.
    (b) Ideal mechanical advantage = 3.7.
        Actual mechanical advantage = 3.3.
49. (a) Ideal mechanical advantage = 7.5.
    (b) Actual mechanical advantage = 5.6.
51. (a) 163 N.
    (b) Actual mechanical advantage = 5.8, Efficiency = 96%.
    (c) Efficiency = 60%.
53. (a) $E = mc^2 + K$. (b) $K = \tfrac{1}{2}mv^2$.
    (c) mc^2 is the total energy of the particle when the particle is at rest.
57. 67.5 MeV.
59. Yes. Creation of an electron-positron pair by a γ-ray.
61. Yes. The hydrogen atom is stable—its nucleus (the proton) does not decay.

65. *B, A, C.*
67. The binding energy of a neutron or a proton to a nucleus is several MeV, while the binding energy of atoms within a molecule is merely a few eV. Rearranging atoms to form different molecules therefore involves only a few eV per molecule, but rearranging neutrons and protons to form different nuclei typically involves several MeV per nucleus.
71. (a) Yes.   (b) Yes.   (c) Yes.   (d) Yes.

S-1. (a) 0.   (b) 2.4×10^3 joules.
 (c) Weight of elevator and the frictional force acting on elevator.
S-3. (a) 1.70 horsepower.   (b) 34.0 sec.
S-5. 46%.
S-9. Speed = $\sqrt{gL}$
S-11. (a) Potential energy = $-Gm^2/r$.
 (b) 0.
 (c) Total mechanical energy = $0 + (-Gm^2/r) = -Gm^2/r$.
S-13. (a) 230 m.   (b) 2.03×10^8 joules.
S-15. (a) 785.   (b) 3.14×10^4 N.
S-17. (a) If no friction, input force would be 40.0 N. Actual mechanical advantage = 25, Efficiency = 25%.
 (b) 3.23%. Efficiency decreases as the load decreases.
S-21. (a) 3.9×10^3 joules.   (b) 3.9×10^3 joules.
 (c) 0.800.   (d) 13 m/sec.
S-23. (Ex 6) $Q = 1.02$ MeV.   (Ex 7) $Q = -1.02$ MeV.
 (Ex 8) $Q = 2.23$ MeV.   (Ex 9) $Q = -9$ joules.
 (Ex 10) $Q = 4.87$ MeV.   (Ex 11) $Q = 33.9$ MeV.
S-25. (a) $_1H^1 + {_3Li^7} \rightarrow {_2He^4} + {_2He^4}$.
 (b) From kinetic energies, $Q = 17.0$ MeV.
  From rest energies, $Q = 17.2$ MeV.
 (c) Within experimental error, the decrease in rest energy (17.2 MeV) is equal to the increase in kinetic energy (17.0 MeV).
S-27. 30°C.
S-29. (a) The law of conservation of energy and the law of conservation of momentum are both violated unless an unseen particle is also emitted with
$$\text{Energy} = c(\text{momentum}).$$
 (b) A neutrino is electrically neutral and interacts only very weakly with nuclei.
 (c) A neutrino possesses energy and momentum [as well as angular momentum (see Section 3.9)].

CHAPTER 5

1. (a) 5.0 m/sec, east. (b) 1.0 m/sec, west.
3. 120 m/sec, east.
5. Current is 2.5 mph, speed of boat relative to water is 17.5 mph.
7. 117 m/sec, 22.6° north of east.
9. 24.1 m/sec, 4.8° south of east.
11. 40 m/sec².
13. When the resultant force acting on a body is zero, the body moves with a constant velocity vector.
15. (a) Not an inertial frame. (b) Inertial frame.
    (c) Not an inertial frame. (d) Not an inertial frame.
17. The frame force, $\vec{F}_{\text{frame}} = -m\vec{a}_{\text{frame}}$, is the quantity which must be added to the resultant of the real forces acting on a body, so that relative to an accelerated frame, we again have

$$\overrightarrow{\text{force}} = (\text{mass}) \times \overrightarrow{\text{acceleration}}.$$

19. 4.9×10^4 N forward.   50:1.
21. 742 N.
23. (a) The centrifuge exerts a force on the test tube directed toward the center of the circular path (centrifuge axle). The acceleration of the test tube is v^2/r directed toward the center.
    (b) In addition to the force exerted by the centrifuge toward the centrifuge axle there is a centrifugal force of equal magnitude in the opposite direction. Relative to this frame the test tube has no acceleration.
25. In the rotating frame there is a frame force, mv^2/r, in the direction opposite to the weight, mg. The acceleration a' is determined by
    $ma' = mg - mv^2/r$.
    If the rotation were to stop, the acceleration would increase to the value given by $ma = mg$.
27. $\tan \theta = v^2/gR$.
29. Relative to laboratory the electron has a speed $(4.0 \times 10^5 + 2.0 \times 10^4)$ m/sec, but the photon speed is again c, not $c + 2.0 \times 10^4$ m/sec.
31. Yes.
33. (a) $\Delta t' = (2/3) \times 10^{-7}$ sec. (b) $\Delta x = 198$ m,
    (c) $\Delta x/\Delta t = 3.0 \times 10^8$ m/sec. $\Delta t = 66 \times 10^{-8}$ sec.
35. (a) $\Delta x' = 27$ m; $\Delta t' = 1.0 \times 10^{-7}$ sec.
    (b) With lab frame as $O'X'$ and other frame as OX (frames vertical with positive direction downward) we have $V = +0.98\ c$ and then find $\Delta x = 282$ m, $\Delta t = 94 \times 10^{-8}$ sec.
    (c) $v/c = \dfrac{0.9 + 0.98}{1 + (0.9 \times 0.98)} = \dfrac{1.88}{1.882}$.   (d) Distance $= 0$.

41. Interval is 36×10^6 m² in both frames.
45. Einstein's law gives 0.983c. $v_x' + V = 1.45c$.
47. 0.988c, east.
49. (a) 1.12 MeV. (b) 0.61 MeV.
51. (a) $E = 1438$ MeV. (b) $p = 1.09 \times 10^3$ MeV/c.
    (c) $v = 0.758c$.
53. (a) 529 MeV. (b) 423 MeV.
    (c) 518 MeV/c (d) 10.2×10^{-6} sec.
55. (a) 6.67×10^{-6} kg-m/sec.
    (b) 1.67×10^{-3} m/sec.
S-1. (a) 10 m/sec, forward.
    (b) 45 m/sec and 55 m/sec.
S-3. Speed relative to car is 11.5 m/sec. Speed relative to ground is 5.77 m/sec.
S-5. 53.1° south of east.
S-7. Acceleration relative to plane is 4.9 m/sec² directed along the downward slope of the plane. Acceleration relative to ground is 11.5 m/sec² at an angle of 12.3° below the horizontal.
S-9. 1.09 sec.

S-11.

Event	Measurements in $O'X'$		Measurements in OX	
	x'	t'	x	t
A	−300 m	1.0×10^{-6} sec	−100 m	0.33×10^{-6} sec
B	300 m	1.0×10^{-6} sec	900 m	3.00×10^{-6} sec

S-21. 3.6×10^2 MeV.
S-23. (b) 2.72×10^{-5} sec.
S-27. $v_y = \dfrac{\Delta y}{\Delta t} = \dfrac{\Delta y'}{\gamma(\Delta t' + \beta \Delta x'/c)}$
    $= \dfrac{v_y'}{\gamma(1 + \beta v_x'/c)}$.

S-29. Examine absorption process from a frame in which the final product is at rest, the initial electron has a total energy $mc^2 + K$, and the photon has an energy E_γ. Then,

Rest energy of final product $= mc^2 + K + E_\gamma$.

Consequently,

Rest mass of final product > rest mass of electron.

Therefore final product cannot be an electron.

CHAPTER 6

1. A push of 1.1×10^3 N perpendicular to tile.
3. 2.9×10^4 N.
5. Mass = 1.0×10^{-3} kg. Weight = 9.8×10^{-3} N.
7. 1.99×10^5 N/m².
9. Force = 2.94×10^2 N. Weight of water = 0.98 N. Force on base is determined by water pressure at base, not by the weight of water above the base. (The water exerts an upward force on the upper walls of the flask. The resultant of all the forces exerted by the water on the flask equals the weight of the water.)
11. 10.2 m.
15. 1.4×10^5 N/m².
17. Because the density of iron is less than the density of mercury, iron floats in mercury. Gold has a greater density than that of mercury, so gold will not float in mercury.
19. (a) 2.00 m.
    (b) 1.96×10^4 N/m².
    (c) $F = PA = 2.35 \times 10^6$ N. Weight of barge = 2.35×10^6 N.
21. (a) 91.9 m³.   (b) 1081 N.
23. 9.57
25. A streamline is a line in a fluid with a direction at each point in the direction of the fluid velocity at that point.
    A tube of flow of a fluid is a tubular region with sidewalls that are streamlines.
29. (b) 6.0×10^5 joules.
31. (a) 0.50 m/sec.   (b) 1.72×10^5 N/m².
33. (a) 7.7 m/sec.   (b) 4.9 m.

S-1. As the fluid density increases, the displaced volume V decreases to maintain the same buoyant force $\rho V g$. The ship therefore rises in salt water.
S-3. (a) 780 mm.
    (b) 743 mm of 5% dextrose solution.
S-5. The volume of water that overflows is equal to the volume of the immersed object and this volume is given by

$$V = M/\rho$$

where the object has a mass M and a density ρ. Consequently the ratio of overflow volumes V_{crown}/V_{gold} determines the ratio of densities of the immersed objects:

$$V_{crown}/V_{gold} = \rho_{gold}/\rho_{crown}.$$

Archimedes' measurements showed that ρ_{crown} was less than ρ_{gold} and this implied that the crown contained some material less dense than gold.

$$\rho_{crown} = 0.900(19.3 \times 10^3 \text{ kg/m}^3)$$
$$= 17.4 \times 10^3 \text{ kg/m}^3.$$

S-7. According to Archimedes' principle, the fluid in the beaker exerts a buoyant force on the immersed finger. By Newton's third law, the finger exerts a force of equal magnitude downward on this fluid. This downward force will depress the pan containing the beaker in which the finger is immersed.

S-9. (a) 1.0×10^5 joules.
(b) Bernoulli's equation predicts no drop in pressure. Frictional effects and turbulence lead to a dissipation of macroscopic mechanical energy and consequent departures from the predictions of Bernoulli's equation.
(c) 1.0×10^5 joules per cubic meter of oil per kilometer of pipeline length.

S-11. 1.4×10^{-4} m^2.

S-15. (a) Bubbles do not move relative to water. In the accelerated frame of the bottle, every portion of water has its weight balanced by an upward frame force. Consequently there is no increase of pressure with depth and no buoyant force is exerted on a bubble.
(b) Balloon moves toward the center of the curve (toward the door on the inside of the turn). In the accelerated frame of the car, there is a centrifugal force acting on every portion of the air in the car. Analogous to the increase in pressure with depth in a fluid because of gravitational force, we will find, associated with this centrifugal force, an increase of pressure in the air of the car as we move from the door on the inside of the turn to the door on the outside of the turn. And as is the case for a helium-filled balloon in the atmosphere, the balloon will be pushed from regions of higher pressure toward regions of lower pressure.

CHAPTER 7

1. 7.5×10^3 N.
3. 6.0×10^{26} molecules.
5. $37.0°$ C, $310.15°$ K.

7. (a) 4.04×10^5 N/m².
   (b) 4.44×10^5 N/m².
9. 33.3%.
11. 788°K.
13. 1.8×10^{24} molecules.
15. 4.87×10^4 N/m².
17. 0.577 kg.
19. The Kelvin temperature is a measure of the average of the translational kinetic energy of a molecule's random thermal motion: (average of $\frac{1}{2}mv^2$) = $\frac{3}{2}kT$.
21. (a) 4.8×10^2 m/sec.
    (b) 1200° K.
23. Knowing k, measure P, V, and T and then calculate N from $N = PV/kT$. Then measure the mass M of the gas and calculate the molecular mass m from $m = M/N$.
25. By measurement of P, V, and T for a measured number of moles n of a gas at low density, the gas constant R can be calculated: $R = PV/nT$. Then Avogadro's number N_0 is given by $N_0 = R/k$.
27. (a) 4.1×10^2 m/sec. (b) 3.4×10^3 joules. (c) 0.
29. $\bar{E} = \frac{3}{2}NkT$
    $= 3.7 \times 10^3$ joules.
31. 2.7×10^2 joules.
33. (a) Heat for process $= \Delta \bar{E} -$ (work for process).
    (b) $\Delta \bar{E} =$ (work for process) $+$ (heat for process).
35. 2.0 kilocalories.
37. (a) Change in thermal energy $\Delta \bar{E}$ depends only on the initial and the final states which are the same as in Question 32. From Question 36 this is $\Delta \bar{E} = 79.7$ kilocalories.
    (b) Work $= (1.0 \times 10^4 \times 1.01 \times 10^5$ N/m²$)(8.6 \times 10^{-5}$ m³$)$
       $= 8.7 \times 10^4$ joules
       $= 20.8$ kilocalories.
    (c) Heat for process $= (79.7 - 20.8)$ kilocalories
       $= 58.9$ kilocalories.
39. 964 C°.
41. 1.67×10^{-2} m³.
43. 26.8° C.
45. Measure the temperature rise ΔT which occurs in a time interval t during which electric power $VI = RI^2$ is supplied to the coil. Calculate the specific heat s of water from $ms\Delta T = VIt$ after measuring ΔT, t, I and V (or R), and the mass m of the water. Energy "losses" to the rest of the equipment, such as the container and the thermometer have been neglected.

47. 0.053 kilocalorie/kg C°.

51.
Temperature range	Hydrogen $\bar{E}$	C_V
Below 70°K	$\frac{3}{2}NkT$	$\frac{3}{2}k$
20°K – 700°K	$\frac{5}{2}NkT$	$\frac{5}{2}k$
Near 3000°K	$\frac{7}{2}NkT$	$\frac{7}{2}k$

S-1. (a) 1.1×10^6 N. (b) 5.5×10^7 N/m².

S-3. 73.43 cm of Hg.

S-5. 1.09 kg/m³.

S-7. (a) 1.84×10^3 m/sec.
(b) Escape speed is 11.2×10^3 m/sec which is approximately six times the thermal speed of hydrogen molecules at 0°C. Nevertheless, there is an appreciable probability that a hydrogen molecule acquire a speed greater than the escape speed and, at high altitudes, escape from the earth before subsequent collisions with slower molecules reduce its speed below the escape speed. Hydrogen will thus escape from the earth.
(c) The rate of escape is greater for hydrogen than for nitrogen or oxygen because the thermal speed for hydrogen is greater ($v_{thermal}$ is inversely proportional to the molecule's mass).

S-9. On the surface of the moon the escape speed is merely 2.37×10^3 m/sec which is nearly the same as the thermal speed at 273°K of molecules of hydrogen, nitrogen, and oxygen. At the moon's surface, at temperatures above a few degrees kelvin, such gas molecules will therefore easily acquire speeds sufficient to escape from the moon.

S-11. Special relativity does not impose an upper limit to temperature. The conclusions of Section 7.2 are not valid in special relativity because Newtonian (rather than relativistic) expressions for momentum and kinetic energy were used.

S-13. $(100 \times \Delta m/m)\% = (100 \times \Delta mc^2/mc^2)\%$
$= 100 \times \left(\dfrac{(501.7 - 3.0) \text{ kilocal/kg } (4.184 \times 10^3 \text{joules/kilocal})}{(3.0 \times 10^8)^2 \text{ joules/kg}} \right) \%$
$= 2.3 \times 10^{-9}\%$.

S-15. (b) The ratio of the heat capacity of water to that of an equal mass of lead is 32.3.
(c) The ratio of the heat capacity of water to that of an equal volume of lead is 2.85.

S-17. The temperature rise is merely 0.96×10^{-3}C°.

CHAPTER 8

1. (a) $\Delta L = (12 \times 10^{-6})(1.00)(10)$ m
   $= 1.2 \times 10^{-4}$ m.
   (b) The coefficient of linear expansion of invar is less than 1/10 of that of steel. Changes of length associated with temperature changes will therefore be much less for invar than for steel.

3. $\Delta T = 64$ C°. Therefore tire must be heated to 84°C.

5. Area at 170°C is 8.03×10^{-3} m².

7. For the benzene $\Delta V = 124 \times 10^{-5}$ m³. The volume of the glass bottle increases by 2.7×10^{-5} m³. Therefore, to three significant figures the overflow of benzene is 121×10^{-5} m³.

9. 4°C. At this temperature water attains its maximum density. Consequently, if there is any water at 4°C, some will be found at the bottom of the lake. After a lake has been cooled to 4°C, further cooling will lead to a distribution of water such that the temperature increases with depth because this corresponds to an increase of density with depth.

11. (a) Radiation.
    (b) Conduction.
    (c) Convection.

13. (a) 2.0 kg.
    (b) 7.3×10^3 m.

15. The hot alloy and cold water of Example 8 in Chapter 7 are not in thermal equilibrium initially but attain thermal equilibrium after sufficient time has elapsed.

17. 307°K.

19. A thermocouple because (1) it can measure temperatures up to 1600°C (platinum-rhodium alloy), (2) it can be read at locations remote from the chimney, and (3) it can follow rapid temperature changes.

21. The mass of ice melted is 5.0 kg. There remains 7.0 kg of ice in equilibrium with 5.0 kg of water. Since the heat supplied was not sufficient to melt all the ice and then warm the water, the temperature remains at 0°C.

23. (a) 40×10^2 kilocalories.
    (b) 80×10^2 kilocalories.
    (c) 33×10^3 sec.

25. $(0.100 \text{ kg})(80 \text{ kilocalories/kg}) + (0.100 \text{ kg}) T$
    $= (0.800 \text{ kg})(1.00 \text{ kilocalorie/kg})(30.0°C - T)$
    $T = 17.8°C$.

27. The pressure of the vapor at which the vapor and the liquid coexist in equilibrium.

29. (a) When a container of solid carbon dioxide (dry ice) is open to the atmosphere, sublimation occurs as gaseous carbon dioxide escapes. (b) The removal of water vapor by sublimation at low pressure.

31. On Fig. 8-14 such a process would be represented by a horizontal line which crosses the sublimation curve (lower curve on graph).

33. Freezing of water at constant pressure is accompanied by an expansion. Hence if the water is confined, there is a tremendous pressure increase when freezing occurs and the confining material is subjected to large forces.

35. 355 mm Hg.

37. The boiling water maintains a constant temperature (100°C when the pressure is one atmosphere) and provides good thermal contact for heat transfer to the egg.

39. Percentage relative humidity $= 100 \times \dfrac{\left(\begin{array}{c}\text{pressure of vapor at}\\ \text{temperature in question}\end{array}\right)}{\left(\begin{array}{c}\text{saturation vapor pressure}\\ \text{at same temperature}\end{array}\right)}$.

The *dew point* is the temperature at which the vapor pressure in the air would become the saturation vapor pressure.

41. Relative humidity $= 100 \times \dfrac{9.21 \text{ mm Hg}}{12.8 \text{ mm Hg}} \%$
$= 72\%$.

S-3. 25°C.

S-5. 0.109 kcal/kgC°. It is assumed that the heat exchanged between the calorimeter and its surroundings is negligible.

S-7. (a) During periods of sunshine, a certain amount of solid sodium sulphate will melt. During cooler periods, a certain amount of liquid sodium sulphate will solidify giving 51.3 kilocalories/kg to heat the house.
(b) 5.85×10^3 kg.

S-9. A path on the phase diagram of Fig. 8-13 which is an "end run" around the critical point (going from the liquid region to the vapor region without crossing the vaporization curve) represents the transformation in question. It can be accomplished by performing the following sequence of transformations:
1. At constant temperature, increase the pressure to a value above

the pressure at the critical point (the critical pressure).

2. At constant pressure, increase the temperature to a value above the temperature at the critical point (the critical temperature).

3. At constant temperature, decrease the pressure to a value below the critical pressure.

4. At constant pressure, decrease the temperature to a value below the critical temperature.

CHAPTER 9

1. (a) Entropy is a measure of molecular disorder.
   (b) The entropy (and molecular disorder) of the liquid is greater than that of the solid.

3. Change of entropy = final entropy − initial entropy
   $= -1.915$ kilocalories/°K.

5. 2.00×10^{-2} kilocalories/°K.

7. 9.3 kilocalories/°K.

9. Entropy decrease of reservoir is 5.36×10^{-3} kilocalorie/°K.

11. (a) 8.5×10^2 m.
    (b) Entropy decrease of water is 2.00×10^{-2} kilocalorie/°K.
    (c) No.
    (d) Yes. No violation of conservation of energy is implied.
    (e) No. This process would lead to a decrease in total entropy.

13. Process is irreversible because it involves a net increase of entropy of 0.30 kilocalorie/°K.

15. (a) 0.058 kilocalorie/°K (increase).
    (b) Yes.

19. Change in total entropy $= \dfrac{2.0 \text{ kilocalories}}{273°K} - \dfrac{2.0 \text{ kilocalories}}{373°K}$
    $= 2.0 \times 10^{-3}$ kilocalorie/°K.

    No. Entropy-increasing processes are irreversible.

21. A hypothetical device which would extract heat from a reservoir and convert this heat entirely into work. Energy would be conserved but the total entropy would decrease.

23. 600°K.

25. Heat should be exhausted to the air in the winter and to the ocean in the summer.

27. Carnot efficiency of the gasoline engine (0.620) exceeds that of the steam engine (0.293) by a factor of 2.1.

29. No. Heat must be exhausted to a cold reservoir to provide an entropy increase that will more than compensate for the entropy decrease of the hot reservoir.

31. The net effect will be a heating of the room since, although the refrigerator extracts a heat Q_{cold}, it gives off a heat $Q_{hot} = Q_{cold} + W$, where W is the energy supplied to operate the refrigerator.

33. In one cycle, the only entropy changes that occur are the entropy increase Q_{hot}/T_{hot} of the hot reservoir and the entropy decrease Q_{cold}/T_{cold} of the cold reservoir. The principle of entropy increase therefore requires that
$Q_{hot}/T_{hot} \geq Q_{cold}/T_{cold}$
or
$Q_{hot} \geq (T_{hot}/T_{cold}) Q_{cold}$.
Since $T_{hot} > T_{cold}$, the above result implies that $Q_{hot} > Q_{cold}$.

35. Electromagnetic radiation from sun
↓
Plant life ————————→ degradation to thermal energy
↓
Fossil fuels ————————→ degradation to thermal energy
↓
Burning in heat engine which ——→ degradation to thermal energy
turns generator and develops
electrical potential energy
↓
Degradation of electric potential
energy to thermal energy in stove
element.

37. $\dfrac{n_2}{n_1} = \dfrac{10^{S_2/ka}}{10^{S_1/ka}} = 10^{S_2/ka - S_1/ka} = 10^{(S_2 - S_1)/ka}$.

39. The entropy change of the hot object is smaller than the entropy change of the cold object.

S-1. Final temperature is 0°C. Change in entropy is an increase of 0.026 kilocalorie/°K.

S-3. 28.5 kilowatts.

S-7. 84 watts.

S-9. Expose the refrigerator interior to the outside environment and its exterior to the inside of the house. The refrigerator will then act as a heat pump, extracting heat from the outside and rejecting a greater amount of heat to the inside of the house.

S-11. $n_2/n_1 = 10^{(S_2 - S_1)/ka} = 10^{2.3 eV/kTa} = 10^{2.3/(1/40)2.3} = 10^{40}$. Notice that a minute increase in energy has increased the number of microscopically different states by a huge factor.

CHAPTER 10

1. The wave speed is the speed at which the wave form travels along the rope. A particle of the rope moves up and down as the wave passes the particle.

3. (a) The wavelength is the distance between adjacent crests. The number of cycles per second is the frequency.
   (b) $T = 1/f$.

5. (a) 0.70 hertz.   (b) 1.8 m/sec.   (c) 2.6 m.

9. Molecules vibrate back and forth in the direction in which the sound waves move. Fig. 10-3(a) shows the displacement at one instant of time of different molecules located along a line in this direction. Fig. 10-3(b) shows the displacement of one molecule at different instants.

11. 1655 m, 2.7%.

13. 113 m.

15. (a) 2.27×10^{-3} sec.
    (b) 0.773 m.

17. Wavelength is approximately 1 millimeter. With $\lambda = 5$ mm, the beam would fan out into a cone of vertex angle about 50°.

19. (a) 258 hertz and 254 hertz.
    (b) 258 hertz.

21. A natural frequency is the frequency of oscillation in a normal mode. The fundamental frequency is the lowest natural frequency. Overtones are higher natural frequencies. Harmonics are overtones that are integral multiples of the fundamental frequency.

23. $f = 1200$ hertz. There are nodes at each end and also at 0.167 m from each end. The loops are halfway between adjacent nodes.

25. 2.01×10^3 watts.

27. The pitch of the voices will be higher because the frequency is increased by the factor 45/33.

29. (a) 0.652 m.   (b) 521 hertz.

31. (a) 0.938 m. Waves reflected from the cliff have the same wavelength.
    (b) 380 m/sec.
    (c) 405 hertz.

35. With observer approaching the source and moving with velocity v_o relative to the air, the observed frequency f_o is

$$f_o = \frac{v + v_o}{\lambda} = \frac{v + v_o}{(v/f)} = \left(1 + \frac{v_o}{v}\right) f.$$

Therefore

$$\frac{\Delta f}{f} = \frac{f_o - f}{f} = \frac{v_o}{v}.$$

37. 60°.

39. 1.22 m/sec.   41. 2.6 × 10⁸ m/sec.

43. Energy = power × time
    = (7.6 × 10³ W/m²)(0.50 m)(0.30 m)(20.0 sec)
    = 2.3 × 10⁴ joules.

45. (a) At a temperature such that v = 340 m/sec, λ = (340 m/sec)/(10 × 10⁶ hertz) = 3.4 × 10⁻⁵ m.
    (b) In body tissue, as in water, the velocity of sound is greater than in air. The frequency is the same. Therefore the wavelength is greater.

S-1. (a) Particles at x = 2.00 m and at x = 4.00 m.
    (b) Particle at x = 1.00 m.

S-5. 65 dB.

S-7. 41 m toward speaker.

S-9. 16%.

S-15. In therapeutic use of ultrasound, such as the destruction of gallstones by the application of ultrasonic mechanical vibrations, general body damage can be minimized and the destructive effect on the gallstones maximized by using a vibration frequency which is a resonant frequency of the stones.

CHAPTER 11

1. The difference of potential, $V_A - V_B$, between points A and B is W/q where W is the work done by the electric force on a charge q that is transported from A to B.

3. (a) The electric field encountered en route from B to C.
   (b) $V_B - V_C$ = −4.0 volts.   (c) C.   (d) −4.0 volts.

5. (a) 400 eV; 6.4 × 10⁻¹⁷ joule.   (b) 400 eV; 6.4 × 10⁻¹⁷ joule.

7. $\vec{E}$ is zero.

9. (a) 185 volts.   (b) 220 volts.   (c) 200 volts.

11. 50 × 10³ volt/m directed 53.1° north of east.

13. (a) Salt-water solution, ionized gases.   (b) Metals.

15. 7.00 coul.

17. 0.020 coul.

19. (a) 4.32 × 10⁵ coul.   (b) 8.64 × 10² sec.

21. 8.0 volts.

23. (a) The resistance R of a circuit element which carries a current I when the potential difference between its terminals is V, is defined by $R = V/I$.
    (b) I = 0.50 amp.

25. Silver, copper, aluminum, tungsten, salt water, glass, oil.

27. In terms of units: $[\rho] = \left[\dfrac{Nm^2\ sec}{coul^2}\right] = \left[\dfrac{kg\ m^3}{coul^2\ sec}\right].$

   In terms of fundamental dimensions: $[\rho] = \left[\dfrac{ML^3}{Q^2T}\right].$

29. (a) 0.050 volt/m.
    (b) 2.9×10^6 amp/m².
    (c) 5.9 amp.

31. Ratio of Cu/kg cost to Al/kg cost is 0.50.

33. (a) The resistance $R(=V/I)$ of a conductor at a given temperature is constant, independent of the value of the current (I) through the conductor.
    (b) 3.0 amp.

35. (a) A circuit element whose resistance depends upon the applied voltage.
    (b) Rectifiers (which conduct current only in one direction) are non-linear circuit elements.

37. 2.4 cents.

39. (a) Power $= I^2R$; 2.5×10^2 watts.
    (b) 7.5×10^4 joules.

41. (a) VI.   (b) I^2R.   (c) V^2/R.

43. 144 ohms.

45. 7.2×10^4 joules.

47. 1.9×10^2 amp.

49. Yes. Connect 20 such bulbs in series.

51. 2.4 ohms.

53. 1/3 ohm.

55. (a) 1.2×10^3 amp.   (b) 2.2×10^5 joules.

57. (a) 2.00×10^{-4} amp.   (b) 4.00 volts.
    (c) 2.18×10^{-4} amp.   (d) 5.45×10^{-5} amp.

59. If a finger inadvertently makes contact with a high-voltage terminal, the electric shock will be particularly severe. The bath water furnishes a low-resistance path to the grounded plumbing. The body resistance itself is abnormally low because wet skin offers much less resistance to the flow of current through it than dry skin.

61. A: 2.0 amp, *a* to *d*.   B: 0.
    C: 2.0 amp, *c* to *b*.

63. 2.7 amp up; 0.55 amp down; 2.2 amp up.

65. 80 volts, 90 volts.

S-1. 6.0 MeV or 9.6×10^{-13} joule; 1.7×10^7 m/sec.

S-3. (a) From Eq.(11.5): $\vec{E} = 0$ implies $\Delta V = 0$.
 (b) For two points on the surface of the conductor, $\Delta V = 0$. Therefore, for Δs parallel to the surface, $\Delta V/\Delta s = 0$, which implies that the component of $\vec{E}$ parallel to the surface is zero. Consequently $\vec{E}$ can have only a normal component.

S-5. The electric current at a given point in the wire will be established as soon as an electric field is set up at that point. The electric field is associated with an electromagnetic wave which travels along the wire with a speed which is almost the speed of light. "The push" ($\vec{E}$) travels much faster than the particles (electrons) which are being pushed.

S-7. (a) $\dfrac{\mathcal{E}^2 R}{(r+R)^2}$.
 (b) The curve goes through the points $0, 0$; $\tfrac{1}{4}r, 0.16\mathcal{E}^2/r$; $\tfrac{1}{2}r, 0.22\mathcal{E}^2/r$; $r, 0.25\mathcal{E}^2/r$; $\tfrac{3}{2}r, 0.24\mathcal{E}^2/r$; $2r, 0.22\mathcal{E}^2/r$; $3r, 0.19\mathcal{E}^2/r$.
 (c) r; $\mathcal{E}^2/4r$.

S-9. (a) 90 volts, polarity opposite to that of the applied EMF.
 (b) 900 watts. (c) 300 watts. (d) 0.75.

S-11. $\dfrac{R_1^2 + 2R_1R_2 + R(R_1 + R_2)}{R + R_1 + R_2}$

S-13. The idea that "30" should last twice as long as "15" is fallacious because the numbers do not refer to some expendable quantity contained in the fuses. The 15-amp fuse is "blowing" because the appliances are drawing current in excess of 15 amp. A 30-amp fuse might not blow but the circuit (wires, connections, etc.) which were presumably designed for a maximum of 15 amp would overheat and constitute a fire hazard.

S-15. (a) 4.0 ohms. (b) 6.0 ohms. (c) 24 volts.
 (d) 4.0 amp. (e) 16 volts. (f) 2.7 amp, 1.3 amp.
 (g) zero. (h) 8.0 volts. (i) 16 volts.
 (j) Through battery: 4.0 amp; through resistor between b and c: zero; through parallel resistor group: 4.0 amp.
 (k) zero.
 (l) $V_a = 16$ volts, $V_b = 0$, $V_c = 0$, $V_d = 0$.
 (m) 21 watts. (n) zero.

S-17. (a) 1.45×10^{-2} amp. (b) 1.61×10^{-3} amp.
 (c) 1.55 volts. (d) $(1.64 - 1.55)$ volt $= 0.09$ volt.
 (e) $0.09/1.55 = 6 \times 10^{-2}$.

S-19. At balance the two circuit points connected through R_G are at the same potential. If I_1 is the current through R_1 and R_3, and I_2 is the current through R_2 and R_4 then

$$I_1 R_1 = I_2 R_2, \quad I_1 R_3 = I_2 R_4.$$

S-21. (a) The relation follows from the general result given in Question S-19.
(b) 7.1×10^2 ohms.

CHAPTER 12

1. Yes. No.
3. The compass needle will be horizontal and perpendicular to the wire. If the current is to your right, the north pole of this compass needle will point toward you.
5. In an unmagnetized iron bar the different domains are aligned in different directions at random. When magnetized, there is a preferred direction of alignment of the domains in the bar.
7. (a) No.
(b) No. The magnetic force is perpendicular to the magnetic field.
(c) The direction of the magnetic field, the direction of the charge's velocity vector, and the sign of the charge.
9. 1.6×10^{-11} N, directed north.
11. $\vec{F}$ and $\vec{v}$, and $\vec{F}$ and $\vec{B}$ are always at right angles. $\vec{v}$ and $\vec{B}$ may have any angle between them.
13. 1.4×10^{-11} N directed south.
17. 2.8×10^9 revolutions/sec.
19. Using Eq. (12.5) one derives $M/M_p = T/T_p$.
21. A child on a swing being periodically pushed.
23. (a) 0.43 m.   (b) 4.4×10^{-8} sec; 2.3×10^7 hertz.
25. 6.0×10^{-3} N directed west.
27. (a) Direction of current.
(b) No. $\vec{F}$ is perpendicular to plane defined by I and $\vec{B}$.
29. (a) 2.8×10^{-3} m-N.   (b) Zero.
31. (a) A low-resistance shunt connected in parallel with a galvanometer.
(b) The shunt is removed and a high resistance is connected in series with the galvanometer.
33. 3.0×10^{-4} ohm.
37. 500 amp.
39. 4.8×10^{-2} weber.
41. (a) 1.6×10^{-3} weber.   (b) Zero.
43. Sense of induced EMF is counterclockwise.
45. 6.0×10^2 volts.
47. 2.94×10^2 volts.

49. 1370/138; step-up.

51. Right to left in wire at magnet gap.

S-1. (a) Superposition principle for magnetic fields.
(b) 8.1×10^{-2} weber/m², 60.3° south of east.

S-3. Closing switch in circuit P energizes electromagnet M which causes the spring-loaded arm C to move to left thereby closing light bulb circuit at X: Light goes on.

S-5. $\dfrac{mv^2}{2} = qV$ and $\dfrac{mv^2}{r} = qvB$, where $r = \tfrac{1}{2}D$. Substituting the expression for v from the first of these equations in the second we obtain the desired result.

S-7. 3.6×10^{-2} weber/m².

S-9. Force is 7.2×10^{-2} N directed north.
Acceleration is 1.8 m/sec², north.

S-11. It takes a time L/v for a charge at one end of wire to reach the other end, a distance L away. Thus Nq is the charge passing an end in the time L/v; that is

$$I = \dfrac{\Delta q}{\Delta t} = \dfrac{Nq}{(L/v)}, \text{ or } v = \dfrac{LI}{Nq}.$$

Now use Eq. (12.1): $F = qvB \sin \theta$. This is the force on a single charge q, so for N charges we have $F = NqvB \sin \theta$. In this expression substitute the result obtained for v, whereupon the desired equation is obtained.

S-13. 0; 3.0×10^{-3} m-N; 6.0×10^{-3} m-N.

S-15. Perpendicular face, 2.4 webers; sloped face, 2.4 webers; horizontal face, 0.

S-17. (a) Induced EMF sets up currents in a direction such that the force exerted by the magnetic field on the conductor carrying these currents is directed so as to oppose the motion of the conductor.
(b) Increase $\vec{B}$ of magnet (make magnet gap as narrow as possible); use material with lower resistivity (e.g., silver) for pendulum plate; make pendulum plate thicker; increase speed of initial pendulum swing.
(c) Increased thermal energy of swinging copper pendulum plate.

S-19. (a) 4.8×10^{-3} volt.
(b) EMF directed along moving cylinder from west to east.

S-21. 1.9×10^{-3} volt.

S-23. F_{ext} must be equal but oppositely directed with respect to the force BIL on the conductor due to its motion through the magnetic field. Hence external power supplied $F_{\text{ext}} v = BILv$. The EMF induced in the circuit is, according to the result of Question S-22,

$$\mathcal{E} = vBL.$$

Thus electrical power converted from mechanical power is given by

$$P = F_{\text{ext}}\, v = BILv = \mathcal{E}I.$$

CHAPTER 13

1. In DC the direction of the drift—that is, the average motion—of the electrons remains the same, whereas in AC this drift reverses direction periodically [see Fig. 13-1(a)].

3. Radio waves, generated by an alternating current, carry away energy.

5. $v = 0$, at these times in seconds: 0, 1/120, 1/60, 1/40, 1/30. Positive peak voltages occur at (1/240) sec, (5/240) sec.

7. (a) (1/15) sec.
   (b) Noticeable flickering of the light.

9. 340 volts.

11. (a) Operation of heater depends only on the Joule heating effect, and for given numerical values of voltage and current, this heating will be the same for DC as for AC.
    (b) 9.5 amp effective (or RMS) value.

13. The real generator has a soft iron core, many turns per armature coil, and usually more coils. The magnetic field is provided by electromagnets rather than permanent magnets. Moreover, it is common practice to keep the armature stationary and to rotate the electromagnets instead.

15. (a) 2.0×10^3 volts.   (b) 2.0×10^4 watts.

17. $C = q/v$, where q is the magnitude of the charge on either of the capacitor plates and v is the voltage across the plates. Farad, or microfarad.

19. 1.4×10^{-2} coul.

21. By moving one set of plates in or out with respect to the other set the effective area of the capacitor is changed.

23. 7.96 Hz.

25. The relation to be plotted is $X_C f =$ constant which is a hyperbola asymptotic to the X_C and the f axes. As f gets very small X_C becomes very large and as f becomes very large X_C becomes very small.

27. 212 amp.

29. (a) The relation to be plotted is $X_L =$ constant $\times f$ which is a straight line. When f is zero X_L is zero and X_L increases in magnitude linearly with f.
    (b) At low frequencies X_C is large while X_L is small, and for high frequencies X_C is small while X_L is large.

31. (a) Increasing the inductance would increase the inductive reactance and hence, in this case, increase the impedance. The current would thus be reduced and the light would be dimmer.
(b) Having an iron core which could be moved in and out of the inductance coil.

33. (a) (i) $Z = V/I$. But if R is the only component, then $R = V/I$, from Eq.(13.1). Thus $Z = R$.
 (ii) If C is the only component, then $X_C = V/I$, from Eq. (13.5). Hence $Z = X_C$; but $X_C = \dfrac{1}{2\pi fC}$, thus $Z = \dfrac{1}{2\pi fC}$.
 (iii) If L is the only component, then $X_L = V/I$, from Eq. (13.6). Hence $Z = X_L$; but $X_L = 2\pi fL$, thus $Z = 2\pi fL$.
 (b) (i) $Z = \sqrt{R^2 + (X_L - X_C)^2}$ [Eq.(13.8)]; with a pure resistance, $X_L = X_C = 0$; hence $Z = \sqrt{R^2 + (0-0)^2} = R$.
 (ii) With a pure capacitance, $R = X_L = 0$; hence
$$Z = \sqrt{0 + (0 - X_C)^2} = X_C = \dfrac{1}{2\pi fC}.$$
 (iii) With a pure inductance, $R = X_C = 0$; hence
$$Z = \sqrt{0 + (X_L - 0)^2} = X_L = 2\pi fL.$$

35. 25 µf.

37. (a) 23 ohms. (b) 1.9×10^2 volts.

39. (a) 1.6×10^3 hertz. (b) 12 volts.

41. $P = 2.2$ kW, given, but $IV = (20\ \text{amp})(110\ \text{volts}) = 2.2$ kW. Thus from $P = IV\ (R/Z)$, one sees that the power factor, $R/Z = 1$. This implies $Z = R$; that is, the circuit is purely resistive.

43. (a) RMS values.
(b) Assuming an ideal capacitor and an ideal inductor, the power dissipated in each of C and L is zero. In R, we get $P = IV = 1.5$ kW.
(c) 1.5 kW.
(d) We know $V_L \neq V_C$. But $V_L = IX_L$ and $V_C = IX_C$. Therefore $X_L \neq X_C$. Consequently the impedance $\sqrt{R^2 + (X_L - X_C)^2}$ is greater than R. Therefore the input voltage $V = IZ$ is greater than IR, the voltage across the resistor.

45. (a) 50.0 kV.
(b) 50.0 amp.
(c) 5.00 kW if power factor is 1.
(d) 100% efficiency; power factor in the secondary circuit is 1.

47. 62.5 volts/amp.

51. (a) Grid and cathode. (b) 15.
(c) About 15 volts.

53. The birds are not part of a complete circuit.

55. About 15 mA. This is larger than is normally likely, but it is quite possible to exceed this value.

S-1. The wave form for the tuning fork would resemble that of Fig. 10-3(b) with a time interval of (1/256) sec between successive peaks.

The different *quality* of the sound of different instruments sounding the same fundamental frequency would have various periodic variations, characteristic of the instrument, superimposed on the basic 256 Hz wave form.

S-3. $V_{total} = V_1 + V_2 + V_3 + \ldots + V_n$, so

$$\frac{q}{C} = \frac{q_1}{C_1} + \frac{q_2}{C_2} + \frac{q_3}{C_3} + \ldots + \frac{q_n}{C_n}, \text{ where } q = q_1 = q_2 = q_3 = q_n,$$

$$\frac{1}{C} = \frac{1}{C_1} + \frac{1}{C_2} + \frac{1}{C_3} + \ldots + \frac{1}{C_n}.$$

S-5. (a) Inductive   (b) 14.0 ohms.   (c) 3.72×10^{-2} henry.

S-7. 10^3 volts.

S-9. With AC, eddy currents are induced in the magnet's core. This energy is ultimately dissipated in heating the core. Also some energy used in continually realigning the magnetic domains in the core is ultimately evidenced in increased core temperature. This latter energy loss is known as the hysteresis loss.

S-11. 6.85 μf.

S-13. 3.00×10^3 volts.

S-15. Transformer supplies two voltages: the voltage used in charging the battery and another voltage used in heating the filament of the vacuum diode. The diode conducts only when transformer voltage polarity is as required for battery. The rheostat in the charging circuit is to control the charging current which is monitored by the ammeter.

S-17. In case of any electrical defect the user is protected in that he will not become part of a circuit carrying AC current to a ground connection.

S-19. (a) 90 volts.
(b) 120 volts.
(c) On a diagram such as that of Fig. 13-34 the 120 volts is indicated by an arrow (phasor) pointing down and the 90-volt phasor points to the right.
(d) 150 volts.   (e) Voltage lags current.

S-21. (a) Inductive   (b) Yes.
(c) Voltage leads current.   (d) 21 ohms.

S-23. (a) 96 volts.
(b) 144 volts.
(c) 204 volts.

S-25. (a) 1.13×10^3 ohms, 442 ohms, 796 ohms, 0.139 amp, 0.503.
(b) 55.6 volts, 157 volts, 61.4 volts.
(c) 7.73 watts.

S-27. (a) 74.0 volts. (b) 90.0 volts. (c) 116 volts.
(d) The voltage across the inductor is represented by a phasor pointing upward while the voltage across the capacitor of Question S-26 is represented by a phasor pointing down.
(e) Voltage leads current. (f) V_L leads V_R.
(g) 39.4°.

S-29. (a) The phasor diagram shows the phasor representing the voltage at 60° above the horizontal axis.
(b) 1.74 ohms. (c) 1.0 ohm.

S-31. (a) No. Yes, because the voltages may be out of phase such that the phasor addition results in a sum of magnitude less than that of one of the component voltages.
(b) 20.

S-33. (b) From the phasor diagram it can be seen that
$$V = 2(v \cos 30°) = 2v \frac{\sqrt{3}}{2} = \sqrt{3}\, v.$$

CHAPTER 14

1. As in Fig. 14-1(c).
3. (a) 8.5 watts. (b) 113 volts/m.
5. (a) 88×10^6 hertz.
   (b) Wavelengths are shorter for FM.
   (c) 3.39 m to 3.40 m.
7. Length of wave train = 3.0 m.
   Number of wavelengths = 5.0×10^6.
9. The reflected light is at least partially polarized in a horizontal direction. Light with this polarization will be absorbed in sunglasses made of polaroid and oriented so that the polaroid transmission axes are vertical.
11. $I_t = 1.33 \times 10^{-3} E_\parallel^2 = 1.33 \times 10^{-3} E_0^2 \cos^2 \theta = I_i \cos^2 \theta$.
13. (a) 1.9 watts/m². (b) 0.
15. $0.21\, I_0$
17. (a) Maximum amplitude = 0.80 volt/m.
    Intensity = 0.85×10^{-3} watt/m².
    (b) Minimum amplitude = 0.20 volt/m.
    Intensity = 0.053×10^{-3} watt/m².

19. Start with:
$$I = kE^2 = k(E_1 + E_2)^2$$
$$= k(E_1^2 + E_2^2 + 2E_1E_2 \cos\theta), \text{ where } \theta = 0, \text{ so } \cos\theta = 1.$$

21. $LL_1 = 2.0 \times 10^2$ m.

23. 5.88×10^{-7} m.

25. 2.8 Å. Very soft x-rays.

27. (a) Light emerging from double slit (Fig. 14-9) if the incident light came from a preceding narrow single slit in front of a light source. (b) Two light bulbs.

29. Approximately 5×10^{-2} degree.

31. 2×10^{-2} m.

33. (a) D(radio waves) = 1.0×10^3 m.
    D(light waves) = 1.7×10^{-6} m.
    (b) Significant diffraction will occur only for apertures and obstacles smaller than the values of D computed in part (a). Since buildings provide obstacles and apertures with a width of the order of 10^2 meters, there will be appreciable diffraction of radio waves (10^2 m < 10^3 m) but negligible diffraction of light waves (10^2 m > 10^{-6} m).

35. Speed of light v in vacuum = c, hence
$$n = \frac{c}{v} = \frac{c}{c} = 1.$$

37. 1.33.

39. 3760 Å.

41. 34.8°. No.

43. Wave theory, giving $n = c/v$, requires that light waves travel more slowly in water than in a vacuum (since $n = 1.33$ for water). But a refractive index greater than 1 would require, for particles obeying Newtonian mechanics, a particle speed greater in water than in a vacuum.

45. 1.0×10^9 light-years.

S-1. $I_t = I_i \cos^2\theta$. Since unpolarized light has $\vec{E}$ randomly orientated we use the average value of $\cos^2\theta$ ($=\frac{1}{2}$), so $I_t = \frac{1}{2}I_i$.

S-3. (a) $\cos 0° = 1$, so $E_0^2 = E_{01}^2 + E_{02}^2 + 2E_{01}E_{02}$
    $= (E_{01} + E_{02})^2$;
    hence $E_0 = E_{01} + E_{02}$.
    (b) $\cos 180° = -1$; following procedure of (a) yields required result.

S-7. 3.4 m.

S-9. (a) $(\Delta f)_1 = (v/c)f$ gives the frequency shift detected by an observer in the car. The signal returned to the monitor originates with a frequency shift of $(\Delta f)_1$ but since the car is moving toward the monitor at speed v, the already frequency-shifted signal is observed to shift an amount $(\Delta f)_2 = (v/c)[f + (\Delta f)_1]$ which is approximately equal to $(v/c)f$. Therefore the total frequency shift with respect to the original signal is $(\Delta f)_1 + (\Delta f)_2 = 2f(v/c)$.

CHAPTER 15

1. The fact that we can't see around corners (without the aid of mirrors) supports the proposition of the rectilinear propagation.
3. (a) Penumbra (see Fig. 15-4).
   (b) Because the light source (sun) is large compared to the obstacle (moon) in the way of the light to the "screen" (earth).
5. The light travels through one medium only, that is, the speed of light is constant.
7. 2.00 m.
9. 0.8 m.
11. *Real image.* Light rays originating from object actually pass through image. Therefore a real image could be captured on a screen.
    Virtual image. Light rays originating from object are bent so as to appear to come from the image, but they do not pass through the image. A virtual image could therefore not be captured on a screen.
13. (b) Virtual image. No. The eye-brain system "projects" the rays backward to give the illusion that the light signal is coming from behind the mirror.
15. See Fig. 15-12. The distance from mirror to point C is 0.50 m, and the distance from the mirror to F is 0.25 m.
17. (a) Real image is formed farther from the mirror than point C.
    (b) No image is formed (one sometimes says that the image is formed at infinity).
19. (b) "Apparent depth" is less than actual depth.
21. Total internal reflection refers to the situation in which light strikes the interface between two media at such an angle that all the incident light is reflected, none is transmitted.
23. *Hint:* See Fig. 15-16 and imagine the candle inside the beaker.
27. Looming causes the sun to be seen before it actually rises above the horizon.
29. Total internal reflection.
31. (a) 30 cm.
    (b) Real and inverted.
33. Since the image is real, it will be inverted with respect to the object. The magnification is therefore -2. Hence $f = 0.67$ m.

35. (b) $q = -4.8$ cm. Image is smaller than object, erect, and virtual.
    (c) 0.72 cm.
    (d) 0.4.
37. No, because the relation between the curvature of the lens and its focal length depends upon the refractive indices of both the lens and that of the medium in which the lens is immersed.
39. 15.6 cm.
41. 1.7 cm.
43. $\frac{1}{f} = \frac{1}{f_1} + \frac{1}{f_2}$ [Eq. (13.4)], but $D = \frac{1}{f}$, hence $D = D_1 + D_2$.
45. 40 cm.
47. *Hint:* See Figs. 15-33(a) and 15-34.
49. Reflection takes place both at air-coating interface and at the coating-glass interface producing two reflected waves of approximately equal amplitude. If the waves reflected from the coating-glass interface travel half a wavelength in going from the air-coating interface to the coating-glass interface and back to air-coating interface, then destructive interference occurs at the air-coating interface with the waves reflected at that interface. Hence if the coating is 1/4 the wavelength of the light in the coating, the net effect is no reflection at the face of the lens and the light normally reflected is instead transmitted.
51. Say we select light with a wavelength of 6000 Å in air. From Eq. (14.18) the wavelength in the coating would be given by
$$\lambda = \frac{\lambda(\text{in air})}{n} = \frac{6000 \text{ Å}}{1.50} = 4000 \text{ Å}.$$
Therefore the required coating thickness is
$$(\lambda/4) = \frac{4000 \text{ Å}}{4} = 1000 \text{ Å}.$$
53. (a) 15 diopters.
    (b) 6.7 cm.
55. (a) 24 mm. (b) 42 diopters.
57. (a) Mental extrapolation based upon past experience with such objects at close-hand, coupled with depth perception.
    (b) Depth perception, i.e., estimation of distances.
59. (a) 5.00 cm.
    (b) About 5.05 cm.
    (c) In spite of the drastically reduced object distance, the image would still be fairly well defined at the film.
    (d) Not normally if a very sharply defined image is required; however, if the light is sufficiently bright and there is no objection to a fairly long exposure time, then the size of the opening in diaphragm could be reduced to the point at which "pinhole optics" would pre-

vail and a fairly well defined image could be obtained even at very short object distances.

61. 9.2×10^{-3} m.

63. $f/4$ is "faster" because it has a larger diameter relative to its focal length than an $f/11$ lens, and hence the $f/4$ lens provides a greater light intensity at the film.

65. The camera with the five-lens system is much more in need of lens coatings than is the simpler camera because reflection losses occur at each of the many interfaces of the multiple-lens system.

67. A continuous spectrum displaying the rainbow colors.

69. 201.

71. (a) The shorter wavelength of ultraviolet light would permit distinguishing small details which would be blurred by diffraction if the longer wavelengths of white light were employed. Therefore ultraviolet light would make higher magnification possible.
(b) As far as vision is concerned, the eye is not sensitive to ultraviolet light, and moreover ultraviolet light is harmful to the eye so some device which transforms the ultraviolet light image into a visible image is required. Further, ultraviolet light is strongly absorbed by the common optical material, crown glass; quartz lenses would have to be employed.

73. 3.

75. (a) Total internal reflection.
(b) The image appears in the same orientation as the object.

77. The relative brightness is given by the square of the ratio obtained by dividing the diameter of the objective by the magnification. Since $\left(\dfrac{50}{7}\right)^2$ is greater than $\left(\dfrac{30}{8}\right)^2$, the 7×50 glasses are better for night vision.

S-1. 3.13×10^6 m, $3.13/12.7$.

S-3. Three images, lying on a circle centered at the junction of the mirrors and having a radius equal to the distance from this junction to the object.

S-5. (a) $\dfrac{x}{d} = \dfrac{x+a}{d} - \dfrac{a}{d} = \tan(\alpha + 2\theta) - \tan \alpha$.

(b) $\dfrac{x}{d} = \tan(\alpha + 2\theta) - \tan \alpha \approx \alpha + 2\theta - \alpha \approx 2\theta$, so $\theta \approx \dfrac{x}{2d}$.

S-7. -15 cm.   S-9. $50°, 31°$.

S-11. As water is added the light rays are sufficiently refracted to permit the viewer in a sense to "see around the corner," that is, over the lip of the container.

S-13. Place a light source in a medium, such as water, with $n > 1$, then determine the critical angle.

S-15. (a) As close as possible to that of water.
(b) Indices are about the same.

S-17. Use the fact that for small θ, $\sin \theta \approx \theta$.

S-19. 0.001.

S-21. The essential safety feature is that the fiber optical coupling provides electrical isolation between the patient and the final monitoring device.

S-23. (a) 30.0 cm.   (b) Fourth case in Fig. 15-26.

S-27. (a) Double convex.
(b) $f(\text{double convex}) = \frac{1}{2} f(\text{plano convex})$.

S-29. (a) 24.0 cm.
(b) Reduction of spherical aberration; reduction of chromatic aberration.

S-31. (a) 1000Å   (b) 5000 Å.

S-33. 71.3 mm.

S-35. (a) The rays of light which would be reflected from the patient's retina such as to enter the examiner's eye must follow approximately the same paths as those which need to enter the eye to illuminate the retina: The examiner's head gets in the way.
(b) Light reflected from the glass plate enters the patient's eye. The plate must be only partially reflecting so some light from the retina will reach the examiner.
(c) Acts as a magnifying glass.

S-37. 1.3×10^2.

S-39. (a) From the definition of angular magnification it can be shown that if the angle subtended by the object when it is 25 cm from eye is θ_1 and the angle subtended when object is placed near focal point of magnifying lens is θ_2 then $M = 25/f$ with f in cm, so $M = D/4$.
(b) 15.

S-41. 680.

S-43. (a) Hyperopia.   (b) See Fig. 15-43(c)   (c) 2.0.

CHAPTER 16

1. (a) Violet, green, red.
   (b) Red, green, violet.
   (c) Red, green, violet.
3. 6200 Å, red.
5. 6.3×10^{19}.
7. 2.1×10^{21} photons/m² sec.
9. (a) No. Red photons have less energy than yellow photons, and we know that the yellow photon energy is insufficient to cause ejection of electrons.
   (b) Yes. Violet photon energy is greater than green photon energy.
11. (a) 4.23 eV. (b) 5.77 eV.
13. (b) 0.124 Å.
15. 0, 0.0032, 0.024, 0.048.
17. 2.53×10^5 eV/c to the right, 1.27×10^5 eV/c to the left, 3.80×10^5 eV/c to the right.
19. 10.2 eV; 1.22×10^3 Å; ultraviolet.
21. 3.12 eV, 3.97×10^3 Å.
23. 1.9 eV.
25. 12.1 eV, 6560 Å is red.
29. The photons arrive at the screen, one at a time, at random intervals. Many photons have struck the photographic plate in the region of an intensity maximum but very few photons have struck near an intensity minimum.
31. 4.31×10^{-6} eV.
33. 1.7×10^{-28} m.
35. The probability of finding an electron in any given region near an atom can be visualized by imagining a charge cloud with a density in each region proportional to this probability; that is, the greater the probability of an electron being in a given region the greater the charge cloud density in that region.
37. $\Delta p \approx 2m\alpha c$,
    $\Delta x \approx 3\hbar^2/mke^2$.
    $\Delta x \Delta p \approx 6\hbar^2 \alpha c/ke^2 = 6\hbar$
    $\qquad\qquad = 6h/2\pi$.
    Therefore
    $$\Delta x \Delta p \approx h,$$
    which is consistent with the Heisenberg uncertainty relation.

41. The values of n, ℓ, m_ℓ, m_s for each state are

$(4, 3, 3, \tfrac{1}{2})$	$(4, 3, 3, -\tfrac{1}{2})$
$(4, 3, 2, \tfrac{1}{2})$	$(4, 3, 2, -\tfrac{1}{2})$
$(4, 3, 1, \tfrac{1}{2})$	$(4, 3, 1, -\tfrac{1}{2})$
$(4, 3, 0, \tfrac{1}{2})$	$(4, 3, 0, -\tfrac{1}{2})$
$(4, 3, -1, \tfrac{1}{2})$	$(4, 3, -1, -\tfrac{1}{2})$
$(4, 3, -2, \tfrac{1}{2})$	$(4, 3, -2, -\tfrac{1}{2})$
$(4, 3, -3, \tfrac{1}{2})$	$(4, 3, -3, -\tfrac{1}{2})$

$(4, 2, 2, \tfrac{1}{2})$	$(4, 2, 2, -\tfrac{1}{2})$
$(4, 2, 1, \tfrac{1}{2})$	$(4, 2, 1, -\tfrac{1}{2})$
$(4, 2, 0, \tfrac{1}{2})$	$(4, 2, 0, -\tfrac{1}{2})$
$(4, 2, -1, \tfrac{1}{2})$	$(4, 2, -1, -\tfrac{1}{2})$
$(4, 2, -2, \tfrac{1}{2})$	$(4, 2, -2, -\tfrac{1}{2})$

$(4, 1, 1, \tfrac{1}{2})$	$(4, 1, 1, -\tfrac{1}{2})$
$(4, 1, 0, \tfrac{1}{2})$	$(4, 1, 0, -\tfrac{1}{2})$
$(4, 1, -1, \tfrac{1}{2})$	$(4, 1, -1, -\tfrac{1}{2})$

$(4, 0, 0, \tfrac{1}{2})$	$(4, 0, 0, -\tfrac{1}{2})$

There are $32 = 2 \times 4^2$ states.

43. (a) The set of states (of an electron in an atom) associated with a given $n\ell$ energy level is called a *subshell*.
(b) Table 16-1 indicates that for each ℓ value there are $2\ell + 1$ different m_ℓ values, and that for each m_ℓ value there are two different m_s values. Therefore for each ℓ value there are $2(2\ell + 1)$ different sets of values of m_ℓ and m_s; that is, for a given ℓ there are $2(2\ell + 1)$ different possible states.

45. $1s^2$ $2s^2$ $2p^6$ $3s^2$.

47.
Helium	$1s^2$	Neon	$2p^6$
Argon	$3p^6$	Krypton	$4p^6$
Xenon	$5p^6$	Radon	$6p^6$

49. *From Question 47*: Except for helium, the inert gases are characterized by a full p-subshell as their outermost subshell. Because this structure is particularly stable, these atoms are relatively inert. All these elements are listed in the last column of the periodic table.
From Question 48: The elements Li, Na, and K have a full p-subshell followed by an s-subshell occupied by a single electron. Since this s-electron is easily detached, these elements are chemically active (valence is $+1$). They are listed in the first column of the periodic table.

Fluorine and chlorine have five outermost electrons in a p-subshell. One more electron is required to form the particularly stable full p-subshell. These elements are chemically active (valence is -1). They are listed in the seventh column of the periodic table.

53. In diamond the empty conduction band lies 6 eV above the full valence band. A visible photon has an energy of less than 3.5 eV and therefore cannot furnish enough energy to promote an electron from the valence to the conduction band. Therefore such photons are not absorbed in diamond.

59. Photoelectric effect, Compton effect, pair production.

61. The water molecule has no resonant frequencies in the band of frequencies corresponding to visible light but has resonant frequencies in both the ultraviolet and the infrared.

63. Yes. The analysis of Section 16.11 shows that range R is related to mass M by $M = h/cR$ or $R = h/Mc$. The smaller the mass of the exchanged particle, the greater the range of the interaction.

S-1. $h = 4.00 \times 10^{-5}$ eV-sec $= 6.40 \times 10^{-34}$ joule-sec; $W = 1.00$ eV.

S-3. (a) 0.012 Å.
(b) 1.03×10^6 eV/c, 3.44×10^5 eV/c at right angles to direction of incident photon.
(c) 1.09×10^6 eV/c at 18.5° with direction of incident photon.

S-7. $R(\text{theory}) = 1.10 \times 10^7$ m^{-1}, i.e. good agreement.

S-11. $\frac{1}{2}$ Bohr radius, or 0.26 Å.

S-13. $mvr = n\hbar = \dfrac{n\lambda mv}{2\pi}$, so $2\pi r = n\lambda$.

S-15. Use relation given in Question S-14. When mc^2 is small compared to K, we have (approximately) $K = E = cp$, so $\lambda = h/p = hc/K$.

S-17. (a) 6.63×10^{-20} kg-m/sec.   (b) 19.9×10^{-12} joule.

S-19. (a) 2.41 eV, 2.54 eV.   (b) blue, green.
(c) Yes. Red objects absorb chiefly wavelengths other than red.
(d) 2.85×10^{-2} joule.   (e) 36.9×10^{15}, 35.0×10^{15}.

Index

Page number in bold denotes major discussion of topic.

Absolute time, **220**
Absolute zero, **278**
AC, 455, 467, **470**
    advantage of, **470**
    production of, **474**
    three-phase, 508
    voltage addition, 480, **496**
Acceleration, 41, **47**
    angular, **127**
    centripetal, **54**, 65, 127
    due to gravity, 49, 63
    vector, **53**
Accelerator, particle, 21, **187**, 222, 442, 449
Accommodation, **582**
Activation analysis, **32**
Activity, **18**, 24
Air molecule, energy of, 157
Algebra, **673**
Algebraic sum, **675**
Alhazen, 517, 553
Alpha decay, **13**
Alpha particle, 5, 8, 12
Alternating current (AC), 455, 467, **470**
Ammeter, 397, 412, **446**
Ampere (unit of current), **397**, **448**
Ampère, A. M., 433
Amplifier, voltage, **491**
Amplitude, **130**, 271, 529, 619
amu, **173**, 176
Angstrom unit (Å), 617, **691**
Angular
    acceleration, **127**
    momentum, **128**, 616; conservation of, 68, **129**; intrinsic (*Sℏ*), 129; natural unit (ℏ), 129, **617**

velocity, **126**
Anode, 77, 486
Antineutrino, **177**, 178
Arago, D., 518
Archimedes, 167, **243**
Archimedes' principle, **251**
Aristotle, 41, 49, 57, 250
Armature, 474
Arrow of time, **333**, 352
Aspirator pump, **258**
Astigmatism, **585**
Astronomy, **542**
Atmosphere, standard, **245**, 250
Atmospheric pressure, 245, 690
Atmospheric refraction, **566**
Atom, **6**, **79**
    bomb, 181, 182, 185
    copper, 398
    hydrogen, 6, 19, 616, 625, **634**
Atomic
    mass number, **5**, 16
    mass unit (amu), **173**, 176
    number, **5**
    representation, quantum mechanical, 19
Atwood's machine, 114
Average life, 633
Avogadro, A., 273
Avogadro's hypothesis, **273**
Avogadro's number, **271**, 272, 273

Back EMF, **481**
Ballistocardiograph, **150**
Balloon, buoyancy of, 252
Balmer, J., 624
Bar, **245**

Bardeen, J., 469
Barometer, **249**, 260
Baryon, 190; number, 191
Basal metabolic rate, **295**
Battery charger, **505**
Battery tester, 253
Beams, J. W., 216
Beats, **365**
Becquerel, H., 1
Békésy, G. von, 371
Berman metal locator, **462**
Bernoulli, D., **244**, 269
Bernoulli's equation, 254, **255**, 257, 259
Beta decay, **13**, 178, 626
Beta particle, 4, 9
Betatron, 449
Bevatron, 155, 157
Binding energy, **178**, 180, 625, 650
    nuclear compared to atomic, 179
    and nuclear stability, **178**
Binocular vision, **583**
Binoculars, **597**; brightness, **598**
Bioelectricity, **430**
Biological
    half-life, 30, **32**
    tissue and ionizing radiation, **26**
Biomicroscopy, **611**
Biot, J. B., 433, 447
Biot-Savart law, **447**, 449; constant (*K*), **447**, 449
Black, J., **297**, 305
Blood pressure, 249, **266**
Blood test, 253
Bohr, N., 11, 542, **615**, 661
Bohr
    atom, 616

731

frequency condition, **625**
quantum hypothesis, **661**
radius, **636**, 690
Boiling, **317**
Boltzmann, L. E., **269**, 270
Boltzmann's constant (k), **272, 273,** 347, 690
Born, M., 239, **633**
Bose-Einstein statistics, 207
Bound
    state, 79
    system, **180**
Boyle, R., 271
Boyle's law, **271**
Bragg, Sir W. H., 534, 535
Bragg, Sir W. L., 534, 535
Brahe, Tycho, **68**, 555
Brattain, W. H., 469
Breeder reactor, 185
Bremsstrahlung, **622**
Brown, Robert, 276
Brown, R. H., 536
Brownian motion, 208, **276**
Bubble chamber, 9
Bunsen, R., 542
Buoyancy, center of, 252
Buoyant force, **251**

Calculus, 43, 47, 55
Calorie, 165
Calorimetry, **286**, 295
Camera, **586**
    f-number, **588**
    pinhole, **580**
    shutter speeds, 587
Cancer therapy, 18, 29, 627
Capacitance, **477**
Capacitive reactance, **479**
Capacitor, **476**; charging and discharging, **478**
Carbon dating, 27
Carbon-14, 27
Carnot, S., 327
Carnot efficiency, **339**
Carnot engine, **339**
Cartesian diver, 263
Cataract cryoextractor, **354**

Cathode, 77, 486; ray oscilloscope, **78**
Cavendish, Lord, 61
Cavitation, 376
Celsius and Fahrenheit temperature scales, **163**
Center
    of buoyancy, 252
    of gravity, **120**
    of mass, **106**, 121, 129; law of motion, **107**; of a system of particles, **106**
Centigrade temperature scale, 163
Centrifugal force, **214**
Centrifuge, **215, 238**
Centripetal
    acceleration, **54**, 65, 127
    force, **66**
Cerebrospinal puncture, 263
Cerenkov, P. A., 374, **375**
Cerenkov radiation, 375, **385**
CERN (European Council for Nuclear Research), 189
Chadwick, J., 21
Chain reaction, **182**
Chance, 19, 626
Change, rate of, **43**
Charge
    cloud, 635
    electric, **3, 70**, 403
    -to-mass ratio of electron (e/m), **77, 441**
    negative, 3
    positive, 3
Chemical behavior, 6
Chladni, E. F., 357
Choke, **481**
Chromatic aberration, **579**
Chromatic dispersion, **564**
Circuit
    breaker, 415, **459**
    electric, **399, 410**
    in motion, **454**
    multiloop, **419**
    parallel, **413**
    series, **412**
    single-loop, **411**
Clausius, R., 269, 327, **328**
Clausius equation, **329**
Clinton pile, 185

Cloud chamber, 9, 20, **319**
Cobalt-60, 18, 24, **626**
Cockcroft, J. D., 21
Coherence, of light, **535**
Colladen, J. D., 358, 363
Compass, magnetic, 433
Component of vector, **52, 108**, 687
Compound nucleus, 20
Compression, 362
Compton, A. H., 154, 624
Compton scattering, **624**, 650
Conant, James, 154
Condensation, 313
Condenser R-meter, **91**
Conductor, electric, 389, 395, **398**, 644
Conic section, 64, 79
Conservation
    of angular momentum, 68, **129**
    of electric charge, **15**, 400
    of energy, 153, 170, **171**
    laws, **191**
    of linear momentum, **102**, 105
    of mechanical energy, 153, **160**
Contact force, 80
Contact lenses, 585
Continuity, equation of, **254**
Coolidge, W. D., 509
Copernicus, 42, 67
Coriolis force, **217**
Correspondence principle, **662**
Cosmic rays, 26, 526
Coulomb, C. A., 61, 74
Coulomb
    (coul), unit of electric charge, **70, 448**
    force, 74, 75, 79, 80
    potential energy, 180
Coulomb's law, **74**; constant (k), **71**, 75, 449
Cowan, C. L., 177
Critical
    angle, 565
    mass, **183**
    point, 314, **317**
Cryosurgery, **352**
Curie, M., **14**, 37
Curie, P., 14
Curie (Ci), unit of radioactivity, **18**
Current

alternating, 455, 467, **470**
conventional direction, **396**
direct (DC), 389, **395, 488**
effective value, **472**
electric, **397, 400, 403**
electricity, **395**
maximum value, **471**, 473
RMS value, **473**
Cycle, **130**, 360
Cyclic process, 337
Cyclotron, 188, **442**

Dating, with radioisotopes, 27
Daughter nucleus, 13
Davisson, C. J., 615, 627
deBroglie, Louis, 615, **627**
deBroglie wavelength, **627**
Decay, exponential, **17**, 479
Decay, radioactive, 13, **17**, 175, 626
Decibel, 368
della Porta, G., 553
Delta, 44
Density, **246;** water, maximum, 302
Depth of focus, **587**
Descartes, R., 541, 595
Deuterium, 7, 186
Deuteron, 7, 173, 186
Dew point, **318**
Dewar flask, 304
Diathermy, **550**
Dielectric, 398
Diffraction, 365, 517, 518, **537**
  grating, 533
  spreading, **365, 538**
Dimensions, **114**
Diode, 77, 486
Diopter, **576**
Dirac, P. A. M., 616, 633, 652
Direct current (DC), 389, **395, 488**
  pulsating, **488**
  steady, 389, 395
Disintegration, probability of, **18**, 174
Disintegration, spontaneous, **174**
Dispersion of light, **541**, 564
Dissipation of mechanical energy, **162**
Dissociation energy of molecules, 197
Domain, magnetic, **437**

Doppler, J. C., 371
Doppler effect
  light, **542**
  sound, **371**
  ultrasonic, in medicine, **384**
Dry cell, 401, 410
Dry ice, 315
Dulong and Petit, 289
Dynamics, **55**, 253
Dynamo, 410, **455, 474**

Ear, **370**
Earth
  centripetal acceleration of, 212
  its ecosystem, **352**
  mass, 690
Echoencephalography, **377**
Eclipse, solar, **556**, 581
Eddington, Sir A. S., 98, 345
Eddy current, 460, 461
Edison, T. A., 467
Effective current, **472**
Effective voltage, **472**
Einstein, A., 94, 170, 203, **207**, 239, 277, 289, 520, 615, 620
Einstein law for addition of velocities, **223, 227**
Elastic limit, **135**
Electric
  analogy, **401**
  and gravitational analogues, table of, **71**
  charge, **3, 70,** 403; circular motion in magnetic field, **440**; conservation, **15,** 400; mechanism of transfer, **395**; momentum measurement, **441**
  circuit, **399,** 410
  conductor, 389, 395, **398, 644**
  current, **397, 400, 403**
  energy, **408,** 454
  field, E, **71,** 73, 391, 394; between parallel plates, **72**
  force, 3, **70,** 72, 80; in atoms, **79**
  generator, 410, **455, 474**
  meter, **445**
  motor, 410, **434, 447**

  potential, **391,** 418; high, 392; low, 392
  potential energy, 159, 180, 391, **392**
  power, **408,** 410, 454
  shock, **415,** 416, **494, 511**
Electrical hazards
  biological damage, **513**
  in patient care areas, **511**
Electricity, current, **395**
Electricity, static, **494**
Electrocardiogram, **429,** 513, 514
Electrode, 76, 486
Electroencephalogram, **429**
Electromagnet, **438**
Electromagnetic
  induction, **434**
  interaction, 80
  radiation, 5, **524**; interaction with matter, **650**
  spectrum, 5, **523**
  waves, 519, **520**
Electromagnetism, **433**
Electrometer, 91
Electromotive force, 401
Electron, **3,** 13, 79
  charge-to-mass ratio, **77, 441**
  cloud, **19**
  configuration, 6, **640**
  free, 398
  microscope, **592**
  -positron annihilation, **172**
  rest energy, 172
  subshells, **639**
  volt (eV), **155, 393**
Electronics, **486**
Electrophoresis, 92
Element, **6, 641, 692**
Elementary particles, **190**
EMF ($\mathcal{E}$) **401,** 403, 409, 411, 471
  equivalent, **412, 414**
  induced, 434, **451, 455**
  source, **400,** 403, 410
  thermal, 310
Energy, 4, 20, **157**
  band; conduction, **644;** forbidden **643;** solids, **642;** valence, **644**
  binding, **178,** 180, 625, 650
  conservation of, 153, 170, **171**
  degradation, **342**

disordered, **269**
electric, **408**, 454
excitation, 182
interaction, **179**
internal, **171**, 284
intrinsic, **171**
kinetic, 153, **156**
level: atomic, 542, 616, **624**, 633;
   nuclear, **626**
and mass, **170**
mechanical, 153, **160**
potential, **158**
primary producers, 352
to remove proton or neutron from nucleus, 157
rest, **170**, 229
thermal, **162**, **280**, **282**
total, 170, 229
transformations, **161**
typical, in eV, **157**
and work, 157
Engineering system of units, 59, 158
Entropy, 327, **343**
   change, **329**
   and disorder, **329**, 332
   increase, heat conduction, **335**
   principle of increase, **332**
   and probability, **344**
   table of, 329
Equations, **679**; and units, **679**
Equilibrium
   chemical, 320
   conditions of, **121**
   mechanical, **125**, 320
   rotational, **122**, 128
   stable, 125, 134
   state, **316**, **319**
   thermal, 297, **305**, 320
Escape velocity, 65, **161**
Euler, L., 55
Evaporation, **313**
Event, as viewed in physics, 209
Excitation energy, 182
Excited state, 79, 625
Exponential decay, **17**, 479
Exponential notation, **676**
Exponents, **675**
Eye, **582**
Eye defects, **585**

Fahrenheit temperature scale, 163, **272**
Fallout, 182
Farad, **478**
Faraday, M., 73, **433**, 444, 520
Faraday's law, **451**, 520
Fermat, P. de., 554
Fermat's principle, **557**
Fermi, E., 21, **153**, 178, 185, 203
Fermion, 154
Ferromagnetism, **437**
Feynman, R. P., 94, 151, 519, 652, **671**
Feynman diagram, **652**
Fiber optics, **565**, **608**
Fibrillation, ventricular, 495
Field
   electric, **71**, 73, 391, 394
   emission, **664**
   gravitational, **59**
   magnetic, 433, **435**, 438, 443, **447**
   theory, **73**, 435
Filament, 486
Filtering, electrical, **488**
Fine structure constant, **634**
Fission
   fragments, 182
   nuclear, 173, **181**
   plutonium, 182
   uranium, 154, 181
Fleming, Sir J. A., 469, 486
Flotation, 252
Fluid
   dynamics, **253**
   irrotational, 257
   statics, **245**
Flux, magnetic, 433, **436**, 451
Focal length, lens, **569**, **575**
Focal length, mirror, **562**
Focus, **561**
Force, 56
   buoyant, **251**
   centrifugal, **214**
   centripetal, **66**
   Coriolis, **217**
   electric, 3, **70**, 72, 80
   external, 102
   frame, 211, **212**
   frictional, 80, **115**, 162

gravitational, **59**, 80, 158
inertial, **213**
internal, 102, 213
magnetic, 434, **438**, **443**, **448**
nuclear, 5, 80, 656
resultant, 56
table of, 80
Foucault, J. B. L., 518, 541, 555
Fourier, J. B., 389
Frame force, 211, **212**
Frame of reference, **44**, 52, **209**, 214
   accelerated, **212**
   inertial, **57**, **210**
   rotating, **214**
Franklin, Benjamin, 395, 396
Fraunhofer, J. von, 518, 533, 537
Free body diagram, 110
Free electron, 398
Freeze-drying, 315
Frequency (f), **131**, 472
   fundamental, **366**
   natural, **366**
   and period, 360
Fresnel, A., 518, 537
Fresnel lens, **586**
Friction, 80, **115**, 162
   at body joints, **150**
   coefficients, 116, **117**
   sliding, **116**
   static, **116**
Frisch, Otto, 185
Fuse, 415
Fusion curve, **316**
Fusion, nuclear, **186**

"g" meter, 137
Galileo, 41, 49, 57, 357, 554, 594
Galvanometer, 397, **446**
Gamma
   camera, 26, 33, **664**
   decay, 14, 626
   ray, 4, 526, 626; photon, energy of, 157
Gas
   ideal, 272
   inert, 7
   kinetic theory, 208, 269, **274**
   law, **271**

thermal energy, **281**
work in compression/expansion, **282**
Gauge pressure, **246**, 249
Gauss, K. F., 606
Gauss (unit of magnetic field $\vec{B}$), **439**
Gedanken Experimente, **219**
Geiger, H., 10, 12
Geiger counter, 10, 11
Gell-Mann, M., 191
General relativity, **231**
Germer, L. H., 615, 627
Gibbs, J. W., 270
Glaser, D. A., **9**
Glass, **563**
GM counter, **11**
Gravitation
    acceleration due to, 49, 63
    Newton's law of, **62**
Gravitational
    constant (G), **60**, 690
    and electrical analogues, table of, **71**
    field ($\vec{g}$), **59**; sources of, **60**
    force, **59**, 80, 158
    potential energy, 159
    torque, 121
Gravity
    center of, **120**
    in medical practice, **90**
Gray, J. A., 624
Gray, S., 389
Gregory, J., 596
Grid, 489; bias, 490
Grimaldi, F. M., 517, 537
Ground state, 79, 625, **636**
Grounding, electric, 392, **416**
Guericke, O. von, 250, 265, 357

Hahn, Otto, 185
Half-life, **17**, 25, 626; biological, 30, **32**
Hall, E. H., 395
Hall effect, 395, 396
Halley's comet, period of, 86
Harmonics, **367**
Hearing, **370**; binaural, 371
Heat, **164**, 283

death of universe, 342
engine, **337**; thermal efficiency, 337
latent; of fusion, 298, **311**; of vaporization, 312, 325
pump, 351
reservoir, **331**
transfer, **302**, 324; conduction, **302**; convection, 303; in humans, **303**, **324**; radiation, 303
Heisenberg, W. K., 616, **632**, 633
Heisenberg uncertainty relation, **631**
Helium, liquid, 348
Helmholtz, H. L. F. von, 153, 358
Henry, J., **434**
Henry (unit of inductance), **481**
Hertz, H. R., 519, **525**, 615
Hertz (unit of frequency), **130**
High-voltage transmission, 409, 454
Holes, 396, 492
Hooke's law, **135**, 160; force constant, 134, 135
Hubble, E. E., **545**
Hubble-Humason law, 545
Humidity, relative, **318**
Huygens, C., 55, 518, 540, 554
Hydraulic brakes, 251
Hydraulic press, **251**
Hydrogen atom, 6, 19, 616, 625, **634**
Hydrogen bomb, 181, **186**
Hydrometer, **253**
Hydrostatics, law of, **248**
Hydrotherapy, 252
Hyperopia, **585**
Hypodermic injection; jet, **90**; needle, **202**
Hypothalamus, 324

Ideal gas law, **271**, 273
Illumination, 546
Image, **559**
    left-right reversal, **560**
    real, **559**
    virtual, **560**
Impedance, **482**
Induced current, 434
Induced EMF, 434, **451**, 455
Inductance, **481**
Inductive reactance, **480**

Inductor, **480**
Inert gases, 7
Inertia, 59; law of, **57**
Inertial force, **213**
Inertial frame of reference, **57**, **210**
Infra-red, 525
Infrasound, 365, 379
Insulator, electric, 395, **398**, **644**
Integrated circuit, 469, 647
Intensity
    electromagnetic wave, **522**, **529**, **619**
    light, **522**, 546, **619**
    sound, **368**
Interaction(s)
    electromagnetic, 80
    energy, **179**
    particle, **5**
    strong, **655**
    weak, 80
    table of, **80**
Interference, 518, **528**, **531**
    constructive, 518, **529**, 530
    destructive, 518, **529**, 530
    particle, **628**
    pattern, **530**, 628
    of probabilities, 615, **629**, 631
    term, **530**, 536, 628, 631
Internal energy, **171**, 284
Intravenous infusions, 90, 262
Invariant interval, **227**
Inverse square law, **683**
Iodine-131, 25, 26
Ion, 8
Ionization, 8
    chamber, 91
    energy, of hydrogen, 179, **637**
IR-drop, 404, 411, 418
I^2R loss, **409**
Irreversible process, **333**
Isolation transformer, **506**
Isotope, **6**, 21
    chart, 21
    tracers; nonradioactive, **34**; radioactive, **25**

Janssen, Z., 517, 553
Joule, J. P., 153, 163, **165**, 269

Joule (unit of energy), **155**
Joule heating, **409**
Joule's law, **409**

Kelvin, Lord (W. Thomson), 327, **328**
Kepler, J., 68, 357, **553**, 555, 595
Kepler's laws, **67**, 129
Kilocalorie, **165**, **282**
Kilogram (kg), **58**
Kilowatt-hour, **157**
Kinematics, **55**
Kinetic energy, 153, **156**; and rest energy, **170**
Kinetic theory of gases, 208, 269, **274**
Kinetic theory of matter, 208
Kirchhoff, G. R., 419, 541
Kirchhoff's rules, **419**
Krönig, A., 269

Land, E., **527**
Laplace, P. S., 98
Laser, 9, 523, 537, 539, **647**, 665; medical applications, **649**, **665**, **667**
Latent heat, fusion, 298, **311**
Latent heat, vaporization, **312**, 325
Laue, M. T. F. von, 534, 535
Laue diffraction, 534, **535**
Lawrence, E., 21, 188, **442**
Lawson, R., 325
Leeuwenhoek, A. van, **554**
Leibnitz, G. W., 98
Length contraction, of moving object, **226**
Lens
  aberrations, **578**
  combinations, **577**
  concave, **573**
  condensing, 586
  converging, **569**
  convex, **569**
  defects, **578**
  diverging, **573**
  equation, **571**
  Fresnel, **586**
  nonreflecting coating, **580**
  objective, **590**
  ocular, **590**
  thin, **568**
Lensmaker's formula, **576**
Lenz, H. F. E., 434
Lenz's law, **452**
Lepton, **190**, 191
Lever, **166**
Libby, W. F., 27
Light, 5, **556**
  absorption, 525, 542, 622, 650
  amplification, 647
  coherence, **535**
  emission and reemission, **536**, 542, 622, 624, 650
  intensity, **522**, 546, **619**; for surgery, **546**
  invariance of speed, 222
  particle nature, 518, 541, 622
  polarized, **526**
  rays, 517, **527**, **562**
  rectilinear propagation, **556**
  unpolarized, 527
  velocity, 218, 449, 517, 520, **690**
  wave nature, 519, **531**, **537**, 541
  wavelength, **523**, **524**, **531**
  year, **545**
Linear accelerator, 188, 222
Linear circuit element, 407
Lippershey, H., 517, 553
Livingston, M. S., 21, 188, **443**
Longitudinal wave, 361, 362
Looming, **566**
Loops (anti-nodes), 366
Lorentz transformations, **222**
Loudness, **368**, 371

Mach, E., 270, 374
Mach number, 374
Machines
  efficiency, **167**, **169**
  mechanical advantage, **166**, 168
  simple, **166**
Macroscopic system, 269
Magdeburg hemispheres, 250, 260
Magnet, 437, 439; ophthalmological, **462**
Magnetic
  domain, **437**
  field ($\vec{B}$), 433, **435**, 438, 443, **447**; typical values, **439**
  flux, 433, **436**, **451**; density, 434, **435**, **436**; line, 433, **436**, **451**
  force, 434, **438**, **448**; and electric current, **443**
Magnification, **573**, 592, 595; angular, **595**, 608
Magnitude, 50
Maiman, T. H., 647
Malus, E., 526
Manometer, **249**
Marius, S., 594
Marsden, E., 12
Mass, 3, 5, **58**, 60
  center of, **106**, 121, 129
  critical, **183**
  and energy, **170**
  number, **5**, 16
  relativistic, 4, **170**
  rest, **170**
  spectrometer, **441**, 459
  standard of, 58
Maxwell, J. C., 270, **517**, 519, 520, 555
Maxwell demon, 336
Maxwell's equations, 519, 528
Mayer, J. R., 153
Mechanical
  advantage, **166**, 168; actual, **168**; ideal, **168**
  energy, 153, **160**; conservation, 153, **160**; conversion into thermal, **163**, **332**; dissipation, **162**
  equilibrium, **125**, 320
  waves, 361
Medical radiography, **509**, **511**, 662, 663, **664**, **667**
Medicine, applications of physics in, **30**
Meitner, L., 185
Mercury hazards, **324**
Mersenne, M., 357
Meson, 176, 177, **190**, **656**
Metastable state, **648**
Meter, electric, **445**

Metric system and equivalents, **691**
Michelson, A. A., **218**, 555
Microscope
    compound, **590**
    electron, **592**
    field ion, 8, **593**
    optical, 554, 577, **590**
    ultraviolet, 551
Microwaves, 524, **550**
Millikan, R. A., **76**, 621
Millikan's oil drop experiment, **75**
Mirage, **566**
Mirror
    concave, **561**
    convex, **561**
    plane, **559**
mks system of units, 59, 158
Moderator, 184
Mole, **271**, 272
Molecule, **7**
Moment arm, 119
Moment of inertia, **128**; of a solid cylinder, 128
Momentum, **100**, 228, 441
    angular, **128**
    conservation law, **102**, 105; present status of, **105**
    and Newton's second law, **100**
    rate of change, 100
    relativistic, **228**, 441
    of a system of particles, **101**
Moon, distance from earth, 139, **690**
Moon, mass of, 139, **690**
Morley, E. W., 218
Motion
    Newton's first law, **57**
    Newton's second law, **56**, 58
    Newton's third law, 62, **99**
Motor, electric, 410, 434, **447**
Müller, E. W., 8
Müller, W., 11
Muon, 235
Myopia, **585**

National Accelerator Laboratory (Fermilab), **189**
Neutrino, 154, **177**; and conservation laws, **177**

Neutron, **4**, 21
    activation analysis, 25, **32**
    capture, 21, **173**, 182
    fast, 182
    mass, 173, 690
    slow, 182
Newton, Sir I., 42, 57, **97**, 243, 517, 537, 554, 596
Newton (N) (unit of force), **56**, 58
Newton's law of gravitation, **62**
Newton's law(s) of motion
    first, **57**
    practical use, **110**
    second, **56**, 58; for accelerated frames of reference, **212**; in terms of rate of change of momentum, 101
    third, 62, **99**; example, **100**
Nodes, 366
Nonconductor, electric, 389, **398**
Nonlinear circuit element, **407**
Nuclear
    chain reaction, first self-sustaining, **185**
    emulsion, 9, 10
    fission, 173, **181**; energy comparison with chemical reaction, 181
    force, 5, 80, 656
    fusion, **186**
    power production in U.S.A., **185**
    reactions, **20**
    reactor, 181, 184, 185
    symbols, 5
    transmutation, **20**
Nucleus, **5, 11**, 79; compound, **20**

Object, in optics, 559
Octave, **370**
Oersted, H. C., 433
Ohm, G. S., **389**
Ohm (Ω) (unit of resistance), **404**
Ohmic circuit element, **406**
Ohm's law, 390, 405, **406**; AC analogue, **484**
Onnes, H. K., 348
Operational definition, **56**
Ophthalmological magnet, **462**

Ophthalmoscope, **609**
Oppenheimer, J. R., 185
Optical
    center, **569**
    diagram, **569**, 570, 574, 577
    pumping, 648
    pyrometer, **308**
Orbit, 63, 65; circular, **65**, 440; speed, 66
Oscilloscope, **78**, 92
Ostwald, W., 270
Overhead projector, **586**
Overtones, **366**, 367

Pacemaker, **489**
Pair annihilation, **172**
Pair production, **172**, 650
Paraboloid, 562
Parallel circuits, **413**
Parallel plates (capacitor), 72
Parent nucleus, 13
Particle
    accelerator, 21, **187**, 222, 442, 449
    detector, **10**
    system, 101
    virtual, **655**
Pascal, B., 243, 251
Pascal (unit of pressure), **245**
Pascal's principle, **251**
Pauli, W., 178, 633, **639**
Pauli exclusion principle, **639**
Pendulum clocks, 137
Pendulum, simple, **136**
Penumbra, **556**
Period, **130**, 360, 472
Periodic table, 640, **641**
Perpetual motion, **337**
Phase, **131**
    angle, **131**
    changes, 310
    diagram, 315
    difference, in AC, **480**
    equilibrium, 312
    relationship, 472, 475, 480, 528
Phases of matter, 310
Phasors, addition of AC voltages, **496**
Photocoagulation, 666

Photoelectric
  effect, 92, 207, **620**, 650
  equation, **621**
  tube, **663**
Photomultiplier, **663**
Photon, **4**, 9, **190**, **615**, **617**
  absorption/emission, 616, 622, 626, 650
  energy, **617**
  exchange, 653
  red light, 157
Phototimer, **663**
Physical constants, **690**
Physical law, **16**
Physics, in medicine, **30**
Physics, theories, **671**
Physiological monitor, **92**
Piezoelectric effect, 378
Pilot blackout, **213**
Pinhole camera, **580**
Pitch, sound, **370**; international standard, 380
Planck, M., **622**
Planck's constant (h), 303, **617**, 622
Plasma, **186**
Plate, 486
P-n junction, 492
Polarization, **526**
Polaroid, 526, **527**
Polonium, 14
Population inversion, 647
Position coordinate, 45
Position vector, **52**
Positron, 21
Postural hypotension, **266**
Potential electric, **391**, 418
Potential difference, electric, **391**, 404
Potential difference, and electric field, 391, **394**
Potential energy, **158**
  electric, 159, **392**
  and force direction, **160**
  gravitational, 159
Powell, C. F., 177, 656
Power, **158**
  AC, **484**
  blackout, **468**
  consumption per capita, 193
  conversion, and EMF, **409**, **455**

electric, **408**, 410, 454
factor, **485**
lens, 577
nuclear in U.S.A., **185**
Pressure, **245**, **271**
  absolute, 249
  atmospheric, 245, 690
  change with depth, **247**
  cooker, 317
  gauge, **246**, **249**
Principal axis, **561**, **569**
Principal focus, **561**, **569**
Probability, **18**, 344, **615**, **628**, **633**; and entropy, **344**
Problem-solving procedure, **110**
Projectiles, **68**
Proportionality, **681**; constant, **682**
Proton, **4**, 70, 173, **690**
Ptolemy, 553
Pulleys, **169**
Pulsating DC, **488**
Pulse sensor, 92
Pumps, peristaltic, **265**
Pumps, piston, **265**
Pythagoras, 357
Pythagorean theorem, **683**

Quantum, 617
  electrodynamics, **652**
  hypothesis, **661**
  mechanical representation of atom, 19
  mechanics, 18, 79, **615**, **627**, **630**, **633**
  number, 129, **625**, **637**
Quark, **191**
Quasar, **545**

Radar waves, 524
Radian measure, **126**, **685**
Radiation
  background, 26
  detectors, 31
  effects on forest ecosystems, **202**
  exposure, dosage and guides, 26, 30, **34**
  hazards, 27

  interaction with matter, 26, **650**
  medical usage, 24, 32, **550**
  standards, **34**
Radio waves, 519, 524, 546, 618
Radioactive
  decay, **13**, **17**, 175, 626,; law, **18**, 24, 25; possibility of, 175
  series, **14**
Radioactivity, **13**
Radioastronomy, **596**
Radiobroadcast bands, **546**
Radiocarbon dating, **27**
Radiography, 39, 509, 511, 652, 662, 663, 664, 667
Radiographic contrast media, **668**
Radioisotopes, **21**, **23**
  in agriculture, 23, 24
  dating, **27**
  in geology, 24
  half-life, 23
  identification, 25
  in industry, 23, 24
  in medicine, 23, 24, **25**, **31**
  radiation source, 23, **32**
  tracers, 25
Radiotelescope, **596**
Radium, 14, 29; decay, **175**
Rainbow, **567**
Random motion, and thermal energy, 162, **276**
Rarefaction, 362
Rate of change, **43**
Rayleigh, Lord (J. W. Strutt), **357**
Rays, light, 517, **527**, **562**
Reactance: capacitive, **479**; inductive, **480**
Rectification, 469, **486**, **487**
  full-wave, **488**
  half-wave, **488**
Rectifier, 407, **486**; solid-state, **492**
Rectilinear scanner, 26, 33
Red shift, 543
Reference, frame of, **44**, **52**, **209**, **214**
Reflection, **539**, **557**; total internal, **564**
Refraction, **540**, **563**; atmospheric, **566**
Refractive index, **540**, 563
Refractive power, **577**

Refrigerator, **340**; coefficient of performance, **351**
Reines, F., 177
Relativistic mass, 4, **170**
Relativistic mechanics, **228**
Relativity
 general theory, **231**
 principle, **222**; in Newtonian mechanics, **211**
 of simultaneity, **220, 225**
 special theory, 153, 170, 218, **228**, 230, 520
Relay, electric, 458
Resistance
 electric, 403, **404, 405, 406, 475**
 equivalent, **412, 413**, 417
 internal, **411**, 418
Resistivity, **405**; table of, **406**
Resistor, **399**
Resonance
 and biological cells, **383**
 electrical, **483**
 mechanical, **368, 383**
 sound, **368**
Resonant frequency, **368, 483, 650**
Rest energy (mc^2), **170**, 180, 229
 and energy conservation, **171**
 uranium, 181
Rest mass, **170**; changes and chemical reactions, **174**
Resultant force, 56, 109
Retina, **582**
Reversible process, **334, 339**
Right-hand rule, **436, 437, 439**
Rigid body, mechanics of, **119, 126**
Ripple, in DC, **488**
RMS current, **473**
RMS voltage, **473**
Rocket
 escape velocity, **161**
 propulsion, **105**
 thrust, **105**
Roemer, O., 517, 554
Roentgen (unit of ionizing radiation), 30
Röntgen, W. C., 1, **526**, 534
Rotating frame of reference, **214**
Rotational motion, **126**
Rotor, 475

Rowland, H. A., 534
Rumford, Count (B. Thompson), 153, 284
Russel, B., 671
Russel traction, **124**
Rutherford, E., **1**, 17, 20
Rutherford scattering, **11**, 79
Rutherford-Soddy law, 1, **18**, 24, 25
Rydberg constant, 661

Satellite, 64
Saturated vapor, 317
Saturation vapor pressure, **313, 317**; water, **314**
Savart, F., 433, 447
Scattering elastic, 20
Scattering, nuclear, **11**
Schrödinger, E., **633**
Schwinger, J., 652
Scintillation counter, 10, 33
Seaborg, G. T., **2**, 38
Secondary emission, **664**
Semiconductor, 396, **398, 492, 644**
Series circuits, **412**
Serpukhov, 189
Shock, electric, **415**, 416, **494, 511**
Shock wave, 374
Shockley, W. B., 469
Short circuit, **414**, 416
Shunt, 446
SI system of units, 59, 94
Siemens, E. W. von, 474
Significant figures, **677**
Simple harmonic motion, **130**, 359
 dynamics, **134**
 force constant, **134**, 135
 phase angle, 131
 total mechanical energy, **160**
 velocity and acceleration, **133**
Simple machines, **166**
Simultaneity, 219, 220, 225
Single-loop equation, **411**
Sinusoidal function, **359**, 471
Smog chamber, **598**
Snell, W., 517, 554
Snell's law of refraction, **540**
Soddy, F., 1, 17
Solid, amorphous, **315**

Solid, crystal, 534
Solid-state rectifier, **492**
Sonar, 375
Sonic boom, 374
Sound, **361**
 frequency of waves, **364**
 intensity, **368**
 pitch, **370**, 371
 pollution, **379**
 quality, **370**
 spectrum, **364, 375**
 speed, **362**
Spark chamber, 9
Special theory of relativity, 153, 170, 218, **228**, 230, 520; assessment of, **230**
Specific gravity, **253**
Specific heat, **165, 286**
 per mole, **292**
 per molecule, **288**
Spectrograph, **589**
Spectroscope, **589**
Spectrum
 characteristic, **541, 642**
 continuous, **541, 623**
 electromagnetic, 5, **523**
 line, 541
 sonic, **364, 375**
 visible light, **523**, 524, **541**
Speedometer, 45
Spherical aberration, **562, 578**
Sphygmomanometer, **249**, 266
Spin quantum number, 129
Spirometer, 295
Stability, **125**
Stable system, 180
Standing wave, **366**
State, of system, 633, 652
Static electricity, **494**
Statics, **119**; fluid, **245**
Statistical mechanics, 320, 343
Statistical physics, 269
Stator, 475
Steinmetz, C. P., **467**
Stereoscopic acuity, **584**
Stevin, S., 56
Stimulated emission, 207, 537, 647
Storage battery, 401, 410
Strassman, F., 185

Streamlines, **254**
Sturm, J. F., 358, 363
Sublimation, **315**; curve, **316**
Subshells, electrons, **639**
Sun, centripetal acceleration of, 212
Sun, energy of, 187
Superconductivity, **348**
Supercritical mass, **183**
Superfluidity, 348
Superposition principle, **61**, **72**, 365, 436, 528; quantum mechanical, **630**
Supersaturated vapor, 317
Supersonic speed, **374**
Surgical gloves, checking, **430**
Svedburg, T., 216
Synchrotron, 177, **188**, 443
System, macroscopic, 269
System of particles, 101, 106, 269

Technitium-99, 26, **33**
Telescope: reflecting, **596**; refracting, **594**
Television picture tube, 78
Temperature, **272**, **278**, **305**, **345**
  Celsius, 163, **272**
  Fahrenheit, 163, **272**
  Kelvin, 272, 275, **305**, **347**
  low, 348
  measurements in medicine, **325**
  and molecular translational kinetic energy, **274**
  typical values, **279**
  warm-blood organisms, **294**
Tension, 124
Terminal voltage, **401**, 418
Tesla, N., 467, 505
Tesla (unit of magnetic field $\vec{B}$), **439**
Tesla coil, **505**
Thermal
  conductivity coefficient, **324**
  energy, **162**, **278**, **280**; and ecosystem of earth, **352**; change ($\bar{E}$), **284**; crystals, **281**; gases, **281**; table of, **282**; and temperature, 163, **278**
  equilibrium, 297, **305**, 320
  excitation, 279

expansion, **299**; anomalous, **301**; linear coefficient, **299**; liquids, **300**; solids, **299**; volume coefficient, **300**; water, **301**
  pollution, **339**
  speed, **276**
Thermionic diode, 486
Thermistor, **309**, 325
Thermocouple, **309**, 325
Thermodynamics, 164, **320**
  in biology, **352**
  first law, **164**, **284**
  second law, **327**, **335**; Clausius statement, **328**, **336**; Kelvin statement, **328**, **337**
  zeroth law, **305**
Thermography, **308**, 325
Thermometer, **305**
  bimetallic, **307**
  clinical, 307; disposal of, **324**
  constant-volume gas, **306**
  electrical resistance, **309**
  liquid-in-glass, **307**
  optical pyrometer, **308**
Thermonuclear reaction, 186
Thermostat, **300**; human, **324**
Thompson, B. (Count Rumford), 153, 284
Thomson, G. P., 615, 627
Thomson, J. J., 77, 627
Thomson, W. (Lord Kelvin), 327, **328**
Time
  absolute, **220**
  arrow of, **333**, 352
  dilation, **225**
TNT, energy release in explosion, 196
Tomonaga, S. I., 652
Tone quality, **370**
Torque, **119**, **127**; gravitational, 121
Torr (unit of pressure), **250**
Torricelli, E., 249
Torsion balance, 74
Townes, C. H., 647
Tracers: nonradioactive, **34**; radioactive, **25**
Track recorder, 9
Traction, **124**
Trajectory of particle in gravitational field, **63**

Transducer, **378**
Transformation, equations, **209**, 210
Transformation, from one frame of reference to another, **209**, 222
Transformer, **452**, **485**, 506
Transistor, 398, 469, 492
Transmission axis, in polarization, **526**
Transverse dimension of moving object, **226**
Transverse wave, **359**, 361
Triangles: congruent, **684**; similar, **683**
Trigonometry, **684**, 693
Triode, **489**
Triple point, **306**, 314, **316**
Tritium, 7
Triton, 7, 187
Tube of flow, **254**
Turbulence, 260
Twiss, R. O., 536

Ultracentrifuge, 216
Ultrasonography, **376**, 378
Ultrasound, 364, **375**
  applications, **375**, 384
  and bats, 385
Ultraviolet microscope, 551
Ultraviolet radiation, 525, **551**
Umbra, **556**
Units, systems of, 59, 158, 691
Uranium
  fission, 154, **181**
  isotopes, 7
  rest energy, 181
  series, **15**, 28
  -235, 7
  -238, 6, 7, 13, 14, 15, 28
Urinalysis, 253

Vacuum tube, 76, 407, **486**, 489
Vapor
  saturated, 317
  supersaturated, 317
  water, in air, **317**
Vaporization curve, **316**
V/c, table of, **222**

Vector, **50**
   addition, **51, 108**
   component, **52, 108**
   magnitude, 50, 109
   multiplication, **51**
   sum, **51**, 109
Velocity, **45**
   addition, 210, 218; Einstein law of, **223, 227**; failure of, **218**
   angular, **126**
   change in, 48
   nature's limit of, 222, 229
   vector, **53**
Venturi meter, 257
Vibration: fundamental mode, 367; normal mode, **366**
Villard, P., 526
Virtual particle, **655**
Viscosity, **255, 259**
Volt, **392**
Volta, A., 389, 391
Voltage
   amplifier, **491**
   effective value, **472**
   maximum value, **471**, 473
   RMS value, **473**
   sinusoidal, **471**
   typical values, 391, 401
Voltmeter, 391, 414, **447**

Walton, E. T. S., 21

Water
   absorption of light, 651
   maximum density, 302
Water beds, **266**
Watt (unit of power), **158, 408**
Wave
   amplitude, 359
   crest, **359**
   electromagnetic, 519, **520**
   frequency, 360
   front, **527**
   function (ψ), 615, **630**, 633
   longitudinal, **361**, 362
   mechanical, 361
   motion, **359**
   shock, **374**
   sinusoidal, **359**, 471
   sound, 361
   speed, 359, **360**
   standing, **366**
   train, 359
   transverse, **359**, 361, 526
   trough, **359**
Wavelength (λ), **359**, 523, 524, **531**; de Broglie, **627**
Weak interaction, 80
Weber (unit of magnetic flux), 439, **451**
Weber/meter^2, **439**, 451
Weight, **60**, 65
Weightlessness, 65, **217**

Whitehead, A. N., 243
Wilson, C. T. R., 9
Work, **155**
   and energy, 157
   function, **621**
   macroscopic mechanical, 337
Wren, C., 55

X-axis, 45
X-ray, 4, 27, 526, **534, 622, 640**
   attenuation, 624, 650, 660, **667**
   circuit, **510**
   photon, energy of, 157, **618**, 623
   spectrum; characteristic, **642**, 664; continuous, **623**, 664
   tube, **201, 509**, 510
Xeroradiography, **665**

Young, T., **518**, 531, 532
Young's double-slit experiment, **531**, 536, 537
Yukawa, H., **656**

Zeno, 630
Zero-point energy, 278
Zero potential, **392**
Zweig, G., 191

ALPHABETICAL LIST OF THE ELEMENTS

Element	Symbol	Atomic number Z	Element	Symbol	Atomic number Z	Element	Symbol	Atomic number Z
Actinium	Ac	89	Hafnium	Hf	72	Praseodymium	Pr	59
Aluminum	Al	13	Helium	He	2	Promethium	Pm	61
Americium	Am	95	Holmium	Ho	67	Protractinium	Pa	91
Antimony	Sb	51	Hydrogen	H	1	Radium	Ra	88
Argon	A	18	Indium	In	49	Radon	Rn	86
Arsenic	As	33	Iodine	I	53	Rhenium	Re	75
Astatine	At	85	Iridium	Ir	77	Rhodium	Rh	45
Barium	Ba	56	Iron	Fe	26	Rubidium	Rb	37
Berkelium	Bk	97	Krypton	Kr	36	Ruthenium	Ru	44
Beryllium	Be	4	Kurchatovium	Ku	104	Samarium	Sm	62
Bismuth	Bi	83	Lanthanum	La	57	Scandium	Sc	21
Boron	B	5	Lawrencium	Lr	103	Selenium	Se	34
Bromine	Br	35	Lead	Pb	82	Silicon	Si	14
Cadmium	Cd	48	Lithium	Li	3	Silver	Ag	47
Calcium	Ca	20	Lutetium	Lu	71	Sodium	Na	11
Californium	Cf	98	Magnesium	Mg	12	Strontium	Sr	38
Carbon	C	6	Manganese	Mn	25	Sulfur	S	16
Cerium	Ce	58	Mendelevium	Md	101	Tantalum	Ta	73
Cesium	Cs	55	Mercury	Hg	80	Technetium	Tc	43
Chlorine	Cl	17	Molybdenum	Mo	42	Tellurium	Te	52
Chromium	Cr	24	Neodymium	Nd	60	Terbium	Tb	65
Cobalt	Co	27	Neon	Ne	10	Thallium	Tl	81
Copper	Cu	29	Neptunium	Np	93	Thorium	Th	90
Curium	Cm	96	Nickel	Ni	28	Thulium	Tm	69
Dysprosium	Dy	66	Niobium	Nb	41	Tin	Sn	50
Einsteinium	Es	99	Nitrogen	N	7	Titanium	Ti	22
Erbium	Er	68	Nobelium	No	102	Tungsten	W	74
Europium	Eu	63	Osmium	Os	76	Uranium	U	92
Fermium	Fm	100	Oxygen	O	8	Vanadium	V	23
Fluorine	F	9	Palladium	Pd	46	Xenon	Xe	54
Francium	Fr	87	Phosphorus	P	15	Ytterbium	Yb	70
Gadolinium	Gd	64	Platinum	Pt	78	Yttrium	Y	39
Gallium	Ga	31	Plutonium	Pu	94	Zinc	Zn	30
Germanium	Ge	32	Polonium	Po	84	Zirconium	Zr	40
Gold	Au	79	Potassium	K	19			